Sustainable Energy Technology

Sustainable Energy Technology

Editors

Wei-Hsin Chen
Aristotle T. Ubando
Chih-Che Chueh
Liwen Jin

MDPI

Basel • Beijing • Wuhan • Barcelona • Belgrade • Novi Sad • Cluj • Manchester

Editors
Wei-Hsin Chen
Department of Aeronautics and Astronautics
National Cheng Kung University
Tainan

Aristotle T. Ubando
Department of Mechanical Engineering
De La Salle University
Manila

Chih-Che Chueh
Department of Aeronautics and Astronautics
National Cheng Kung University
Tainan

Liwen Jin
School of Human Settlements and Civil Engineering
Xi'an Jiaotong University
Xi'an

Editorial Office
MDPI
St. Alban-Anlage 66
4052 Basel, Switzerland

This is a reprint of articles from the Topic published online in the open access journals *Energies* (ISSN 1996-1073), *Journal of Marine Science and Engineering* (ISSN 2077-1312), *Processes* (ISSN 2227-9717), and *Wind* (ISSN 2674-032X) (available at: https://www.mdpi.com/topics/sustainable).

For citation purposes, cite each article independently as indicated on the article page online and as indicated below:

Lastname, A.A.; Lastname, B.B. Article Title. *Journal Name* **Year**, *Volume Number*, Page Range.

ISBN 978-3-7258-0197-8 (Hbk)
ISBN 978-3-7258-0198-5 (PDF)
doi.org/10.3390/books978-3-7258-0198-5

Contents

About the Editors

Wei-Hsin Chen

Professor Wei-Hsin Chen is a Distinguished Professor at the Department of Aeronautics and Astronautics and the Director of the Energy Industry Talent Education Center, National Cheng Kung University. He was the visiting professor of Princeton University, University of New South Wales, University of Edinburg, University of British Columbia, University of Lorraine, and RWTH Aachen University. He was also a visiting Professor at the University of Lorraine in 2019–2022. His research topics include bioenergy (combustion, torrefaction, pyrolysis, gasification, liquefaction, bioethanol, etc.), hydrogen production and purification, clean energy (clean coal technology, wind power, thermoelectric generation, carbon capture and utilization, etc.), energy system analysis (optimization, evolutionary computation, machine learning, etc.), and atmospheric science (aerosol absorption and PM2.5). He has published around 600 papers with an h-index of 82 (Web of Science). He is the Editor, Associate Editor, and Editorial Board Member of several prestigious international journals and the Guest Editor of over 15 international journals. He has published 14 books and chapters concerning energy science and air pollution.

Aristotle T. Ubando

Dr. Aristotle T. Ubando is the Assistant Dean for Research and Advanced Studies at the Gokongwei College of Engineering at De La Salle University. He is a Full Professor and Research Fellow at the Mechanical Engineering Department at the same University. He is the head of the Thermomechanical Analysis Laboratory at De La Salle University—Laguna Campus. He has published more than 180 Scopus-indexed journal papers at national and international levels with an h-index of 26. Dr. Ubando was listed among the top 2% of the world's researchers based on composite metrics by Stanford University and Elsevier statisticians in 2021 and 2022. He was selected as one of the 2017 Asian Scientists 100 for his contribution to the field of sustainability. He is a multi-awarded scientist in the Philippines, and among his national prizes is the award for Outstanding Young Scientist, 2016, bestowed by the National Academy of Science and Technology. He is a recipient of the Achievement Award, 2020, for his contribution to the Engineering Division of the National Research Council of the Philippines. His research includes finite element analysis, computational fluid dynamics, process integration and optimization, and energy system modeling.

Chih-Che Chueh

Dr. Chih-Che Chueh's academic journey is a fantastic tale, shaped by educational milestones across diverse landscapes. Beginning in Taipei and later becoming immersed in the academic milieu of Tainan, Dr. Chueh earned a Bachelor's degree from the esteemed National Cheng Kung University in 2002. This was merely the prelude to their deeper scholarly pursuit, leading Dr. Chueh to Taipei again, where their Master's degree from National Taiwan University was attained in 2004. Their quest for knowledge then took a transcontinental turn, guiding Dr. Chueh to Canada, where they earned a Ph.D. from the University of Victoria, British Columbia, in 2011. Dr. Chueh embraced a research-intensive role as a postdoctoral fellow at Queen's University in Canada, specializing in fuel cell simulations. Subsequently, they were entrusted with a second postdoctoral position at the Research Center for Environmental Changes in Academia Sinica. Dr. Chueh is now an Assistant Professor at the Department of Aeronautics and Astronautics in National Cheng Kung University.

Liwen Jin

Dr. Liwen Jin is currently a Professor working at the School of Human Settlements and Civil Engineering of Xi'an Jiaotong University. He obtained his B.Sc. degree from the School of Energy and Power Engineering of Xi'an Jiaotong University, and his Ph.D. degree in the School of Mechanical and Aerospace Engineering from Nanyang Technology University, Singapore. Prior to his current affiliation, he worked at the National University of Singapore as a research fellow. His research interests include geothermal energy applications, multiple-scale heat-transfer enhancement, and thermal management technology. He is the author and coauthor of more than 120 papers in referred international journals. He has co-authored seven books/chapters published by Taylor & Francis, Springer Science, Bentham Science, Wiley-VCH, etc. He has also been invited to speak at various international conferences related to sustainable energy and green technology.

Preface

This Topic encompasses diverse research papers covering various aspects of renewable energy, energy systems, and environmental sustainability.

Here is a preface summarizing the key themes and contributions of the mentioned papers: The first set of papers focuses on tidal and wind energy. Hu et al. (2021) analyze the coupled interactions of a floating tidal current power station, considering uniform flow conditions. Wang et al. (2021) investigate the thrust characteristics of a downstream offshore floating wind turbine under yawed inflow conditions. Lin et al. (2022) delve into the equivalent aerodynamic design of blades for offshore floating wind turbine models. Ma et al. (2022) contribute blade-resolved computational fluid dynamics (CFD) simulations of a periodic array of NREL 5 MW rotors, both with and without towers. The second set of papers explores solar energy applications and thermoelectric-based systems. Seyednezhad and Najafi (2021) conduct a parametric study on a solar-powered thermoelectric-based cooling and heating system for building applications. Shakya et al. (2021) focus on thermal comfort and an energy analysis of a hybrid cooling system combining natural ventilation with radiant and indirect evaporative cooling. Wang et al. (2022) examine the energy-saving potential of thermal diode tank-assisted refrigeration and air-conditioning systems. The third set of papers addresses life cycle assessment and recycling methods. Zhou et al. (2021) present a comparative life cycle assessment of merging recycling methods for spent lithium-ion batteries. Belliardi et al. (2021) propose a method to analyze the performance of geocooling systems with borehole heat exchangers in a monitored residential building. Zdun and Uhl (2022) contribute to improving the properties of an insulated wall for refrigerated trailers through numerical and experimental studies. The fourth set of papers focuses on optimizing and monitoring energy systems. Hu et al. (2021) provide a review of multi-objective optimization in organic Rankine cycle (ORC) system design. Nunes et al. (2021) explore the use of learning mechanisms to enhance the condition monitoring of wind turbine generators. Abdalla et al. (2022) investigate biomass-integrated air gasification regenerative gas turbine power plants. The final set of papers covers diverse topics such as the evaluation of LNG supply chain resilience (Al-Haidous et al., 2022), economic feasibility of green hydrogen production (Osman Awaleh et al., 2022), and a production and delivery model of hydrogen from solar thermal energy (Joubi et al., 2022). Additionally, research on electric vehicles (Dik et al., 2022), waste-plastic pyrolysis products (Belbessai et al., 2022), and advanced concepts in ocean thermal energy conversion (OTEC) (Shi et al., 2023) further enrich the collection.

Overall, these papers contribute their valuable insights to the ongoing research in renewable energy, sustainable technologies, and environmental impact assessment.

Wei-Hsin Chen, Aristotle T. Ubando, Chih-Che Chueh, and Liwen Jin
Editors

Journal of
Marine Science and Engineering

MDPI

Article

Coupled Interactions Analysis of a Floating Tidal Current Power Station in Uniform Flow

Chao Hu [1,2], Chenxuan Tang [3], Chenyang Yuwen [4] and Yong Ma [2,3,*]

1 College of Shipbuilding Engineering, Harbin Engineering University, Harbin 150001, China; huchao@hrbeu.edu.cn
2 Southern Marine Science and Engineering Guangdong Laboratory (Zhuhai), Zhuhai 519000, China
3 School of Marine Engineering and Technology, Sun Yat-sen University, Guangzhou 518000, China; tangchenxuan126@163.com
4 Jiangnan Shipbuilding (Group) Co., Ltd., Shanghai 201913, China; 18845634625@163.com
* Correspondence: mayong3@mail.sysu.edu.cn; Tel.: +86-136-8462-3478

Abstract: For a floating tidal current power station moored in the sea, the mutual interactions between the carrier and the turbine are pretty complex. Current simulation methods based on potential flow theory could not consider the complicated viscous effects between the carrier motion and rotor rotation. To accurately account for the viscous effect, developing a different numerical simulation method based on computational fluid dynamics is necessary. This paper deals with a moored FTCPS (floating tidal current power station) with 6-degree-of-freedom motion in uniform flow based on dynamic fluid body interactions (DFBI) method. Results showed that the blockage effect caused by the columns would increase the average power output of the turbine, while the power output fluctuation also increased. When the carrier is individually moored in the sea, the motion response of the carrier is pretty small, and the carrier is obviously trimming by the bow. However, when the turbine is mounted on the carrier, the carrier motion response is simple harmonic. The motion response frequency of the carrier is in relation to the rotation frequency of the turbine.

Keywords: tidal current energy; vertical axis twin-rotor turbine; hydrodynamic performance; dynamic fluid boundary interaction

Citation: Hu, C.; Tang, C.; Yuwen, C.; Ma, Y. Coupled Interactions Analysis of a Floating Tidal Current Power Station in Uniform Flow. *J. Mar. Sci. Eng.* **2021**, *9*, 958. https://doi.org/10.3390/jmse9090958

Academic Editor: Elisabetta Tedeschi

Received: 12 August 2021
Accepted: 28 August 2021
Published: 3 September 2021

1. Introduction

The tidal current energy is one of the most abundant clean renewable energy resources worldwide [1]. Due to its high energy density, strong predictability, and energy stability, tidal current energy has become a hot topic in ocean energy research in recent years [2]. The tidal current turbine is the core component that converts the ocean tidal current kinetic energy into electricity. Tidal current turbines can be classified into fixed turbine and floating turbine based on their supporting platforms [3]. Figure 1 shows some typical fixed and floating tidal current turbine worldwide. Most of them are still in the demonstration stage and far from commercialized applications. Compared to fixed turbines, floating tidal current turbines are suitable for different kinds of complex marine environments. Floating tidal current turbines are easy to install [4]. The large deck area is convenient for daily operation and maintenance. The turbines can better make use of the strong ocean surface tidal current kinetic energy. The floating platform can be easily removed and redeployed when the platform experience extreme weather conditions. The floating platform has no requirements for the submarine topography and ocean depth [5]. Thus, the FTCPS possesses broad development prospects in the future.

(a) SeaGen (b) AK-1000 (c) "Hai Ming" I

(d) BlueTec (e) Evopod (f) "Hai Neng" II

Figure 1. Fixed and floating tidal current turbine worldwide. (**a**) SeaGen; (**b**) AK-1000; (**c**) "Hai Ming" I; (**d**) BlueTec; (**e**) Evopod; (**f**) "Hai Neng" II.

The floating platform can be easily removed and redeployed when the platform experience extreme weather conditions. The floating platform has no requirements for the submarine topography and ocean depth. Thus, the FTCPS possesses broad development prospects in the future [6]. However, the FTCPS is moored to the sea in the long-term. The turbine's operation would generate a high working load, which would heavily influence the motion response of the carrier. Since the flow field is highly changed due to the motion, the motion response of the carrier would, in turn, affect the turbine's power output. To enhance the reliability and safety of such a kind of device, it is necessary to thoroughly study the mutual and complex interactions between the carrier and the turbine before commercial development and application.

Experiments about the coupled interactions between the platform motion and turbine's hydrodynamics are seldom seen in the literature. Firstly, it is difficult to simulate the load condition of the turbine mounting on a floating platform. Secondly, scaled-down FOWT models have underlying limitations for simultaneously satisfying essential scaling laws (Froude and Reynolds number scaling) [7]. That is, when the Froude scaling law is applied, the Reynolds number scaling law is not guaranteed, and vice versa. Usually, an experimental test costs much more than ordinary numerical analysis. Currently, there is no systematic theoretical framework or mature commercial software to analyze the coupled interactions between the carrier and the turbine for such a system. The turbine and the carrier are often considered as two independent parts. The mutual interactions between the turbine and carrier are usually ignored or simplified. Since the structure of FTCPS is similar to the floating offshore wind turbine (FOWT), some researchers suggest that the analysis method of FTCPS could learn from the OFWT [8]. Thanhtoan et al. studied a FOWT experiencing the prescribed pitching motion of a supporting platform as a sine function [9]. Results showed that the unsteady dynamic loads of the turbine were dramatically changed by the frequency and amplitude of the platform motion. Shifeng et al. carried out a wind-tunnel experiment for a horizontal axis wind turbine with prescribed rolling and pitching motion to study the wake and power fluctuations [10]. Results showed that the power output of the turbine could increase with moderate tower oscillations. The cube of the relative incoming flow velocity impinging the rotor in the pitching, and a momentum replenishing in the rolling motion could lead to the power increasing of the rotor. Thanh et al. established a computation fluid dynamics framework based on dynamic fluid body interaction approach to inspect the turbine's power fluctuations and the platform's motion responses [11]. Three different methods (DFBI, FAST-BEM, GDW) have been conducted and detailed compared for the simulation of FOWT. Compared to the conventional techniques, such as FAST-BEM and generalized dynamic wake (GDW) methods, the maximum oscillating amplitude of turbine power output could be an approximate four-fold difference by the proposed DFBI methods. Some researchers have used similar methods to study the tidal current turbine learnt from wind turbine research. Guo et al. established a framework based on a modified blade element momentum to investigate the power output and rotor loads of a horizontal axis tidal current turbine with prescribed floater motions [12,13]. The floater motions were calculated in the time domain by potential flow theory. Results showed that floater-motion-induced contributions dominate the rotor load. They suggested that the rotor should be mounted close to the floater center to avoid the floater-motion-induced velocity on the blade section. Brown et al. developed a computational HATT model for generalized incident flow conditions using the actuator theory [14]. The arbitrary movement of the turbine was modeled based on the 'weighted body force implementation'. This method requires no constraints in the local structure of the mesh around the turbine, which allows for a quick and efficient simulation of the turbine. Shuqi et al. established a numerical model based on the dynamic mesh technique to study the turbine's hydrodynamics with prescribed platform oscillating motions [15,16]. Results showed that the turbine's loads and power output were highly affected by the oscillating frequency and oscillating amplitude. The damping coefficient and added mass of the turbine were obtained by the least square method. They explained that the fully coupled motion prediction of the platform could be realized by adding the damping and added mass of the turbine to the platform's motion equation in time domain. However, the technique about the motion prediction of the platform are not reported yet in the literature. Xu et al. studied the coupled motion of a catamaran by linearly adding the wave excitation force of the turbine to the platform's motion equation [17]. However, the motion response of the catamaran tends to be underestimated.

As described above, there is currently no systemic method for analyzing the fully coupled characteristics of floating tidal current power stations. The turbine's hydrodynamics and platform's motion response are always investigated separately. Therefore, a fully coupled method which can take the platform's motion, turbine's hydrodynamics, and mooring line tension into consideration is required for the FTCPS.

Therefore, this study establishes a six-degree-of-freedom motion model of a platform-turbine-mooring coupled system under the gas-water two-phase flow based on the dynamic fluid-solid interaction and volume fraction methods. The complicated coupling mechanism between the turbines and the platform is studied. The dual-rotor turbines and the power station carrier model are simplified to reduce the interference of other complex components in the investigation of the coupling effect between the turbines and the platform. The force and moment are transmitted between the carrier and hydraulic turbines through a rotating hinge to reduce interference with connecting construction on hydraulic turbine-carrier coupling. Meanwhile, a double overlapping grid will ensure the convergence of the computational results to prevent the large-scale transient movement of the carrier and the rotating motion of the turbines from causing the calculation results to diverge. Based on the above methods, this article first compares the simulation results with the test results to verify the reliability of the research method. Then, the interactions between the motion of the carrier and the rotating motion of turbines under the uniform flow condition are computed and the mutual influence law and the flow field are obtained by using the aforementioned numerical simulation method. Finally, the influence of the speed ratio on the hydrodynamic characteristics of the turbine under the uniform flow condition is discussed. This research not only outputs the information of turbine load, power and flow field in real-time by the CFD prediction method for the coupled characteristics of a tidal current power station but also predicts the motion response of the power station carrier. The research results can provide specific theoretical guidance and technical support for the design and engineering of floating tidal current power plants.

2. Floating Tidal Current Power Station Design

In this paper, a four-column semi-submersible tidal current power station is designed by adopting a vertical-axis counter-rotating double-rotor turbine. The main structure of the FTCPS includes the main floating bodies on the left and right sides, two symmetrical lower beams, four columns, the platform deck, and the upper buildings. Figure 2 shows the sketch of the floating tidal current power station model. The twin-rotor turbine is symmetrically deployed.

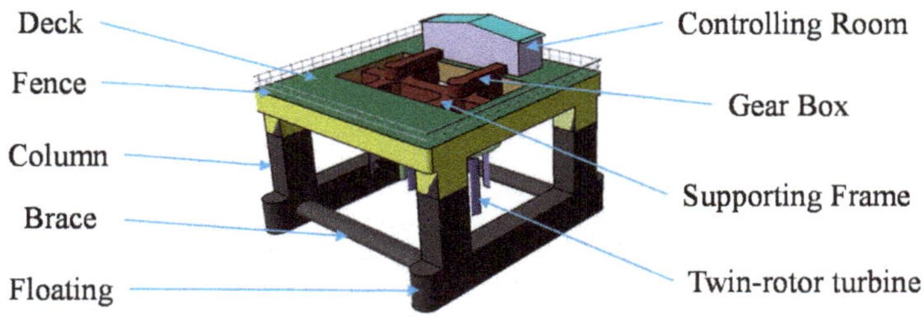

Figure 2. Effect picture of floating tidal current power station.

The airfoil of the turbine is NACA-0018, as shown in Figure 3a. The designed capacity of a single turbine is 120 kW, the designed flow velocity is 2.5 m/s, the turbine's diameter is 8 m, and the airfoil chord length is 1 m. A sketch of the turbine is shown in Figure 3b. To increase the computation efficiency and minimize the influence of unnecessary components of the FTCPS, the FTCPS has been simplified in the numerical simulation process. A simplified model of the FTCPS is shown in Figure 4.

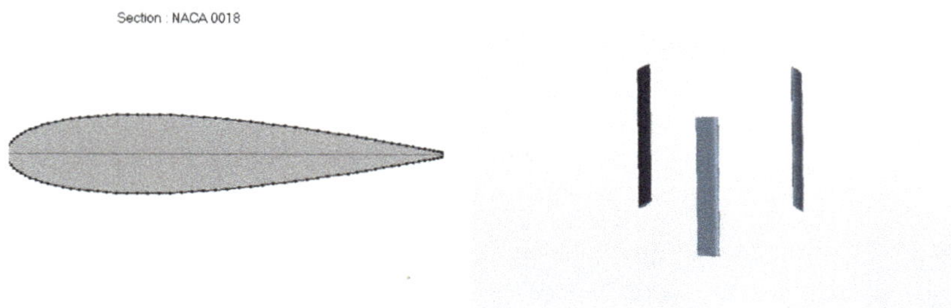

(**a**) Blade airfoil (**b**) sketch of a single turbine.

Figure 3. Twin-rotor turbine. (**a**) Blade airfoil; (**b**) sketch of a single turbine.

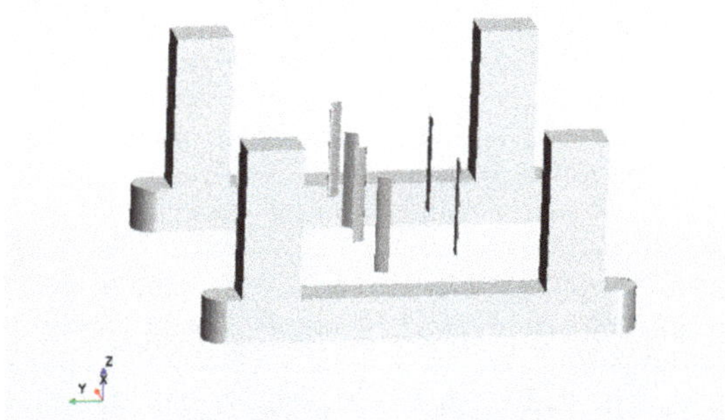

Figure 4. Simplified numerical model of floating tidal current power station.

3. Numerical Research on the Performance of Vertical-Axis Twin-Rotors Tidal Current Turbine

3.1. Numerical Verification of CFD Simulation

In the present research, all the simulations employed a semi-implicit method for pressure-linked equations (SIMPLE) solution algorithms. A second-order up-wind scheme was used for the convection terms. In the unsteady simulation of the floating power station, a second-order central difference scheme was used for the time discretization. Meanwhile, the overset method was used to avoid the computation divergence caused by the significant platform's motion coupled with turbine rotation. In general, the computational is composed of a background domain, an overset refined turbine rotation domain, and a platform motion domain. Detailed discussions have been discussed in previously published literature [18,19].

The overall computational process for the DFBI analysis used in this study is well described in References [11,20]. The turbine rotation is coupled in the 6-DOF motions of the platform, which superimposes a body rotation in addition to the platform motion. The volume of fluid (VOF) and the 6-DOF motion solver is used to solve the fluid-induced motion of the FTCPS in a two-phase flow composed of water and air. The catenary mooring line method is used to model the four mooring lines attached to the FTCPS in time domain.

3.2. Computational Domain Settings

In the present numerical simulations, a global fixed coordinate system (XYZ), one 6-DOF coordinate system ($X_0Y_0Z_0$), and two local rotating coordinate systems ($X_1Y_1Z_1$ and $X_2Y_2Z_2$) has been established in the computational domain. As shown in Figure 5a,b, in the simulation process, the left turbine rotates clockwise in $X_1Y_1Z_1$ by the Z_1-axis, while the right turbine rotates counterclockwise in $X_2Y_2Z_2$ by the Z_2-axis. This internal counter-rotation method can maximize the power output of the twin-rotor system, which has been proved by many researchers [12]. As has been mentioned in Section 3.1, the rotation coordinate ($X_1Y_1Z_1$ and $X_2Y_2Z_2$) of the twin-rotor turbine is coupled with the platform's body motion coordinate $X_0Y_0Z_0$. The carrier is set as a DFBI body. The carrier and the turbine are connected by a virtual hinge. The virtual hinge can transmit the force and momentum between the carrier and the turbine at each time step.

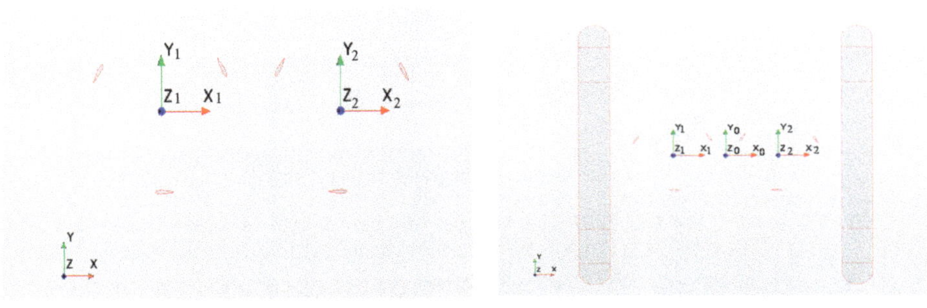

Figure 5. Coordinate system establishment. (**a**) Coordinate system of the twin-rotor turbine; (**b**) Coordinate system of the FTCPS.

As Figure 6 shows, the four catenary mooring lines are symmetrically attached to the platform, forming an angle of 30 degrees with the incoming flow direction (the Y-axis direction). The mooring line simulation is based on quasi-static theory. The net weight of the mooring line is 10 kg/m, and the overall length of each line is 7 m.

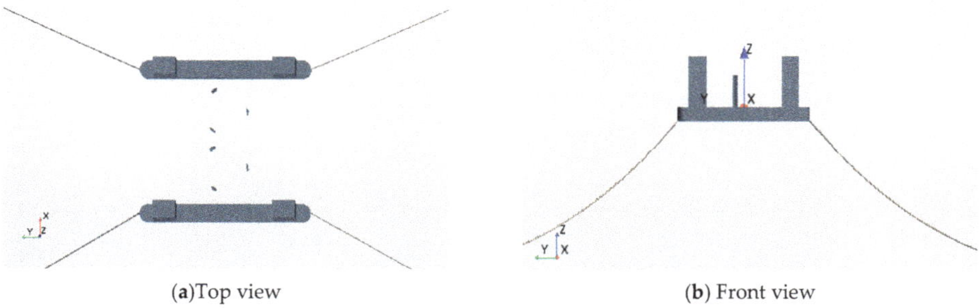

Figure 6. Mooring design. (**a**)Top view; (**b**) Front view.

When calculating the coupled motion of the carrier, turbine and mooring system are under uniform flow conditions. There is a two-phase flow in the numerical model due to free surfaces. VOF (Volume of Fluid) is an effective numerical technique for free surface modelling based on the Euler method [21]. Therefore, the VOF fifth-order wave model is used to simulate the numerical tank.

As Figure 7 shows, the whole computational domain is divided into three sub-domains: background domain, rotating domain, and moving domain. The background domain and the moving domain are connected by overset. The implementation of overset method can reduce the risk of computational divergence resulted from the large movement of the carrier. The rotation domain and the background domain are also connected by the overset method.

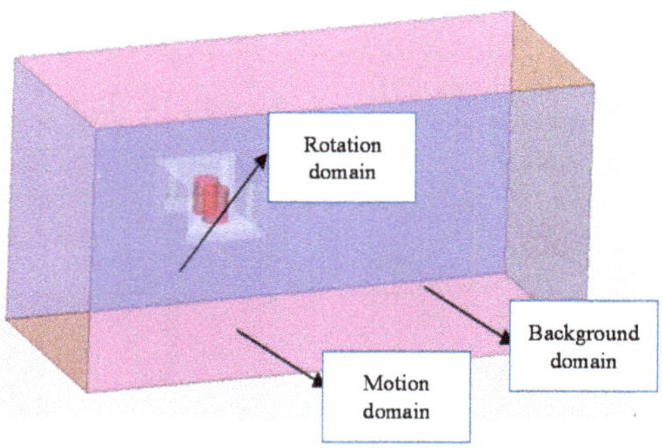

Figure 7. Schematic diagram of the computational domain.

The size of the whole computational domain is shown in Figure 8. The numerical tank measures 20D × 10D × 10D (D is the diameter of the turbine). The entrance of the computational domain is set as velocity inlet, and the incoming flow speed is 1.5 m/s. The outlet and the top of the computational domain is set as pressure outlet, and the relative pressure is set as 0 Pa. The bottom is set as a solid wall. Both sides of the computational domain are set as symmetry planes.

The trimmed cell mesh technique was used in the mesh generation for the whole numerical simulation cases. Due to the complex rotation and 6-DOF movement of the turbine, the mesh around the turbine blade has been refined to capture the flow separation and accurately simulate the mesh around the turbine blade, and the free surface has been refined to accurately simulate the physical scene and capture the flow separation near the blade surface. The mesh around the interface between the water and the air has also been refined to maintain the stability and accuracy of free surface in space and time. The generated computational domain grid, boundary layer grid, and blade surface grid are shown in Figure 9a–c. During the mesh generation around the blade surface, there are 8 layers inside the boundary, with a first layer of 0.0004 m and a progression ratio of 1.5.

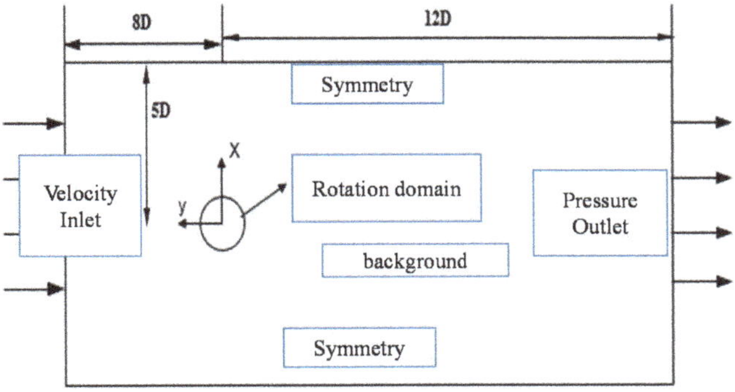

Figure 8. Geometry layout of the computational domain.

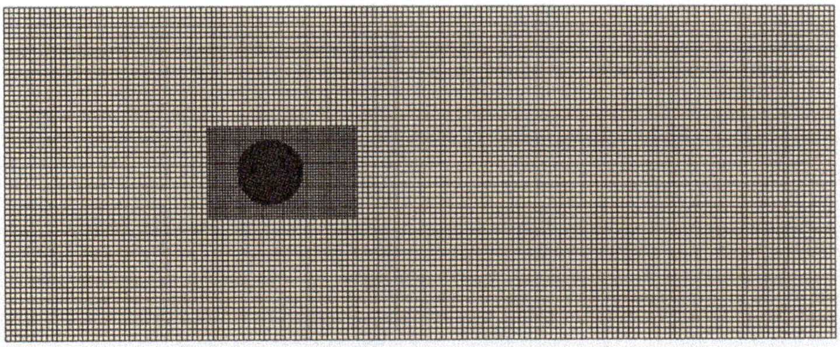

(**a**) Computational domain grid of the turbine.

(**b**) Boundary layer grid.

(**c**) Blade surface mesh.

Figure 9. Grid settings. (**a**) Computational domain grid of the turbine; (**b**) Boundary layer grid; (**c**) Blade surface mesh.

For the numerical simulation under uniform flow conditions, due to the existence of the free surface, the motion of the carrier and the turbine will lead to the mesh deformation of the free surface. Therefore, the surrounding grid needs to be encrypted to maintain the stability of the free surface in space and time. At the same time, to reduce the error caused by the iterative transmission of data between grids, this paper encrypts the grid from the rotation domain to the background domain layer by layer, and the overall grid division of the computational domain is shown in Figure 10.

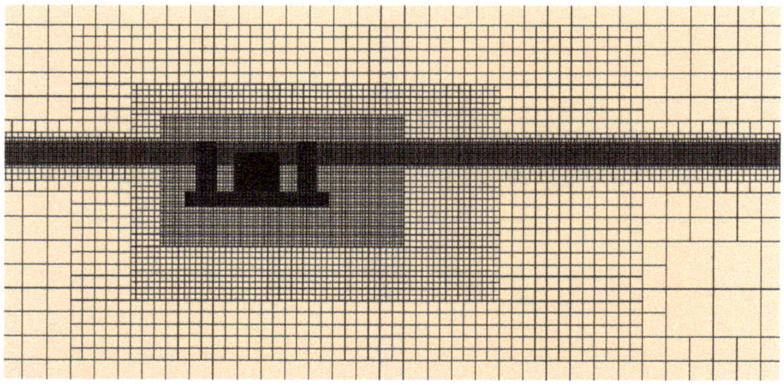

Figure 10. Refined grid of the FTCPS model.

Notably, a grids independency test has been carried out to keep a balance between computation efficiency and computation resources. Three different refined meshes (coarse mesh: 170 W, medium mesh: 250 W, fine mesh: 340 W) have been simulated, and the results are shown in Figure 11. The difference between the medium mesh and the fine mesh is relatively small, while the coarse mesh tends to underestimate the power output efficiency of the turbine. Thus, the fine mesh has been selected in the following numerical simulations.

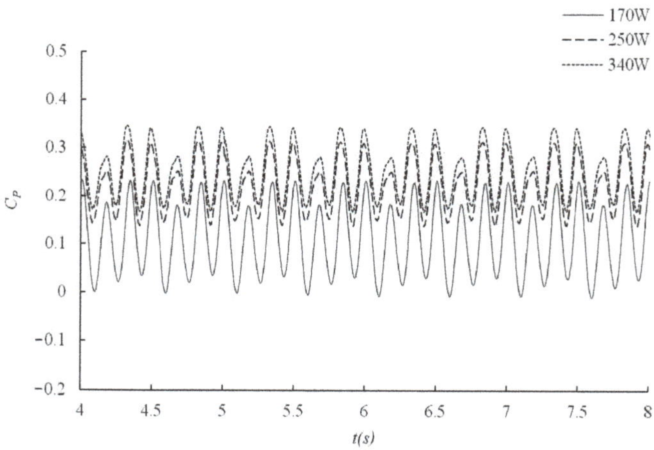

Figure 11. Power output efficiency.

3.3. Validation of the Computational Model

Figure 12 shows the comparison between the numerical simulation results and the corresponding model test results [22]. The test model parameters are consistent with the turbine parameters designed in this paper. It can be seen that, as the speed ratio increases, the turbine power output efficiency (C_p) first increases and then decreases. The value difference between the experiment and numerical simulation at the best speed ratio is small, and the overall trend is in good agreement. However, since the mechanical friction, free liquid surface, and waves are not considered in the numerical simulation, the computational results are a little higher than the experimental results. The simulation error does not exceed 10% (within the error range) in the low-speed ratio area. In general, the simulation results of the CFD numerical simulation method for the vertical axis turbine are in good agreement with the test results, which verifies the reliability of the CFD numerical simulation method for computing the hydrodynamic performance of the vertical axis turbine.

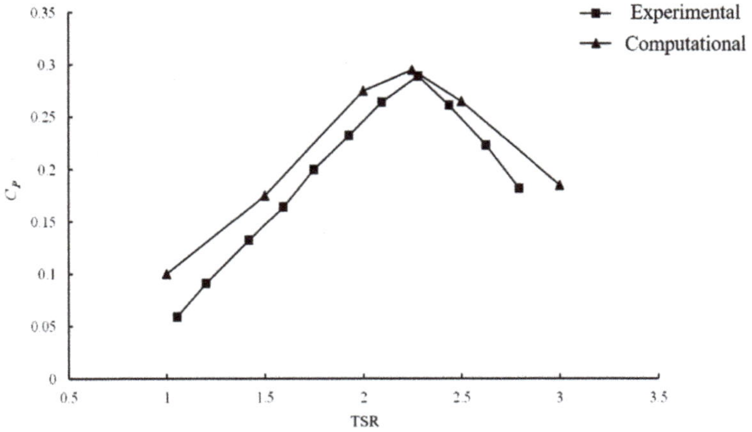

Figure 12. Comparison of computational results and experiment results.

4. Results and Discussions

4.1. Influence of Speed Ratio on the Hydrodynamic Characteristics of Hydraulic Turbine

During the operation of a floating tidal current power station, the tidal current speed is not always constant. In order to maximize the power output of the turbine, it is necessary to adjust the rotation speed of the turbine. This section discusses the changing tendency of the hydrodynamic characteristics of the floating turbines under different tip speed ratios (1, 1.5, 2, 2.5, 3). The incoming flow velocity is 2.5 m/s, and the direction is along the negative direction of the y-axis.

4.1.1. Power Output

Figure 13 shows the time history curve of turbine power output efficiency at different tip speed ratios (TSRs). It can be seen that, under the condition of uniform flow, the power output efficiency changes periodically, and its change period is consistent with the rotation period of the turbine. Sawtooth line sections could be observed at the time history curve of the turbine power output curve when the TSR is low. When the turbine is operating at low TSR, flow separations occurs around the blade surfaces, which caused the blade force fluctuations.

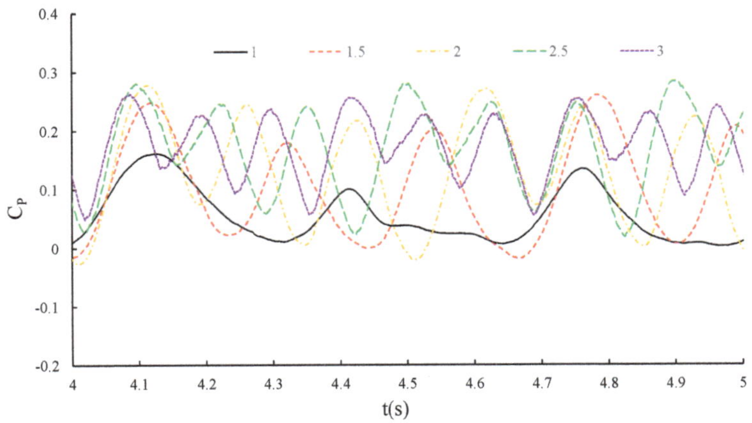

Figure 13. Time history curves of power output efficiency at different speed ratios.

Figure 14 shows the fluctuation amplitude ($\hat{C}_p$) and average value ($\overline{C}_p$) of power output efficiency of the turbine at different speed ratios. It can be seen that, as the speed ratio increases, both $\overline{C}_p$ and $\hat{C}_p$ first increase and then decrease. When the speed ratio approaches 2, both $\overline{C}_p$ and $\hat{C}_p$ reach their maximum values. Meanwhile, when the speed ratio is increasing from 1 to 2, the increasing trend of $\hat{C}_p$ is faster than that of $\overline{C}_p$, and, when the speed ratio is from 2 to 3, the decreasing trend of $\overline{C}_p$ is slower than $\hat{C}_p$.

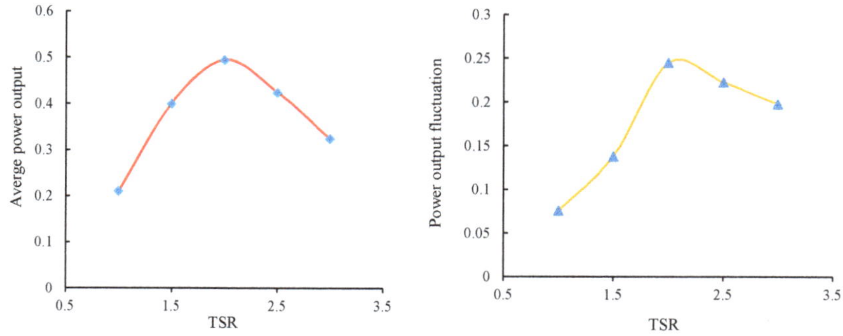

(**a**) Average power output (**b**) Fluctuation of power output

Figure 14. The value of C_p and $\overline{C}_p$ at different speed ratios. (**a**) Average power output; (**b**) Fluctuation of power output.

4.1.2. Thrust Coefficient

Figure 15 shows the time history curve of the thrust coefficient of one of the turbine at different TSRs. It can be seen that, as the TSR increases, the fluctuation rate of the thrust coefficient becomes more intense. It can be seen from Figure 16 that the average thrust coefficient ($\overline{C}_T$) is positively correlated with the change of the speed ratio because the thrust of the hydraulic turbine increases with the increase of the speed ratio.

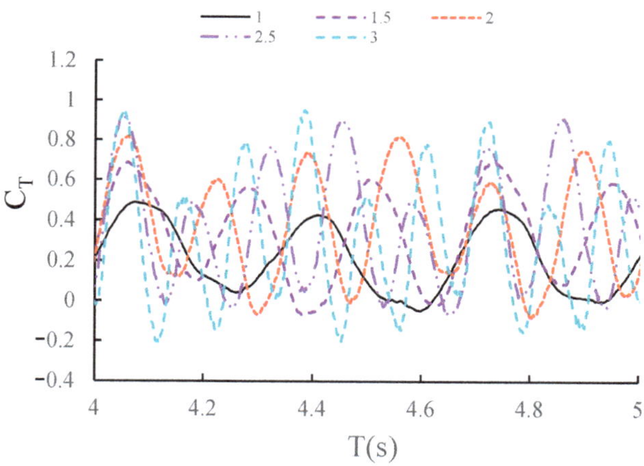

Figure 15. Time history curve of thrust coefficient at different TSRs.

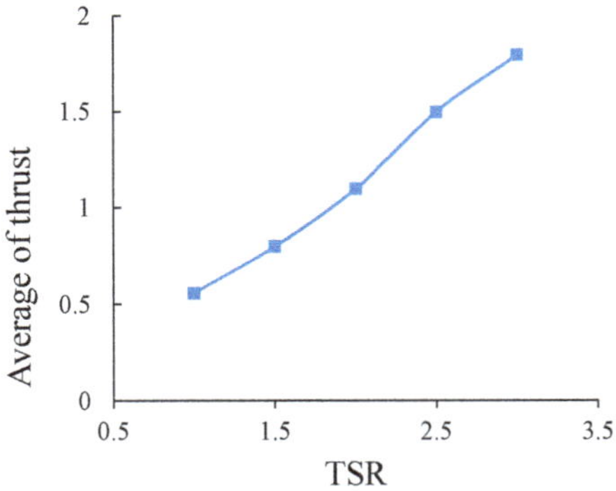

Figure 16. Average of thrust coefficient at different TSRs.

4.1.3. Lateral Force Coefficient

Figure 17 is the time history curve of the lateral force coefficient of a single turbine at different speed ratios. It can be seen that the lateral force coefficient of the hydraulic turbine changes periodically, and its change period is consistent with the rotation period of the hydraulic turbine. When the speed ratio is low, the lateral force fluctuation of the hydraulic turbine is small. As the speed ratio increases, the lateral force coefficient of the turbine increases, and the fluctuation become more intense. Figure 18 shows that the amplitude of the lateral force coefficient of the turbine (C_L) increases with the increasing speed ratio, but the increasing trend gradually slows down. As the speed ratio increases, the average value of the lateral force coefficient ($\overline{C}_L$) first increases and then decreases. The maximum lateral force coefficient at the best speed ratio of 2 is 0.35. At this time, the hydraulic turbine continues to receive positive lateral force. This is because changing the direction of rotation will change the direction of the side force. In order to offset the influence of lateral force on the rolling and swaying motion of the turbine and its carrier, the usual method is to arrange an even number of inwardly rotating turbines side by side.

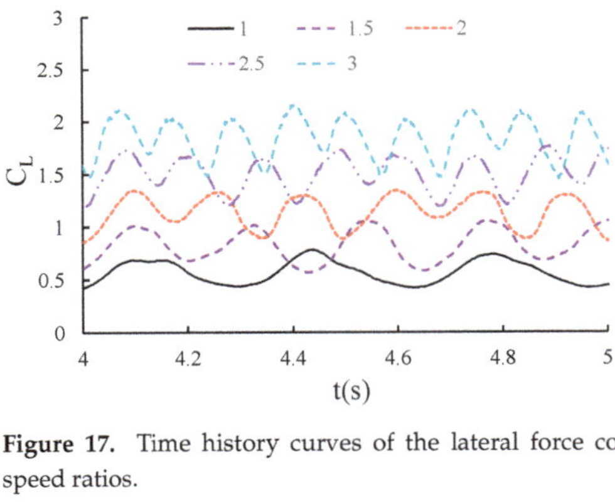

Figure 17. Time history curves of the lateral force coefficient of a single turbine at different speed ratios.

4.2. Influence of Carrier Motion on the Hydrodynamic Characteristics of Hydraulic Turbine

This section compares the hydrodynamic performance difference of the twin-rotor turbine at two different operation conditions, e.g., fixed and floating conditions. In the fixed condition, the computational domain only includes the twin-rotor turbine without any motion. In the floating condition, the computational domain includes the twin-rotor turbine and the carrier with 6-DOF motion. The TSR of the turbine is 2, and the incoming flow velocity is 2.5 m/s. The specific design parameters of the computational domain are shown in Table 1.

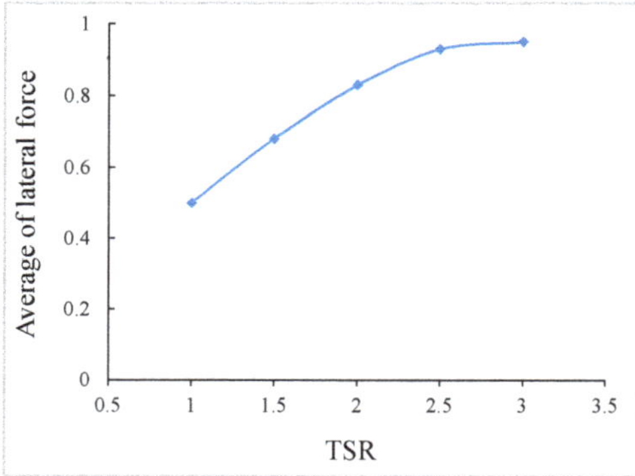

Figure 18. Average of lateral force at different TSRs.

Table 1. Specific design parameters of the computational model.

Case	Velocity (m/s)	TSR	Computational Domain
A	2.5	2.0	Stand-alone turbine
B	2.5	2.0	FTCPS

4.2.1. Power Output

Figure 19 shows the comparison of power output efficiency with respect to time between case A and case B due to the carrier 6-DOF motion.

It can be seen that the fluctuation rate of the turbine with 6-DOF motion is bigger than the fixed turbine. The motion response of the carrier caused higher flow fluctuations for the incoming flow acting on the turbine disk, leading to larger power output fluctuations for the turbine. The average power output of the turbine in case A is 0.24, while the average power output of the fixed turbine is 0.2. The power output increasing of the floating turbine is caused by the two front columns of the carrier. The submerged part of the column changed the boundary conditions of the flow field around the turbine. The two front columns are similar to a flow deflector, which increases the flow velocity passing through the turbine. A flow acceleration zone is clearly observed in Figure 26. The flow velocity increasing effect would be thoroughly discussed in Section 4.4.1.

4.2.2. Thrust Coefficient

Figure 20 shows the comparison of thrust coefficient with respect to time for the twin-rotor turbine. The maximum value of thrust for the floating turbine is 1.3 times that of the fixed turbine. The fluctuation rate of thrust for the floating turbine is 1.13 that of the fixed turbine. The deflector effect of the columns increased the power output of the floating turbine, but it also increased the thrust. As for the fluctuation rate difference between the two cases, the motion response of the carrier increased the flow velocity changing rate acting on the turbine, which definitely would increase the fluctuation rate. It can also be observed that the fluctuation frequency of the thrust for the floating turbine is almost equal to the fixed turbine, which implies that the motion response of the carrier would not influence the thrust frequency response characteristics.

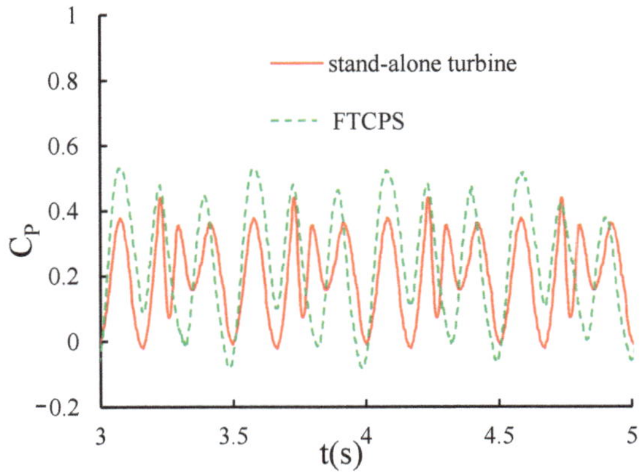

(a) Time history curves of turbine power output efficiency.

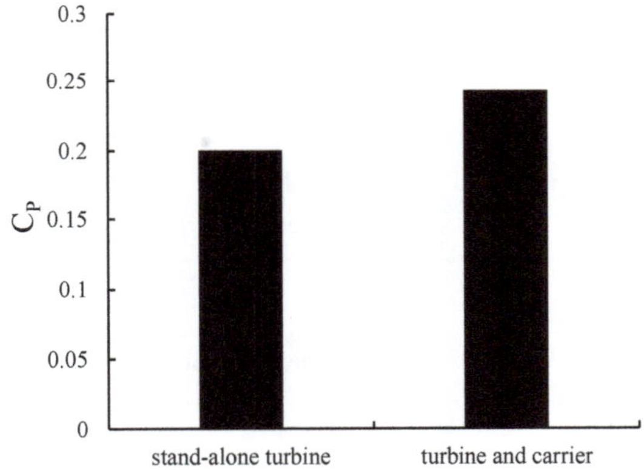

(b) Average power output efficiency.

Figure 19. Influence of the carrier motion on the power output efficiency. (a) Time history curves of turbine power output efficiency; (b) Average power output efficiency.

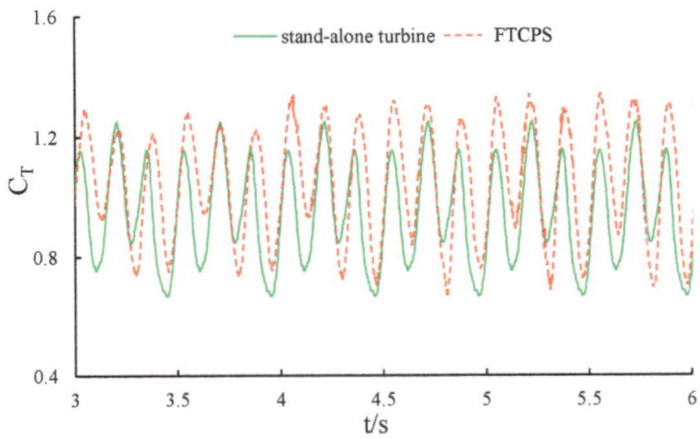

Figure 20. Time history curve of thrust coefficient.

4.2.3. Lateral Force Coefficient

The change of the lateral force of the twin-rotor turbine will affect the rolling and swaying motion of the vertical-axis tidal current turbine. Figure 21 is a time history curve of the lateral force coefficient of the vertical axis turbine. It can be seen that the fluctuation range and maximum value of the lateral force coefficient of the turbine with a carrier are 1.125 times and 1.3 times that of the turbine without a carrier. The reason is that the carrier increases the lateral force acting on the turbine. However, the minimum values of the lateral force coefficients are all-around 0. It can be seen that the lateral force coefficient is much smaller than thrust coefficient. The lateral force between the two rotor is opposite leading to the resultant of lateral force decreasing. Meanwhile, the carrier is equivalent to shortening the distance from the turbine to the boundary. The reflected wave formed by the fluid passing through the surface of the carrier is also one reason for the increase in the lateral force coefficient.

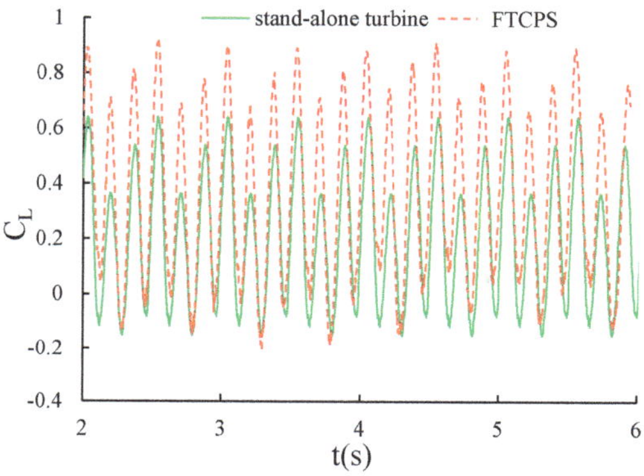

Figure 21. Time history curves of the lateral force coefficient.

4.3. Influence of the Turbine on the Hydrodynamic Response of the Carrier

This section compares the motion response difference of the carrier at two different scenarios, e.g., a stand-alone carrier (case C) and a carrier with the twin-rotor turbine (case D). Only three difference directions have been chosen in the discussion e.g., surge, pitch and heave. The motion response at these three directions is the most essential for the safe operation of such a kind of device. TSR of the twin-rotor turbine is set as 2, and the incoming flow velocity is 2.5 m/s. The specific design parameters of the computational domain are shown in Table 2.

Table 2. Specific design parameters of the computational domain.

Case	Velocity (m/s)	TSR	Computational Domain
C	2.5	2.0	Stand-alone carrier
D	2.5 m/s	2.0	FTCPS

4.3.1. Surge Motion of the Carrier

Figure 22 shows the comparison of surge motion response with respect to time due to the operation of twin-rotor turbine. It can be seen that the stand-alone carrier would move towards the negative axis of X first and then move downstream. The negative value of surge motion response is caused by the pretension of the mooring line. In the end, the surge motion of the carrier is almost a constant value when the carrier moves to the equivalent position. However, for the FTCPS scenario, the carrier would directly move downstream, which is different from the stand-alone carrier. This is caused by the thrust generated by the turbine. The surge motion of the carrier is almost sinusoidal when the carrier moves to the equivalent position. The frequency of the surge motion response is equal to the rotation period of the twin-rotor turbine. It can also be seen that the surge motion response of case D is much bigger than case C. The thrust of the twin-rotor dramatically increased the motion response of the carrier and changed the equivalent position of the carrier.

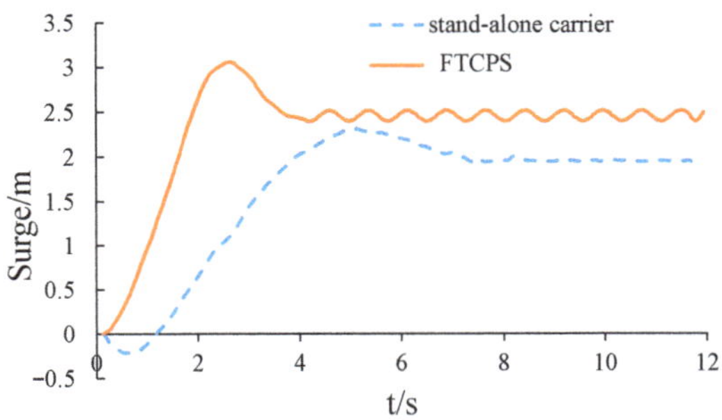

Figure 22. Time history curve of surge motion.

4.3.2. Pitch Motion of the Carrier

Figure 23 shows the comparison of pitch motion response with respect to time due to the operation of twin-rotor turbine. In the initial stage, the motion response of the two scenarios is almost the same. However, the pitch motion of the stand-alone carrier is almost a constant value when the carrier moves to the equivalent position. The average pitch angle is $-1.77°$, which implies that the carrier is trimming by head. Compared to the stand-alone carrier, the pitch motion of the FTCPS is also trimming by head. However, the pitch motion of the FTCPS is almost sinusoidal when the carrier reaches equivalent position. The pitch motion frequency of the FTCPS is almost equal to the rotation period of the twin-rotor turbine. The average pitch angle is $-2.25°$, which implies that the thrust of the twin-rotor turbine would increase the trimming by head situation, which increases the risk of wave running-up.

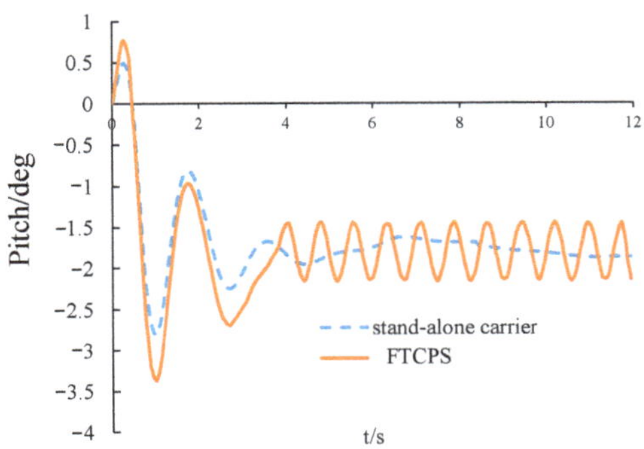

Figure 23. Time history curves of pitch motion.

4.3.3. Heave Motion of the Carrier

Figure 24 shows the time history curve of the heave motion response of the carrier. It can be seen that the heave motion of the carrier with turbine is much bigger than the stand-alone carrier. The stand-alone carrier would stay steady when the carrier moves to the equivalent position. The heave motion of the stand-alone turbine can be neglected compared to the carrier with the twin-rotor turbine. For the FTCPS, the heave motion is also sinusoidal. The heave motion frequency is also linked with the turbine rotation period. Due to the existence of the twin-rotor turbine, the full system would reach a new equivalent position.

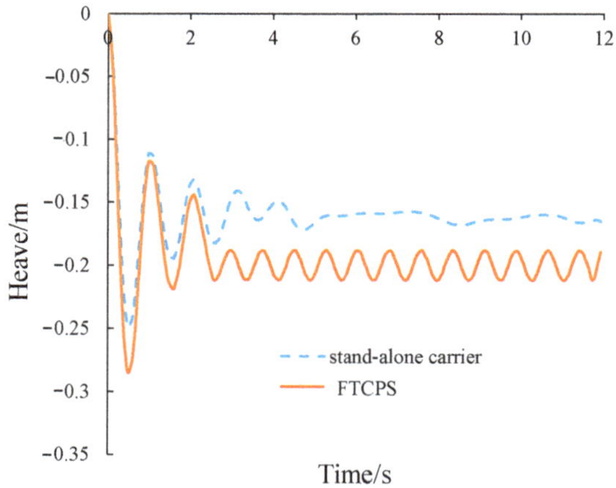

Figure 24. Time history curves of heave motion.

It can be seen that the heave motion response value difference is pretty small compared to the surge motion and roll motion. Since the turbine is a straight-bladed turbine, the horizontal cross-sectional area of the blade is small, which has little influence on the vertical heave damping and added mass of the turbine.

4.4. *Flow Field Analysis*

4.4.1. Flow Velocity

To investigate the hydrodynamics of the turbine and carrier dynamic motion response in detail, the flow field chosen from different sections in the computational domain is analyzed. Figures 25–30 show the instantaneous iso-velocity contours of the turbine with rotating blades and the carrier with 6-DOF motions in the computational domain.

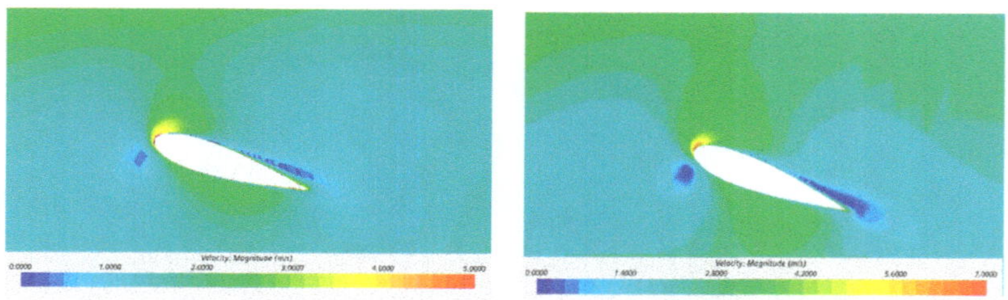

(**a**) Stand-alone turbine (**b**) Turbine with 6-DOF motion

Figure 25. Comparison of velocity field between the fixed stand-alone twin-rotor turbine and the twin-rotor turbine with 6-DOF motion. (**a**) Stand-alone turbine; (**b**) Turbine with 6-DOF motion.

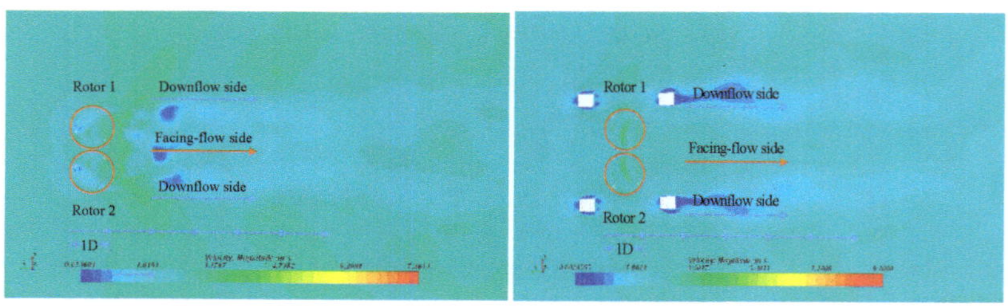

(**a**) Fixed stand-alone turbine (**b**) Turbine with 6-DOF motions

Figure 26. Comparison of horizontal iso-velocity field for the fixed stand-alone turbine and the turbine with 6-DOF motions. (**a**) Fixed stand-alone turbine; (**b**) Turbine with 6-DOF motions.

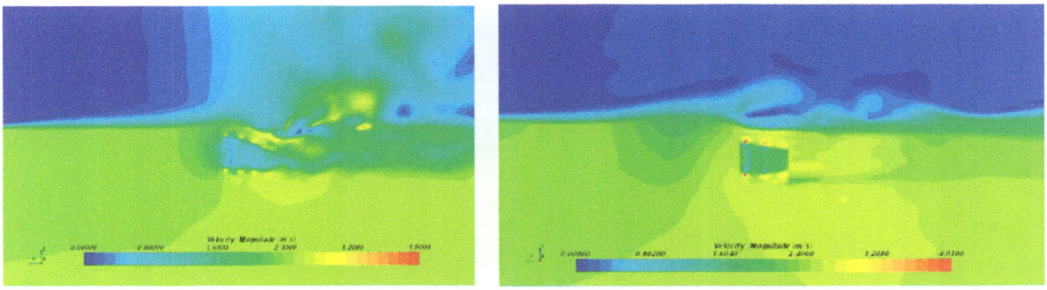

(**a**) Fixed stand-alone turbine (**b**) Turbine with 6-DOF motions

Figure 27. Comparison of vertical iso-velocity field for the fixed stand-alone turbine and the turbine with 6-DOF motions. (**a**) Fixed stand-alone turbine; (**b**) Turbine with 6-DOF motions.

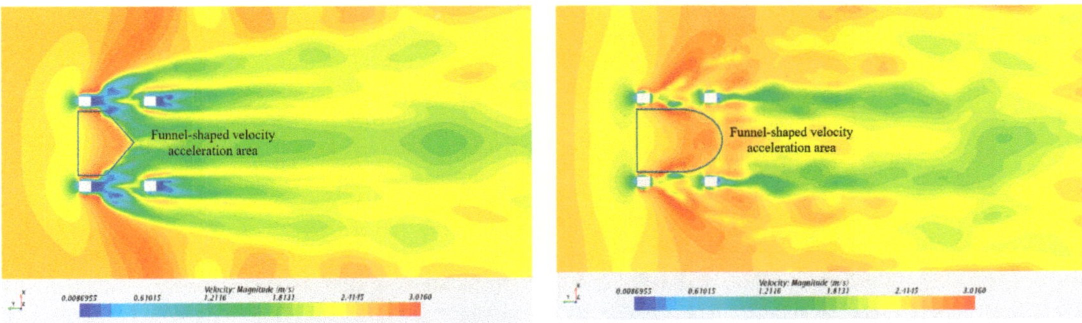

(**a**) Stand-alone carrier (**b**) FTCPS

Figure 28. Comparison of the vertical section iso-velocity flow field for the stand-alone carrier and the carrier with twin-rotor turbine. (**a**) Stand-alone carrier; (**b**) FTCPS.

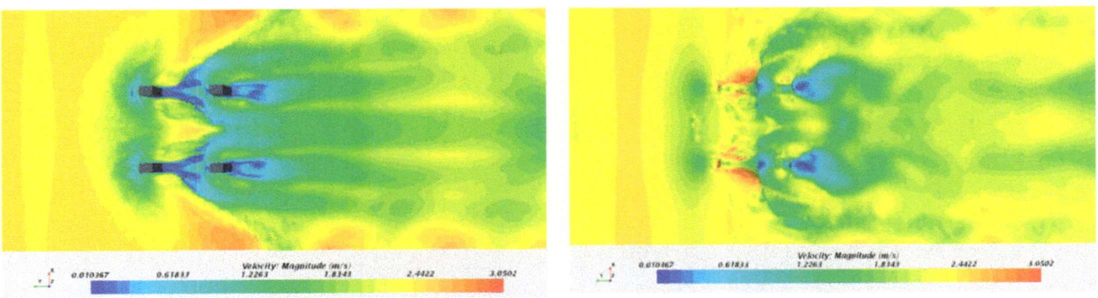

(**a**) Stand-alone carrier (**b**) FTCPS

Figure 29. Comparison of the free surface iso-velocity flow field for the stand-alone carrier and the carrier with twin-rotor turbine. (**a**) Stand-alone carrier; (**b**) FTCPS.

Figure 25 shows the comparison of velocity field around the blade cross section between the fixed stand-alone twin-rotor turbine and the twin-rotor turbine with 6-DOF motion. It can be seen that the velocities around the leading edge of the blade cross section are all less than 2 m/s, forming an obvious velocity stagnation area, which produces the pressure difference and makes the turbine rotate. The velocity difference for the turbine with 6-DOF motion is much bigger than the fixed turbine, which explains that the thrust and lateral force of turbine with 6-DOF motion is larger than the other.

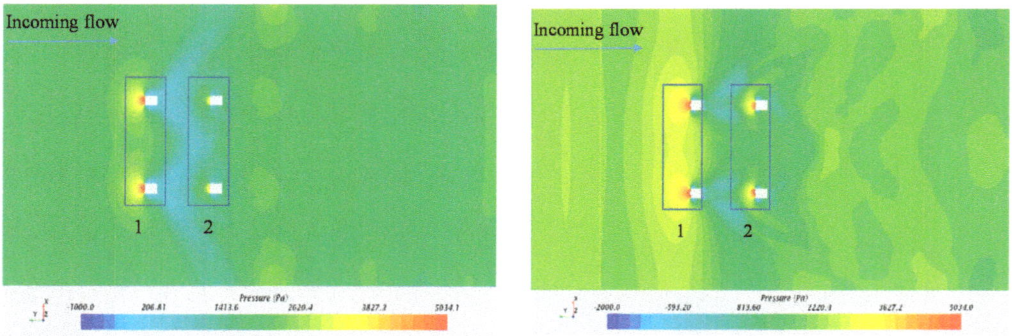

(a) Stand-alone carrier (b) FTCPS

Figure 30. Flow pressure for the stand-alone carrier and FTCPS. (a) Stand-alone carrier; (b) FTCPS.

Figure 26 shows the comparison of horizontal section iso-velocity field for the fixed stand-alone turbine and the turbine with 6-DOF motions. For the fixed stand-alone turbine, the velocity flow field is almost symmetrical, and a minimum velocity area could be observed at about 1D (D is the diameter of the rotor) downstream. The velocity flow field would recover at about 5D downstream. An obvious velocity stagnation area could be seen between 2D~5D downstream in the facing-flow side. Meanwhile, the facing-flow side velocity recovery is faster than the downflow side. For the twin-rotor turbine with 6-DOF motion, due to the blockage effect of the carrier, the flow velocity in the facing-flow side has been accelerated, which led to the increasing of power output [23]. Compared to the fixed stand-alone turbine, the facing-flow side velocity recovery for the twin-rotor turbine with 6-DOF motion is much faster, and there is no velocity stagnation area in the facing-flow side. Therefore, more fluid could pass across the turbine in a period, and more kinetic energy could be captured by the turbine.

Figure 27 shows the comparison of vertical section iso-velocity field of the fixed turbine and the turbine with 6-DOF motion. It can be seen that flow fields around carriers are distributed symmetrically. There is a clear speed-dividing line of the flow field around the middle section of the carrier without turbines. The backstream velocity around the carrier support structure has a clear stagnation zone. With the development of the flow field, the carrier wake field gradually recovers at a slower speed. However, for the carrier with turbines, the rotational motion of the turbines makes the fluid velocity flowing through the carrier have no clear dividing line. The backflow velocity of the carrier support structure has a smaller stagnation zone. The velocity increases, and the internal wake field velocity gradually decreases, forming a clear deceleration zone and, finally, mixing with the surrounding flow field.

Figure 28 shows the comparison of the vertical section iso-velocity flow field for the stand-alone carrier and the carrier with twin-rotor turbine. The velocity flow field for the stand-alone turbine at the vertical section is almost symmetrical. A funnel-shaped velocity acceleration area could be both observed for the two scenarios. However, the funnel-shaped velocity acceleration area for the stand-alone carrier is smaller and the funnel shape is more cuspidal than the other. The low velocity area of the carrier with the rotating turbine is smaller than the other, which implies that the rotation of the turbine accelerates the velocity recovery of the flow field.

It can be seen that the wave run-up effect is more obvious for the carrier with the rotating turbine in Figure 29. Compared to the symmetrical velocity flow field of the stand-alone turbine, the free surface flow field of the carrier with the rotating turbine is more disorganized than the other. Meanwhile, the rotation of the turbine enlarged the influenced flow velocity recovery area at the free surface area.

4.4.2. Flow Pressure

Figure 30 shows the flow pressure contour field for the stand-alone carrier and the carrier with twin-rotor turbine. It can be seen that both contours are symmetric. Due to the blockage effect of the twin-rotor turbine, the pressure field in front of the FTCPS is much larger than the stand-alone turbine. The fluid velocity passing across the turbine decreased due to turbine energy capturing, which leads to the pressure field increasing for the FTCPS. Meanwhile, the pressure field before the second row columns for the FTCPS is much increased due to the operation of the turbine. Therefore, the operation of the turbine can dramatically change the flow pressure field around the turbine, and the structure strength of the FTCPS should be given more consideration during the construction stage.

5. Conclusions

This research establishes a numerical simulation method for the fully coupled viscous turbine-carrier-mooring system simulation based on body fluid boundary interaction approach. The STAR-CCM+ software was used to compute the hydrodynamic performance of the vertical axis turbine and compared with the experimental results, which verified the reliability of the CFD numerical simulation method. Then, the DFBI approach was used to compute the coupled motion of the carrier, turbine, and mooring system under uniform flow conditions. The hydrodynamic characteristics, flow field characteristics, and speed ratio effects of the carrier and turbine were analyzed. According to the numerical results, the following conclusions were obtained:

(1) The average value of power output efficiency for the twin-rotor turbine reaches the maximum value when the TSR is around 2. Lateral force coefficient and the fluctuation amplitude of turbine thrust coefficient increase with the increase of speed ratio.

(2) The blockage effect caused by the columns of the carrier can increase the power output efficiency of the turbine. In the meantime, the complex flow field change also increases the fluctuation range of the thrust and lateral force of the turbine.

(3) The motion response of the carrier with the turbine in swaying and pitching will increase, but the influence of turbine rotation on the carrier heaving motion response is relatively small.

(4) The turbine output efficiency benefits from the blockage effect of the carrier. However, the existence of the carrier would increase the pressure acting on the columns of the carrier, which should be given more consideration during the construction stage of the carrier.

Author Contributions: C.H. Conceptualization, methodology, software, review and editing; C.T. Original draft preparation, formal analysis; C.Y. Data curation, software, investigation; Y.M. Supervision, project administration, funding acquisition. All authors have read and agreed to the published version of the manuscript.

Funding: This paper was funded by the National Natural Science Foundation of China grant number 51779062, 51879064,52071348, the General Project of Guangdong Province Department of Natural Resources to Promote High-quality Economic Development grant number GDOE [2020] 026), the Innovation Group Project of Southern Marine Science and Engineering Guangdong Laboratory (Zhuhai) grant number 311021014.

Institutional Review Board Statement: Not applicable.

Informed Consent Statement: Not applicable.

Data Availability Statement: Not applicable.

Conflicts of Interest: The authors declare no conflict of interest.

References

1. Khan, N.; Kalair, A.; Abas, N.; Haider, A. Review of ocean tidal, wave and thermal energy technologies. *Renew. Sustain. Energy Rev.* **2017**, *72*, 590–604. [CrossRef]
2. Nachtane, M.; Tarfaoui, M.; Goda, I.; Rouway, M. A review on the technologies, design considerations and numerical models of tidal current turbines. *Renew. Energy* **2020**, *157*, 1274–1288. [CrossRef]
3. Zhang, L.; Li, X.; Jing, G.; Zhang, X. Tidal enery updates 2013. *Adv. New Renew. Energy* **2013**, *1*, 53–68.
4. Zhou, Z.; Benbouzid, M.; Charpentier, J.-F.; Scuiller, F.; Tang, T. Developments in large marine current turbine technologies—A review. *Renew. Sustain. Energy Rev.* **2017**, *71*, 852–858. [CrossRef]
5. Ma, Y.; Li, B.; Xu, Y.; Dong, Y. Design and structural strength analysis of the gate-type tidal current energy converter. In *International Conference on Offshore Mechanics and Arctic Engineering*; American Society of Mechanical Engineers: New York, NY, USA, 2016.
6. Uihlein, A.; Magagna, D. Wave and tidal current energy—A review of the current state of research beyond technology. *Renew. Sustain. Energy Rev.* **2016**, *58*, 1070–1081. [CrossRef]
7. Jing, F.; Sheng, Q.; Zhang, L. Experimental research on tidal current vertical axis turbine with variable-pitch blades. *Ocean Eng.* **2014**, *88*, 228–241. [CrossRef]
8. Zhang, Y.; Zhang, J.; Lin, X.; Wang, R.; Zhang, C.; Zhao, J. Experimental investigation into downstream field of a horizontal axis tidal stream turbine supported by a mono pile. *Appl. Ocean Res.* **2020**, *101*, 102257. [CrossRef]
9. Tran, T.; Kim, D.; Song, J. Computational Fluid Dynamic Analysis of a Floating Offshore Wind Turbine Experiencing Platform Pitching Motion. *Energies* **2014**, *7*, 5011–5026. [CrossRef]
10. Fu, S.; Jin, Y.; Zheng, Y.; Chamorro, L. Wake and power fluctuations of a model wind turbine subjected to pitch and roll oscillations. *Appl. Energy* **2019**, *253*, 113605. [CrossRef]
11. Tran, T.; Kim, D.-H. Fully coupled aero-hydrodynamic analysis of a semi-submersible FOWT using a dynamic fluid body interaction approach. *Renew. Energy* **2016**, *92*, 244–261. [CrossRef]
12. Guo, X.; Yang, J.; Lu, W.; Li, X. Dynamic responses of a floating tidal turbine with 6-DOF prescribed floater motions. *Ocean Eng.* **2018**, *165*, 426–437. [CrossRef]
13. Guo, X.; Yang, J.; Li, X.; Lu, W.; Peng, T.; Li, J. Dynamic responses of a floating tidal turbine considering prescribed wave-frequency motion. In *International Conference on Offshore Mechanics and Arctic Engineering*; American Society of Mechanical Engineers: New York, NY, USA, 2018; Volume 10.
14. Brown, S.; Ransley, E.; Greaves, D. Developing a coupled turbine thrust methodology for floating tidal stream concepts: Verification under prescribed motion. *Renew. Energy* **2020**, *147*, 529–540. [CrossRef]
15. Wang, S.-Q.; Sun, K.; Zhang, J.-H.; Zhang, L. The effects of roll motion of the floating platform on hydrodynamics performance of horizontal-axis tidal current turbine. *J. Mar. Sci. Technol.* **2017**, *22*, 259–269. [CrossRef]
16. Wang, K.; Sun, K.; Sheng, Q.-H.; Zhang, L.; Wang, S.-Q. The effects of yawing motion with different frequencies on the hydrodynamic performance of floating vertical-axis tidal current turbines. *Appl. Ocean Res.* **2016**, *59*, 224–235. [CrossRef]
17. Xu, G.; Sun, S.; Liu, F. Motion response analysis of floating tidal current power station. *Zhong Guo Shui Yun. J.* **2015**, *15*, 137–140.
18. Lynch, C.; Smith, M. Unstructured overset incompressible computational fluid dynamics for unsteady wind turbine simulations. *Wind. Energy* **2012**, *16*, 1033–1048. [CrossRef]
19. Rogers, S.E.; Suhs, N.E.; Dietz, W.E. PEGASUS 5: An Automated Preprocessor for Overset-Grid Computational Fluid Dynamics. *AIAA J.* **2003**, *41*, 1037–1045. [CrossRef]
20. Park, S.; Jeun, G. Coupling of rigid body dynamics and moving particle semi-implicit method for simulating isothermal multi-phase fluid interactions. *Comput. Methods Appl. Mech. Eng.* **2011**, *200*, 130–140. [CrossRef]
21. Hirt, C.W.; Nichols, B.D. Volume of fluid (VOF) method for the dynamics Free Boundaries. *J. Comput. Phys.* **1981**, *39*, 201–225. [CrossRef]
22. Xie, Y. *Hydrodynamic Performance Research and Experiment of Vertical Axis Tidal Current Turbine*; Dalian University of Technology: Dalian, China, 2016.
23. Zhang, Y.; Zhang, Z.; Zheng, J.; Zhang, J.; Zheng, Y.; Zang, W.; Lin, X.; Fernandez-Rodriguez, E. Experimental investigation into effects of boundary proximity and blockage on horizontal-axis tidal turbine wake. *Ocean Eng.* **2021**, *225*, 108829. [CrossRef]

energies

MDPI

Article

Solar-Powered Thermoelectric-Based Cooling and Heating System for Building Applications: A Parametric Study

Mohadeseh Seyednezhad and Hamidreza Najafi *

Department of Mechanical and Civil Engineering, Florida Institute of Technology, Melbourne, FL 32901, USA; mseyednezhad2015@my.fit.edu
* Correspondence: hnajafi@fit.edu; Tel.: +1-321-674-8408

Abstract: Thermoelectric (TE) based cooling and heating systems offer significant advantages over conventional vapor compression systems including no need for refrigeration or major moving parts, high controllability, and scalability. The purpose of the present study is to provide an energy and economic assessment of the performance of a TE-based radiant cooling and heating system for building applications. It is considered that TE modules are integrated in the ceiling to lower/increase the ceiling temperature through the Peltier effect during the hot/cold season to provide thermal comfort for the occupants via radiation and convection. The study explores the possibility of using rooftop PV panels to produce electricity required for the operation of TE modules. An actual office building located in Melbourne, FL, USA is considered for a test study, and the hourly cooling and heating loads of the building are calculated through building energy simulation in eQuest. Various operating conditions, including different input voltages and temperature gradient across TE modules, are considered, and the system is sized to properly address the year-around cooling/heating demand. It is shown that a nominal cooling capacity of 112.8 W and a nominal PV capacity of 31.35 W per unit area of the building is required to achieve the target goal when the system operates at the optimal condition. An economic analysis is also performed, and estimated cost, as well as potential savings, are calculated for each operating condition. The optimal operating condition with minimum cost is selected accordingly. The results demonstrated that the initial cost of the proposed system is considerably higher than conventional heating/cooling systems. However, the system offers other benefits that can potentially make it an attractive option for building cooling/heating applications.

Keywords: building energy; Peltier effect; photovoltaic panels; and sustainable buildings

Citation: Seyednezhad, M.; Najafi, H. Solar-Powered Thermoelectric-Based Cooling and Heating System for Building Applications: A Parametric Study. *Energies* **2021**, *14*, 5573. https://doi.org/10.3390/en14175573

Academic Editor: Wei-Hsin Chen

Received: 8 August 2021
Accepted: 1 September 2021
Published: 6 September 2021

1. Introduction

The growth of the worldwide energy demand in light of the depletion of conventional sources of energy have accelerated the urge to pursue energy-efficient systems and use of alternative energies. Buildings are responsible for nearly 40% of total electricity consumption and one-third of global gas emissions [1]. Cooling and heating systems are among the largest energy end-users in buildings [2,3] and, therefore, innovative cooling and heating techniques that potentially lead to energy savings and reducing emissions are of great significance. Thermoelectric (TE) systems, as an alternative cooling/heating system, through the Peltier effect, offer several attractive characteristics including no need for refrigeration or major moving parts, quiet operation, high controllability, stability, and minimum maintenance [4–6]. TE modules consist of n-type and p-type doped semiconductor elements that are connected thermally in parallel and electrically in series [7]. TE modules are capable of generating power as the result of having a temperature gradient between the two sides of them (Seebeck effect) and producing a temperature gradient between the sides when supplied by DC electricity (Peltier effect). TE modules have been used in a variety of applications including cooling electronic devices [8,9], heat recovery [10,11], water treatment [12], solar stills [13], and thermal management of solar panels [14,15].

Applications of TE systems are thoroughly reviewed by Zhao and Tan [5] and Twaha et al. [7].

One of the emerging applications of TE systems is the use of Peltier modules for building cooling and heating, either through integration into building envelope or as separate units [16]. Various configurations have been proposed and studied for integrated TE modules in windows [17–20] and walls [17,21–25] demonstrated promising results. A few previous studies have also been focused on incorporating TE modules in the ceiling.

The TE modules integrated into the ceiling provide means for radiant cooling. The concept is basically making a cold/hot radiating surface by lowering/increasing the surface temperature of the ceiling. The radiant ceiling panels provide thermal comfort for the occupants through radiant heat exchange between the radiating surface and the human body. Based on ASHRAE's definition [26], radiant ceiling panel systems are the systems for which at least half of the total heat exchange is occurring through radiation. Radiant cooling systems have received a lot of attention in recent years, owing to the advantages such as improved thermal comfort and reduced energy consumption [27–29]. The energy performance of radiant systems in comparison with conventional systems is investigated in several studies [30–32]. Most of the studies that have been done on radiant cooling/heating systems were focused on applications of chilled/hot water tubing behind the ceiling, and the study of thermoelectric based radiant cooling and heating systems is still in the early stages.

Lertsatitthanakorn et al. [33,34] investigated the performance of a TE system integrated in the ceiling for cooling purposes. They used a constant temperature heat sink and studied the impact of heat sink temperature on the performance of a TE based cooling system in a 1.5 m × 1.5 m × 2 m chamber. As expected, a lower temperature heat sink resulted in a better performance of the system. Lertsatitthanakorn et al. [35] also looked into the thermal comfort aspects of their proposed system in the test chamber and reported that the system was capable of maintaining an indoor temperature of 27 °C with a *COP* of 0.75. In a similar study by Cheng et al. [36], a constant temperature heat sink (copper water channel) was used for the hot side of TE modules and a copper plate was used as the ceiling attached to the cold side of the TE modules. They studied the variation of *COP* with water temperature by maintaining a constant input current. They showed that their new design is capable of providing an acceptable thermal comfort for the occupants. Bhargava and Najafi [6] conducted a study to evaluate the performance of a solar-powered TE system integrated in the ceiling for space cooling and demonstrated that the system is capable of maintaining a comfort level temperature for the simulated room. The performance of a solar-powered TE based cooling and heating system integrated into the ceiling is studied by He et al. [37]. They developed a model for both cooling and heating modes of operation and validated the results experimentally via a small apparatus for the summer season, where a *COP* of 0.45 was achieved. Liu et al. [38] proposed a solar-powered TE based system for cooling, heating, and dehumidification purposes. The TE modules were integrated into the ceiling to provide cooling for the building. They assessed the performance of the system under various input voltage, ambient temperature, and indoor temperature conditions and reported a *COP* of 0.9 under operating voltage 5 V in the cooling mode and a *COP* of 1.9 with an input voltage of 4 V in the heating mode. An analysis of a building envelope integrated with thermoelectric modules and radiative sky cooler is performed by [39]. They optimized the system and achieved cooling capacity of 25.49 W/m^2 and a *COP* of 2 in the presence of 1000 W/m^2 of solar irradiation and ambient temperature of 35 °C. An investigation of the dynamic thermal characteristics of thermoelectric radiant cooling panel system is conducted by Luo et al. [40]. They developed a new system model by combining finite difference method and state-space matrix. They also used an artificial neural network for fast evaluation under dynamic conditions.

Lim et al. [41] studied application of a phase change material layer as energy storage for a thermoelectric ceiling radiant cooling panel through numerical modeling and experiment. They showed that a 10-mm-thick PCM layer when combined with fins (>five per unit length of the panel) was the most desirable case among all cases that they studied. Duan et al. [42] developed a model and performed a sensitivity analysis to provide a clear vision on the optimization of thermoelectric materials and how they can be improved to increase energy performance of thermoelectric cooling systems. They considered three major properties of thermoelectric modules including Seebeck coefficient, electrical conductivity, and thermal conductivity, and the results showed that *COP* of cooling is a stronger function of the Seebeck coefficient.

A comprehensive review of the application of TE systems for building cooling, heating, and ventilation is conducted by Zuazua-Ros et al. [16]. They reviewed TE based systems, integrated and non-integrated in a building envelope. More recently, a detailed review of radiative cooling systems (integrated in buildings), including thermoelectric systems, is presented by [43]. They summarized and discussed recent applications of thermoelectric cooling systems for building applications along with other radiative cooling technologies.

Most of the previous studies indicated a strong correlation between the performance of the TE system, the input electricity, and the heat sink in use. A detailed parametric study is much needed, however, to provide insights regarding the optimal operating conditions. This becomes more important particularly when a full scale building model is in use, which is scarce in the literature.

The purpose of the present paper is to conduct a feasibility study on using TE-based cooling/heating systems for the real-world scale from an energy and economic standpoint. In this regard, the cooling and heating load of an office building located in Melbourne, FL, USA is determined via building energy simulation in eQuest. A TE-based cooling and heating system, integrated into the ceiling, is sized to satisfy the thermal loads for various operating conditions accordingly. It is assumed that the system is powered by solar PV panels. The total number of TE modules, as well as the required number of PV panels to supply the system, is evaluated for multiple input voltages and temperature gradients between the two sides of the TE modules associated with different thermal resistances. The energy performance of the system as well as the estimated savings and cost for each operating condition, are evaluated for both heating and cooling modes of operations. The results are thoroughly discussed to provide a picture regarding the energy, and cost aspects of TE based cooling and heating systems. The major practical challenges are also discussed.

2. System Description and Method

Figure 1 shows the schematic of the solar powered TE-based system integrated in the ceiling. It is considered that TE modules from the bottom side are attached to an aluminum sheet that serves as the ceiling. In hot and cold seasons, the bottom side of the TE modules cools down and heats up respectively, forming a cool/warm ceiling surface to allow the transfer of heat via radiation and convection to the occupants and provide the desired comfort level. It should be noted that by switching the direction of the electrical current supplied to TE modules the hot and cold sides of the module are switched as well. Therefore, a set of TE modules can potentially be used for both heating and cooling purposes, even though this will introduce practical challenges which will be discussed in the next section.

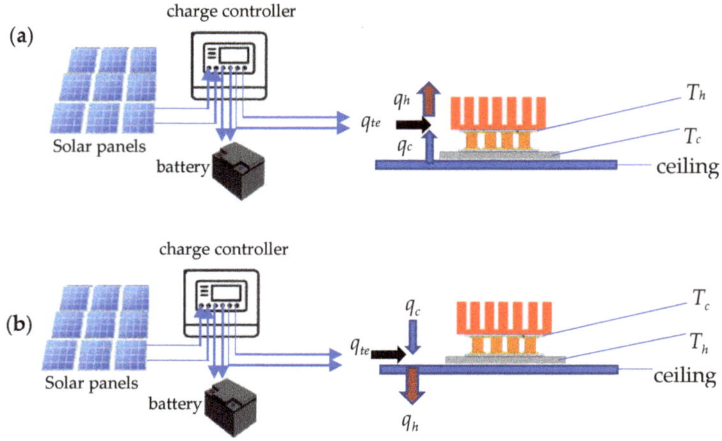

Figure 1. Schematic of the solar-powered TE-based system, (**a**) cooling (hot season); (**b**) heating (cold season).

It is assumed that the TE modules are solar powered. PV panels are used to generate electricity. The schematic in Figure 1 shows a possible configuration of the system which involves battery storage and charge controller. When properly sized, the system will be able to supply the TE modules with adequate electricity year-around. A more typical system with lower cost includes a charge controller, an inverter, and a net-meter that allows exporting the PV power to the grid and receive the utility power for building use (including the TE modules, after inversion to DC) simultaneously.

2.1. Methodology

The system must be sized properly to ensure the year-around occupant's comfort. This includes the number of TE modules and PV panels to provide adequate cooling/heating and electrical power, respectively. An office building in Melbourne, FL, USA is considered for the case study and a building energy model is developed in eQuest accordingly to determine the peak hourly cooling/heating loads ($Q_{c,max}$ and $Q_{h,max}$), as well as the yearly amount of thermal energy that must be added ($Q_{h,y}$) or removed ($Q_{c,y}$) to the building for maintaining desired set-point temperature.

For a given operating condition (V and ΔT_{te}), the rate of heat removal, q_c (or addition q_h) and the coefficient of performance of the TE module (COP) are determined from the performance curves provided by the TE manufacturer. The number of required TE modules (N_{te}) to meet the peak cooling/heating loads can be evaluated accordingly:

$$N_{te,c} = \frac{Q_{c,\max}}{q_c},\ N_{te,h} = \frac{Q_{h,\max}}{q_h} \quad (1)$$

The total annual thermal energy that needs to be removed, $Q_{c,y}$, from (or added, $Q_{h,y}$, to) the building in order to maintain the desired set-point temperature are found from the building energy simulation. The total annual energy consumption by the TE modules for cooling and heating can be found as:

$$Q_{te,y,c} = \frac{Q_{c,y}}{COP_c},\ Q_{te,y,h} = \frac{Q_{h,y}}{COP_h} \quad (2)$$

For simulation purposes, it is assumed that the PV panels are placed south facing (azimuth angle of 180 degrees) and at 28 degrees tilt angle (equivalent to the latitude) to maximize electricity production [43] on the surface of the roof of a canopy that is right next to the building (south side). The PVWatts calculator [44] is used to calculate the monthly electricity production by one panel. The number of PV panels are then calculated to cover the energy required to operate the TE modules, as described in the previous section. The number of required PV panels to provide adequate energy for the TE modules in cooling and heating modes can be then calculated as:

$$N_{pv,c} = \frac{Q_{te,y,c}}{P_{PV}},\ N_{pv,h} = \frac{Q_{te,y,h}}{P_{PV}} \tag{3}$$

The performance of the proposed system and the required number of TE and PV panels are strong functions of input voltage (V) and temperature gradient between the hot and cold sides of the TE modules (ΔT_{te}). Therefore, a parametric study is performed to compare the energy performance, required size, and cost of the system under various operating conditions.

2.2. Building Energy Model

An office building on the campus of the Florida Institute of Technology is considered for the case study. The floor area of the building is approximately 200 m^2. The exterior walls are constructed from concrete block (with a U-factor of 0.89 W/m^2 K). The roof is made of a white lacquer color on weathered asphalt pavement on a standard wood frame (U-factor 1.14 W/m^2 K). The building is occupied by six people between 8:00 a.m. to 5:00 p.m. every day except for weekends and the standard US holiday. The building energy audit was performed in order to gather all the necessary data for the building model. Table 1 shows the list of all doors and windows and their characteristics, and Table 2 demonstrates the list of lights and plug-loads.

Table 1. Windows and doors data.

Windows					
Orientation	Num	Glass Type	Frame Type	Size (cm)	Dist. to Ground (cm)
East	1	Clear	Wood/AL	88.9 × 124.5	83.8
East	2	Clear	Wood/AL	182.9 × 91.4	121.9
South	2	Clear	Wood/AL	91.4 × 94.0	129.5
West	2	Clear	Wood/AL	91.4 × 94.0	116.8
West	1	Clear	Wood/AL	182.9 × 94.0	116.8
North	2	Clear	Wood/AL	182.9 × 91.4	129.5
Doors					
Orientation	Num	Type		Size (cm)	
East	1	Glass (Metal Frame)		76.2 × 181.6	
West	1	Sliding Glass		152.4 × 213.4	
North	1	Wood		91.4 × 213.4 × 4.5	

Table 2. Plug loads and lights.

Plug Loads				
Type	Location	Num	Wattage	Hours/year
Computer	Offices	4	130	2000
Refrigerator	Kitchen	1	180	8742
Microwave	Kitchen	1	1200	125
TV	Kitchen	1	300	1250
Coffee Pot	Kitchen	1	750	500
Lights				
Type	Location	Num	Wattage	Hours/year
T8	All	56	32	1821
T2	Bath	9	13	4500
A19	Closet	1	75	15
A15	Closet	1	55	15
GU24	Garage	1	13	15

A cooling and heating set-point temperature of 298 K and 292 K are considered, respectively, for the occupied hours. The building model was developed in eQuest, and the hourly cooling/heating loads are determined for a one-year period. The highest hourly cooling and heating loads are found as 10.4 kW ($Q_{c,max} \approx 3$ tons of regrigeration) for 3:00–4:00 p.m. on 8 July and 6.74 kW ($Q_{h,max} \approx 2$ tons of refrigeration) for 7:00–8:00 a.m. on 26 January for the hot and cold seasons, respectively.

Table 3 demonstrates the characteristics of the PV panels (Panasonic 330 W panel) and thermoelectric modules (TE Technology-HP-127-1.4-1.15-71 [45]) that are considered for this study.

Table 3. Characteristics of the PV panel and the thermoelectric module.

PV Panel		Thermoelectric Module	
Dimensions of the PV panel	1038 × 533 × 35 mm	$V_{TE,max}$	16.1 V
Rated power	330 W	$I_{TE,max}$	8 A
Nominal cell efficiency	19.7%	$\Delta T_{TE,max}$	71 K
Maximum power voltage, V_{mp}	58 V	$Q_{TE,max}$	80 W
Maximum power current, I_{mp}	5.7 A	Dimensions	40 × 40 mm
Open circuit voltage, V_{oc}	69.7 V		
Short circuit current, I_{sc}	6.07		

For a given operating condition, including input voltage (V_{TE}) and the temperature gradient between hot and cold sides of the module (ΔT_{TE}), the cooling or heating effect from each module (q_c or q_h) as well as the corresponding coefficient of performance (COP_c or COP_h) can be obtained from the manufacturer performance curves. The calculation with regard to the system sizing is already described in the previous section.

3. Results and Discussion

In this section, the building energy modeling results, system sizing as well as energy and cost analysis of the system are presented and discussed in detail.

3.1. Weather Data, Cooling and Heating Load, and Solar Power

The building that is considered for simulation is in Melbourne, FL, USA. Melbourne, situated along the Atlantic coastline in central Florida, has a humid subtropical climate (climate zone 2A, according to ASHRAE). The typical meteorological year (TMY3) is used as the source of weather data for the simulation. The TMY3 contains 12-months of hourly data that represent median weather conditions in a multiyear period and therefore is considered as a typical data set that can properly represent the weather data for a given location. The schematic of the building layout as well as the 3D model built in eQuest are shown in Figure 2.

The maximum hourly cooling and heating loads were found to occur on 8 July at 3:00 p.m.–4:00 p.m. (10.4 kW) and 26 January 7:00 a.m.–8:00 a.m. (6.74 kW), respectively. The solar radiation data and outside dry-bulb temperature for 8 July and 26 January are plotted in Figure 3a,b. The calculated hourly cooling load and heating load for these two days are plotted in Figure 4. The total monthly heating and cooling demand of the building are calculated and demonstrated in Figure 5.

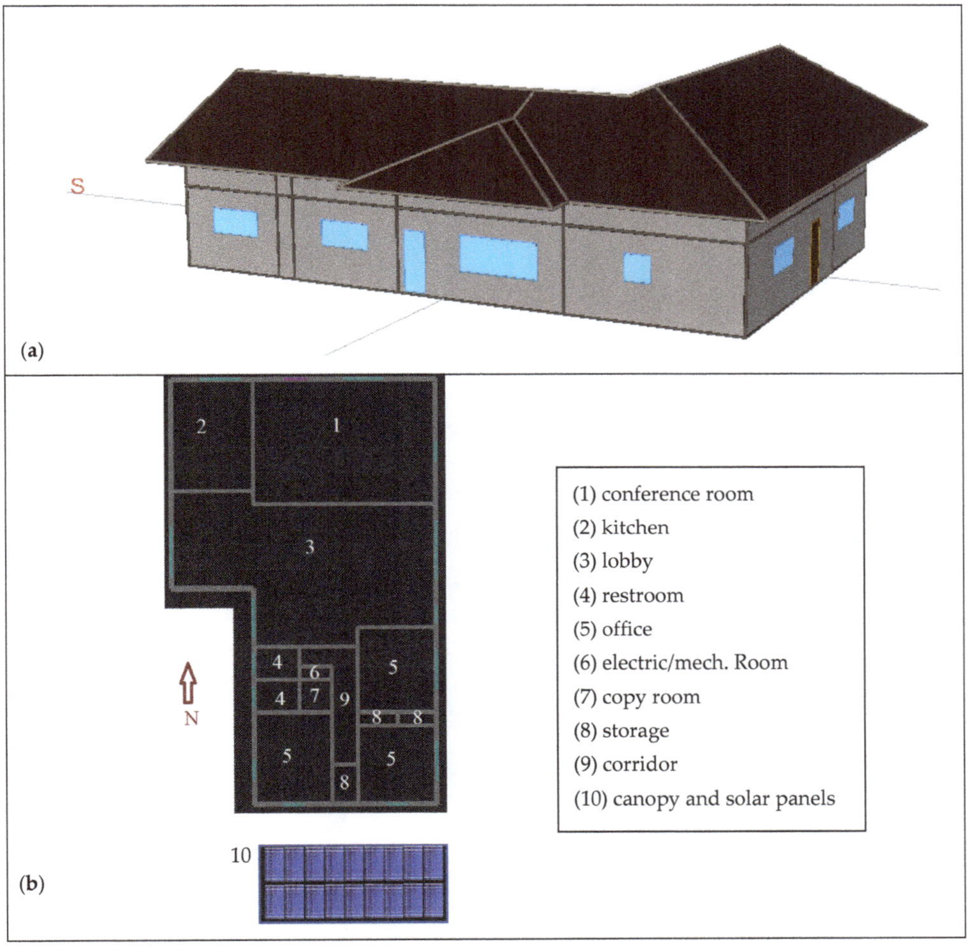

Figure 2. (**a**) 3D building model developed in eQuest; (**b**) building layout.

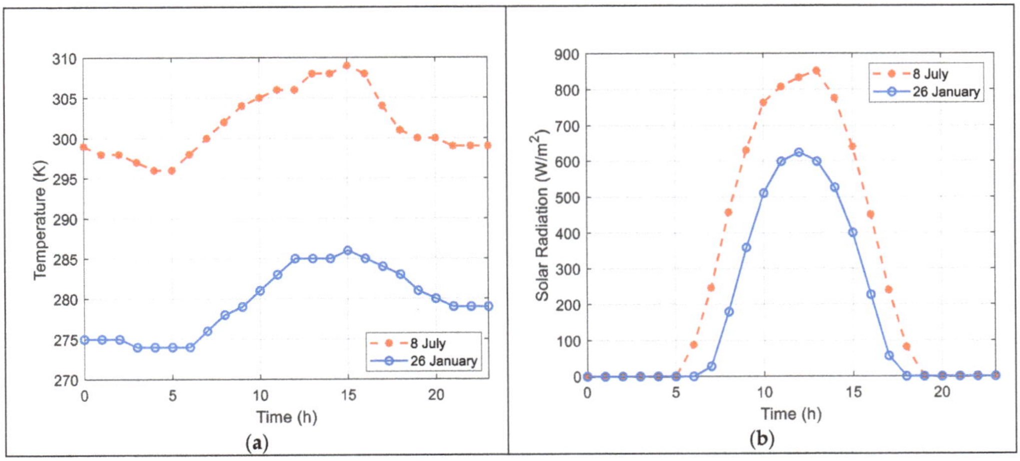

Figure 3. Weather Data for Melbourne, FL, USA: (**a**) ambient temperature; (**b**) solar irradiation.

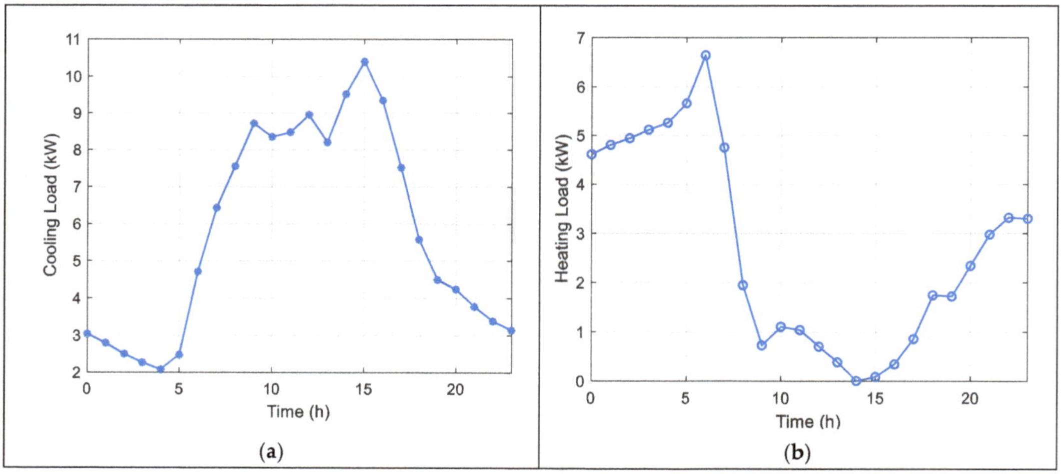

Figure 4. Hourly (**a**) cooling load for 8 July and (**b**) heating load for 26 January.

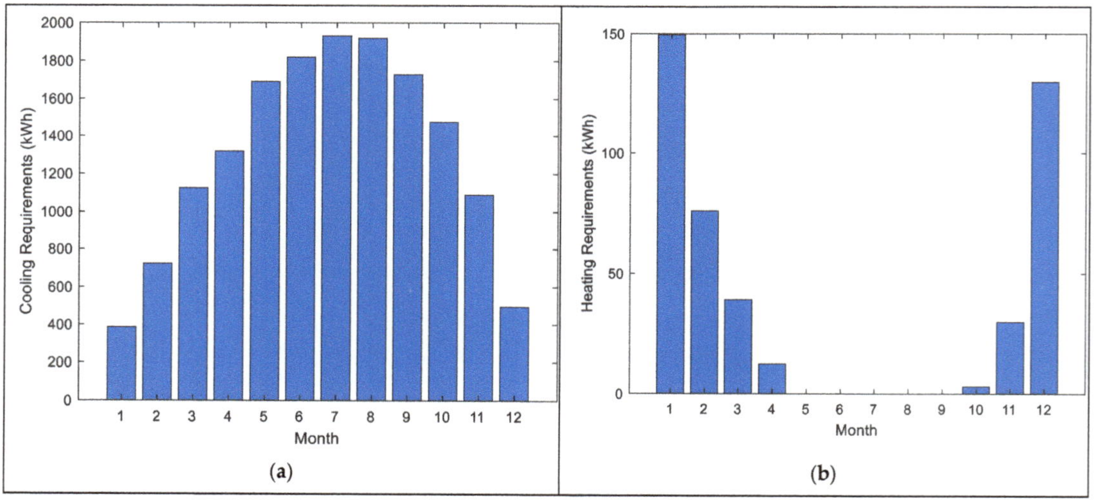

Figure 5. Monthly cooling (**a**) and heating (**b**) requirements for the building.

As expected for the given hot and humid climate, the cooling load is significantly larger than the heating load, particularly for an office building with operating hours during the daytime. The highest cooling and heating demands occur in August and January, respectively. The total annual thermal energy that needs to be added/removed to/from the building in heating/cooling season to maintain the set-point temperature is found as 440 kWh and 15,707 kWh, respectively. It should be noted that this number is associated with occupied hours only, and holidays and weekends, as well as other unoccupied hours, are excluded.

The monthly electricity production by a PV panel is calculated via PVWatts calculator [40]. It is assumed that the solar panel is placed horizontally on the roof with a tilt angle equivalent to the latitude (28 degrees). The number of required PV panels for each operating condition is found using Equation (3).

As shown in Figure 6, the maximum electricity production is associated with May in the amount of 50.4 kWh, and the total annual electricity production is calculated as 522.7 kWh/yr.

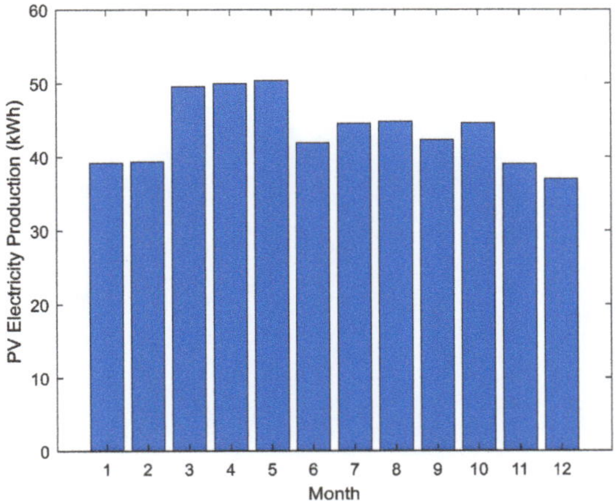

Figure 6. Summation of monthly energy produced by a PV panel.

3.2. *Thermoelectric Module Performance*

The performance of TE modules under various operating conditions, including input voltage (V) and temperature gradients between hot and cold sides (ΔT_{te}) is evaluated. The coefficient of performance for a TE module in cooling mode of operation may be given as:

$$COP_c = \frac{q_c}{q_{te,c}} = \frac{\alpha T_c I - \frac{1}{2} I^2 R - k(T_h - T_c)}{\alpha I (T_h - T_c) + R I^2} \tag{4}$$

where α, k, R, T_c, and T_h represent the Seebeck coefficient, thermal conductance, electrical resistance, and cold and hot side temperature of the TE modules, respectively. In addition, q_c is the heat removal rate and $q_{te,c}$ is the input electricity supplied to the TE module (for cooling), which may be defined as:

$$q_{te,c} = q_h - q_c \tag{5}$$

The coefficient of performance for the heating mode of operation (COP_h) assuming similar V and ΔT_{te} can be found as:

$$COP_h = \frac{q_h}{q_{te}} = COP_c + 1 \tag{6}$$

It should be noted that, as seen in Equation (4), the COP of TE modules is a strong function of the temperature gradient between the hot and cold sides of the module (ΔT_{te}). The larger ΔT_{te} results in a lower coefficient of performance. Maintaining a reasonably low ΔT_{te} requires effective thermal management and selection of an appropriate heat sink that provides minimum thermal resistance. The relationship between the thermal resistance on the hot/cold side of the TE module and the corresponding temperatures can be given as:

$$T_h = T_a + q_h R_{th,h} \tag{7}$$

$$Tc = T_a - q_c R_{th,c} \tag{8}$$

where T_a is the ambient/surrounding temperature, and $R_{th,h}$, and $R_{th,c}$ are the thermal resistances on the hot and cold sides of the TE module, respectively. The effect of the thermal resistances on the COP of the TE module is further explored in Figure 7. In this analysis, it is assumed that the thermal resistances on both sides of the TE modules are equal.

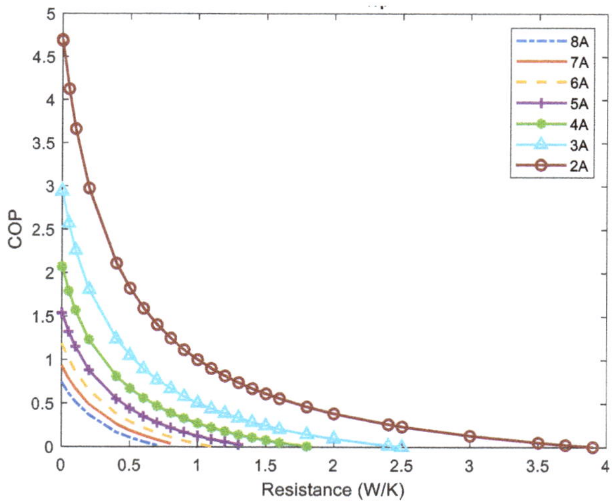

Figure 7. Variation of *COP* with a thermal resistance of the two sides of the TE module.

As seen, an increase in thermal resistance for a given input electrical current results in a significant reduction of *COP*. It is also observed that higher electricity input results in lower *COP* for the TE modules. These results further emphasize the importance of heat sink design and the adjustment of electricity input to the TE modules for achieving optimal performance. The detailed design of the heat sink, however, is outside the scope of the present study.

3.3. Parametric Study

The procedure for system sizing is already described in Section 2.1. For the selected TE module, five different ΔT_{te} are considered, including 9.6, 12, 19.2, 28.8, and 38.3 °C. For each particular ΔT_{te}, a number of possible input voltages are considered, and the corresponding *COP*, heat removal/addition rate (q_c or q_h), and energy consumed by a TE module (q_{te}) operating under these conditions are evaluated and listed in Table 4. As an example, consider the first row of Table 4, in the cooling mode operation, for ΔT_{te} = 9.6 °C and V = 2.9 V, the values of COP_c, q_c, and $q_{te,c}$ are determined as 3, 14 W, and 4.67 W, respectively. Similarly, for the first row, in the heating mode operation, the values of COP_h and q_h are evaluated as 4 and 18.67 W, respectively.

Table 4. Summary of the results for various operating conditions.

						Cooling						Heating	
Case Num	ΔT (°C)	V	COP_c	q_c (W)	$q_{te,c}$ (W)	N_{tec}	$Q_{tec,y}$ (kWh/yr)	$N_{pv,c}$	$C_{init,c}$ ($)	COP_h	q_h (W)	N_{teh}	$Q_{teh,y}$ (kWh/yr)
1	9.6	2.9	3	14	4.67	743	5236	11	22,700	4	18.67	358	110
2	9.6	5.9	1.7	37	21.76	282	9239	18	13,800	2.7	58.76	114	163
3	9.6	8.8	1	45	45.00	232	15,707	31	17,425	2	90.00	75	220
4	9.6	11.7	0.8	52	65.00	200	19,634	38	19,250	1.8	117.00	58	244
5	12	5.9	1.62	38	23.46	274	9696	19	13,975	2.62	61.46	109	168
6	12	8.8	1.14	45	39.47	232	13,778	27	15,925	2.14	84.47	79	206
7	12	11.7	0.78	52	66.67	200	20,137	39	19,625	1.78	118.67	57	247
8	19.2	5.9	1.3	23	17.69	453	12,082	24	20,325	2.3	40.69	164	191
9	19.2	8.8	0.9	36	40.00	289	17,452	34	19,975	1.9	76.00	88	232
10	19.2	11.7	0.6	44	73.33	237	26,178	51	25,050	1.6	117.33	57	275
11	28.8	5.9	0.8	10	12.50	1040	19,634	38	40,250	1.8	22.50	297	244
12	28.8	8.8	0.7	26	37.14	400	22,439	43	26,125	1.7	63.14	106	259
13	28.8	11.7	0.5	34	68.00	306	31,414	61	30,525	1.5	102.00	66	293
14	38.3	8.8	0.5	16	32.00	650	31,414	61	39,125	1.5	48.00	139	293
15	38.3	11.7	0.4	24	60.00	434	39,268	76	39,350	1.4	84.00	80	314

Using the building energy model, the maximum cooling and heating loads (sensible and latent load) are found as 10.4 kW and 6.74 kW, respectively. The required number of TE modules for the hot and cold seasons are then calculated using Equation (1) for each operating condition. For the ΔT_{te} = 9.6 °C and V = 2.9 V (first row in Table 4), these values are determined as 743 and 358, respectively. Note that, since the building under study is located in a hot and humid climate zone, the cooling load is much more significant than the heating load. Therefore, assuming each TE module can operate in both cooling and heating mode by switching the direction of input electricity, the system must be sized based on the cooling needs.

As mentioned, the total annual heat that must be removed and added to the building for maintaining the comfort conditions during the hot and cold seasons are found as 15,707 kWh and 440 kWh, respectively. These values are substituted in Equation (3) to evaluate the number of PV panels that are needed to provide sufficient electricity for the modules. For the first row of Table 4 (ΔT_{te} = 9.6 °C and V = 2.9 V), 11 PV panels are needed to provide sufficient electricity for TE modules in the cooling mode operation. In contrast, only four PV panels are required to cover the electrical energy for the TE modules in the heating mode of operation. Note that the sizing must be performed based on the cooling demands since the cooling load is significantly higher than the heating load.

It can be seen from Table 4 that *COP* values decline as the temperature gradient between the hot and cold sides increases. The performance of the TE modules can be improved if they are integrated with proper heat sinks in order to have effective heat dissipation. A closer look at the trend of the results in Table 4 also shows that, at a given temperature gradient, for most cases, each module operates more efficiently (higher *COP*) when supplied with lower input voltages. However, at lower input voltages, the heat removal/addition rate by the TE module is smaller. As a result, a more efficient cooling/heating system requires a larger number of TE modules to satisfy the load. In other words, a lower operating cost requires a higher initial investment in TE modules.

In order to provide a clearer picture of the system's performance, the variation of N_{pv} and N_{te} versus *COP* at different temperature gradients and for various operating voltages are plotted in Figure 8 for cooling mode of operation. It is shown that, as *COP* value increases, each module provides less cooling/heating effect, and, as a result, the required number of TE modules to address the load increases, whereas the required number of PV panels decreases as the result of more efficient operation of each TE module.

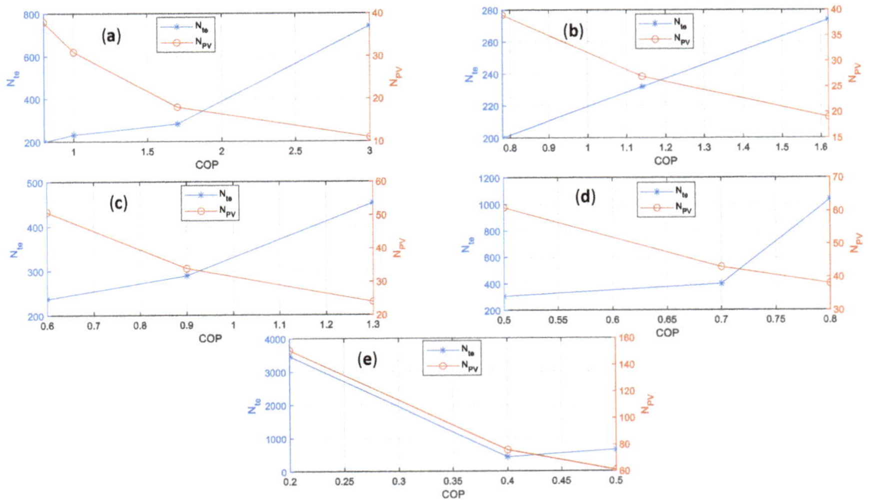

Figure 8. Variation of the number of the TE modules and PV panels versus *COP* for cooling mode of operation: (**a**) ΔT = 9.6 °C; (**b**) ΔT = 12 °C; (**c**) ΔT = 19.2 °C; (**d**) ΔT = 28.8 °C; and (**e**) ΔT = 38.3 °C.

An important observation can be made by reviewing the range of *COP* values that are obtained. For the majority of operating conditions, the *COP* is very small and not comparable with conventional high-efficiency heating (heat pump) and air conditioning systems. As previously discussed, a well-designed heat sink would allow efficient operation by maintaining a low ΔT_{te}. The cases with reasonably high *COP* are associated with the lowest ΔT_{te} values (9.6 °C and 12 °C). It should be noted that the *COP* should not be the only parameter when it comes to the energy efficiency of the proposed system. The TE based cooling/heating system allows individual control over each module, which facilitates cooling and heating based on occupancy of a certain area within each room. This will reduce the overall energy used for cooling/heating since not all the areas are always occupied.

3.4. Economic Analysis

A basic cost model is developed to provide a perspective on the economic aspects of the proposed system. The initial cost of the system, assuming the cost of instrumentation and installation is roughly 25% of the major parts, can be estimated as:

$$C_{init} = (N_{te}C_{te} + N_{pv}C_{pv})\,(1 + 0.25) \quad (9)$$

where C_{init} refers to the total initial cost of the system and C_{te} and C_{pv} are the price of one TE module and one PV panel, respectively. The estimated price for each TE module is $20, and the cost of each PV panel is obtained from the manufacturer at $300.

As mentioned, the system must be sized based on the cooling mode of operation as the cooling load is significantly higher than the heating load. The estimated initial cost of the system for each mode of operation is shown in Figure 9.

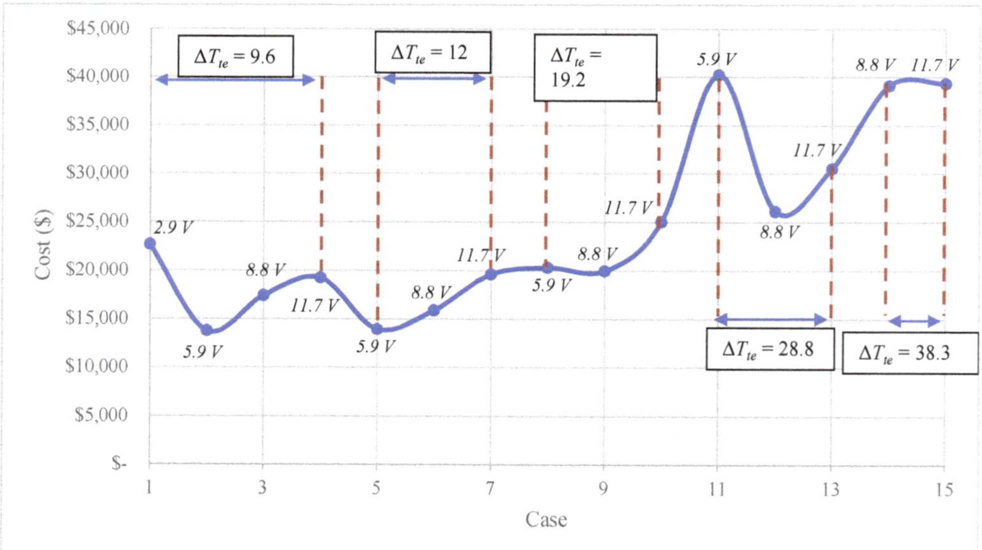

Figure 9. Initial cost of the proposed hybrid PV-TE system.

It can be seen from Figure 9 that case 2 (closely followed by case 5) offers the lowest initial cost of $13,800. Case 2 demonstrated the second-highest *COP* (1.7) among all the cases. A higher *COP* lowers the number of required PV panels, which significantly brings the initial cost down. The second case is, therefore, selected as the optimal operating condition. It should be noted that, although the Case 1 conditions show a higher *COP*, it is not cost-effective since it requires a much large number of TE modules to provide adequate cooling.

For Case 2, it is determined that 282 TE modules and 18 PV panels are needed. This is nearly 1.41 TE module (rated at 80 W) and 0.09 PV panel (rated at 330 W) per unit of area of the building, when the system operates at an approximate *COP* of 1.7. In other words, a nominal cooling capacity of 112.8 W and a nominal PV capacity of 31.35 W per unit area of the building is required to achieve the target goal when the system operates at the optimal condition.

Since the system is solar-powered, cost savings will be realized as the result of a reduction in the electricity bill. Assuming an average $COP_{HP,c}$ of 3 and $COP_{HP,h}$ of 4 for cooling and heating by a heat pump, the electricity consumption by such a system is evaluated as 5,346 kWh using Equation (10):

$$EC_{HP} = \frac{Q_{c,y}}{COP_{HP,c}} + \frac{Q_{h,y}}{COP_{HP,h}} \tag{10}$$

Assuming an average usage rate of \$0.12/kWh, the annual cost savings is found as \$641.5 using:

$$CS = EC_{HP} \times UR \tag{11}$$

The installed cost of a comparable size commercial heat pump system (excluding duct-work) is approximately \$5000–\$6000. This indicates that, at first look, the proposed system is significantly more expensive than conventional heat pump systems with an estimated simple payback period of over 12 years. However, there are other cost considerations that need to be discussed. It is expected that the maintenance cost of the proposed system is significantly lower than the conventional air conditioning systems and they also offer a longer lifecycle. The lifespan of a conventional air conditioning unit is about 12–15 years while this value is over 20 years for the solar panels and TE modules. Accounting for the lifecycle, the payback period of using the proposed system versus conventional units can further reduce to about 8.6 years.

Additional cost savings may be achieved through the reduction of regular maintenance costs and accurate control that allows for operating the system only when the area within the coverage is occupied. Given the function of the considered building (office), many areas remain unoccupied during working hours. By automatic shut-down of the system in the unoccupied areas, significant energy/cost savings can be realized. Considering Case 2, a conservative estimate of a 25% reduction in operating hours will result in a decrease in the number of required PV panels down to 14, which will reduce the initial cost by \$1875. This will bring down the payback period to just below six years. Additionally, for the economic analysis, the cost savings is calculated using an average usage rate of \$0.12 which is estimated based on electricity cost in Florida. This number could be much larger in other areas which will significantly decrease the payback period.

4. Prospects and Challenges

The TE-based cooling/heating system is a promising technology owing to several major advantages that it offers. In comparison with conventional heat pump systems, besides the green nature of the TE system due to eliminating the need for refrigerant and ability to use DC electricity (i.e., through solar PV), the system does not require any major moving parts and requires minimum maintenance. The individual control over TE modules provides opportunities for smart control based on occupancy in each area, which holds the potential for significant improvement in energy efficiency and occupants' thermal comfort. Additionally, with the continuous improvement in material sciences, any increase in the efficiency of thermoelectric materials will result in a more cost-effective system. It should be also noted that optimization of the building and hybridizing the TE technology with passive techniques [46,47] can also significantly reduce the energy demands, prior implementation of the TE-bases system and as a result reduces the overall system cost.

There are also several challenges that must be addressed to pave the road in commercializing the TE-based cooling/heating systems. A major challenge is the optimal design of heat sinks for both hot and cold sides of the module to ensure a low-temperature gradient between the two sides and maintain a reasonably high *COP*. Harvesting the waste heat from the TE modules and using it for applications such as preheating water or power generation via thermoelectric generators (Seebeck effect) is another challenge that must be thoroughly studied. Optimal control of the system is also a critical task to maintain high energy efficiency and ensure occupants' comfort. The installation of the system, including modules, wiring, and control system, must also be explored. Additionally, the proper ventilation system must be incorporated in the design to provide fresh air. The overall payback period of the system may be reduced by improving thermoelectric materials and the use of a well-designed system with optimal control.

5. Conclusions

In the present paper, the feasibility of using solar powered TE-based cooling and heating system for an office building is investigated. The system consists of solar-powered TE modules that are integrated into the ceiling and lower/increase the temperature of the ceiling in the hot/cold season to provide year-around thermal comfort for the occupants via radiation and convection. A test study is performed for a building in Melbourne, FL, USA with a hot and humid climate. The building model is developed in eQuest, the hourly cooling and heating load values are evaluated, and an algorithm is developed for sizing the system. A parametric study is performed, and the size and performance of the system under various operating conditions are evaluated. A basic cost model is also developed, and the initial cost of the system, as well as the resulted savings, are estimated and compared against conventional heat pump systems. The conditions which led to the lowest initial cost of the system are selected as the optimal conditions and thoroughly discussed. It was shown that, while the system offers major benefits including no need for refrigeration or major moving parts, high controllability, and reasonably high *COP*, the cost of the system is significantly higher than conventional heat pumps with a simple payback period of about 9–12 years, assuming a usage rate of $12/kWh. Considering the much longer lifespan of TE and PV systems and their relatively low maintenance cost in comparison with conventional air conditioning systems, the payback period may reduce down to about six years. Further improvement in thermoelectric materials with a higher figure of merit and the use of smart control based on occupancy can significantly decrease the cost of the system and introduce it as a green and cost-effective alternative cooling and heating system.

Author Contributions: Formal analysis, writing—original draft preparation, M.S.; Conceptualization, methodology, investigation, writing—review and editing, H.N. All authors have read and agreed to the published version of the manuscript.

Funding: This research was supported by the Florida Institute of Technology.

Conflicts of Interest: The authors declare no conflict of interest.

References

1. Ruparathna, R.; Hewage, K.; Sadiq, R. Improving the energy efficiency of the existing building stock: A critical review of commercial and institutional buildings. *Renew. Sustain. Energy Rev.* **2015**, *53*, 1032–1045. [CrossRef]
2. Global Energy Assessment Writing Team. Energy End-Use-Buildings. In *Global Energy Assessment Writing Team*; Cambridge University Press: Cambridge, UK, 2012; Chapter 10-IIASA.
3. Betharte, O.; Najafi, H.; Nguyen, T. Towards Net-Zero Energy Buildings: A Case Study in Humid Subtropical Climate. In Proceedings of the ASME 2018 International Mechanical Engineering Congress and Exposition, Pittsburgh, PA, USA, 9–15 November 2018. [CrossRef]
4. Riffat, S.; Ma, X. Thermoelectrics: A review of present and potential applications. *Appl. Therm. Eng.* **2003**, *23*, 913–935. [CrossRef]
5. Zhao, D.; Tan, G. A review of thermoelectric cooling: Materials, modeling and applications. *Appl. Therm. Eng.* **2014**, *66*, 15–24. [CrossRef]

6. Bhargava, A.; Najafi, H. Photovoltaic-Thermoelectric Systems for Building Cooling Applications: A Preliminary Study. In Proceedings of the ASME 2016 International Mechanical Engineering Congress and Exposition, Phoenix, AZ, USA, 11–17 November 2016. [CrossRef]
7. Twaha, S.; Zhu, J.; Yan, Y.; Li, B. A comprehensive review of thermoelectric technology: Materials, applications, modelling and performance improvement. *Renew. Sustain. Energy Rev.* **2016**, *65*, 698–726. [CrossRef]
8. Liu, D.; Zhao, F.-Y.; Yang, H.; Tang, G.-F. Thermoelectric mini cooler coupled with micro thermosiphon for CPU cooling system. *Energy* **2015**, *83*, 29–36. [CrossRef]
9. Ahammed, N.; Asirvatham, G.; Wongwises, S. Thermoelectric cooling of electronic devices with nanofluid in a multiport minichannel heat exchanger. *Exp. Therm. Fluid Sci.* **2016**, *74*, 81–90. [CrossRef]
10. Zhang, H.; Kong, W.; Dong, F.; Xu, H.; Chen, B.; Ni, M. Application of cascading thermoelectric generator and cooler for waste heat recovery from solid oxide fuel cells. *Energy Convers. Manag.* **2017**, *148*, 1382–1390. [CrossRef]
11. Najafi, H.; Woodbury, K.A. Modeling and Analysis of a Combined Photovoltaic-Thermoelectric Power Generation System. *J. Sol. Energy Eng.* **2013**, *135*, 031013. [CrossRef]
12. Tan, Y.Z.; Han, L.; Chew, N.G.P.; Chow, W.H.; Wang, R.; Chew, J.W. Membrane distillation hybridized with a thermoelectric heat pump for energy-efficient water treatment and space cooling. *Appl. Energy* **2018**, *231*, 1079–1088. [CrossRef]
13. Rahbar, N.; Esfahani, J.A. Experimental study of a novel portable solar still by utilizing the heatpipe and thermoelectric module. *Desalination* **2012**, *284*, 55–61. [CrossRef]
14. Najafi, H.; Woodbury, K.A. Optimization of a cooling system based on Peltier effect for photovoltaic cells. *Sol. Energy* **2013**, *91*, 152–160. [CrossRef]
15. Najafi, H. *Evaluation of Alternative Cooling Techniques for Photovoltaic Panels*; University of Alabama: Tuscaloosa, AL, USA, 2012.
16. Zuazua-Ros, A.; Martín-Gómez, C.; Ibañez-Puy, E.; Vidaurre-Arbizu, M.; Gelbstein, Y. Investigation of the thermoelectric potential for heating, cooling and ventilation in buildings: Characterization options and applications. *Renew. Energy* **2018**, *131*, 229–239. [CrossRef]
17. Liu, Z.; Zhang, L.; Gong, G.; Luo, Y.; Meng, F. Evaluation of a prototype active solar thermoelectric radiant wall system in winter conditions. *Appl. Therm. Eng.* **2015**, *89*, 36–43. [CrossRef]
18. Birthwright, R.-S.; Messac, A.; Harren-Lewis, T.; Rangavajhala, S. Heat Compensation in Buildings Using Thermoelectric Windows: An Energy Efficient Window Technology. In Proceedings of the ASME 2008 International Design Engineering Technical Conferences and Computers and Information in Engineering Conference, Brooklyn, NY, USA, 3–6 August 2008; Volume 1, pp. 977–987. [CrossRef]
19. Van Dessel, S.; Foubert, B. Active thermal insulators: Finite elements modeling and parametric study of thermoelectric modules integrated into a double pane glazing system. *Energy Build.* **2010**, *42*, 1156–1164. [CrossRef]
20. Zhang, J.; Messac, A.; Zhang, J.; Chowdhury, S. Adaptive optimal design of active thermoelectric windows using surrogate modeling. *Optim. Eng.* **2013**, *15*, 469–483. [CrossRef]
21. Liu, Z.; Zhang, L.; Gong, G.; Han, T. Experimental evaluation of an active solar thermoelectric radiant wall system. *Energy Convers. Manag.* **2015**, *94*, 253–260. [CrossRef]
22. Luo, Y.; Zhang, L.; Liu, Z.; Wang, Y.; Meng, F.; Wu, J. Thermal performance evaluation of an active building integrated photovoltaic thermoelectric wall system. *Appl. Energy* **2016**, *177*, 25–39. [CrossRef]
23. Martín-Gómez, C.; Ibáñez-Puy, M.; Bermejo-Busto, J.; Fernández, J.A.S.; Ramos, J.C.; Rivas, A. Thermoelectric cooling heating unit prototype. *Build. Serv. Eng. Res. Technol.* **2015**, *37*, 431–449. [CrossRef]
24. Irshad, K.; Habib, K.; Basrawi, F.; Saha, B. Study of a thermoelectric air duct system assisted by photovoltaic wall for space cooling in tropical climate. *Energy* **2017**, *119*, 504–522. [CrossRef]
25. Piantanida, P. PV & Peltier Façade: Preliminary Experimental Results. *Energy Procedia* **2015**, *78*, 3477–3482. [CrossRef]
26. American Society of Heating, Refrigerating and Air-Conditioning Engineers. *2012 ASHRAE Handbook: Heating, Ventilating, and Air-Conditioning Systems and Equipment*; ASHRAE: Atlanta, GA, USA, 2012.
27. Rhee, K.-N.; Kim, K.W. A 50 year review of basic and applied research in radiant heating and cooling systems for the built environment. *Build. Environ.* **2015**, *91*, 166–190. [CrossRef]
28. Zhao, K.; Liu, X.-H.; Jiang, Y. Application of radiant floor cooling in large space buildings—A review. *Renew. Sustain. Energy Rev.* **2016**, *55*, 1083–1096. [CrossRef]
29. Chiang, W.-H.; Wang, C.-Y.; Huang, J.-S. Evaluation of cooling ceiling and mechanical ventilation systems on thermal comfort using CFD study in an office for subtropical region. *Build. Environ.* **2012**, *48*, 113–127. [CrossRef]
30. Kazanci, O.B.; Shukuya, M.; Olesen, B.W. Theoretical analysis of the performance of different cooling strategies with the concept of cool exergy. *Build. Environ.* **2016**, *100*, 102–113. [CrossRef]
31. Bojić, M.; Cvetković, D.; Marjanović, V.; Blagojević, M.; Djordjević, Z. Performances of low temperature radiant heating systems. *Energy Build.* **2013**, *61*, 233–238. [CrossRef]
32. Li, R.; Yoshidomi, T.; Ooka, R.; Olesen, B.W. Field evaluation of performance of radiant heating/cooling ceiling panel system. *Energy Build.* **2015**, *86*, 58–65. [CrossRef]
33. Lertsatitthanakorn, C.; Tipsaenprom, W.; Srisuwan, W.; Atthajariyakul, S. Study on the Cooling Performance and Thermal Comfort of a Thermoelectric Ceiling Cooling Panel System. *Indoor Built Environ.* **2008**, *17*, 525–534. [CrossRef]

34. Lertsatitthanakorn, C.; Srisuwan, W.; Atthajariyakul, S. Experimental performance of a thermoelectric ceiling cooling panel. *Int. J. Energy Res.* **2008**, *32*, 950–957. [CrossRef]
35. Lertsatitthanakorn, C.; Wiset, L.; Atthajariyakul, S. Evaluation of the Thermal Comfort of a Thermoelectric Ceiling Cooling Panel (TE-CCP) System. *J. Electron. Mater.* **2009**, *38*, 1472–1477. [CrossRef]
36. Cheng, T.-C.; Cheng, C.-H.; Huang, Z.-Z.; Liao, G.-C. Development of an energy-saving module via combination of solar cells and thermoelectric coolers for green building applications. *Energy* **2011**, *36*, 133–140. [CrossRef]
37. He, W.; Zhou, J.; Hou, J.; Chen, C.; Ji, J. Theoretical and experimental investigation on a thermoelectric cooling and heating system driven by solar. *Appl. Energy* **2013**, *107*, 89–97. [CrossRef]
38. Liu, Z.; Zhang, L.; Gong, G. Experimental evaluation of a solar thermoelectric cooled ceiling combined with displacement ventilation system. *Energy Convers. Manag.* **2014**, *87*, 559–565. [CrossRef]
39. Su, X.; Zhang, L.; Liu, Z.; Luo, Y.; Chen, D.; Li, W. Performance evaluation of a novel building envelope integrated with thermoelectric cooler and radiative sky cooler. *Renew. Energy* **2021**, *171*, 1061–1078. [CrossRef]
40. Luo, Y.; Yan, T.; Zhang, N. Study on dynamic thermal characteristics of thermoelectric radiant cooling panel system through a hybrid method. *Energy* **2020**, *208*, 118413. [CrossRef] [PubMed]
41. Lim, H.; Kang, Y.-K.; Jeong, J.-W. Application of a phase change material to a thermoelectric ceiling radiant cooling panel as a heat storage layer. *J. Build. Eng.* **2020**, *32*, 101787. [CrossRef]
42. Duan, M.; Sun, H.; Lin, B.; Wu, Y. Evaluation on the applicability of thermoelectric air cooling systems for buildings with thermoelectric material optimization. *Energy* **2021**, *221*, 119723. [CrossRef]
43. Chen, J.; Lu, L. Development of radiative cooling and its integration with buildings: A comprehensive review. *Sol. Energy* **2020**, *212*, 125–151. [CrossRef]
44. National Renewable Energy Laboratory. PVWatts Calculator. Available online: https://pvwatts.nrel.gov/ (accessed on 20 April 2021).
45. TE Technology. Hp-127-1.4-1.15-71. 2018. Available online: https://tetech.com/product/hp-127-1-4-1-15-71/ (accessed on 1 April 2021).
46. Ford, B.; Schiano-Phan, R.; Vallejo, J.A. *The Architecture of Natural Cooling*, 2nd ed.; Routledge: London, UK, 2019. [CrossRef]
47. Widera, B. Bioclimatic Architecture. *J. Civ. Eng. Archit. Res.* **2015**, *2*, 567–578.

energies

Review

Comparative Life Cycle Assessment of Merging Recycling Methods for Spent Lithium Ion Batteries

Zhiwen Zhou [1,2], Yiming Lai [1,2], Qin Peng [1,2] and Jun Li [1,2,*]

[1] Key Laboratory of Low-Grade Energy Utilization Technologies and Systems, Chongqing University, Ministry of Education, Chongqing 400044, China; zhouzhiwenglzx@163.com (Z.Z.); YMlai@cqu.edu.cn (Y.L.); pengqin@cqu.edu.cn (Q.P.)

[2] Institute of Engineering Thermophysics, Chongqing University, Chongqing 400044, China

* Correspondence: lijun@cqu.edu.cn; Tel./Fax: +86-23-65102474

Abstract: An urgent demand for recycling spent lithium-ion batteries (LIBs) is expected in the forthcoming years due to the rapid growth of electrical vehicles (EV). To address these issues, various technologies such as the pyrometallurgical and hydrometallurgical method, as well as the newly developed in-situ roasting reduction (in-situ RR) method were proposed in recent studies. This article firstly provides a brief review on these emerging approaches. Based on the overview, a life cycle impact of these methods for recovering major component from one functional unit (FU) of 1 t spent EV LIBs was estimated. Our results showed that in-situ RR exhibited the lowest energy consumption and greenhouse gas (GHG) emissions of 4833 MJ FU^{-1} and 1525 kg CO_2-eq FU^{-1}, respectively, which only accounts for ~23% and ~64% of those for the hydrometallurgical method with citric acid leaching. The H_2O_2 production in the regeneration phase mainly contributed the overall impact for in-situ RR. The transportation distance for spent EV LIBs created a great hurdle to the reduction of the life cycle impact if the feedstock was transported by a 3.5–7.5 t lorry. We therefore suggest further optimization of the spatial distribution of the recycling facilities and reduction in the utilization of chemicals.

Keywords: spent lithium-ion batteries; recycling; life cycle analysis; pyrometallurgical method; hydrometallurgical method; in-situ roasting reduction; energy consumption; greenhouse gas emission

Citation: Zhou, Z.; Lai, Y.; Peng, Q.; Li, J. Comparative Life Cycle Assessment of Merging Recycling Methods for Spent Lithium Ion Batteries. *Energies* **2021**, *14*, 6263. https://doi.org/10.3390/en14196263

Academic Editors: Mark Laser and Joshua M. Pearce

Received: 2 September 2021
Accepted: 28 September 2021
Published: 1 October 2021

1. Introduction

The utilization of lithium-ion batteries (LIBs) has increased dramatically due to the accelerated adaptation of electric vehicles (EVs) and portable electronics. It is predicted that 11 million tons of spent LIBs will be produced worldwide by 2030 [1]. However, to satisfy the operation safety requirement and to ensure the road haul of EV, the LIBs need to be replaced once their capacity decays to below 80% [2]. Consequently, there is an increasing demand for the disposal of spent LIBs in the forthcoming years. However, only less than 5% of spent LIBs are recycled currently [1]. Direct disposal of spent LIBs leads to serious release of toxins such as heavy metals and organic chemicals [1]. On the other hand, valuable metals (e.g., Ni, Co, Li) present in the spent LIBs are at very high levels, even higher than those found in natural ores. Therefore, the recycling of major components from spent LIBs is regarded as an extremely important way to prevent environmental pollution and to meet the requirement of sustainable utilization of valuable metals.

To address the issue of recycling spent LIBs, pyrometallurgical or hydrometallurgical approach, which derived from cobalt or nickel extraction metallurgy have been adopted by many companies. Specifically, pyrometallurgical methods like the Umicore, Inmetco, Accurec, and Glencore processes have been commercialized at an industrial scale [3–7], but the high energy requirement and hazardous gas emissions are the main drawbacks of these technologies. As for hydrometallurgical methods such as GEM High-Tech, Brunp, Retriev, and Recupyl processes [3–7], large amount of leachants are required to ensure a high leaching efficiency, posing a key challenge for waste water treatment and waste acid recovery [6]. To solve these problems, many novel methods, such as ultra-high temperature (UHT) method [3], hydrometallurgical methods with organic or inorganic acid leaching [8–14], and in-situ reduction roasting (in-situ RR) method [15–24] has been developed [25]. However, most of the studies are dedicated to developing processing method and/or optimizing operational conditions at bench scale, but the scaled-up industrial application is still absent. Although some excellent reviews [5,26–28] provided insightful suggestions and development orientation on recycling spent LIBs, a systematic quantitative estimation of these emerging methods in industrial application was still lacking, causing a huge gap between academic research and commercialization.

Life cycle assessment (LCA) is a "cradle to grave" approach to evaluate the life cycle impacts generated during the entire life cycle of products, processes and systems. It also helps the decision maker to find out the optimal design and the critical step required for improvements [29]. Although valuable insights have been gained by the LCA study of the recycling processes like Umicore, BIT, Toxco and EcoBat processes in previous studies [30–35], the energy consumption and the environmental burden of the emerging methods in an industrial scale was still unknown. Therefore, a systematical and comprehensive evaluation on these methods is needed for the development and application of LIB recycling technology and lowering the risk of commercial failures.

In this work, the emerging methods for LIB recovery, namely UHT, hydrometallurgical methods with organic or inorganic acid leaching, and in-situ RR method were briefly reviewed. Then, an industrial-scaled spent LIB recycling system based on these processes was proposed. The life-cycle impacts (i.e., energy and material consumption and greenhouse gas (GHG) emissions) of these methods were quantified by the process-based LCA approach. Additionally, the sensitivity analysis of each method was evaluated based on the uncertainties caused by changing the key parameters. The reasonable opportunities for reducing the life cycle impacts were also analyzed in this paper. To our best knowledge, this paper is the first comparative study of these methods by LCA approach and may provide direction guidance for the industrial application of spent LIB recycling technology in the future.

2. Description of the Merging Recycling Methods for LIBs

To establish a basis for the modeling of spent LIB recycling systems for LCA, the three merging recycling methods proposed in previous studies were provided in this section [7–24].

2.1. UHT Method

Traditional pyrometallurgical method usually involve UHT smelting and purification steps [6,27]. During these steps, spent LIBs was usually smelted with other types of batteries (e.g., NiMH), or ores and industrial wastes. In some cases (e.g., Umicore), batteries are directly smelted with process slag in the furnace at a temperature above 1450 °C to optimize valuable metal recovery efficiencies. The main products of UHT smelting is Co or Ni-based alloys. Then, the alloys undergo a series of hydrometallurgical processes (i.e., leaching and solvent extraction) to obtain purified products.

Recently, a LIB-dedicated UHT method with a component recovery efficiency of ca. 50% was developed based on the traditional pyrometallurgical method [3]. Figure 1 depicts the LIB dedicated UHT process according to refs. [3,36,37]. In this scale-up process, an electric arc furnace is replaced by a shaft furnace to avoid the agglomeration of the electrode materials and the excess graphite in the electrodes is used as fuel. Prior to the smelting step, spent LIB cells undergo a serials of pretreatment steps, (i.e., discharging, crushing, and material separating) to obtain the Cu, Fe, and Al metal scraps, plastic, and a mixture of anode and cathode electrode materials. Then, along with process slag (e.g., limestone sand and slag), the mixture are fed into the UHT furnace where the Co compounds are converted to Co alloy, a majority of Li enters into flue dust, and Al and Mn go into the slag. The smelted products then undergo a series hydrometallurgical processes to regenerate $LiCoO_2$ (LCO). Li in the dust is then extracted by sequential leaching and chemical precipitation steps to obtain Li_2CO_3. And the alloy is firstly leached by H_2SO_4 at 55–85 °C [38,39], and then is oxidized to Co_3O_4 by H_2O_2 at room temperatures [3,36]. Finally, the recovered Li_2CO_3 and Co_3O_4 are sintered to regenerate LCO. During the regeneration step, additional virgin Li_2CO_3 are usually needed to compensate the considerable Li loss in flue dust.

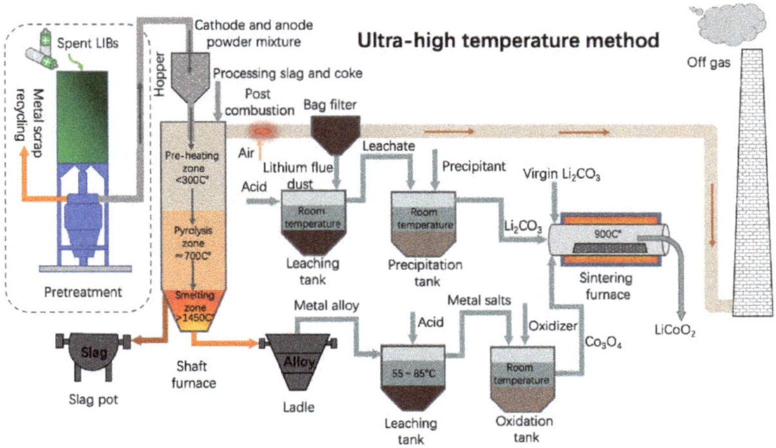

Figure 1. Schematic of the Ultra-high treatment method of spent LIB recycling system.

In general, the LIB dedicated UHT method offers a number of attractive advantages compared to the traditional pyrometallurgical one such as a higher productivity, a substantial improvement in component recovery efficiency, and a marketable or directly utilizable products for LIBs [3]. However, the UHT method also has distinct disadvantages of high energy consumption, hazardous gas emission, need of adding additional Li sources, and its economic feasibility being highly dependent on the content of valuable metals in LIBs.

2.2. *Hydrometallurgical Method*

The hydrometallurgical method typically includes a leaching step that dissolves the metallic components in spent LIBs, and the subsequent purification, separation and recovery process [8–14]. To ensure a high recovery efficiency, strong acids, such as H_2SO_4, HCl, HNO_3, were usually adopted at an excess dosage [26], but these chemicals also lead to an extra environmental burden. To alleviate this issue, organic acids like citric, oxalic, and malic acids with a high biodegradability and strong acidity were proposed as the leachant in recent studies [26]. After the sequential purification, separation and regeneration steps, the dissolved metallic ions can be recovered in different forms (e.g., Co, Li salts, or metal oxides) [6,27].

Figure 2 illustrates a typical hydrometallurgical process modified from refs. [8–14,40–42]. To reduce the leachant dosage, pretreatment process is needed to separate the anode and cathode materials so that the cathode active materials can be leached individually. Then, the valuable metals in the powdery mixture of electrode materials are leached by inorganic or organic acid with the help of reducing agents (such as H_2O_2, glucose) at elevated temperatures of 40–90 °C to enhance the leaching efficiency. Then, the residue (mainly graphite) is filtered, and the metal ions in the solution were precipitated as $CoCO_3$ and Li_2CO_3, which was solid-state sintered to regenerate LCO.

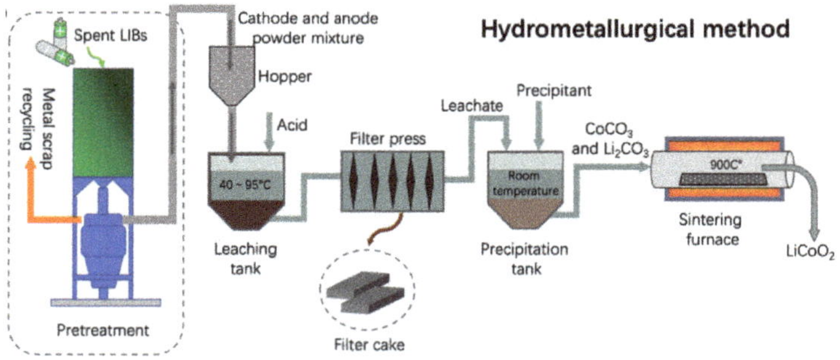

Figure 2. Schematic of the hydrometallurgical method of spent LIB recycling system.

Compared with the LIB-dedicated UHT method, the hydrometallurgical method has many advantages, including higher purity products (95.9~99.9%), a higher leaching efficiency (90~100%), and a lower operation temperature (40~95 °C) [28]. However, the disadvantages of the hydrometallurgical method is also obvious, such as the need for manual sorting, complex operation steps for purification and separation, and serious environmental issues caused by the harmful waste acids [26–28].

2.3. *In-Situ RR Method*

In-situ RR method was firstly proposed by Xu et al. in 2016 [16]. Similar to the methods mentioned above, the spent LIBs are pretreated to obtain the powdery mixture of the electrode materials. Then the mixture was roasted at a mild temperature lower than 1000 °C under oxygen-free conditions. In in-situ RR process, the cathode materials are converted to Li_2CO_3, metal and metal oxide by carbothermal reduction. Then Li_2CO_3 is recovered from the roasted products by water [16] or carbonated water leaching [38,39], while the magnetic products (i.e., Co or Ni metal) can be recovered by magnetic separation.

Figure 3 illustrates a typical in-situ RR process. Like the UHT and hydrometallurgical method, the spent LIBs are pretreated to obtain the powdery mixture of the electrode materials for further processing. Then, the mixture of the electrode active materials was subjected to reduction roasting in an electrical furnace. Subsequently, Li_2CO_3 in the roasted products are separated by carbonated water leaching [38,39]. Like in the UHT method, Co metal in the residue is then magnetically separated and leached with H_2SO_4 to produce $CoSO_4$. In a following step, it is oxidized by H_2O_2 to produce Co_3O_4 [38]. Finally, the recovered Li_2CO_3 and Co_3O_4 are solid-state sintered to regenerate LCO.

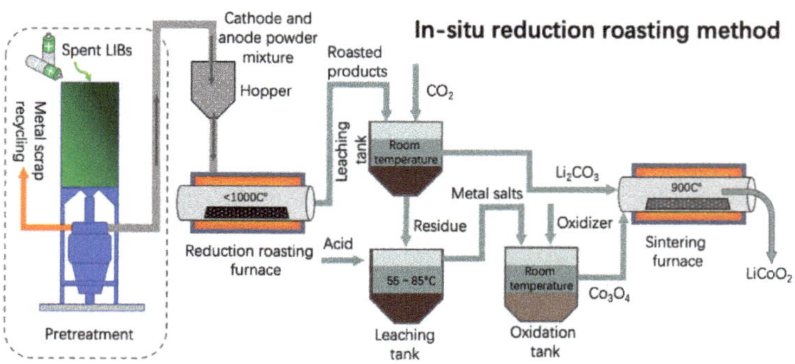

Figure 3. Schematic of the In-situ reduction roasting method of spent LIB recycling system.

This method has the advantages of: (1) a lower processing temperature (≤1000 °C) that is lower than the boiling point of the transition metals, (2) no need of extra reductants, and (3) a much simplified separation processes for the roasted products. However, the emission of hazardous gases during roasting process is obvious [26].

3. Materials and Methods

3.1. Goal and Scope Definition

The system boundary of this study is shown in Figure 4. In this system, we defined 4 phases: collection and transportation of spent LIB packs (LIBPs), pretreatment of spent LIBPs, pyrometallurgical or hydrometallurgical conversion, and regeneration phases. The functional unit (FU) was 1 t spent EOL (end-of-life) EV LIBPs. In the collection and transportation phase, we a transportation distance of 500 km by a lorry of the size class >32 t gross vehicle weight and Euro III emissions class was taken as the base case. To provide a more precise estimation on the effect of transportation on the spent LIB recovery in China, the supply chain of spent LIBs was also established based on a network dataset was created in Excel. The collection points of the spent LIBPs was assumed to be the EOL vehicle dismantling plants according to the measures on Management of EOL Vehicle Recycling. Data on the location of 769 listed EOL vehicle disassembly plants and 26 listed domestic LIB treatment facilities in China were sourced from the website of Ministry of Commerce of the People's Republic of China and Minstry of Industry and Information Technology of the People's Republic of China [32,43,44]. The transportation distance from the EOL vehicle disassembly plant to the closest LIB treatment facilities was estimated by using Baidu Map software. The LIBPs was assumed to be transported by 3.5–7.5 t lorry due to the lorry restriction rule in the urban area of most Chinese cities. The energy consumption for transportation (E_t, MJ FU^{-1}) could be obtained according to Equation (1):

$$E_t = \overline{E_t} \times L \quad (1)$$

where $\overline{E_t}$ is the energy consumption of transportation per kilometer (MJ$_{\text{-eq}}$ (t·km)$^{-1}$), L is transport mileage (km). The energy consumption with different types of lorry were listed in Table S1.

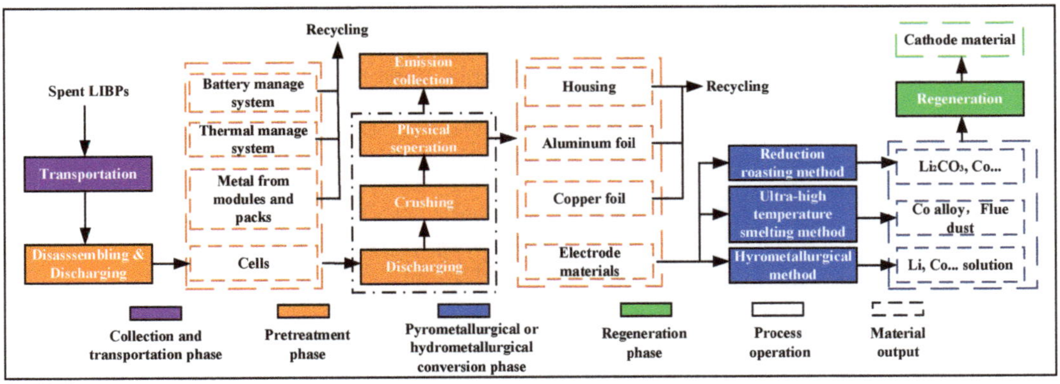

Figure 4. System boundary of spent LIB recycling used in this study.

The spent LIBs were transported by road, and the distance from the collection site to the recycling plant was 500 km, and the energy consumption of the lorry was 1.56 MJ (t·km)$^{-1}$ (lorry > 32 t, Ecoinvent v3.5 (Sphera, Hauptstrasse, Germany)). As shown in Figure S1, when one FU of spent LIBPs was inputted in the system, the total energy consumption for the collection and transportation phase was 781 MJ. But in the case that the

LIBPs was transported by 3.5–7.5 t lorry, a transportation distance of 500 km was resulted in an energy consumption of 4120 MJ due to the much higher $\overline{E_t}$ value of a 3.5–7.5 t lorry.

In the pretreatment phase, the spent LIBPs underwent a series of pretreatment to separate the components for further processing. In the conversion phase, the mixture of the electrode materials was processed by pyrometallurgical or hydrometallurgical method. To investigate the effect of leachants on the life-cycle impacts, LIB recycling systems using the hydrometallurgical method with two typical types of leachants (sulfuric acid (HM-SA) and citric acid (HM-CA)) were compared. To study the role of roasting condition on the life-cycle impact of the in-situ RR method, the LIB recycling systems in which the electrode materials were roasted under N_2 atmosphere (RR-N_2) and vaccine conditions (RR-Vac) were also established. In regeneration phase, valuable materials recovered in the previous phases were heat-treated to regenerate LCO.

Although many types of cathode materials (such as nickel cobalt manganese oxide, lithium manganate, and LCO) has been used in commercialized LIBs, LCO is still the mainstream of cathode materials in spent LIBs due to its excellent performances and prolonged usage [45] Therefore, LIBs with LCO were used in this study. It should be mentioned that although EV LIBPs (material inventories for EV LIBs and the carbon-containing components in lithium-ion battery cells listed in Tables S2 and S3)) were used as the analysis scenario, the recycling methods investigated in this study were also valid for other scenarios (such as portable electronics and energy storage system with different battery chemistries) except for a slight difference in the disassembling process.

3.2. Evaluation Methodology and Data Sources

This work was carried out using an OpenLCA software (GreenDelta, Berlin, Germany) based on the methods introduced by ISO 14041 and 14044 [26]. According to the regulations, four basic processes (i.e., goal and scope defined, inventory analysis, impact assessment, and result interpretation and recommendations) were involved in this study. The ReCiPe Midpoint (H) Method was adopted to evaluate the environmental impacts of different recycling methods based on the input/output of materials, energy and environmental emissions within the life cycle inventories. In this work, the data on the energy flow and material flow for different recycling methods were mainly derived from our experimental results, patents, academic literature, similar industrial processes, the Ecoinvent v3.5 database or theoretical calculations due to the limited industrial data arising from the immaturity of the methods investigated in this study. The process-level calculation included collection and transportation phase, pretreatment phase as well as conversion phase were exhibited in detail in Supplementary Files (Figures S1–S7).

The GHG emissions generated from the carbonaceous materials containing in spent LIBs were calculated based on the assumption that they were converted to CO_2 during recycling. To achieve a reasonable comparison, it was also assumed that the carbonaceous residues of spent LIBs during hydrometallurgical conversion were converted into CO_2 and accumulated in the total GHG emissions. For the in situ-RR method, the carbonaceous materials were completely converted to CO_2, except for the fixed carbon in Li_2CO_3. The energy consumption and the related GHG emissions derived from the LIBP transportation and the production of chemicals used in LIB recycling processes also came from the same database and summarized in Tables S1 and S4, respectively. The GHG emissions generated by electricity production were also calculated based on the electricity grid structure of Guangdong, Shandong, and Sichuan province (data from Ecoinvent v3.5, Table S5). Unless mentioned, the material consumption in this study was calculated based on the theoretical stoichiometric data, and the loss of substances during the processing was not considered. The reagent consumptions during sulfuric acid and citric acid leaching were estimated based on the average laboratory data shown in Tables S6 and S7.

3.3. *Life Cycle Impact Assessment*

The life-cycle impacts were evaluated in terms of total energy consumption and GHG emissions. The primary contribution to the GHG emissions is associated with CO_2, CH_4 and N_2O. The three GHG emissions are integrated into the CO_2 equivalence (CO_2-eq) of 1, 25, and 298, respectively, in accordance with a time horizon of 100 years [46].

The reduction rate of energy consumption (ξ_E,%) and GHG emissions (ξ_G, %) of various methods were defined based on the following equations:

$$\xi_E = \frac{E_v - E_r}{E_v} \quad (2)$$

$$\xi_G = \frac{G_v - G_r}{G_v} \quad (3)$$

where E_v is energy consumption for producing virgin materials (MJ FU^{-1}), E_r is energy consumption for recovering electrode materials from LIBPs (MJ FU^{-1}), G_v is GHG emissions for producing virgin materials (kg $CO_{2\text{-eq}}$ FU^{-1}), and G_r is GHG emissions for recovering electrode materials from LIBPs (kg $CO_{2\text{-eq}}$ FU^{-1}).

To investigate the effect of the change in the recovery rate of Co and Li as well as the electricity structure on the energy consumption and GHG emissions of the system, sensitivity analysis were carried out according to the following equations:

$$\Delta E = \frac{E^{\circ} - E^{i}}{E^{\circ}} \quad (4)$$

$$\Delta G = \frac{G^{\circ} - G^{i}}{G^{\circ}} \quad (5)$$

where ΔE is the changes in energy consumption (MJ FU^{-1}), E° is the initial energy consumption (MJ FU^{-1}), E^i is the altered energy consumption (MJ FU^{-1}), ΔG is the changes in GHG emissions (kg $CO_{2\text{-eq}}$ FU^{-1}), G° is the initial GHG emissions (kg $CO_{2\text{-eq}}$ FU^{-1}), and G^i is the altered GHG emissions.

4. Results

4.1. *Energy Consumption Analysis*

Our calculation (Figures 5 and S4–S7) shows that 284.2, 275.5, 261.0, 273.0, 274.3 kg LCO can be recovered from one FU by the UHT, HM-SA, HM-CA, RR-N_2 and RR-Vac method, respectively. Figure 5 also shows that, if only LCO recovery was considered, the energy consumption of all recycling methods were significantly lower than that for producing an equivalent weight virgin LCO in industry (38,367~41,777 MJ [45]). Among the five methods, HM-CA was the most energy intensive method which had an energy consumption of 20,892 MJ FU^{-1} during the whole process, while RR-N_2 consumed the least amount of energy (4833 MJ FU^{-1}). The ξ_E value for UHT, HM-SA, HM-CA, RR-N_2 and RR-Vac was 71%, 71%, 46%, 88% and 87%, respectively. All methods exhibited the same energy consumption for the collection & transportation and pretreatment phase of 781 and 474 MJ FU^{-1}, respectively, accounting to only a small faction (6% (HM-CA)~26% (RR-N_2) of the total energy consumed. From Figure 5, it is clear that the main contributor of the total energy consumption for each method was different. For the UHT method, the conversion and regeneration phases were mainly responsible to the total energy consumption, and the energy consumption of these two phases was rather similar. For the hydrometallurgical methods (HM-SA and HM-CA), the major part of the total energy consumption was the conversion phase, while the regeneration phase was identified as the main contributor for the total energy consumption in the case of in-situ RR methods (RR-N_2 and RR-Vac) due to the much low processing temperature (800~850 °C).

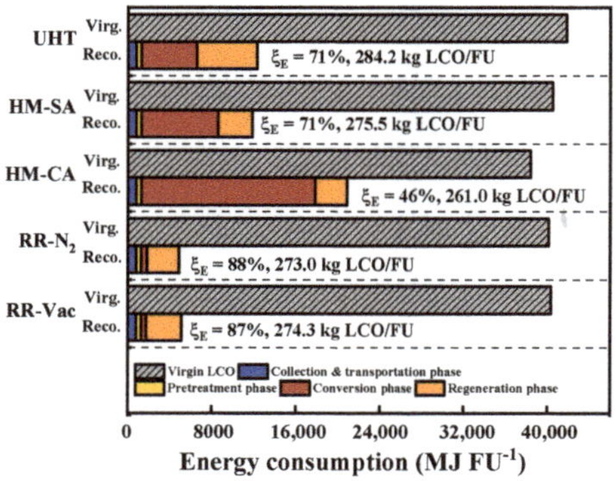

Figure 5. Contribution analysis of the total energy consumption for the five different methods. Each pair of bars represents a method, where the bottom bar represents the energy consumption distribution of each phase for 1 FU, the top bar is the energy consumption for producing an equivalent weight virgin LCO in industry, and the numbers indicate the reduction rate of energy consumption and recovered LCO from 1 FU for each method.

A detailed analysis of the energy consumption in the conversion and regeneration phase for the five different recycling methods are displayed in Figure 6.

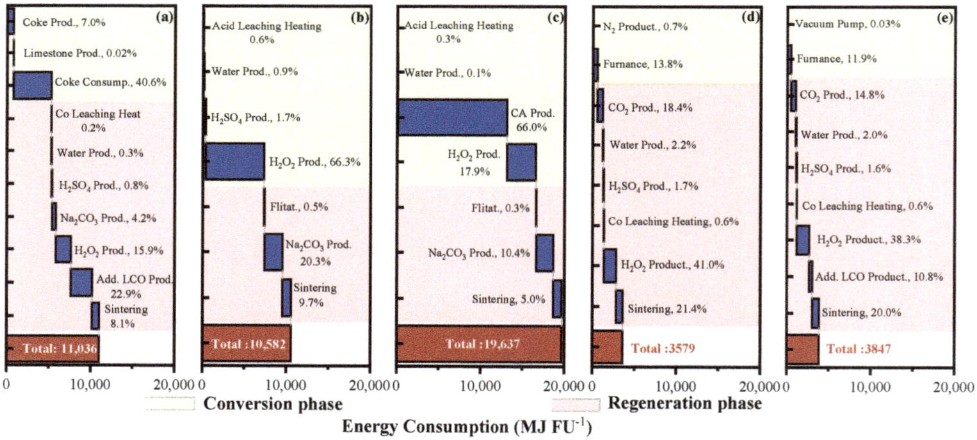

Figure 6. Contribution analysis of the energy consumption in the conversion and regeneration phase for the five different recycling methods. (**a**) UHT method, (**b**) HM-SA, (**c**) HM-CA, (**d**) RR-N_2, and (**e**) RR-Vac. The percentage labeling indicates the proportion of the energy consumption associated with the manufacture of the materials or active electrode material processing needed for the treatment of 1FU in the conversion and regeneration phase. The red bar indicates the total energy consumption of the conversion and regeneration phase for the five different recycling methods.

The total energy consumption of HM-CA in the conversion and regeneration phase is 19,637 MJ FU^{-1}, which was more than five times higher than that of the in-situ RR methods and about 1.7 times higher than that of UHT. The upstream production of citric acid and H_2O_2 are the main contributor of the total energy consumption for the conversion and regeneration phase of HM-CA, accounting for 66.0% and 17.9% of the energy consumption of these phases, respectively (Figure 6c). UHT (11,036 MJ FU^{-1}) consumed a slightly higher energy as compared to HM-SA (10,582 MJ t^{-1} FU) in these phases. The coke consumption and H_2O_2 production in the conversion phase is the main contributors of the UHT and HM-SA methods, accounting for 40.6% and 66.3% of the total energy consumption, respectively (Figure 6a,b). From Figure 6, it is also derived that although the energy consumption composition of the regeneration phase for various methods were different, the energy consumption of HM-SA, HM-CA, RR-N_2 and RR-Vac is rather close, ranging from 3059 MJ FU^{-1} for RR-N_2 to 3390 MJ FU^{-1} for RR-Vac, UHT method possess the highest energy consumption (5780 MJ FU^{-1}) in the regeneration phase as compared with other methods because of the extra energy consumption in the production of virgin Li_2CO_3 for the compensation of the significant Li loss in the conversion phase (Figure 6a).

4.2. GHG Emission Analysis

From Figure 7, it is clear that the reduction in the GHG emission of different recycling methods for recovering an equivalent weight LCO from one FU is evident compared to virgin LCO production. The ξ_G values for UHT, HM-SA, HM-CA, RR-N_2 and RR-Vac were 40%, 42%, 31%, 57%, and 57%, respectively. RR- N_2 and RR-Vac has a lowest total GHG emissions of 1525 kg $CO_{2\text{-eq}}$ FU^{-1}, while the HM-CA method exhibited the highest amount of GHGs (2351 kg $CO_{2\text{-eq}}$ FU^{-1}) among the five methods. Consistent with the energy consumption data (Figure 6), the GHG emissions of both collection and transportation and pretreatment phase were the same and only contribute to a small fraction of the total GHG emissions, ranging from 7% (HM-CA) to 10% (RR-N_2). Figure 7 demonstrates that the conversion phase dominated the GHG emissions during the recycling processes. The proportion of conversion phase in the total GHG emission for UHT, HM-SA, HM-CA, RR-N_2 and RR-Vac are 66%, 64%, 70%, 65%, and 64%, respectively. A detailed analysis of the GHG emissions in this phase (Figure 8) shows that they were mainly caused by the conversion of carbonaceous components in spent LIB cells, which were 45.2%, 49.3%, 42.8%, 64.0%, and 63.6%, for UHT, HM-SA, HM-CA, RR-N_2, and RR-Vac, respectively. The GHG emission from the regeneration phase of UHT, HM-SA, HM-CA, RR-N_2 and RR-Vac were 584, 580, 550, 377, and 400 kg $CO_{2\text{-eq}}$ FU^{-1}, respectively, indicating that it was also an important contributor to the total GHG emissions. The GHG emissions of this phase were mainly derived from the electricity consumption for sintering, chemical production and the decomposition of Li_2CO_3.

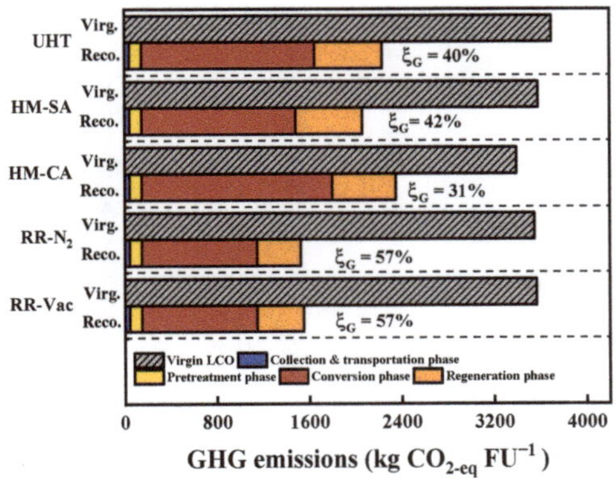

Figure 7. Contribution analysis of total GHG emissions for the five different recycling methods. Each pair of bars represents a method, where the bottom bar indicated the GHG emissions distribution of each phase for 1 FU, the top bar is the GHG emissions for producing an equivalent weight virgin LCO in industry, and the difference represents the GHG emission reduction. The numbers indicate the reduction rate of GHG emission for each method.

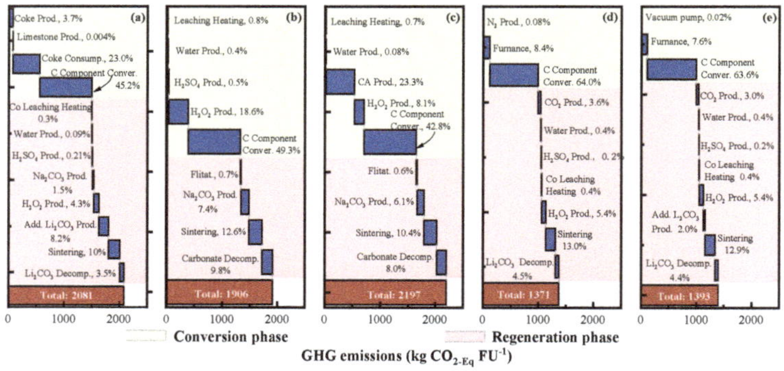

Figure 8. Contribution analysis of the GHG emissions in the conversion and regeneration phase for the five different recycling methods. (**a**) UHT method, (**b**) HM-SA, (**c**) HM-CA, (**d**) RR-N_2, and (**e**) RR-Vac. The percentage labeling indicates the proportion of the GHG emission resulting from the manufacture of the materials or active electrode material processing needed for the recycling 1FU in the conversion and regeneration phase. The red bar indicates the total GHG emissions of the conversion and regeneration phase for the five different recycling methods.

4.3. Benefits of Resource Recovering

When the recovering of Cu and Al were considered, the energy consumption and GHG emissions of Cu and Al recovering processes were calculated based on the similar process flows in Ecoinvent v3.5 and a Chinese case [47] (relevant data shown in Table S4) by assuming a recovery rate of 85%. The production of virgin Cu and Al with raw materials was also estimated based on the same data sources. The energy and environmental benefits of recovering Cu, Al together with LCO were quantified by ξ_E and ξ_G (Figure 9). It can be seen that when Cu, Al and LCO are recovered, ξ_E can reach 68% (HM-CA)~88% (RR-N_2/Vac) and ξ_G increased to 59% (HM-CA)~72% (RR-N_2/Vac). As shown in Figure 9a,b, although Cu and Al recovery has a different impact on the ξ_E and ξ_G values of the methods, RR-N_2 and RR-Vac still exhibited the higher reduction in energy consumption and GHG emissions as compared with UHT, HM-SA, and HM-CA. These results suggest that in-situ RR method (RR-N_2 and RR-Vac) is more suitable for recycling spent LIBs than the UHT and hydrometallurgical methods (HM-CA and HM-SA).

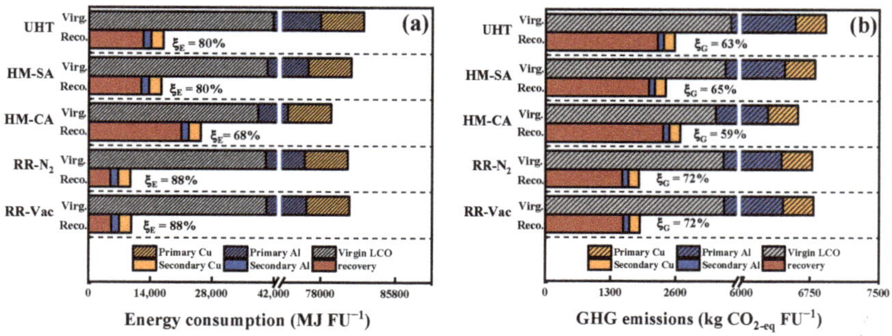

Figure 9. Contribution analysis of energy consumption (**a**) and GHG emissions (**b**) for the five different recycling methods. Each pair of bars is related to a method, where the bottom bar represents the energy consumption (**a**) or GHG emissions (**b**) required for each recycling method and Cu and Al recovering processes for 1 FU, the top bar is the energy consumption (**a**) or GHG emissions (**b**) for an equivalent weight virgin LCO and primary Cu and Al production in industry, and the difference represents the energy saving. The numbers indicate the reduction rate of energy consumption (**a**) or GHG emissions (**b**) from 1 FU for each method.

4.4. Sensitivity Analysis

Since the industrial application of the methods evaluated in this study is still in infancy, its energy consumption and GHG emission faces significant uncertainties. Therefore, a sensitivity analysis (Figures 10 and 11) is carried out to assess the impact of the key parameters on the evaluation results. It is seen from Figure 10 that both energy consumption and GHG emissions of the methods were sensitive to the recovery rate of Li and Co. However, changes in the recovery rate of Li and Co have a diverse impact on different methods. For instance, when a recovery rate of 70% was taken as the benchmark, changing the recovery rate of Li and Co from 50% to 90% had a substantial impact on the energy consumption of the HM-CA and HM-SA (from 45% to −44%), whereas this parameter only causes a variation of 8% to −8% in the energy consumption of UHT. Changing this parameter had a similar influence on the GHG emissions of the methods. That is, the influence degree of HM-CA and HM-SA (19%~−19% and 16%~−16%, respectively) was significantly higher than that of the rest methods (from 5% to −5%).

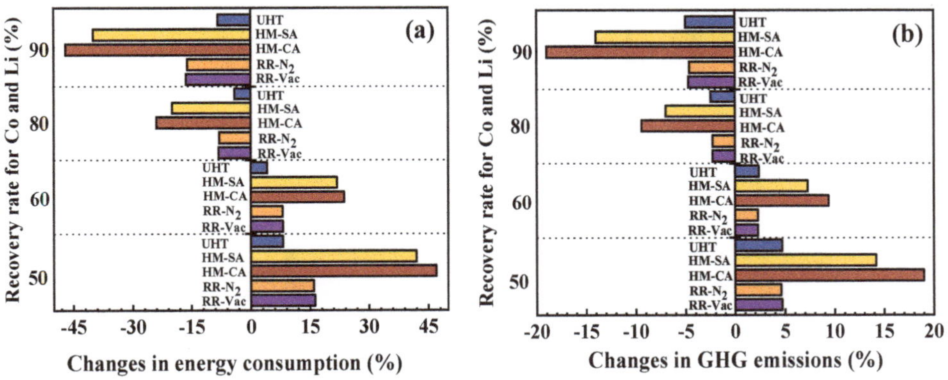

Figure 10. Sensitivity analysis of the total energy consumption (**a**) and GHG emissions (**b**) for different recycling methods by changing the recovery rate of Li and Co.

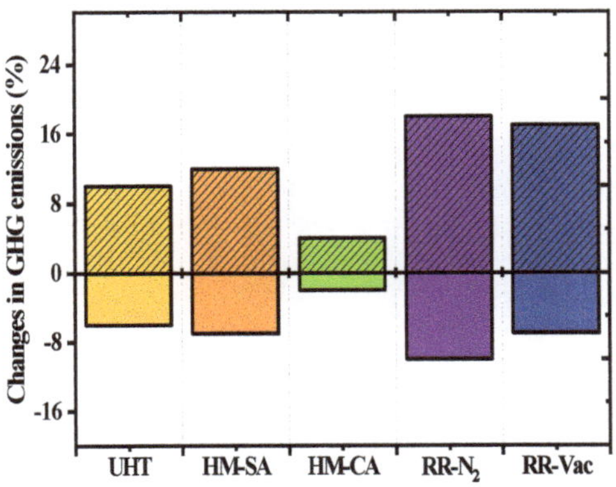

Figure 11. Sensitivity analysis of the GHG emissions for the recycling methods with different electricity sources.

Figure 11 shows the sensitivity analysis of the GHG emissions in different recycling methods when using different electricity structures.

It can be seen from Figure 11 that when the electricity structure was changed from a hydropower-dominated grid (Sichuan electricity grid mixes) to a coal-dominated one

(Shandong electricity grid mixes), the GHG emissions of in-situ RR method showed the most obvious impact of −10% to 18%. By contract, HM-CA exhibits the least change in GHG emissions of only −2% to 4%. These results are understandable because in the case of RR-N_2 and RR-Vac, the electricity consumption accounts for a considerable large proportion of the total energy consumption (Figure 6), while the electricity consumption for HM-CA was the least within our system boundary.

5. Discussion

5.1. Collection and Transport Phase

As shown in Figures 5 and 7, the proportion of the energy consumption and GHG emissions for the collection and transport phase only accounts for 4~16% and 2~3% of the total energy consumption and GHG emissions for different recycling methods, respectively, based on the assumption of a transportation distance of 500 km by a 32 t lorry. Considering that there is usually a heavy-duty lorry restriction rule in the urban area of most Chinese cities, it is more realistic to transport the spent EV LIBPs by a 3.5–7.5 t lorry in China. Thus, the effect of transportation distance by a 3.5–7.5 t lorry on the energy consumption (dark red circles) and GHG emission (dark blue squares) of the collection and transportation phase was estimated and compared to those of the rest part of the recycling phases (Figure 12). It is seen that the transportation distance can make a significant contribution to the energy consumption of the LCO recovery if a 3.5–7.5 t lorry was used to transport the spent EV LIBPs. For instance, the energy consumption for a transportation distance of above 300 km by a 3.5–7.5 t lorry alone would exceed half of that for the rest part of the in-situ RR process.

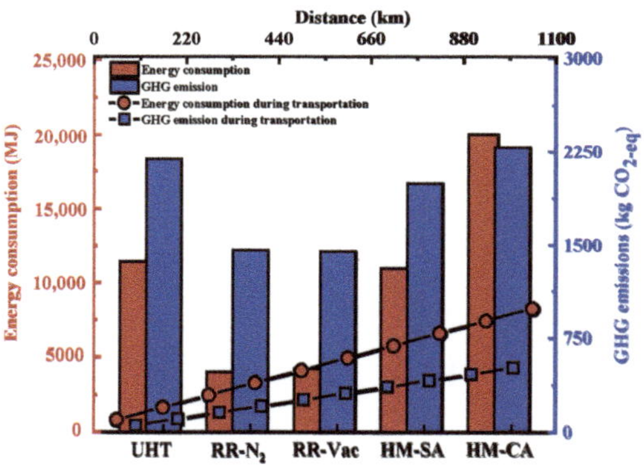

Figure 12. Effect of transportation distance (using lorry 3.5–7.5 t) in spent LIB recycling systems on the energy consumption (dark red circles) and GHG emission (dark blue squares) of the collection and transportation phase as compared to those of the rest part of the recycling methods.

Based on public data [43–45], we also estimated the spatial distribution of the transportation distance of spent EV LIBs from the vehicle dismantling plant to the closest LIB treatment facilities in China. It is found that the transportation distance was strongly correlated with the distribution of the spent LIB treatment facilities, as demonstrated in Figure 13. The provinces with a high distribution density of LIB treatment facilities, such as Guangdong, Fujian, Zhejiang, Jiangxi has a reasonable transportation distance of less than 200 km, but in the case of Sichuan province, a transportation distance of above 200 km is needed due to the low distribution density of LIB treatment facilities in Southwestern China. Based on the fact that Shandong, Shanxi, Chongqing, Guizhong, Guangxi, Yunnan, and Hainan significantly contribute to the national EV parc, additional treatment facilities are suggested to be established in these provinces.

From Figure 13, it is also seen that due to the poor cold-temperature performance of LIBs, the EV parc of the provinces in northeast and northwest China is limited, resulting in the absence of the treatment facilities in these provinces, and therefore an unacceptable long transportation distance of above 1000 km.

Figure 13. Spatial distribution of the transportation distance of spent EV LIBs from the vehicle dismantling plant to the closest LIB treatment facilities in China. The numbers indicate the average closest transportation distance of spent EV LIBs for each province.

It also should be noted that the collection and sorting of variety spent LIBs according to the battery chemistry also has a considerable impact on the manpower and other resource investment during this phase. Thus, a safe and efficient collection system is urgently needed to be established so as to reduce the recycling costs and corresponding environmental impacts. Here we can refer to the currently mature solution of lead-acid batteries [48]. To achieve safe and efficient collection and sorting of spent LIBs by using existing infrastructure, the government can formulate corresponding incentive policies to encourage consumers to return spent LIBs, and manufacturers can establish corresponding measures through their existing sales or after-sales channels.

5.2. Pretreatment Phase

Although the pretreatment of spent LIBPs only consumed 2% (HM-CA)~10% (RR-N_2) of the total energy consumption, corresponding to 5% (HM-CA)~7% (RR-N_2) of the total GHG emissions, most of the LIB packs is manually disassembled at present due to the diversity of cell types, cell chemistries, and pack structures produced by various manufacturers. This would greatly increase labor costs and reduce the benefits for recycling spent LIBs. Therefore, it is necessary to standardize the production of LIBs for facilitating automatic disassembly in the future. It should be noted that although it is assumed that the components (e.g., Al, Cu, separator) are separated from the powdery mixture of electrode materials by physical separation in our study, there is still a big challenge to achieve such a high-purity enrichment of electrode materials in practical operation [28]. These impurities will have a notable impact on the subsequent treatment, and thus additional purification steps are usually required, which increase the cost and complexity of the process. In addition, the emissions generated must be treated effectively during this phase due to its flammability and toxicity [49,50].

5.3. Conversion Phase

5.3.1. UHT Method

The LIB dedicated UHT method is still facing the problem of excessive energy consumption and GHG emissions due to its high operating temperature. Moreover, this method is not conducive to the Li recovery due to the high Li loss in the flue dust and slag [6]. When the method is used to process LIBs with electrode chemistries like lithium manganate and lithium phosphate, its feasibility will be greatly reduced due to the limited Li recovery. Nevertheless, compared with the conventional pyrometallurgical methods that LIBs are directly smelted in a shaft furnace, only the powdery electrode mixture are smelted in the furnace, the LIB dedicated UHT method substantially reduce the processing materials by 40%, and thus a decrease of 20% in energy consumption (43 MJ (kg recovered-LCO)$^{-1}$ of this study) as compared with 54 MJ (kg recovered-LCO)$^{-1}$ for the Umicore Process [51]. These results also highlight the importance of pretreatment phase on spent LIB recycling and valuable component enrichment. Thus, our estimation indicated that UHT method has a potential for solving the excessive energy consumption problem of the traditional pyrometallurgical method.

5.3.2. Hydrometallurgical Method

It is well known that the leachant dosage play a crucial role on the leaching efficiency of the valuable metals. Li et al. reported [52] that the leaching efficiency of Co and Li in HM-CA was only ~34% and 58% when the dosage of citric acid and H_2O_2 were about 4 and 2 times of stoichiometric requirement. While the leaching efficiency of Co and Li can increase to 90% and 99% when the dosage of citric acid and H_2O_2 were about 6 and 3 times of stoichiometric requirement, respectively [52]. Therefore, to ensure an acceptable leaching efficiency, the latter stoichiometric dose of citric acid and H_2O_2 were used in this study. Our calculation indicates that HM-CA consumed the most energy for producing an equivalent weight LCO (80 MJ (kg recovered-LCO)$^{-1}$) among the five methods even based on the assumption that that 90% consumed citric acid was recovered. This value is nearly twice higher than the results obtain by Dunn (42 MJ (kg recovered-LCO)$^{-1}$ [51]). These results are understandable because their estimation was based on 1.1 times the stoichiometric dosage [36,51,53]. Our calculation also demonstrated that HM-CA consumed a 2.3-fold of energy and emitted a 1.9-fold of GHGs compared to HM-SA for reagent production during this phase. Thus, in terms of life-cycle impacts, organic acid leaching is not a suitable solution for migrating the environmental issues of strong acid leaching.

In LIBs, the valuable metals are basically at the cathode active material. Therefore, to solve the problem of excessive leachant consumption, the separation of cathode and anode materials before hydrometallurgical processes is highly recommended. However, it is still rather challenging to achieve a complete separation of anode and cathode materials in an industrial scale [22,54]. Fortunately, He et al. demonstrated that the effective separation between anode and cathode materials can be achieved by Fenton assisted flotation. However, the impacts of its application on the downstream processing have not yet to be studied.

5.3.3. In-Situ RR Method

Our results indicated the in-situ RR method is a promising method for spent LIB recycling. For example, the total energy consumption of RR-N_2 only accounted for 39% of UHT and 23% of HM-CA, corresponding to 68% and 65% of the total GHG emissions. This reduction can be mainly attributed to the considerable reduction in processing temperature from >1450 to 800 °C. Additionally, the in-situ RR method can significantly reduce the amount of electrode active materials by converting the cathode active materials into Li_2CO_3, Co and other metal oxides (depending on the chemistries of cathode). This is important to the following regeneration phase because it can greatly reduce the reagent consumption.

However, this method still has some shortcomings, such as a low Li recovery rate of ~70%, and the emission of the F-containing gases during the roasting step. In addition, although our estimation was based on the assumption that graphite is completely converted to CO_2, transition metals existed in metallic form, and Li is completely converted to Li_2CO_3 during the roasting step, previous studies showed that considerable carbonaceous materials remained after roasting [17] and part of the transition metals exists in high valence oxides. To overcome these issues, roasting the electrode active materials in a reductive atmosphere can be an attractive solution. In such an atmosphere, the carbon content in electrode materials can be converted into gaseous fuels (such as H_2, CO) or the energy required for roasting. Besides, cathode materials can be reduced into metals or low valence metal oxides (e.g., Co or CoO), which consumes much less reductant in the sequential leaching steps.

It should be pointed out that the profitability of the merging methods for LIBs is also one of the most important issues affecting the commercialization of these methods. However, the profitability of the LIBs recycling method may vary because of the diverging emission standards for pollutants and the labor costs of different regions and countries. Therefore, it is extremely difficult to compare the profitability of the merging methods for LIBs, and only energy consumption and GHG emissions were compared in this study. Recently, many novel methods with for industrial spent LIBs recycling also had been developed [55,56]. For example, the company TOXCO in Canada has adopted liquid nitrogen freezing crushing and physical separation technology to recycle metals such as Cu, Fe and Al from spent LIBs, and are reported to be capable of dismantling of 4500 tons of spent LIBs per year. Duesenfeld GmbH, a company in Germany, proposed a combination of mechanical, thermodynamic, and hydrometallurgical treatment processes to recycle Ni and Co from spent LIBs, and the recycling rates of main metals can reach 75%. In addition, Contemporary Amperex Co. Ltd. (CATL, Fujian, China), one of the biggest lithium-ion battery manufacturers in China, reported a novel hydrometallurgical process combined with solvent extraction for recovering cathode materials from spent LIBs, and the recovery rate of Ni, Co and Mn can reach 99.3%. Lastly, it should be noted that our estimation is based on our experimental and basic industrial data. Additional steps are still needed to process the recovered electrode material to achieve the same battery performance as the virgin materials. Nevertheless, the results can still provide the information on whether the emerging recycling methods can provide energy and environmental benefits, and which method is more commercially competitive.

6. Conclusions

In this work, five merging technologies for spent LIB recycling were reviewed and a quantitative analysis to evaluate the life cycle impacts of these technologies was conducted. It is shown that the five emerging methods for spent LIB recycling have significant potential for reducing energy consumption and GHG emissions when only LCO recovery was considered. Among the five different methods, in-situ RR method consumed the least energy coupled with the least GHG emissions, showing significant advantages. While, HM-CA exhibited significant disadvantages, which has about 3.3 times higher energy consumption than that for in-situ RR method, corresponding to 1.4 times higher GHG emissions. Further reduction in energy consumption and GHG emission can be achieved if the recycling of Cu and Al are also taken into account. It is also found that transportation distance has a significant effect of life cycle impacts of the spent LIB recycling technologies in China. For a province without recycling facilities, the energy consumption for collection and transportation would become unacceptably high. We also proposed that further investigation on the in-situ RR method should focus on the removal of carbonaceous residues and complete conversion of valuable content into metals or low valence metal oxides.

Supplementary Materials: The following are available online at https://www.mdpi.com/1996-1073/14/19/6263/s1, Figure S1. Energy and material flows for the collection and transportation phase of spent LIB recycling system. Figure S2. Energy and material flows in the pretreatment phase of spent LIB recycling system. Figure S3. Flow diagram of the UHT method. Figure S4. Flow diagram of the HM-SA method. Figure S5. Flow diagram of the HM-CA method. Figure S6. Flow diagram of the RR-N2 method. Figure S7. Flow diagram of the RR-Vac method. Table S1: Energy consumption and GHG emissions associated with different transportation modes. Table S2: Material invento-ries for EV LIBs. Table S3: Carbon-containing components in lithium-ion battery cells and their carbon contents. Table S4: Energy consumption and GJG emissions associated with chemicals used in LIB recycling. Table S5: GHG emissions associated with different electricity structure. Table S6: Sulfuric acid leaching conditions for recycling of valuable metals from LIBs. Table S7: Citric acid leaching conditions for recycling of valuable metals from LIBs.

Funding: This research was funded by the National Natural Science Foundation of China (No. 52076022), and Program for Back-up Talent Development of Chongqing University (No. cqu2018CDHB1A03).

Institutional Review Board Statement: Not applicable.

Informed Consent Statement: Informed consent was obtained from all subjects involved in the study.

Data Availability Statement: Not applicable.

Conflicts of Interest: The authors declare no conflict of interest.

References

1. Natarajan, S.; Aravindan, V. Burgeoning prospects of spent lithium-ion batteries in multifarious applications. *Adv. Energy Mater.* **2018**, *8*, 1802303. [CrossRef]
2. Wang, W.; Wu, Y. An overview of recycling and treatment of spent $LiFePO_4$ batteries in China. *Resour. Conserv. Recycl.* **2017**, *127*, 233–243. [CrossRef]
3. Georgi-Maschler, T.; Friedrich, B.; Weyhe, R.; Heegn, H.; Rutz, M. Development of a recycling process for Li-ion batteries. *J. Power Sour.* **2012**, *207*, 173–182. [CrossRef]
4. Al-Thyabat, S.; Nakamura, T.; Shibata, E.; Iizuka, A. Adaptation of minerals processing operations for lithium-ion (LiBs) and nickel metal hydride (NiMH) batteries recycling: Critical review. *Miner. Eng.* **2013**, *45*, 4–17. [CrossRef]
5. Lv, W.; Wang, Z.; Cao, H.; Sun, Y.; Zhang, Y.; Sun, Z.H. A critical review and analysis on the recycling of spent lithium-ion batteries. *ACS Sustain. Chem. Eng.* **2018**, *6*, 1504–1521. [CrossRef]
6. Pinegar, H.; Smith, Y.R. Recycling of end-of-life lithium ion batteries, part I: Commercial processes. *J. Sustain. Met.* **2019**, *5*, 402–416. [CrossRef]
7. Harper, G.; Sommerville, R.; Kendrick, E.; Driscoll, L.; Slater, P.; Stolkin, R.; Walton, A.; Christensen, P.; Heidrich, O.; Lambert, S.; et al. Recycling lithium-ion batteries from electric vehicles. *Nature* **2019**, *575*, 75–86. [CrossRef]
8. Shin, S.M.; Kim, N.H.; Sohn, J.S.; Yang, D.H.; Kim, Y.H. Development of a metal recovery process from Li-ion battery wastes. *Hydrometallurgy* **2005**, *79*, 172–181. [CrossRef]
9. Xu, J.; Thomas, H.R.; Francis, R.W.; Lum, K.R.; Wang, J.; Liang, B. A review of processes and technologies for the recycling of lithium-ion secondary batteries. *J. Power Sour.* **2008**, *177*, 512–527. [CrossRef]
10. Kang, J.; Senanayake, G.; Sohn, J.; Shin, S.M. Recovery of cobalt sulfate from spent lithium ion batteries by reductive leaching and solvent extraction with Cyanex 272. *Hydrometallurgy* **2010**, *100*, 168–171. [CrossRef]
11. Chen, X.; Luo, C.; Zhang, J.; Kong, J.; Zhou, T. Sustainable recovery of metals from spent lithium-ion batteries: A green process. *ACS Sustain. Chem. Eng.* **2015**, *3*, 3104–3113. [CrossRef]
12. Chen, X.; Fan, B.; Xu, L.; Zhou, T.; Kong, J. An atom-economic process for the recovery of high value-added metals from spent lithium-ion batteries. *J. Clean. Prod.* **2016**, *112*, 3562–3570. [CrossRef]
13. Almeida, J.R.; Moura, M.N.; Barrada, R.V.; Barbieri, E.M.S.; Carneiro, M.T.; Ferreira, S.A.D.; Lelis, M.D.F.F.; Freitas, M.; Brandão, G.P. Composition analysis of the cathode active material of spent Li-ion batteries leached in citric acid solution: A study to monitor and assist recycling processes. *Sci. Total Environ.* **2019**, *685*, 589–595. [CrossRef]
14. Yu, M.; Zhang, Z.; Xue, F.; Yang, B.; Guo, G.; Qiu, J. A more simple and efficient process for recovery of cobalt and lithium from spent lithium-ion batteries with citric acid. *Sep. Purif. Technol.* **2019**, *215*, 398–402. [CrossRef]
15. Li, J.; Lai, Y.; Zhu, X.; Liao, Q.; Xia, A.; Huang, Y.; Zhu, X. Pyrolysis kinetics and reaction mechanism of the electrode materials during the spent $LiCoO_2$ batteries recovery process. *J. Hazard. Mater.* **2020**, *398*, 122955. [CrossRef]
16. Li, J.; Wang, G.; Xu, Z. Environmentally-friendly oxygen-free roasting/wet magnetic separation technology for in situ recycling cobalt, lithium carbonate and graphite from spent $LiCoO_2$/graphite lithium batteries. *J. Hazard. Mater.* **2016**, *302*, 97–104. [CrossRef]
17. Xiao, J.; Li, J.; Xu, Z. Recycling metals from lithium ion battery by mechanical separation and vacuum metallurgy. *J. Hazard. Mater.* **2017**, *338*, 124–131. [CrossRef]
18. Xiao, J.; Li, J.; Xu, Z. Novel approach for in situ recovery of lithium carbonate from spent lithium ion batteries using vacuum metallurgy. *Environ. Sci. Technol.* **2017**, *51*, 11960–11966. [CrossRef]
19. Mao, J.; Li, J.; Xu, Z. Coupling reactions and collapsing model in the roasting process of recycling metals from $LiCoO_2$ batteries. *J. Clean. Prod.* **2018**, *205*, 923–929. [CrossRef]
20. Huang, Z.; Ruan, J.; Yuan, Z.; Qiu, R.-L. Characterization of the materials in waste power banks and the green recovery process. *ACS Sustain. Chem. Eng.* **2018**, *6*, 3815–3822. [CrossRef]

21. Zhang, Y.; Wang, W.; Fang, Q.; Xu, S. Improved recovery of valuable metals from spent lithium-ion batteries by efficient reduction roasting and facile acid leaching. *Waste Manag.* **2020**, *102*, 847–855. [CrossRef]
22. Wang, W.; Zhang, Y.; Zhang, L.; Xu, S. Cleaner recycling of cathode material by in-situ thermite reduction. *J. Clean. Prod.* **2020**, *249*, 119340. [CrossRef]
23. Wang, W.; Han, Y.; Zhang, T.; Zhang, L.; Xu, S. Alkali metal salt catalyzed carbothermic reduction for sustainable recovery of $LiCoO_2$: Accurately controlled reduction and efficient water leaching. *ACS Sustain. Chem. Eng.* **2019**, *7*, 16729–16737. [CrossRef]
24. Wang, W.; Zhang, Y.; Liu, X.; Xu, S. A simplified process for recovery of li and co from spent $LiCoO_2$ cathode using al foil as the in situ reductant. *ACS Sustain. Chem. Eng.* **2019**, *7*, 12222–12230. [CrossRef]
25. Liu, C.; Lin, J.; Cao, H.; Zhang, Y.; Sun, Z. Recycling of spent lithium-ion batteries in view of lithium recovery: A critical review. *J. Clean. Prod.* **2019**, *228*, 801–813. [CrossRef]
26. Zhang, X.; Li, L.; Fan, E.; Xue, Q.; Bian, Y.; Wu, F.; Chen, R. Toward sustainable and systematic recycling of spent rechargeable batteries. *Chem. Soc. Rev.* **2018**, *47*, 7239–7302. [CrossRef]
27. Chen, M.; Ma, X.; Chen, B.; Arsenault, R.; Karlson, P.; Simon, N.; Wang, Y. Recycling end-of-life electric vehicle lithium-ion batteries. *Joule* **2019**, *3*, 2622–2646. [CrossRef]
28. Xiao, J.; Li, J.; Xu, Z. Challenges to future development of spent lithium ion batteries recovery from environmental and technological perspectives. *Environ. Sci. Technol.* **2020**, *54*, 9–25. [CrossRef]
29. Wang, H.; Wu, J.-J.; Zhu, X.; Liao, Q.; Zhao, L. Energy–environment–economy evaluations of commercial scale systems for blast furnace slag treatment: Dry slag granulation vs. water quenching. *Appl. Energy* **2016**, *171*, 314–324. [CrossRef]
30. Gaines, L.; Sullivan, J.; Burnham, A.; Belharouak, I. Paper No. 11-3891. Life-cycle analysis for lithium-ion battery production and recycling. In Proceedings of the Transportation Research Board 90th Annual Meeting, Washington, DC, USA, 23–27 January 2011.
31. Dewulf, J.; van der Vorst, G.; Denturck, K.; van Langenhove, H.; Ghyoot, W.; Tytgat, J.; Vandeputte, K. Recycling rechargeable lithium ion batteries: Critical analysis of natural resource savings. *Resour. Conserv. Recycl.* **2010**, *54*, 229–234. [CrossRef]
32. Dunn, J.B.; Gaines, L.; Sullivan, J.; Wang, M.Q. Impact of recycling on cradle-to-gate energy consumption and greenhouse gas emissions of automotive lithium-ion batteries. *Environ. Sci. Technol.* **2012**, *46*, 12704–12710. [CrossRef]
33. Dunn, J.B.; Gaines, L.; Kelly, J.C.; James, C.; Gallagher, K.G. The significance of Li-ion batteries in electric vehicle life-cycle energy and emissions and recycling's role in its reduction. *Energy Environ. Sci.* **2015**, *8*, 158–168. [CrossRef]
34. Ciez, R.E.; Whitacre, J.F. Examining different recycling processes for lithium-ion batteries. *Nat. Sustain.* **2019**, *2*, 148–156. [CrossRef]
35. Xiong, S.; Ji, J.; Ma, X. Environmental and economic evaluation of remanufacturing lithium-ion batteries from electric vehicles. *Waste Manag.* **2020**, *102*, 579–586. [CrossRef]
36. Dunn, J.B.; Gaines, L.; Barnes, M.; Wang, M.; Sullivan, J. *Material and Energy Flows in the Materials Production, Assembly, and End-of-Life Stages of the Automotive Lithium-Ion Battery Life Cycle*; Argonne National Lab. (ANL): Argonne, IL, USA, 2012.
37. Cheret, D.; Santen, S. Battery Recycling. U.S. Patent No. 71692062007, 18 April 2005.
38. Hu, J.; Zhang, J.; Li, H.; Chen, Y.; Wang, C. A promising approach for the recovery of high value-added metals from spent lithium-ion batteries. *J. Power Sour.* **2017**, *351*, 192–199. [CrossRef]
39. Zhang, J.; Hu, J.; Zhang, W.; Chen, Y.; Wang, C. Efficient and economical recovery of lithium, cobalt, nickel, manganese from cathode scrap of spent lithium-ion batteries. *J. Clean. Prod.* **2018**, *204*, 437–446. [CrossRef]
40. Haibo, Y.; Hui, L.; Jicheng, H.; Ping, L. Research on structure and properties of $LiCoO_2$ prepared from spent lithium ion batteries. *Rare Met. Mater. Eng.* **2006**, *35*, 836–840.
41. Granata, G.; Moscardini, E.; Pagnanelli, F.; Trabucco, F.; Toro, L. Product recovery from Li-ion battery wastes coming from an industrial pre-treatment plant: Lab scale tests and process simulations. *J. Power Sour.* **2012**, *206*, 393–401. [CrossRef]
42. Wang, B.; Lin, X.-Y.; Tang, Y.; Wang, Q.; Leung, M.K.; Lu, X.-Y. Recycling $LiCoO_2$ with methanesulfonic acid for regeneration of lithium-ion battery electrode materials. *J. Power Sour.* **2019**, *436*, 226828. [CrossRef]
43. Heelan, J.; Gratz, E.; Zheng, Z.; Wang, Q.; Chen, M.; Apelian, D.; Wang, Y. Current and prospective Li-ion battery recycling and recovery processes. *JOM* **2016**, *68*, 2632–2638. [CrossRef]
44. Sun, C.; Xia, A.; Liao, Q.; Fu, Q.; Huang, Y.; Zhu, X. Life-cycle assessment of biohythane production via two-stage anaerobic fermentation from microalgae and food waste. *Renew. Sustain. Energy Rev.* **2019**, *112*, 395–410. [CrossRef]
45. Ning, D.; Gao, F.; Wang, Z.; Gong, X. Comparative analysis of primary aluminum and recycled aluminum on energy consumption and greenhouse gas emission. *Chin. J. Nonferrous Met.* **2012**, *22*, 2908–2915.
46. Gaines, L. The future of automotive lithium-ion battery recycling: Charting a sustainable course. *Sustain. Mater. Technol.* **2014**, *1*, 2–7. [CrossRef]
47. Diaz, F.; Wang, Y.; Weyhe, R.; Friedrich, B. Gas generation measurement and evaluation during mechanical processing and thermal treatment of spent Li-ion batteries. *Waste Manag.* **2019**, *84*, 102–111. [CrossRef] [PubMed]
48. Yang, T.; Lu, Y.; Li, L.; Ge, D.; Yang, H.; Leng, W.; Zhou, H.; Han, X.; Schmidt, N.; Ellis, M.; et al. An effective relithiation process for recycling lithium-ion battery cathode materials. *Adv. Sustain. Syst.* **2020**, *4*. [CrossRef]
49. Li, L.; Ge, J.; Wu, F.; Chen, R.; Chen, S.; Wu, B. Recovery of cobalt and lithium from spent lithium ion batteries using organic citric acid as leachant. *J. Hazard. Mater.* **2010**, *176*, 288–293. [CrossRef] [PubMed]
50. Li, L.; Dunn, J.B.; Zhang, X.X.; Gaines, L.; Chen, R.J.; Wu, F.; Amine, K. Recovery of metals from spent lithium-ion batteries with organic acids as leaching reagents and environmental assessment. *J. Power Sour.* **2013**, *233*, 180–189. [CrossRef]

51. Zhang, T.; He, Y.; Wang, F.; Ge, L.; Zhu, X.; Li, H. Chemical and process mineralogical characterizations of spent lithium-ion batteries: An approach by multianalytical techniques. *Waste Manag.* **2014**, *34*, 1051–1058. [CrossRef]
52. Zhang, T.; He, Y.; Wang, F.; Li, H.; Duan, C.; Wu, C. Surface analysis of cobalt-enriched crushed products of spent lithium-ion batteries by X-ray photoelectron spectroscopy. *Sep. Purif. Technol.* **2014**, *138*, 21–27. [CrossRef]
53. He, Y.; Zhang, T.; Wang, F.; Zhang, G.; Zhang, W.; Wang, J. Recovery of $LiCoO_2$ and graphite from spent lithium-ion batteries by Fenton reagent-assisted flotation. *J. Clean. Prod.* **2017**, *143*, 319–325. [CrossRef]
54. Chen, Y.; Liu, N.; Jie, Y.; Hu, F.; Li, Y.; Wilson, B.P.; Xi, Y.; Lai, Y.; Yang, S. Toxicity identification and evolution mechanism of thermolysis-driven gas emissions from cathodes of spent lithium-ion batteries. *ACS Sustain. Chem. Eng.* **2019**, *7*, 18228–18235. [CrossRef]
55. Jung, J.C.-Y.; Sui, P.-C.; Zhang, J. A review of recycling spent lithium-ion battery cathode materials using hydrometallurgical treatments. *J. Energy Storage* **2021**, *35*, 102217. [CrossRef]
56. Diekmann, J.; Rothermel, S.; Nowak, S.; Kwade, A. The LithoRec process. In *Recycling of Lithium-Ion Batteries*; Kwade, A., Diekmann, J., Eds.; Sustainable Production, Life Cycle Engineering and Management; Springer: Cham, Switzerland, 2018; pp. 33–38.

energies

Review

A Review of Multi-Objective Optimization in Organic Rankine Cycle (ORC) System Design

Shuozhuo Hu, Zhen Yang, Jian Li and Yuanyuan Duan *

Key Laboratory for Thermal Science and Power Engineering of MOE, Key Laboratory for CO_2, Utilization and Reduction Technology, Tsinghua University, Beijing 100084, China; hsz17@mails.tsinghua.edu.cn (S.H.); zhenyang@tsinghua.edu.cn (Z.Y.); lijian20@mail.tsinghua.edu.cn (J.L.)
* Correspondence: yyduan@tsinghua.edu.cn

Abstract: Organic Rankine cycle (ORC) is considered a promising heat-to-power technology to utilize waste heat and renewable energy, including solar, biomass and geothermal. However, since the thermodynamic, economic and environmental performance is usually conflict, the single objective design could no longer meet the requirements of the ORC system, putting forward urgent requirements for multi-objective optimization, which has attracted increasing attention with lots of papers published. However, due to these different decision variables, optimization objectives and approaches, existing research is significantly different from each other and is difficult to compare without a systematic summary. Therefore, this paper provides an overview of ORC multi-objective research from three perspectives: optimization objective, method and optimization parameters. Based on the classification of different objectives, this work summarizes the involved variables and provides a recommendation for selecting appropriate objectives in different scenarios. For the optimization method, this work compares different approaches and reveals their advantages and disadvantages. Finally, the decision variables are reviewed and classified into four levels. Then the integrated design approach considering "system-process-component-fluid" is proposed and recommended for further development.

Keywords: multi-objective optimization; organic Rankine cycle; system design; waste heat recovery; carbon emission; intelligent algorithm; superstructure; fluid design; off-design

Citation: Hu, S.; Yang, Z.; Li, J.; Duan, Y. A Review of Multi-Objective Optimization in Organic Rankine Cycle (ORC) System Design. *Energies* **2021**, *14*, 6492. https://doi.org/10.3390/en14206492

Academic Editors: Wei-Hsin Chen, Aristotle T. Ubando, Chih-Che Chueh, Liwen Jin and Andrea De Pascale

Received: 29 August 2021
Accepted: 4 October 2021
Published: 11 October 2021

1. Introduction

Energy and climate crises are the common challenges facing mankind. Reducing fossil fuel consumption, increasing the proportion of renewable energy and energy efficiency are beneficial to realizing carbon neutrality in the middle of this century, which has become the consensus of the world. Organic Rankine cycle is a widely used power system in utilizing medium-to-low temperature thermal energy, which could effectively use renewable energy, including solar energy, geothermal energy, biomass energy and ocean energy [1]. Furthermore, ORC could also recover waste heat resources, including industrial waste heat and engine waste heat [2]. Therefore, ORC could improve the proportion of renewable energy, increase energy efficiency and reduce carbon emissions, thereby showing huge application potential in a low-carbon energy system in the future [3].

In the past decade, ORC has attracted widespread interest from worldwide scholars with an amount of research conducted. These studies mainly focus on system optimization design, working fluid selection, cycle configuration improvement, component design and control strategies, which have made significant progress [4]. Particularly, as the basis of fluid selection, component design and configuration comparison, system optimization has attracted the most attention [5]. Early research only aims to optimize a single objective, usually the maximum thermal efficiency or net output power. As the research develops, the economic and environmental performance of ORC systems such as the total cost, payback period (PBP), levelized cost of energy (LCOE), net present value (NPV), total carbon emission and sustainability indicator (SI) have gradually attracted increasing attention.

Results show that different optimization objectives are often contradictory and cannot be optimal simultaneously [6]. Therefore, the multi-objective optimization of ORC systems was proposed and developed. Up to date, the multi-objective optimization method has become increasingly popular in ORC system design.

The multi-objective optimization (MOO) means that the number of optimized objectives is more than two, as shown in Equation (1). The N is the number of target variables, and M is the number of optimized variables, namely the decision variable. Here, $g(x)$ is the inequality constraint, and $h(x)$ is the equality constraint. Essentially different from the single-objective optimization, MOO will not result in a single solution, but a series of trade-off solutions, which is called the Pareto frontier, meaning that no objective could be improved without sacrificing any other objectives [7,8]. From the mathematic perspective, all trade-off solutions are equally important in theory. However, a single solution should be selected to guide the practical engineering according to the preference information, which is usually characterized as the weight.

$$\begin{array}{c} \min y = f(x) = [f_1(x), f_2(x), f_3(x) \ldots, f_n(x)],\ n = 1,2,3,\ldots,N \\ g(x) \leq 0,\ i = 1,2,3,\ldots,I \\ h(x) = 0,\ j = 1,2,3,\ldots,J \\ x = [x_1, x_2, x_3, \ldots, x_m],\ m = 1,2,3,\ldots,M \end{array} \tag{1}$$

Recently, there have been many researches discussing the MOO in ORC system design. For instance, Wang et al. focused on maximizing the thermal efficiency and minimizing the heat exchanger area per power output of the subcritical ORC using the simulated annealing algorithm [9]. Yang et al. optimized the evaporator/condenser pressure and superheat degree to minimize the total investment cost and maximize the W_{net}, using the multi-objective Genetic algorithm [10]. Furthermore, the MOO method has also been used in working fluid selection [11]. Feng et al. carried out a thermo-economic comparison between pure and mixture fluids in ORC, in which the exergy efficiency and Levelized energy cost (LEC) are selected as the objective variables [12,13]. Xi et al. proposed a graphical criterion, namely the Pareto frontier, to distinguish the optimal fluid for waste heat recovery, in which the exergy efficiency and annual cash flow were considered [14]. Ghasemian et al. compared eight fluids from the perspective of thermal, exergy and cost. Results indicated that dry fluids have a better performance of costs and exergy efficiency than wet fluids [15]. Moreover, the MOO method could be used to compare different ORC architectures. Lecompte et al. explored the specific investment cost (SIC) and total power output (W_{net}) of subcritical and transcritical ORC using the Non-dominated Sorting Genetic Algorithm II (NSGA-II). Results indicated that subcritical ORC shows lower W_{net} but with better economics [6,16]. Sadeghi et al. compared the simple ORC, parallel two-stage ORC and series two-stage ORC from the viewpoints of turbine size and $W_{net.}$ Results indicated that STORC outperforms other configurations under the same conditions [17]. Song et al. compared the subcritical, transcritical, superheated and recuperated ORC considering the exergy efficiency and payback period Results indicated that the superheating is more suitable for the fluids with lower critical parameters, while the recuperation is not attractive under most operating conditions [18].

In general, MOO has been widely applied in ORC, from system design to cycle configuration and to fluid selection. The involved objectives are also numerous, including the conventional thermodynamic indicators such as exergy efficiency, power output and also the economic indicators such as the LCOE and SIC. Moreover, there are also multiple optimization methods such as the NSGA-II and weighted sum method (WSM). Due to these different decision variables, optimization objectives and approaches, existing researches are significantly different with each other and are difficult to compare or evaluate. Therefore, this work is carried out with two main purposes. The first is to review the optimization objectives, methods and variables involved in the existing research and summarize the main findings. The second is to refine further the important issues that deserve additional attention to provide valuable references for future work.

Authors searched on the Web of Science using the keywords of "multi-objective" and "organic Rankine cycle (Including other expressions: multi objective, multi criteria, bi-objective, tri-objective, Pareto, ORC.)". Then, based on the searched results, the literature is further judged whether its focus is on ORC, removing the combined systems such as CCHP or CHP. Finally, a total of 122 papers are obtained and analyzed in this work, which are directly related to multi-objective optimization of ORC system design. As Figure 1 shows, most papers are published after 2015, revealing the popular trend of MOO in system design. There are mainly six chapters in this work. The second section discusses the optimization objectives from thermodynamic, economic, environmental and safe indicators. The third section summarizes the popular optimization methods and further compares their advantages/disadvantages. The fourth section classifies main decision variables from the system, process, component and fluid level. This work has two main contributions: (1) Summarize the main progress made in the existing multi-objective research and provide specific suggestions for selecting optimization objectives and methods in ORC optimization. (2) Propose a four-level design approach considering "system-process-component-fluid" aspects, which could design ORC architecture, component structure and working fluid molecules simultaneously. This approach is expected to improve the overall performance further and is worthy of future discussion.

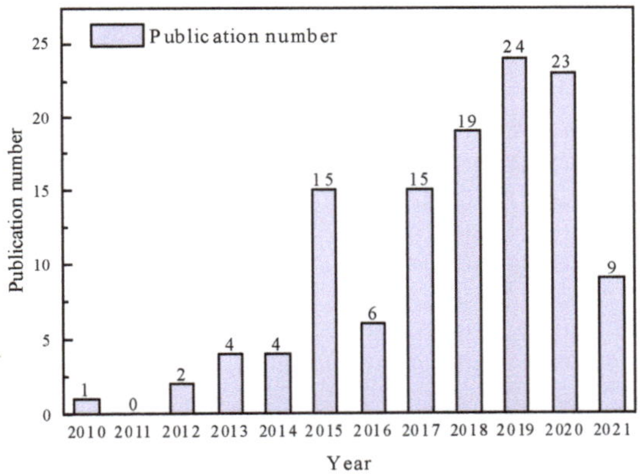

Figure 1. Timeline of publications from 2010 to 2021.

2. Optimization Objective

Optimization objective is also called the target variable in MOO, referring to the variables to be maximized or minimized. The earliest and most studied evaluation indicator is the thermodynamic criterion, including the power output, thermal efficiency and exergy efficiency. Gradually, the economics attract increasing attention, including the total costs, area power ratio (APR), electricity production cost (EPC) and the specific investment cost (SIC) [19,20]. As the climate crisis deepens, the environmental performance has also been quantified and evaluated in ORC, including carbon emission and sustainability indicators. Moreover, due to the diversification of ORC application scenarios, some special needs should be considered. For instance, ORCs deployed on mobile devices to recover exhaust heat should simultaneously consider the space limitations, equipment weight and safety. In this section, different indicators in ORC will be sorted out and analyzed.

According to the optimization objective, all 122 papers are counted and classified. A list of all reviewed papers could be found in the Supplementary Material, and this section mainly summarizes the key findings. Figure 2a shows the paper number involving different indicators. Results indicate that the thermodynamic criterion is the most popular indicator, especially the exergy efficiency, accounting for almost half of all papers. The economic criterion is followed, including the LCOE, SIC and total cost. In contrast, there are very few studies on environmental performance, volume, weight and safety. Only nine papers calculate the total carbon emission with four papers discussing the sustainability index. Figure 2b shows the number of objective indicators. The majority of studies (75%) only focus on bi-objective optimization. 16% of studies focus on tri-objective optimization. In contrast, only one study discusses the optimization with over five objectives, due partly to the immaturity of high-dimensional optimization methods, which will be discussed in Section 3.

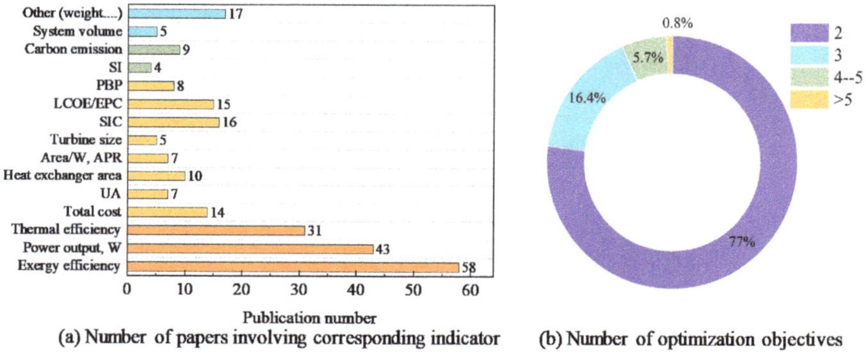

Figure 2. Classification and statistical results of optimization objectives.

2.1. Thermodynamic

As the most basic index in ORC, the thermodynamic criterion indicates the system's ability to utilize heat sources, which is usually quantified by the power output, thermal efficiency and exergy efficiency. The basic concepts and equations are listed in Appendix A and will not be discussed in the main part.

Usually, the designer only chooses one of the power output and thermal efficiency as the optimization objective, according to the specific application scenario. In terms of thermal efficiency, it is more popular in solar or biomass ORC since the thermal oil cycle is usually included. This oil cycle belongs to the closed system defined by Zhai et al. [21], which is more suitable for efficiency evaluation. In contrast, the power output is more suitable for an open system, including the industrial waste heat, engine waste heat, geothermal and other scenarios without oil cycle, as shown in Figure 3. In these open systems, the outlet heat is directly discharged without being recycled.

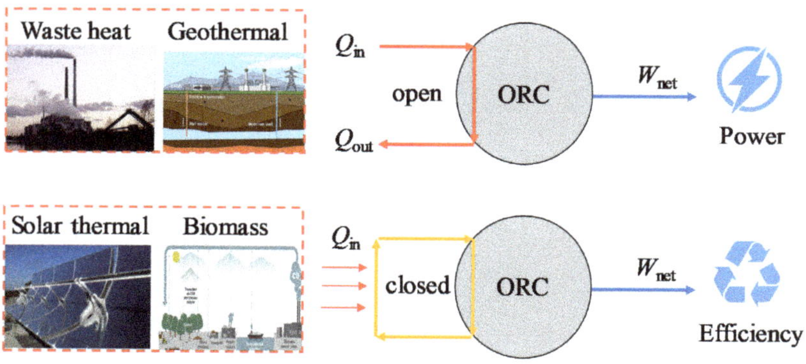

Figure 3. Selection of thermodynamic criteria for different heat sources cases.

The traditional exergy loss analysis for each component helps facilitate the optimal design of the component and ORC system [22]. Furthermore, the advanced exergy analysis methods have become increasingly popular in recent years, which decomposes the exergy loss into endogenous and exogenous parts, thereby determining the possibility of reducing the loss through technological improvement [23]. Therefore, if the endogenous loss of all components is summed up, the efficiency improvement limit of the ORC system could be obtained. For instance, Wang et al. used advanced exergy analysis to estimate the improvement potential of dual-loop ORC. Results indicated that the high-temperature turbine is the first component that needs technical modifications [23].

2.2. Economic

In addition to the thermodynamic performance, the economic criterion is also an important indicator to evaluate ORC performance, which has become a necessary option for ORC optimization. Up to date, the popular indicators include the indirect indicator, and direct indicators (total cost, SIC, PBP, etc). Particularly, exergoeconomic analysis is different from traditional economic analysis and deserves additional discussion.

2.2.1. Indirect Indicator

Indirect indicators do not directly measure the cost or economics of the whole ORC sysem, but use the parameters such as heat exchanger area to characterize the component cost qualitatively [24]. The commonly used index includes heat exchanger internal heat transfer requirement (UA) [25,26], heat exchanger area, turbine size, area to power ratio (APR) [27] and other indexes.

Specifically, the UA is simple to use since only the heat capacity and temperature difference are needed without complicated iterative calculation for U. Feng et al. used UA as an economic evaluation index to evaluate the cost of a supercritical-subcritical ORC in waste heat utilization [28]. Tiwari et al. optimized the exergy efficiency and the total UA of the evaporator and condenser. For a known working fluid, the total UA will exhibit the required area [29]. Further, the heat transfer coefficient U could be calculated to determine the heat exchanger area based on various heat exchange correlations [30]. Gotelip et al. designed the heat exchanger to minimize the total heat exchanger area and maximize the power of ORC in floating product storage offload [31]. The turbine diameter is also selected as the optimization objective to obtain a smaller turbine size [32]. Bahadormanesh et al. defined an objective function by dividing the turbine inlet radius by power output [33]. Rahbara et al. minimized the overall turbine size ($d_{\max}$) for a highly compact ORC layout applied in diesel truck engines [34]. However, the smaller area or diameter does not necessarily represent lower costs since the impacts of material, pressure level, component structure are not considered [35].

These indirect indicators could only characterize the cost of a certain component. However, they could not represent the overall economy of the overall ORC system. For different system sizes, the results are difficult to compare, either. Therefore, more direct indicators are required to represent the costs of the whole ORC system.

2.2.2. Direct Indicator

Direct indicators are much more frequently used in existing research since they are strictly defined in engineering or economics, including the Total cost, Specific investment cost (SIC), Payback period (PBP), Levelized cost of electricity (LCOE), Net present value (NPV) and Internal Rate of Return (IRR), as shown in Table 1. Total cost is the most basic index to evaluate ORC economics [36], which are mainly calculated using empirical correlations, fitted from the equipment cost on the market [37]. The two most popular correlations are from Turton [38] and Smith [39]. In Sun's work, the carbon emission costs and water consumption costs were also considered [40]. Based on total cost, SIC describes the unit cost per power output [41] and has the advantage of easy use and intuitive comparison between different cases. Sun et al. maximized the exergy efficiency and minimized the specific power cost, revealing the better performance of double pressure ORC over traditional ORC [42]. However, SIC does not consider the depreciation, operation costs or discount rate [43]. PBP measures the number of years required to recover the total cost, including the static and dynamic ones. Generally, the dynamic PBP is more usual since it considers the time value over multiple years. For instance, Wang et al. used dynamic PBP to evaluate the superiority of the ORC investment plan in engine waste heat recovery [44]. LCOE denotes the cost of unit electricity and could be directly compared with the local electricity price to represent the profitability. In addition, the comparison between different energy generating technologies is also convenient, as suggested by Lecompte [6]. It's worth noting that many existing researches also use other abbreviations such as EPC or LEC to represent the cost per power generation, especially in the research of industrial/engine waste heat [45–47]. In this review, these abbreviations are unified as LCOE.

Table 1. Comparison of different economic indicators.

Indicator	Definition	Advantage	Disadvantage
C_{tot}	Total investment cost	Easy to use;	Not consider time value or technical criterion
SIC	Specific investment cost	Easy to use; simultaneously consider cost and power;	Not consider the time value
PBP	Payback period	Consider time value	Not consider system lifetime
LCOE	Levelized cost of energy	Compared with electricity price to determine the feasibility	Need to presume multiple parameters; significantly affect the results.
NPV	Net present value	Intuitively shows the profitability	Difficult to presume return rate
IRR	Internal rate of return	Not affected by external parameters; dependent on the cash flow	Calculation is complex

The net present value (NPV) intuitively shows the profitability during the lifetime operation. Hu et al. used NPV to evaluate the total profits of a hybrid geothermal-solar power system during the 30-year operation [48]. Pirerobon et al. used NPV to evaluate the profitability of three waste heat recovery units for offshore platforms [49]. However, the expected discount rate should be defined in advance, which is related to the company's lowest investment rate of return and is usually difficult to determine. In contrast, IRR has an outstanding advantage in that it depends entirely on investment cash flow and is not affected by external parameters, including the discount rate [50]. However, IRR only represents a ratio rather than an absolute value. When the project scale is large, a low IRR may also result in large profits. Therefore, it is suggested that IRR and NPV should be simultaneously considered when comparing different schemes. In general, authors recommend the direct indicators in ORC design, especially the LCOE, NPV and IRR, which describe the lifetime economy of the entire system, rather than individual component or static periods.

2.2.3. Exergoeconomic Analysis

Exergoeconomic analysis combines advanced exergy analysis and economic analysis, aiming to explore the relative importance of each component in the design of the ORC system [51]. Different from the normal economic analysis, exergoeconomic analysis also calculates the exergy cost rate when considering the equipment cost and operation cost [52,53]. Özahi et al. calculated the capital cost rate of components using the SPECO method to determine the optimal fluid from toluene, MDM, D4 and n-decane [54]. Behzadi et al. identified the relative cost importance of each component and designed a cost-effective ORC system for waste recovery. Results indicated that R123 is the best working fluid from exergy efficiency and total production cost [55].

2.3. *Environmental*

The research on environmental performance started later than thermodynamic and economic performance. The most studied environmental performance is the carbon emission. Moreover, there are also some studies on sustainability indicators (SI) [56].

2.3.1. Carbon Emission

Carbon emission is the most intuitive and commonly used environmental indicator. The emissions of ORC system involve all processes through the whole life cycle, including production, transportation, operation and recycling. Specifically, the emission sources include: (1) the process of manufacturing the components, (2) leakage of organic fluids in operating process, (3) replacement of coal/natural gas using ORC. Some researches also discuss the impact of recycling water on carbon emission [57]. Moreover, the methane also results in the greenhouse effect, which could be converted into carbon emissions using coefficients. Common analysis method includes the Total equivalent warming impact (TEWI) method, Life cycle climate performance (LCCP), life cycle analysis (LCA) and carbon emission coefficient method, which are described in detail:

(1) Total equivalent warming impact (TEWI)

TEWI method is a measure of the greenhouse gas emissions, with the advantages of simple calculation and easy application. TEWI is mainly composed of two parts [58]: (1) direct emission: emissions caused by fluid leakage and end-of-life recovery. (2) Indirect emission: energy saving due to ORC power. Yang et al. introduced an extended TEWI method to evaluate the impact of refrigerants on climate change, considering that the generated power of ORC could compensate for the warming relief [58].

(2) Life cycle climate performance (LCCP)

LCCP is developed based on the TEWI method, which considers chemical manufacturing and end-of-life disposal compared with TEWI, as shown in Zhang's work [36]. The diagram of LCCP is shown in Figure 4.

(3) Carbon coefficient

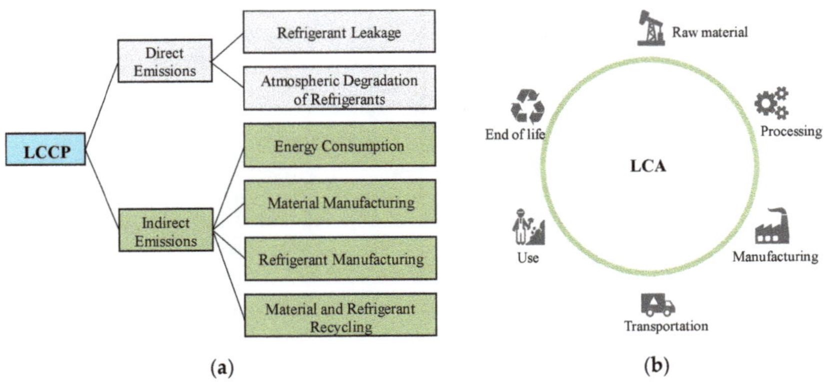

Figure 4. Diagram of LCCP (**a**) and LCA (**b**) method.

The carbon coefficient method is a general method for direct multiplication coefficient to calculate carbon emissions. For instance, power generation could multiply a coefficient to estimate carbon emissions [23]. Xia et al. calculated the warming value considering the conversion factor of CO_2, CO and CH_4 based on coal-fired power plants [59].

(4) Life cycle analysis (LCA)

LCA is a standardized method based on ISO 14,044 [60], which could evaluate the environmental effects in each stage during the life cycle, including the construction, operation, maintenance and disposal [61]. Herrera et al. assessed the exergo-environmental performance of the waste heat recovery systems with simple and regenerative ORC. Results indicated that toluene presented the best efficiency and lower climate effects with 0.00181 kg CO_2 eq·kWh^{-1} [62].

Table 2 summarizes related researches on carbon emission. Results show that existing studies usually calculate emission from two or three categories. However, no research has considered all factors simultaneously, which may underestimate the ORC's environmental performance.

Table 2. Typical references involving carbon dioxide emission.

Reference	**Method**	**CO_2 Source**				
		CH_4 Equivalence	**Component Construction**	**Working FLUID Leakage**	**Power Reduction**	**Other**
Wang et al. [23]	Carbon coefficient	√	√		√	
Herrera et al. [62]	LCA		√	√	√	
Xia et al. [59]	Carbon coefficient	√	√		√	
Zhang et al. [36]	LCCP		√	√	√	
Yi et al. [63]	LCA	√	√	√		
Martinez et al. [57]	TEWI		√	√	√	Water consumption
Kalikatzarakis et al. [64]	Carbon coefficient	√			√	

2.3.2. Exergoenvironmenal Analysis

Exergoenvironmenal analysis has a similar principle to advanced exergy analysis by linking the environmental impact with the energy flow and internal inefficiency, which is different from the direct calculation of carbon emission [53]. Fergani et al. compared three fluids in cement industry ORC by taking the exergy efficiency, cost and environmental impact per exergy unit of the net produced power as optimization objectives. Results indicated that benzene is the best from an exergo-environmental point of view [65].

2.3.3. Sustainability Index (SI)

The definition of SI is the ratio of exergy loss rate to the exergy decline of heat source (water, exhaust gas), which is used to evaluate the impact of ORC on the environment [66]. The key idea is to improve the exergy efficiency, reduce the emitted heat and alleviate the damage to the environment [67]. A lower SI means a smaller impact [68].

2.4. Other

In addition to the thermodynamic, economic and environmental indicators, there are also other evaluation indicators, including system volume, weight and safety.

2.4.1. Volume

ORC could be used on mobile vehicles and marine scenarios. Due to the limited space, ORC volume is a very important factor in these scenarios. Many researchers focus on the volume of the heat exchanger since it is the largest component in the ORC system. For instance, Xu et al. [69], Baldasso et al. [70], Liu et al. [71], Barbazza et al. [72] optimized the system parameters and structure by taking the heat exchanger volume as the optimization objective. Moreover, some researchers also consider the volume of turbines, feed pumps and auxiliary equipment [73]. Based on each component's structure, related empirical correlations could also be used to calculate the volume [74].

2.4.2. Weight

Similar to the volume, the weight of the ORC system is also an important factor in mobile vehicles. Imran et al. [73] and Pierobon et al. [49] used correlation equations to calculate the weights of turbines, heat exchangers, feed pumps, pipes, instrumentation/control systems and other auxiliary components. Then an additional mass of 10% is supplemented to represent the mass of storage tanks. Finally, the total weight is selected as one of the optimization objectives. More detailed equations could refer to these references.

2.4.3. Safety

Safety is a very important index in all application scenarios but is seldom discussed. Safety mainly refers to the risk of workers when exposed to the ORC environment, which is usually measured by the individual risk [57]. These risks are mainly caused by the leakage of flammable and toxic working fluids and high-pressure pipes. There are special methods for quantitative safety assessment, such as the Quantitative Risk Analysis (QRA).

Lee et al. simplified the QRA method and focused on the riskiest part, namely the pipeline between the feed pump and evaporation, since the pressure is the highest in this process [75]. The authors focused on source modeling, dispersion modeling and fire event modeling, and used the concept of risk distance to characterize the ORC safety. In addition, some scholars directly optimize the ORC system from the perspective of fluid design to minimize the flammability, thereby improving ORC safety. Papadopoulos et al. researched the molecular design of mixed fluids by minimizing the flammability of two pure fluids using the group-contribution method [76].

2.4.4. Stability

Most of the aforementioned studies only discuss a single working condition. During the practical operation, the ORC system always deviates from the design condition due to the influence of solar radiation, wind speed and ambient temperature. Therefore, some scholars take the power fluctuation as the optimization objective, and minimize the fluctuation range as small as possible by parameter optimization.

Li et al. focused on solar-based ORC and selected the fluctuation of output ($W \cdot min^{-1}$) as the optimization objective [77]. Results indicated that a larger energy storage capacity could reduce power fluctuation, but will significantly increase the costs. Bufi et al. focused on maximizing the thermal efficiency and minimizing its variance [78]. Zhang et al. proposed a multi-objective estimation of distribution algorithm to keep superheat following a target value by controlling the pump speed [79].

3. Optimization Method

Multi-objective optimization method is essentially different from single-objective optimization. A single optimal solution could be obtained in single-objective optimization. However, different indicators compete with each other, and there is no unique optimal solution in multi-objective optimization (MOO), which is also more complex and time-consuming to converge. MOO is usually divided into the Priori method and No preference method. Further, the Priori method could be divided into the Apriori method, interactive method and Aposteriori method, according to whether the preference information is determined before, during or after the optimization process, as shown in Figure 5. At present, the Apriori method and evolutionary algorithm method are widely used in ORC, including the linear weighted sum method (WSM), ε-constraint method and smart algorithms such as NSGA-II, MOPSO and etc.

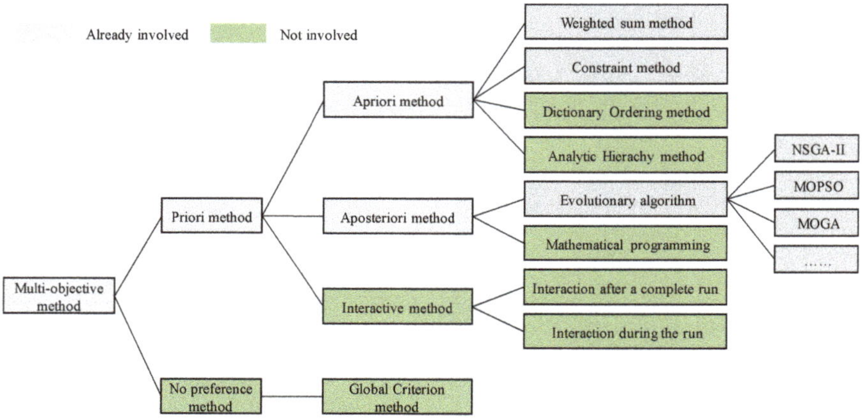

Figure 5. Multi-objective optimization methods.

This work has summarized the application of these methods in the ORC MOO application, as shown in Figure 6 [7,80]. Results show that, from the perspective of optimization methods, many interesting methods have not been applied in ORC, including the interactive methods that could feedback the decision makers' preferences during the design process. Applying these methods may make the system design more in line with the needs of designers and engineering projects, thus worth future exploration.

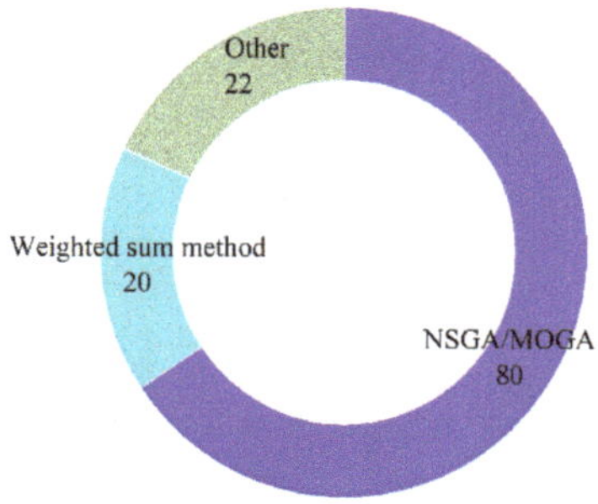

Figure 6. Statistical results of optimization methods.

In particular, MOGA and NSGA-II are both developed from the single-objective Genetic algorithm and are not distinguished in many previous researches. Therefore, this review uses NSGA-II to represent these two methods. Results show that NSGA-II is the most popular algorithm, accounting for about 66% of all existing studies. The second popular method is WSM, which accounts for 16%. Other methods such as MOPSO and ε-constraint method only account for 18%. Therefore, this work will take WSM, ε-constraint and intelligent algorithm as examples to introduce the principle and application in detail, and compare the advantages and disadvantages of each method.

3.1. Weighted Sum Method (WSM)

3.1.1. Principle

WSM combines multiple target variables into a single one according to a certain weight ratio, thereby transforming the multi-objective optimization problem into a simpler single-objective optimization problem [81], as shown in Equation (2).

$$f(x) = \omega_1 f_1(x) + \omega_2 f_2(x) + \ldots + \omega_N f_N(x) \quad (2)$$

where w represents the weight factor, ranging from 0 to 1. The sum of all factors is 1.

WSM has a simple principle and is easy to use. There is no theoretical upper limit to the number of optimization objectives. Thus 2, 3, 5 or even more than 10 objectives could be combined into one [82]. For instance, Arasteh et al. combined the thermal efficiency and exergy efficiency into one objective function with each factor's contribution of 0.5. Then the Genetic Algorithm is used to solve this optimization problem [83]. Zhu et al. combined the exergy efficiency and the heat exchanger area per power output into one function. Then the optimization is conducted to determine the optimal evaporation temperature, condensation temperature and working fluid [84]. In addition to the Genetic algorithm, the PSO could also be used to solve this single-objective problem [71].

3.1.2. Methods to Determine the Weight

WSM is a priori method with the weight and preference being determined before optimization. Thus a consequent problem is: how to determine the weight factor of each target variable? In many previous studies, the weight factor is directly assumed. For instance, the weight is usually set as 0.5:0.5 [83,85] or 0.6:0.4 [56] when two target variables are used. When four target variables are considered, the weight is usually set as 0.1:0.2:0.3:0.4 [86]. This direct assumption usually only considers typical weight scenarios but is too subjective and does not consider the characteristics of each target variable. Therefore, in some researches, the weight factor is not subjectively determined by the designer, but using other mathematical models to calculate the coefficient according to each variable's characteristics. The typical methods include: the α-method, Analytic Hierarchy Process (AHP), and Grey relational analysis (GRA) [87].

(1). α-method

α-method calculates the weight factor by considering the magnitude and the range of each optimization objective. A variable with a larger range will have a larger weight to ensure that two indicators have similar magnitudes after multiplying the weight [68,88]. This method is usually used for bi-objective optimization, in which the weight factor could be calculated by:

$$\omega_1 = \frac{f_2^{max} - f_2^{min}}{(f_1^{max} - f_1^{min}) + (f_2^{max} - f_2^{min})}, \ \omega_2 = \frac{f_1^{max} - f_1^{min}}{(f_1^{max} - f_1^{min}) + (f_2^{max} - f_2^{min})} \tag{3}$$

α-method is simple and easy to use. However, the maximum and minimum value of two optimization objectives should be determined in advance, which could be calculated through four single-objective optimizations, thereby significantly increasing the calculation complexity. For instance, Kazemi et al. firstly determines the weight coefficients of exergy efficiency and SIC using the α-method, and then explores the effects of system parameters [89].

(2). Analytic Hierarchy Process method (AHP)

The AHP method is a kind of multi-criteria decision-making method, which divides complex problems into orderly levels to make them organized. Then the importance of each element is quantitatively compared and described, which could be used to calculate the weight factor of each variable, as shown in Figure 7 [90].

Zhang et al. discussed four target variables, including power, thermal efficiency, exergy efficiency and carbon emission by dividing these variables into four levels: energetic, exergetic, economic and environmental criteria. Then the author used nine absolute numbers (1–9) to indicate the importance intensity (equal, moderate, strong, extreme importance and etc.) and construct the judgment matrix [36]. It is worth noting that different indicators should be dimensionless.

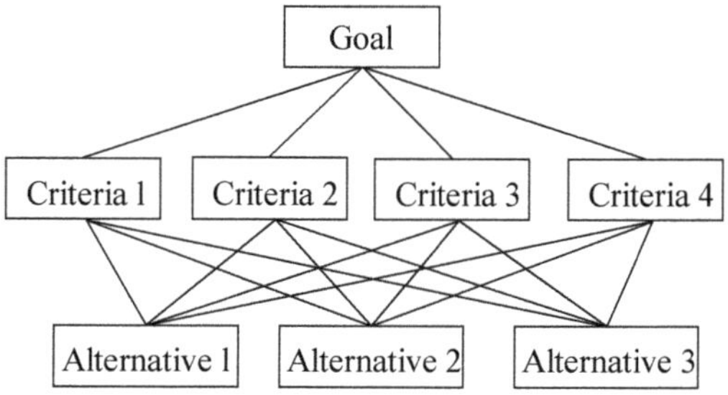

Figure 7. Diagram of AHP method.

(3). Taguchi method

Taguchi method is a statistical method to obtain the importance of different factors to the objective function, which could be used in the experiment, theory and numerical simulation [91]. Bademlioglu used this method to explore the impact of changes in different ORC parameters on system thermal efficiency and exergy efficiency. The author first used this method to obtain an orthogonal array of optimization objectives (thermal and exergy efficiency). Then this array is converted into a signal-to-noise (S/N) ratio in 27 cases. Consequently, the weight factors could be calculated as 52.61% and 47.38% [91].

3.2. ε-Constraint

The core idea of the ε-constraint method is to regard a single indicator as the optimization objective and convert other objectives into constraints, thereby transforming the multi-objective optimization problem into a single-objective optimization problem. Different from the WSM method, ε-constraint could deal with the non-convex problems [63].

$$\begin{gathered} \min f_i(x) \\ f_j(x) \in \varepsilon_j (j = 1, 2, \ldots, M, j \neq i) \end{gathered} \tag{4}$$

Multiple calculations could result in a relatively complete Pareto frontier. However, Figure 8 shows that if the ε is too small, f_1 may exceed the feasible region, resulting in non-convergence. If ε is too large, there will be more feasible solutions, resulting in lower convergence speed. Therefore, some prior knowledge is often needed to determine a reasonable range of ε. For instance, Yi et al. firstly conducted two single-objective optimizations to determine the lower and upper bounds. Then authors implemented the ε-constrained method to solve this multi-objective optimization problem in GAMS 23.6 [63].

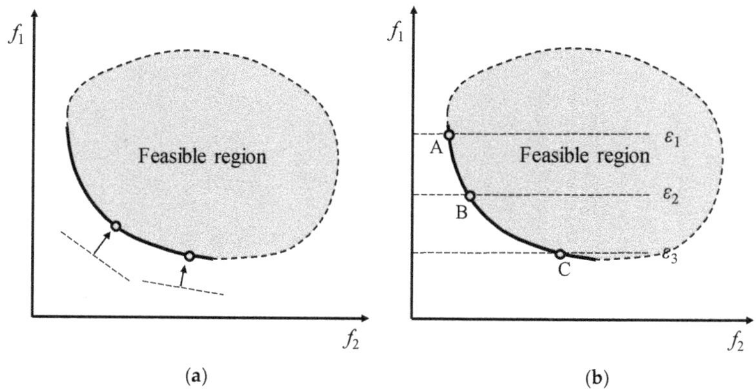

Figure 8. Schematic diagram of the weighted sum method (**a**) and the ε-constraint method (**b**).

3.3. Intelligent Algorithm

Regardless of the WSM or the ε-constraint method, there is either the invalidity of the non-convex problem or the requirement for prior information, resulting in limitations to practical application. As the algorithm develops, some intelligent optimization algorithms with wider applicability have been gradually developed and improved, which have been widely used in different fields.

Popular intelligent algorithms include the NSGA-II [33], MOPSO [92], MOEA [93]. Non-dominated Sorting Genetic Algorithm II (NSGA-II) is an improved algorithm for NSGA based on GA's selection, crossover and mutation ideas, which was proposed by Deb in 2001 [94] It is worth mentioning that the *gamultiobj* function embedded in the Matlab toolbox is also a modified version of NSGA-II. Therefore, this review uses NSGA-II to simultaneously characterize the method of self-programming or calling the Matlab toolbox. The multi-objective particle swarm optimization (MOPSO) algorithm was proposed by Carlos A. Coello in 2004 for multi-objective optimization based on the PSO algorithm [95], which simplifies the crossover and mutation process and shortens the convergence time. The disadvantage of PSO is that it is easy to fall into local optimization, resulting in low convergence accuracy and poor solution diversity. Multiobjective Evolutionary Algorithm Based on Decomposition (MOEA/D) transforms the multi-objective optimization into a single-objective problem with the advantage of lower computational complexity [96]. The disadvantage is that the weight vectors need to be set artificially, which will determine the quality of the final solution [96].

In addition to the intelligent algorithms mentioned above, there are also other algorithms applied in ORC, including the multi-objective heat transfer search (MOHTS) [97], Artificial Cooperative Search (ACS) [98], multi-objective grey wolf optimizer (MOGWO) [99], multi-objective firefly algorithm (MOFA) [33], artificial bee colony algorithm (ABC) [100] and simulated annealing (SA) [101]. Even though these methods are rarely used, it will still be a very interesting topic to compare these different methods. However, for high-dimensional optimization with 4 or more objectives, these intelligent algorithms are currently ineffective since the calculation time will increase significantly and the solution is not accurate, either. Therefore, WSM method is recommended for three or more optimization objectives, as shown in Table 3.

Table 3. Comparison of different multi-objective optimization methods.

Optimization Method	Advantages	Disadvantages	Recommended Scenario	Case
Weighted sum method	·Simple, easy to use ·could include multiple objectives (>10)	·Pareto is not uniform ·cannot tackle the nonconvex problem ·need normalization for objectives	$Ns \geq 4$	[20]
ε-constraint	·could tackle the nonconvex problem	·calculation time varies for different formulations ·Pareto is not uniform ·epsilon is difficult to determine	-	[63]
Intelligent algorithm	·could tackle the nonconvex problem ·Pareto is uniform	·only include several objectives (<4) ·time consuming ·multiple adjustable parameters	$Ns \leq 3$	[44,102]

3.4. Decision Making

The multi-criteria decision-making method (MCDM) develops from scheme sorting in operations research and is also used to select the optimal solution from the Pareto in the MOO of ORC. This decision-making process selects the optimal solution for design guidance according to the decision maker's preference [103]. In ORC optimization, the typical MCDM methods include TOPSIS, LINMAP, Shannon entropy, GRA, fuzzy set theory, etc. The differences between these methods lie in the definition of the optimal solution. For instance, LINMAP only requires the solution closest to the ideal one [104], while TOPSIS requires the solution closest to the ideal solution and the farthest from the non-ideal one at the same time [105,106]. The Shannon entropy method could measure the uncertainty with information sources using the probability theory [107]. The key point is that the objective with a sharp distribution will have lower importance compared with that following the biased distribution [13]. As a part of grey system theory, grey relational analysis defines the black area, white area and grey area [108]. Theoretically, this analysis proposes a dependence to measure the correlation degree of factors, meaning that more similarity leads to more factor correlation [40].

Statistical results in Figure 9 show that nearly half of the multi-objective optimization studies use the MCDM method to determine the optimal solution. TOPSIS is the most popular approach, accounting for over 60%. LINMAP comes next, accounting for about 35%, while Shannon entropy and GRA methods are relatively less used. Since these MCDM methods have various concepts and usually lead to different final solutions, some researchers propose to apply multiple methods simultaneously and then determine the final solution using the aggregation method, which may improve the robustness of the decision-making process [13,90,103]. Detailed descriptions are shown in Table 4.

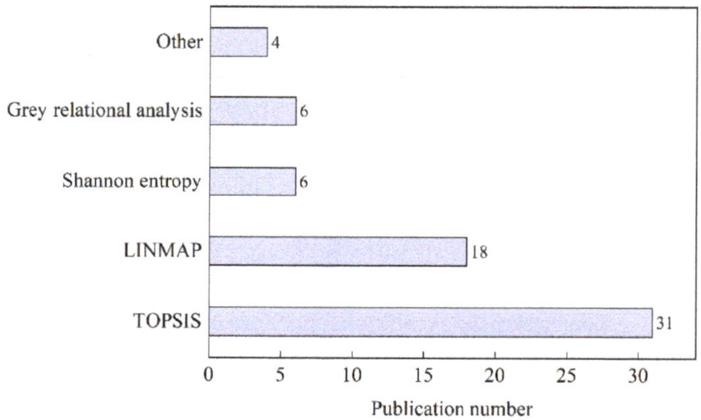

Figure 9. Statistical results of MCDM applied in ORC.

Table 4. Descriptions of different MCDM methods.

Refs.	Method	Principle	Calculation
[109–111]	LINMAP	Closest to the ideal solution	$d_{i+} = S_i^* = \sqrt{\sum_{j=1}^{N} (V_{ij} - V_j^*)^2}$
[112–114]	TOPSIS	Closest to the ideal solution. Furthest to the non-ideal solution.	$d_{i-} = S_i^- = \sqrt{\sum_{j=1}^{N} (V_{ij} - V_j^-)^2} C_i^* = \frac{d_{i-}}{d_{i-}+d_{i+}}$
[13,107]	Shannon entropy	A sharp distribution leads to lower importance	$SE_j = -\frac{1}{\ln(n)} \sum_{i=1}^{n} P_{ij} \ln P_{ij}$
[40,44,59,108,115]	Grey relational analysis	More similarity leads to more factor correlation	$\xi_i(k) = \frac{\min(\Delta_i(\min))+\rho\max(\Delta_i(\max))}{\lvert x_0(k)-x_i(k)\rvert+\rho\max(\Delta_i(\max))}$

4. Optimization Parameter

During the MOO process, many ORC parameters could be optimized. The most popular ones are evaporation pressure, superheat, condensation pressure and other parameters, which all belong to the system level. In addition, there are also some parameters at the component level and working fluid level that could be optimized, which will be discussed in detail in this section.

4.1. System Level

System-level optimization parameters mainly include the evaporating pressure/ temperature, condensing pressure/temperature, subcooling and superheating. Under given conditions of heat source, the efficiency and net power output of ORC could be calculated according to the above system-level parameters. For the transcritical ORC, the superheat degree is not required, but the evaporation pressure and turbine inlet temperature should be determined at the same time [6,52]. For other new architectures such as dual-pressure evaporation ORC and two-stage ORC, the optimized parameters are more but are similar to the simple cycle [116]. System-level parameters are the most basic parameters of ORC, which are involved in almost all ORC optimization researches and will not be discussed in detail.

4.2. Process Level

Process-level design mainly refers to the design of cycle processes and system configurations, such as the conventional subcritical cycles, transcritical cycles, two-stage cycles, multi-pressure evaporation cycles. Most of the existing researches select the configuration by directly comparing the Pareto frontier of different forms through multi-objective optimization. However, this comparison could only study simple and several configurations. When there are multiple possible configurations, the computational complexity will increase sharply. Superstructure optimization could discuss various alternative configurations by analyzing the process stream, thereby parameterizing the ORC process design. Then the intelligent algorithms could be used to quickly solve the problem and obtain the best system structure and process, as shown in Figure 10. Kermani et al. [117] conducted a superstructure modeling for ORC systems driven by industrial waste heat, including regenerative, superheating, turbine-bleeding, reheating, multi-stage and transcritical cycles, etc. The multi-objective optimization is carried out with the net power output and total cost as the objective. Results indicate that the coupling of multiple forms results in better economics and thermodynamic performance.

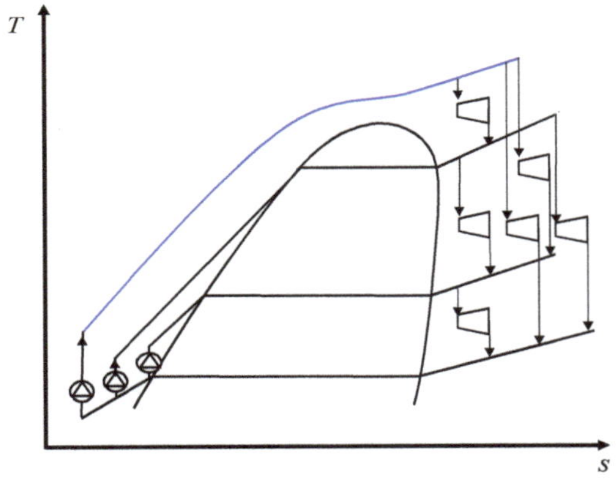

Figure 10. Multi-objective optimization of the ORC superstructure.

4.3. Component Level

In addition to the system-level and process-level parameters, the parameters at the component level also have a very significant impact on ORC performance. Key ORC components include the turbine, evaporator, condenser and feed pump. Up to date, most existing studies focus on the heat exchanger or the expansion device. The heat exchangers mainly include the shell-and-tube heat exchanger, plate heat exchanger and tube-fin heat exchanger. The expansion device mainly includes the radial turbine, axial turbine and scroll expander [118]. It is worth mentioning that no multi-objective optimization on the feed pump is discovered in this literature review. Part of the reason may be that the cost of the feed pump is much lower than that of the turbine or heat exchanger. However, as one of the key components in ORC, the efficiency of the feed pump has a significant impact on system efficiency and operational stability. Therefore, the structural design of the feed pump could be further discussed in the future.

(1). Heat exchanger

The heat exchanger is an important and also the largest component in ORC, including the evaporator, condenser, regenerator, preheater and superheater. The structure design of the heat exchanger is beneficial to reducing the exergy loss, improving system efficiency and increasing the safety and stability of ORC operation. Generally, the heat exchanger includes shell-and-tube, plate, tube-fin and plate-fin type [106,119,120]. The structure and optimization parameters of different heat exchanger types are also different, as shown in Table 5.

Table 5. Comparison of different heat exchangers.

Type	Advantages	Disadvantages	Application Scenario
Shell-and-tube	Mature, lower cost, long life	Large volume, low heat transfer coefficient	For liquid-liquid heat exchange
Plate	High heat transfer coefficient, low heat loss, compact structure, small volume	Poor sealing, only low pressure, higher flow resistance	For liquid-liquid and liquid-gas exchange
Tube-finned	High heat transfer coefficient, compact structure	Complex structure, expensive	For liquid-liquid and liquid-gas exchange
Plate-finned	High heat transfer coefficient, compact structure	Easy to block, poor resistance to corrosion	For liquid-liquid, liquid-gas and gas-gas exchange

In terms of the shell-and-tube heat exchanger, the optimization parameters mainly include the tube diameter, shell diameter, baffles spacing, tube number and tube length. In terms of the plate exchanger, the popular decision variables include the plate number, plate length, plate thickness, plate spacing, channel length, channel width and pass number. More detailed parameters are shown in Table 6.

(2). Expansion device

The expansion device is the key component in ORC, which could be divided into the turbine, scroll expander and screw expander according to the power range. Ref. [118] introduces the categories and applications of different expansion devices in ORC. This work aims to reveal the parameters and turbine types in multi-objective optimization. According to the literature, most existing researches focus on radial turbine but pay little attention to other turbine types. The popular decision variables include the expansion ratio, specific speed, blade angle, blade height, radius ratio and flow coefficient. For an axial turbine, only Schilling et al. [121] have optimized the turbine stages but not discuss other structural parameters such as inlet angle and turbine size. For the positive displacement shown in

Figure 11, both scroll expander and screw expander are more suitable for medium and small-scale ORC systems. However, there is currently no published work on incorporating expander structure parameters into the multi-objective optimization of ORC, which is worthy of further research.

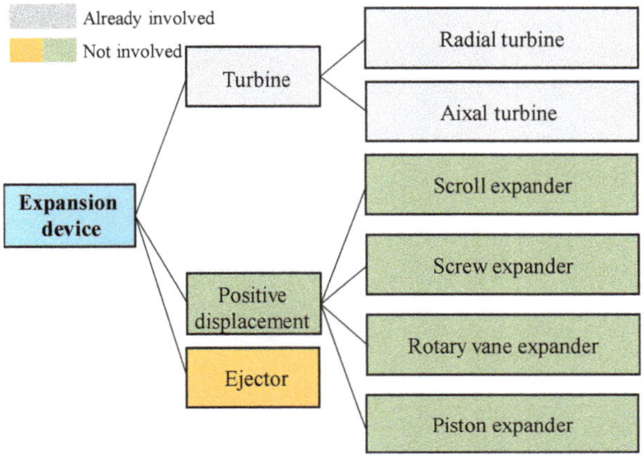

Figure 11. Type of expansion devices in ORC multi-objective optimization.

Table 6. List of the reference focusing on the component level.

Component	Type	Year	Author	Optimized Parameter
Heat exchanger	Shell-and-tube	2021	Turgut et al. [22,98]	Outer tube diameter, shell diameter, baffles spacing, number of tube passes, tube arrangement.
	double-pipe tubular	2019	Van Kleef et al. [122]	Working fluid velocities
	Tube-finned, plate-finned	2019	Holik et al. [123]	Length, width, height.
	Tube-finned	2019	Baldasso et al. [70]	Inner tube diameter, tube length, fin height, fin thickness, fin spacing, transversal pitch
	Shell-and-tube	2019	Baldasso et al. [70]	Inner tube diameter, tube length, baffle spacing
	Tube-finned	2017	Liu et al. [71]	Inlet radius on tube, Inlet radius on shell, fin eight, fin thickness, fin spacing.
	Shell-and-tube	2016	Andtreasen et al. [11]	Inner tube diameter, number of tubes, Baffle spacing
	Plate	2015	Lecompte et al. [6]	Number of passes
	Plate	2015	Kalikatzarakis et al. [64]	Number of plates, plate thickness, plate length, channel length, channel width
	Plate	2015	Imran et al. [124]	Channel length, channel width, plate spacing
	Shell-and-tube	2014	Pierbon et al. [49]	Inner tube diameter, tube length, number of tubes, baffle spacing
	plate-finned	2014	Pierbon et al. [49]	Fin height, fin frequency, fin length, number of plates, flow length
	Plate	2014	Barbazza et al. [72]	Plate width, channel spacing, number of channels, number of passes
	Plate	2013	Wang et al. [125]	Plate length, plate width, channel distance
	Shell-and-tube	2013	Pierbon et al. [74]	Outer diameter, tube pitch, baffle spacing

Table 6. *Cont.*

Component	Type	Year	Author	Optimized Parameter
Turbine	radial	2021	Li et al. [126]	Expansion ratio, specific speed
	radial	2021	Alshammari et al. [127]	Rotor blade angle, exit vane angle, blade thickness, vane thickness
	radial	2020	Jankowski et al. [128]	Specific speed
	radial	2019	Palagi et al. [129]	Specific speed, specific diameter
	radial	2019	Bekiloglu et al. [32]	Specific speed, radius ratio
	axial	2017	Schilling et al. [121]	Turbine stages
	radial	2017	Bahadormanesh et al. [33]	Velocity ratio, rotational speed, inlet flow angle, radius ratio, meridional velocity ratio
	radial	2017	Al Jubori et al. [130]	Flow coefficient, nozzle radius ratio, rotor radius ratio, Rotational speed
	radial	2015	Rahbara et al. [34]	Expansion ratio, rotational speed, flow coefficient, radius ratio
	radial	2015	Erbas et al. [131]	Flow coefficient, inlet blade height, inlet flow angle, exit flow angle, number of blades, meridional speed ratio

4.4. *Fluid Level*

Different organic working fluids significantly affect ORC performance. Therefore, determining a suitable working fluid from diverse types of fluids is particularly important for the practical application of ORC. In general, the current research on the selection of working fluids is divided into two levels. The first level is to compare the ORC performance when using different working fluids directly. The second level is to design the working fluid, that is, the working fluid is parameterized using the equation of state or the molecular group. Therefore, the second level is not limited to existing organic fluids, but can also calculate the ORC performance using potential working fluids in the future. This section will introduce the research status of these two fluid levels.

4.4.1. Selection from Limited Fluids

Most researches are based on comparing existing working fluids, including the pure fluids or mixtures. The typical approach is to construct the ORC model and calculate system performance, in which the physical property is obtained from software such as REFPROP or CoolProp. This method has the advantages of simplicity and high accuracy since most of the physical parameters in REFPROP have been verified and fitted by experiments.

(1) Pure fluids

The optimization of pure fluid is mainly reflected in the comparison of ORC performance using different fluids but does not discuss or design the fluid itself. Therefore, the pure fluid will not be expanded in detail in this work. For more discussion, please refer to the existing literature [132].

(2) Mixture

Due to the temperature slide during the phase change process, zeotropic mixtures could obtain a better temperature matching with heat/cold source, thereby reducing the heat loss in the heat exchanger and improving system performance. Different composition ratios could result in different temperature slides. Thus the design of zeotropic mixture mainly focuses on optimizing the component ratio, especially the ratio of binary mixtures [133]. However, the application potential of ternary or quaternary mixture is less discussed and could be explored in the near future.

4.4.2. Fluid Design

Even though the comparison of existing fluids is easy to conduct, it is usually difficult to determine the optimal fluid from diverse working fluids since the calculation process needs to be repeated many times, which significantly increases calculation time. Moreover, the direct selection of working fluids can only be based on existing working fluids. These diverse fluids usually have different critical temperatures, critical pressures, flammability and toxicity, which are difficult to quantify and compare with each other. Therefore, some researchers propose to incorporate the fluid characteristics into the design optimization of the ORC system, which is usually called computer-aided molecular design (CAMD) [122]. Currently, there are two popular molecular design methods. One is the contribution group design method, which calculates the physical properties of fluid by determining its chemical structure through the contribution group design. The other approach is to use the general equation of state (EOS) to parameterize the working fluid by characterizing the physical properties of the working fluid with critical temperature, critical pressure and eccentricity factor [134]. Consequently, the fluid characteristics could be included in the optimization process. In this section, the two methods will be introduced.

(1) Group-contribution method (GCM)

The principle of GCM is that through the combination of different structural groups (—CH_3, =CH-), any kind of organic fluid can be constructed theoretically. As shown in Figure 12, popular organic fluids in ORC such as R601a and R134a can be represented using the combination of common structural groups [135]. Using this principle, the physical properties of the working fluid are determined by the number of each group and its contribution ratio. In this GCM method, the contribution of each group is considered the same, regardless of the specific spatial structure and location. Therefore, the GCM method can predict the physical properties of both existing working fluids and also unknown fluids. GCM has a much faster calculation speed than CCM but has less accuracy.

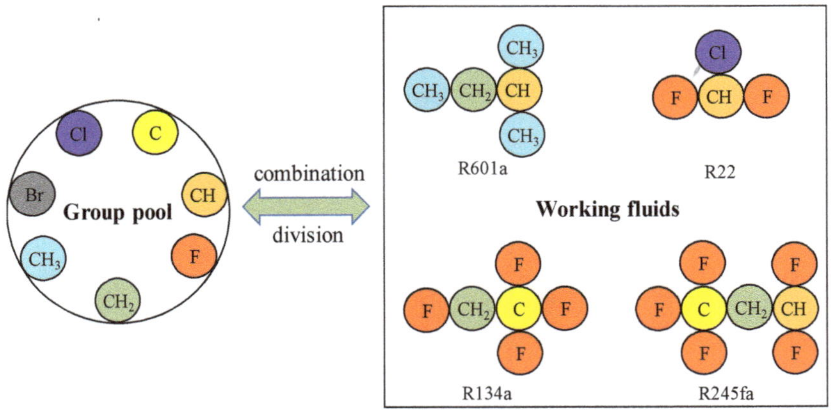

Figure 12. Diagram of the group-contribution method (GCM).

Papadopoulos et al. used the GCM method to generate a series of ORC candidate fluids, including existing fluids and also novel fluids. Then the optimal fluid is selected considering the requirements of safety and environmental performance [76,136,137]. Van Kleef et al. used the GCM method to estimate the critical physical properties and transport properties of the working fluid. Then the authors used a multi-objective optimization method to optimize the thermodynamic and economic performance simultaneously, and explored the optimal working fluid design in different heat source conditions [122]. Moreover, Su et al. reviewed the application of the GCM method in ORC and summarized the commonly used equations for predicting different macroscopic parameters of working fluids, such as boiling point, critical physical properties, transport properties, flammability and toxicity [135].

(2) Equation-of-State method (EOS)

The equation of state method uses a few parameters, including the critical parameters and eccentricity factors, to calculate the basic and transport properties of a certain working fluid. In this way, the fluid properties and the thermodynamic and economic performance of the ORC system could be calculated. In traditional physical property software such as REFPROP, although high-precision multi-parameter equations of state can be used to accurately obtain the thermophysical properties of the working fluid, these equations have many adjustable parameters and require a large number of experimental fittings. The forecast for novel fluids is not precise enough for ORC performance prediction. In contrast, the general-purpose state function is simpler and has fewer parameters, which can be more reliably extended to unknown working fluids, as shown in Figure 13 [134,138].

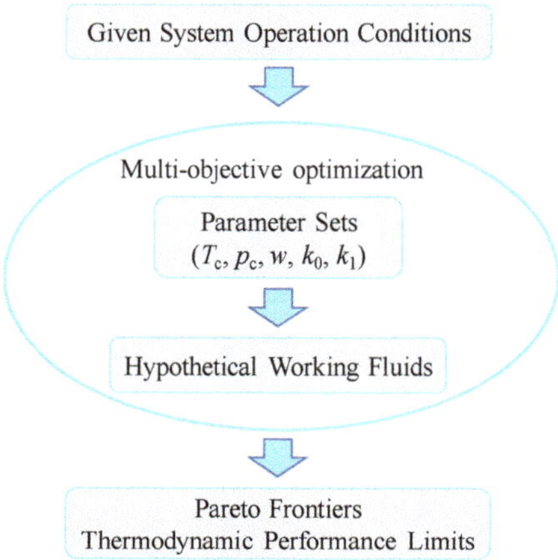

Figure 13. Schematic diagram of EOS method in ORC optimization.

Yang et al. proposed a corresponding states model to predict the fluid properties using five parameters: critical temperature, critical pressure, acentric factor, molar ideal gas isobaric heat capacity and temperature gradient [134]. In this way, different working fluids could be parameterized. The authors compared 13 common existing fluids and obtained the thermodynamic performance limit, which characterizes the best performance that can be obtained by improving the working fluid under ideal conditions.

(3) Computational chemistry method (CCM) method

In addition, there is a computational chemistry method (CCM) from the view of autom, which has a too long convergence time (usually from hours to days for a molecular structure). Therefore, this CCM method is not currently suitable for fluid design in the ORC system [135]. Comparison of three CAMD methods is shown in Table 7.

Table 7. Comparison of three computer-aided method for property prediction.

Method	Advantages	Disadvantages	Appropriate for Cycle Optimization?	References
CCM	Accurate	Time consuming, (hours to days for one molecule)	×	[135,139]
GCM	Wide range, efficient for new fluids	Less accurate	√	[76,122,136]
EOS	Quite accurate	May be inefficient for new fluids	√	[134]

5. Discussion and Future Outlook

The single-objective optimization could no longer meet the requirements from the increasingly serious climate crisis, different technical requirements and rich application scenarios. Instead, the thermodynamic, economic, environmental and other performance should be considered simultaneously, which, however, usually conflict with each other. Thus it is difficult to obtain a unique optimal solution, proposing challenges to ORC multi-objective optimization. In recent years, there have been increasing studies involved in multi-objective optimization and discussing the ORC design. Correspondingly, multi-objective optimization has also become increasingly popular and gradually become a typical method for system design.

5.1. Discussion

In general, existing research has made significant progress, which could promote the development of many new indicators or methods and lay the basis for the practical application of the ORC system. The progress could be divided into the following areas:

5.1.1. Optimization Objective

The optimization objectives of the ORC system have been greatly enriched, including the thermodynamic, economic, environmental indicators, as well as the system weight, volume, safety and stability, which provide valuable guidance for the design and operation of ORC system. Different indicators have their advantages, disadvantages and application scopes, which are also summarized in this article. However, most of the existing research focuses on thermodynamic performance, followed by economic performance, and less attention to other properties. Especially when the climate issue becomes increasingly important, the research on the environmental protection performance of the system, the role of emission reduction, and the positioning of the ORC system in low-carbon energy scenarios are still insufficient.

In terms of selecting the optimization objective, unless specific preferences are required, authors are recommended to consider the thermodynamic, economic and environmental performance simultaneously. Other indicators such as safety could be represented by the working fluids. For each category, a single indicator could be selected according to the practical application scenario, as shown in Table 8. The power output and thermal efficiency are recommended for thermodynamic performance. Particularly, the power output is more suitable for geothermal, waste heat scenarios, which does not include the thermal oil cycle. In contrast, the thermal efficiency is more suitable for solar, biomass scenarios with oil cycles. In terms of the economic criterion, LCOE and IRR are recommended since these two indexes are independent of system size. The calculation of IRR is based on electricity price, thereby being more suitable for the region with a benchmark price. In contrast, LCOE describes the electricity price and is more suitable for the region with the electricity market. Particularly, LCOE is applicable for engine waste heat recovery in cars or marine applications since the electricity is usually consumed locally and not sold to the grid. Moreover, the system volume and weight should also be considered in these waste heat scenarios since the volume is very limited.

Table 8. Recommended optimization objectives for different ORC applications.

Application	Electric Market	Electricity Benchmark Price
Geothermal, industrial waste heat	W + LCOE + Carbon emission	W + IRR + Carbon emission
Solar, biomass (with oil cycle)	η_{th} + LCOE + Carbon emission	η_{th} + IRR + Carbon emission
Engine waste	W + LCOE + Carbon emission + Volume + Weight	

Specifically, for calculating the carbon emission, the combination of LCA and the carbon coefficient method is recommended. This combined method could calculate the carbon emission during production and transportation, and also consider the emission reduction by replacing other plants using ORC, thereby improving the accuracy over separate LCA or carbon coefficient method.

5.1.2. Optimization Method

As the algorithm develops, ORC optimization approaches are also gradually improved. The number of optimization objectives increases from a single one to two, three and multiple. The approach varies from the linear weighting method to the ε-constraint method, and the intelligent algorithms, which could obtain the Pareto frontier more conveniently. Furthermore, based on the Pareto frontier, many studies introduce the multi-criteria decision theory to select the desired solution with corresponding weights. Thus the unique solution could be determined to guide engineering design according to the decision maker's preference. However, most of the studies are based on optimization with three or fewer objectives. When the objective number exceeds three, the current intelligent algorithms could not converge very well. Up to date, only the traditional weighted method could be used to deal with high-dimensional optimization (>3) by combining multiple indicators into one. However, this method only considers a special scenario, and the weight is difficult to determine.

Based on this review, the authors recommend the intelligent algorithms in the optimization with 2-3 objectives, such as the NSGA-II, which is relatively simple to use and could effectively solve non-convex problems. These non-convex problems are very common in the design of configuration and system planning in complex thermal systems [140], which cannot be effectively solved by the WSM method. However, when the optimization objectives are more than or equal to 4, these intelligent algorithms such as NSGA-II could not work effectively with an explosive computation time and poor convergence at this stage. Therefore, the authors recommend the WSM method for this high-dimensional optimization. The specific weights can be calculated according to the AHP or α-method discussed in Section 3.1.

In addition, when using intelligent algorithms for optimization, it is usually necessary to adopt a decision-making method to obtain a unique solution for engineering applications. The authors recommend using multiple decision-making methods simultaneously by comparing the decision-making results to determine the final solution, which could effectively improve the robustness of decision-making.

5.1.3. Optimization Parameter

According to Chapter 4, the optimization parameters involved in existing research cover four levels: working fluid, components, processes and systems. Particularly, the system level is the most studied, focusing on the optimization design of evaporation/condensing pressure, superheat degree and etc. The research on process mainly aims to design the system structure and architecture by establishing a superstructure model. The research on components focuses on heat exchangers (tube and shell type, plate type and etc.) and turbines (mainly radial), while little attention is paid to feed pumps. For the working fluid, the general approach of using the group contribution method or the equation of state method to determine the optimal working fluid has been widely applied. However, the current research only focuses on one of these levels, and the research on multi-objective optimization of ORC systems considering multiple (four) levels is still scarce.

Authors recommend the collaborative design approach by considering multiple levels, especially the system, fluid and component levels. At the fluid level, the GCM or the EOS method could be used to parameterize the physical properties of the fluid and incorporate them into decision variables. At the component level, component size and structural parameters could also be included in the optimization parameters. Finally, a comprehensive design method considering four levels of system-process-component-fluid will be developed.

5.2. Future Outlook

In view of the above research status and problems, the following directions need to be developed.

5.2.1. Discussion on Carbon Emissions

The climate issue has become a major challenge threatening human survival. To achieve carbon neutrality, it is necessary to make concessions or even sacrifice certain economics. Therefore, future attention to environmental performance needs to be further strengthened.

Even though ORC could use turn heat resources into treasure (power), it is not completely carbon-free. Existing studies have used various methods such as LCA, LCCP, TEWI to measure its carbon emissions during the manufacturing process. However, there are some important issues that need further exploration, including: (1) Under the existing energy structure, how much room for emission reduction could be achieved by large-scale promotion of ORC? (2) In the future energy structure with a high proportion of renewable energy, how does ORC's emission reduction compare to its lifetime carbon emissions? Should ORC be promoted on a large scale? (3) With the continuous improvement of the carbon trading market, could ORC independently or in conjunction with other units benefit from the carbon trading market?

5.2.2. High-Dimension Optimization

When the number of optimization objectives exceeds 3, the results and convergence of existing intelligent algorithms, including NSGA-II and MOPSO, cannot meet the requirements. At present, only a few studies have focused on high-dimensional multi-objective optimization [36,64,86,141], and the methods used are mainly linear weighted method or multi-criteria decision-making method. However, as mentioned above, the weights should be determined in advance, which makes their application limited. Therefore, the following issues need to be further discussed: (1) How to simultaneously consider multiple indicators such as thermodynamic, economic, environmental and safety index for high-dimension Pareto optimization? (2) How to reasonably convert the preferences of decision-makers and the constraints of practical projects such as limited funds and limited equipment space into weights and reflect them in the objective function?

5.2.3. Consider the Practical Operation

Most of the existing multi-objective optimizations are based on design conditions. That is, the heat source and environmental conditions are assumed the same. However, this assumption is inconsistent with practical operating conditions, resulting in significant differences between expected operation and practical operation [111], or even leading to completely opposite conclusions [142,143]. Therefore, some research proposed to integrate long-term off-design operation into the system design process. Specifically, the typical heat source conditions and ambient temperature are selected as boundary conditions to guide the multi-objective design, resulting in a more robust design scheme [116,144]. Even though this method is closer to practical engineering, it also has limitations such as time-consuming calculations and uncertain weather and heat source conditions. Even though Hu et al. proposed an approach to reduce the calculation time using the ANN method, a duration of 1–2 days is still required [48]. Yang et al. and Feng et al. proposed a BPNN model based on experimental data [145,146], but the mapping between environmental fluctuation and ORC output needs to be further established. To improve the practicability of this off-design approach, some novel methods need to be improved to reduce the calculation time and improve precision. The following aspects could be explored in the future: (1) How to integrate the dynamic response of ORC into system design? Especially the unit response speed (for heat exchanger or heat storage) and the loss during the start-stop process. (2) How much impact does the fluctuation of user-side load have on ORC design? (3) How much influence will participation in the carbon trading market have on system design?

(4) How to improve the design method to speed up the optimization process based on a large number of off-design conditions and dynamic responses? Possible methods may include deep learning methods such as CNN, and the specific approach still needs to be developed.

5.2.4. Multi-Level Optimization

At present, most researches focus on the optimization of the system level, or the coupling of the two levels such as "system-component" and "system-fluid". There is still a lack of collaborative design of "system-process-component-fluid", which may further improve the results over conventional system optimization or component design, as shown in Figure 14. In this integrated design process, there are numerous decision variables, including the physical parameters of the working fluid, the structural parameters of components, the process parameters and the system parameters. Therefore, the optimization method and calculation capabilities need certain requirements. In general, the following aspects need further exploration: (1) Existing studies paid little attention to fluid pumps. How does the structural design of the fluid pump affect the thermodynamic and economic performance of the ORC system? (2) How much could the proposed "system-process-component-fluid" method improve system performance compared with conventional methods? What is the computational cost? (3) What's the difference between optimization results of four levels, three levels and two levels? Which level of optimization is most appropriate considering the improvements and computation cost?

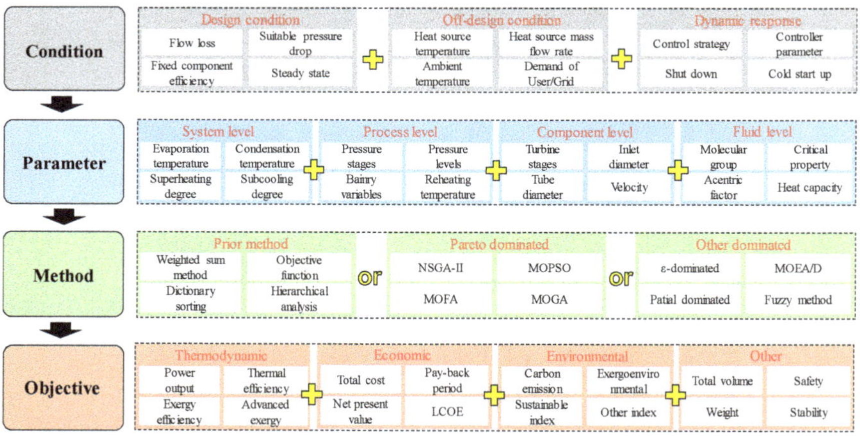

Figure 14. The multi-level optimization approach considering "system-process-component-fluid".

6. Conclusions

A single evaluation indicator could no longer meet the development needs of power systems such as ORC. Collaborative optimization considering multiple indicators such as thermodynamic, economic and environmental indexes has become an inevitable trend in system design. This article summarizes the research on multi-objective optimization of ORC system design. From the perspective of optimization objective, optimization methods and optimization parameters, this work classifies and statistically summarizes the existing research, and prospects the future development. Main conclusions are as follows

1. For the optimization objectives, multiple aspects should be considered, including the thermodynamic, economic and environmental indicators, which should be selected according to specific application scenarios and the local electricity market. For the thermodynamic indicator, the output power is recommended for geothermal and waste heat. Thermal efficiency is recommended for solar energy and biomass. For

economics, LCOE is recommended for benchmark electricity price, and IRR is recommended for the electricity market. Furthermore, for the waste heat recovery in mobile transportation, additional attention should be paid to the ORC volume and weight.

2. In terms of the optimization method, intelligent algorithms including NSGA-II and MOPSO are recommended for low-dimension optimization (objective is less than 4). In contrast, the WSM method is recommended for high-dimension optimization (the objective number is more than 3), which should determine the weight in advance and then convert the complex problem into single-objective optimization.
3. For optimization parameter, the researches are mainly carried out from four levels: system, process, component and fluid. However, the existing researches are relatively independent. A comprehensive design method that couples multiple levels simultaneously is expected to improve the system performance further.
4. For future development, the following aspects could be further explored: (1) Exploring the emission reduction potential of ORC in the future high-proportion renewable energy system and its economic value in the carbon trading market. (2) Transform the preferences and engineering constraints into boundary constraints in multi-objective optimization. (3) Propose and expand the comprehensive optimization methods considering four levels "system-process-component- fluid".

Supplementary Materials: The following are available online at https://www.mdpi.com/1996-1073/14/20/6492/s1, Table S1: List of the reviewed papers.

Author Contributions: Conceptualization, S.H. and Y.D.; methodology, S.H.; software, S.H.; validation, S.H. and J.L.; formal analysis, S.H. and Z.Y.; writing—original draft preparation, S.H.; writing—review and editing, S.H. and Y.D.; visualization, S.H.; supervision, Y.D.; project administration, S.H. and Y.D.; funding acquisition, Y.D. and J.L. All authors have read and agreed to the published version of the manuscript.

Funding: This research was funded by the National Natural Science Foundation of China, grant number 51736005 and 52106017, the National Postdoctoral Program for Innovative Talents, grant number BX20200178 and the China Postdoctoral Science Foundation, grant number 2020M680548.

Conflicts of Interest: The authors declare no conflict of interest.

Abbreviations

Nomenclature

C	Cost [$]
d	Distance
E	Electricity generation [kW]
Ex	Exergy [kJ]
h	Enthalpy [kJ·kg^{-1}]
$\dot{m}$	Mass flow rate [kg·s^{-1}]
Ns	Number of optimization objectives
pr	Price [$]
Q	Heat
r	Discount rate
T	Temperature [°C]
V	Volume [m^3]
W	Power output [kW]
Greek Symbols	
β	Indirect emission factor [kg·kWh^{-1}]
ω	Weight
η	Efficiency
ξ	Grey relational value

Sub- or Superscripts

con	Condenser
eva	Evaporator
H	Heat source
in	Inlet
out	Outlet
pum	Pump
tot	Total
tur	Turbine

Abbreviations

ABC	Artificial bee colony
ACS	Artificial Cooperative Search
AHP	Analytic Hierarchy Process
ANN	Artificial Neural Network
APR	The heat exchanger area to power ratio
BPNN	Back Propagation Neural Network
CEPCI	Chemical Engineering Plant Cost Index
DEP	Annual depreciation
EPC	Electricity production cost
GA	Genetic algorithm
GRA	Grey relational analysis
IRR	Internal Rate of Return
LCA	Life cycle analysis
LCOE	Levelized cost of electricity
LEC	Levelized electricity cost
LINMAP	Linear Programming Technique for Multidimensional Analysis of Preference
MOGWO	Multi-objective grey wolf optimizer
MOGA	Multiple objective genetic algorithm
MOO	Multiple objective optimization
MOPSO	Multiple objective particle swarm optimization
MPO	Mass flow rate of heat source per net power output
NPV	Net present value
NSGA-II	Non-dominated sorting genetic algorithm II
ORC	Organic Rankine cycle
PBP	Payback period
SA	Simulated annealing algorithm
SI	Sustainability Index
SIC	Specific investment cost
TOPSIS	Technique for Order of Preference by Similarity to Ideal Solution
UA	The product of the overall heat transfer coefficient and the total area
WSM	Weighted sum method

Appendix A

This appendix describes the definition of various objective variables in the reviewed papers and provides general formulas for reference.

Appendix A.1. Thermodynamic Index

Appendix A.1.1. Power Output

The power output is the most fundamental indicator in the power system. It is usually the turbine power minus the power consumption of the fluid pump and cooling pump. If the heat source is geothermal, the downhole pump should be considered. Alternatively, if the heat source is solar heat transfer oil, the power consumption of the oil pump should be considered, as shown in Equation (A1).

$$W_{\text{net}} = W_{\text{tur}} - W_{\text{pum,f}} - W_{\text{pump,c}} - W_{\text{other}} \quad \text{(A1)}$$

Some studies further use the total power generation to represent electricity generation over a period (usually a single year or 20 years). The difference lies in considering the generator and motor efficiency, although they are usually 95% [20].

$$E_{\text{tot}} = W_{\text{tur}}\eta_{\text{tur}} - \frac{W_{\text{pum,f}}}{\eta_{\text{pum,f}}} - \frac{W_{\text{pump,c}}}{\eta_{\text{pum,c}}} - \frac{W_{\text{other}}}{\eta_{\text{other}}} \quad \text{(A2)}$$

where η is the efficiency of turbine or pump. E is the total electricity generation.

Appendix A.1.2. Thermal Efficiency

Different from the power output, thermal efficiency measures the system's external output and input energy simultaneously, characterizing the utilization level of the input heat by ORC. Thermal efficiency could be calculated by Equation (A3).

$$\eta_{\text{th}} = \frac{W_{\text{net}}}{Q_{\text{in}}} = \frac{W_{\text{net}}}{\dot{m}_{\text{H}}(h_{\text{H,in}} - h_{\text{H,out}})} \quad \text{(A3)}$$

where Q_{in} denotes the absorbed heat by ORC, m_{H} denotes the mass flow rate of the heat source. $h_{\text{H,in}}$ and $h_{\text{H,out}}$ represents the inlet and outlet enthalpy of the heat source.

Appendix A.1.3. Exergy Efficiency

Exergy efficiency further considers the energy grade and describes the effective utilization of exergy by ORC. In particular, the exergy loss analysis for each component helps facilitate the optimal design of the component and ORC system. The calculation of exergy efficiency is as follows:

$$\eta_{\text{ex}} = \frac{W_{\text{net}}}{Ex_{\text{in}}} = \frac{W_{\text{net}}}{W_{\text{net}} + Ex_{\text{out}} + Ex_{\text{loss}}} \quad \text{(A4)}$$

where the Ex_{in}, Ex_{out}, Ex_{loss} represent the exergy at the heat source inlet, outlet and exergy loss in ORC, respectively.

Appendix A.2. Economic Index

Appendix A.2.1. UA

UA could evaluate the heat exchanger cost according to the log mean temperature difference (LMTD) method [28,29]. A lower UA indicates lower costs and better economic performance [30], which could be calculated by:

$$\text{UA} = \sum \frac{Q_{\text{eva}}}{\Delta T_{\text{eva}}} + \frac{Q_{\text{con}}}{\Delta T_{\text{con}}} \quad \text{(A5)}$$

UA has the advantage of simple calculation. However, UA does not consider the impact of different working fluids and heat transfer capabilities, resulting in a relatively large cost deviation.

Appendix A.2.2. Total Cost

Total cost is the most basic index to evaluate ORC economics. Almost all direct economic indicators are calculated based on the total cost. The component costs are mainly calculated using empirical correlations, fitted from the cost of different types and sizes of equipment on the market. The two most popular correlations are from Turton [38] and Smith [39]:

$$C_{\mathrm{tot}} = \sum_i C_i \frac{\mathrm{CEPCI}_{2020}}{\mathrm{CEPCI}_m} \quad \text{(A6)}$$

where C_i denotes the total investment cost of each component, including the turbine, heat exchanger, pump. CEPCI denotes the correction to inflation or deflation. m denotes the benchmark year when fitting the correlations [37].

Appendix A.2.3. Specific Investment Cost (SIC)

SIC is a very common index to evaluate the thermo-economic performance of ORC, which describes the unit cost per power output and could be calculated by [41]:

$$\mathrm{SIC} = \frac{C_{\mathrm{tot}}}{W_{\mathrm{net}}} \quad \text{(A7)}$$

SIC has the advantage of easy use and intuitive comparison between different cases. The disadvantage is that SIC is too simplified and does not consider the depreciation, operation costs or discount rate [43].

Appendix A.2.4. Payback Period (PBP)

PBP measures the number of years required to recover the total cost, mainly including the static and dynamic PBP [44]. The calculation processes are shown in Eqs. A8 and A9, respectively. The dynamic PBP is more frequently used since it considers the time value and has higher accuracy than static PBP.

$$\mathrm{PBP}_{\mathrm{sta}} = \frac{C_{\mathrm{tot}}}{C_{\mathrm{profit}}} \quad \text{(A8)}$$

$$\mathrm{PBP}_{\mathrm{dyn}} = -\frac{\ln\left(1 - i\frac{C_{\mathrm{tot}}}{C_{\mathrm{profit}}}\right)}{\ln(1+i)} \quad \text{(A9)}$$

Appendix A.2.5. Levelized Cost of Electricity (LCOE)

LCOE denotes the cost of unit electricity considering the project construction, operation and maintenance, depreciation and residual value [6]. This indicator could be directly compared with the local electricity price to represent the profitability. If LCOE is lower than the electricity price, then this project will be economically feasible. The calculation process is:

$$\mathrm{LCOE} = \frac{\sum_{t=1}^{LT}\left(\frac{C_{\mathrm{OM}}}{(1+r)^t} - \frac{DEP}{(1+r)^t}\right) + C_{\mathrm{tot}} - \frac{C_{\mathrm{residual}}}{(1+r)^{LT}}}{\sum_{t=1}^{LT}\frac{E_{\mathrm{yr}}}{(1+r)^t}} \quad \text{(A10)}$$

where C_{resudual} represents the residual value of the asset at the end. The denominator represents the discount of the annual power generation.

Appendix A.2.6. Net Present Value (NPV)

The net present value (NPV) represents the difference between the discounted cash flow of an investment in the future and the total cost [48]. The expected discount rate is determined according to the company's lowest investment rate of return, which is the lowest acceptable limit. A positive NPV represents a feasible project, and a larger NPV represents a better return on investment. NPV could be calculated by:

$$\mathrm{NPV} = -C_{\mathrm{tot}} + \sum_{t=1}^{LT}\frac{(pc \times E_{\mathrm{yr}} - C_{\mathrm{OM}} - DEP) \times (1 - tax) + DEP}{(1+r)^t} \quad \text{(A11)}$$

where the pc represents the electricity price. E_{yr} represents annual electricity generation. LT is the operating years of ORC. C_{OM} is the operation and maintenance cost. DEP is the annual depreciation with linear or accelerated depreciation. tax is the tax rate of electricity sales, which varies in different countries or regions. r denotes the discount rate, which could be taken as the company's expected return on investment.

IRR is defined as the discount rate when the total present values of capital inflow and the total outflow are equal to the net present value of 0. IRR could also represent the ability to resist inflation, the calculation of which usually requires relatively complex iteration. A higher IRR indicates better economic performance. Generally, the project is considered feasible if IRR is higher than the benchmark value of the return rate [50].

Appendix A.2.7. Exergoeconomic Analysis

Under the nominal condition, the input and output costs should be balanced, and the total cost in exergoeconomic analysis could be calculated by [53]:

$$\sum_{output} \dot{C}_{out,tot} = \sum_{input} \dot{C}_{in,tot} + \dot{Z}_{CA} + \dot{Z}_{OM} \tag{A12}$$

$$F_{eco} = \sum \dot{C}_k + \dot{Z}_k \tag{A13}$$

where C_k denotes the exergy cost of component k. Z_k represents the total investment and operation cost of component k.

Appendix A.3. Environmental Index

Appendix A.3.1. Total Equivalent Warming Impact (TEWI)

$$\text{TEWI} = \text{GWP} \cdot M \cdot L \cdot n + \text{GWP} \cdot M \cdot (1 - a_{recovery}) + E_{annual} \cdot \beta \cdot n \tag{A14}$$

where L is the annual leakage rate, kg [58]. n is operating life, year. M is the refrigerant charge, kg. $a_{recovery}$ is the recycling factor. E_{annual} is the annual energy consumption, kWh. β is the indirect emission factor, kg·kWh^{-1}.

Appendix A.3.2. Life Cycle Climate Performance (LCCP)

$$\text{LCCP} = \text{TEWI} + MM{\cdot}m + REM{\cdot}M + n{\cdot}L{\cdot}RFM{\cdot}M{\cdot} \tag{A15}$$

where n is the lifetime [36]. MM is the CO_2 production of material, kg·kg^{-1}. m is the mass of unit material, kg. RFM is the refrigerant manufacturing emissions, kg·kg^{-1}.

Appendix A.3.3. Life Cycle Analysis (LCA)

The environmental balance in LCA is shown below, which helps in exploring the origin of pollution in each process [62]:

$$\sum_{i=1}^{N} M_i \cdot v_{M_i} + \sum_{i=1}^{N} E_i \cdot v_{E_i} - \sum_{i=1}^{N} W_i \cdot v_{W_i} = \sum_{i=1}^{N} P_i \cdot v_{P_i} \tag{A16}$$

where M_i denotes the mass input. E_i is the energy input. P_i denotes the outlet stream. W_i represents the residues. v is the synergy matrix of mass and energy.

Appendix A.3.4. Exergoenvironmenal Analysis

Main process is divided into three steps [53]. The first step is to perform exergy analysis on the overall system to identify the exergy loss in each component. The second step is to use the LCA method to analyze each component and corresponding energy input. The third step is to allocate the environmental impact to exergy flow in ORC using LCA.

For each component, the balanced equation of environmental impact could be calculated by:

$$\dot{B}_{P,k} = \dot{B}_{F,k} + \dot{Y}_k \tag{A17}$$

where $\dot{B}_{P,k}$ represents the output or product of component k. $\dot{B}_{F,k}$ represents the input or fuel consumption of component k. $\dot{Y}_k$ denotes the environmental impact during the production, transportation and installation. By establishing the balance equation of each component, the evaluation index F_{env} of exergoenvironmental performance could be calculated by:

$$F_{\text{env}} = \sum B_{D,k} + \dot{Y}_k \tag{A18}$$

where $B_{D,k}$ represents the environmental impact caused by exergy loss in component k.

Appendix A.3.5. Sustainability Index (SI)

SI could be calculated by:

$$\text{SI} = \frac{Ex_{\text{loss}}}{m_{\text{H}}(h_{\text{in}} - h_{\text{out}} - T_{\text{amb}} \cdot \Delta s_{\text{H}})} \tag{A19}$$

where Ex_{loss} represents the exergy loss in ORC. m_{H} denotes the heat source flow rate. h_{in}, h_{out} represents the input and outlet enthalpy of the heat source. Δs denotes the entropy change of heat source.

Appendix A.4. Other Index

Appendix A.4.1. Volume

The total volume could be calculated by:

$$V_{\text{sys}} = V_{\text{comp}} + V_{\text{aux}} = V_{\text{hx,eva}} + V_{\text{hx,con}} + V_{\text{exp}} + V_{\text{pump}} + V_{\text{aux}} \tag{A20}$$

where V_{comp} represents the volume of system components, including the heat exchanger, feed pump and turbine. V_{aux} represents the volume of auxiliary equipment.

Appendix A.4.2. Safety

The evaluation index in Quantitative Risk Analysis (QRA) is shown below [75]:

$$R_{overall} = \sum_i (outcome frequency_i \cdot probability\ of\ death\ from\ the\ outcome_i) \tag{A21}$$

$$d_{risk} = maximum\ distance\ (10^{-4} < R_{overall}) \tag{A22}$$

where $R_{overall}$ is the process risk. d_{risk} is the risk distance.

References

1. Roumpedakis, T.C.; Loumpardis, G.; Monokrousou, E.; Braimakis, K.; Charalampidis, A.; Karellas, S. Exergetic and economic analysis of a solar driven small scale ORC. *Renew. Energy* **2020**, *157*, 1008–1024. [CrossRef]
2. Loni, R.; Najafi, G.; Bellos, E.; Rajaee, F.; Said, Z.; Mazlan, M. A review of industrial waste heat recovery system for power generation with Organic Rankine Cycle: Recent challenges and future outlook. *J. Clean. Prod.* **2020**, *287*, 125070. [CrossRef]
3. Imran, M.; Haglind, F.; Asim, M.; Zeb Alvi, J. Recent research trends in organic Rankine cycle technology: A bibliometric approach. *Renew. Sust. Energ. Rev.* **2018**, *81*, 552–562. [CrossRef]
4. Lecompte, S.; Huisseune, H.; van den Broek, M.; Vanslambrouck, B.; De Paepe, M. Review of organic Rankine cycle (ORC) architectures for waste heat recovery. *Renew. Sust. Energ. Rev.* **2015**, *47*, 448–461. [CrossRef]
5. Bao, J.; Zhao, L. A review of working fluid and expander selections for organic Rankine cycle. *Renew. Sust. Energ. Rev.* **2013**, *24*, 325–342. [CrossRef]
6. Lecompte, S.; Lemmens, S.; Huisseune, H.; van den Broek, M.; De Paepe, M. Multi-Objective Thermo-Economic Optimization Strategy for ORCs Applied to Subcritical and Transcritical Cycles for Waste Heat Recovery. *Energies* **2015**, *8*, 2714–2741. [CrossRef]
7. Cui, Y.; Geng, Z.; Zhu, Q.; Han, Y. Review: Multi-objective optimization methods and application in energy saving. *Energy* **2017**, *125*, 681–704. [CrossRef]

8. Wang, J.; Yan, Z.; Wang, M.; Li, M.; Dai, Y. Multi-objective optimization of an organic Rankine cycle (ORC) for low grade waste heat recovery using evolutionary algorithm. *Energy Convers. Manag.* **2013**, *71*, 146–158. [CrossRef]
9. Wang, Z.Q.; Zhou, N.J.; Guo, J.; Wang, X.Y. Fluid selection and parametric optimization of organic Rankine cycle using low temperature waste heat. *Energy* **2012**, *40*, 107–115. [CrossRef]
10. Yang, F.; Zhang, H.; Song, S.; Bei, C.; Wang, H.; Wang, E. Thermoeconomic multi-objective optimization of an organic Rankine cycle for exhaust waste heat recovery of a diesel engine. *Energy* **2015**, *93*, 2208–2228. [CrossRef]
11. Andreasen, J.; Kærn, M.; Pierobon, L.; Larsen, U.; Haglind, F. Multi-Objective Optimization of Organic Rankine Cycle Power Plants Using Pure and Mixed Working Fluids. *Energies* **2016**, *9*, 322. [CrossRef]
12. Feng, Y.; Hung, T.; Greg, K.; Zhang, Y.; Li, B.; Yang, J. Thermoeconomic comparison between pure and mixture working fluids of organic Rankine cycles (ORCs) for low temperature waste heat recovery. *Energy Convers. Manag.* **2015**, *106*, 859–872. [CrossRef]
13. Feng, Y.; Hung, T.; Zhang, Y.; Li, B.; Yang, J.; Shi, Y. Performance comparison of low-grade ORCs (organic Rankine cycles) using R245fa, pentane and their mixtures based on the thermoeconomic multi-objective optimization and decision makings. *Energy* **2015**, *93*, 2018–2029. [CrossRef]
14. Xi, H.; Li, M.-J.; He, Y.-L.; Tao, W.-Q. A graphical criterion for working fluid selection and thermodynamic system comparison in waste heat recovery. *Appl. Therm. Eng.* **2015**, *89*, 772–782. [CrossRef]
15. Ghasemian, E.; Ehyaei, M.A. Evaluation and optimization of organic Rankine cycle (ORC) with algorithms NSGA-II, MOPSO, and MOEA for eight coolant fluids. *Int. J. Energy Environ. Eng.* **2017**, *9*, 39–57. [CrossRef]
16. Lecompte, S.; Lemmens, S.; Verbruggen, A.; van den Broek, M.; De Paepe, M. Thermo-economic comparison of advanced Organic Rankine Cycles. *Energy Procedia* **2014**, *61*, 71–74.
17. Sadeghi, M.; Nemati, A.; Ghavimi, A.; Yari, M. Thermodynamic analysis and multi-objective optimization of various ORC (organic Rankine cycle) configurations using zeotropic mixtures. *Energy* **2016**, *109*, 791–802. [CrossRef]
18. Song, J.; Loo, P.; Teo, J.; Markides, C.N. Thermo-Economic Optimization of Organic Rankine Cycle (ORC) Systems for Geothermal Power Generation: A Comparative Study of System Configurations. *Front. Energy Res.* **2020**, *8*, 6. [CrossRef]
19. Dimitrova, Z.; Lourdais, P.; Marechal, F. Performance and economic optimization of an organic rankine cycle for a gasoline hybrid pneumatic powertrain. *Energy* **2015**, *86*, 574–588. [CrossRef]
20. Samadi, F.; Kazemi, N. Exergoeconomic analysis of zeotropic mixture on the new proposed organic Rankine cycle for energy production from geothermal resources. *Renew. Energy* **2020**, *152*, 1250–1265. [CrossRef]
21. Zhai, H.; An, Q.; Shi, L.; Lemort, V.; Quoilin, S. Categorization and analysis of heat sources for organic Rankine cycle systems. *Renew. Sust. Energ. Rev.* **2016**, *64*, 790–805. [CrossRef]
22. Turgut, M.S.; Turgut, O.E. An oppositional Salp Swarm: Jaya algorithm for thermal design optimization of an Organic Rankine Cycle. *Sn Appl. Sci.* **2021**, *3*, 224. [CrossRef]
23. Wang, Z.; Xia, X.; Pan, H.; Zuo, Q.; Zhou, N.; Xie, B. Fluid selection and advanced exergy analysis of dual-loop ORC using zeotropic mixture. *Appl. Therm. Eng.* **2021**, *185*, 116423. [CrossRef]
24. Gimelli, A.; Luongo, A.; Muccillo, M. Efficiency and cost optimization of a regenerative Organic Rankine Cycle power plant through the multi-objective approach. *Appl. Therm. Eng.* **2017**, *114*, 601–610. [CrossRef]
25. Noriega Sanchez, C.J.; Gosselin, L.; da Silva, A.K. Designed binary mixtures for subcritical organic Rankine cycles based on multiobjective optimization. *Energy Convers. Manag.* **2018**, *156*, 585–596. [CrossRef]
26. Bufi, E.A.; Camporeale, S.; Fornarelli, F.; Fortunato, B.; Pantaleo, A.M.; Sorrentino, A.; Torresi, M. Parametric multi-objective optimization of an Organic Rankine Cycle with thermal energy storage for distributed generation. In Proceedings of the Ati 2017—72nd Conference of the Italian Thermal Machines Engineering Association, Lecce, Italy, 6–8 September 2017; Fortunato, B., Ficarella, A., Torresi, M., Eds.; Volume 126, pp. 429–436.
27. Feng, Y.; Zhang, Y.; Li, B.; Yang, J.; Shi, Y. Sensitivity analysis and thermoeconomic comparison of ORCs (organic Rankine cycles) for low temperature waste heat recovery. *Energy* **2015**, *82*, 664–677. [CrossRef]
28. Feng, Y.-q.; Zhang, W.; Niaz, H.; He, Z.-x.; Wang, S.; Wang, X.; Liu, Y.-z. Parametric analysis and thermo-economical optimization of a Supercritical-Subcritical organic Rankine cycle for waste heat utilization. *Energy Convers. Manag.* **2020**, *212*, 112773. [CrossRef]
29. Tiwari, D.; Sherwani, A.F.; Kumar, N. Optimization and thermo-economic performance analysis of organic Rankine cycles using mixture working fluids driven by solar energy. *Energy Sources Part A Recovery Util. Environ. Eff.* **2019**, *41*, 1890–1907. [CrossRef]
30. Tiwari, D.; Sherwani, A.F.; Arora, A.; Haleem, A. Thermo-economic and multiobjective optimization of saturated and superheated organic Rankine cycle using a low-grade solar heat source. *J. Renew. Sustain. Energy* **2017**, *9*, 054701. [CrossRef]
31. Gotelip Correa Veloso, T.; Sotomonte, C.A.R.; Coronado, C.J.R.; Nascimento, M.A.R. Multi-objective optimization and exergetic analysis of a low-grade waste heat recovery ORC application on a Brazilian FPSO. *Energy Convers. Manag.* **2018**, *174*, 537–551. [CrossRef]
32. Bekiloglu, H.E.; Bedir, H.; Anlas, G. Multi-objective optimization of ORC parameters and selection of working fluid using preliminary radial inflow turbine design. *Energy Convers. Manag.* **2019**, *183*, 833–847. [CrossRef]
33. Bahadormanesh, N.; Rahat, S.; Yarali, M. Constrained multi-objective optimization of radial expanders in organic Rankine cycles by firefly algorithm. *Energy Convers. Manag.* **2017**, *148*, 1179–1193. [CrossRef]

34. Rahbara, K.; Mahmoud, S.; Al-Dadah, R.K.; Moazami, N. Integrated modelling and multi-objective optimization of organic Rankine cycle based on radial inflow turbine. In Proceedings of the ASME Turbo Expo 2015: Turbine Technical Conference and Exposition. Volume 3: Coal, Biomass and Alternative Fuels; Cycle Innovations; Electric Power; Industrial and Cogeneration, Montreal, QC, Canada, 15–19 June 2015; ASME: New York, NY, USA, 2015.
35. Galindo, J.; Climent, H.; Dolz, V.; Royo-Pascual, L. Multi-objective optimization of a bottoming Organic Rankine Cycle (ORC) of gasoline engine using swash-plate expander. *Energy Convers. Manag.* **2016**, *126*, 1054–1065. [CrossRef]
36. Zhang, C.; Liu, C.; Xu, X.; Li, Q.; Wang, S. Energetic, exergetic, economic and environmental (4E) analysis and multi-factor evaluation method of low GWP fluids in trans-critical organic Rankine cycles. *Energy* **2019**, *168*, 332–345. [CrossRef]
37. Shu, G.; Liu, P.; Tian, H.; Wang, X.; Jing, D. Operational profile based thermal-economic analysis on an Organic Rankine cycle using for harvesting marine engine's exhaust waste heat. *Energy Convers. Manag.* **2017**, *146*, 107–123. [CrossRef]
38. Turton, R.; Bailie, R.C.; Whiting, W.B.; Shaeiwitz, J.A. *Analysis, Synthesis and Design of Chemical Processes*; Pearson Education: London, UK, 2008.
39. Smith, R. *Chemical Process: Design and Integration*; John Wiley & Sons: Hoboken, NJ, USA, 2005.
40. Sun, Z.; Liu, C.; Xu, X.; Li, Q.; Wang, X.; Wang, S.; Chen, X. Comparative carbon and water footprint analysis and optimization of Organic Rankine Cycle. *Appl. Therm. Eng.* **2019**, *158*, 113769. [CrossRef]
41. Yang, M.-H.; Yeh, R.-H.; Hung, T.-C. Thermo-economic analysis of the transcritical organic Rankine cycle using R1234yf/R32 mixtures as the working fluids for lower-grade waste heat recovery. *Energy* **2017**, *140*, 818–836. [CrossRef]
42. Sun, Q.; Wang, Y.; Cheng, Z.; Wang, J.; Zhao, P.; Dai, Y. Thermodynamic and economic optimization of a double-pressure organic Rankine cycle driven by low-temperature heat source. *Renew. Energy* **2020**, *147*, 2822–2832. [CrossRef]
43. Oyewunmi, O.A.; Markides, C.N. Thermo-Economic and Heat Transfer Optimization of Working-Fluid Mixtures in a Low-Temperature Organic Rankine Cycle System. *Energies* **2016**, *9*, 448. [CrossRef]
44. Wang, Z.; Hu, Y.; Xia, X.; Zuo, Q.; Zhao, B.; Li, Z. Thermo-economic selection criteria of working fluid used in dual-loop ORC for engine waste heat recovery by multi-objective optimization. *Energy* **2020**, *197*, 117053. [CrossRef]
45. Laouid, Y.A.A.; Kezrane, C.; Lasbet, Y.; Pesyridis, A. Towards improvement of waste heat recovery systems: A multi-objective optimization of different Organic Rankine cycle configurations. *Int. J.* **2021**, *11*, 100100.
46. Fang, Y.; Yang, F.; Zhang, H. Comparative analysis and multi-objective optimization of organic Rankine cycle (ORC) using pure working fluids and their zeotropic mixtures for diesel engine waste heat recovery. *Appl. Therm. Eng.* **2019**, *157*, 113704. [CrossRef]
47. Feng, Y.; Zhang, Y.; Li, B.; Yang, J.; Shi, Y. Comparison between regenerative organic Rankine cycle (RORC) and basic organic Rankine cycle (BORC) based on thermoeconomic multi-objective optimization considering exergy efficiency and levelized energy cost (LEC). *Energy Convers. Manag.* **2015**, *96*, 58–71. [CrossRef]
48. Hu, S.; Yang, Z.; Li, J.; Duan, Y. Thermo-economic optimization of the hybrid geothermal-solar power system: A data-driven method based on lifetime off-design operation. *Energy Convers. Manag.* **2021**, *229*, 113738. [CrossRef]
49. Pierobon, L.; Benato, A.; Scolari, E.; Haglind, F.; Stoppato, A. Waste heat recovery technologies for offshore platforms. *Appl. Energy* **2014**, *136*, 228–241. [CrossRef]
50. Moreira, L.F.; Arrieta, F.R.P. Thermal and economic assessment of organic Rankine cycles for waste heat recovery in cement plants. *Renew. Sust. Energ. Rev.* **2019**, *114*, 109315. [CrossRef]
51. Tsatsaronis, G. Definitions and nomenclature in exergy analysis and exergoeconomics. *Energy* **2007**, *32*, 249–253. [CrossRef]
52. Xia, J.; Wang, J.; Zhang, G.; Lou, J.; Zhao, P.; Dai, Y. Thermo-economic analysis and comparative study of transcritical power cycles using CO_2-based mixtures as working fluids. *Appl. Therm. Eng.* **2018**, *144*, 31–44. [CrossRef]
53. Nasruddin, N.; Saputra, I.D.; Mentari, T.; Bardow, A.; Marcelina, O.; Berlin, S. Exergy, exergoeconomic, and exergoenvironmental optimization of the geothermal binary cycle power plant at Ampallas, West Sulawesi, Indonesia. *Therm. Sci. Eng. Prog.* **2020**, *19*, 100625. [CrossRef]
54. Özahi, E.; Tozlu, A.; Abuşoğlu, A. Thermoeconomic multi-objective optimization of an organic Rankine cycle (ORC) adapted to an existing solid waste power plant. *Energy Convers. Manag.* **2018**, *168*, 308–319. [CrossRef]
55. Behzadi, A.; Gholamian, E.; Houshfar, E.; Habibollahzade, A. Multi-objective optimization and exergoeconomic analysis of waste heat recovery from Tehran's waste-to-energy plant integrated with an ORC unit. *Energy* **2018**, *160*, 1055–1068. [CrossRef]
56. Jankowski, M.; Borsukiewicz, A.; Szopik-Depczynska, K.; Ioppolo, G. Determination of an optimal pinch point temperature difference interval in ORC power plant using multi-objective approach. *J. Clean. Prod.* **2019**, *217*, 798–807. [CrossRef]
57. Martinez-Gomez, J.; Pena-Lamas, J.; Martin, M.; Ponce-Ortega, J.M. A multi-objective optimization approach for the selection of working fluids of geothermal facilities: Economic, environmental and social aspects. *J. Environ. Manag.* **2017**, *203*, 962–972. [CrossRef]
58. Yang, J.; Gao, L.; Ye, Z.; Hwang, Y.; Chen, J. Binary-objective optimization of latest low-GWP alternatives to R245fa for organic Rankine cycle application. *Energy* **2021**, *217*, 119336. [CrossRef]
59. Xia, X.X.; Wang, Z.Q.; Zhou, N.J.; Hu, Y.H.; Zhang, J.P.; Chen, Y. Working fluid selection of dual-loop organic Rankine cycle using multi-objective optimization and improved grey relational analysis. *Appl. Therm. Eng.* **2020**, *171*, 115028. [CrossRef]
60. Valencia, G.; Fontalvo, A.; Forero, J.D. Optimization of waste heat recovery in internal combustion engine using a dual-loop organic Rankine cycle: Thermo-economic and environmental footprint analysis. *Appl. Therm. Eng.* **2021**, *182*, 116109. [CrossRef]

61. Matuszewska, D.; Kuta, M.; Gorski, J. Multi-objective optimization of ORC geothermal conversion system integrated with life cycle assessment. In Proceedings of the 17th International Conference Heat Transfer and Renewable Sources of Energy, Międzyzdroje, Poland, 2–5 September 2018; Kujawa, T., Stachel, A.A., Zapalowicz, Z., Eds.; Volume 70.
62. Herrera-Orozco, I.; Valencia-Ochoa, G.; Duarte-Forero, J. Exergo-environmental assessment and multi-objective optimization of waste heat recovery systems based on Organic Rankine cycle configurations. *J. Clean. Prod.* **2021**, *288*, 125679.
63. Yi, Z.; Luo, X.; Yang, Z.; Wang, C.; Chen, J.; Chen, Y.; Ponce-Ortega, J.M. Thermo-economic-environmental optimization of a liquid separation condensation-based organic Rankine cycle driven by waste heat. *J. Clean. Prod.* **2018**, *184*, 198–210. [CrossRef]
64. Kalikatzarakis, M.; Frangopoulos, C.A. Multi-criteria Selection and Thermo-economic Optimization of an Organic Rankine Cycle System for a Marine Application. *Int. J. Thermodyn.* **2015**, *18*, 133–141. [CrossRef]
65. Fergani, Z.; Touil, D.; Morosuk, T. Multi-criteria exergy based optimization of an Organic Rankine Cycle for waste heat recovery in the cement industry. *Energy Convers. Manag.* **2016**, *112*, 81–90. [CrossRef]
66. Jankowski, M.; Borsukiewicz, A. Multi-objective approach for determination of optimal operating parameters in low-temperature ORC power plant. *Energy Convers. Manag.* **2019**, *200*, 112075. [CrossRef]
67. Rosen, M.A.; Dincer, I.; Kanoglu, M. Role of exergy in increasing efficiency and sustainability and reducing environmental impact. *Energy Policy* **2008**, *36*, 128–137. [CrossRef]
68. Xiao, L.; Wu, S.-Y.; Yi, T.-T.; Liu, C.; Li, Y.-R. Multi-objective optimization of evaporation and condensation temperatures for subcritical organic Rankine cycle. *Energy* **2015**, *83*, 723–733. [CrossRef]
69. Xu, G.; Zhu, P.; Quan, Y.; Dong, B.; Jin, R. Multi-objective optimization design of plate-fin vapor generator for supercritical organic Rankine cycle. *Int. J. Energy Res.* **2019**, *43*, 2312–2326. [CrossRef]
70. Baldasso, E.; Andreasen, J.G.; Mondejar, M.E.; Larsen, U.; Haglind, F. Technical and economic feasibility of organic Rankine cycle-based waste heat recovery systems on feeder ships: Impact of nitrogen oxides emission abatement technologies. *Energy Convers. Manag.* **2019**, *183*, 577–589. [CrossRef]
71. Liu, H.; Zhang, H.; Yang, F.; Hou, X.; Yu, F.; Song, S. Multi-objective optimization of fin-and-tube evaporator for a diesel engine-organic Rankine cycle (ORC) combined system using particle swarm optimization algorithm. *Energy Convers. Manag.* **2017**, *151*, 147–157. [CrossRef]
72. Barbazza, L.; Pierobon, L.; Mirandola, A.; Haglind, F. Optimal design of compact organic rankine cycle units for domestic solar applications. *Therm. Sci.* **2014**, *18*, 811–822. [CrossRef]
73. Imran, M.; Haglind, F.; Lemort, V.; Meroni, A. Optimization of organic rankine cycle power systems for waste heat recovery on heavy-duty vehicles considering the performance, cost, mass and volume of the system. *Energy* **2019**, *180*, 229–241. [CrossRef]
74. Pierobon, L.; Tuong-Van, N.; Larsen, U.; Haglind, F.; Elmegaard, B. Multi-objective optimization of organic Rankine cycles for waste heat recovery: Application in an offshore platform. *Energy* **2013**, *58*, 538–549. [CrossRef]
75. Lee, Y.; Kim, J.; Ahmed, U.; Kim, C.; Lee, Y.-W. Multi-objective optimization of Organic Rankine Cycle (ORC) design considering exergy efficiency and inherent safety for LNG cold energy utilization. *J. Loss Prev. Process. Ind.* **2019**, *58*, 90–101. [CrossRef]
76. Papadopoulos, A.I.; Stijepovic, M.; Linke, P.; Seferlis, P.; Voutetakis, S. Molecular Design of Working Fluid Mixtures for Organic Rankine Cycles. In *23 European Symposium on Computer Aided Process Engineering*; Kraslawski, A., Turunen, I., Eds.; Elsevier: Amsterdam, The Netherlands, 2013; Volume 32, pp. 289–294.
77. Li, S.; Li, W. Thermo-economic optimization of solar organic Rankine cycle based on typical solar radiation year. *Energy Convers. Manag.* **2018**, *169*, 78–87. [CrossRef]
78. Bufi, E.A.; Camporeale, S.M.; Cinnella, P. Robust optimization of an Organic Rankine Cycle for heavy duty engine waste heat recovery. *Energy Procedia* **2017**, *129*, 66–73. [CrossRef]
79. Zhang, J.; Ren, M.; Xiong, J.; Lin, M. Multi-objective Optimal Temperature Control. for Organic Rankine Cycle Systems. In Proceedings of the World Congress on Intelligent Control and Automation (WCICA), Shenyang, China, 29 June–4 July 2014; pp. 661–666.
80. Xin, B.; Chen, L.; Chen, J.; Ishibuchi, H.; Hirota, K.; Liu, B. Interactive Multiobjective Optimization: A Review of the State-of-the-Art. *IEEE Access* **2018**, *6*, 41256–41279. [CrossRef]
81. Wang, Z.-q.; Zhou, N.-j.; Zhang, J.-q.; Guo, J.; Wang, X.-y. Parametric optimization and performance comparison of organic Rankine cycle with simulated annealing algorithm. *J. Cent. South. Univ.* **2012**, *19*, 2584–2590. [CrossRef]
82. Liu, X.; Wei, M.; Yang, L.; Wang, X. Thermo-economic analysis and optimization selection of ORC system configurations for low temperature binary-cycle geothermal plant. *Appl. Therm. Eng.* **2017**, *125*, 153–164. [CrossRef]
83. Arasteh, A.M.; Akbari, H.H.A. Waste heat recovery from data centers using Organic Rankine Cycle (ORC), and Multi-objective energy and exergy optimization of the system in marine industries. *Res. Mar. Sci.* **2020**, *5*, 645–654.
84. Zhu, Y.; Li, W.; Sun, G.; Li, H. Thermo-economic analysis based on objective functions of an organic Rankine cycle for waste heat recovery from marine diesel engine. *Energy* **2018**, *158*, 343–356. [CrossRef]
85. Jankowski, M.; Wisniewski, S.; Borsukiewicz, A. Multi-objective analysis of an influence of a brine mineralization on an optimal evaporation temperature in ORC power plant. In Proceedings of the 17th International Conference Heat Transfer and Renewable Sources of Energy, Międzyzdroje, Poland, 2–5 September 2018; Kujawa, T., Stachel, A.A., Zapalowicz, Z., Eds.; Volume 70.
86. White, M.; Sayma, A.I. System and component modelling and optimisation for an efficient 10kWe low-temperature organic Rankine cycle utilising a radial inflow expander. *Proc. Inst. Mech. Eng. Part. A J. Power Energy* **2015**, *229*, 795–809. [CrossRef]

87. Karimi, S.; Mansouri, S. A comparative profitability study of geothermal electricity production in developed and developing countries: Exergoeconomic analysis and optimization of different ORC configurations. *Renew. Energy* **2018**, *115*, 600–619. [CrossRef]
88. Zhang, M.-G.; Zhao, L.-J.; Xiong, Z. Performance evaluation of organic Rankine cycle systems utilizing low grade energy at different temperature. *Energy* **2017**, *127*, 397–407. [CrossRef]
89. Kazemi, N.; Samadi, F. Thermodynamic, economic and thermo-economic optimization of a new proposed organic Rankine cycle for energy production from geothermal resources. *Energy Convers. Manag.* **2016**, *121*, 391–401. [CrossRef]
90. Sun, Z.; Liu, C.; Wang, S. Multi-objective decision framework for comprehensive assessment of organic Rankine cycle system. *J. Renew. Sustain. Energy* **2020**, *12*, 014702. [CrossRef]
91. Bademlioglu, A.H.; Canbolat, A.S.; Kaynakli, O. Multi-objective optimization of parameters affecting Organic Rankine Cycle performance characteristics with Taguchi-Grey Relational Analysis. *Renew. Sust. Energ. Rev.* **2020**, *117*, 109483. [CrossRef]
92. Wang, M.; Jing, R.; Zhang, H.; Meng, C.; Li, N.; Zhao, Y. An innovative Organic Rankine Cycle (ORC) based Ocean Thermal Energy Conversion (OTEC) system with performance simulation and multi-objective optimization. *Appl. Therm. Eng.* **2018**, *145*, 743–754. [CrossRef]
93. Rosset, K.; Mounier, V.; Guenat, E.; Schiffmann, J. Multi-objective optimization of turbo-ORC systems for waste heat recovery on passenger car engines. *Energy* **2018**, *159*, 751–765. [CrossRef]
94. Wang, W.; Deng, S.; Zhao, D.; Zhao, L.; Lin, S.; Chen, M. Application of machine learning into organic Rankine cycle for prediction and optimization of thermal and exergy efficiency. *Energy Convers. Manag.* **2020**, *210*, 112700. [CrossRef]
95. Coello, C.C.; Lechuga, M.S. *MOPSO: A Proposal for Multiple Objective Particle Swarm Optimization, Proceedings of the 2002 Congress on Evolutionary Computation. CEC'02 (Cat. No. 02TH8600), Honolulu, HI, USA, 12–17 May 2002*; IEEE: Piscataway, NJ, USA, 2002; pp. 1051–1056.
96. Zhou, A.; Qu, B.-Y.; Li, H.; Zhao, S.-Z.; Suganthan, P.N.; Zhang, Q. Multiobjective evolutionary algorithms: A survey of the state of the art. *Swarm Evol. Comput.* **2011**, *1*, 32–49. [CrossRef]
97. Prajapati, P.P.; Patel, V.K. Thermo-economic optimization of a nanofluid based organic Rankine cycle: A multi-objective study and analysis. *Therm. Sci. Eng. Prog.* **2020**, *17*, 100381. [CrossRef]
98. Turgut, M.S.; Turgut, O.E. Multi-objective optimization of the basic and single-stage Organic Rankine Cycles utilizing a low-grade heat source. *Heat Mass Transf.* **2019**, *55*, 353–374. [CrossRef]
99. Li, P.; Han, Z.; Jia, X.; Mei, Z.; Han, X.; Wang, Z. An Improved Analysis Method for Organic Rankine Cycles Based on Radial-Inflow Turbine Efficiency Prediction. *Appl. Sci. Basel* **2019**, *9*, 49. [CrossRef]
100. Sadeghi, S.; Maghsoudi, P.; Shabani, B.; Gorgani, H.H.; Shabani, N. Performance analysis and multi-objective optimization of an organic Rankine cycle with binary zeotropic working fluid employing modified artificial bee colony algorithm. *J. Therm. Anal. Calorim.* **2019**, *136*, 1645–1665. [CrossRef]
101. Han, F.; Wang, Z.; Ji, Y.; Li, W.; Sunden, B. Energy analysis and multi-objective optimization of waste heat and cold energy recovery process in LNG-fueled vessels based on a triple organic Rankine cycle. *Energy Convers. Manag.* **2019**, *195*, 561–572. [CrossRef]
102. Li, P.; Mei, Z.; Han, Z.; Jia, X.; Zhu, L.; Wang, S. Multi-objective optimization and improved analysis of an organic Rankine cycle coupled with the dynamic turbine efficiency model. *Appl. Therm. Eng.* **2019**, *150*, 912–922. [CrossRef]
103. Hu, S.; Li, J.; Yang, F.; Yang, Z.; Duan, Y. Multi-objective optimization of organic Rankine cycle using hydrofluorolefins (HFOs) based on different target preferences. *Energy* **2020**, *203*, 117848. [CrossRef]
104. Han, Z.; Mei, Z.; Li, P. Multi-objective optimization and sensitivity analysis of an organic Rankine cycle coupled with a one-dimensional radial-inflow turbine efficiency prediction model. *Energy Convers. Manag.* **2018**, *166*, 37–47. [CrossRef]
105. Yao, S.; Zhang, Y.; Yu, X. Thermo-economic analysis of a novel power generation system integrating a natural gas expansion plant with a geothermal ORC in Tianjin, China. *Energy* **2018**, *164*, 602–614. [CrossRef]
106. Yang, F.; Cho, H.; Zhang, H.; Zhang, J. Thermoeconomic multi-objective optimization of a dual loop organic Rankine cycle (ORC) for CNG engine waste heat recovery. *Appl. Energy* **2017**, *205*, 1100–1118. [CrossRef]
107. Jankowski, M.; Borsukiewicz, A.; Hooman, K. Development of Decision-Making Tool and Pareto Set Analysis for Bi-Objective Optimization of an ORC Power Plant. *Energies* **2020**, *13*, 5280. [CrossRef]
108. Zhang, X.; Bai, H.; Zhao, X.; Diabat, A.; Zhang, J.; Yuan, H.; Zhang, Z. Multi-objective optimisation and fast decision-making method for working fluid selection in organic Rankine cycle with low-temperature waste heat source in industry. *Energy Convers. Manag.* **2018**, *172*, 200–211. [CrossRef]
109. Habibi, H.; Zoghi, M.; Chitsaz, A.; Shamsaiee, M. Thermo-economic performance evaluation and multi-objective optimization of a screw expander-based cascade Rankine cycle integrated with parabolic trough solar collector. *Appl. Therm. Eng.* **2020**, *180*, 115827. [CrossRef]
110. Nematollahi, O.; Hajabdollahi, Z.; Hoghooghi, H.; Kim, K.C. An evaluation of wind turbine waste heat recovery using organic Rankine cycle. *J. Clean. Prod.* **2019**, *214*, 705–716. [CrossRef]
111. Li, P.; Han, Z.; Jia, X.; Mei, Z.; Han, X.; Wang, Z. Comparative analysis of an organic Rankine cycle with different turbine efficiency models based on multi-objective optimization. *Energy Convers. Manag.* **2019**, *185*, 130–142. [CrossRef]

112. Tian, Z.; Yue, Y.; Zhang, Y.; Gu, B.; Gao, W. Multi-Objective Thermo-Economic Optimization of a Combined Organic Rankine Cycle (ORC) System Based on Waste Heat of Dual Fuel Marine Engine and LNG Cold Energy Recovery. *Energies* **2020**, *13*, 1397. [CrossRef]
113. Tian, Z.; Yue, Y.; Gu, B.; Gao, W.; Zhang, Y. Thermo-economic analysis and optimization of a combined Organic Rankine Cycle (ORC) system withLNGcold energy and waste heat recovery of dual-fuel marine engine. *Int. J. Energy Res.* **2020**, *44*, 9974–9994. [CrossRef]
114. Jankowski, M.; Borsukiewicz, A.; Wisniewski, S.; Hooman, K. Multi-objective analysis of an influence of a geothermal water salinity on optimal operating parameters in low-temperature ORC power plant. *Energy* **2020**, *202*, 117666. [CrossRef]
115. Wang, Y.Z.; Zhao, J.; Wang, Y.; An, Q.S. Multi-objective optimization and grey relational analysis on configurations of organic Rankine cycle. *Appl. Therm. Eng.* **2017**, *114*, 1355–1363. [CrossRef]
116. Hu, S.; Li, J.; Yang, F.; Yang, Z.; Duan, Y. How to design organic Rankine cycle system under fluctuating ambient temperature: A multi-objective approach. *Energy Convers. Manag.* **2020**, *224*, 113331. [CrossRef]
117. Kermani, M.; Wallerand, A.S.; Kantor, I.D.; Maréchal, F. Generic superstructure synthesis of organic Rankine cycles for waste heat recovery in industrial processes. *Appl. Energy* **2018**, *212*, 1203–1225. [CrossRef]
118. Zhao, Y.; Liu, G.; Li, L.; Yang, Q.; Tang, B.; Liu, Y. Expansion devices for organic Rankine cycle (ORC) using in low temperature heat recovery: A review. *Energy Convers. Manag.* **2019**, *199*, 111944. [CrossRef]
119. Valencia, G.; Núñez, J.; Duarte, J. Multiobjective Optimization of a Plate Heat Exchanger in a Waste Heat Recovery Organic Rankine Cycle System for Natural Gas Engines. *Entropy* **2019**, *21*, 655. [CrossRef] [PubMed]
120. Wei, M.; Fu, L.; Zhang, S.; Zhao, X. Experimental investigation on vapor-pump equipped gas boiler for flue gas heat recovery. *Appl. Therm. Eng.* **2019**, *147*, 371–379. [CrossRef]
121. Schilling, J.; Tillmanns, D.; Lampe, M.; Hopp, M.; Gross, J.; Bardow, A. From molecules to dollars: Integrating molecular design into thermo-economic process design using consistent thermodynamic modeling. *Mol. Syst. Des. Eng.* **2017**, *2*, 301–320. [CrossRef]
122. van Kleef, L.M.T.; Oyewunmi, O.A.; Markides, C.N. Multi-objective thermo-economic optimization of organic Rankine cycle (ORC) power systems in waste-heat recovery applications using computer-aided molecular design techniques. *Appl. Energy* **2019**, *251*, 112513. [CrossRef]
123. Holik, M.; Zivic, M.; Virag, Z.; Barac, A. Optimization of an organic Rankine cycle constrained by the application of compact heat exchangers. *Energy Convers. Manag.* **2019**, *188*, 333–345. [CrossRef]
124. Imran, M.; Usman, M.; Park, B.-S.; Kim, H.-J.; Lee, D.-H. Multi-objective optimization of evaporator of organic Rankine cycle (ORC) for low temperature geothermal heat source. *Appl. Therm. Eng.* **2015**, *80*, 1–9. [CrossRef]
125. Wang, J.; Wang, M.; Li, M.; Xia, J.; Dai, Y. Multi-objective optimization design of condenser in an organic Rankine cycle for low grade waste heat recovery using evolutionary algorithm. *Int. Commun. Heat Mass Transf.* **2013**, *45*, 47–54. [CrossRef]
126. Li, Y.; Li, W.; Gao, X.; Ling, X. Thermodynamic analysis and optimization of organic Rankine cycles based on radial-inflow turbine design. *Appl. Therm. Eng.* **2021**, *184*, 11627. [CrossRef]
127. Alshammari, F.; Pesyridis, A.; Elashmawy, M. Turbine optimization potential to improve automotive Rankine cycle performance. *Appl. Therm. Eng.* **2021**, *186*, 116559. [CrossRef]
128. Jankowski, M.; Borsukiewicz, A. A Novel Exergy Indicator for Maximizing Energy Utilization in Low-Temperature ORC. *Energies* **2020**, *13*, 1598. [CrossRef]
129. Palagi, L.; Sciubba, E.; Tocci, L. A neural network approach to the combined multi-objective optimization of the thermodynamic cycle and the radial inflow turbine for Organic Rankine cycle applications. *Appl. Energy* **2019**, *237*, 210–226. [CrossRef]
130. Al Jubori, A.M.; Al-Dadah, R.; Mahmoud, S. Performance enhancement of a small-scale organic Rankine cycle radial-inflow turbine through multi-objective optimization algorithm. *Energy* **2017**, *131*, 297–311. [CrossRef]
131. Erbas, M.; Biyikoglu, A. Design and multi-objective optimization of organic Rankine turbine. *Int. J. Hydrog. Energy* **2015**, *40*, 15343–15351. [CrossRef]
132. Scardigno, D.; Fanelli, E.; Viggiano, A.; Braccio, G.; Magi, V. A genetic optimization of a hybrid organic Rankine plant for solar and low-grade energy sources. *Energy* **2015**, *91*, 807–815.
133. Tiwari, D.; Sherwani, A.F.; Arora, A. Thermodynamic and multi-objective optimisation of solar-driven Organic Rankine Cycle using zeotropic mixtures. *Int. J. Ambient. Energy* **2019**, *40*, 135–151. [CrossRef]
134. Yang, F.; Yang, F.; Chu, Q.; Liu, Q.; Yang, Z.; Duan, Y. Thermodynamic performance limits of the organic Rankine cycle: Working fluid parameterization based on corresponding states modeling. *Energy Convers. Manag.* **2020**, *217*, 113011. [CrossRef]
135. Su, W.; Zhao, L.; Deng, S. Group contribution methods in thermodynamic cycles: Physical properties estimation of pure working fluids. *Renew. Sust. Energ. Rev.* **2017**, *79*, 984–1001.
136. Papadopoulos, A.I.; Stijepovic, M.; Linke, P. On the systematic design and selection of optimal working fluids for Organic Rankine Cycles. *Appl. Therm. Eng.* **2010**, *30*, 760–769. [CrossRef]
137. Su, W.; Zhao, L.; Deng, S. Simultaneous working fluids design and cycle optimization for Organic Rankine cycle using group contribution model. *Appl. Energy* **2017**, *202*, 618–627. [CrossRef]
138. White, M.T.; Sayma, A.I. Simultaneous Cycle Optimization and Fluid Selection for ORC Systems Accounting for the Effect of the Operating Conditions on Turbine Efficiency. *Front. Energy Res.* **2019**, *7*, 50.
139. Klamt, A. Conductor-like screening model for real solvents: A new approach to the quantitative calculation of solvation phenomena. *J. Phys. Chem.* **1995**, *99*, 2224–2235.

140. Yang, J.; Yang, Z.; Duan, Y. Optimal capacity and operation strategy of a solar-wind hybrid renewable energy system. *Energy Convers. Manag.* **2021**, *244*, 114519. [CrossRef]
141. Bina, S.M.; Jalilinasrabady, S.; Fujii, H. Thermo-economic evaluation of various bottoming ORCs for geothermal power plant, determination of optimum cycle for Sabalan power plant exhaust. *Geothermics* **2017**, *70*, 181–191. [CrossRef]
142. Van Erdeweghe, S.; Van Bael, J.; Laenen, B.; D'haeseleer, W. Design and off-design optimization procedure for low-temperature geothermal organic Rankine cycles. *Appl. Energy* **2019**, *242*, 716–731. [CrossRef]
143. Wang, L.; Bu, X.; Li, H. Multi-objective optimization and off-design evaluation of organic rankine cycle (ORC) for low-grade waste heat recovery. *Energy* **2020**, *203*, 117809. [CrossRef]
144. Petrollese, M.; Cocco, D. A multi-scenario approach for a robust design of solar-based ORC systems. *Renew. Energy* **2020**, *203*, 117809. [CrossRef]
145. Feng, Y.-q.; Liu, Y.-Z.; Wang, X.; He, Z.-X.; Hung, T.-C.; Wang, Q.; Xi, H. Performance prediction and optimization of an organic Rankine cycle (ORC) for waste heat recovery using back propagation neural network. *Energy Convers. Manag.* **2020**, *226*, 113552. [CrossRef]
146. Yang, F.; Cho, H.; Zhang, H. Performance prediction and optimization of an orgaic Rankine cycle (ORC) using back propagation neural network for diesel engine waste heat recovery. In *Energy Sustainability*; American Society of Mechanical Engineers: Lake Buena Vista, FL, USA, 2018.

energies

MDPI

Review

Use of Learning Mechanisms to Improve the Condition Monitoring of Wind Turbine Generators: A Review

Ana Rita Nunes [1], Hugo Morais [1,2,*] and Alberto Sardinha [1,2]

1 Instituto Superior Técnico, Universidade de Lisboa, Av. Rovisco Pais, 1049-001 Lisboa, Portugal; ana.rita.nunes@tecnico.ulisboa.pt (A.R.N.); jose.alberto.sardinha@tecnico.ulisboa.pt (A.S.)
2 INESC-ID, Universidade de Lisboa, 1049-001 Lisboa, Portugal
* Correspondence: hugo.morais@tecnico.ulisboa.pt

Abstract: The main goal of this paper is to review and evaluate how we can take advantage of state-of-the-art machine learning techniques and apply them in wind energy operation conditions monitoring and fault diagnosis, boosting wind turbines' availability. To accomplish this, we focus our work on analysing the current techniques in predictive maintenance, which are aimed at acting before a major failure occurs using condition monitoring. In particular, we start framing the predictive maintenance problem as an ML problem to detect patterns that indicate a fault on turbine generators. Then, we extend the problem to detect future faults. Therefore, this review will consist of analysing techniques to tackle the challenges of each machine learning stage, such as data pre-processing, feature engineering, and the selection of the best-suited model. By using specific evaluation metrics, the expected final result of using these techniques will be an improvement in the early prediction of a future fault. This improvement will have an increase in the availability of the turbine, and therefore in energy production.

Keywords: condition monitoring; fault detection; machine learning; wind farm

Citation: Nunes, A.R.; Morais, H.; Sardinha, A. Use of Learning Mechanisms to Improve the Condition Monitoring of Wind Turbine Generators: A Review. *Energies* **2021**, *14*, 7129. https://doi.org/10.3390/en14217129

Academic Editor: Wei-Hsin Chen

Received: 10 August 2021
Accepted: 19 October 2021
Published: 1 November 2021

1. Introduction

As a consequence of increasing climate change awareness, research on topics such as renewable energy are of extreme importance. An event that played a significant role in rapidly finding practical solutions to stop those changes was the Paris Agreement. The long term goal of that agreement was to limit temperature rises; therefore, countries needed to adapt their current practices to reduce carbon emissions (https://unfccc.int/process-and-meetings/the-paris-agreement/the-paris-agreement (accessed on 17 October 2021)).

Since the power sector is one of the main contributors to global greenhouse gas emissions (https://www.c2es.org/content/international-emissions (accessed on 17 October 2021)), new technologies for cleaner energy production have been developed to replace conventional production (e.g., production based on fossil fuels) [1]. To create incentives for developers to invest in cleaner energy technologies, countries (e.g., China, Russia, India, Japan, Brazil, and European countries) have proposed national action plans. By analysing those national plans, we can observe that some of them highlighted that wind energy can have a major contribution as a zero-emission energy source. Consequently, the "European Green Deal" (https://ec.europa.eu/info/strategy/priorities-2019-2024/european-green-deal_en (accessed on 17 October 2021)) emerged as an extension at the continent level. The main objective of the EU Green Deal is for the EU to become the first climate-neutral continent by 2050 [2]. Regarding the energy sector, these are some of the main goals: build interconnected energy systems and better integrated grids to support renewable energy sources, decarbonize the gas sector, promote smart integration across sectors, and develop the full potential of Europe's off-shore wind energy [3].

As a consequence of the previous initiative, the future of energy will profoundly depend on renewable energies, such as wind and photovoltaic energy. Therefore, this review will focus on wind technology due to its increasing importance in the last few years. The cumulative installed capacity of both off- and on-shore wind energy has been growing, with global yearly new installations rounding 50 GW. In 2020, it even surpassed 90 GW, a 53% growth compared to 2019, bringing the total installed capacity to 743 GW. (https://gwec.net/wp-content/uploads/2021/03/GWEC-Global-Wind-Report-2021.pdf (accessed on 17 October 2021)) According to [4], the global installed capacity of on-shore wind power would increase (taking 2018 as reference) three-fold by 2030 (to 1787 GW) and ten-fold by 2050 (to 5044 GW). Concerning off-shore installations, the predictions are similar, with the global off-shore wind capacity rising 228 GW in 2030 and nearly 1000 GW in 2050, compared with the 23 GW installed in 2018 [4].

Wind farms are usually connected in remote areas far from the big cities and large consumption centers. A key reason for energy producers to select remote locations, such as off-shore, is the better wind conditions. Near big cities, wind suffers a lot of interference, which is not ideal for wind generators. Adding to that, we avoid visual impact and land use issues. However, some disadvantages arise from this location decision, such as connecting the wind farm to the transmission networks, transportation and installation of giant turbines, and operation and maintenance (OM) costs. Nevertheless, knowing that previously mentioned benefits might reveal that it is worth investing in transmission infrastructure to access them [5]. To reinforce the need for OM, wind farms in general and wind turbines in particular are exposed to unpredictable and harsh weather conditions, which result in highly variable and volatile operational conditions, leading to intense mechanical stress (the description of more specific problems of wind turbine generators is covered in Section 2.1).

Since it is complicated to reduce the initial costs of transportation and installation, most of the research has focused on maintenance, more specifically in the use of condition monitoring (CM). CM consists of monitoring the components of a wind turbine to identify changes in operation that can be indicative of a developing fault and preventing it through maintenance. CM increases the availability of the wind farm and, as a result, the production of electricity, decreasing the global cost of the project. There are different types of maintenance, as we will see in Section 2.1. However, the one that uses CM is called predictive maintenance. This type of maintenance has two main advantages: (i) increasing the availability of wind generators and (ii) reducing the costs of wind farms maintenance by reducing or delaying corrective maintenance actions. The first advantage plays quite an important role. For example, a wind turbine (WT) of 2.0 MW can generate 48 MWh during a day (in maximum) and generate a revenue of 3600€ when considering an average FiT (feed-in tariff) (Policy mechanism designed to accelerate investment in renewable energy technologies by offering long-term contracts to renewable energy producers) of 75 €/MWh.

Until recently, monitoring has only relied on manual and straightforward analyses of specific measurements and aspects of operation [6]. However, this type of analysis is inefficient when detecting electrical (including power electronics), mechanical, and hydraulic problems. To date, developments in sensors and signal processing systems have improved the quality and quantity of data obtained [7]. These data, when combined with machine learning (ML) and big data analysis and management, have opened up a new world of possibilities for reliable, cost-effective, and more accurate decision-making in CM.

The key goal of this review is to take advantage of these previously mentioned advances, applying them in wind turbines generator condition monitoring. In particular, we can frame the predictive maintenance problem as an ML problem, finding the best tools and methods for each one of the sub-tasks in Figure 1. Using real data obtained from SCADA (Supervisory Control And Data Acquisition) systems, the first task, Data acquisition and pre-processing, will deal with the variety and quantity of data obtained. In other words, this task does the necessary pre-processing of the data to remove outliers and treat missing values. The second task, feature selection, selects the features that better represent the patterns in the data, removing unnecessary noise. According to the obtained data, the third stage aims to select the best ML model to detect and predict faults. Finally, in the validation stage, we assess the accuracy of the ML model on classifying new data as representative of a failure in the turbine or not.

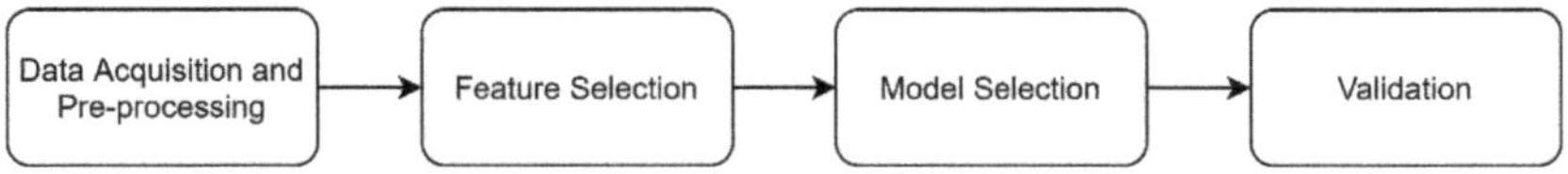

Figure 1. Machine Learning Phases.

The main difference between this review and the existing ones [8–13] is that it tries to gather the strengths of each article from a ML point of view. It will not be a general review on wind turbine CM, but a tailored version comprising only ML approaches. Moreover, it will be organized as an ML problem, focusing on each of the stages mentioned in Figure 1. In addition, we will trim the review to turbine generator problems. We will start by giving a broader view on fault detection through models that can predict production related problems, models that focus on multi-target prediction, others that use transfer learning and ending with federated deep learning. Subsequently, by focusing on generator faults, we review the prediction of more specific problems, giving additional information regarding that fault that can help to prevent it. In conclusion, the present review discusses the most important ML techniques used in condition monitoring, considering a structured methodology used in data analytics. Therefore, this review will help to structure the implementation of each of the ML phases on turbine generators CM.

The remainder of this paper is organized as follows. Section 2 presents a brief background on CM of wind turbines and a state-of-the-art review on machine learning. Section 3 provides the relevant related work on each of the machine learning phases. Finally, in Section 4, we present a discussion of measures to validate the quality of the proposed models, and Section 5 summarizes the findings with conclusions, and discusses future work.

2. Background

To better understand the following sections, in Section 2.1, we present an introduction regarding condition monitoring of wind turbines. Secondly, in Section 2.2, we review state-of-the-art machine learning to help substantiate the critical analysis. Additional information on how a wind turbine works or on its key components can be found in several references in the literature, such as in Pao et al. [14].

2.1. Condition Monitoring of Wind Turbines

This section shows how to apply CM on wind turbines and presents the different approaches.

The maintenance cost of components of a wind turbine strongly depends on how we address the problem. Reactive maintenance consists of only replacing the component when it fails and does not use CM, which is the most expensive approach. Small wind farms typically rely on this approach because they do not have a permanent maintenance team. On the other hand, predictive maintenance, through CM, enables an operator to know when to replace a component before a fault occurs. This approach prevents major failures, decreasing the costs and saving up to 20–25% of maintenance costs of wind turbines [15]. Consequently leading us to another aspect: how do we choose the turbine's element(s) to be monitored? A good strategy would be to prioritize components that are more likely to fail or lead to long down periods. Components such as the rotor and transmission system tend to have a higher rate of failure [16]. Generators tend to have a higher rate of failure in off-shore wind turbines than in on-shore ones [8].

There are also various methods of performing CM, ones more intrusive (wear out the component) than others, including acoustic emission measurement, power quality (harmonics measurements) and temperature monitoring, oil debris monitoring, and vibration analysis [17].

Finally, we can use CM for diagnosis, i.e., fault detection in real-time, or we can use it for prognosis, i.e., fault prediction. For instance, for generator faults, we have some common problems as wearing, electrical problems, rotor asymmetries, overheating and overspeed. Some of the respective current CM techniques are temperature, vibration, torque, current, voltage and power signal analysis, performance monitoring and thermography [9].

Tchakoua et al. [9] discussed some of the limitations and possible improvements on current CM techniques:

- Select a cost-effective monitoring method;
- Automate the diagnosis made by experts;
- Focus on developing precise prognostic techniques;
- Optimize the use of SCADA data for automatic monitoring;
- Improve sensors making them wireless, being easy to place in locations hard to reach;
- Use signal processing techniques for feature extraction useful for predicting WT's components health.

Although these future research areas may appear challenging to address, they also represent great opportunities for CM to boost the wind industry's success by reducing the cost of energy (COE) and increasing its competitiveness.

2.2. State-of-the-Art Machine Learning

To better understand how machine learning can be helpful, we present a brief introduction to the subject and the state-of-the-art regarding condition monitoring to address fault detection and diagnosis issues.

Recently, the field of condition monitoring has moved from the use of conventional techniques to artificial intelligence (AI) techniques [18]. The conventional methods consisted of sensing technologies or analysing physical quantities, as seen in the previous section, having the major problem of needing an expert to do the diagnosis. AI tries to automate this diagnosis, removing human error while handling more data in real-time. AI through machine learning techniques has been widely used to improve the accuracy and efficiency of fault detection and diagnosis, as we will see in Section 3.

ML models can follow two different approaches: supervised learning that predicts an output variable using labelled input data, or unsupervised learning that can learn from unlabelled data. In addition, for supervised learning, we have two different models; one predicts a numeric variable (regression) and the other a categorical variable (classifiers) [19].

The ML model selection phase is the most important as it is the main tool that learns from past data and generalizes into the future. For example, neural networks (NNs) and support vector machines (SVMs) are two popular models that have been used in ML for diagnostics and prognostics [8].

A NN is the adaptation of the learning ability of neurons in the brain to a computational architecture. We arrange NN in layers, and each layer is composed of a set of artificial neurons. Each neuron receives an input signal, manipulates it, and then the output is forwarded to the next layer of neurons [20]. NNs have been evolving rapidly over time. In the beginning, these models could only solve linear classification problems, which in the majority of the cases, we cannot apply in fault detection. Then, NNs evolved to multi-layered architecture that could solve non-linear problems, such as the feed-forward multi-layered method [21,22], in which no feedback from the previous signal is provided to the next. Another example is recurrent neural networks (RNNs) [23], which have feedback connections, and past signals are used to identify new features. Long short-term memory networks (LSTM) [24,25] are a type of RNNs, but instead of taking as input a single data point, they can process entire sequences of data.

In this review, self-organizing maps (SOMs), another type of NN, are also considered [21]. SOMs are trained using unsupervised learning. They produce a low-dimensional (usually two-dimensional) and discretized representation of the input training space. For this reason, we call it a map, and we typically use this method for dimensionality reduction. SOMs differ from other NNs, as they apply competitive learning as opposed to error-correction learning.

We also present two more types of NNs used for fault detection: generalized mapping regressor (GMR) and general regression neural network (GRNN) [22]. GMR is a self-supervised incremental neural network. This method can approximate every multidimensional function or relation that presents any discontinuity. The goal of GMR is to turn the function approximation problem into an unsupervised problem capable of pattern recognition. Therefore, it uses a coarse-to-fine strategy mapping. GRNN is an incremental self-organizing competitive neural network. GRNNs belong to the family of kernel neural networks. The typical GRNN training procedure minimizes the mean square error (MSE) and uses a cross-validation (leave-one-out) approach. Finally, adaptive neuro-fuzzy inference system (ANFIS) is also an important type of NN in the field of condition monitoring [23,26]. ANFIS integrates both neural networks and fuzzy logic principles. Therefore, it captures the strengths of both in a single framework. Its inference system is based on a set of fuzzy IF-THEN rules, with the learning ability to approximate non-linear functions.

The availability of larger datasets, variety of activation functions, and stronger computational power made it possible to add hidden layers (layers that allow a NN to filter/transform the data). This approach is called deep learning, and has started to be used in the wind energy field. NNs can be used for a variety of tasks, such as control (e.g., wind turbine power control) and fault diagnosis and forecasting (e.g., wind speed forecasting) [27,28], as we will describe in Section 3.

As for SVMs, they are often used in fault detection [23,29–31]. SVMs work by finding decision boundary hyperplanes that best separate classes samples; more specifically, the ones that leave the widest possible margin to the samples closest to the hyperplanes. They evolved from performing only linear classification or regression to non-linear problems by adding polynomial features created from existing ones. This method makes the problem linearly separable in a higher-dimensional space. They have recently gained significant importance because of their superior ability to accurately represent the relationship between the input and the output from a small amount of training information.

3. Machine Learning Techniques Applied to Condition Monitoring

In the previous section, we ended by summarizing the evolution of the use of ML in CM. In this section, we will cover the recent research on the subject of this review, including possible limitations and suggested improvements.

Before using ML methods, we typically use pre-processing techniques on the data, such as feature selection. Hence, it might be helpful to first look into work related to those initial tasks. After that, we will cover models for specific tasks. In Section 3.3, we start with models for more general issues, such as "Turbine performance assessment" and "Power curve monitoring", that are not specific to a turbine component. Moving to "Multi-target normal behaviour models", where we cover models that can be used for establishing the normal behaviour of multiple turbine components. In "Transfer Learning models", we introduce models that can be adapted to other datasets. This ends with "Federated Deep Learning", where we cover collaborative learning between multiple wind farm local data centers. Then, in Section 3.4, we focus only on the generator; i.e., "Fault detection, diagnosis, and prediction of generator faults". Finally, in the same section, we present specific generator faults: "Generator bearing failure prediction", "Generator temperature monitoring", "Generator Brush Failure prediction", and "Generator speed anomaly".

3.1. Data Pre-Processing

We obtain data for most existing CM models through SCADA (Supervisory Control And Data Acquisition) systems. This is an advantage, because using data from SCADA turns out to be a cheap alternative (e.g., does not require any extra hardware investment) [32]. This type of system has been integrated into wind farms and wind turbines by using sensors, controlling electricity generation, and providing time-series signals in regular intervals. Unfortunately, there is still a high non-conformity between sets of SCADA signals and taxonomies [33] used by different turbine manufacturers, which makes it challenging to compare existing research.

Another challenge to be faced is that typically a wind farm has hundreds of sensors in each turbine, all of them producing signals at a high rate; this results in "big data" problem [8].

Canizo et al. [34] present an efficient solution to data processing. They suggested a big data framework to manage the data, observing an increase in speed, scalability, automation, and reliability, but also better results in overall accuracy and sensitivity rate.

After dealing with the previously mentioned problems, we can start by looking into the raw SCADA data collected, and perform pre-processing. Peng et al. [35] proposed a novel approach to deal with data loss problems in remote CM. By the use of wireless data transmission, remote CM systems solve the local limited data and computational resources problem of onsite CM. Remote CM grants access to additional computational resources, allowing advanced algorithms to process data from multiple wind turbines, however, it has drawbacks regarding data loss. Therefore, the authors [35] proposed a compressive sensing (CS)- based missing-data-tolerant fault detection method to solve this drawback. The CS technique can reconstruct sparse signals; hence, the original signals are converted to a sparse frequency domain. Then, the signals are sampled by a compressive-sensing-based signal algorithm before being transmitted wirelessly. Hence, the proposed method adds the novelty of treating the signals before transmission. CS technique relies on a small number of sparse signals containing most of the salient information. Therefore, it is possible to reconstruct the signals with loss transmission problems. The reconstruction error is rounded to 0.3 for losses close to 95%, indicating a high tolerance to missing data.

Data provided by SCADA are influenced by structural problems, but also can take into account other important factors. For example, temperature spikes can occur due to external temperatures and not due to an internal problem in the wind turbine components. This type of event can be removed using outlier identification and removal techniques. At first, one could expect that a simple outlier removal technique might solve the problem, but Marti-Puig et al. [36] showed that this was not the case. Although these methods can decrease the training dataset' errors, they also can increase the test dataset errors. Meaning that most of the values considered outliers by the simpler methods are true failures. Consequently, Marti-Puig et al. suggest the aid of an expert on the subject to define absolute and relative ranges.

Lapira et al. [21] applied three changes to the SCADA data obtained to filter outlier samples:

- Remove the samples when the output power is negative or wind speed is below the rated cut-in wind speed;
- Segmenting the data into week bins. In this way, the health value can be computed every week;
- Normalizing the data.

W. Yang et al. [37] also developed a method to pre-process raw SCADA data based on expected value calculation. The advantage of that method is that the expected value reduces the statistical error caused by outliers. Additionally, methods based on the average value, as previously mentioned, may fail to consider the probability distributions of outliers.

We presented the previous papers by order of complexity and efficiency. Hence, if we want to guarantee an increase in accuracy, the last approach could surpass the unavailability of expert knowledge.

Another approach covered in existent literature is the over or under-sampling of the data. It is challenging for the classifier to learn abnormal behaviour when the representative part of the data consists of non-fault samples. Therefore, an additional experience can be removing normal samples or oversampling a certain failure. In terms of oversampling, we have techniques such as synthetic minority over-sampling technique (SMOTE) [25,30], which generates synthetic samples from the minor class instead of creating copies, and random oversampling [38].

Conversely, for undersampling, we have methods to choose samples to keep, to delete, or a combination of both. To keep the samples from the majority class with the smallest average distance from examples of the minority class, near miss undersampling can be used. Another technique is condensed nearest neighbors (CNN), that seeks a subset of a collection of samples that results in no loss in model performance, referred to as a minimal consistent set. On the other hand, random undersampling [30], Tomek links [30,31] or the edited nearest neighbors (ENN) rule [30] can be used to select which samples should be deleted. Tomek links use Euclidean distance information of input data points to identify borderline and noisy data. Therefore, these procedures only remove points along the class boundary, yielding better performance when combined with another undersampling method. Combinations that can be tried out are one-sided selection, which combines Tomek links, and the CNN rule and neighborhood cleaning rule, which combines CNN and the ENN rule.

Another two distinct methods are penalized classification and cluster centroids (CC) [30]. Penalized classification tries to impose an additional cost on the model for making classification mistakes on the minority class during training. These penalties can bias the model to pay more attention to the minority class. CC is another undersampling method that splits all the samples of the majority class into clusters using the k-means algorithm. The centroids of these clusters are then used instead of considering all the samples from that cluster.

Huaikuan et al. [39] proposed an improvement to SMOTE that also uses clustering. Classical SMOTE uses linear interpolation to generate more samples from adjacent samples of the minority class. Therefore, if the data are unevenly distributed, i.e., has sparse regions containing few samples, the interpolation method may fail in those cases. Since the minority class is characterized as having few samples, these situations tend to occur. Hence, the paper [39] developed a method called minority clustering-SMOTE (MC-SMOTE), which replaces interpolation for clustering. Samples from the minority class are divided into several clusters. Then, new samples are created by adjacent clusters in combination with SMOTE, reaching a uniform new minority class distribution, since clustering will produce new samples incorporating sparse areas.

Jiang et al. [40] also proposed a method using SMOTE, however, combined with dependent wild bootstrap (DWB), which they entitled synthetic and dependent wild bootstrapped over-sampling technique (SDWBOTE). The SMOTE does not take into consideration temporal dependence, which is important for time-series, being the case of SCADA data. Additionally, it is not prepared to deal with missing data causing unfixed length inputs. Therefore, they start by modifying SMOTE to allow unfix length data, aligning and slicing samples, as described in detail in the paper. Afterwards, they add DWB to resample the data, capturing the time dependence of the sample. These two modifications combined can solve the mentioned SMOTE disadvantages. As will be seen in Section 3.3.4, transfer learning can also be used to solve this problem by transferring the knowledge from a balanced dataset to one suffering from data imbalance. Qun et al. [41] also proposed a different approach to deal with imbalanced data. Instead of using cross-entropy as the loss function, they used focal loss (FL). FL is an extension of cross-entropy, being dynamically scaled, reducing the weight of samples from the majority class during training.

In addition, performance metrics that can deal with imbalanced data will be covered in Section 4.

At last, an uncommon step of pre-processing was broached by Xu et al. [42], selecting the data corresponding to the normal periods of operation. This pre-processing is useful for the normal behaviour models, which normally combine the status data and historical data to label the data. However, due to the remote location of WTs and the consequent unavailability of regular maintenance, some fault information may be ignored. This means that the data which is supposed to be normal contains faulty behaviour, which presents an issue to the normal behaviour models. The paper [42] proposed the use of quantile regressions combined with NN structure to obtain a nonlinear quantile regression. The quantile regression neural network (QRNN) receives, as inputs, the variables of the normal behaviour models and outputs the conditional quantiles. They considered the range of 0.4 to 0.6 of quantiles levels as representative of normal behaviour, since, according to the significance of median in statistics, a one-to-one mapping rounds 0.5. The method showed good results on constructing intervals of normal behaviour data which are robust against outliers.

3.2. Feature Selection

There is no conventional method for feature selection when using ML on CM, because it depends on the component being monitored. However, it can be as simple as asking an expert if it is more useful to focus on the acoustic sensor or the generator's vibration or going beyond that, and using an automatic method.

Auto-encoders, or principal component analysis (PCA), can reduce the extracted features or combine them. An autoencoder is a type of NN used to learn efficient codings in unsupervised data. They are useful for dimensionality reduction, since they learn a representation (encoding) for a set of data by training the network to ignore signal "noise" [43]. PCA is the process of computing the principal components and creating projections of each data point onto only the first few principal components to obtain lower-dimensional data, while trying to preserve the data's variation [44]. Auto-encoders can perform similarly when the activation functions are linear, and the cost function is the mean squared error. However, when compared to techniques that use dimension reduction, non-linear techniques rarely outperform traditional linear techniques.

Y. Wang et al. [45] proposed a feature selection algorithm based on PCA, with multiple selection criteria, selecting a set of features that better identify fault signals without altering the variety of data in the original dataset. Moreover, it also has the advantage of reducing the number of sensors installed by removing the variables that are not relevant. More specifically, the selection method proposed in the paper is the T selection method, which targets a specific fault signal [46]. This algorithm maximizes variance and maintains the independence among the selected variables, while preserving underlying features regarding the fault. Once a set of features is selected, three performance metrics were used to evaluate the selection algorithm: cumulative percentage partial variance (CPPV), the average correlation coefficient ($\bar{r}$), and the percentage information entropy (ηe).

W. Zhang and X. Ma [47] proposed a model that uses parallel factor analysis (PARAFAC) for fault detection and sensor selection of wind turbines based on SCADA data. PARAFAC, in resemblance with other decomposition methods, such as Tucker3 or unfolded PCA, is part of the family of bi-linear or multi-linear decomposition methods of multi-way data into a set of loading and score matrices [48]. The difference is in PARAFAC, using fewer degrees of freedom than the other mentioned methods. This fact presents an advantage since it leads to simpler models, while excluding noise and insignificant or redundant information. PARAFAC has gained importance because it is a processing technique capable of simultaneously optimizing the factors and selecting the relevant contributions to the dataset in trilinear systems. This method has firstly been applied to the condition monitoring of wind turbines by [47]. More recently, in [49], they proposed the use of PARAFAC and sequential probability ratio test for multi-source and multi-fault condition monitoring; nevertheless, this is not specific to the wind farms domain.

Peng et al. [50] proposed a method called Mahalanobis distance (MD) to reduce the input variable number of the prediction model. MD tries to reduce redundancy while keeping relevant features. MD analyses the effects of using different units to measure the distance between a point and a distribution, thus, detecting correlations between variables. In addition, the MD method computes the univariate distance containing the main features of multivariate data. This advantage plays an important role in reducing the number of input variables of the prediction model. Furthermore, most wind farms are in remote locations, and the data collected are usually transmitted to an analysis center by wireless or optical fibre networks. Therefore, fewer input variables decrease the communication load.

Fernando P. G. de Sá et al. [51] proposed a framework for automatic feature selection called non-dominated sorting genetic algorithm II (NSGAII). NSGAII is a multi-objective genetic algorithm, gaining the name since it adopts a search method that employs concepts from natural genetics. It uses Pareto dominance relationships to rank solutions, simultaneously optimizing each objective without being dominated by other solutions. NSGAII was used to select simultaneously a subset of features and hyperparameters to increase the performance of fault detection. Since we have a codependent relation between the optimal subset of features and the model's hyperparameters, this approach appears to have a great advantage. By using this algorithm, they were able to find the optimal balance between the number of features and the model's ability to detect faults. Additionally, they also determine the hyperparameters that allowed the detection of the fault before it happens.

A. Stetco et al. [52] suggested a featureless approach using convolutional neural networks (CNN). CNNs are NNs, however, they can filter and pool the input data to create a feature map that summarizes the important features in the input. Therefore, they do not need feature engineering. They also used class activation maps (CAMs) to investigate the features selected by the CNN, and to identify the discriminative patterns in signals. By doing this, they can inform engineers which time segments are useful to determine the normal behaviour of operation or failure pattern.

Qun et al. [41] addressed the problem of spatio-temporal correlations between features. They used two modules in parallel, multi-scale deep echo state network (MSDeepESN) to deal with temporal multi-scale features, and the multi-scale residual network (MSResNet) module for the spatial multi-scale features. MSDeepESN is a type of RNN that rapidly and efficiently captures temporal correlations. To prove its effectiveness, it was compared with the LSTM model, presenting better results. MSResNet consisted of an optimized (one-dimensional) CNN for spatial correlation detection. Surpassing the ordinary CNN model. They also found that using the spatio-temporal fusion yielded better results than using them isolated.

Kong et al. [53] also addressed the spatio-temporal issue. They combined the ability of spatial feature extraction of one one-dimensional CNN with the temporal feature extraction of the gated recurrent unit (GRU). Primarily, they reduced the number of features by using Pearson prod-moment correlation to select the most important variables. Pearson weights the degree of association between variables, excluding the ones with small correlation with most others. Afterwards, CNN extracts the spatial features, for each point in time. Subsequently, and not in parallel, as in the previous paper, temporal features are extracted by the GRU. GRU is an RNN with improved state information storage capacity, being the hidden units replaced by gating units.

The results from [41] surpassed the ones from [53], as the authors [41] stated, due to CNN-GRU extracting single-scale features instead of multi-scale.

As previously mentioned, we do not have a conventional method for feature selection, which can be proved by the number of different approaches in the literature. With that in mind, a good approach would be, as a starting point, to test different algorithms, beginning with simpler methods such as PCA. Taking into consideration the ground truths of all wind farm data, there will be non-linear signal relations, tremendous variations in signals, and negative values.

3.3. ML Models for Wind Turbine Condition Monitoring

3.3.1. Turbine Performance Assessment

Lapira et al. [21] used the SCADA data from a large-scale on-shore wind turbine to assess which of the three selected models better captures the turbine's performance and degradation. The methods used to pre-process and filter outlier samples were already mentioned in Section 3.1.

The important SCADA parameters were chosen to model the wind turbine's system performance (wind speed and the average active power), splitting them into two steps: multi-regime (dynamic-wind turbine operating regimes) partitioning and baseline comparison. Finally, a confidence value was computed during the baseline comparison step, which describes the health state of the wind turbine. The multi-regime models being tested were SOM and gaussian mixture model (GMM). GMM is a probabilistic model which assumes that all the data points are generated from a mixture of a finite number of Gaussian distributions. Finally, feed-forward NNs used an approach based on residuals greater than a given threshold during a given time segment. A comparison between the first two, unsupervised models and the last one, a regression model, was a major conclusion of the paper.

They found that the GMM model presents a more gradual health change, being more suitable in performance prediction. Nevertheless, the other two methods can be used for fault or anomaly detection. The suggested future work was to predict the progression of the degradation using predictive techniques, computing the remaining operational time before a future downtime.

The most interesting feature of this paper is the use of unsupervised methods, since most datasets composed of SCADA signals are not labelled as fault or not. As the paper states, an interesting approach is to use SOM and NNs on fault detection to label the data. The paper's addition to the existing literature is to produce a standard for manufacturers to compare performance.

3.3.2. Power Curve Monitoring

The predicted power usually does not meet reality due to various reasons. For instance, the wind speed on a wind farm is not uniform and the air density is different than during the calibration. Additionally, the wind data available are not always measured at the height of the turbine's hub [54]. This fact is true both for a single turbine or for a whole wind farm, making it hard to assess a prediction of the energy output of a wind farm. An efficient wind power forecasting model is important for energy management. Wind power forecasting and prediction techniques allow better scheduling, and unit commitment of thermal generators, hydro and energy storage plants. Thus, this reduces the risk of uncertainty of wind power production for all electricity market sellers and clients. Even though this is not why this tool is helpful for CM, it was probably a good reason for investing in it in terms of the market.

Marvuglia, A. et al. [22] present a data-driven approach for building a steady-state model of a wind farm's power curve under normal operating conditions. This approach allows the creation of quality control charts that can be used as a reference profile for detecting anomalous functioning conditions of the wind farm and power forecasts.

The paper compares three different machine learning models to estimate the relationship between the wind speed and the generated power in a wind farm: GMR, GRNN and a feed-forward multi-layer perceptron (MLP).

This paper has the novelty of applying power curve models to an entire wind farm, and is focused on GMR. When looking into the results, the first two non-parametric methods provided more accurate results when compared with the classical parametric MLP.

Regarding future work, the paper states that labelled data classified as normal or abnormal could lead to various improvements. One of those possible improvements is the utilization of this type of algorithm to perform the prediction and diagnosis of wind turbine faults. In this case, the ML approach should be used to build a steady-state model of the reference power curve of the wind farm under normal operating conditions, and through deviations from that behaviour, detect future faults.

The paper [22] also covers a problem already mentioned; the lack of labelled data, being the learning focused on determining what are normal behaviours and abnormal behaviours (fault detection) and not on fault prediction. Nevertheless, the approach of considering the wind farm as a whole, instead of specific turbines or components, could be extended to other tasks (e.g., obtaining more general statistics that could indicate a possible fault not detected by a single turbine). The fact that it focuses on the whole wind farm is one of the points that was added by this paper; the other point is that it uses GMR, a novel incremental self-organizing competitive neural network.

When modelling power curves, wind speed may not be the only dependent variable used. For example, Schlechtingen et al. [26] compared two models: one using only wind speed as the dependent variable, and another also using wind direction and ambient temperature. After searching among the several existent comparative works in literature, they selected the models that presented the best results for WT power curve monitoring and applied them for their study cases. Those models were cluster center fuzzy logic (CCFL), k-nearest neighbor (K-NN) and ANFIS. the K-NN model predicts the values for new points based on feature similarity with the points in the training set.

Schlechtingen et al. [26] proved that by adding wind direction and ambient temperature, the models fit the data better, reducing the variance in the prediction errors. This finding made it possible for the earlier detection of abnormal turbine performance. Specifically, for the used dataset, the anomaly was detected with the addition of up to five days notice from the models using only the wind speed. The ANFIS model showed the best performance in terms of prediction and in terms of abnormal power output detection, whereas the K-NN model performed worst. The paper's explanation for the poor performance of the K-NN model was that the number of considered neighbors decreased by increasing the dimension of the space by adding wind direction and ambient temperature. Consequently, this makes the predictions more sensitive to outliers.

In contrast, with the first paper [22], the previous used the presence of labelled data to predict errors having best results using the ANFIS model, which allows the incorporation of a priori knowledge in the form of rules. In addition to the previously mentioned model, another novelty added to the literature was including wind direction as an input variable. This addition would be a good approach to be followed, since it improves the detection of abnormal turbine performance. The goal of assessing the power curve's normal behaviour is to detect anomalies when the power deviates from the expected. As will be seen, this approach can be followed for other wind turbine variables.

3.3.3. Multi-Target Normal Behaviour Models

A common approach to CM is to define models for the normal behaviour of a specific component. Then, from that model, detect deviations from the normal operation that can indicate a failure. A disadvantage of this approach is that each of the models needs to be updated and maintained. A. Meyer [24] suggested multi-target regression models in order to deal with this problem. A multi-target regression model receives, as input, a set of features, and outputs multiple target values simultaneously. This means that, for example, instead of having two separate models for predicting the power and the generator temperature, we could have only one model. This technique decreases the time and work of having to do the pre-processing tasks, train and select the thresholds for multiple models. They developed six multi-target regression models, some using deep neural networks, and others classical ML algorithms. Secondly, they compared the model's prediction error with the single target models. They also investigate if using models that take into consideration past observations, such as CNN and LSTM, leaves us with better results than the ones considering only present observations (K-NN and MLP). The results showed that the multi-target models achieved similar, and in some cases, even smaller, predictive errors, than single-target models. Another interesting conclusion was that taking into consideration past observations as input did not improve the performance of the model when the target variables were strongly correlated. Even though it is a novel approach, it is a promising one, since we can reach the same performance as when multiple models are used.

3.3.4. Transfer Learning Models

The goal of transfer learning is to ensure that knowledge from one domain can generalize in a different domain, being used in cases where there is a lack of labelled training data or small training sets. Therefore, transfer learning can bring multiple advantages for WT CM. In that case, we can use it to transfer knowledge to small data sets, or to deal with imbalanced data.

W. Chen et al. [55] suggested using transfer learning for fault diagnosis between two wind turbines. The covered transfer learning algorithms were Inception V3 and TrAdaBoost. Inception V3 is based on a deep NN and is formed by units called inceptions. Each inception unit includes nonlinear convolution modules, being the last layer, a Softmax classifier. TrAdaBoost uses a small amount of data to build a classifier, part of the abundant data from the original dataset, and the remaining data from the target dataset, both probably having different distributions and feature spaces. TrAdaBoost iteratively updates its weights based on each sample from both datasets. These two transfer learning models are then compared with two conventional ML algorithms, K-NN and random forest. Random forest is an ensemble of unpruned classification or regression trees, trained from bootstrap samples of the training data. Additionally, they created a new metric to compare the performance between these algorithms, called comprehensive index (CI). CI takes into account two metrics, Sensitivity and Specificity, both with equal weight. Sensitivity and Specificity represent the percentage of correctly classified normal and faulty data, respectively. The use of this new metric tries to dim the effect of imbalanced data and emphasise the role of correctly classified data. TrAdaBoost showed the best results, dealing with imbalanced data and different distributions.

J. Chatterjee et al. [25] also proved the appeal of using transfer learning. They combined the classification accuracy of an RNN with the transparency of the XGBoost decision tree classifier. RNNs can predict a failure, however, they are not able to provide a detailed diagnosis on which components were affected and what caused it. This type of detail could help the process of OM of the affected component. They use LSTM, an already mentioned type of RNN, and they combined it with XGBoost. XGBoost is a supervised learning method that produces optimal results from the combination of multiple decision tree classifiers. The model computes the importance of the features in a transparent way, giving us insight into which ones play an important role in the deep learning model. Additionally, they use SMOTE to oversample the minority samples. Finally, and as the major conclusion of this paper, they use transfer learning to use the knowledge from the model trained on an offshore WT to an onshore WT. The original model had an accuracy of 97%, as the target model had 65%, and was able to detect 85% of the anomalies. Taking into consideration that it was an unseen dataset, the results were encouraging.

Ren et al. [56] covered the use of transfer learning for fault diagnosis under variable working conditions. The same fault may present different working conditions with dissimilar distributions, decreasing the fault detection accuracy. They added the lack of labelled samples to the aforementioned problem, proposing a method to solve the two issues. The paper [56] proposed a novel method based on composite variational mode entropy (CVME) and weighted distribution adaptation (WDA). Primarily, the original signals presenting various working conditions are used to obtain intrinsic mode function (IMF) components by performing variational mode decomposition (VMD). A low correlation between source and target domain affects the ability of transfer learning. Therefore, multi-scale analysis of the IMF components is carried out to filter noise, selecting the components with a larger correlation with the original vibration signal for feature extraction, with the feature set with the highest correlation with the target feature set being selected. This correlation under different working conditions is used as transferability evaluation for effective transfer to the target domain. Feature extraction results in CVME feature vectors with different frequency bands, which are input into WDA. The WDA decreases the data distribution discrepancy between the labelled source and unlabelled target domain by constructing a transformation matrix to adapt the marginal distribution and conditional distribution, and reduces the class imbalance between domains. At last, the trained classifier is applied to the target samples to identify the fault types. The CVME-WDA method is compared with traditional machine learning methods, yielding better accuracy in fault diagnosis under variable working conditions.

3.3.5. Federated Deep Learning

The state-of-the-art for CM has relied on deep learning models, which typically require a great amount of data. Federated deep learning allows collaborative learning between spatially distributed data, sharing only the prediction model parameters among participants, and not the training data. This characteristic solves the problems of security and privacy related to data sharing, allowing the collection of a greater amount of data to train the deep learning model. Collecting data from multiple WTs will also add fault diversity that is not usually present on only local data, boosting fault diagnosis. This approach has been applied in energy systems for energy demand forecast, preserving consumers' privacy [57,58]. In terms of maintenance, it is starting to be applied in industry, collecting labelled data from multiple devices or machines to help detect and diagnose an anomaly [59,60]. Wang et al. [61] have proposed a novel collaborative deep learning framework for fault diagnosis of renewable energy systems, using three of the four case studies related to wind farm datasets. For all the cases, they considered a distributed network of five local data centers, which they called agents. First, each agent initializes their model's parameters and uses the model to obtain a prediction error, more specifically, the chosen model was LSTM. Next, comes the key of collaborative learning; each agent needs to exchange parameters information to minimize the model's loss. Therefore, a communication layer was used for synchronization, collecting and averaging all the agents' parameters. The first two case studies used different wind farm datasets to prove that the framework can generalize for different datasets. Both showing better results when using the distributed scheme in comparison with using a local strategy. The third case study represented some agents having the imbalanced data issue, also achieving better results for the distributed scheme. Due to agents suffering from the imbalance problem being able to learn information from the other agents, the fourth case study was not specific of WTs, however, it showed the scheme's ability to deal with data with different distributions.

3.4. ML Models for Wind Turbine Generator Condition Monitoring

3.4.1. Fault Detection, Diagnosis, and Prediction of Generator Faults

Looking into literature that covers conditions monitoring and fault prediction, the prediction of more than a half-hour notice is currently very weak for minor faults. Even though they are minor, they occur quite often, contributing to power system-related failures. A study carried out by the EU FP7 ReliaWind project (https://cordis.europa.eu/project/id/212966/reporting (accessed on 17 October 2021)), states that under 40% of overall turbine downtime can be attributed to power system failures [62].

Leahy et al. [30] focused on fault detection, fault diagnosis, and fault prediction of generator minor faults. The first classification level, fault detection, is distinguishing between two classes: "fault" and "no-fault". Fault diagnosis is a more advanced level of classification than fault detection. Fault diagnosis aims to detect specific faults from the rest of the data. Faults were labelled in five classes, including generator heating, power feeder cable, generator excitation, air cooling malfunction faults, and others. The last level was fault prediction/prognosis, which has the objective of predicting the fault before it occurs. The predictions focused only on generator heating and excitation faults, as these showed the most promising results for early detection. The data used came from a SCADA system, and 29 features were selected to be used in classification, using SVM as the ML classification model. Several scoring metrics were used to evaluate final performance. The precision score is one of them, as many false positives can lead to unnecessary checks or rectifications carried out on the turbine. Conversely, many false negatives can lead to failure of the component with no detection taking place, and the recall score captures this.

For fault detection, the recall was high (78% to 95%), but precision was low (2–4%), indicating a high number of false positives. For the diagnostic and prognostic, high recall and low precision were also found. For fault diagnosis, generator heating faults showed few false positives and correctly predicted 89% of faults. In fault prediction, the best performance was achieved with SVM trained with the addition of class weight, using a linear kernel. In general, for fault detection and diagnosis, the recall scores were above 80%, and prediction up to 24 h notice of specific faults, representing a significant improvement over previous techniques.

Possible improvements, excluding adding more data, are using feature selection methods to find only the relevant features, speeding up training time. In addition, a possible avenue for future research is determining whether trained models would still be accurate after a significant change in the turbine, e.g., after replacing a major component.

The most interesting feature in this article was how they use operational and status data to label the data. For example, they considered an operational data point as faulty if it occurred in a time frame of 10 min, before or after a fault present in the status data. Conversely, as the authors stated, a technique that could be improved is the feature selection, as it was based on a personal judgment that is always prone to error. In general, the paper presents simple yet efficient solutions for the three different levels of fault monitoring.

3.4.2. Generator Bearing Failure Prediction

Schlechtingen et al. [63] compared the performance of two artificial intelligence approaches (autoregressive NNs and full signal reconstruction (FSRC) NNs (non-linear NNs)) to a regression-based approach, when learning to approximate the normal bearing temperature. In order to learn regression models, the work used SCADA input signals, such as power output, nacelle temperature, generator speed, and generator stator temperature. This task also used data smoothing techniques in combination with the learning techniques. By using a smoothing filter, the variations of high order can be filtered and the model's prediction error can be reduced.

Although NNs can deal with fuzzy or incomplete data, they perform poorly with invalid data. Therefore, one must typically use a pre-processing technique, which is particularly important when training a network. The network might not give an optimal generalization otherwise. The principal pre-processes applied were: Validity check—ranges and consistency are checked by filtering outliers and data with irregular high gradients; data scaling; missing data processing; and lag removal—WT signals usually do not respond immediately to changes of operational conditions. Many wind turbine signals can be correlated to other measured signals, and only some are related to the output signal (bearing temperature). We can use cross-correlation to find these related signals and their lag to the desired signal.

In [63], the authors found that the non-linear NN approaches outperform the regression models. However, they are more challenging to interpret. In comparison to the regression model, NN had an averaged error with reduced amplitude and was more accurate, leading to reduced alarm limits. An alarm is triggered 30 days before the bearing breaks. The autoregressive model has a very high accuracy, due to the large heat capacity. Thus, this model can detect minor changes in the autoregression of the temperature signal (50 days in advance).

Kusiak and Verma [64] estimated an expected behaviour model of a generator bearing by training an MLP to predict generator bearing temperature. The model is trained on high-frequency (10 s) SCADA data from 24 wind turbines of the same type and location. Two turbines that showed high-temperature faults were used for testing and model validation. Some of the input variables were selected by domain knowledge (selecting 50 out of 100), and subsequently by applying three different data-mining algorithms: wrapper with genetic search (WGS), wrapper with best-first search (WBFS), and boosting tree algorithm (BTA). The residuals were smoothed with a moving average filter (window size of 1 h). If these residuals exceeded two standard deviations, an alarm was triggered. The authors find that their method can predict a high-temperature fault with an average of 1.5 h notice.

Both papers [63,64] used NNs to detect faults on the generator bearing. However, the first paper [63] used more complex approaches, resulting in an earlier prediction of the fault when compared with the second paper. Nevertheless, the authors [64] presented interesting ways of pre-processing the data and three different feature selection algorithms. Before training either of the different approaches of NNs, a combination of the previously mentioned strengths of both articles could be interesting.

Lastly, D. Yang et al. [65] used a vibration CM system to detect generator bearing faults. Wind turbine vibration signals are subjected to high noise disturbance; therefore, they use a noise suppression method for feature frequency extraction. This method was supplemented by a multi-point data fusion. The method for denoising and feature extraction consists of using empirical mode decomposition (EMD)—correlation. EMD decomposes signals into the sum of IMFs of different frequencies. Afterwards, the IMFs containing the relevant fault feature frequencies are selected and used to reconstruct a new signal. Then, autocorrelation is applied to remove noise, and wavelet package transform (WPT) is used to extract features. Secondly, this method is supplemented with multi-point data fusion using adaptive resonance theory-2 (ART-2). The ART-2 is an unsupervized neural network that recognizes the patterns of feature frequency, indicating a possible fault. The results showed that the proposed method reduces the noise and extracts clearer fault features. This is due to the ART-2 ability to strengthen the recurrent patterns in a sequence and remove low amplitude noise by using normalization and non-linear functions. The developed method was implemented in an actual WT to prove that the CM system was able to identify the fault for the generator bearing and that the analysis of the vibration signals successfully diagnosed the fault.

Chen et al. [66] addressed the problem of defining a threshold for unsupervised normal behaviour models that need to establish boundaries representative of that behaviour. The authors proposed a self-setting threshold method using a deep convolutional generative adversarial network (DCGAN) applied to monitor generator bearings. DCGAN are the integration of CNN into the vanilla generative adversarial network (VGAN). VGAN consists of two competing networks—a generator (G) and a discriminator (D). G and D will be replaced by deep CNNs in DCGAN. Each of the networks optimizes their loss function until reaching the Nash equilibrium, where regardless of G/D behaviour, the other is not affected. At this point, the threshold is self-defined based on the discriminator output of the DCGAN. A fault sample will move that output away from the Nash equilibrium; therefore, the DCGAN model is capable of self-defining anomalous samples, not requiring the human intervention or manual setting of a threshold. Thus, a monitoring indicator function (MIF) is computed based on the sample discrepancy analysis of DCGAN output to quantify the health condition of the generator bearing. Finally, the method is compared with other techniques used by regression models such as autoencoders, yielding a more stable and reliable choice of threshold.

3.4.3. Generator Temperature Monitoring

Most of the generator high-temperature failures occur in spring and autumn, especially in spring. This fact is due to the increase in the ambient temperature in springtime and high wind speeds. If this causes a fault on the generator that leads to a shut down in the wind turbine, significant energy generation will be lost, due to the time required to change/repair the generator.

P. Guo et al. [67] proposed a new condition monitoring method, consisting of a temperature trend analysis method based on the non-linear state estimation technique (NSET). NSET is used to model the normal operating behaviour for each wind turbine generator temperature, and then, is used to predict it. In addition, a new and improved memory matrix construction method is used to better cover the generator's normal operational space.

The time series of residuals between the real measured temperature and the predicted is smoothed using a moving average window. This reduces the method's sensitivity to isolated model errors, improving its robustness. The average and standard deviations computed by that moving window are used to detect potential faults early, when significant changes occur, exceeding predefined thresholds, a future failure is pointed out.

The model uses SCADA data from a wind farm that records all wind turbine parameters every 10 s; in total, 47 parameters are recorded for each turbine. At the same time, the SCADA system keeps logs of wind turbine operation and fault information. Nevertheless, only five variables were considered relevant (stored in an observation vector): power, ambient temperature, nacelle temperature, and the generator cooling air temperature.

The results showed that the new approach to the memory matrix increased the model's accuracy. The model can identify dangerous generator high temperature before damage has occurred, which would result in a shutdown of the turbine. In order to compare with the NSET method, a NN was developed and then used to model the normal behaviour of the same wind turbine. Results showed that NSET achieves considerably higher accuracy in modelling the normal behaviour of the wind turbine generator temperature. Moreover, NSET has another benefit compared with the neural network; it can more easily adapt to a new normal working condition.

The level of specificity in terms of fault detection will depend on the information available in the dataset, therefore, determining if it is possible to focus on generator temperature monitoring or not. If that is the case, the approach followed in the paper, NSET, can be used. Regardless, using the sliding window to detect failures is an interesting approach that can be added to any coarse detection fault.

Tautz-Weinert et al. [23] compare different approaches to normal behaviour modelling of bearing and generator temperature, based on 6 months of 10-min SCADA data from 100 turbines. The different approaches were: linear regression, SVMs, an MLP with one hidden layer of six neurons, and an RNN with two recurrence steps, ANFIS and Gaussian process regression (GPR). GPR is a non-parametric Bayesian approach to regression. The input variables are found by analysing cross-correlations between SCADA variables and the target variables.

The authors used only two input variables in their baseline configuration, and added further ones for a sensitivity analysis. They concluded that the performance of RNN was close to the MLP, with both NN types usually outperforming other approaches. GPR and SVM, however, were not as accurate as the other models. SVM and ANFIS tend to have larger errors with more inputs. GPR worked well for the generator temperature prediction, but not that well for the bearing temperature prediction. The authors stated that adding interactions to linear models was advantageous—conversely, the use of recurrence in the NN model was only helpful for some turbines.

An important resemblance can be found in both papers, the small number of variables taken into account when modelling the normal behaviour of generator temperature. This fact reinforces the need for a good technique for feature selection. However, the approach followed by the first paper, inference based on knowledge, cannot always be followed due to the lack of expert insight. Conversely, as in the second paper, doing cross-correlation is a simple technique that can, and should, always be tried out.

3.4.4. Generator Brush Failure prediction

Carbon brushes are one of the critical components of the WT generator. Malfunctions on these components can lead to reduced performance and unnecessary shut-downs, because WTs are taken out of service, so that brushes can be replaced or cleaned.

Verma et al. [31]developed generator brush failure classification models based on SCADA data sampled every ten minutes. Both status and operational parameters are used in this paper. Snapshot files, operational data files that are automatically generated whenever some critical fault occurs in the turbine, were analyzed.

In order to improve prediction and avoid the curse of dimensionality, irrelevant features were removed. Using domain knowledge provided by experts, the initial 100-dimensional data were reduced to 50 dimensions. Three known parameter selections were used to determine the best subset of parameters for the prediction, namely: chi-square, a statistical test of independence to determine the dependency of two variables, in order to select parameters (filter technique); boosting tree (embedded method), which uses a gradient boosting machine approach to rank the parameters and a wrapper algorithm with genetic search used as a black box to rank/score subsets of features according to their importance. The feature selection approach has reduced 50 features to 14 (nacelle revolution, drive train acceleration, etc.).

Considering the quantity of data, for a typical fault, the ratio between normal and fault samples can be as large as 1000:1. Verma et al. [31] used a combination of Tomek links and a random forest algorithm as the data sampling approach. Four data-mining algorithms were studied to evaluate the quality of the models for predicting generator brush faults: MLP, boosting tree, K-NN (K = 10), and SVM. The boosting tree algorithm is an ensemble learning algorithm that combines many weak classifiers to produce a powerful one.

Results of three cases, (1) the original dataset; (2) the sampled dataset based on Tomek links only; and (3) the sampled dataset using Tomek links and random forest algorithm, were obtained. The prediction accuracy using Tomek links and random forest algorithm was in the range of 82.1–97.1% for all timestamps. The significant improvement in accuracy indicates the effectiveness of data sampling methods. In case (2), the initial imbalance in the output class was reduced to 80%:20%. By also applying random forest-based data sampling, it reduced the class imbalance ratio to 65%:35%.

The data-mining model that presented better prediction results was the boosting tree. The results presented in this paper [31] offer an early prediction of future faults. This allows engineers to schedule maintenance and minimize OM costs.

As described, Verma et al. [31] suggest many algorithms for data pre-processing, some for feature selection, but also for data sampling, that as the authors stated, improved the performance of the model. A similar approach should be followed when working with an imbalanced dataset, since it is hard to detect patterns in the data if they are almost not represented among the normal status data.

3.4.5. Generator Speed Anomaly

Jiang et al. [68] used a new fault detection technique based on a recently developed unsupervized learning method, denoising autoencoder (DAE), using SCADA data. This study selected two different fault scenarios that occurred in different turbines, generator speed sensor fault, and gearbox filter blocking fault.

To include the relation between time series of the SCADA data, they use a sliding-window approach which inputs sequences of values into the DAE training model. Thus, a sliding window denoising autoencoder (SW-DAE) for WT fault detection is proposed [68]. The main advantage of the proposed technique is the capability to capture non-linear correlations among sensor signals. Additionally, it also captures the temporal dependency of each sensor variable, consequently improving fault detection performance.

DAE is able to build a multivariate reconstruction model from multiple sensors. Afterwards, the DAE's reconstruction error trained with normal data is used for fault detection. The main characteristic of DAE is its ability to, from a corrupted signal, reconstruct the original one. Therefore, DAE can learn from corrupted data, improving its generalization capability and achieving state-of-the-art performance on feature learning chores [69].

Another particularity of the approach proposed in [68], is that they use the Mahalanobis distance instead of the usual squared error to compute the reconstruction error of the autoencoder. For evaluating the performance of the different fault detection methods, they used the receiver operating characteristics (ROC) curve and the resulting quantification metric area under the ROC curve (AUC). Compared with the static approaches (DAE,

AE, and PCA), the proposed method achieved better fault detection performance in terms of AUC metric.

Normally, in WT, the control actions can be affected by sensor faults. So, as future work, they suggested the introduction of fault tolerance control (FTC). The FTC allows reconfiguration of the control action based on real-time information about the state of the WT. This information includes the fault detection and diagnosis scheme for sensors, actuators, or the system itself.

The main contribution [68] was that by using an SW-DAE, they were able to capture non-linear correlations among variables combined with the time dependency, being the last part something that may lack on some approaches. We also believe that adding time dependency will increase the prediction of the model. Therefore, a sliding window technique should also be used. The evaluation metrics used in the paper can be used, even for an imbalanced dataset.

4. Validation

In this section, we present metrics that can be used to validate if the objectives of the previously mentioned models were accomplished. The integration of these metrics takes place in the validation stage, where we test if the model is capable to be generalized to new data. We can also use this stage to prevent overfitting, adjusting the hyperparameters of the model to new data, avoiding only making good predictions to known data. We can choose a random part of the dataset as a validation set, or we can do cross-validation, doing multiple runs and alternating the fraction of the dataset; more details can be found in [70].

Metrics similar to the ones that Verma et al. [31] used, accuracy and f1 measure with the addition of recall and precision are commonly used for classification problems. As for regression problems, we can use metrics such as the ones that Marvuglia, A. et al. [22] mentioned. For instance, mean absolute error (MAE), mean square error (MSE), or root mean square error (RMSE).

Nevertheless, when dealing with an imbalanced dataset [71], if over or undersampling the data only improves the balance to a certain extent, some of these metrics may not be suited. For instance, the accuracy can have a high value because the classifier is able to detect normal samples, and those are the representative part of the training and test sets. For those cases, accuracy is only reflecting the underlying class distribution. Even though recall, precision, and F1 give more insight into the accuracy of the model than traditional classification accuracy, some other metrics will help to better represent the performance of the model:

- Kappa (or Cohen's kappa): Accuracy normalized by the imbalance of the classes in the dataset;
- Receiver operating characteristic (ROC) curves: Accuracy is divided into sensitivity and specificity, so that the models can be chosen based on the balance thresholds of these values.

5. Conclusions

In this paper, we entailed the state-of-the-art regarding the review topic, i.e., the use of machine learning to improve turbine generators operation conditions monitoring. With this insight, some of the strengths present in the existing literature were pointed out. More specifically, techniques regarding the pre-processing of the data (dealing with missing data, outliers, and data labelling), feature engineering, and finally, knowledge about the models that proved to be more accurate in predicting failures, helping the increase in the availability of the turbine generators. In each of the subsections of Section 3, we gather possible methods to deal with each stage of an ML problem. Data pre-processing presented solutions of increased complexity to give the readers the choice of selecting the one that best suits their problem. In Feature Selection, we concluded that we do not have a conventional method that leads to the best results; it will depend on the specific model and dataset. As for the models, we gave some general insights on how to achieve fault detection in Section 3.3.

For instance, monitoring the turbine production, to capture the turbine's degradation or building a model for the steady-state of a wind farm's power curve. Any deviation of the normal behaviour of production can be indicative of a fault. On the other hand, we also covered a novel approach that uses multi-target normal behaviour models, using only one model for fault diagnosis, being able to monitor multiple variables simultaneously. If we are dealing with a small or unlabelled dataset, it is suggested that we could also use transfer learning to transfer the knowledge from a model used in a different dataset. Federated deep learning opened the doors for combining information from multiple wind farms, surpassing the problem of data privacy, sharing only the model's parameters. Subsequently, we focus only on generator problems, gathering models used for specific problems. Since the majority opted for regression models of the normal behaviour of that specific generator component, we believe that the future of this type of fault detection could be using a multi-target normal behaviour model for the generator. Finally, we introduce metrics on how to validate the previously discussed techniques and models.

In Table 1, we present a summary of all the methods covered for each of the ML stages. In the present analysis, we screened around one hundred papers, and we selected 23 papers. These papers were selected mainly based on the description and importance of ML techniques applied to condition monitoring in wind turbines. This detailed analysis allows a deeper understanding of the used ML techniques when applied to CM problems. In our analysis, we started reading the references mentioned in previous reviews [8–13], and we did an update, with some references published after the last reviews.

Table 1. Summary of methods.

ML Stage	Task/Method	Paper
Data acquisition and pre-processing	Big Data framework	[34]
	Data Loss	[35]
	Outlier Removal	[21,36,37]
	Data Sampling	[25,30,31,38–41]
	Normal Behaviour Data Selection	[42]
Feature Selection	Autoencoder	[43]
	T-Selection	[45]
	PARAFAC	[47]
	Mahalanobis Distance	[50]
	Wrapper Techniques and Boosting Tree	[31,64]
	Chi-square	[31]
	NSGAII	[51]
	Featureless Approach	[52]
	Spatio-Temporal Features	[41,53]
Model Selection	SOM	[21]
	GMM	[21]
	NN	[21–25,31,61,63,64]
	CCFL	[26]
	K-NN	[26,31,55]
	ANFIS	[23,26]
	SVM	[23,30,31]
	NSET	[67]
	GPR	[23]
	Boosting Tree	[31]
	Denoising Auto-Encoder	[68]
	Decision Tree	[24,25]
	Random Forest	[24,55]
	ART-2	[65]
	Inception V3 and TrAdaBoost	[55]
	CVME-WDA	[56]
	DCGAN	[66]
Validation	Accuracy, F1, Recall, Precision	[31]
	ROC	[68]

Author Contributions: A.R.N. developed most of the study and wrote a first draft of the paper. H.M. and A.S. contributed to the definition of the initial structure of the paper, in the revision of the final version, and in the supervision of A.R.N. work. All authors have read and agreed to the published version of the manuscript.

Funding: This work was supported by national funds through FCT, Fundação para a Ciência e a Tecnologia, under projects UIDB/50021/2020 and PTDC/CCI-COM/7203/2020.

Institutional Review Board Statement: Not applicable.

Informed Consent Statement: Not applicable.

Data Availability Statement: Not applicable.

Conflicts of Interest: The authors declare no conflict of interest.

Abbreviations

Abbreviations

The following abbreviations are used in this manuscript:

AI	Artificial Intelligence
CM	Condition Monitoring
ML	Machine Learning
NN	Neural Network
OM	Operation and Maintenance
RNN	Recurrent Neural Network
ROC	Receiver Operating Characteristic
SCADA	Supervisory Control And Data Acquisition
SOM	Self-Organizing Map
SVM	Support Vector Machine
WT	Wind Turbine

References

1. Al Essa, M. The Integration of Distributed Energy Resources into Electric Power Systems. Ph.D. Thesis, Cardiff University, Cardiff, UK, 2017.
2. Tsiropoulos, I.; Nijs, W.; Tarvydas, D.; Ruiz, P. Towards net-zero emissions in the EU energy system by 2050. In *Insights from Scenarios in Line with the 2030 and 2050 Ambitions of the European Green Deal*; Publications Office of the European Union: Luxembourg, 2020.
3. Akerman, P.; Cazzola, P.; Christiansen, E.S.; Van Heusden, R.; Kolomanska-van Iperen, J.; Christensen, J.; Crone, K.; Dawe, K.; De Smedt, G.; Keynes, A.; et al. *Reaching Zero with Renewables*; Hydrogen Knowledge Centre: Derby, UK, 2020.
4. Irena, I. *Future of Wind: Deployment, Investment, Technology, Grid Integration and Socio-Economic Aspects*; International Renewable Energy Agency: Abu Dhabi, United Arab Emirates, 2019.
5. Nick, M.; Riahy, G.; Hosseinian, S.; Fallahi, F. Wind power optimal capacity allocation to remote areas taking into account transmission connection requirements. *IET Renew. Power Gener.* **2011**, *5*, 347–355. [CrossRef]
6. Leahy, K.; Gallagher, C.; O'Donovan, P.; O'Sullivan, D.T. Issues with data quality for wind turbine condition monitoring and reliability analyses. *Energies* **2019**, *12*, 201. [CrossRef]
7. Ranchin, T.; Furevik, B.; Stette, M.; Wensink, H.; Van Hulle, F.; Hasager, C.; Johnsen, H.; Fichaux, N.; Christensen, L.C.; Soerensen, P.B.; et al. Obtaining data for wind farm development and management: The EO-WINDFARM project. In Proceedings of the Energies Renouvelables en Mer, Brest, France, 20–21 October 2004.
8. Stetco, A.; Dinmohammadi, F.; Zhao, X.; Robu, V.; Flynn, D.; Barnes, M.; Keane, J.; Nenadic, G. Machine learning methods for wind turbine condition monitoring: A review. *Renew. Energy* **2019**, *133*, 620–635. [CrossRef]
9. Tchakoua, P.; Wamkeue, R.; Ouhrouche, M.; Slaoui-Hasnaoui, F.; Tameghe, T.A.; Ekemb, G. Wind turbine condition monitoring: State-of-the-art review, new trends, and future challenges. *Energies* **2014**, *7*, 2595–2630. [CrossRef]
10. Lu, B.; Li, Y.; Wu, X.; Yang, Z. A review of recent advances in wind turbine condition monitoring and fault diagnosis. In Proceedings of the 2009 IEEE Power Electronics and Machines in Wind Applications, Lincoln, NE, USA , 24–26 June 2009; pp. 1–7.
11. de Azevedo, H.D.M.; Araújo, A.M.; Bouchonneau, N. A review of wind turbine bearing condition monitoring: State of the art and challenges. *Renew. Sustain. Energy Rev.* **2016**, *56*, 368–379. [CrossRef]
12. Tautz-Weinert, J.; Watson, S.J. Using SCADA data for wind turbine condition monitoring—A review. *IET Renew. Power Gener.* **2016**, *11*, 382–394. [CrossRef]

13. Hameed, Z.; Hong, Y.; Cho, Y.; Ahn, S.; Song, C. Condition monitoring and fault detection of wind turbines and related algorithms: A review. *Renew. Sustain. Energy Rev.* **2009**, *13*, 1–39. [CrossRef]
14. Pao, L.Y.; Johnson, K.E. Control of Wind Turbines. *IEEE Control Syst. Mag.* **2011**, *31*, 44–62. [CrossRef]
15. Godwin, J.; Matthews, P. Classification and detection of wind turbine pitch faults through SCADA data analysis. *Int. J. Progn. Health Manag.* **2013**, *4*, 016.
16. Pérez, J.M.P.; Márquez, F.P.G.; Tobias, A.; Papaelias, M. Wind turbine reliability analysis. *Renew. Sustain. Energy Rev.* **2013**, *23*, 463–472. [CrossRef]
17. Trojan, F.; Marçal, R. Proposal of maintenance-types classification to clarify maintenance concepts in production and operations management. *J. Bus. Econ.* **2017**, *8*, 560–572.
18. Ali, Y.H. Artificial intelligence application in machine condition monitoring and fault diagnosis. In *Artificial Intelligence: Emerging Trends and Applications*; BoD–Books on Demand: Norderstedt, Germany, 2018; p. 275.
19. Berry, M.W.; Mohamed, A.; Yap, B.W. *Supervised and Unsupervised Learning for Data Science*; Springer: Berlin/Heidelberg, Germany, 2019.
20. Guresen, E.; Kayakutlu, G. Definition of artificial neural networks with comparison to other networks. *Procedia Comput. Sci.* **2011**, *3*, 426–433. [CrossRef]
21. Lapira, E.; Brisset, D.; Davari, H.; Siegel, D.; Lee, J. Wind turbine performance assessment using multi-regime modeling approach. *Renew. Energy* **2012**, *45*, 86–95. [CrossRef]
22. Marvuglia, A.; Messineo, A. Monitoring of wind farms' power curves using machine learning techniques. *Appl. Energy* **2012**, *98*, 574–583. [CrossRef]
23. Tautz-Weinert, J.; Watson, S.J. Comparison of different modelling approaches of drive train temperature for the purposes of wind turbine failure detection. *J. Phys. Conf. Ser.* **2016**, *753*, 072014. [CrossRef]
24. Meyer, A. Multi-target normal behaviour models for wind farm condition monitoring. *Appl. Energy* **2021**, *300*, 117342. [CrossRef]
25. Chatterjee, J.; Dethlefs, N. Deep learning with knowledge transfer for explainable anomaly prediction in wind turbines. *Wind Energy* **2020**, *23*, 1693–1710. [CrossRef]
26. Schlechtingen, M.; Santos, I.F.; Achiche, S. Using data-mining approaches for wind turbine power curve monitoring: A comparative study. *IEEE Trans. Sustain. Energy* **2013**, *4*, 671–679. [CrossRef]
27. Kusiak, A.; Li, W. The prediction and diagnosis of wind turbine faults. *Renew. Energy* **2011**, *36*, 16–23. [CrossRef]
28. Ibrahim, R.; Weinert, J.; Watson, S. *Neural Networks for Wind Turbine Fault Detection via Current Signature Analysis*; Loughborough University: Loughborough, UK, 2016.
29. Tang, B.; Song, T.; Li, F.; Deng, L. Fault diagnosis for a wind turbine transmission system based on manifold learning and Shannon wavelet support vector machine. *Renew. Energy* **2014**, *62*, 1–9. [CrossRef]
30. Leahy, K.; Hu, R.L.; Konstantakopoulos, I.C.; Spanos, C.J.; Agogino, A.M.; O'Sullivan, D.T. Diagnosing and predicting wind turbine faults from SCADA data using support vector machines. *Int. J. Progn. Health Manag.* **2018**, *9*, 1–11.
31. Verma, A.; Kusiak, A. Fault monitoring of wind turbine generator brushes: A data-mining approach. *J. Sol. Energy Eng.* **2012**, *134*, 021001. [CrossRef]
32. Wilkinson, M.; Darnell, B.; Van Delft, T.; Harman, K. Comparison of methods for wind turbine condition monitoring with SCADA data. *IET Renew. Power Gener.* **2014**, *8*, 390–397. [CrossRef]
33. Shafae, M.S.A. Advancing the Utility of Manufacturing Data for Modeling, Monitoring, and Securing Machining Processes. Ph.D. Thesis, Virginia Polytechnic Institute, Blacksburg, VR, USA, 2018.
34. Canizo, M.; Onieva, E.; Conde, A.; Charramendieta, S.; Trujillo, S. Real-time predictive maintenance for wind turbines using Big Data frameworks. In Proceedings of the 2017 IEEE International Conference on Prognostics and Health Management (ICPHM), Dallas, TX, USA, 19–21 June 2017; pp. 70–77.
35. Peng, Y.; Qiao, W.; Qu, L. Compressive Sensing-Based Missing-Data-Tolerant Fault Detection for Remote Condition Monitoring of Wind Turbines. *IEEE Trans. Ind. Electron.* **2021**. [CrossRef]
36. Marti-Puig, P.; Blanco-M, A.; Cárdenas, J.J.; Cusidó, J.; Solé-Casals, J. Effects of the pre-processing algorithms in fault diagnosis of wind turbines. *Environ. Model. Softw.* **2018**, *110*, 119–128. [CrossRef]
37. Yang, W.; Court, R.; Jiang, J. Wind turbine condition monitoring by the approach of SCADA data analysis. *Renew. Energy* **2013**, *53*, 365–376. [CrossRef]
38. Velandia-Cardenas, C.; Vidal, Y.; Pozo, F. Wind Turbine Fault Detection Using Highly Imbalanced Real SCADA Data. *Energies* **2021**, *14*, 1728. [CrossRef]
39. Yi, H.; Jiang, Q.; Yan, X.; Wang, B. Imbalanced Classification Based on Minority Clustering SMOTE with Wind Turbine Fault Detection Application. *IEEE Trans. Ind. Inform.* **2021**, *17*, 5867–5875. [CrossRef]
40. Jiang, N.; Li, N. A wind turbine frequent principal fault detection and localization approach with imbalanced data using an improved synthetic oversampling technique. *Int. J. Electr. Power Energy Syst.* **2021**, *126*, 106595. [CrossRef]
41. He, Q.; Pang, Y.; Jiang, G.; Xie, P. A spatio-temporal multiscale neural network approach for wind turbine fault diagnosis with imbalanced SCADA data. *IEEE Trans. Ind. Inform.* **2020**, *17*, 6875–6884. [CrossRef]
42. Xu, Q.; Fan, Z.; Jia, W.; Jiang, C. Quantile regression neural network-based fault detection scheme for wind turbines with application to monitoring a bearing. *Wind Energy* **2019**, *22*, 1390–1401. [CrossRef]
43. Jiang, G.; He, H.; Xie, P.; Tang, Y. Stacked multilevel-denoising autoencoders: A new representation learning approach for wind turbine gearbox fault diagnosis. *IEEE Trans. Instrum. Meas.* **2017**, *66*, 2391–2402. [CrossRef]

44. Ringnér, M. What is principal component analysis? *Nat. Biotechnol.* **2008**, *26*, 303–304. [CrossRef]
45. Wang, Y.; Ma, X.; Qian, P. Wind turbine fault detection and identification through PCA-based optimal variable selection. *IEEE Trans. Sustain. Energy* **2018**, *9*, 1627–1635. [CrossRef]
46. Zhou, N.; Wang, L. A modified T-test feature selection method and its application on the HapMap genotype data. *Genom. Proteom. Bioinform.* **2007**, *5*, 242–249. [CrossRef]
47. Zhang, W.; Ma, X. Simultaneous fault detection and sensor selection for condition monitoring of wind turbines. *Energies* **2016**, *9*, 280. [CrossRef]
48. Rajih, M.; Comon, P.; Harshman, R.A. Enhanced line search: A novel method to accelerate PARAFAC. *SIAM J. Matrix Anal. Appl.* **2008**, *30*, 1128–1147. [CrossRef]
49. Yang, L.; Chen, H.; Ke, Y.; Li, M.; Huang, L.; Miao, Y. Multi-source and Multi-fault Condition Monitoring Based on Parallel Factor Analysis and Sequential Probability Ratio Test. *EURASIP J. Adv. Signal Process.* **2021**, *37*, 1–33.
50. Qian, P.; Tian, X.; Kanfoud, J.; Lee, J.L.Y.; Gan, T.H. A novel condition monitoring method of wind turbines based on long short-term memory neural network. *Energies* **2019**, *12*, 3411. [CrossRef]
51. de Sá, F.P.; Brandão, D.N.; Ogasawara, E.; Coutinho, R.d.C.; Toso, R.F. Wind Turbine Fault Detection: A Semi-Supervised Learning Approach With Automatic Evolutionary Feature Selection. In Proceedings of the 2020 International Conference on Systems, Signals and Image Processing (IWSSIP), Niterói, Brazil, 1–3 July 2020; pp. 323–328.
52. Stetco, A.; Mohammed, A.; Djurović, S.; Nenadic, G.; Keane, J. Wind Turbine operational state prediction: Towards featureless, end-to-end predictive maintenance. In Proceedings of the 2019 IEEE International Conference on Big Data (Big Data), Los Angeles, CA, USA, 9–12 December 2019; pp. 4422–4430.
53. Kong, Z.; Tang, B.; Deng, L.; Liu, W.; Han, Y. Condition monitoring of wind turbines based on spatio-temporal fusion of SCADA data by convolutional neural networks and gated recurrent units. *Renew. Energy* **2020**, *146*, 760–768. [CrossRef]
54. European Wind Energy Association. *Wind Energy-The Facts: A Guide to the Technology, Economics and Future of Wind Power*; Routledge: Oxfordshire, UK, 2012.
55. Chen, W.; Qiu, Y.; Feng, Y.; Li, Y.; Kusiak, A. Diagnosis of wind turbine faults with transfer learning algorithms. *Renew. Energy* **2021**, *163*, 2053–2067. [CrossRef]
56. Ren, H.; Liu, W.; Shan, M.; Wang, X.; Wang, Z. A novel wind turbine health condition monitoring method based on composite variational mode entropy and weighted distribution adaptation. *Renew. Energy* **2021**, *168*, 972–980. [CrossRef]
57. Briggs, C.; Fan, Z.; Andras, P. Federated Learning for Short-term Residential Energy Demand Forecasting. *arXiv* **2021**, arXiv:2105.13325.
58. Briggs, C.; Fan, Z.; Andras, P. Privacy Preserving Demand Forecasting to Encourage Consumer Acceptance of Smart Energy Meters. *arXiv* **2020**, arXiv:2012.07449.
59. Liu, Y.; Garg, S.; Nie, J.; Zhang, Y.; Xiong, Z.; Kang, J.; Hossain, M.S. Deep anomaly detection for time-series data in industrial iot: A communication-efficient on-device federated learning approach. *IEEE Internet Things J.* **2020**, *8*, 6348–6358. [CrossRef]
60. Zhang, W.; Li, X.; Ma, H.; Luo, Z.; Li, X. Federated learning for machinery fault diagnosis with dynamic validation and self-supervision. *Knowl. Based Syst.* **2021**, *213*, 106679. [CrossRef]
61. Wang, H.; Liu, C.; Jiang, D.; Jiang, Z. Collaborative deep learning framework for fault diagnosis in distributed complex systems. *Mech. Syst. Signal Process.* **2021**, *156*, 107650. [CrossRef]
62. Gayo, J. *ReliaWind Project Final Report*; Technical Report; Gamesa Innovation and Technology: Madrid, Spain, 2011.
63. Schlechtingen, M.; Santos, I.F. Comparative analysis of neural network and regression based condition monitoring approaches for wind turbine fault detection. *Mech. Syst. Signal Process.* **2011**, *25*, 1849–1875. [CrossRef]
64. Kusiak, A.; Verma, A. Analyzing bearing faults in wind turbines: A data-mining approach. *Renew. Energy* **2012**, *48*, 110–116. [CrossRef]
65. Yang, D.; Li, H.; Hu, Y.; Zhao, J.; Xiao, H.; Lan, Y. Vibration condition monitoring system for wind turbine bearings based on noise suppression with multi-point data fusion. *Renew. Energy* **2016**, *92*, 104–116. [CrossRef]
66. Chen, P.; Li, Y.; Wang, K.; Zuo, M.J.; Heyns, P.S.; Bagger'ohr, S. A threshold self-setting condition monitoring scheme for wind turbine generator bearings based on deep convolutional generative adversarial networks. *Measurement* **2021**, *167*, 108234. [CrossRef]
67. Guo, P.; Infield, D.; Yang, X. Wind turbine generator condition-monitoring using temperature trend analysis. *IEEE Trans. Sustain. Energy* **2011**, *3*, 124–133. [CrossRef]
68. Jiang, G.; Xie, P.; He, H.; Yan, J. Wind turbine fault detection using a denoising autoencoder with temporal information. *IEEE Asme Trans. Mechatronics* **2017**, *23*, 89–100. [CrossRef]
69. Majumdar, A. Blind denoising autoencoder. *IEEE Trans. Neural Netw. Learn. Syst.* **2018**, *30*, 312–317. [CrossRef] [PubMed]
70. Berrar, D. Cross-Validation. In *Encyclopedia of Bioinformatics and Computational Biology*; Elsevier: Amsterdam, The Netherlands, 2019; pp. 542–545.
71. Kotsiantis, S.; Kanellopoulos, D.; Pintelas, P. Handling imbalanced datasets: A review. *GESTS Int. Trans. Comput. Sci. Eng.* **2006**, *30*, 25–36.

Journal of
Marine Science and Engineering

MDPI

Article

Investigation on Thrust Characteristics of a Downstream Offshore Floating Wind Turbine under Yawed Inflow Conditions

Yangwei Wang, Jiahuan Lin, Huawei Duan and Jun Zhang *

School of Mechanical Engineering and Automation, Fuzhou University, Fuzhou 350116, China; m170210011@fzu.edu.cn (Y.W.); n190220050@fzu.edu.cn (J.L.); 200227065@fzu.edu.cn (H.D.)
* Correspondence: zhang_jun@fzu.edu.cn

Abstract: In the natural marine environment, offshore floating wind turbines (OFWTs) inevitably experience yawed inflow conditions, which will make their aerodynamics more complicated than those experiencing uniform inflow conditions and difficult to understand. In the present study, the thrust characteristics of a wake-influenced OFWT under dynamic, static, and coupled yawed inflow conditions are investigated thoroughly. Analytical characterizations of yawed inflow and upstream wake are integrated into the blade element momentum (BEM) method to achieve the investigation. Based on this method, simulations by the FAST code have been conducted, and the results are analyzed. It is shown that the three inflow conditions have considerable influences on the thrust coefficient of the wind rotor or the normal force at the blade section, especially in the wake case where the downstream OFWT is located at a specific offset from the central line of a single upstream wake. In order to validate the analyses of simulation results, experimental tests by a set of dedicated apparatus are conducted. The comparison results are good, proving the reliability of simulation results. This work can provide some theoretical contributions to the aerodynamic design and control of OFWTs.

Keywords: thrust characteristics; offshore floating wind turbine; wake; yawed inflow conditions

Citation: Wang, Y.; Lin, J.; Duan, H.; Zhang, J. Investigation on Thrust Characteristics of a Downstream Offshore Floating Wind Turbine under Yawed Inflow Conditions. *J. Mar. Sci. Eng.* **2021**, *9*, 1215. https://doi.org/10.3390/jmse9111215

Academic Editor: Aristotle T. Ubando

Received: 11 October 2021
Accepted: 29 October 2021
Published: 3 November 2021

1. Introduction

In recent years, with the rapid development of the offshore wind power industry, the offshore floating wind turbine (OFWT) has been widely studied and applied [1–4]. When operating in natural marine environmental conditions, an OFWT will inevitably experience yawed inflow [5,6]. Under the yawed inflow conditions, the aerodynamics of the OFWT become rather complicated and difficult to obtain, which will bring great challenges to its control in practical engineering [6,7]. As one of the most crucial aerodynamics, the thrust characteristic of the wind rotor is also significantly influenced because the normal force at each blade section used to calculate the thrust varies with the dynamic angle of attack at different azimuths in the yawed inflow conditions [8]. The thrust characteristic is important for two main reasons. On the one hand, the thrust characteristic of an OFWT is closely related to its wake's expansion and wind velocity distribution [9–12]. As the wake is disturbed, not only the aerodynamics of the OFWT itself but also the aerodynamics of its downstream OFWTs can be deeply influenced [13]. On the other hand, the thrust characteristic of an OFWT is also related to its dynamic characteristics because the thrust is a critical external excitation of the wind turbine system [14,15]. Unfortunately, investigations on the thrust characteristic are still scarce, especially under complex conditions like the yawed inflow.

The yawed inflow condition refers to that there is an included angle between the inflow wind direction and the vertical direction of the wind rotor plane, that is, the yawed angle θ_{yaw}. This kind of inflow conditions can be divided into dynamic yawed inflow condition and static yawed inflow condition. The dynamic yawed inflow condition is caused by the platform yaw motions. When the OFWT operates with platform yaw motions, the yawed angle varies in the time domain [16–18]. To clarify the influences of the yaw motions, Qiu et al. [19] used the free-vortex method to predict the unsteady aerodynamic loads of wind turbines. They found that the shaft torque of the blade could be significantly affected by the yaw motions. Besides, Tran et al. [20] applied the blade element momentum (BEM) and the computational fluid dynamics (CFD) methods to investigate the aerodynamics of an OFWT. They detected that the blade-tip vortices were generated in the process of yaw motions, which could make the aerodynamic loads complicated. However, the knowledge about the influences of the dynamic yawed inflow condition on the aerodynamics of an OFWT is still limited. As for the static yawed inflow condition, it is caused by the changes of the incoming wind direction [21,22] or the wind rotor facing direction [23–25]. When these phenomena happen, a fixed yawed angle is generated. Extensive investigations about the static yawed inflow condition have been done [8]. Greco et al. [26] used a free-wake panel method to investigate the aerodynamic performance of a horizontal wind turbine in the static yawed inflow condition. Bangga et al. [27] used CFD and BEM methods to predict the aerodynamic loads on a 2.3 MW wind turbine rotor in three different inflow cases considering the static yawed inflow condition. Both the above two investigations indicated that the static yawed inflow condition has significant influences on the aerodynamics of an OFWT. In addition, Wen et al. [16] pointed out that the static yawed inflow condition could reduce the energy conversion efficiency due to the reduction of effective wind area. Other investigations about the static yawed inflow condition focus on its influences on the OFWT's wake. Lee et al. [25] conducted a numerical simulation on the National Renewable Energy Laboratory (NREL) Phase VI wind turbine model based on the nonlinear vortex lattice method. They found that the static yawed inflow condition could make the wake skewed, which would finally affect the aerodynamic behavior of the wind turbine. Bastankhah et al. [28–30] conducted wind tunnel measurements to study the wake characteristics of a wind turbine with different static yawed angles. They found that the static yawed angles could reduce the wake velocity deficit while increase the wake deflection.

Based on the above literature reviews, there are two major limitations of the existing investigations. Firstly, the existing investigations on the aerodynamics under the yawed inflow conditions are mostly based on an isolated solo OFWT, without considering the influences of the upstream wake in the wind farm. With the gradual large-scale applications of the OFWT, the upstream wake conditions, especially the typical wake conditions under specific wind farm layouts, should be considered in the aerodynamics calculations of a downstream OFWT. Secondly, most existing investigations only focus on the aerodynamics under the dynamic yawed inflow condition or the static yawed inflow condition, and seldom consider the coupled condition of the two yawed inflow. It should be noted that due to the randomness of wave and wind directions in the environment and the hysteresis of yaw control system of the OFWT, the coupled condition of the two yawed inflow is quite likely to come into being. The drawbacks mentioned above result in the motivation of the present study: to investigate the thrust characteristics of a wake-influenced OFWT under dynamic yawed inflow, static yawed inflow, and coupled yawed inflow conditions. To achieve this investigation, the yawed inflow and the wake characterizations are integrated into the numerical method based on the BEM theory, and then simulations applying the FAST code (an open source simulation code defined by "Fatigue", "Aerodynamic", "Structure", and "Turbulence") are carried out for a typical semi-submersible OFWT. According to the simulation results, the thrust characteristics in typical wake flow under the three yawed inflow conditions are analyzed and discussed. Finally, a series of experimental tests are conducted to validate the analyses of simulations. The innovation of this method lies in that it can achieve the correct calculation of yawed inflow aerodynamics for the downstream OFWT.

The rest of this paper is organized as follows. In Section 2, the characterization of the yawed inflow conditions, the calculation method of aerodynamics in wake cases, the application of FAST and the simulation setups are described. In Section 3, the results of the thrust characteristics under different yawed inflow conditions from the simulations are analyzed and discussed. In Section 4, the experimental tests are set up to validate the analyses of simulations. Finally, in Section 5, the main conclusions of this paper are summarized.

2. Methodology

The object analyzed in the present study adopts the NREL 5-MW wind turbine [31] which is installed on the Offshore Code Comparison Collaboration Continuation (OC4) DeepCwind semi-submersible floater [32], as shown in Figure 1. The basic parameters of the OFWT are listed in Tables 1 and 2, and more details can be found in Refs. [31,32].

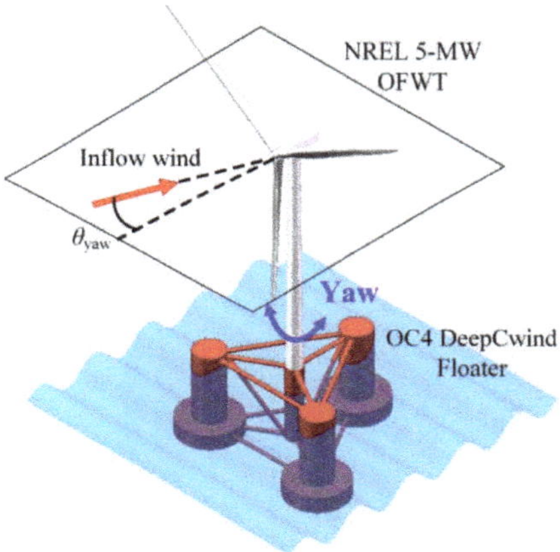

Figure 1. Object OFWT under the yawed inflow.

Table 1. Basic parameters of the NREL 5-MW wind turbine.

Terms	Value
Rated power	5 MW
Rotor type	3 Blades, upwind
Drive-train	Multiple-stage gearbox, high speed
Rotor diameter, hub diameter and hub height	126 m, 3 m, 90 m
Shaft tilt, pre-cone and overhang	5°, 2.5°, 5 m
Tower length, tower top/bottom diameter	77.6 m (for OC4), 3.87 m/6 m
Cut-in, rated, cut-out wind speed	3 m/s, 11.4 m/s, 25 m/s
Rotor, nacelle and tower mass	1.1×10^5 kg, 2.4×10^5 kg, 3.4746×10^5 kg

Table 2. Basic parameters of the OC4 DeepCwind semi-submersible floater.

Terms	Value
Draft	20 m
Elevation of platform top/offset columns	10 m/12 m
Spacing between offset columns	50 m
Length of upper columns/base columns	26 m/6 m
Depth to top of base columns	14 m
Diameter of main column/upper columns/base columns	6.5 m/12 m/24 m
Platform mass/Displacement	13,473,000 kg/13,986.8 m^3
Platform roll inertia	6.827×10^9 kg m^2
Platform pitch inertia	6.827×10^9 kg m^2
Platform yaw inertia	1.226×10^{10} kg m^2

2.1. Modeling of the Yawed Inflow Condition

The yawed inflow condition of an OFWT is shown in Figure 2. In this figure, the yawed angle and the yawed induced wind velocity can be clearly demonstrated. Note that *oxyz* represents the local coordinate system which is fixed to the geometric center of wind rotor. Its *x*-direction is along the nacelle's direction; *z*-direction is along the vertical direction; *y*-direction is along the horizontal direction.

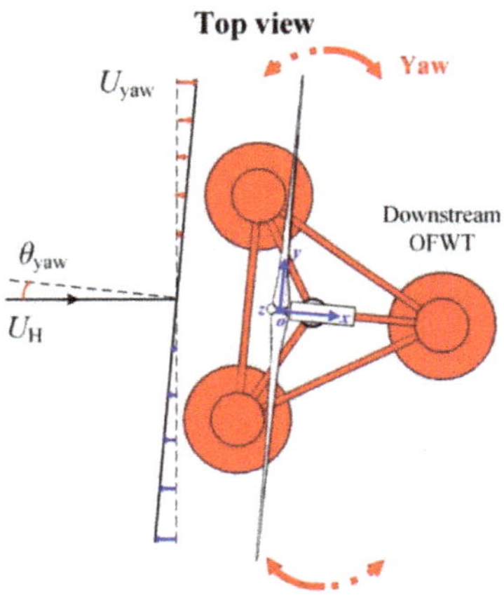

Figure 2. Schematic diagram of the yawed inflow condition from the top view.

As done by prior investigations [8], the rotation center of the yaw motion is assumed to be the origin of the *oxyz* system, the static, dynamic, and coupled yawed angle can be expressed as follows:

$$\begin{cases} \theta_{yaw} = \overline{\theta}_{yaw} & \text{static condition} \\ \theta_{yaw} = \underline{\theta}_{yaw}(t) = A_{yaw}\cos(2\pi f_{yaw}t + \phi_{yaw,0}) & \text{dynamic condition} \\ \theta_{yaw} = \overline{\theta}_{yaw} + \theta_{yaw}(t) & \text{coupled condition} \end{cases}, \quad (1)$$

where $\overline{\theta}_{yaw}$, $\theta_{yaw}(t)$ and $\overline{\theta}_{yaw} + \theta_{yaw}(t)$ denote the static, dynamic and coupled yawed angles, respectively; A_{yaw} and f_{yaw} denote amplitude and frequency of the yaw motion (resulting in dynamic yawed angle); $\phi_{yaw,0}$ denotes the initial phase, and it is assumed to be 0 in this paper.

Then, the corresponding yawed induced angular velocity can be expressed as follows:

$$\omega_{yaw}(t) = -\frac{d\theta_{yaw}}{dt} = \begin{cases} 0 & \text{static condition} \\ 2\pi A_{yaw} f_{yaw}\sin(2\pi f_{yaw}t) & \text{dynamic condition} \\ 2\pi A_{yaw} f_{yaw}\sin(2\pi f_{yaw}t) & \text{coupled condition} \end{cases}. \quad (2)$$

In order to convert the yawed angular velocity into the yawed induced wind velocity in the wind flow direction, an equivalent wind shear model is adopted, which can be expressed as the follow:

$$U_{\text{yaw}}(y,t) = \omega_{\text{yaw}}(t) y \cos\theta_{\text{yaw}}(t), \tag{3}$$

where $U_{\text{yaw}}(y,t)$ denotes the yawed induced wind velocity. Note that $\theta_{\text{yaw}}(t)$ is usually small and therefore the value of $\cos\theta_{\text{yaw}}(t)$ is set as 1 in this paper. Inserting Equation (2) into Equation (3), the yawed induced wind velocity can be finally expressed as:

$$U_{\text{yaw}}(y,t) = \omega_{\text{yaw}}(t) y = \begin{cases} 0 & \text{staticcondition} \\ 2\pi A_{\text{yaw}} f_{\text{yaw}} y \sin(2\pi f_{\text{yaw}} t) & \text{dynamic condition} \\ 2\pi A_{\text{yaw}} f_{\text{yaw}} y \sin(2\pi f_{\text{yaw}} t) & \text{coupling condition} \end{cases}. \tag{4}$$

2.2. Modeling of the Aerodynamics

The yawed inflow condition of a blade airfoil is shown in Figure 3. In this figure, $W_{\text{r,y}}$ denotes the relative wind velocity of the blade airfoil considering the yawed inflow; Ω denotes the rotational angular speed; r denotes the distance from the rotor axis to the airfoil; a and b denote the axial flow induction coefficient and the tangential induction coefficient, respectively; dF denotes the resultant force acting on the airfoil; dL and dD denote the lift force and drag force, respectively; dT and dQ denote the axial force (thrust) and tangential force, respectively; γ denotes the twist angle; β denotes the angle of attack; c denotes the chord length.

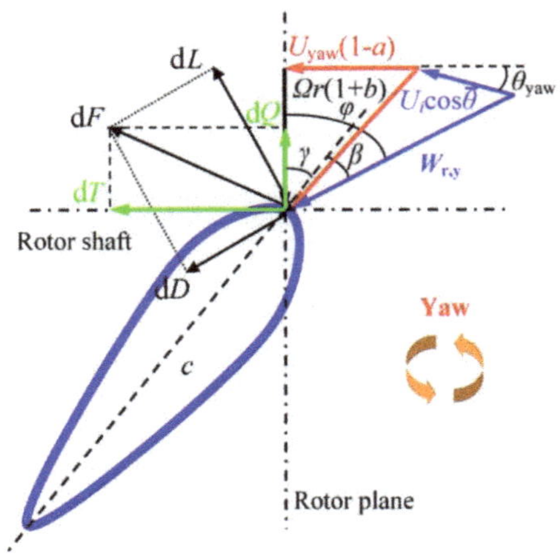

Figure 3. Schematic diagram of the yawed inflow condition of a blade airfoil.

As can be seen from Figure 3, due to the yawed inflow the magnitude and direction of relative wind velocity has been changed. Based on the geometry theory, the relative wind velocity $W_{\text{r,y}}$ and the corresponding angle of attack β can be calculated by the following Equations:

$$W_{\text{r,y}} = \sqrt{[U_{\text{yaw}}(1-a) + U_i \cos\overline{\theta} \cos\theta_{\text{yaw}}]^2 + [\Omega r(1+b) - U_i \cos\overline{\theta} \sin\theta_{\text{yaw}}]^2}, \tag{5}$$

$$\beta = \tan^{-1}[\frac{U_{\text{yaw}}(1-a) + U_i \cos\overline{\theta} \cos\theta_{\text{yaw}}}{\Omega r(1+b) - U_i \cos\overline{\theta} \sin\theta_{\text{yaw}}}] - \gamma, \tag{6}$$

where U_i denotes the incoming wind velocity at the position where the airfoil locates.

For a downstream OFWT in a wind farm, the incoming wind velocity is actually the wake velocity of the upstream OFWT. Considering this situation, the distribution of the incoming wind velocity of a downstream OFWT is expressed by a widely used Gaussian wake model as the follow [10,33,34]:

$$U_i(X,Y,Z) = U_{\mathrm{In}j}(Z) - \sqrt{\sum_{j=1}^{n}[\frac{A_j(X)}{2\pi\sigma_j(X)^2}e^{\frac{-Y^2-(Z-H_0)^2}{2\sigma_j(X)^2}} + B_j(X)]^2}, \quad (7)$$

where the subscript i and j denote the downstream OFWT and the upstream OFWT, respectively; $U_{\mathrm{In}j}(Z)$ denotes the incoming wind velocity of upstream OFWTs considering the wind shear effect as expressed in Equation (8) [35]; $\sigma_j(X)$ denotes the parameter characterizing wake expanding as expressed in Equation (9) [9]; $A_j(X)$ and $B_j(X)$ denote the two key parameters determining Gaussian-shaped wake deficit as expressed in Equation (10) and Equation (11) [10], respectively. Note that X, Y, and Z here belong to the global coordinate system $OXYZ$. The origin O is located at the central position of the upstream OFWTs at the still water level; X-direction is the stream-wise direction; Z-direction is the vertical upward direction; Y-direction is the horizontal direction determined by the right-hand rule.

$$U_{\mathrm{In}j}(Z) = U_{\mathrm{H}}(\frac{Z}{H_0})^{\alpha}, \quad (8)$$

$$\sigma_j(X) = \frac{R_0 I_0 + k_0 X\sqrt{I_0^2 + [0.73a^{0.8325} I_0^{0.0325}(\frac{X}{D_0})^{-0.32}]^2}}{C I_0}, \quad (9)$$

$$A_j(X) = \frac{\pi R_0^2 V_0 - 2\int_{H_0-C\sigma_j(X)}^{H_0+C\sigma_j(X)} U_{\mathrm{In}j}(Z)\sqrt{C^2\sigma_j(X)^2 - (Z-H_0)^2}\mathrm{d}Z}{1 - e^{-\frac{C^2}{2}} - \frac{C^2}{2}e^{-\frac{C^2}{2}}}, \quad (10)$$

$$B_j(X) = -\frac{C^2 A_j(X)}{2\pi\sigma_j(X)^2}e^{-\frac{C^2}{2}}, \quad (11)$$

where U_{H} denotes the wind velocity at the hub height H_0; α denotes the wind shear exponent empirical constant; R_0 and D_0 denotes the radius and the diameter of wind rotor, respectively; I_0 denotes the turbulence intensity of the ambient wind; k_0 denotes the wake expansion rate; C is an empirical constant; V_0 denotes the wind velocity just behind the wind rotor.

After solving the relative wind velocity and the angle of attack, the axial force (thrust) $\mathrm{d}T$ and tangential force $\mathrm{d}Q$ can be calculated by the following Equations:

$$\mathrm{d}T = \frac{1}{2}\rho W_{\mathrm{r,y}}^2[C_{\mathrm{L}}\cos(\beta+\gamma) + C_{\mathrm{D}}\sin(\beta+\gamma)]c\mathrm{d}r, \quad (12)$$

$$\mathrm{d}Q = \frac{1}{2}\rho W_{\mathrm{r,y}}^2[C_{\mathrm{L}}\sin(\beta+\gamma) - C_{\mathrm{D}}\cos(\beta+\gamma)]c\mathrm{d}r, \quad (13)$$

where ρ denotes the air density; C_{L} and C_{D} denote the lift coefficient and the drag coefficient. Then only C_{L} and C_{D} are not solved.

Based on the momentum theory in BEM, $\mathrm{d}T$ and $\mathrm{d}Q$ have the following Equation relationships [36]:

$$\mathrm{d}T = \frac{4}{3}\pi r^2\rho U_i^2(1-a)a\mathrm{d}r, \quad (14)$$

$$\mathrm{d}Q = \frac{4}{3}\pi r^2\rho U_i^2\Omega(1-a)b\mathrm{d}r, \quad (15)$$

By combining Equations (12)–(15), a and b can be derived as follows:

$$a = \{1 + \frac{8\pi r \sin^2(\beta + \gamma)}{3c[C_L \cos(\beta + \gamma) + C_D \sin(\beta + \gamma)]}\}^{-1}, \tag{16}$$

$$b = \{-1 + \frac{8\pi r \sin(\beta + \gamma)\cos(\beta + \gamma)}{3c[C_L \sin(\beta + \gamma) - C_D \cos(\beta + \gamma)]}\}^{-1}, \tag{17}$$

As seen from the above equations, C_L and C_D can be obtained after a and b are determined. By the BEM iteration algorithm as introduced in Ref. [36], a and b can be accurately calculated.

After all the necessary parameters are ready, the aerodynamic force of each blade section can be obtained based on Equations (12) and (13). Then the thrust of the wind rotor T_{rotor} can be calculated by integrating Equation (12) as the follow:

$$T_{\text{rotor}} = \sum_{n=1}^{3} \int_{R_h}^{R_0} \frac{1}{2} \rho W_{r,y}{}^2 [C_L \cos(\beta + \gamma) + C_D \sin(\beta + \gamma)] c dr, \tag{18}$$

where n denotes the number of the blades; R_h denotes the radius of the hub.

Finally, the thrust coefficient of the wind rotor can be obtained as the follow:

$$C_{T,\text{rotor}} = \frac{T_{\text{rotor}}}{0.5 \rho \pi R_0{}^2 U_H{}^2}. \tag{19}$$

2.3. Application of the FAST

The simulation in this study applies the FAST code [37–39] developed by the NREL. However, the wind file code TurbSim of the FAST cannot provide the input wind velocity file under the required conditions such as the wake condition. Thus, the authors proposed a specialized converter which contains an algorithm supporting conversion of the decimal data (from the wake model and the yawed inflow model) to the binary data (compatible in FAST). The application method of this converter in FAST is shown in Figure 4.

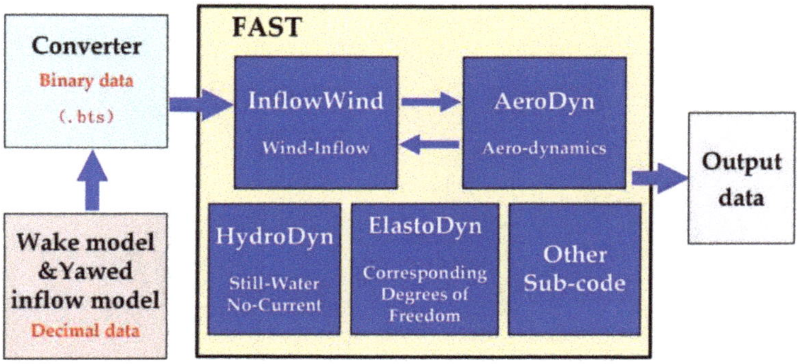

Figure 4. Application of the proposed converter in FAST.

2.4. Simulation Setting up

To obtain the aerodynamics of the analyzed OFWT, two main input conditions need to be set.

The first one is the wake condition. As seen in Figure 5, the downstream OFWT in a normal wind farm usually experiences three wake cases. Figure 5a shows a single-wake case where the downstream OFWT is located just along the central wake region of one upstream OFWT; Figure 5b shows another single-wake case where the downstream OFWT is offset from the central wake region of one upstream OFWT; Figure 5c shows a multiple-wake case where the downstream OFWT is located just behind two upstream OFWTs. In this study, the three wake cases are adopted in the simulations. The input parameters of the wake conditions are as follows: the downstream distance ΔX is set to be $5D_0$ and $10D_0$; the lateral offset ΔY is set to be equal to R_0; the relative distance ΔL of the two upstream OFWTs is set to be 144 m; the incoming wind velocity is set to be 11.4 m/s; the tip speed ratio (TSR) is set to be 7; the wind shear exponent empirical constant is set to be 0.1; the parameter C is set to be 2.12; the turbulent intensity of the incoming wind is set to be 0.08; the wake expansion rate is set to be 0.02.

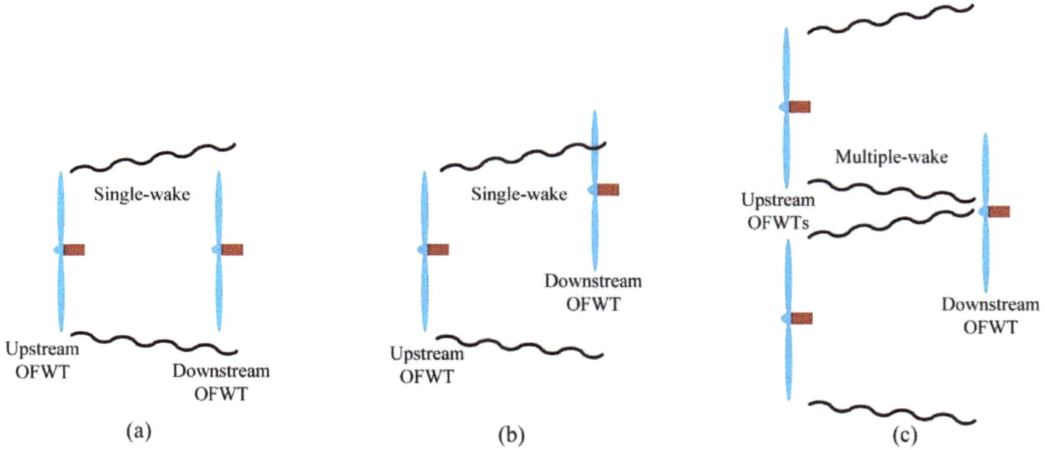

Figure 5. Schematic diagram of three upstream wake cases: (**a**) A single-wake case; (**b**) Another single-wake case; (**c**) A multiple-wake case from two upstream OFWTs.

The second one is the yawed inflow condition. In this study, three yawed inflow conditions, i.e., the dynamic, static, and coupled yawed inflow conditions, are adopted. Specifically, the detailed parameters of each yawed inflow case are listed in Table 3. The platform fixed case (i.e., FC) is regarded as a reference in the comparisons. A 300 s simulation for each case is conducted, and only the data from 100 s~200 s (101 s in total) are taken for analysis to avoid the start and stop effects of the simulation.

Table 3. Yawed inflow parameters in the three wake cases.

Yaw	Wake Case (a)		Wake Case (b)		Wake Case (c)	
	A_{yaw} [°]	f_{yaw} [Hz]	A_{yaw} [°]	f_{yaw} [Hz]	A_{yaw} [°]	f_{yaw} [Hz]
FC [1]	0	0	0	0	0	0
YC1 [2]	2	0.04	2	0.04	2	0.04
YC2 [2]	2	0.08	2	0.08	2	0.08
YC3 [2]	5	0.08	5	0.08	5	0.08
SYC1 [3]	2	0	2	0	2	0
SYC2 [3]	10	0	10	0	10	0
SYC3 [3]	20	0	20	0	20	0
YC3 + SYC1	5 + 2	0.08	5 + 2	0.08	5 + 2	0.08
YC3 + SYC2	5 + 10	0.08	5 + 10	0.08	5 + 10	0.08
YC3 + SYC3	5 + 20	0.08	5 + 20	0.08	5 + 20	0.08
YC3 − SYC3	5 − 20	0.08	5 − 20	0.08	5 − 20	0.08

[1] FC denotes the platform fixed case; [2] YC1~3 denote three cases of yaw motion, i.e., dynamic yawed inflow cases; [3] SYC1~3 denote three static yawed inflow cases.

3. Results and Discussion

3.1. Wake Cases

In this subsection, the thrust coefficients in the three wake cases are analyzed. For clarity, the YC3, SYC3, and YC3 + SYC3, as described in Table 3, are taken into account as case studies. The results at $\Delta X = 5D_0$ and $\Delta X = 10D_0$ are shown in Figures 6 and 7.

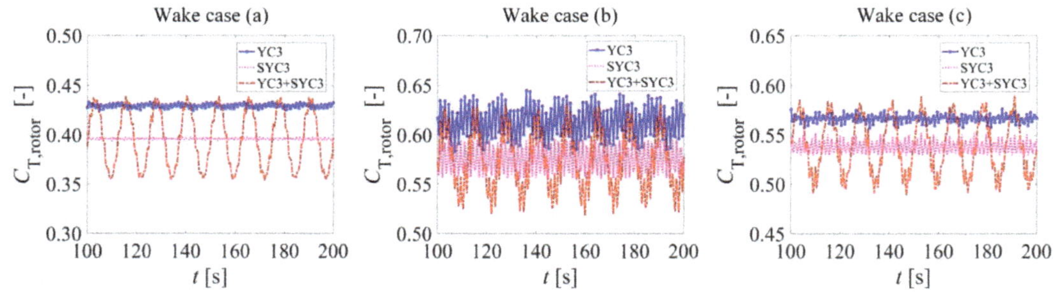

Figure 6. Thrust coefficients of the wind rotor in the three wake cases at $\Delta X = 5D_0$.

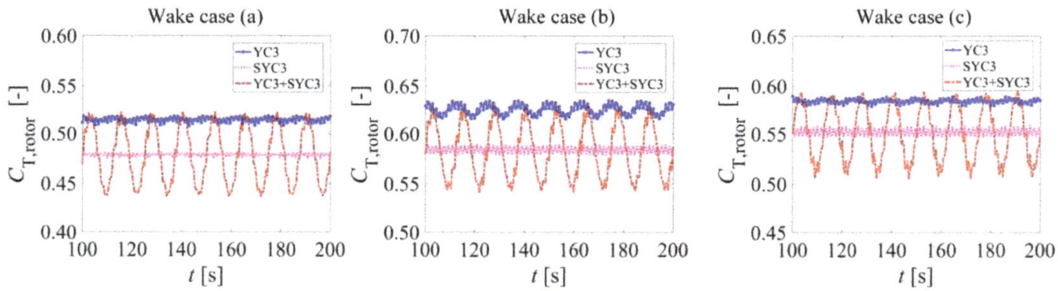

Figure 7. Thrust coefficients of the wind rotor in the three wake cases at $\Delta X = 10D_0$.

As seen in Figure 6, one finding at $\Delta X = 5D_0$ is that the thrust coefficients of SYC3 and YC3 + SYC3 seem to be lower than that of YC3 in all the wake cases. This may be due to that the static yawed inflow conditions can reduce the equivalent windward area of the wind rotor resulting in the loss of the aerodynamic load. Another finding is that the fluctuations of thrust coefficient of YC3 + SYC3 are more violent than the other two cases. Due to the violent fluctuations, some thrust coefficient values of YC3 + SYC3 at the maximum points even exceed the values of YC3 marginally. This may be explained by that the combination of the two yawed inflow make the actual inflow condition more complex, which could lead to greater fluctuations of aerodynamic load. The third finding is that the fluctuations of thrust coefficient of the three yawed inflow conditions in wake case (b) are more complex than those in wake cases (a) and (c), which makes the corresponding time varying ripples of thrust coefficient curve more complex. The main reason may be that the wake case (b) has higher turbulent intensity and more unsteady wind velocity distribution, which could aggravate the fluctuations of aerodynamic load.

As seen in Figure 7, similar findings can be found at $\Delta X = 10D_0$. The thrust coefficients of YC3 are still the largest and the thrust coefficient fluctuations of YC3 + SYC3 are still the most violent in the three wake cases. However, although the fluctuation amplitudes and frequencies of the three yawed inflow conditions are still the largest in wake case (b), the trends are not presented as significant as those at $\Delta X = 5D_0$. This could be explained by the recovery of the wake flow with the increasing downstream distance at the far wake region.

In order to analyze the aerodynamics of yawed inflow conditions under different wake cases more specifically, the mean values, the variations, and the variances of the thrust coefficients ($\overline{C}_T$, ξ_{C_T}, $\delta^2_{C_T}$) are defined and calculated as follows:

$$\overline{C}_T = \sum_{k=1}^{m} C^k_{T,rotor}/m, \tag{20}$$

$$\xi_{C_T} = \frac{C_{T,rotor}(max) - C_{T,rotor}(min)}{2\overline{C}_T}, \tag{21}$$

$$\delta^2_{C_T} = \sqrt{\sum_{k=1}^{m} \frac{\left(C^k_{T,rotor} - \overline{C}_T\right)^2}{m-1}}, \tag{22}$$

where m denotes the number of analyzed points during the selected time period; $C^k_{T,rotor}$ denotes the thrust coefficient at the kth point (k = 1~m); $C_{T,rotor}(max)$ and $C_{T,rotor}(min)$ denote the maximum and minimum values of the thrust coefficient, respectively.

The results are listed in Table 4. As seen from the results of $\overline{C}_T$, all the values under YC3 are maximums in each wake case, as highlighted in bold. For example, in wake case (a), the values of $\overline{C}_T$ under YC3 (0.4296 at $\Delta X = 5D_0$ and 0.5143 at $\Delta X = 10D_0$) are larger than the values under SYC3 (0.3961 at $\Delta X = 5\ D_0$ and 0.4787 at $\Delta X = 10D_0$) and the values under YC3 + SYC3 (0.3963 at $\Delta X = 5D_0$ and 0.4789 at $\Delta X = 10D_0$). As seen from the results of ξ_{C_T}, the values under YC3 + SYC3 are maximums in each wake case, as highlighted in bold. For example, in wake case (b), the values of ξ_{C_T} under YC3 (0.0978 at $\Delta X = 5D_0$ and 0.0751 at $\Delta X = 10D_0$) are obviously larger than the values under YC3 (0.0337 at $\Delta X = 5D_0$ and 0.0085 at $\Delta X = 10D_0$) and the values under SYC3 (0.0165 at $\Delta X = 5D_0$ and 0.0091 at $\Delta X = 10D_0$). As for the results of $\delta^2_{C_T}$, compared with the values of all the yawed inflow conditions, the values in wake case (b) are larger than the other wake cases, as highlighted in bold. Take YC3 at $\Delta X = 5D_0$ for instance, the value of $\delta^2_{C_T}$ in wake case (b) is 0.0160, while the value in wake case (a) is 0.0018 and the value in wake case (c) is 0.0038. This phenomenon indicates that the fluctuations of thrust coefficient of the yawed inflow conditions are more unstable in wake case (b) which should be further studied. The specific data shown in Table 4 can fully validate the findings from Figures 6 and 7.

Table 4. Mean values, variations, and variances of the thrust coefficient under three wake cases.

Wake Case	**Yawed Inflow Conditions**	$\overline{C}_T$		ξ_{C_T}		$\delta^2_{C_T}$	
		5D_0	**10D_0**	**5D_0**	**10D_0**	**5D_0**	**10D_0**
Wake case (a)	YC3	**0.4296**	**0.5143**	0.0098	0.0101	0.0018	0.0023
	SYC3	0.3961	0.4787	0.0048	0.0064	0.0009	0.0015
	YC3 + SYC3	0.3963	0.4789	**0.1069**	**0.0894**	0.0282	0.0284
Wake case (b)	YC3	**0.6139**	**0.6252**	0.0499	0.0156	**0.0160**	**0.0053**
	SYC3	0.5757	0.5840	0.0337	0.0085	**0.0138**	**0.0034**
	YC3 + SYC3	0.5755	0.5849	**0.0978**	**0.0751**	**0.0300**	**0.0273**
Wake case (c)	YC3	**0.5667**	**0.5841**	0.0165	0.0091	0.0038	0.0022
	SYC3	0.5383	0.5520	0.0207	0.0106	0.0066	0.0030
	YC3 + SYC3	0.5407	0.5528	**0.0914**	**0.0805**	0.0283	0.0218

According to the discussions in this subsection, the aerodynamics of a downstream OFWT under the three yawed inflow conditions in wake case (b) will be further analyzed in the following subsections.

3.2. Dynamic Yawed Inflow

In this subsection, the thrust coefficients of the wind rotor in wake case (b) at $\Delta X = 5D_0$ under different dynamic yawed inflow conditions are firstly demonstrated, as shown in Figure 8. As seen from the figure, with the increment of the yaw motion frequency and amplitude, the fluctuations of the thrust coefficient become more violent. However, this trend is not significant. To understand the influences of the yaw motion on the aerodynamics of a downstream OFWT more deeply, the normal forces at $0.2R_0$ section and $0.8R_0$ section of the blade are calculated and included in the analyses as shown in Figure 9.

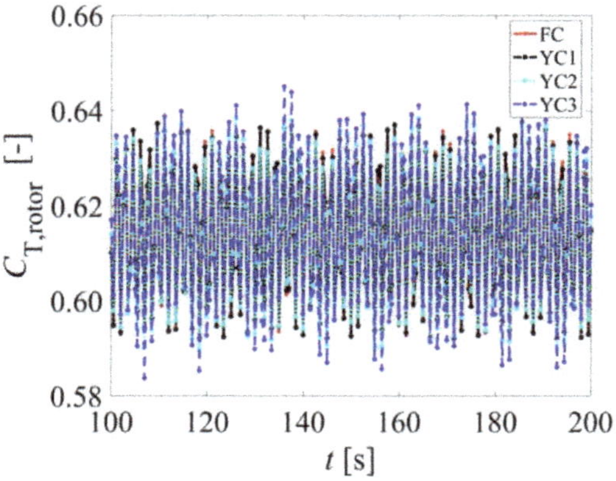

Figure 8. Thrust coefficients of the wind rotor under dynamic yawed inflow.

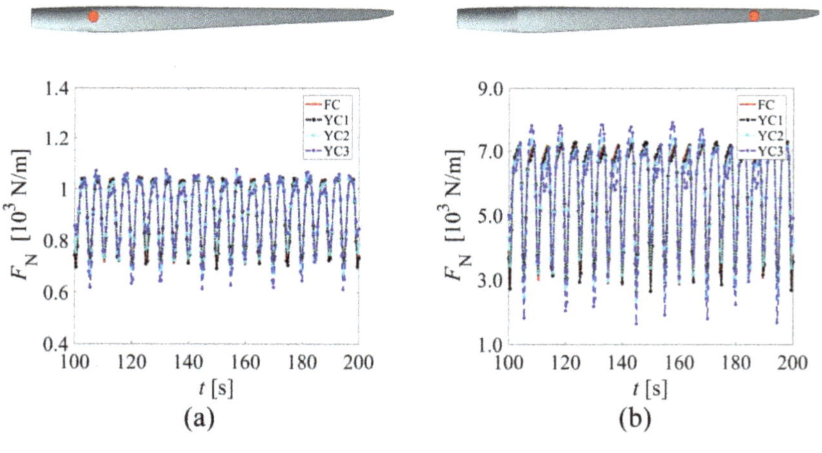

Figure 9. Normal forces at $0.2R_0$ section of the blade (**a**) and at $0.8R_0$ section (**b**) of the blade.

As seen in Figure 9, the normal forces on the blade sections present similar trends with the thrust coefficient, i.e., with the increment of the yaw motion frequency and amplitude, the fluctuations of the normal force become more violent. This is because the thrust coefficient is calculated from the normal force at all the blade sections, as seen in Equations (12), (18) and (19). However, the normal force at $0.8R_0$ section of the blade seems to be larger and fluctuating more violently than the normal force at $0.2R_0$ section of the blade. Specifically, the mean values, variations, and variances of normal force at the two sections of the blade are listed in Table 5. The mean values of normal forces at each section are nearly the same under different inflow conditions. However, with the increment of the yaw motion frequency and amplitude from FC to YC3, both the variations and the variance of normal forces increase, which indicates that the normal forces fluctuate more violently and complicatedly. Especially at $0.8R_0$ section of the blade, the mean values of normal forces in all the inflow conditions are quite larger than those at $0.2R_0$ section of the blade, and the variations and the variances of normal forces increase more obviously from FC to YC3. This can be explained by the more complex wind velocity distribution experienced by the blade section near the outer edge of the wind rotor (i.e., near the blade tip).

Table 5. Mean values, variations and variances of normal force at sections $0.2R_0$ and $0.8R_0$.

Normal Force	Mean Value [10^3 N/m]				Variation				Variance [10^3 N/m]			
	FC	YC1	YC2	YC3	FC	YC1	YC2	YC3	FC	YC1	YC2	YC3
$0.2R_0$	0.914	0.914	0.915	0.915	0.182	0.195	0.211	0.254	0.121	0.122	0.122	0.125
$0.8R_0$	5.836	5.834	5.840	5.843	0.375	0.398	0.432	0.537	1.477	1.484	1.489	1.576

To further illustrate the harmful influences of the higher frequency and larger amplitude of the yaw motion, the fast Fourier transform (FFT) of the time series normal force at $0.8R_0$ section of the blade is conducted, and the results are shown in Figure 10. As seen from the figure, the yaw motions do not significantly affect the 1P (once per revolution, introduced by the wind rotor rotation) or 3P (thrice per revolution, introduced by the blade rotation) component. However, they result in several other components (around 0.12 Hz and 0.28 Hz) which can be observed on both sides of the 1P component. With the increment of the yaw motion frequency and amplitude, these components are strengthened. This phenomenon will increase the possibility of the wind rotor resonance, which is not conducive to the safety of the OFWT structure.

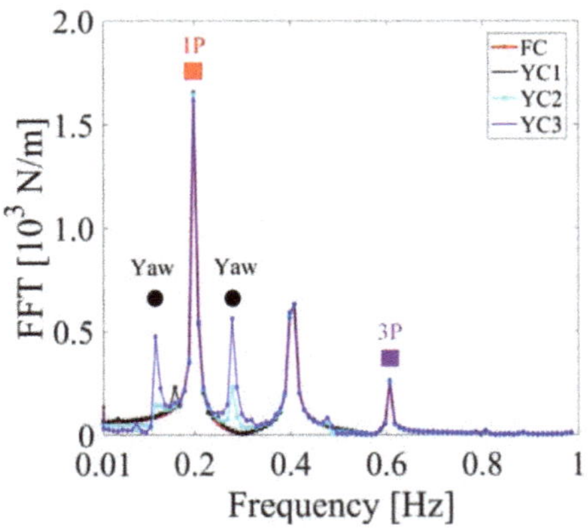

Figure 10. FFT of the normal force at $0.8R_0$ section of the blade.

It can be concluded from the above analysis that higher frequency and larger amplitude of yaw motion can take harmful influences on the blade, especially on the position near the blade tip. Thus this condition should be carefully controlled in actual engineering in considerations of the fatigue and service life of the OFWT blade.

3.3. Static Yawed Inflow

In this subsection, the thrust coefficients of the wind rotor in wake case (b) at $\Delta X = 5D_0$ under different static yawed inflow angles are demonstrated. As seen from Figure 11, with the increment of the static yawed angle (from 2° in SYC1 to 20° in SYC3), the mean value of the thrust coefficient decreases. To be more specific, in the platform fixed case (FC, 0°), the mean value of the thrust coefficient is 0.6143, while from SYC1 to SYC3 (2°, 10°, 20°), the mean values of the thrust coefficient are 0.6131, 0.6024 and 0.5757, respectively.

However, the changes in the fluctuations of these cases are not significant. It indicates that the static yawed inflow angle has greater influences on the mean value characteristic of the aerodynamics and less on the fluctuation characteristic of the aerodynamics. This finding can be explained by that the static yawed inflow can reduce the effective wind capturing area of the wind rotor, and this behavior is actually static. To fully prove this finding, the normal forces at $0.2R_0$ section and $0.8R_0$ section of the blade are also calculated, and the results are shown in Figure 12.

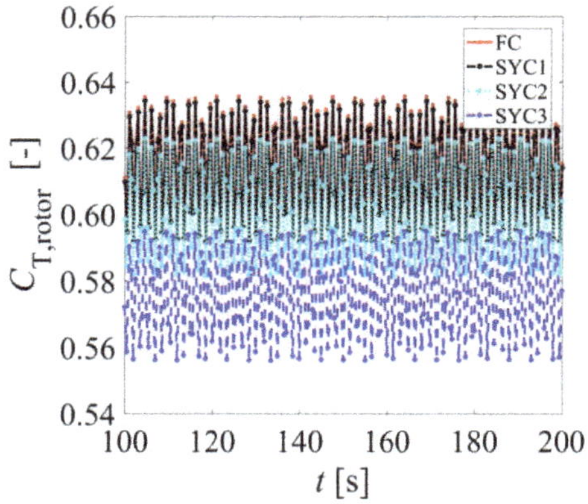

Figure 11. Thrust coefficients of the wind rotor under different static yawed inflow angle.

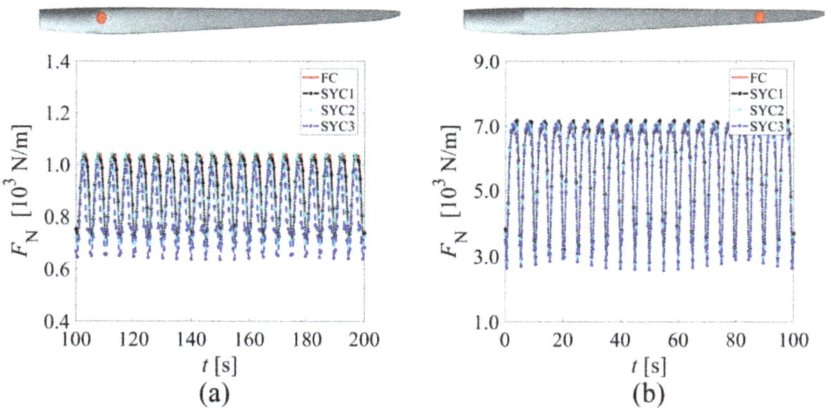

Figure 12. Normal forces at $0.2R_0$ section of the blade (**a**) and at $0.8R_0$ section (**b**) of the blade.

As seen from Figure 12, similar findings can be obtained in both the normal forces at $0.2R_0$ section and $0.8R_0$ section of the blade. However, from 0° to 20° static yawed angle, the mean normal force at $0.2R_0$ section of the blade is reduced by 11.82% (0.914×10^3 N/m to 0.806×10^3 N/m), and the mean normal force at $0.8R_0$ section of the blade is only reduced by 4.39% (5.836×10^3 N/m to 5.580×10^3 N/m). That means the static yawed angle has a greater influence on the forces near the blade root. The finding is quite different from the dynamic yawed inflow conditions in Section 3.2 which have a greater influence on the forces near the blade tip. The main reason for the results of Section 3.2 may be that the aerodynamic force near the blade tip is more susceptible to dynamic motions in the wake because it is farther from the center of rotation. However, for the static yawed inflow conditions here, the aerodynamic force near the blade tip is not as susceptible as that near the blade root, because the blade root may be closer to the wake boundary area which has severe turbulence.

From the perspective of physical intuition, the directions of static yawed inflow will also have considerable influences on the aerodynamics in the complex asymmetric wake flow field such as the wake case (b). To analyze the influences, the static yawed inflow direction is defined, as shown in Figure 13. The positive value of static yawed inflow angle is defined as leaving away from the central wake area counterclockwise, as seen in Figure 13(left); the negative value is defined as closing to the central wake area clockwise, as seen in Figure 13(right). The static yawed angle is set to be 0°, 20° and −20 °, respectively. The thrust coefficients of the wind rotor and the normal forces at $0.2R_0$ section of the blade under these static yawed angles are calculated and the results are shown in Figure 14.

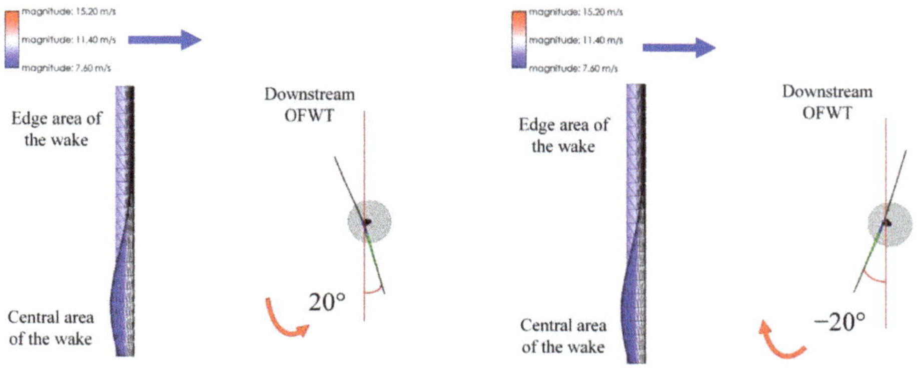

Figure 13. Definition of the signs of static yawed angle from the top view: positive value (**left**) and negative value (**right**).

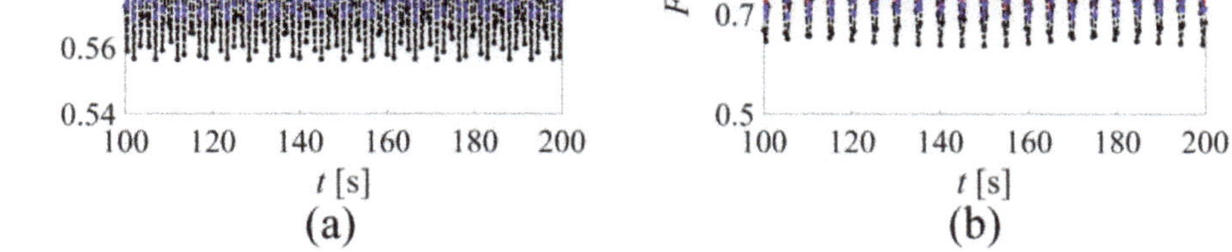

Figure 14. Thrust coefficients of the wind rotor (**a**) and the normal forces at $0.2R_0$ section of the blade (**b**) under different static yawed inflow directions.

As seen from Figure 14a, the thrust coefficient is reduced due to the existence of the static yawed angle. Nevertheless, the reduction of thrust coefficient under static yawed angle −20° is smaller than that under static yawed angle 20°. In other words, the mean value of thrust coefficient under static yawed angle −20° is larger than that under static yawed angle 20°. Specifically, compared with the mean value of thrust coefficient under no yawed condition (0.6143), the mean values of thrust coefficient under static yawed angle −20° and 20° are 0.5922 (reduced by 3.60%) and 0.5757 (reduced by 6.28%), respectively.

This phenomenon may be caused by "time lag", that is, the longer downstream distance along the central wake means that the wind reaches the wind rotor later, not to mention that the wind velocity in the central wake area is small inherently. In addition, the directions of static yawed inflow, i.e., the signs of static yawed angle, have less influence on the fluctuation characteristic of thrust coefficient. For the normal force at $0.2R_0$ section of the blade, as shown in Figure 14b, the signs of static yawed angle have significant influences. The mean normal force under static yawed angle $-20°$ is greater while the mean normal force under static yawed angle $20°$ is smaller than that under no static yawed inflow condition (FC, static yawed angle $0°$). Similarly, the signs of static yawed angle also have less influence on the fluctuation characteristic of the normal force.

3.4. Coupled Yawed Inflow

In order to analyze the influences of coupled yawed inflow conditions on the aerodynamics of an OFWT, the thrust coefficients of wind rotor and normal forces of the blade in YC3 coupled with SYC1~SYC3 as listed in Table 3 are calculated and analyzed in this subsection. The results of thrust coefficients are shown in Figure 15, and the results of normal forces at $0.2R_0$ section of the blade and at $0.8R_0$ section of the blade are shown in Figure 16.

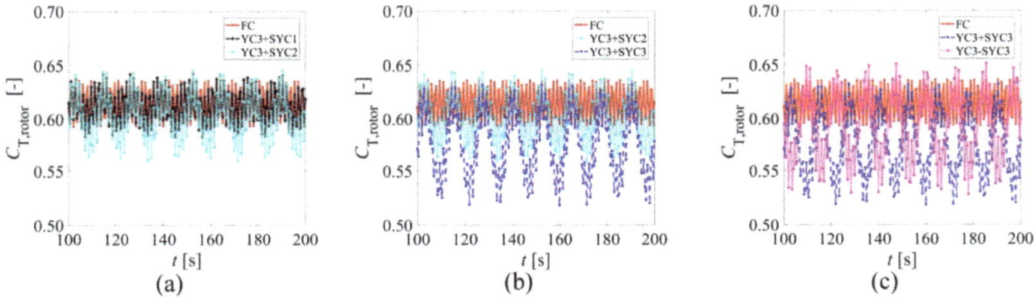

Figure 15. Thrust coefficients of the wind rotor under different coupled yawed inflow conditions: (**a**) YC3 + SYC1 and YC3 + SYC2; (**b**) YC3 + SYC2 and YC3 + SYC3; (**c**) YC3 + SYC3 and YC3 − SYC3. Note that FC is regarded as reference.

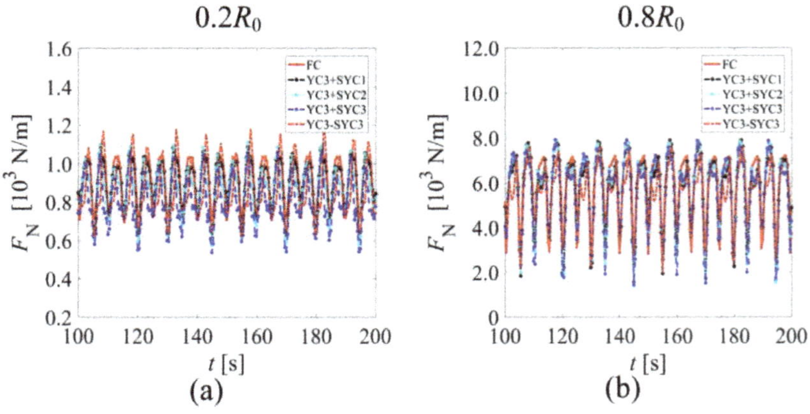

Figure 16. Normal forces at $0.2R_0$ section of the blade (**a**) and at $0.8R_0$ section of the blade (**b**).

As seen in Figure 15a, the fluctuation of thrust coefficient in YC3 + SYC2 is more violent than those in YC3 + SYC1 and FC. Specifically, the variations of the thrust coefficient in YC3 + SYC2, YC3 + SYC1, and FC are 0.0719, 0.0488, and 0.0351, respectively. When the static yawed angle is smaller than the amplitude of yaw motion (dynamic yawed angle), the fluctuation of thrust coefficient in the coupled yawed inflow conditions is marginally slighter than that in the single dynamic yawed inflow condition. For example, the variation of thrust coefficient in YC3 + SYC1 is 0.0488 while the variation of thrust coefficient in YC3 is 0.0499. However, when the static yawed angle is larger than the dynamic yawed angle, the fluctuation of thrust coefficient in the coupled yawed inflow conditions is more violent than that in the single dynamic yawed inflow condition. It could be clearly seen that compared with 0.0499 in YC3, the variation of thrust coefficient in YC3 + SYC2 is 0.0719. As seen in Figure 15b, with the increment of the static yawed angle, the mean value of the thrust coefficient keeps decreasing and the fluctuation of thrust coefficient becomes more violent. Specifically, the mean values of the thrust coefficient in YC3 + SYC2 and YC3 + SYC3 are 0.6015 and 0.5745, while the variations of thrust coefficient in them are 0.0719 and 0.0984. As seen in Figure 15c, when the value of static yawed angle is negative (YC3 − SYC3), the mean value of thrust coefficient in the coupled yawed inflow condition increases, which is quite similar to that in single static yawed inflow condition. The fluctuation of the thrust coefficient under YC3 − SYC3 seems to be more violent than that under YC3 + SYC3. In addition, as clearly seen in the figure, when the thrust coefficient curve of YC3 + SYC3 is at the peak, the curve of YC3 − SYC3 is at the trough.

As seen in Figure 16, with the increment of the static yawed angle in the coupled yawed inflow conditions, both the fluctuations of normal forces at $0.2R_0$ and $0.8R_0$ sections of the blade become more violent and complex. Although the mean values of the normal forces seem to decrease, the trends are not significant. However, for the coupled yawed inflow condition when the static yawed angle is negative (YC3 − SYC3), things are quite different. At $0.2R_0$ section of the blade, the fluctuation of the normal force in YC3 − SYC3 is more violent and more complex than that in other conditions, while the trend at $0.8R_0$ section of the blade is just opposite. Specifically, the detailed data of mean value, variation, and variance of normal force at the two sections of the blade are listed in Table 6.

Table 6. Mean values, variations and variances of normal force at sections $0.2R_0$ and $0.8R_0$.

Normal Force	Mean Value [10^3 N/m]				Variation				Variance [10^3 N/m]			
	YC3 + SYC1	YC3 + SYC2	YC3 + SYC3	YC3 − SYC3	YC3 + SYC1	YC3 + SYC2	YC3 + SYC3	YC3 − SYC3	YC3 + SYC1	YC3 + SYC2	YC3 + SYC3	YC3 − SYC3
$0.2R_0$	0.909	0.875	0.804	0.871	0.266	0.306	0.323	0.327	0.125	0.124	0.127	0.143
$0.8R_0$	5.833	5.762	5.576	5.664	0.544	0.568	0.583	0.498	1.592	1.628	1.619	1.314

The FFT of the time series normal force at $0.2R_0$ section and $0.8R_0$ section of the blade is conducted and the results are shown in Figure 17. As seen from the figure, the coupled yawed inflow conditions from YC3 + SYC1 to YC3 − SYC3 result in multiple components which can be clearly observed around 1P and 3P components as indicated by the red arrows in the figure. However, the coupled yawed inflow conditions do not significantly affect the 1P and 3P components except YC3 − SYC3. The condition of YC3 − SYC3 at $0.2R_0$ section of the blade enhances the 1P component which can be seen in Figure 17a, while that at $0.8R_0$ section of the blade weakens the 1P component which can be seen in Figure 17b. These phenomena are also not conducive to the safety of the OFWT structure. Thus the coupled yawed inflow conditions should also be carefully controlled in actual engineering.

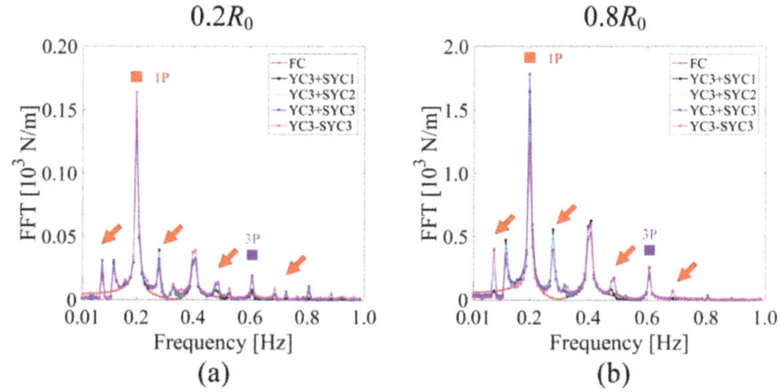

Figure 17. FFT of the normal forces at 0.2R_0 section of the blade (**a**) and at 0.8R_0 section (**b**) of the blade.

4. Experimental Validation

4.1. Experiments Set up

In order to validate the analysis results of the simulation in the above section, experimental tests are conducted on a set of dedicated apparatus. As shown in Figure 18, the apparatus consists of a wind generating system, an upstream scale model wind turbine (SMWT), a downstream SMWT, a yawed conditions generating mechanism, and a load cell. The wind generating system can provide a maximum 8 m/s wind flow field with a turbulent intensity about 0.22 in an area of 15.75 m $\times$ 3.2 m $\times$ 2.52 m (Length, width, height). The SMWTs are designed at a 1/80 scaling ratio with reference to the NREL 5-MW wind turbine, and the major components such as the blades and the towers are manufactured of carbon fiber by 3D printing technology. The upstream SMWT, mounted on a fixed base, is used to provide the wake conditions (the wake case (b) in this section). The downstream SMWT is mounted on the yawed conditions generating mechanism which is actually a Stewart platform. The Stewart platform can provide yaw motions or static yawed poses. The six-component load cell is installed between the nacelle and the tower to measure the thrust characteristics of the wind rotor.

During the tests, the downstream SMWT and the yawed conditions generating mechanism are located at a downstream distance of $\Delta X = 5D_0$ (7.875 m) and an offset of $\Delta Y = R_0$ (0.7875 m) behind the upstream SMWT. The experimental wind speed is set to be 1.63 m/s and the tip speed ratios of the two SMWTs are both controlled to be 7. The yawed conditions in the tests are listed in Table 7. The data acquisition time is 60 s.

Table 7. Yaw parameters in the experimental tests.

Yaw	FC	YC	SYC	YC + SYC
A_{yaw} [°]	0	5, 10	10, 20, −20	5 + 10, 5 + 20, 5 − 20
f_{yaw} [Hz]	0	0.36, 0.72	0	0.72

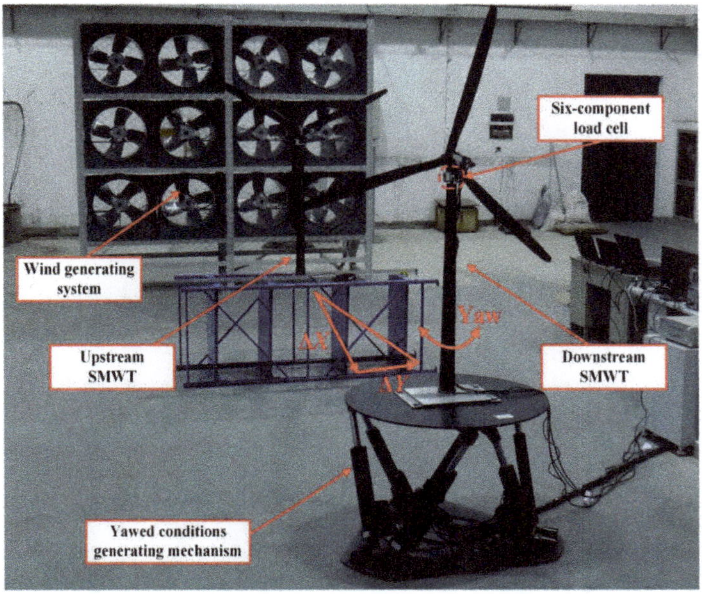

Figure 18. Images of the apparatus for experimental tests.

4.2. Validation

In the experiments, the thrust can be measured by the six-component load cell and then the thrust coefficient of the wind rotor can be calculated based on Equation (19).

The results of the thrust coefficients of wind rotor under dynamic yawed inflow conditions are presented in Figure 19. As seen from Figure 19a, due to the existence of the dynamic yawed condition, a clear increment of the fluctuation of thrust coefficient can be observed. As seen from Figure 19b, with the increment of yaw motion amplitude (from 5° to 10°), an increment of the fluctuation also can be observed. This phenomenon is clearer when the yaw motion frequency is larger (0.72 Hz), as observed from Figure 19c. Besides, as seen from Figure 19d, with the increment of yaw motion frequency (from 0.36 Hz to 0.72 Hz), a more violent fluctuation can be observed. For the dynamic yawed inflow conditions, the influence trends of the experimental data are in consistent with those of the simulation data as discussed in Figure 8 of Section 3.2. In order to illustrate the reliability of this validation, the mean values of the thrust coefficient of the experimental data in FC and YC are calculated and compared with the simulation data. The value of the experimental data in FC (0°, 0 Hz) is 0.5725, and the relative error is −6.28% compared with the value 0.6143 of the simulation data. The value of the experimental data in YC (5°, 0.72 Hz) is 0.5918, and the relative error is −3.60% compared with the value 0.6139 of the corresponding simulation data in YC3 (5°, 0.08 Hz).

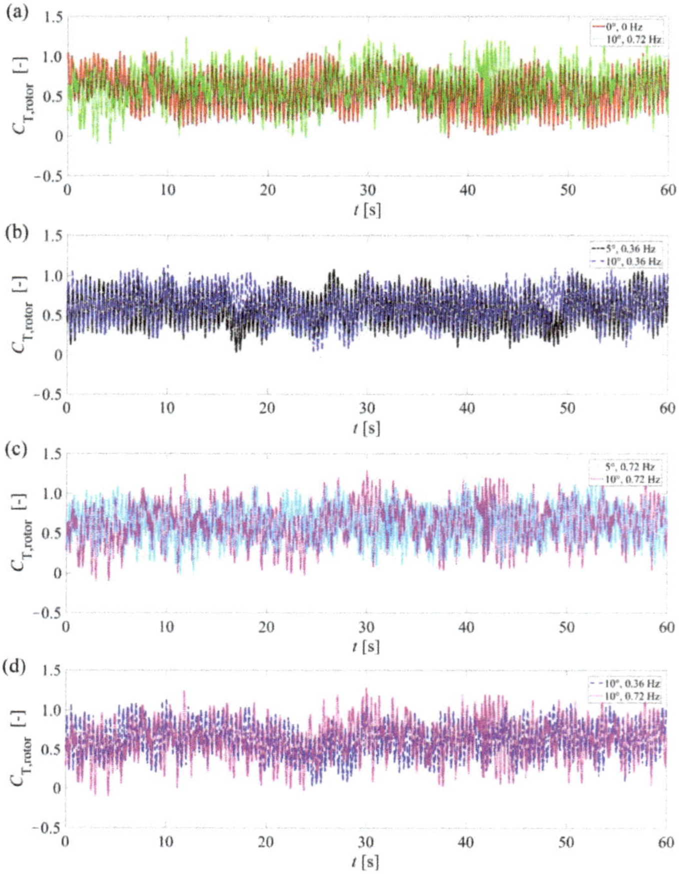

Figure 19. The thrust coefficients by the experimental tests under the dynamic yawed inflow conditions: (**a**) 0°, 0 Hz and 10°, 0.72 Hz; (**b**) 5°, 0.36 Hz and 10°, 0.36 Hz; (**c**) 5°, 0.72 Hz and 10°, 0.72 Hz; (**d**) 10°, 0.36 Hz and 10°, 0.72 Hz.

The results of the thrust coefficients of wind rotor under static yawed inflow conditions are presented in Figure 20. As seen from figure, with the increment of the static yawed angle (from 0° to 20°), the thrust coefficient gradually decreases. However, when the static yawed angle is negative (−20°), the thrust coefficient increases. It is quite similar to the trends of the simulation results discussed in Figures 11 and 14 of Section 3.3. The comparisons of mean thrust coefficients between the experimental data and the simulation data are also conducted. The experimental data values under static yawed angles 10°, 20° and −20° are 0.5615, 0.5319 and 0.5497, respectively. The corresponding relative errors are −6.79%, −7.61% and −7.18% compared with the values 0.6024, 0.5757 and 0.5922 of the simulation data, respectively.

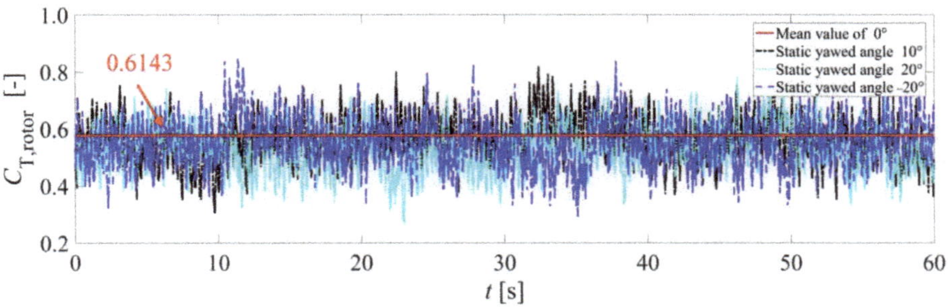

Figure 20. Thrust coefficients by the experimental tests under the static yawed inflow conditions.

The thrust coefficients of the wind rotor under the coupled yawed inflow conditions are presented in Figure 21. As seen from the figure, with the increment of static yawed angle (from 0° to 20°) in the coupled yawed condition, the thrust coefficient keeps decreasing and its fluctuation becomes more violent. However, when the static yawed angle is negative (−20°), the thrust coefficient increases which is also similar to that in the single static yawed inflow condition as discussed in Figure 20. The fluctuation of thrust coefficient under the condition (5° − 20°, 0.72 Hz) is more violent than that under the condition (5° + 20°, 0.72 Hz). Thus, for the coupled yawed inflow conditions, the influence trends of the experimental data are in consistent with those of the simulation data as discussed in Figure 15 of Section 3.4. The values of the experimental data under coupled yawed conditions (5° + 10°, 0.72 Hz), (5° + 20°, 0.72 Hz) and (5° − 20°, 0.72 Hz) are 0.5488, 0.5090 and 0.5205, respectively, and the corresponding relative errors are −8.76%, −11.40% and −11.99% compared with the values 0.6015, 0.5745 and 0.5914 of the simulation data.

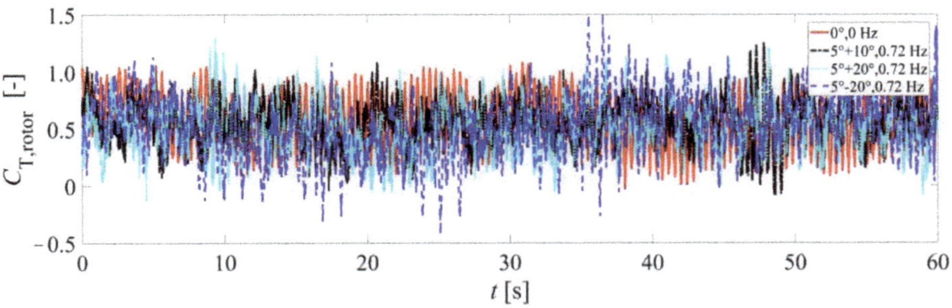

Figure 21. Thrust coefficients by the experimental tests under the coupled yawed inflow conditions.

From the above discussions, it can be roughly concluded that the thrust characteristics of a downstream OFWT under dynamic, static, and coupled yawed inflow conditions from the simulations are validated by the experimental testing data.

5. Conclusions

In this study, the aerodynamics of a downstream NREL 5-MW OFWT in the wake flow under dynamic, static, and coupled yawed inflow conditions are investigated based on the FAST simulations. According to the simulation results, the three yawed inflow conditions have more significant influences on the thrust characteristics in wake case (b) than in wake cases (a) and (c). Thus the following discussions are all conducted in wake case (b).

For the dynamic yawed inflow conditions, the thrust coefficient fluctuates more violently with the increment of the yaw motion frequency and amplitude. The normal forces present similar trends, but the normal force near the blade tip fluctuates more violently and complicatedly than that near the blade root. Significant components can be observed around the 1P of FFT of normal force, which indicates higher yaw motion frequency and amplitude are not conducive to the safety of the OFWT structure.

For the static yawed inflow conditions, the thrust coefficient decreases with the increment of the yawed angle, but the fluctuation does not change much. The normal forces present similar trends, but the normal force near the blade root is reduced more than that near the blade tip. Both the thrust coefficient and the normal force under the negative static yawed angle are larger than those under the corresponding positive one.

For the coupled yawed inflow conditions, although the thrust coefficient decreases with the increment of static yawed angle of the coupled inflow condition, it fluctuates more violently. The normal forces present similar trends but are not as significant as the thrust coefficient. Both the thrust coefficient and the normal force under the negative static yawed angle are larger, and they fluctuate more violently than those under the corresponding positive one.

Experimental tests are conducted on a set of dedicated apparatus. According to the analysis of the experimental data of thrust coefficient and comparisons with the simulation results, it can be roughly validated that the analyses of thrust characteristics of a downstream OFWT under the dynamic, static, and coupled yawed inflow conditions from the simulations are correct. More experimental studies will be done to further verify the accuracy of the simulation results in our future work.

Author Contributions: Conceptualization, Y.W.; methodology, Y.W.; software, Y.W. and J.L.; validation, Y.W., J.L. and H.D.; investigation, Y.W., J.L. and H.D.; writing—original draft preparation, Y.W.; writing—review and editing, J.Z.; supervision, J.Z.; funding acquisition, J.Z. All authors have read and agreed to the published version of the manuscript.

Funding: This research was funded by [National Natural Science Foundation of China] grant number [51875105].

Institutional Review Board Statement: Not applicable.

Informed Consent Statement: Not applicable.

Data Availability Statement: Not applicable.

Conflicts of Interest: The authors declare no conflict of interest.

Abbreviations

Abbreviations	
OFWT	Offshore Floating Wind Turbine
BEM	Blade Element Momentum
FAST	Fatigue, Aerodynamic, Structure, and Turbulence
CFD	Computational Fluid Dynamics
NREL	National Renewable Energy Laboratory
OC4	Offshore Code Comparison Collaboration Continuation
TSR	Tip Speed Ratio
FC	Platform fixed case
YC	Dynamic yawed inflow case
SYC	Static yawed inflow case
1P	Once per revolution
3P	Thrice per revolution

References

1. Fernández-Guillamón, A.; Das, K.; Cutululis, N.A.; Molina-García, Á. Offshore Wind Power Integration into Future Power Systems: Overview and Trends. *J. Mar. Sci. Eng.* **2019**, *7*, 399. [CrossRef]
2. Kosasih, K.M.A.; Suzuki, H.; Niizato, H.; Okubo, S. Demonstration Experiment and Numerical Simulation Analysis of Full-Scale Barge-Type Floating Offshore Wind Turbine. *J. Mar. Sci. Eng.* **2020**, *8*, 880. [CrossRef]
3. Zhao, Z.; Li, X.; Wang, W.; Shi, W. Analysis of Dynamic Characteristics of an Ultra-Large Semi-Submersible Floating Wind Turbine. *J. Mar. Sci. Eng.* **2019**, *7*, 169. [CrossRef]

4. Feng, X.; Lin, Y.; Zhang, G.; Li, D.; Liu, H.; Wang, B. Influence of Combined Motion of Pitch and Surge with Phase Difference on Aerodynamic Performance of Floating Offshore Wind Turbine. *J. Mar. Sci. Eng.* **2021**, *9*, 699. [CrossRef]
5. Sebastian, T.; Lackner, M.A. Characterization of the unsteady aerodynamics of offshore floating wind turbines. *Wind. Energy* **2013**, *16*, 339–352. [CrossRef]
6. Sant, T.; Cuschieri, K. Comparing three aerodynamic models for predicting the thrust and power characteristics of a yawed floating wind turbine rotor. *J. Sol. Energy Eng.* **2016**, *138*, 031004. [CrossRef]
7. Ebrahimi, A.; Sekandari, M. Transient response of the flexible blade of horizontal-axis wind turbines in wind gusts and rapid yaw changes. *Energy* **2018**, *145*, 261–275. [CrossRef]
8. Wen, B.; Tian, X.; Dong, X.; Peng, Z.; Zhang, W.; Wei, K. A numerical study on the angle of attack to the blade of a horizontal-axis offshore floating wind turbine under static and dynamic yawed conditions. *Energy* **2019**, *168*, 1138–1156. [CrossRef]
9. Gao, X.; Yang, H.; Lu, L. Optimization of wind turbine layout position in a wind farm using a newly-developed two-dimensional wake model. *Appl. Energy* **2016**, *174*, 192–200. [CrossRef]
10. Sun, H.; Yang, H. Study on an innovative three-dimensional wind turbine wake model. *Appl. Energy* **2018**, *226*, 483–493. [CrossRef]
11. Bastankhah, M.; Porté-Agel, F. A new analytical model for wind-turbine wakes. *Renew. Energy* **2014**, *70*, 116–123. [CrossRef]
12. Ge, M.; Wu, Y.; Liu, Y.; Li, Q. A two-dimensional model based on the expansion of physical wake boundary for wind-turbine wakes. *Appl. Energy* **2019**, *234*, 975–984. [CrossRef]
13. Kanev, S. Dynamic wake steering and its impact on wind farm power production and yaw actuator duty. *Renew. Energy* **2020**, *146*, 9–15. [CrossRef]
14. Tian, X.L.; Xiao, J.R.; Liu, H.X.; Wen, B.R.; Peng, Z.K. A Novel Dynamics Analysis Method for Spar-Type Floating Offshore Wind Turbine. *China Ocean Eng.* **2020**, *34*, 99–109. [CrossRef]
15. Cai, X.; Wang, Y.; Xu, B.; Feng, J. Performance and effect of load mitigation of a trailing-edge flap in a large-scale offshore wind turbine. *J. Mar. Sci. Eng.* **2020**, *8*, 72. [CrossRef]
16. Wen, B.; Tian, X.; Dong, X.; Peng, Z.; Zhang, W. Influences of surge motion on the power and thrust characteristics of an offshore floating wind turbine. *Energy* **2017**, *141*, 2054–2068. [CrossRef]
17. Xu, B.F.; Wang, T.G.; Yuan, Y.; Cao, J.F. Unsteady aerodynamic analysis for offshore floating wind turbines under different wind conditions. *Philos. Trans. R. Soc. A Math. Phys. Eng. Sci.* **2015**, *373*, 20140080. [CrossRef] [PubMed]
18. Aird, J.; Gaertner, E.; Lackner, M. Dynamic prescribed-wake vortex method for aerodynamic analysis of offshore floating wind turbines. *Wind. Eng.* **2019**, *43*, 47–63. [CrossRef]
19. Qiu, Y.X.; Wang, X.D.; Kang, S.; Zhao, M.; Liang, J.Y. Predictions of unsteady HAWT aerodynamics in yawing and pitching using the free vortex method. *Renew. Energy* **2014**, *70*, 93–106. [CrossRef]
20. Tran, T.T.; Kim, D.H. The aerodynamic interference effects of a floating offshore wind turbine experiencing platform pitching and yawing motions. *J. Mech. Sci. Technol.* **2015**, *29*, 549–561. [CrossRef]
21. Li, X.; Zhu, C.; Fan, Z.; Chen, X.; Tan, J. Effects of the yaw error and the wind-wave misalignment on the dynamic characteristics of the floating offshore wind turbine. *Ocean Eng.* **2020**, *199*, 106960. [CrossRef]
22. Li, X.; Qiu, Y.; Feng, Y.; Wang, Z. Wind turbine power prediction considering wake effects with dual laser beam LiDAR measured yaw misalignment. *Appl. Energy* **2021**, *299*, 117308. [CrossRef]
23. Wei, D.; Zhao, W.; Wan, D.; Xiao, Q. A new method for simulating multiple wind turbine wakes under yawed conditions. *Ocean Eng.* **2021**, *239*, 109832. [CrossRef]
24. Ma, H.; Ge, M.; Wu, G.; Du, B.; Liu, Y. Formulas of the optimized yaw angles for cooperative control of wind farms with aligned turbines to maximize the power production. *Appl. Energy* **2021**, *303*, 117691. [CrossRef]
25. Lee, H.; Lee, D.J. Wake impact on aerodynamic characteristics of horizontal axis wind turbine under yawed flow conditions. *Renew. Energy* **2019**, *136*, 383–392. [CrossRef]
26. Greco, L.; Testa, C. Yaw-Wind turbine unsteady aerodynamics and performance by a free-wake panel method. *Renew. Energy* **2021**, *164*, 444–459. [CrossRef]
27. Bangga, G.; Lutz, T. Aerodynamic modeling of wind turbine loads exposed to turbulent inflow and validation with experimental data. *Energy* **2021**, *223*, 120076. [CrossRef]
28. Bastankhah, M.; Porté-Agel, F. Experimental and theoretical study of wind turbine wakes in yawed conditions. *J. Fluid Mech.* **2016**, *806*, 506–541. [CrossRef]
29. Bastankhah, M.; Porté-Agel, F. A new miniaturewind turbine for wind tunnel experiments. Part I: Design and performance. *Energies* **2017**, *10*, 908. [CrossRef]
30. Bastankhah, M.; Porté-Agel, F. A new miniaturewind turbine for wind tunnel experiments. Part II: Wake structure and flow dynamics. *Energies* **2017**, *10*, 923. [CrossRef]
31. Jonkman, J.; Butterfield, S.; Musial, W.; Scott, G. *Definition of a 5-MW Reference Wind Turbine for Offshore System Development*; National Renewable Energy Laboratory (NREL): Golden, CO, USA, 2009.
32. Robertson, A.; Jonkman, J.; Masciola, M.; Song, H.; Goupee, A.; Coulling, A.; Luan, C. *Definition of the Semisubmersible Floating System for Phase II of OC4*; National Renewable Energy Laboratory (NREL): Golden, CO, USA, 2014.
33. Sun, H.; Yang, H. Numerical investigation of the average wind speed of a single wind turbine and development of a novel three-dimensional multiple wind turbine wake model. *Renew. Energy* **2020**, *147*, 192–203. [CrossRef]

34. Shao, Z.; Wu, Y.; Li, L.; Han, S.; Liu, Y. Multiple wind turbine wakes modeling considering the faster wake recovery in overlapped wakes. *Energies* **2019**, *12*, 680. [CrossRef]
35. Thiringer, T.; Dahlberg, J.-A. Periodic Pulsations from a Three-Bladed. *IEEE Trans. Energy Convers.* **2001**, *16*, 128–133. [CrossRef]
36. Wen, B.; Wei, S.; Wei, K.; Yang, W.; Peng, Z.; Chu, F. Power fluctuation and power loss of wind turbines due to wind shear and tower shadow. *Front. Mech. Eng.* **2017**, *12*, 321–332. [CrossRef]
37. Jonkman, B.J.; Jonkman, J.M. *FAST v8.16.00a-bjj User's Guide*; National Renewable Energy Laboratory (NREL): Golden, CO, USA, 2016.
38. Jonkman, J.M.; Hayman, G.J.; Jonkman, B.J.; Damiani, R.R. *AeroDyn v15 User's Guide and Theory Manual*; National Renewable Energy Laboratory (NREL): Golden, CO, USA, 2015.
39. Platt, A.; Jonkman, B.; Jonkman, J. *Inflow Wind User's Guide*; National Renewable Energy Laboratory (NREL): Golden, CO, USA, 2016.

energies

Article

A Method to Analyze the Performance of Geocooling Systems with Borehole Heat Exchangers. Results in a Monitored Residential Building in Southern Alps

Marco Belliardi [1,2,*], Nerio Cereghetti [1], Paola Caputo [2] and Simone Ferrari [2]

1 ISAAC (Institute for Applied Sustainability to the Built Environment), SUPSI (University of Applied Sciences and Arts of Southern Switzerland), CH-6850 Mendrisio, Switzerland; nerio.cereghetti@supsi.ch
2 Department of Architecture, Built Environment and Construction Engineering (ABC), Politecnico di Milano, IT-20133 Milano, Italy; paola.caputo@polimi.it (P.C.); simone.ferrari@polimi.it (S.F.)
* Correspondence: marco.belliardi@supsi.ch

Abstract: Geothermal heat is an increasingly adopted source for satisfying all thermal purposes in buildings by reversible heat pumps (HP). However, for residential buildings located in moderate climates, geocooling, that implies the use of geothermal source for cooling buildings without the operation of HP, is an efficient alternative for space cooling not yet explored enough. Geocooling allows two main benefits: to cool the buildings by high energy efficiencies improving summer comfort; to recharge the ground if space heating is provided by HP exploiting the geothermal source (GSHP). In these cases, geocooling allows to avoid the decreasing of the performances of the GSHP for space heating over the years. To explore these issues, a method has been developed and tested on a real case: a new residential building in Lugano (southern Switzerland) coupled with 13 borehole heat exchangers. The system provides space heating in winter by a GSHP and space cooling in summer by geocooling. During a 40 months monitoring campaign, data such as temperatures, heat flows and electricity consumptions were recorded to calibrate the model and verify the benefits of such configuration. Focusing on summer operation, the efficiency of the system, after the improvements implemented, is above 30, confirming, at least in similar contexts, the feasibility of geocooling. Achieved results provides knowledge for future installations, underlining the replication potential and the possible limits.

Keywords: geothermal energy; geocooling; borehole heat exchangers; residential space cooling; radiant underfloor systems

Citation: Belliardi, M.; Cereghetti, N.; Caputo, P.; Ferrari, S. A Method to Analyze the Performance of Geocooling Systems with Borehole Heat Exchangers. Results in a Monitored Residential Building in Southern Alps. *Energies* **2021**, *14*, 7407. https://doi.org/10.3390/en14217407

Academic Editors: Wei-Hsin Chen, Aristotle T. Ubando, Chih-Che Chueh and Liwen Jin

Received: 3 November 2021
Accepted: 4 November 2021
Published: 7 November 2021

1. Introduction

The role of buildings in relation to energy consumption is widely known on a global and local scale. To satisfy the growing energy needs, while reducing the related greenhouse gases (GHG) emissions, academic research and policies have been supported for decades the diffusion of different solutions. Relevant results can be obtained through the implementation of high efficient measures at envelope and HVAC (Heating, Ventilation and Air Conditioning) level. These measures allow decreasing the energy needs and increasing the integration of local RES for electric and thermal purposes, according to goals defined for 2030 and 2050 (reported at www.roadmap2050.eu), as described in [1] and, referring to Switzerland, in [2–4].

According to [5], geothermal energy is a renewable energy suitable for energy conversion to electricity as well as heating and cooling of different buildings and other facilities. The most popular technology to extract the shallow geothermal resources are GSHP (ground source heat pump) systems, which can be properly adopted for low temperature distribution systems such as floor heating and/or cooling systems.

In [6] Ground Source Heat Pumps (GSHP) are defined as energy efficient and environmental friendly systems for space heating and cooling in buildings, underlining the recent remarkable growth of the installed capacity worldwide and the most important role in the direct uses of geothermal energy. Moreover, they observe that the introduction of the Nearly Zero Energy Buildings (NZEB) concept and standard is expected to further increase the penetration of GSHP in Europe.

A deep review of the worldwide applications of geothermal energy for direct utilization, updated at 2020, is presented in [7], underlining that this technology is one of the oldest, most versatile and most common forms to exploit this source. The paper presents information on direct applications of geothermal heat based on country update papers involving 113 countries and regions. Among these, Switzerland appears in the "top five" countries for installed capacity in terms of thermal power for population; in terms of installed capacity for land area; and in terms of annual energy use for area.

Indeed, in the European area, Switzerland, with more than 100,000 ground source heat pumps (GSHP), is a leader in the use of geothermal energy for heating and domestic hot water and recently has given insights about the appropriateness of using this source also for summer cooling.

In [7] it is included a complete description of geothermal applications in Switzerland, underlining that borehole heat exchangers with heat pumps are still the most spread application and that the number of GSHP, now increasingly used for heating and cooling, are growing constantly. In fact borehole heat exchangers amount to 1843.8 thermal MW and 10,733.8 TJ per year, more than the 80% of the total.

Reversible GSHP can have high performances during the whole year when conventional cooling systems are adopted, i.e., systems for removing sensible and latent heat and with a level of temperature of the carrier around 7 °C. Conversely, when reversible GSHP are coupled with radiant systems for providing heat and cool and the cooling mode is devoted at removing only sensible heat, the EER (Energy Efficiency Ratio, defined as the ratio of cooling capacity provided to electricity consumed and adopted in rating system air conditioners, allowing for straightforward comparisons of different units) can decrease due to the levels of temperature. In fact, the ground temperature is around 14–16 °C while the design temperature of the carrier is around 20–21 °C. In these cases different solutions can be searched, as reported also in [8] that, starting from the temperature comparison among indoor air, envelopes, fresh air and undisturbed soil, suggested an integrated system aimed at combining pipe-embedded walls, pipe-embedded windows, and fresh air pre-handling system with the conventional GSHP system with the aim to utilize shallow geothermal energy more efficiently.

By a different perspective, among the different options for promoting renewable district heating and cooling systems, in a deep review [9] considered also geothermal energy. In particular, the benefit of the geothermal energy used in district cooling system (DCS) worldwide are mentioned because this source is considered sustainable, cheap, and safe.

Among the solutions cited above, geocooling, i.e., "geothermal free cooling", allows removing heat from the buildings in summer exploiting the difference of temperature between indoor temperature and ground temperature, without the adoption of the GSHP. This technology has undergone a fair expansion in the last decade, in particular in Switzerland where there are hundreds of cases, according also to the current regulation that limits the adoption of traditional air conditioning systems. However, geocooling has not been yet very deepened in the technical literature. A state of the art for geocooling technology is presented in [10], summarizing results related to a dozen of existing buildings, underlining the need to design these solutions as an integral part of the whole building and that even if theoretical and technical knowledge is available, practical applications are few. Other contributions were provided by [11–13], dealing with geocooling potential for administrative buildings. More recently, [14] studied the effect of geocooling in combination with thermal energy storage (TES) to decrease the energy and carbon impact of cooling in a small, lightweight commercial building located in a Mediterranean climate. The study addressed the effectiveness of this solution, even if with a seasonal performance lower than that achieved by the experience documented in the present work. An interesting update is reported in [15], with an investigation about the impacts of free cooling supply using ground-source heat pumps on the basis of the available pertinent literature. On their opinion, according to the few available literature, geothermal systems could also play an important role in meeting cooling demand through free and active cooling. In the comparison to other technologies, they stress the lowest impacts of free cooling configurations as a consequence of negligible electricity consumption and no additional equipment. However, the statement that geothermal free cooling has the lowest impacts cannot be generalized.

Focusing on the methods available for designing geocooling systems, insights are reported in [16] about PILESIM2 and in the recent [17], by a deep review of the modeling aspects and practices of shallow geothermal systems, and [18], which detail the comparison among active chilled beams and thermally activated building systems. In particular, [17] evidenced the lack of systematic design guidelines and reviewed the available modeling options (including PILESIM2) presenting the main analytical and numerical models and methods and discussing the most important supplementary factors affecting such modeling. At country level, the Swiss standards regarding geocooling are [19], the one more considered in the present work, and [20,21], where the term "geocooling" is used in combination with "freecooling" and "natural cooling".

In this framework, the contribution here presented is aimed at defining a method for analyzing the effectiveness of geocooling at least for residential buildings located in climate without extreme summer conditions. The method, described in Section 2 and Figure 1, includes energy simulation, the thermal characterization of the underground and the accurate design and simulation of the BHE field. This allows overcoming the lacks of a proper design of geocooling systems. The method is then tested on a real case and the work benefits from the direct participation of part of the authors in the interventions for improving and monitoring the performance of the same real case for which a large amount of monitored data are available. The described real case application has been promoted by the Swiss Federal Office of Energy and the Renewable Energy Fund of the Canton of Ticino, as a demonstration project deeply analyses in [22].

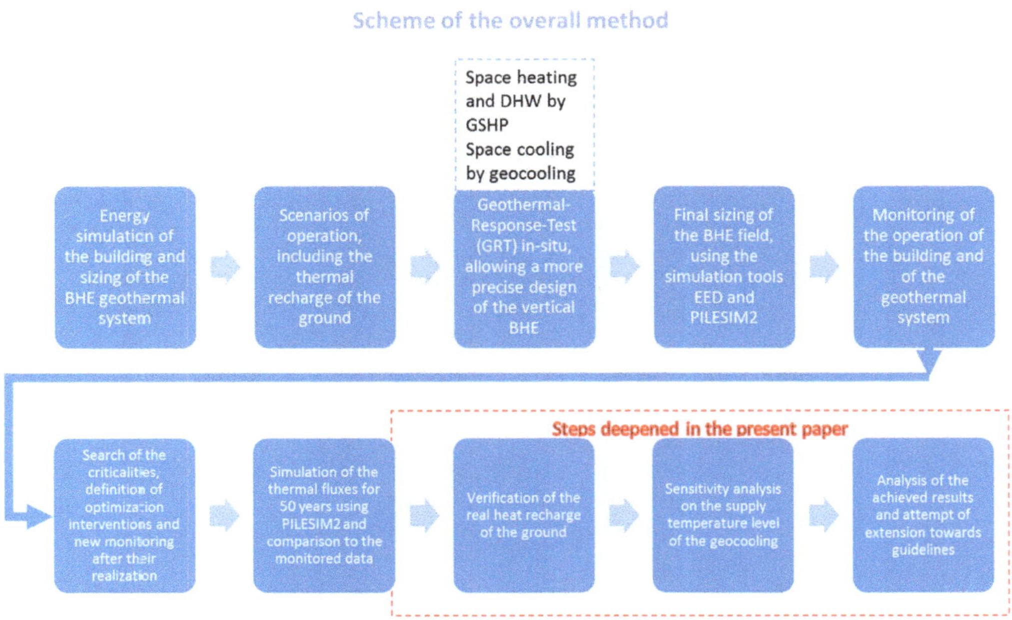

Figure 1. Scheme of the overall method.

2. Materials and Method

A method, able to include simulations and calibration, for analyzing geocooling systems has been developed and the following activities for testing the method in a real case, a new residential building in Lugano (southern Switzerland) have been accomplished:

1. Preliminary energy simulation of the building and dimensioning of the BHE geothermal systems. The total length and the number of BHE and their configuration have been defined to satisfy the building thermal needs. In this preliminary stage, some scenarios has been drawn, suggesting also the thermal recharge of the ground, showing its importance from the economic and energy point of view and stressing the need of deepening the investigation about the thermal behavior of the overall system;
2. Carrying out of the Geothermal-Response-Test (GRT) in-situ for a detailed study of the thermal characteristics of the ground according to [23]; the GRT allows a more precise evaluation of the length of the vertical BHE. When it is properly determined, supplementary costs are avoided;
3. Final dimensioning of the BHE field, using updated thermal needs and ground properties found by the GRT, by the simulation tools EED [24] and PILESIM2 [16]. The system was realized for providing space heating and DHW by GSHP, together with a solar thermal plant. The cooling demand of the building was expected to be satisfied with geocooling technology that can as well as recharge the ground.
4. Monitoring of the operation of the building and of the geothermal system, search of the criticalities, definition of optimization interventions and new monitoring after their realization;
5. Simulation of the thermal fluxes from and to the geothermal field for 50 years using the PILESIM2 program and comparison to the monitored data. This step is mainly aimed at verifying the real heat recharge of the ground in order to show the long-term effects; a sensitivity analysis on the supply temperature level of the geocooling was also set;
6. Analysis of the achieved results and attempt of extension.

The present paper deepens in particular the last steps of the method described and sketched in Figure 1, starting from the output of the simulations by PILESIM2 and on the comparison to monitored data. Further details about each steps of the method are reported in [22], the already mentioned project report, taking into account the same case study.

In relation to steps 1 and 2, the need of a detailed design of geothermal system has been stressed in the technical literature already in the first decade of 2000, as reported by [25] that provided a review of systems, models and applications about ground heat exchangers and, in the conclusions, referred to a typical residential house in Switzerland. More recently, the same need is stressed also by [26] that underlined the importance of the GRT and reported the results of an experimental analysis carried out in France referring to the heating and cooling performance. Within the framework of the IEA-ECES Annex 31, [27] provided a review of configurations, thermal response, analytical and numerical models and, analyzing the seasonal performances, illustrated the effect of proper design and sizing of such systems. In particular they underlined the need to develop general procedures for early stage sizing and with reasonable accuracy for energy piles.

PILESIM2 program has been cited in the technical literature, e.g., recently in the review by [28] that assessed of the impacts of system geometric configuration, pipe material, working fluid, and depth of ground heat exchanger on heat flux, heat transfer coefficient, outlet temperature, thermal resistance, and pressure drop, as a help in overtaking the lacks of knowledge that still affect this technology.

In relation to step 4, in order to verify the real operation of the system, several temperature sensors (thermocouples), flow meters and power meters were installed and connected to a data-logger. In addition, the indoor air temperature and humidity of a sample flat were constantly monitored, paying attention to the control of the heating and cooling system and to the indoor thermal comfort conditions.

It has to be noted that the optimizations have been discussed and then implemented by technical staff (not ever aware) and that each measure requires an entire season to be monitored and eventually corrected only the next year. Main optimization options were:

- Flow rate regulation of the geothermal BHE circuit (underground circuit);
- Adjustment of the intermediate circulation pumps;
- Flow regulation of the circulation pump of the underfloor circuit;
- Increase of the supply temperature of the geocooling circuit.

Further details about the optimization strategies implemented and monitored in relation to the present real case are reported in [22].

Finally, several exchanges with the technicians involved have been organized together with the realization of a survey with the tenants of the building by a questionnaire.

In relation to step 5, the importance of long-term evaluation is stressed e.g., in [29] that developed a detailed, effective and validated numerical model for transient simulation of borehole heat exchangers. PILESIM2, based on TRNSYS (a flexible software environment used to simulate the behavior of transient systems, in particular for assessing the performance of thermal energy systems as described at see trnsys.com), uses hourly thermal energy values (thermal energy actually exchanged with the ground) and then calculates the respective fluid temperature over the years. For the first three years, the simulated supply temperatures have been compared with the monitored ones. Several simulations have been run setting different supply temperature for geocooling, looking at the related effect on the cooling potential of the building.

The present article focuses mainly on steps from 4 and 6, as described in the following sections.

3. Results

According to the steps reported in Section 2, in the following sections are described the real case of study and the achieved results in terms of predesign evaluation and monitoring before and after optimization.

3.1. *Case of Study*

The building called "City Residence" is located in Lugano (southern Switzerland). It is a Minergie® building in operation since 2014, with 46 flats on 7 floors (see Figure 2) and an energy surface of 5700 m^2. Heating and cooling are provided by an underfloor radiant systems and a mechanical ventilation system with air-to air heat exchangers is also in operation.

Following the Minergie® standard referred to the age of the building, as average thermal transmittance, a U value of about 0.18 W/m^2K was reached for the opaque façade and a U value of about 1.2 W/m^2K was reached for windows.

In the beginning of the project and before the development of the study here reported, the energy behavior of the building has been simulated in steady state way, according to the local rules in force, i.e., [21,30], resulting in a design thermal power for space heating and DHW of about 130 kW and for cooling of about 80 kW.

In order to characterize the ground, a geothermal response test (GRT) was carried out in-situ on a BHE. The test revealed a thermal conductivity of the ground of 3.3 W/mK, a ground temperature on the surface of 13.7 °C, and a geothermal temperature gradient of 11 K/km.

Based on the preliminary thermal demands and geological literature data, a system of 17 BHE of 200 m depth was evaluated. The final dimensioning of the BHE field was achieved using updated thermal needs (estimated again in steady state way and before the development of the study here reported) and ground properties and taking into account a proper value of regeneration of the ground, i.e., the ratio between energy injected over energy extracted from the ground.

Figure 2. Rendering (**top**) and photograph (**bottom**) of the building located in the densely populated centre of Lugano (source: [22]).

These design conditions result in 13 BHE, with a depth of 200 m each, i.e., saving 4 BHE in comparison to the initial estimation (see Figure 3).

The thermal demands (heating and DHW) are expected to be satisfied by 3 HP coupled to the BHE field. A solar thermal plant of about 100 m^2 contributes to the supply of the DHW and of the space heating on yearly basis. Solar thermal heat in excess is not injected into the ground to avoid an unbalance of the temperature levels, during the geocooling mode.

According to the expertise of the authors, the thermal recharging of the ground is considered essential to ensure a constant efficiency during the operation of the system in the long term. To that end and based on the features of the building and of the emission system (underfloor radiant panels), a geocooling system coupled to the same BHE field has been implemented. This solution allows also to avoid the use of traditional cooling machines in summer and therefore to lower summer electricity needs for cooling.

As anticipated, the BHE field has been designed based on the winter needs satisfied by the HP. During winter season, the ground is thermally discharged, while during summer season geocooling can recharge the ground.

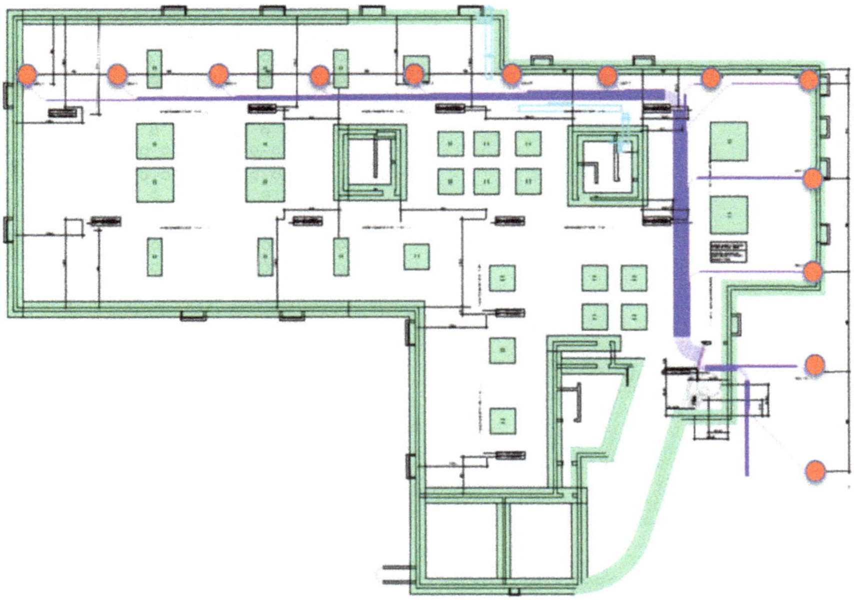

Figure 3. Position of 13 BHE under the building (source: [22]).

As a result of the sizing carried out by the design team, for a steady long-term operation of the system, it was suggested to set a recharge of the ground in the range between a minimum of 17% and a maximum of 52%. Indeed, if summer recharge is lower than 17%, the extraction of heat from the ground could be excessive compared to the injection; therefore this can cause a rapid decrease in the temperatures of the fluid entering the ground. While, if summer recharge is higher than 52%, the operation would limit the expected potential for geocooling. In addition, according to the expertise of the authors and the local thermo-hygrometric conditions, it was suggested to inlet water into the underfloor radiant system of the flats at temperatures equal or higher than 21 °C, in order to prevent too high relative humidity and vapor condensation on the indoor surfaces, but this suggestion was not considered by the technicians during the first period of operation.

During winter operation the geocooling circuit is switched off, the HP charge the heat storage tank and the system allows to heat the flats.

Figure 4 shows a simplified scheme of the analyzed system with the detail of the main hydraulic components in accordance to during the summer operation.

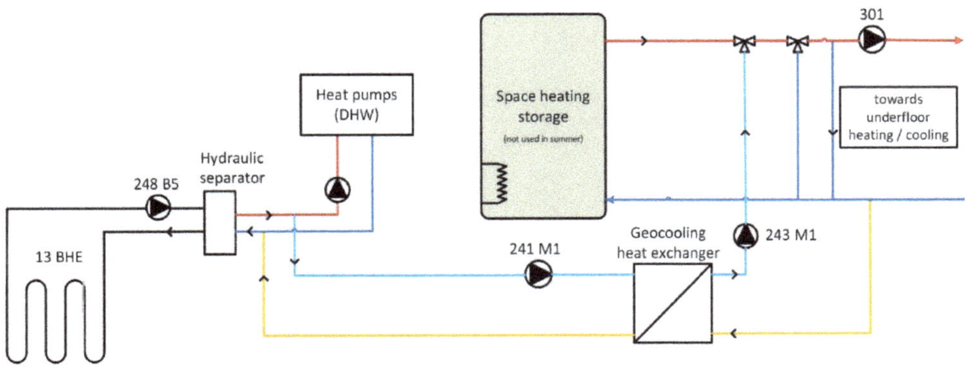

Figure 4. Scheme of the summer operation using geocooling (source: [22]).

3.2. Monitoring System and Campaign

The method is calibrated and tested according to a complete monitoring campaign. In order to analyze and verify geocooling benefits over several seasons, the years 2016 to 2019 have been monitored, starting at the end of June 2016.

For characterizing the outside environment, the weather data adopted are those recorded in the nearest meteo-station in Lugano.

The monitoring system provides the values of temperatures and water flow rates at different points of the circuit. The main ones are positioned upstream of the BHE field to record the thermal exchanges with the ground (in correspondence of pump 248 B5, see Figure 4) and another one downstream of the underfloor circuit (in correspondence of pump 301, see Figure 4). In order to evaluate the efficiency of the geocooling, a monitoring of all the circulation pumps (electricity) necessary for the summer operation mode was carried out.

The monitoring system includes the following components:

- 1 technical cabinet with a data-logger inside;
- 5 volume flow meters (electromagnetic);
- 15 thermocouple sensors–Type T (grade insulated wires);
- 3 electric power-meters installed on the heat pumps (electric metering of heat pumps consumptions);
- 4 electric power-meters on the circulation pumps dedicated to the geocooling circuit (electric metering of circulations pumps consumptions);
- 1 indoor weather station in one flat (temperature and humidity).

The heat meters installed in the technical room are described in Figure 5. In particular:

- Q1: heat meter of the BHE field. This is necessary for the monitoring of the thermal exchange with the ground. In this counter, water with glycol circulates;
- Q2: geocooling heat meter after the hydraulic separator and before the heat exchanger. In this counter, water with glycol circulates;
- Q3: heat meter after heat pumps, only for DHW production. In this counter circulates simply water;
- Q4: heat meter towards the underfloor heating (heating and cooling of flats). In this counter simply water circulates;
- Q5: heat meter after the hydraulic separator towards heat pumps. In this counter flows water with glycol.

The four pumps dedicated to geocooling have been monitored for the evaluation of its efficiency, as the pumps represent its only electricity consumption.

Conversely, the monitoring of the solar thermal pump is outside the scope of the study. In the present real case, solar thermal collectors could contribute to satisfy heating needs, after ensuring DHW production. This contribution was monitored through temperature sensors and without a mass flow sensor. According to the collected data, the energy contribution was negligible, considering the overall and annual thermal balance of the case of study.

The system for data acquisition, visualization and supply runs due to the GSM Internet connection with a dedicated SIM card in-situ. The acquired data have been stored on a server and can be viewed and downloaded via Grafana, an open source platform for the organization and graphic display of data.

The accuracy of the monitoring and related results depends on the measurement methods, the temperature sensors, the data logger, and the measurement circuits used. Considering the technical information available for the sensors and the data logger and the experience acquired by other thermal data acquisition systems, a total accuracy of flow measurement of 2.4% has been noticed, while the total error on temperature measurements is less than 0.3 K. Although all devices installed for the monitoring have been accurately calibrated, these levels of accuracy affects the thermal measurements, depending on the entity of the temperature difference observed on the carrier for providing heating and cooling. Further details and clarifications about the equipment of the monitoring system are reported in [22].

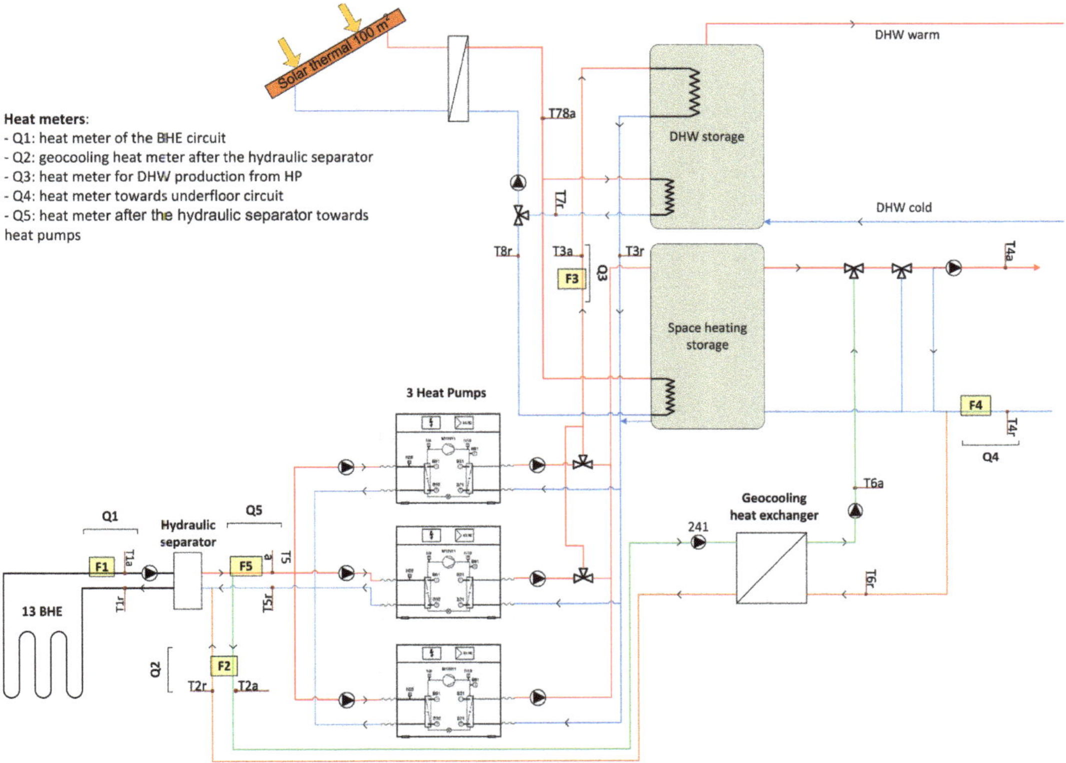

Figure 5. Scheme of the system and of the monitoring concept with the detail of the sensors and of the meters (source: [22]).

3.3. Monitored Seasons

In the technical literature is very rare to find out results of monitoring campaign as long-lasting as in the real case here presented. In fact, the whole monitored period from 2016 to 2019 has been studied in order to obtain more detailed and aggregate results. Table 1 reports the monitored seasons including their duration in days and hours and the number of hours in which the system was in operation (about the 80% of the total hours of the season). It has to be explained that in Switzerland winter seasons are not defined by law depending on climatic data; heating and cooling are provided when needed during a year. In the analyzed real case, the switch between heating and cooling operation mode is done manually by the person responsible for maintenance at the request of the administrator of the building because there is not an automatic control logic. Based on Table 1, the winter season has a duration of about 240 days, while summer season of about 120 days.

Based on the results of the monitoring of the first 15 months, some measures for the optimization of the system were suggested. In autumn 2017, energy efficiency optimization measures were agreed and fully implemented at the end of July 2018. The results of the monitoring of August and September 2018 were representative to show the effects of the optimizations implemented.

Table 1. Duration of the summer (cooling by geocoding) and winter (heating by GSHP) seasons monitored (elaborations based on [22]).

Season	Start Date	End Date	Season Duration [days]	Seasonal Duration [h]	Working Period [h]
Summer 2016 *	23 June 2016	28 September 2016	97	2328	1858
Winter 2016–17	28 September 2016	23 May 2017	237	5664	4523
Summer 2017	23 May 2017	15 September 2017	115	2760	2203
Winter 2017–18	15 September 2017	23 May 2018	250	5808	4637
Summer 2018	23 May 2018	3 October 2018	133	3190	2571
Winter 2018–19	3 October 2018	4 June 2019	244	5880	4737
Summer 2019 **	4 June 2019	7 October 2019	125	2952	2378

* cooling started later than in the other summer seasons due to technical reasons in the manual switching of the system; ** technical problems occurred affecting the management of the system.

3.4. Energy Balance of the System

The following Figures 6 and 7 show the monthly values of the energy exchanged with the building through the underfloor radiant system (heat meter Q4 in Figure 5) and the monthly energy exchanged with the ground through the BHE field (heat meter Q1 in Figure 5), during the entire monitored period.

3.5. Energy Performance

The overall performance has been evaluated at building level by investigating the heating and cooling demands and at system level investigating the seasonal coefficient of performance of the GSHP, the recharge of the ground and the efficiency of geocooling.

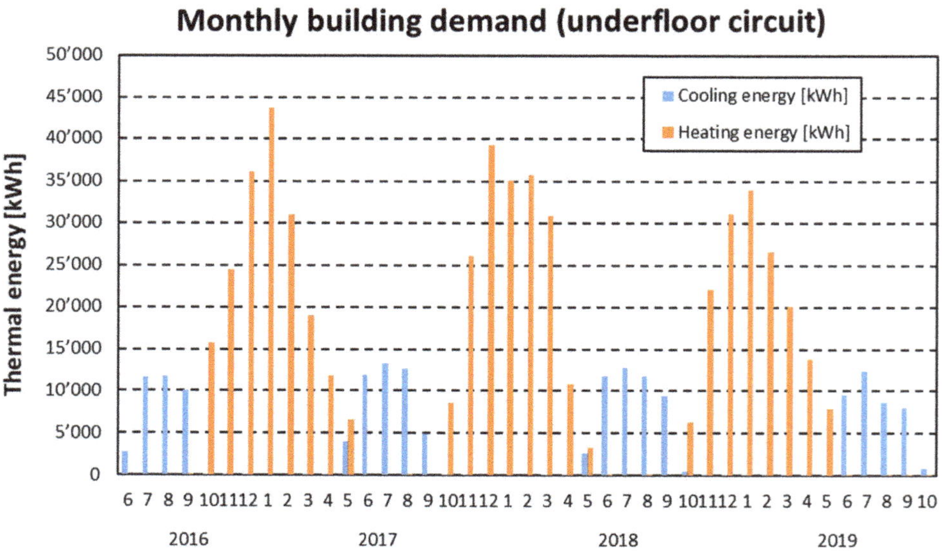

Figure 6. Monthly values of the energy demands (at the heat meter "Q4") during the monitoring (years and months in the abscissa labels; source: [22]). This measure is called demand to differentiate it from the thermal needs at envelope level.

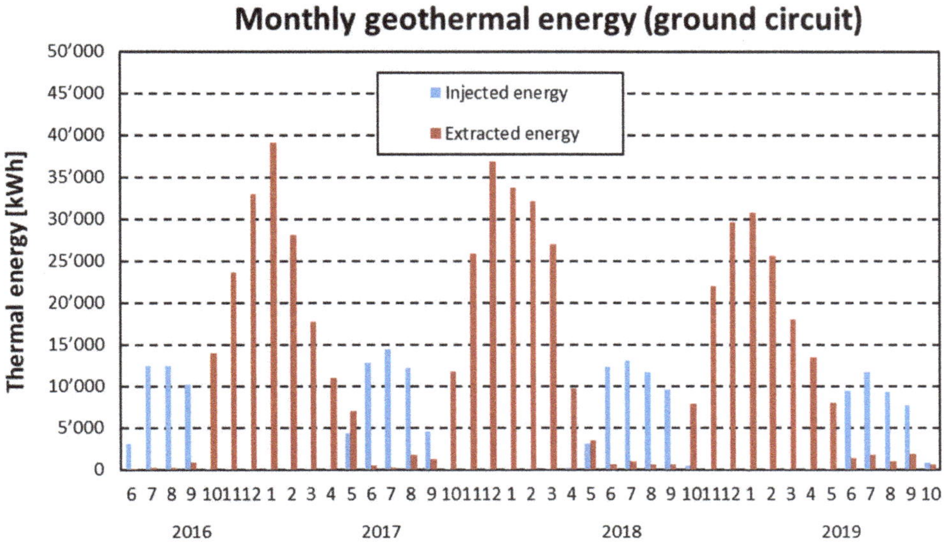

Figure 7. Monthly energy exchanged with the ground (at the heat meter "Q1") during the monitoring (years and months in the abscissa labels; source: [22]).

Based on monitored data and geometric features of the building, such as the reference energy surface, i.e., the surface of the thermal zones to be heated and cooled, the following indicators have been calculated (Table 2). In order to consider the impact of climatic conditions on the monitored results, the heating and cooling degree-days (in Table 2, HDD and CDD respectively, [°Cd]) were taken into account and specific indexes, aimed at overcoming this impact, were calculated (as the ratio between Heating index and HDD and between Cooling index and CDD, respectively) and named normalized heating and cooling indexes.

Table 2. Space heating and cooling demands and related indicators (elaborations based on [22]).

Monitored Season	Heating Demand MWh	Heating Index kWh/m^2y	HDD	Normalised Heating Index $Wh/m^2/°Cd$	Monitored Season	Cooling Demand MWh	Cooling Index kWh/m^2y	CDD	Normalised Cooling Index $Wh/m^2/°Cd$
					Summer 2016	**36.4**	**6.4**	**73**	**87.9**
Winter 2016–17	188.4	33.1	2279	14.5	Summer 2017	47.8	8.2	106	77.5
Winter 2017–18	190.2	33.4	2350	14.2	Summer 2018	49.1	8.6	112	76.6
Winter 2018–19	162.0	28.4	2041	13.9	Summer 2019	39.3	6.9	113	61.1

The HDD were calculated according to [31], as the sum of differences between the daily mean external air temperature and the set indoor temperature (20 °C), where only days with daily mean external air temperature minor or equal to 12 °C are taken into account. The CDD were calculated analogously but based on a reference internal temperature of 25 °C, taking into account that summer is not defined based on meteo-climatic data, but it corresponds to the period during which geocooling is witched on manually.

The winter normalized heating values (always about 14 $Wh/m^2/°Cd$) describe coherent results, while the summer ones (from 61 to 88 $Wh/m^2/°Cd$) are more variable, mainly depending on the occupants' behavior and on technical management practices.

As anticipated, it has to be taken into account that during summers in 2016 and 2019, the geocooling system has encountered some operative problems that lightly reduced the hours of operation.

Furthermore, the thermal demands monitored differ to the thermal needs estimated in the first stage of the project attesting, a part the difference related to the heat losses along the water distribution system from the technical room to the flats, an initial underestimation of the heating needs and overestimation of the cooling needs.

A realistic index of the efficiency of a HP over the year can be expressed by the Seasonal Coefficient of Performance (*SCOP*), i.e., the ratio between the thermal energy supplied and the electricity required, in a season. In summer, the GSHP of the building provide only DHW, when heat from solar collectors is not sufficient, at higher supply temperature compared to space heating needs in winter. Therefore, due to the different levels of temperature, the *SCOP* in summer is expected to be lower than in winter.

In order to better understand these results, the main features of the three identical GSHP adopted in the present case are described in Table 3, resulting in a nominal thermal power of 133.2 kW. In Table 3, B0W35 and B0W60 are two operative conditions of the heat pump, it means the brine inlet temperature is 0 °C, and the water outlet temperatures are 35 °C and respectively 60 °C.

Table 3. Characteristics of the GSHP provided by the manufacturing company.

	B0W35	**B0W60**
Thermal Power supplied at condenser [kW]	44.4	41.3
Electrical Power absorbed by compressor [kW]	10.0	13.5
COP [–]	4.4	3.1

Due to the "Q5" heat meter before the GSHP evaporators (see Figure 4) and by the electric monitoring of the GSHP compressors, it is possible to estimate the *SCOP* by the following equation:

$$SCOP = \left(\frac{E_{th\ evaporator}[\mathrm{kWh}]}{E_{el\ compressor}[\mathrm{kWh}]} \right) + 1 \quad (1)$$

where $E_{th\ evaporator}$ is the geothermal heat exchanged to the evaporator and $E_{el\ compressor}$ is the electricity provided to the compressor, while the electricity for the operation of the circulation pumps and other devices has not been considered in the calculation of the *SCOP*.

The obtained results are reported in Table 4, where it is possible to note an intensive growing of the heat exchanged to the evaporator in summer. This condition is mainly due to a different contribution of the solar thermal sysyem devoted to provide DHW because of technical and meteo-climatic issues.

Table 4. Results of the calculation of the *SCOP* (source: [22]).

Seasons	$E_{th\ evaporator}$	$E_{el\ compressor}$	***SCOP***
Winter 2016–17	167,306	59,741	3.8
Winter 2017–18	178,260	63,625	3.8
Winter 2018–19	157,069	55,517	3.8
Summer 2016	1695	887	2.9
Summer 2017	6794	3387	3.0
Summer 2018	8705	4186	3.1
Summer 2019	14,315	6667	3.1

Further, it is important to observe that the seasonal performances of the GSHP (3.8 in winter) are lower than the instantaneous or nominal ones (B0W35 in Table 3). This is mainly due to the DHW production even in winter (at about 60 °C) and to the non optimized control and management system, i.e., to the many switching on and off of heating supply. This result underlines the importance of real monitored data to evaluate the performance of GSHP coupled to low energy buildings, considering that, in general, higher values are reported, often because the different operative conditions and system management options are not taken into account.

As anticipated, another index, related to the global operation of the geothermal system, is the recharge of the ground. According to [16], this parameter affects the long-term dynamics behavior of the ground as heat source and then affects the global performance of geothermal systems.

Indeed, the real thermal recharge must be verified and confirmed over the years and geothermal district heating systems with huge BHE fields must be sized and designed carefully to avoid serious problems (e.g., ground freezing, system shutdown) during their operation.

According to [12], to evaluate the thermal recharge of the ground, in terms of heat injection over heat extraction it is necessary to consider an entire year starting from the winter season. Therefore, the first summer monitored (2016) was neglected. The results of the calculations based on monitored data are reported in Table 5.

Table 5. Annual values of the energy extracted and injected and related thermal recharge (source: [22]).

Operation Period	Energy Extracted	Energy Injected	Thermal Recharge
Winter 2016–2017 + Summer 2017	173.3 MWh/y	44.6 MWh/y	25.7%
Winter 2017–2018 + Summer 2018	181.8 MWh/y	46.9 MWh/y	25.8 %
Winter 2018–2019 + Summer 2019	154.3 MWh/y	32.4 MWh/y	21.0 %

The same calculation has been provided also for all the monitored period (4 summers, considering also summer 2016, and 3 winters) by the following equation, where I (i from 1 to 4) is the injection in the i summer season and E_j (j from 1 to 3) is the extraction from the ground in the j winter season, attesting an average value for "R" of 23.7%.

$$R = \frac{\frac{(I_1+I_2+I_3+I_4)}{4}}{\frac{(E_1+E_2+E_3)}{3}} \tag{2}$$

The last performance index elaborated is the geocooling efficiency, defined as the ratio between the cooling demand of the building (all the flats) and the electricity needed for the pumps, following the equation:

$$\varepsilon = \frac{E_{th\_cool}\ [\text{kWH}]}{E_{el}\ [\text{kWH}]} \tag{3}$$

Based on monitored data, the values of the monthly demand, of the monthly electricity consumption and of the geocooling efficiencies are reported in Figure 8.

In Figure 8 is reported an efficiency of 30 achieved in August 2018, after the implementation of the improvement measures. Nevertheless, in September 2018 the electricity consumption remained almost unchanged, while cooling demand was reduced resulting and this means a worse efficiency. Probably, without a better regulation of the pump 301 (Figure 3), it will not be possible to further reduce electricity consumption, especially when cooling demand decreases. Indeed summer 2019 shows the negative effects of an inappropriate management of the system, with the worst efficiencies of the monitoring campaign recorded in August and September.

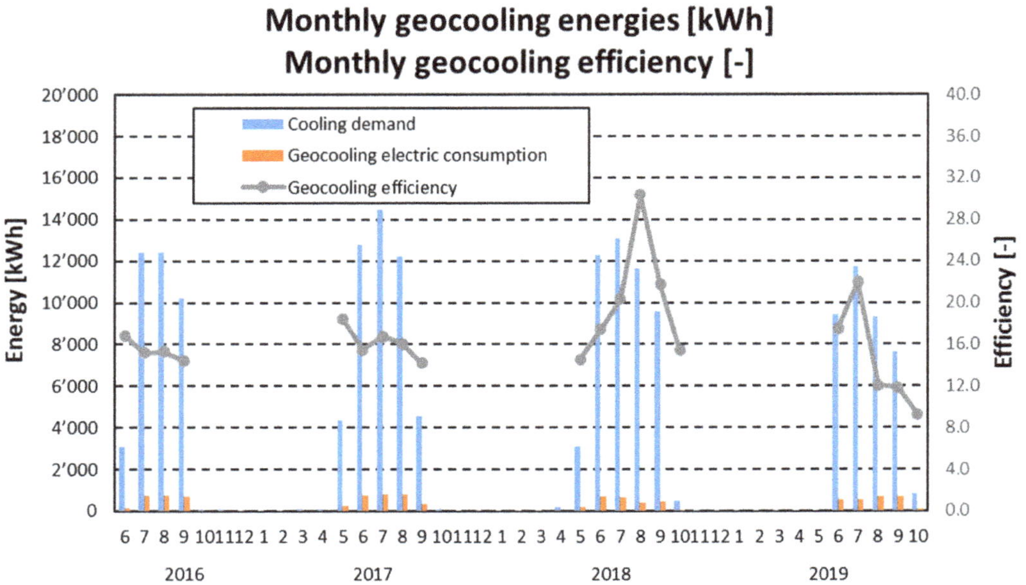

Figure 8. Monthly cooling demand, electricity consumption and related geocooling efficiency (years and months in the abscissa labels; source: [22]).

However, based on results achieved during August 2018, it can be stated that efficiency values of 30 are easily approachable with an optimal regulation of the system. Further, with a regulation of all circulation pumps, it would be possible to reach efficiencies around 35–40.

4. Discussion

The issues explored in the work, bringing interesting insights for the involved stakeholders, can be summarized as it follows:

- The efficiency and potential of geocooling technology;
- The real quantification of a thermal recharging of the ground by geocooling;
- The limits of the system and the performance achievable by an underfloor radiant distribution system.

About the first point, according to the mentioned theoretical studies, the measured geocooling efficiency (defined as the ratio between the total cooling demand and the electricity used by the circulation pumps at monthly level) is in the range between a maximum of 30 and a minimum of 12, depending on pumps settings and other issues. In particular, the circulation pump that set warm/cool water to the flats does not modulate the flow rate setting a temperature difference but by taking into account the pressure difference. Indeed, a different control system would allow a lower electricity consumption and a better modulation of the temperature of the supplied warm/cool water. Values higher than 30 can be also achieved by optimized control and management systems.

About the second point, the value achieved for the thermal recharge of the ground over the monitored seasons was about 24%, in accordance to the recommendations given during the first stages of the project (preliminary analysis, sizing and design). Indeed, to recharge by geocooling allows a proper thermal balance of the ground, in addition to the possibility to provide cooling to the occupants. Further, from the economic point of view, an issues that will be better described in further developments, geocooling can help reducing the initial investment, i.e., saving meters of drilling because less BHE can be sufficient, and developing a new approach about thermal tariffs. This is a critical issue because, according to [5,6], despite the described pros, the economic barrier of the initial capital cost still affect GSHP. This is mainly due to the excavation cost that represent about the half of the total cost. This condition stimulates research and practice in finding new solutions to reduce or avoid drilling cost such as those proposed by [6] as thermo-active foundations or other energy geostructures. To that end, geocooling represent another effective alternative.

The assessment of the technical and economic factors that influence the design and performance of vertical GSHP is deepened in [5], with a focus on the spatial correlation that these factors have with geographic features such as geology and climatic conditions. By the way they indicated that Switzerland is one of the country with the highest capital costs for vertical GSHP systems at least 10 years ago. Further, they stressed that, as yet subsurface characteristics are not adequately considered in the planning and design phases, under or oversizing can happen, with a long-term effect on the maintenance costs and payback time of these systems.

About the third point, during the first two years of monitoring, some critical issues emerged in relation to geocooling efficiency and thermal comfort in summer. These criticalities are mainly caused by an uncontrolled management of the system. However, some optimizations were implemented to overcome these problems such as the modulation of the circulation pumps and the increasing of about 2 °C of the temperature of the water circulating in the radiant underfloor system in summer. The monitoring in the sample flat demonstrate that cooling was requested by the occupants even with external temperature lower than 25–26 °C. According to thermal comfort theory, the monitored indoor temperature were too low (22–24 °C) and relative humidity too high (70–85%) in many summer hours, underlining again the need of a better regulation of the cooling system, e.g., to switch on when really needed, to supply temperature not lower than 20–21 °C in the underfloor circuit, to consider also the effect of the mechanical ventilation equipped by the air to air heat exchanger. These results demonstrate also that the obtained thermal recharge is the maximum achievable for the monitored building and respective boundary conditions (energy standard, solar gains, typology, loads, and climate). Therefore, it is possible to guess that better performance of the systems can be achieved in terms of comfort conditions in summer in climates more warm and dry.

5. Conclusions and Further Developments

The research here presented is based on a long-lasting research supported by validated tools and expertise and on data collection and elaboration related to a monitoring campaign of about 40 months, involving thermal and electric measurements in the geothermal system and in one sample flat (users' side), collaboration with the technicians involved and with the tenants of the building by a questionnaire survey.

The obtained results prove that geocooling through underfloor radiant distribution is an efficient and applicable technology, at least for residential buildings and moderate climatic conditions. Indeed, it is evident that geocooling cannot be assimilated to conventional air conditioning systems, based on chillers. In the analyzed real case, heat is extracted from the buildings by radiant underfloor systems able to act only on sensible heat. Therefore, if there are not air treatment units, this kind of technology is in general not recommended in case of building with high loads in terms of latent heat (non residential buildings with a high density of occupation). The heat exchange due to the radiant system depends on the temperature level of the water inside. The best temperatures from the point of view of indoor thermal comfort are equal or greater than 21°C, but this condition limits the heat exchange. In addition, to avoid surface condensation problems, it is advisable to satisfy the internal comfort keeping the upper levels of set point temperature inside, according to the reference standards. The survey by questionnaire was carried out exactly to better understand the perception of cooling and to instruct tenants in setting indoor temperatures, underlining the role of the involvement of the occupants in such systems. The relatively high temperature levels suggested for underfloor radiant cooling, compared to conventional air conditioning, imply a careful preliminary design and management during the operation to guarantee a proper control of the global system.

In addition, since the thermal recharge of the ground is a key parameter for an efficient long-term operation of the system, it has to be designed by accurate energy simulations. If the designed thermal recharge cannot be achieved over the years, the possibility to modify the installation or its operation should be provided, since long-term thermal storage problems can seriously affect the whole effectiveness of the system. Therefore, the achieved results can support strategies for new and existing geothermal systems, giving solutions for effective space heating and cooling systems.

Furthermore, the results reported confirm that large geothermal systems, with a huge extension of BHE, should be planed and sized very carefully in order to reduce the risk of technical troubles and decreasing of performance. In this framework, the work presented gives also an important contribution to the definition of methods and tools needed to design and monitor properly such systems.

The conclusions of this work can be framed also in the recent progress of rules and regulations about geothermal systems (e.g., the [19]) that consider the issue of the thermal interference in case of geothermal plants in the same area, underlining the promising role of properly sized geocooling systems in the next future.

First estimations referred to the real case here presented, which will be better detailed in further developments of the work, indicate that geocooling allows saving in the range 10–25% of the investment cost for installing BHE thanks to the recharge of the ground that allows a minor drilling length (considering the long-term performance of the overall system, at least in similar boundary conditions). Taking into account the overall operation of the system all over the year, ref. [22] estimated that the costs for geocooling devices (about 20,000 euros) are comparable to the economic saving due to the recharge. In addition, the building real estate value would certainly further increase thanks to the better indoor comfort conditions and lower annual operative costs. An economic saving for the final users can be achieved also in comparison to other conventional technologies such as cooling machines that, due to drastically lower efficiencies, imply higher operative costs for final users.

Further efforts are necessary to assess experimentally the energy performance of these systems, to model and analyze the influence of the boundary condition and to optimize the heat exchanger layout, even in the framework of assessment related to geothermal potential in wide areas such as that proposed by [32]. In this paper all these aspects are investigated, starting from the results of the monitoring campaign of the analyzed real case.

The adopted method can be replicated and the results can be generalized to new residential buildings, well insulated and with underfloor radiant systems for providing heat and cool. In fact, it could be possible to calibrate and estimate space cooling needs for residential buildings in different climatic conditions, according to the indexes described in Table 2.

Finally, although the analyzed geothermal system serves only one building of 46 units, the method and part of the achieved results can be extended not only to similar cases but also to geothermal district heating network, where thermal needs of several buildings must be considered for the correct sizing of the BHE field. Following developments of the work will face this issue in the framework of the evolution of district thermal systems to low temperature applications able to provide both heat and cool to the final users.

Author Contributions: Conceptualization, P.C.; methodology, M.B., P.C.; software, M.B.; validation, M.B.; formal analysis, P.C., S.F.; investigation, M.B.; resources, M.B., N.C., data curation, M.B., P.C., S.F.; writing—original draft preparation, M.B.; writing—review and editing, P.C.; Visualization, P.C.; supervision, M.B., P.C., S.F., project administration, M.B.; Funding acquisition, M.B., N.C. All authors have read and agreed to the published version of the manuscript.

Funding: This research received no external funding.

Acknowledgments: The authors thank the Swiss Federal Office of Energy, the Renewable Energy Fund of the Canton of Ticino, Suisse Promotion Immobiliare SA, for funding the project and the monitoring activities and for realizing and managing the buildings; Andrea Andreoli for his precious technical advices and consulting; the occupants and other people involved for their collaboration.

Conflicts of Interest: The authors declare no conflict of interest.

Abbreviations

Abbreviations	
BHE	Borehole Heat Exchanger/Exchangers
CDD	Cooling Degree Days
COP	Coefficient of Performance
DHW	Domestic Hot Water
EER	Energy Efficiency Ratio
GRT	Geothermal-Response-Test
GSHP	Ground Source Heat Pump/Pumps or Geothermal Pump/Pumps
HDD	Heating Degree Days
HP	Heat pump/pumps
HVAC	Heating, Ventilation and Air Conditioning
RES	Renewable Energy Source/Sources
SCOP	Seasonal Coefficient of Performance
SIA	Swiss society of Engineers and Architects (www.sia.ch/en/the-sia/)

References

1. Agora Energiewende. European Energy Transition 2030: The Big Picture. Ten Priorities for the Next European Commission to Meet the EU's 2030 Targets and Accelerate towards 2050. 2019. Available online: https://www.agora-energiewende.de/fileadmin2/Projekte/2019/EU_Big_Picture/153_EU-Big-Pic_WEB.pdf (accessed on 5 November 2021).
2. BFE–Swiss Federal Office of Energy. Energy Strategy 2050. 2018. Available online: https://www.bfe.admin.ch/bfe/en/home/policy/energy-strategy-2050.html (accessed on 5 November 2021).
3. Confederation Suisse. *Message Relatif a la Revision Totale de la Loi Sur le CO2 Pour la Periode Posterieure a 2020*; Confederation Suisse: Bern, Switzerland, 2017.
4. Narula, K.; Chambers, J.; Streicher, K.N.; Patel, M.K. Strategies for decarbonising the Swiss heating system. *Energy* **2019**, *169*, 1119–1131. [CrossRef]
5. Blum, P.; Campillo, G.; Kölbel, T. Techno-economic and spatial analysis of vertical ground source heat pump systems in Germany. *Energy* **2011**, *36*, 3002–3011. [CrossRef]
6. Sterpi, D.; Tomaselli, G.; Angelotti, A. Energy performance of ground heat exchangers embedded in diaphragm walls: Field observations and optimization by numerical modelling. *Renew. Energy* **2020**, *147*, 2748–2760. [CrossRef]
7. Lund, J.W.; Toth, A.N. Direct utilization of geothermal energy 2020 worldwide review. *Geothermics* **2021**, *90*, 101915. [CrossRef]
8. Lyu, W.; Li, X.; Yan, S.; Jiang, S. Utilizing shallow geothermal energy to develop an energy efficient HVAC system. *Renew. Energy* **2020**, *147*, 672–682. [CrossRef]
9. Inayat, A.; Raza, M. District cooling system via renewable energy sources: A review. *Renew. Sustain. Energy Rev.* **2019**, *107*, 360–373. [CrossRef]
10. Hollmuller, P.; Lachal, B.; Pahud, D. *Rafraîchissement par Géocooling. Bases pour un Manuel de Dimensionnement*; Rapport Final; Office Fédéral de l'Energie: Berne, Switzerland, 2005. Available online: https://repository.supsi.ch/2808/1/105-Pahud-2005-Geocooling.pdf (accessed on 5 November 2021).
11. Pahud, D.; Caputo, P.; Branca, G.; Generelli, M. *Etude du Potentiel D'utilisation de "Geocooling" D'une Installation Avec Sondes Géothermiques Verticales Appliqué à un Bâtiment Administratif Minergie à Chiasso*; Rapport Final; Office Fédéral de L'énergie: Berne, Switzerland, 2008. Available online: https://www.aramis.admin.ch/Default?DocumentID=63151 (accessed on 5 November 2021). (in French)
12. Pahud, D.; Belliardi, M. *Geocooling Handbook-Cooling of Buildings using Vertical Borehole Heat Exchangers*; Final Report; Swiss Federal Office of Energy: Bern, Switzerland, 2011. Available online: https://repository.supsi.ch/2803 (accessed on 5 November 2021).
13. Pahud, D.; Belliardi, M.; Caputo, P. Geocooling potential of borehole heat exchangers' systems applied to low energy office buildings. *Renew. Energy* **2012**, *45*, 197–204. [CrossRef]
14. McKenna, P.; Turner, W.J.N.; Finn, D.P. Geocooling with integrated PCM thermal energy storage in a commercial building. *Energy* **2018**, *144*, 865–876. [CrossRef]
15. Pratiwi, A.S.; Trutnevyte, E. Life cycle assessment of shallow to medium-depth geothermal heating and cooling networks in the State of Geneva. *Geothermics* **2021**, *90*, 101988. [CrossRef]

16. Pahud, D. PILESIM2: Simulation Tool for Heating/Cooling Systems with Energy Piles or Multiple Borehole Heat Exchangers. 2007. Available online: https://repository.supsi.ch/3067 (accessed on 5 November 2021).
17. Christodoulides, P.; Vieira, A.; Lenart, S.; Maranha, J.; Vidmar, G.; Popov, R.; Georgiev, A.; Aresti, L.; Florides, G. Reviewing the Modeling Aspects and Practices of Shallow Geothermal Energy Systems. *Energies* **2020**, *13*, 4273. [CrossRef]
18. Arghand, T.; Javed, S.; Trüschel, A.; Dalenbäck, J.O. A comparative study on borehole heat exchanger size for direct ground coupled cooling systems using active chilled beams and TABS. *Energy Build.* **2021**, *240*, 110874. [CrossRef]
19. SIA 384/6. *Borehole Heat Exchangers*; SIA: Zurich, Switzerland, 2010; (In French, German and Italian).
20. SIA 384/7. *Use of Heat of Underground Water*; SIA: Zurich, Switzerland, 2015; (In French, German and Italian).
21. SIA 382/1. *Plants of Ventilation and Air Conditioning. General Basis and Needs*; SIA: Zurich, Switzerland, 2014. (In French, German and Italian)
22. Belliardi, M. Applied Analysis of Geocooling Technology for a Residential Building in Lugano, Final Report, Swiss Federal Office of Energy (SFOE), Energy Research and Cleantech. 2020. Available online: https://www.aramis.admin.ch/Default?DocumentID=66659 (accessed on 5 November 2021).
23. Sanner, B.; Hellström, G.; Spitler, J.; Gehlin, S. Thermal response test–current status and world-wide application. In *Proceedings of the World Geothermal Congress 2005, Antalya, Turkey, 24–29 April 2005*; International Geothermal Association: Bonn, Germany, 2005; Volume 1436. Available online: https://www.geothermal-energy.org/pdf/IGAstandard/WGC/2005/1436.pdf (accessed on 5 November 2021).
24. Hellström, G.; Sanner, B. EED-Earth Energy Designer, User Manual, Version 2.0. Borehole Heat Exchangers. Available online: https://buildingphysics.com/manuals/EEDONTHEWEB.pdf (accessed on 5 November 2021).
25. Florides, G.; Kalogirou, S. Ground heat exchangers—A review of systems, models and applications. *Renew. Energy* **2007**, *32*, 2461–2478. [CrossRef]
26. Sivasakthivel, T.; Philippe, M.; Murugesanc, K.; Vermad, V.; Pingfang, H. Experimental thermal performance analysis of ground heat exchangers for space heating and cooling applications. *Renew. Energy* **2017**, *113*, 1168–1181. [CrossRef]
27. Fadejev, J.; Simson, R.; Kurnitski, J.; Haghighat, F. A review on energy piles design, sizing and modelling. *Energy* **2017**, *122*, 390–407. [CrossRef]
28. Javadi, H.; Mousavi Ajarostaghi, S.S.; Rosen, M.A.; Pourfallah, M. Performance of ground heat exchangers: A comprehensive review of recent advances. *Energy* **2019**, *178*, 207–233. [CrossRef]
29. Biglarian, H.; Abbaspour, M.; Saidi, M.H. A numerical model for transient simulation of borehole heat exchangers. *Renew. Energy* **2017**, *104*, 224–237. [CrossRef]
30. SIA 380/1. *Thermal Energy in Building Construction*; SIA: Zurich, Switzerland, 2009. (In French, German and Italian)
31. SIA 381/3. *Heating Degree-Days in Switzerland*; SIA: Zurich, Switzerland, 1982. (In French, German and Italian)
32. Casasso, A.; Sethi, R. Assessment and mapping of the shallow geothermal potential in the province of Cuneo (Piedmont, NW Italy). *Renew. Energy* **2017**, *102 Pt B*, 306–315. [CrossRef]

energies

MDPI

Article

Thermal Comfort and Energy Analysis of a Hybrid Cooling System by Coupling Natural Ventilation with Radiant and Indirect Evaporative Cooling

Pradeep Shakya [1], Gimson Ng [1], Xiaoli Zhou [1], Yew Wah Wong [2], Swapnil Dubey [3] and Shunzhi Qian [4,*]

1 SJ-NTU Corporate Lab, Nanyang Technological University, 50 Nanyang Avenue, Singapore 639798, Singapore; pradeep.s@ntu.edu.sg (P.S.); gimson.ng@ntu.edu.sg (G.N.); xzhou006@163.com (X.Z.)
2 S-Lab for Advanced Intelligence, Nanyang Technological University, 50 Nanyang Avenue, Singapore 639798, Singapore; mywwong@ntu.edu.sg
3 Energy Research Institute, Nanyang Technological University, 50 Nanyang Avenue, Singapore 639798, Singapore; SDubey@ntu.edu.sg
4 School of Civil and Environmental Engineering, Nanyang Technological University, 50 Nanyang Avenue, Singapore 639798, Singapore
* Correspondence: SZQian@ntu.edu.sg

Citation: Shakya, P.; Ng, G.; Zhou, X.; Wong, Y.W.; Dubey, S.; Qian, S. Thermal Comfort and Energy Analysis of a Hybrid Cooling System by Coupling Natural Ventilation with Radiant and Indirect Evaporative Cooling. *Energies* **2021**, *14*, 7825. https://doi.org/10.3390/en14227825

Academic Editor: Wei-Hsin Chen

Received: 19 October 2021
Accepted: 19 November 2021
Published: 22 November 2021

Abstract: A hybrid cooling system which combines natural ventilation with a radiant cooling system for a hot and humid climate was studied. Indirect evaporative cooling was used to produce chilled water at temperatures slightly higher than the dew point. With this hybrid system, the condensation issue on the panel surface of a chilled ceiling was overcome. A computational fluid dynamics (CFD) model was employed to determine the cooling load and the parameters required for thermal comfort analysis for this hybrid system in an office-sized, well-insulated test room. Upon closer investigation, it was found that the thermal comfort by the hybrid system was acceptable only in limited outdoor conditions. Therefore, the hybrid system with a secondary fresh air supply system was suggested. Furthermore, the energy consumptions of conventional all-air, radiant cooling, and hybrid systems including the secondary air supply system were compared under similar thermal comfort conditions. The predicted results indicated that the hybrid system saves up to 77% and 61% of primary energy when compared with all-air and radiant cooling systems, respectively, while maintaining similar thermal comfort.

Keywords: natural ventilation; radiant cooling; indirect evaporative cooling (IEC); sustainable building; energy saving; thermal comfort

1. Introduction

Energy efficiency in buildings has received great interest due to the increasing energy demand and significant depletion of natural resources. Owing to its hot and humid tropical climate, in Singapore, more than 60% of the total energy consumed by residential and non-residential building is by air-conditioning for thermal comfort [1]. As an effective strategy for reducing energy use in buildings, natural ventilation utilizes the natural forces of wind pressure and stack effects to direct the movement of air through openings [2]. Natural ventilation is becoming increasingly popular, owing to the following significant advantages: (1) low capital and maintenance costs [3]; (2) environmental sustainability [4]; reduced risk of airborne contagions [5]; and (3) positive perception as there is evidence that occupants in buildings prefer to have control over their environment [6]. However, natural ventilation often fails to provide enough cooling capacity for buildings in a hot and humid area or with dense occupancy [7], such as offices where the high-level thermal comfort is required for intellectual productivity [8]. Moreover, natural ventilation performance is easily limited by building design and shape, both of which are difficult to modify after construction [6].

In order to overcome the barriers to the adoption of natural ventilation, numerous efforts have been made in the literature to enhance its wind utilization [9–12] or buoyancy effect [13,14] and thereby achieve adequate air flow rate and desired thermal comfort. In heat-intensive scenarios, however, natural ventilation alone still cannot meet the requirement of thermal comfort, leading to the increased adoption of mixed-mode or hybrid systems [6]. The basic concept of hybrid ventilation is to maintain a satisfactory indoor environment while minimizing the energy use and operating costs associated with air conditioning by alternating between and combining natural and mechanical systems [15].

Recently, radiant cooling systems, utilizing temperature-controlled indoor surfaces as the cooling source [16], have been gaining much popularity due to good thermal comfort, reduced energy consumption, quiet operation, and space-saving features [17]. It would be very promising to couple radiant cooling with natural ventilation as a hybrid system for both thermal comfort and energy saving in buildings. The literature reveals that, so far, few studies have been conducted on such a hybrid system, mainly because of the condensation issue on radiant surfaces, especially in hot and humid area like Singapore, where dew-point temperature typically exceeds the surface temperature of a chilled ceiling panel (CCP) [18]. A possible solution to the condensation issue is the application of an indirect evaporative cooling (IEC) system to produce chilled water at a temperature between the wet-bulb temperature and dew point of the outdoor air [19]. Moreover, IEC shows the significant advantages of a high coefficient of performance (*COP*), energy-saving, and environmental sustainability over other conventional air conditioning systems.

Most previous studies have mainly focused on hybrid systems in enclosed environment by coupling radiant cooling with mechanical ventilation (e.g., airbox convector [18,20], personalized ventilation [21,22], underfloor/displacement ventilation [23–26]), which usually precool the supplied outdoor air to prevent condensation on a chilled surfaces. Yu et al. [27,28] proposed a novel system combining natural ventilation with diffuse ceiling inlet and thermally activated building systems (TABS) and investigated its cooling performance experimentally in Denmark, which has a relatively cold climate even in summer, and hence the condensation issue on the TABS was not considered. Meggers et al. [29] experimentally analyzed the thermal comfort of an open air pavilion using radiant cooling and indirect evaporation under the temperate climate of Princeton. Vangtook and Chirarattananon [30] employed water cooled by a cooling tower for radiant cooling and found that it was sufficient to achieve thermal comfort in night-time application, while the ventilation air precooled by chilled water at 10 °C was required to treat the latent load in day-time application. The effect of natural ventilation on enhancing the cooling capacity of radiant cooling was not studied in their work. To the authors' knowledge, no studies in the literature have evaluated the thermal comfort of a hybrid system coupling natural ventilation with radiant and IEC for a tropical climate, as well as compared their energy performance with conventional all-air and radiant cooling systems.

Therefore, the present work aims (i) to investigate a hybrid cooling system which couples natural ventilation together with radiant cooling, and which uses IEC to supply chilled water, and (ii) to study the thermal comfort and energy-saving potential of the hybrid cooling system in an office-sized, well-insulated environmental chamber.

2. Proposed Hybrid Cooling System

The proposed hybrid system mainly consists of three parts, namely, the chilled ceiling for radiant cooling, IEC, and a test room with cross ventilated openings for natural ventilation, as shown in Figure 1a,b, where the return water from the CCP is cooled down by IEC first to the temperature between the wet bulb temperature and dew point, and is then circulated into the CCP as the cooling source to remove the intensive sensible loads caused by solar radiation, occupants, lighting, and equipment in the working space. At the same time, natural ventilation removes the latent and a part of the sensible load and provides abundant fresh air.

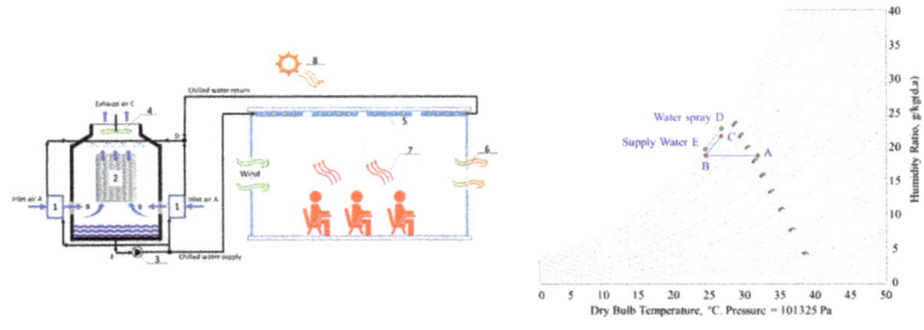

(**a**) (**b**)

Figure 1. (**a**) Configuration of proposed hybrid cooling system: (1) air–water counterflow heat exchanger; (2) air–water counterflow padding tower; (3) water pump; (4) fan; (5) chilled ceiling; (6) window opening; (7) internal heat load; (8) solar flux. (**b**) Psychrometric process to produce chilled water in IEC.

The psychrometric process to produce a chilled water supply in IEC is shown in Figure 1a. Firstly, the inlet air *A* is cooled to the state *B* through the air–water counterflow heat exchanger, while its humidity ratio keeps constant as there is no air–water contact during the indirect heat exchanging process. The air at the state *B* is then blown into the air–water counterflow padding tower for the air–water direct evaporative cooling process. After being heated and humidified by the water spray, the air is exhausted from the top of the IEC at the state *C*. Meanwhile, the water spray *D* is cooled by the air to the state *E* and is divided into two parts: a smaller portion of water is pumped into the air–water counterflow heat exchanger to pre-cool the inlet air, and the larger portion is pumped into the CCPs, where it removes the heat in the working space together with natural ventilation and then returns to the IEC.

3. CFD Model

To determine the thermal comfort and energy consumption of the proposed hybrid system, the values of key parameters are required, such as the air temperature, surface temperature, airflow rate, and cooling loads. Using CFD, the variation of these key parameters under the effects of radiant heat transfer between two surfaces at different temperature and convective heat transfer induced by the air motion can be analyzed. For this reason, CFD simulation using the FLUENT program was performed. In the present work, CFD simulation was performed to evaluate the thermal conditions in the office-sized space with both radiant cooling and natural ventilation, and to provide the required parameters for thermal comfort and energy calculations. The following assumptions were used for the simulation:

- Turbulent flow was obtained using the realizable (k-ε) model with enhanced wall treatment including the effects of gravity for flow analysis.
- An incompressible ideal gas model was assumed.
- A surface-to-surface radiation heat transfer model was used.
- A solution method employing a pressure-based coupled scheme for pressure-velocity coupling was used.
- A body force weighted method for the pressure interpolation scheme was used as it resulted in a stable solution for buoyancy calculations.

In accordance with the BS EN 14240 testing standard of chilled ceilings [31], the test room was developed using FLUENT, as schematically shown in Figure 2a. Twelve cylindrical dummies were uniformly deployed to simulate the internal heat load and ceiling panels were installed at 2.6 m above the floor. In order to reduce the computational cost of the CFD model, a simplified geometry of the ceiling panels as a flat rectangular parallelopiped with uniform temperatures on the lower and upper surfaces was employed instead of a fully detailed model, which may reduce the accuracy unless all the input parameters are well known [32].

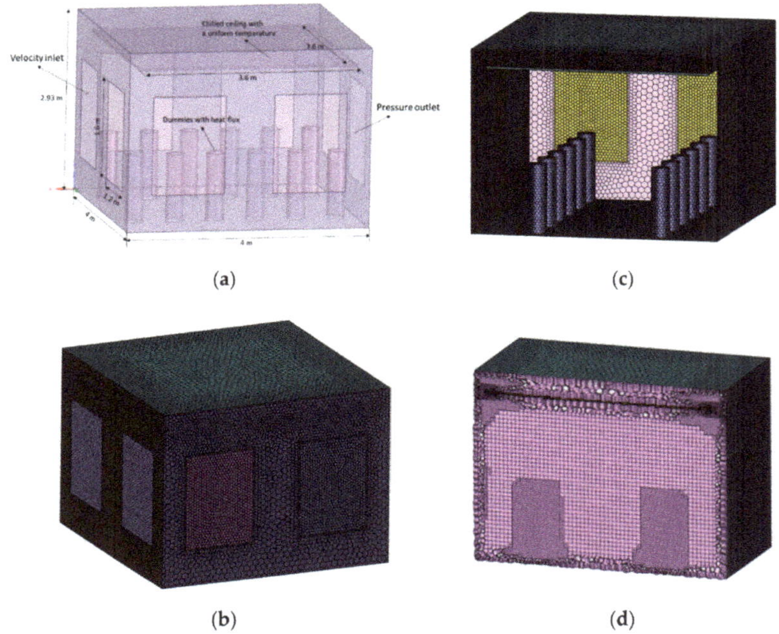

Figure 2. (**a**) Computational domain and dimensions for office model; (**b**) exterior surface mesh; (**c**) interior surface mesh; and (**d**) volume mesh.

Regarding the boundary conditions, the 'velocity inlet' was imposed to the openings in the upstream area of prevailing wind (detailed in Appendix A), while other openings were assigned as 'pressure outlets'. Internal surfaces were assumed to be 'no slip walls', while the internal load and solar flux were assigned to 12 dummies and wall surfaces, respectively, as heat flux in the unit of W/m^2, and uniform temperature was applied to the ceiling panel surfaces. Moreover, the floor and roof surfaces were taken as well-insulated walls in a building; hence, their boundary conditions can be assumed to be adiabatic.

It was necessary to adopt the minimum outdoor supply air, solar heat flux, and internal heat loads properly based on relevant standards, which are shown in Table 1. The unit heat flux assigned to the dummy surface was then achieved by dividing the total internal loads over the dummy surface area.

Table 1. Heat loads and fresh air rate for the test room model.

Heat from Envelope [33], W/m^2	Internal Loads [34,35]				Minimum Outdoor Supply Air [36], l/s/Person
	Occupant Density, /100 m^2	Occupant Activity, W/Person	Lighting, W/m^2	Equipment, W/m^2	
40	5	75 (Sensible), 55 (Latent)	12	16	8.5

4. Mathematical Models

In this section, the mathematical models to determine the thermal comfort parameters and the power input required for various equipment are presented.

4.1. Thermal Comfort

In general, there are two different categories for thermal comfort evaluation, namely, the heat balance approach for mechanical ventilation and the adaptive approach for natural ventilation [37]. For the hybrid system, with both mechanical and natural ventilation as proposed in the present work, the predicted mean vote (*PMV*) and predicted percentage of dissatisfied (*PPD*) models [38] based on the heat balance approach were adopted to quantify the thermal comfort. This is because both indices have been used with success during the past decades and their validity extends to the range of the proposed hybrid system [39]. *PMV* is an index that predicts the mean value of the votes of a large group of persons on the seven-point thermal sensation scale (−3 cold, −2 cool, −1 slightly cool, 0 neutral, +1 slightly warm, +2 warm, and +3 hot), while *PPD* is determined from *PMV* to predict the percentage of thermally dissatisfied people [40], and they are given by [37]

$$PMV = f(t_a, t_{mrt}, v, p_a, M, I_{cl}) \tag{1}$$

$$PPD = 100 - 95 \times \exp\left(-0.03353 \times PMV^4 - 0.219 \times PMV^2\right) \tag{2}$$

where t_a is the air temperature, t_{mrt} is the mean radiant temperature (MRT), v is the relative air velocity, p_a is the vapor pressure, M the activity level, and I_{cl} the clothing insulation. t_{mrt} is derived from the temperature of the surrounding surfaces and their positions with respect to the occupant, as described in Equation (3) [41]:

$$t_{mrt}^4 = t_1^4 F_{o-1} + t_2^4 F_{o-2} + \ldots + t_N^4 F_{o-N} \tag{3}$$

where t_N is the surface temperature of a surface N and $F_{o\text{-}N}$ is the angle factor between a surface N and an occupant; the defining method is detailed in [41]. The ranges of *PPD* and *PMV* for an acceptable thermal environment are defined as $PPD < 10\%$ and $-0.5 < PMV < +0.5$, respectively [40].

4.2. Energy Calculation

Obviously, a cooling system with both good thermal comfort and minimal energy consumption is much preferred. In the present work, energy consumption for the proposed hybrid system is compared with the that of the conventional all-air and radiant cooling systems, in order to evaluate its potential energy saving contributed to by these factors:

1. The supply fresh air comes from natural ventilation, leading to a reduced fan energy;
2. Higher supply water temperature means higher *COP* of IEC;
3. The hybrid system can achieve the same thermal comfort with a higher room temperature, because of the elevated air speed and lower MRT.

The primary energy-consuming equipment in the cooling systems include the chiller, fan, and pump. The required electric power input P_{ch} for a chiller was calculated using Equation (4).

$$P_{ch} = \frac{Q_c}{COP} \tag{4}$$

where Q_c is the cooling energy (W) and *COP* is the coefficient of performance of the water chiller and is dependent on the required chilled water temperature, ambient temperature, and humidity, as well as the operation capacity.

Fan and pump energy, as essential factors in the energy consumption of cooling system, were determined using Equation (5).

$$P_f = \frac{\dot{V}\Delta p}{3600\eta_f} \quad (5)$$

where $\dot{V}$ is the air (water) volumetric flow rate (m^3/h), Δp is the total pressure rise (P_a), and η_f is the fan (pump) efficiency. Using the standard heat load presented in Table 1, Equation (6) was used to calculate the air (or water) volumetric flow rate.

$$\dot{Q} = \rho \dot{V} \Delta h / 3600 \quad (6)$$

where $\dot{Q}$ is the rate of heat supplied to the system (W), ρ is the density (kg/m^3), and Δh is the specific enthalpy change (J/kg).

5. Results and Discussion

In this section, the results obtained using the CFD, thermal comfort, and energy modeling are presented.

5.1. *Effect of CCP Surface Temperature*

To simulate the office model with natural ventilation and radiant cooling, it is important to assume an appropriate temperature for the chilled ceiling surfaces. The CCP surface temperature will depend upon the chilled water supply temperature and the heat load in the test room. According to the Singapore weather station data, throughout the year, the dew point and wet bulb temperature in Singapore generally varies from 22 °C to 24 °C and 24 °C to 26 °C, respectively [42]. IEC can produce chilled water at a temperature between the wet bulb temperature and dew point of the outdoor air, which are assumed to be 26 °C and 24 °C, respectively, in Singapore as cooling design parameters to avoid the condensation on the chilled ceiling surfaces [43]. Therefore, the chilled water produced by IEC can be assumed as 25 °C for the present work. The panel surface temperature is usually non-uniform due to the temperature difference of the outlet and inlet section. This difference is estimated to be about 1–2 °C [32]; hence, average surface temperature in the model was assumed to be 26 °C and higher. In addition, the wind temperature through the openings was set as 29.5 °C according to the BCA standard [33]. The air flow through the ventilated openings was obtained from the velocity contours determined using the macro CFD model, which is presented in Appendix A, Figure A3.

To evaluate the thermal comfort of the hybrid system, the test room was firstly simulated with only natural ventilation, that is, with the CCP turned off. Secondly, with both CCP and natural ventilation on, three variations of the CCP surface temperature were considered (Table 2). The variation of the CCP surface from 26 °C to the upper threshold of 28 °C was considered due to the reason that CCP surface temperatures higher than 28 °C result in failure to comply with thermal comfort requirements.

Table 2. Test designs for thermal comfort evaluation of the hybrid system.

S. No.	System	Heat through Envelope (W/m^2)	Heat Output of Dummy (W/m^2)	Wind Temperature (°C)	CCP Surface Temperature (°C)
1	Natural ventilation				/
2	The hybrid system	40	38	29.5	26
3					27
4					28

Using the CFD simulation, the sensible cooling load obtained for the CCP and ventilation was 842 W and 358 W, respectively. Figure 3a–d shows the variation of temperature on the vertical plane 1.6 m from the wall inside the room. The MRT was calculated using Equation (3) based on the wall surface temperatures obtained from the CFD. For the test scenarios considered, it is evident that as the CCP surface temperature increased, the air temperature and MRT became higher. It was also found that the pure natural ventilation system results in slightly a higher air velocity inside the room, followed by Test no. 4, 3, and 2, possibly because the zone air with the higher temperature usually introduces more drastic air movement.

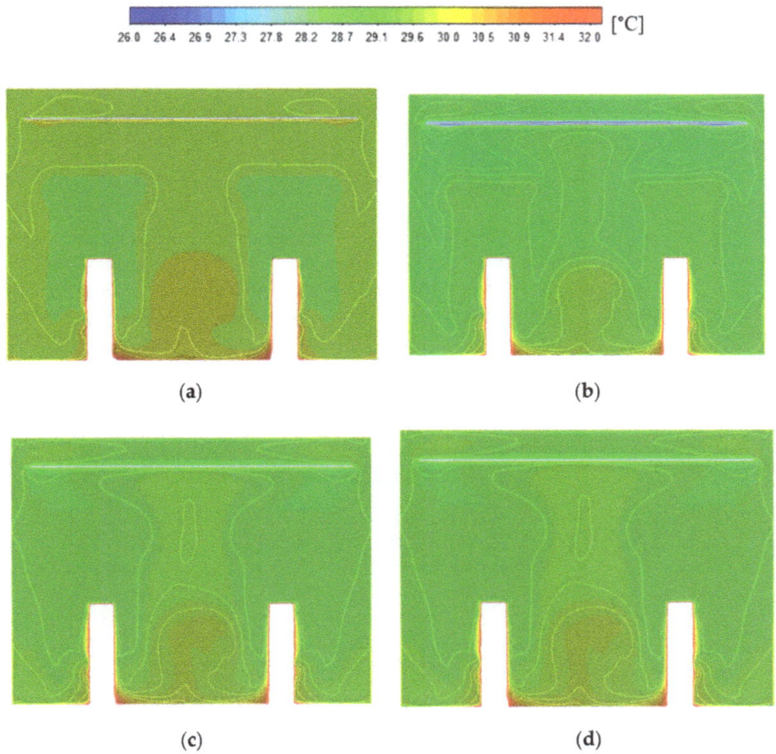

Figure 3. Temperature contours (1.6 m from the wall) for the four different scenarios: (**a**) natural ventilation with radiant cooling, (**b**) hybrid system with CCP surface temperature of 26 °C, (**c**) hybrid system with CCP surface temperature of 27 °C, and (**d**) hybrid system with CCP surface temperature of 28 °C.

The standard effective temperature (SET) as a comprehensive metric of outdoor thermal comfort was introduced to quantify the thermal comfort enhancement by the developed hybrid system. The SET for the four cases is shown in Figure 4, wherein the SET of the natural ventilation system was reduced by 0.5 °C using the proposed hybrid system compared with a CCP surface temperature of 26 °C, and the SET increased almost linearly with respect to the increase in the CCP surface temperature.

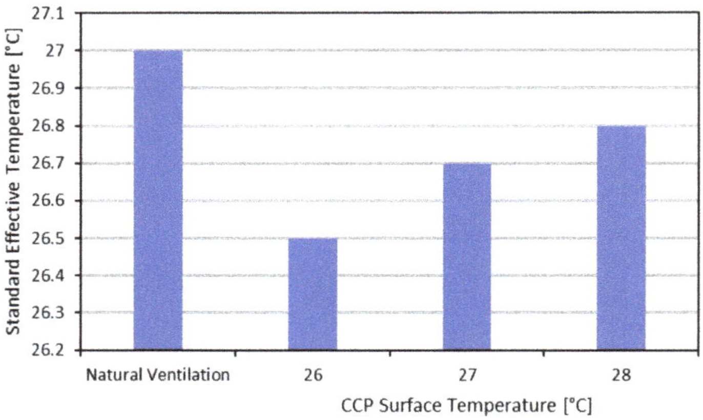

Figure 4. Standard effective temperatures of the four cases.

Furthermore, together with proper assumptions of metabolic rate (1 met) and clothing level (0.5 clo), simulation results on air speed, air temperature, relative humidity, and MRT from Table 3 were fed into the *PMV-PPD* model to compare the thermal comfort of the four cases, which is shown in Table 4. Obviously, it was found that the thermal comfort of the pure natural ventilation system (Case 1) can be significantly enhanced by coupling with the radiant cooling and IEC. For the reference wind velocity of 2.9 m/s, the developed hybrid system with the CCP surface temperature below 28 °C can only satisfy the thermal comfort and cooling design of office space in the humid and hot climate of Singapore.

Table 3. Simulation results for the various test cases.

S. No.	System	CCP Surface Temperature (°C)	Simulation Results		
			Air Velocity (m/s)	Air Temperature (°C)	MRT (°C)
1	Natural ventilation	\	0.66	29.7	30.9
2	The hybrid system	26	0.64	29.5	29.9
3		27	0.64	29.6	30.2
4		28	0.65	29.6	30.4

Table 4. Thermal comfort comparison between the four cases.

	1	2	3	4
Compliance to thermal comfort	No	Yes	Yes	No
PMV	0.57	0.43	0.46	0.5
PPD	12%	9%	9%	10%
Sensation	Slightly warm	Neutral	Neutral	Slightly warm

As the thermal comfort is dependent upon the prevalent wind speed outside the test room, the variation of *PMV* and *PPD* with respect to air speed in the test room was studied. As shown in Figure 5, the percentage dissatisfied was higher for lower air speeds. The allowable air speed for the outdoor air conditions selected was within the narrow range of 0.6–0.8 m/s. The upper threshold of 0.8 m/s is the allowable maximum air speed according to the ASHRAE 55 standard. According to the local weather station data, the prevalent wind condition in Singapore can vary from 0.2 to over 6 m/s [42]. This implies that there will be conditions, such as during rainfall when the outdoor humidity is high and ventilation might be limited, where the thermal comfort compliance will not be satisfied when the air speed in the room decreases below 0.6 m/s. Therefore, to meet the thermal comfort compliance for all climatic conditions in Singapore, a secondary fresh air delivery system, such as using a desiccant dehumidification system (DDS), is required. A DDS can be used for both fresh air supply and latent heat removal.

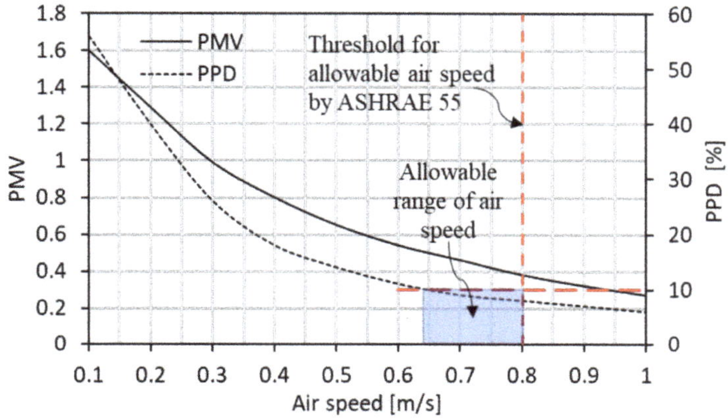

Figure 5. Effect of air speed on *PMV* and *PPD*.

In a DDS, desiccants dry the air by adsorbing moisture without the requirement to cool the air below its dew point. Once the desiccants are saturated with moisture, heat is used to desorb the moisture and regenerate it back to its dry state. This means the electrical energy demand in a conventional dehumidification system can be replaced by thermal energy supplied using sustainable and clean energy sources such as solar thermal systems. In a hot and humid climate, such as in Singapore, where the solar energy is abundant, this kind of desiccant system is highly relevant. Therefore, during recent years, a desiccant dehumidification and cooling system has been considered for air-conditioning due to its energy saving and environmentally friendly features [34,44,45].

5.2. Energy Analysis

As the thermal comfort from the hybrid system including a CCP combined with natural ventilation alone was not always satisfactory, DDS was also integrated as the secondary source of fresh air supply. The issue with the hybrid system with DDS is that this system will require higher initial capital for installation. However, the energy savings from this kind of hybrid system may be substantial. In this section, the proposed hybrid system is compared with conventional all-air and radiant cooling systems for energy consumption. To achieve the same comfort level ($PMV = 0.43$, $PPD = 9\%$) for the hybrid system with the CCP surface temperature of 26 °C, the designed indoor air conditions were a dry bulb temperature of 27 °C and a relative humidity of 50% for the all-air and radiant cooling systems.

The air handling process of the conventional all-air system is shown in Figure 6 and corresponding air state parameters are shown in Table 5. In the all-air system, outdoor fresh air *O* is mixed with the return air *R* in the ratio of 1:6 to obtain the mixture at state *M*. The air at state *M* is then cooled and dehumidified by the cooling coil to the state *S*. The supply air *S* then absorbs the sensible and latent heat of the space and returns it as the air at state *R*. In the cooling coil, the chilled water supply and return temperature (CHWS/RT) was assumed at 5 °C and 12 °C, respectively. Based on the total heat load of 1244 W and Equation (6), the supply volume flow rates were computed as 207 m^3/h and 0.179 m^3/h for air and water, respectively.

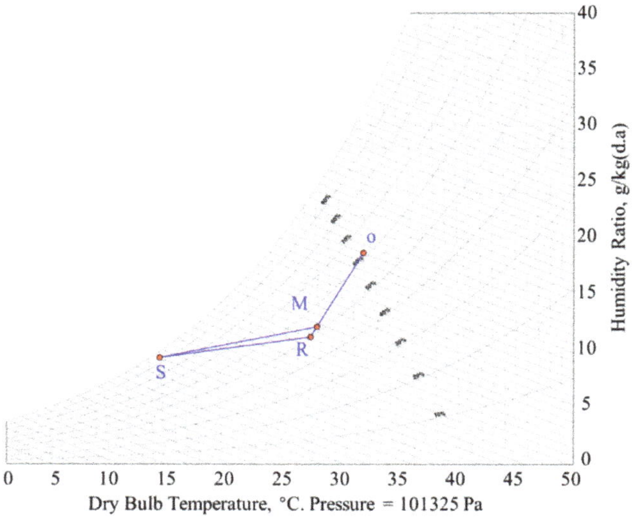

Figure 6. Psychrometric chart for air handling process.

Table 5. Air state parameters of conventional all-air system.

State Point	Dry Bulb Temperature (°C)	Humidity Ratio (g/kg)	Specific Enthalpy (kJ/kg)	Relative Humidity (%)
O	32	18.729	80.111	62
M	27.8	12.3	59.4	52
R	27.2	11.324	56.237	50
S	14	9.502	38.088	95

For the radiant cooling system, the air handling process and corresponding air state parameters were similar to those in Figure 6 and Table 5, respectively. To maintain indoor air quality and remove the moisture load, the fresh air at *O* is cooled and dehumidified by a cooling coil to state *S* and is then transported by a fan to the office space to absorb the latent load and a small portion of the sensible load. Afterwards, the return air *R* is exhausted to the atmosphere. The main sensible load is treated by CCP. The CHWS/RT was assumed to be 15/17 °C and 5/12 °C for the chilled ceiling and air handling unit, respectively. By computation, the total supply volume flow rates were 28.2 m^3/h and 0.4 m^3/h for air and water, respectively.

For the hybrid system, the air state parameters in IEC corresponding to the air handling process in Figure 1b are shown in Table 6. Using the developed model of the test room, the sensible heat removed by CCP was computed as 1200 W. Based on the air state parameters, the volume flow rates were calculated as 248.8 m^3/h and 0.36 m^3/h for air and water, respectively.

Table 6. Air state parameters for the hybrid system.

State Point	Dry Bulb Temperature (°C)	Humidity Ratio (g/kg)	Specific Enthalpy (kJ/kg)	Relative Humidity (%)
A	32	18.7	80.111	63
B	25	18.7	72.89	93.5
C	27	21.2	81.16	93.5

In the case of the hybrid system, ventilation alone is not able to remove the latent heat, especially when the outdoor humidity is high during rainfall. Therefore, a desiccant dehumidification system (DDS) was proposed for latent heat removal and fresh air supply. The psychrometric process for the dehumidification part of DDS is presented in Figure 7 and Table 7. The fresh air at state '*1*' is drawn through the desiccant wheel using a fan. This results in air to be heated and dehumidified to state '*2*' [46]. An air–water heat exchanger, using the chilled water from IEC, is used to sensibly cool the air to state '*3*'. The supply air at state '*3*' removes the latent heat and a small portion of sensible heat. The air state of the room is at state '*R*'. Based on the air state parameters provided in Table 7, the required air and water flowrates of 28.6 and 0.093 m^3/h, respectively, were obtained. For the regeneration, the return air at air state 'R' will be passed to the solar thermal collectors. The air will then be heated up to temperature of 65 °C before passing through the desiccant wheel for regeneration. The total air flowrate required is approximately 28.6 m^3/h. For the operation of CCP and IEC, the total air and water flowrates of 438.8 and 0.5 m^3/h, respectively, were obtained.

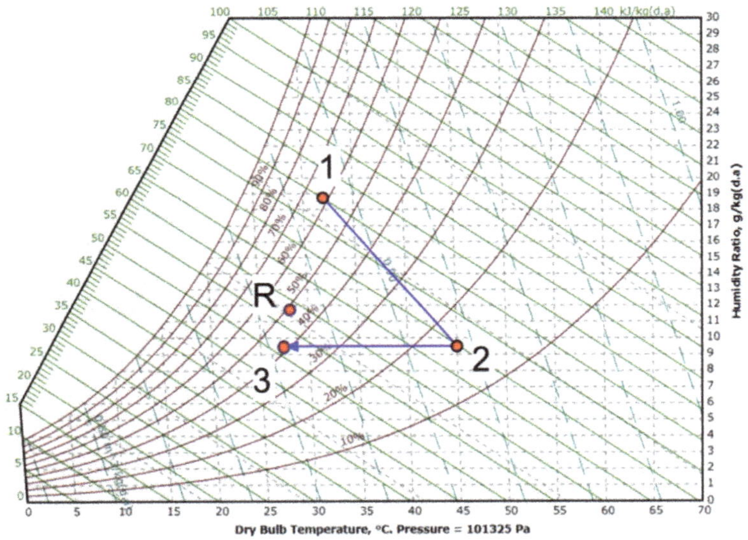

Figure 7. Psychrometric chart for dehumidification in DDS.

Table 7. Air state parameters for dehumidification in DDS.

State Point	Dry Bulb Temperature (°C)	Humidity Ratio (g/kg)	Specific Enthalpy (kJ/kg)	Relative Humidity (%)
1	32	18.729	80.111	62
2	45	9.5	69.730	15.827
3	27.2	9.5	51.585	42
R	27.2	11.324	56.237	50

Energy consumed by the primary equipment is computed with constant indices of performance [19,47,48]: (i) the *COP* of the chiller is 4.39 and 3.31 for evaporating temperatures of 15 °C and 5 °C, respectively; (ii) the pump and fan efficiencies are 0.6; (iii) the pressure drop (ΔP) of the fan is 1400 Pa, 1600 Pa, and 422 Pa, respectively, for conventional all-air, radiant, and hybrid cooling systems; and (iv) the ΔP of the pump is 0.3 bar, 0.5 bar, and 0.89 bar, respectively, for conventional all-air, radiant, and hybrid cooling systems. The fan electricity consumption was calculated assuming that the pressure drops across the airflow path to the desiccant wheel for both the dehumidification and regeneration are equal.

Using Equations (4) and (5), electricity consumptions were obtained and then converted into the equivalent primary energy by assuming a primary energy factor of 3 [49]. The resulting primary and total energy consumptions are illustrated in Figure 8 for the three systems. It can be seen from Figure 8 that, in terms of fan power, the radiant cooling system consumes the least amount of energy, followed by the hybrid and all-air systems, because most of the loads are treated by the chilled water in radiant cooling system. The all-air system consumes the least pump power among the three systems, followed by the radiant and hybrid systems. However, the chiller plant used to supply chilled water, which is the major source of energy consumption in all-air and radiant cooling systems, is replaced by IEC in the proposed hybrid system. As a result, when compared with conventional all-air and radiant cooling systems, total energy savings of 77% and 61%, respectively, can be achieved by the proposed hybrid system. Therefore, it is promising to couple natural ventilation with CCP, IEC, and DDS for thermal comfort and energy saving.

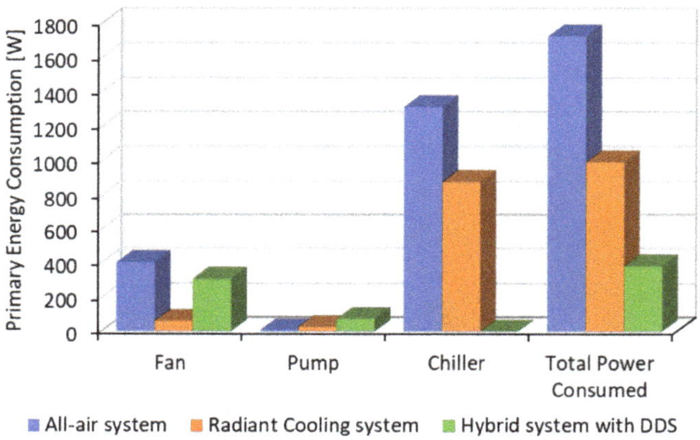

Figure 8. Comparison of primary energy consumption.

6. Conclusions

A hybrid cooling system which combines radiant cooling with natural ventilation was studied. CFD modelling was used to determine the parameters required for thermal comfort and energy analysis. Furthermore, energy consumptions of conventional all-air, radiant cooling, and hybrid systems were calculated and compared under similar thermal comfort. Finally, the results from this study were summarized to draw the following conclusions:

(1) Compared to natural ventilation, cooling capacity and thermal comfort can be significantly enhanced by the proposed hybrid system with minimal energy increase due to the coupled high-temperature radiant cooling system.
(2) While condensation issue can be avoided using the IEC to produce chilled water with a temperature slightly higher than dew point, in hot and humid climate countries such as Singapore, the hybrid system with natural ventilation alone cannot satisfy the thermal comfort requirement of heat-intensive office spaces in all climatic conditions, especially when the air supply from ventilation is low or when the outdoor dew point is higher than 24 °C. For a higher reliability of such a system, a secondary fresh air supply system such as DDS should be installed.
(3) The proposed hybrid system with DDS can save up to 77% and 61% of primary energy in comparison with conventional all-air and radiant cooling systems, respectively.
(4) The hybrid system by coupling natural ventilation with radiant cooling, IEC, and DDS is a feasible alternative to conventional air-conditioning systems for desired

thermal comfort and minimal energy consumption in tropical areas with a hot and humid climate.

As for future work, realistic building designs will be included and an experimental approach to compare these cooling systems in the dynamic cooling load condition will be implemented.

Author Contributions: Conceptualization, X.Z., P.S. and Y.W.W.; methodology: X.Z., P.S. and G.N.; writing, X.Z. and P.S.; review and editing, P.S., S.D., Y.W.W. and S.Q.; software, G.N.; formal analysis, P.S.; supervision, S.D.; project administration, S.D. and S.Q.; funding acquisition, S.D. and S.Q. All authors have read and agreed to the published version of the manuscript.

Funding: This research is supported under RIE 2020 grant (Grant number: I1801E0020).

Data Availability Statement: The data presented in this study are available on request from the corresponding author.

Acknowledgments: This study is supported under the RIE2020 Industry Alignment Fund—Industry Collaboration Projects (IAF-ICP) Funding Initiative, as well as cash and in-kind contribution from Surbana Jurong Pte, Ltd.

Conflicts of Interest: The authors declare no conflict of interest.

Appendix A

Appendix A.1 Macro Simulation

The computational domain and boundary conditions are schematically shown in Figure A1 for the simulation of natural ventilation. As seen from Figure A1a, the office model, located in the center of the air domain, was sized 4 m × 4 m × 2.93 m (length × width × height) with two cross-ventilated openings on each wall. By taking the height of the office model as the reference length scale H_r, the upstream and downstream were both sized 7 H_r while the height was 5.5 H_r for the computational domain, to ensure the airflow was fully developed in the CFD model [33].

The boundary conditions are shown in Figure A1b,c for side and top views, respectively. The inbound vertical wind profile as an inlet boundary condition is described by the logarithmic law and given by Equation (A1) [33]:

$$U(z) = \frac{u^*_{ABL}}{\kappa} \ln\left(\frac{z+z_0}{z_0}\right) \tag{A1}$$

$$k(z) = \frac{u^{*2}_{ABL}}{\sqrt{C_\mu}} \tag{A2}$$

$$\varepsilon(z) = \frac{u^{*3}_{ABL}}{\kappa(z+z_0)} \tag{A3}$$

$$u^*_{ABL} = \frac{U_{ref}\kappa}{\ln\left(\frac{h+z_0}{z_0}\right)} \tag{A4}$$

where $U(z)$ is the velocity at height z, u^*_{ABL} is the atmospheric boundary layer friction velocity, κ is the von Kármán constant (0.42), C_μ is a constant (0.09), z_0 is the aerodynamic roughness length, and U_{ref} is the specified velocity at the reference height of 15 m. The standard k-ε model was employed as the turbulence model. Moreover, grid independence analysis was necessarily carried out to ensure that the obtained results were independent on the grid density. In this regard, the grid convergence index (GCI) for the different cell numbers (1.8 million, 3.2 million, 4.3 million) was evaluated using the Richardson extrapolation method [50]. The average wind speed through the upstream openings was adopted as the quantity of interest. The GCI decreased from 6.8% (1.8 million cells) to 0.7% (3.2 million cells), suggesting that the total cell number of 3.2 million can produce a grid independent solution.

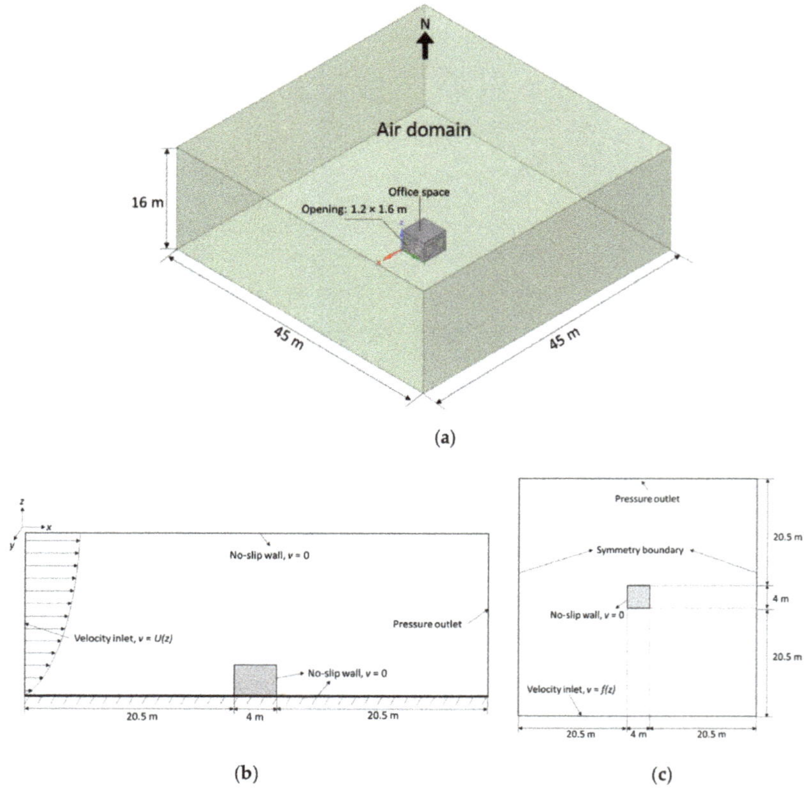

Figure A1. Design conditions for simulation of natural ventilation illustrating various domains: (**a**) Computational domain, (**b**) Boundary conditions: side view, and (**c**) Boundary conditions: top view.

Appendix A.2 Results

The prevailing wind conditions in Singapore are shown in Table A1. According to the relevant standard [33], the simulation can be conducted based on two best prevailing wind directions, which are NE and SE winds as adopted in the present study, to evaluate the performance of natural ventilation. By applying the selected wind profiles to the developed model of natural ventilation, the results are shown in Figures A2 and A3a,b for the velocity contours on the horizontal and vertical planes, respectively. From the results, the average wind velocities through the openings of the office model were 1.5 m/s and 1.3 m/s, respectively, for SE and NE winds. To evaluate the thermal comfort of the office space with the hybrid cooling system, the lower average wind velocity of 1.3 m/s was adopted for the micro simulation.

Table A1. Prevailing wind conditions in Singapore [33].

Wind Direction	Reference Velocity (U_{ref})
North	2.0
North-East (NE)	2.9
South	2.8
South-East (SE)	3.2

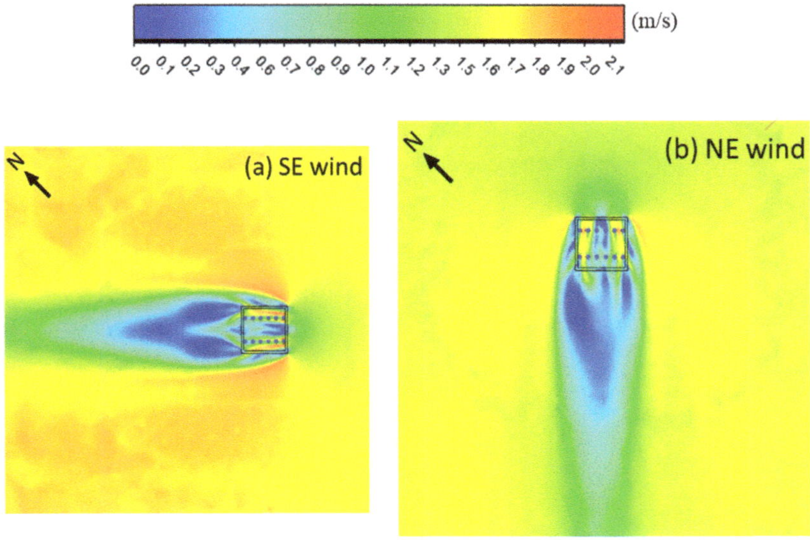

Figure A2. Velocity contours on the horizontal (x–y) plane: z = 1.1 m for (**a**) south-east wind, (**b**) north-east wind.

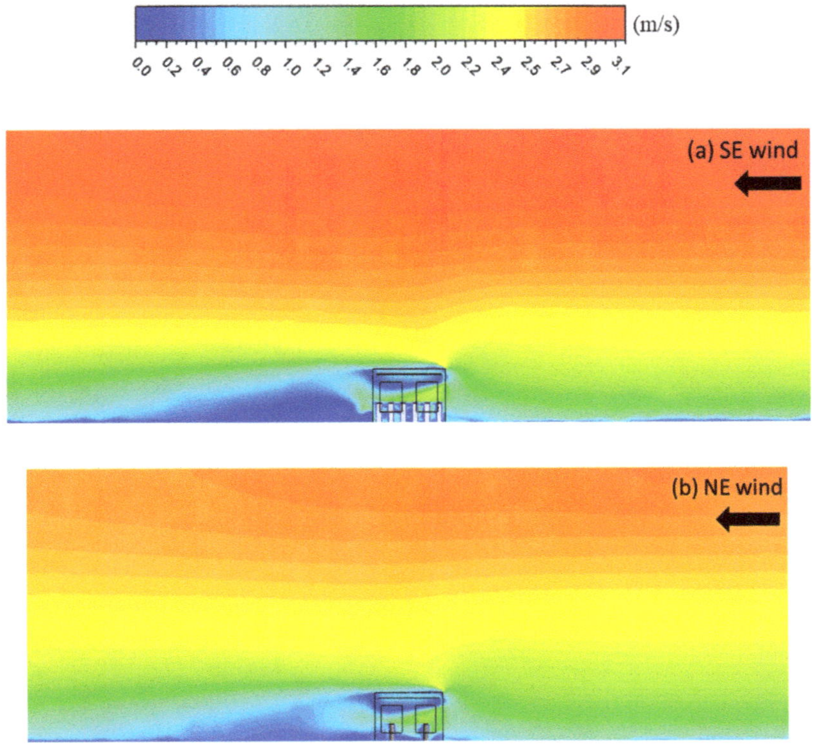

Figure A3. Velocity contours on the vertical plane through middle of window opening for (**a**) south-east wind and (**b**) north-east wind.

Appendix A.3 Model Validation

In this section, comparisons between simulation and experimental results are made to verify the macro and micro models separately.

Appendix A.4 Macro Simulation

In order to verify the standard *k*-ε turbulence model of natural ventilation, our geometry was reconfigured to replicate the cross-ventilation model in the experimental study of Ohba [51], as shown in Figure A4. Assuming the model height as the reference length scale H_r, the upstream and downstream were both sized as 7 H_r while the height was 5.5 H_r for the computational domain. Based on the measurement, the wind profile as the inlet velocity was described in the form of the power law as shown in Equation (A5) [52].

$$U(z) = 7\left(\frac{z}{0.15}\right)^{0.25} \quad \text{(A5)}$$

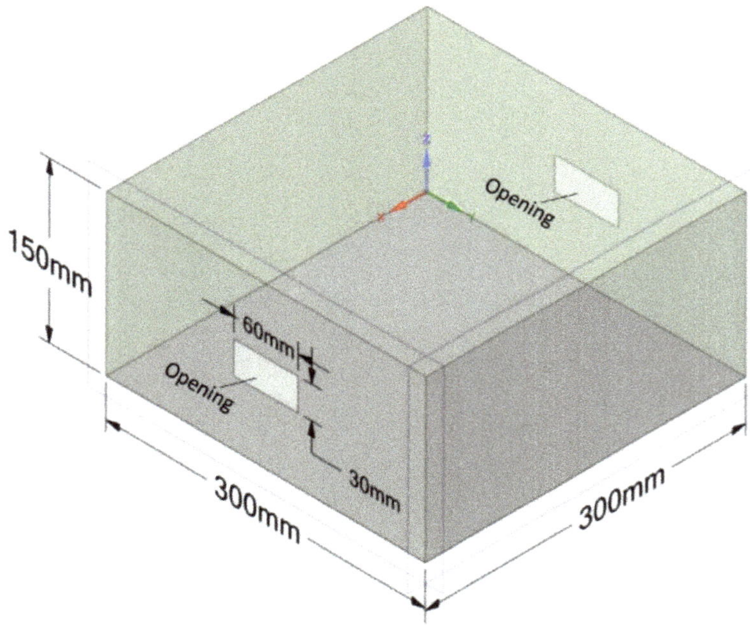

Figure A4. Cross-ventilation model [51].

Figure A5 compares the simulated and measured air velocity through the opening in various wind directions. To quantitively evaluate the model accuracy, the cumulative variation of root mean square error (*CVRMSE*) was introduced and given by Equation (A6) [53].

$$CVRMSE = \frac{\sqrt{\sum_{i=1}^{j}\left(\left(E_i - S_i\right)^2/j\right)}}{\sum_{i=1}^{j} E_i/j} \quad \text{(A6)}$$

where E_i and S_i are the *i*th experimental and simulated results, respectively. Using the equation, the calculated *CVRMSE* between the simulated and measured air velocity was equal to 1.3%, which is acceptably within engineering tolerance. Therefore, the developed model proved sufficiently accurate in assessing the natural ventilation performance.

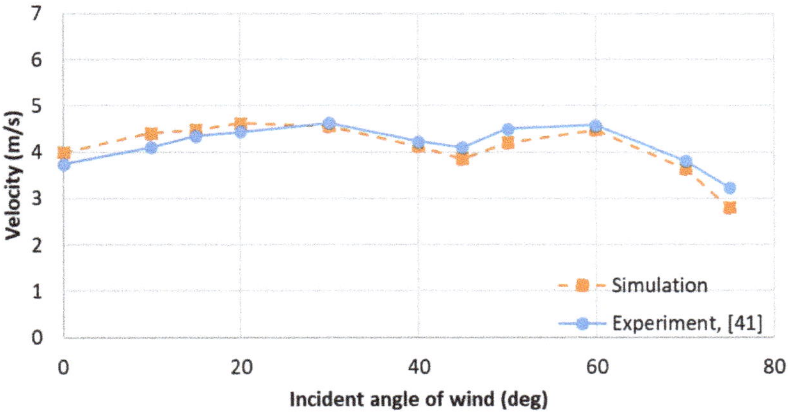

Figure A5. Comparison of simulated and experimental results for air velocity through the opening.

Appendix A.5 Micro Simulation

To validate the test room model, the simulation results were compared accordingly with the mockup test conducted by Shin [32], as the simulation model and mockup test are constructed based on the same testing standard of chilled ceilings [31]. For model validation under various load conditions, two cases of experimental design are shown in Table A2 by adjusting the cooling load and supply water temperature.

The results are shown in Figure 6a,b for the two cases, by comparing the simulated and measured air temperature and cooling capacity. Using Equation (12), the *CVRMSE* for the air temperature was achieved as 2.55% and 2.08% for Cases 1 and 2, respectively. For the grid independence analysis, a poly-hexcore mesh was prepared using ANSYS Fluent meshing, where 797.759 cells were generated and the value of 0.5 was achieved for the minimum orthogonal quality. Moreover, the difference of the cooling capacity between the simulation and experiment was limited, indicating that good agreement was achieved between the proposed CFD model and the published experimental results.

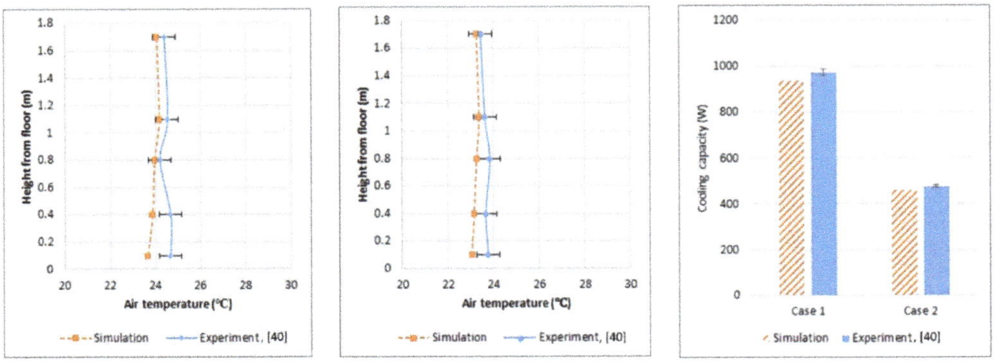

(**a**) Case 1. (**b**) Case 2. (**c**) Cooling capacity.

Figure 6. Comparison of simulated and experimental results for air temperature and cooling capacity.

Table A2. Case design of experiments for model validation [32].

No.	Case 1	Case 2
Heating capacity of dummy, W	939.8	469.9
Cooling water flow rate, lpm	7	7
Cooling water flow rate, g/s	117	117
Supply water temperature, °C	15	17

References

1. Oh, S.J.; Ng, K.C.; Thu, K.; Chun, W.; Chua, K.J.E. Forecasting long-term electricity demand for cooling of Singapore's buildings incorporating an innovative air-conditioning technology. *Energy Build.* **2016**, *127*, 183–193. [CrossRef]
2. Emmerich, S.; Dols, W.; Axley, J. *Natural Ventilation Review and Plan for Design and Analysis Tools*; NIST Interagency/Internal Report (NISTIR); National Institute of Standards and Technology: Gaithersburg, MD, USA, 2001. Available online: https://tsapps.nist.gov/publication/get_pdf.cfm?pub_id=860854 (accessed on 8 August 2021).
3. Omrani, S.; Garcia-Hansen, V.; Capra, B.; Drogemuller, R. Natural ventilation in multi-storey buildings: Design process and review of evaluation tools. *Build. Environ.* **2017**, *116*, 182–194. [CrossRef]
4. Gil-Baez, M.; Barrios-Padura, Á.; Molina-Huelva, M.; Chacartegui, R. Natural ventilation systems to enhance sustainability in buildings: A review towards zero energy buildings in schools. *E3S Web Conf.* **2017**, *22*. [CrossRef]
5. Escombe, A.R.; Oeser, C.C.; Gilman, R.H.; Navincopa, M.; Ticona, E.; Pan, W.; Martínez, C.; Chacaltana, J.; Rodríguez, R.; Moore, D.A.J.; et al. Natural Ventilation for the Prevention of Airborne Contagion. *PLoS Med.* **2007**, *4*, e68. [CrossRef]
6. Etheridge, D. *Natural Ventilation of Buildings: Theory, Measurement and Design*; John Wiley & Sons: Hoboken, NJ, USA, 2011.
7. Khanal, R.; Lei, C. Solar chimney—A passive strategy for natural ventilation. *Energy Build.* **2011**, *43*, 1811–1819. [CrossRef]
8. Nomura, M.; Hiyama, K. A review: Natural ventilation performance of office buildings in Japan. *Renew. Sustain. Energy Rev.* **2017**, *74*, 746–754. [CrossRef]
9. Mochida, A.; Yoshino, H.; Takeda, T.; Kakegawa, T.; Miyauchi, S. Methods for controlling airflow in and around a building under cross-ventilation to improve indoor thermal comfort. *J. Wind Eng. Ind. Aerodyn.* **2005**, *93*, 437–449. [CrossRef]
10. Karava, P.; Stathopoulos, T.; Athienitis, A.K. Wind-induced natural ventilation analysis. *Sol. Energy* **2007**, *81*, 20–30. [CrossRef]
11. Aldawoud, A.; Clark, R. Comparative analysis of energy performance between courtyard and atrium in buildings. *Energy Build.* **2008**, *40*, 209–214. [CrossRef]
12. Khan, N.; Su, Y.; Riffat, S.B. A review on wind driven ventilation techniques. *Energy Build.* **2008**, *40*, 1586–1604. [CrossRef]
13. Cheng, X.; Shi, L.; Dai, P.; Zhang, G.; Yang, H.; Li, J. Study on optimizing design of solar chimney for natural ventilation and smoke exhaustion. *Energy Build.* **2018**, *170*, 145–156. [CrossRef]
14. Lei, Y.; Zhang, Y.; Wang, F.; Wang, X. Enhancement of natural ventilation of a novel roof solar chimney with perforated absorber plate for building energy conservation. *Appl. Therm. Eng.* **2016**, *107*, 653–661. [CrossRef]
15. Deuble, M.P.; de Dear, R.J. Mixed-mode buildings: A double standard in occupants' comfort expectations. *Build. Environ.* **2012**, *54*, 53–60. [CrossRef]
16. ASHRAE Handbook. *HVAC SYSTEMS and Equipment*; American Society of Heating, Refrigerating and Air-Conditioning Engineers, Inc.: Atlanta, GA, USA, 1996; Volume 39.
17. Rhee, K.N.; Olesen, B.W.; Kim, K.W. Ten questions about radiant heating and cooling systems. *Build. Environ.* **2017**, *112*, 367–381. [CrossRef]
18. Kim, M.K.; Leibundgut, H. A case study on feasible performance of a system combining an airbox convector with a radiant panel for tropical climates. *Build. Environ.* **2014**, *82*, 687–692. [CrossRef]
19. Jiang, Y.; Xie, X. Theoretical and testing performance of an innovative indirect evaporative chiller. *Sol. Energy* **2010**, *84*, 2041–2055. [CrossRef]
20. Kim, M.K.; Liu, J.; Cao, S.-J. Energy analysis of a hybrid radiant cooling system under hot and humid climates: A case study at Shanghai in China. *Build. Environ.* **2018**, *137*, 208–214. [CrossRef]
21. Lipczynska, A.; Kaczmarczyk, J.; Melikov, A.K. Thermal environment and air quality in office with personalized ventilation combined with chilled ceiling. *Build. Environ.* **2015**, *92*, 603–614. [CrossRef]
22. Al Assaad, D.; Ghali, K.; Ghaddar, N. Effectiveness of intermittent personalized ventilation assisting a chilled ceiling for enhanced thermal comfort and acceptable indoor air quality. *Build. Environ.* **2018**, *144*, 9–22. [CrossRef]
23. Fernández-Hernández, F.; Fernández-Gutiérrez, A.; Martínez-Almansa, J.J.; del Pino, C.; Parras, L. Flow patterns and heat transfer coefficients using a rotational diffuser coupled with a radiant floor cooling. *Appl. Therm. Eng.* **2020**, *168*, 114827. [CrossRef]
24. Hao, X.; Zhang, G.; Chen, Y.; Zou, S.; Moschandreas, D.J. A combined system of chilled ceiling, displacement ventilation and desiccant dehumidification. *Build. Environ.* **2007**, *42*, 3298–3308. [CrossRef]
25. Rees, S.J.; Haves, P. An experimental study of air flow and temperature distribution in a room with displacement ventilation and a chilled ceiling. *Build. Environ.* **2013**, *59*, 358–368. [CrossRef]
26. Hernández, F.F.; López, J.M.C.; Gutiérrez, A.F.; Muñoz, F.D. A new terminal unit combining a radiant floor with an underfloor air system: Experimentation and numerical model. *Energy Build.* **2016**, *133*, 70–78. [CrossRef]

27. Yu, T.; Heiselberg, P.; Lei, B.; Pomianowski, M.; Zhang, C.; Jensen, R. Experimental investigation of cooling performance of a novel HVAC system combining natural ventilation with diffuse ceiling inlet and TABS. *Energy Build.* **2015**, *105*, 165–177. [CrossRef]
28. Yu, T.; Heiselberg, P.; Lei, B.; Zhang, C.; Pomianowski, M.; Jensen, R.L. Experimental study on the dynamic performance of a novel system combining natural ventilation with diffuse ceiling inlet and TABS. *Appl. Energy* **2016**, *169*, 218–229. [CrossRef]
29. Meggers, F.; Guo, H.; Teitelbaum, E.; Aschwanden, G.; Read, J.; Houchois, N.; Pantelic, J.; Calabrò, E. The Thermoheliodome—'Air conditioning' without conditioning the air, using radiant cooling and indirect evaporation. *Energy Build.* **2017**, *157*, 11–19. [CrossRef]
30. Vangtook, P.; Chirarattananon, S. Application of radiant cooling as a passive cooling option in hot humid climate. *Build. Environ.* **2007**, *42*, 543–556. [CrossRef]
31. *British Standard—Ventilation for Buildings—Chilled Ceilings—Testing and Rating*; BS EN 14240; British Standards Institution: London, UK, 2004.
32. Shin, M.S.; Rhee, K.N.; Park, S.H.; Yeo, M.S.; Kim, K.W. Enhancement of cooling capacity through open-type installation of cooling radiant ceiling panel systems. *Build. Environ.* **2019**, *148*, 417–432. [CrossRef]
33. Building and Construction Authority (BCA). Green Mark for New Non-Residential Buildings: Technical Guide and Requirements. 2021. Available online: https://www1.bca.gov.sg/buildsg/sustainability/green-mark-certification-scheme (accessed on 2 September 2021).
34. Janssen, J. Ventilation for acceptable indoor air quality. *ASHRAE J.* **1989**, *31*, 40–42.
35. *ASHRAE Standard 55-2004: Thermal Environmental Conditions for Human Occupancy*; American Society of Heating, Refrigerating and Air-Conditioning Engineers, Inc.: Atlanta, GA, USA, 2004.
36. *Singapore Standard: Code of Practice for Air-Conditioning and Mechanical Ventilation in Buildings*; SS 553: 2009; Enterprise: Singapore, 2009.
37. Yang, L.; Yan, H.; Lam, J.C. Thermal comfort and building energy consumption implications—A review. *Appl. Energy* **2014**, *115*, 164–173. [CrossRef]
38. Fanger, P.O. *Thermal Comfort: Analysis and Applications in Environmental Engineering*; McGraw-Hill: New York, NY, USA, 1970.
39. Catalina, T.; Virgone, J.; Kuznik, F. Evaluation of thermal comfort using combined CFD and experimentation study in a test room equipped with a cooling ceiling. *Build. Environ.* **2009**, *44*, 1740–1750. [CrossRef]
40. *ASHRAE Handbook: Fundamentals*; American Society of Heating, Refrigerating and Air-Conditioning Engineers, Inc.: Atlanta, GA, USA, 2009.
41. Fanger, P.O. *Thermal Comfort*; Robert E. Krieger Publishing Company: Malabar, FL, USA, 1982.
42. National University of Singapore, Singapore, Geography Weather Station. 2021. Available online: https://inetapps.nus.sg/fas/geog/ (accessed on 22 September 2021).
43. *ASHRAE Energy Standard for Buildings Except Low-Rise Residential Buildings*; American Society of Heating, Refrigerating and Air-Conditioning Engineers, Inc.: Atlanta, GA, USA, 2004; Volume 90.
44. Narayanan, R.; Saman, W.Y.; White, S.D.; Goldsworthy, M. Comparative study of different desiccant wheel designs. *Appl. Therm. Eng.* **2011**, *31*, 1613–1620. [CrossRef]
45. Katejanekarn, T.; Chirarattananon, S.; Kumar, S. An experimental study of a solar-regenerated liquid desiccant ventilation pre-conditioning system. *Sol. Energy* **2009**, *83*, 920–933. [CrossRef]
46. Xu, M.M.; Li, H. *Solar Desiccant Cooling System BT—Handbook of Energy Systems in Green Buildings*; Wang, R., Zhai, X., Eds.; Springer: Berlin/Heidelberg, Germany, 2018; pp. 301–323.
47. Niu, J.L.; Zhang, L.Z.; Zuo, H.G. Energy savings potential of chilled-ceiling combined with desiccant cooling in hot and humid climates. *Energy Build.* **2002**, *34*, 487–495. [CrossRef]
48. Porumb, B.; Bălan, M.; Porumb, R. Potential of Indirect Evaporative Cooling to Reduce the Energy Consumption in Fresh Air Conditioning Applications. *Energy Procedia* **2016**, *85*, 433–441. [CrossRef]
49. Coz, T.D.; Kitanovski, A.; Poredos, A. Primary energy factor of a district cooling system/Faktor primarne energije sistema daljinskega hlajenja. *Stroj. Vestn. J. Mech. Eng.* **2016**, *62*, 717–731. Available online: https://link.gale.com/apps/doc/A474041498/AONE?u=tacoma_comm&sid=googleScholar&xid=45003155 (accessed on 17 November 2021). [CrossRef]
50. Celik, I.B.; Ghia, U.; Roache, P.J.; Freitas, C.J. Procedure for Estimation and Reporting of Uncertainty Due to Discretization in CFD Applications. *J. Fluids Eng.* **2018**, *130*, 078001.
51. Ohba, M.; Irie, K.; Kurabuchi, T. Study on airflow characteristics inside and outside a cross-ventilation model, and ventilation flow rates using wind tunnel experiments. *J. Wind Eng. Ind. Aerodyn.* **2001**, *89*, 1513–1524. [CrossRef]
52. Cheung, J.O.P.; Liu, C.-H. CFD simulations of natural ventilation behaviour in high-rise buildings in regular and staggered arrangements at various spacings. *Energy Build.* **2011**, *43*, 1149–1158. [CrossRef]
53. Measurement of energy, demand, and water savings. In *ASHRAE Guideline14-2014*; American Society of Heating, Refrigerating and Air-Conditioning Engineers, Inc.: Atlanta, GA, USA, 2014.

energies

Article

Improvement of Properties of an Insulated Wall for Refrigerated Trailer-Numerical and Experimental Study

Konrad Zdun * and Tadeusz Uhl

Faculty of Mechanical Engineering and Robotics, University of Science and Technology AGH, 30-059 Krakow, Poland; tuhl@agh.edu.pl
* Correspondence: zdun@agh.edu.pl

Abstract: In the paper, we report our research on the improvement of thermal efficiency of refrigerated trailers by modification of their wall structure by placing a layer of phase change material inside them. The research was carried out in the field of transport, meeting the requirements of all classes provided for in the ATP agreement for refrigerated trailers. As part of the research, we formulated a numerical model of the proposed design of the refrigerator walls, which was subsequently validated by comparing the modeling results with the results of experimental tests carried out on a test bench designed specifically for this purpose. Based on the validated simulation conditions, we formulated the numerical model of a full-scale refrigerated semi-trailer, which was numerically tested under the conditions specified in the ATP Agreement. The results proved that adding a 6 mm layer of the SP-24 phase change material in each of the walls of the cold store allows the temperature inside the trailer to be kept below −20 °C for a period of 24 h without the need to supply cold from the outside during operation. The passive refrigerated semi-trailer system implemented in this manner with 6 mm PCM layer allows for a reduction in primary energy consumption by up to 86% in a period of 22 h. The mentioned percentage did not take into account the efficiency of the cooling system of the phase change material.

Keywords: phase change materials (PCMs); refrigerated trailer; cold storage; latent heat storage

Citation: Zdun, K.; Uhl, T. Improvement of Properties of an Insulated Wall for Refrigerated Trailer-Numerical and Experimental Study. *Energies* **2022**, *15*, 51. https://doi.org/10.3390/en15010051

Academic Editors: Wei-Hsin Chen, Aristotle T. Ubando, Chih-Che Chueh and Liwen Jin

Received: 14 October 2021
Accepted: 16 December 2021
Published: 22 December 2021

1. Introduction

Low-temperature cargo transports is an important link of the cold supply chain. The rapid development of mankind requires the transport of more and more products, especially food and medicines, which are subject to strictly defined temperature conditions. This, on a growing scale, is carried out with the use of refrigerated trailers, each of which have an individual refrigeration unit driven by low-efficiency diesel engines. It is estimated that worldwide transport of goods in low temperatures is responsible for some 14% of the total amount of CO_2 emitted [1], and diesel-driven refrigeration systems are responsible for up to 40% of these emissions [2].

The scale of the issue of emissions connected to the cold chain is so substantial that it has attracted the attention of numerous groups of researchers, who have taken steps to reduce the environmental footprint of this industry [3–8]. The replacing of diesel units by the use of central, highly efficient refrigeration units in cooperation with cold thermal energy storage in semi-trailers is particularly promising and is currently enjoying great interest among technologies that seem promising in this respect [3,5,6]. This technology is in line with the general trend of searching for solutions reducing greenhouse gas emissions from industry and energy plants with the use of thermal energy storage [9].

The aforementioned cold thermal energy storage can be implemented in cold storage in two ways [10]: (a) they can be integrated with the walls to improve their insulation [4] or (b) in the form of the cooling source for the cooling system [3]. It is also possible to combine both solutions, where the cold storage is implemented as a layer in the cold store wall [5,6]. The solution proposed by the authors of the present research provides for the use of a combination of both solutions.

What is particularly interesting among the available cold thermal energy storage (CTES) technologies is the energy storage using latent heat in phase change materials (PCMs) [11]. Phase change materials are characterized by their ability to accumulate significant amounts of thermal energy in the process of phase change, i.e., during an isothermal thermodynamic process caused by a change in temperature or pressure. The phase transformation most often consists of a change of the aggregate state or a polymorphic transformation. Currently, many research teams around the world are working on the use of phase change materials to reduce primary energy consumption and reduce CO_2 emissions to the atmosphere [12,13]. Owing to the use of latent heat, it is possible to obtain high energy storage density, both per unit of mass and volume [14]. This aspect is particularly significant when applied in refrigerated vehicles because one of the key aspects of transport is reducing the weight of the trailer—the increased weight would negatively affect the amount of fuel consumed and the overall energy efficiency of the solution, as well as reduce the load capacity of the vehicle [15]. What is also a crucial feature of phase change materials is also the fact that the phase transition takes place at a specific temperature, thanks to which such material enables the transfer or absorption of thermal energy under constant temperature conditions, thus allowing passive cooling [14].

When we consider the applicability of phase change materials to improve the efficiency of refrigerated transport, it should be stressed that for most materials, phase change can be considered at two different points, related to different phenomena: their phase transformation from solid to liquid and from liquid to gas. Despite a much larger change in enthalpy during the liquid–gas transformation, from the practical point of view of the use of PCM for storing cold in the walls of refrigerated trailers, the particularly interesting type of phase transformation, in the context of thermal energy storage, is the solid–liquid transformation, which is not connected with significant volume changes, and thus an increase in pressure, which is characteristic for the liquid–gas transformations [14].

The research carried out so far on the effectiveness of the use of phase change materials to improve the energy efficiency of refrigerated transport was characterized by different approaches to implement cold thermal energy storage and usage to ensure the climatic conditions required during transport.

One of the first research works carried out in this area was the work of the team of Liu et al. [3]. They developed a cooling system based on a central tank with a PCM bed, from which the cold was distributed by means of forced internal air circulation inside the trailer. This solution allowed for the elimination of the requirement to use a refrigeration unit installed on the semi-trailer, but it cannot be considered a fully passive system due to the forced air circulation. Research demonstrated that this method is justified, as it allows maintenance of the required temperature inside the trailer for the required time, and the weight of the PCM tank was comparable to that of a refrigeration unit. The application of the described solution reduced the energy expenditure to maintain the temperature by 50%.

Similar work was carried out by the team of Mousazade et al. [16], who placed the tanks filled with the phase change material in 6 tons refrigerated trailer without additional, forced, internal air circulation. The tests were carried out by them during the actual transport (vehicle movement). The cooling panels used by them allowed to maintain the required temperature inside the trailer for up to 4 h 46 min while the vehicle is in motion and up to 5 h 7 min for the stationary truck. Such approach can also be applied to containers intended for rail transport, as demonstrated by Tong et al. [17].

Another way to implement the phase change material in the construction of a refrigerated semi-trailer is to locate it in walls. The walls of typically designed refrigerated semi-trailers or refrigerated containers are made of a sandwich-structure panels, filled with polyurethane foam, which is characterized by a very low thermal conductivity of the range of 0.023 to 0.026 $Wm^{-1}K^{-1}$ [18], allowing, depending on the thickness of this partition, to achieve the value of the insulation parameter K, in the range of 0.22 to 0.27 $Wm^{-2}K^{-1}$.

The phase change material can be placed in the walls of refrigerated trailers in several different manners. The first is to place PCM in the wall by using a composite made of polyurethane foam and PCM itself. Research in this area was carried out by Michel et al. [4], who created two composites dedicated for the application of insulating the walls of refrigerated vehicles, differing in their content of the phase change material. The walls manufactured on the basis of these composites were tested for their thermal conductivity. The results obtained by the team proved the potential for doubling the thermal insulation of the walls that are currently used in refrigerated trailers.

A different approach to integrating PCM into the walls of refrigerated trailers was presented by Fioretti et al. [5], who located the phase change material in the wall in the form of a layer of encapsulated material. This solution was dedicated to refrigerated containers, however, in this case, the method of storing the cold thermal energy can be analogically used in refrigerated trailers. It was based on placing a layer of phase change material enclosed in a specially prepared structure on the outer side of the container. This was aimed at having the thermal energy penetrating inside the container through the material layer absorbed. To ensure that the internal temperature was maintained, the team used a phase change material with a phase transition temperature of 35 °C. The advantage of this solution is in its versatility—it can be used to maintain any temperature inside a trailer or container, albeit it has one major disadvantage, which is the lack of effectiveness during these days when the temperature does not exceed 35 °C and there is no direct insolation.

Huang and Piontek [6], who, similarly to Fioretti [5], applied layers of encapsulated phase change material, but placed it inside of the container, present a very similar approach. In order to maintain the temperature at an appropriate level, they used a phase change material with a phase transition temperature of 5 °C. The team was able to demonstrate that by placing a layer of frozen phase change material on the internal side of the wall of a trailer or a container, it is possible to maintain the required temperature for up to 80 h, which allows for the goods to be transported at low temperatures without the need to install refrigeration units on the trailer.

The research carried out for the purposes of this paper uses the concept of building a cold storage wall, similar to presented by Huang and Piontek [6], but its objective is to verify the effectiveness of using this form of improving the energy efficiency of refrigerated transport, in the case of temperatures below 0 °C. The ATP agreement [19], which defines the conditions to be met by vehicles with regulated internal temperature, depending on the method of implementing temperature and transport temperature regulation, was what motivated us to conduct the present research. According to the ATP agreement adopted in 1970, and ratified by most countries of the world, four types of temperature-controlled means of transport are distinguished:

- Thermally insulated means of transport, for which there are minimum requirements for the wall heat transfer coefficient of $K = 0.7\ Wm^{-2}K^{-1}$ for measures with ordinary thermal insulation and $K = 0.4\ Wm^{-2}K^{-1}$ for measures with reinforced thermal insulation;
- Means of transport with a non-mechanical refrigeration system (ice room) that can be cooled by means of ice, eutectic plates, or dry ice. In this category, the agreement distinguishes four temperature classes, depending on the maximum permissible temperature inside the trailer: Class A—up to +7 °C, class B—up to −10 °C, class C—up to −20 °C, and class D—up to 0 °C. The agreement does not specify the minimum time for which the required temperature is to be maintained for this type of means of transport, but it can be assumed that it is meant to be maintained throughout the duration of the transport;
- Means of transport with a mechanical refrigeration device (cold store)—A thermally insulated means of transport, equipped with a refrigeration device that allows for lowering and maintaining the temperature inside such means of transport. The ATP agreement distinguishes three classes of this type of transport, depending on the temperature range that can be maintained in a given means of transport: Class A—temperature range of +12 °C to 0 °C, Class B—temperature range of +12 °C to

−10 °C, Class C—temperature range of +12 °C to −20 °C, and three classes depending on the ability to keep the maximum temperature below a certain level: class D—up to 0 °C, class E—up to −10 °C, and class F—up to −20 °C;

- Heated means of transport—means of transport equipped with a heating device that allows the temperature inside such means of transport to be maintained at a minimum level of 12 °C for a minimum of 12 h. This transport was also divided into classes, depending on the outside temperature. There are 4 classes in total: class A—heating at an outside temperature of −10 °C, class B—heating at an outside temperature of −20 °C, class C—heating at an outside temperature of −30 °C, and class D—heating at an outside temperature of −40 °C.

The objective of the project is to build a refrigerator wall, using cold thermal energy storage in the form of a phase change material layer that is charged with cold thermal energy from a central cold production unit, dedicated to a vehicle that meets the requirements of all four classes for refrigerated means of transport with a non-mechanical system, as specified in the ATP agreement. This action will allow for the verification of the possibility of using cold storage based on phase change materials also in the transport of deep-frozen goods, which has not been subjected to scientific research so far.

The authors of the research, striving to create a solution adequate for use in the case of transporting materials in all classes defined in the ATP agreement for non-mechanically cooled means of transport, i.e., for transporting goods at a maximum temperature of −20 °C, propose to combine the traditional wall of a refrigerated semi-trailer made of polyurethane foam with a layer of phase change material with a phase transition temperature equal to −24 °C.

The paper contains five chapters and a list of the quoted bibliography. The first chapter presents a review of the subject literature, including typical solutions for cold store walls and the possibility of applying phase change materials, storing cold for the construction of a cold store wall. The second chapter contains the elaborated numerical models to simulate the properties of walls containing cold thermal energy storage phase change materials. The third chapter presents the results of laboratory tests carried out to validate the developed models, while the fourth chapter presents the concept of a refrigerated container with walls filled with phase change materials along with the prediction of its properties. The fifth chapter contains conclusions and final remarks on the feasibility of designing a cold store without a refrigerating unit.

2. Modeling and Simulation of Wall Properties

2.1. *Refrigerated Trailer Wall Concept*

The proposed structure is based on a fragment of a conventional refrigerated trailer wall, made in the form of a sandwich panel made of two steel sheets, with 86 mm thick filling, made of polyurethane foam, with a thermal conductivity coefficient of $\lambda = 0.025$ $Wm^{-1}K^{-1}$. The macro-encapsulated phase change material was added to the inside of the wall. A structure made of cellular polycarbonate was used to contain the phase change material. The developed structure of the tested walls is presented in Figure 1a. Figure 1b shows the theoretical heat transfer through the proposed wall in the state when the phase change material remains frozen.

The walls were made in two variants of the layer thickness of the phase change material. The first variant utilizes a 6 mm thick PCM layer, while the second one—a 10 mm. Both tested walls were sized 1000 mm × 1000 mm and had a thickness of 92 mm to 96 mm, depending on the variant of a wall design.

The use of cellular polycarbonate for encapsulation purposes a reduced volume of phase change material accumulated in the layers. In Variant I, 153 chambers with a capacity of 32 mL each were available, resulting in a total PCM volume of 4896 mL, weighing 6.36 kg. In Variant II, 93 chambers were filled with 93 mL each. The total volume and weight of PCM in this variant was 8649 mL and 11.24 kg.

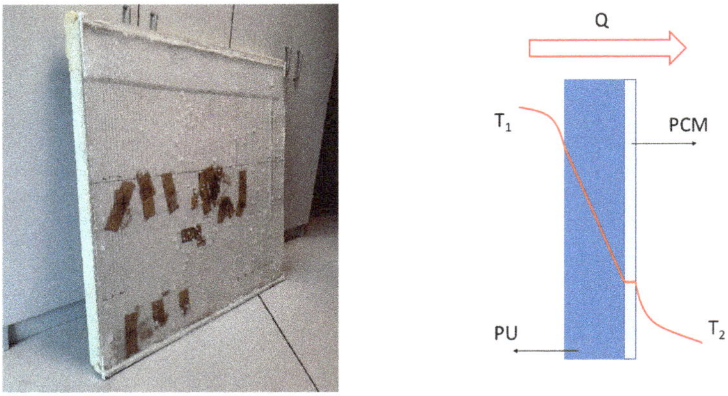

(**a**) (**b**)

Figure 1. (**a**) The tested wall with a layer of phase change material; (**b**) Heat flow through the proposed wall structure.

2.2. *Phase Change Material*

The selection of the phase change material for the tests was a significant aspect of the wall structure. The main assumption in the material selection was its ability to maintain the required temperature inside the chamber and/or the trailer at the maximum level of −20 °C, so the phase change material had to have a phase change temperature below this point. From the practical point of view, this temperature must also not be too low, due to the need to cool it down to very low temperatures, which could occur more expensive and more energy-consuming.

After the analysis of the phase change materials available on the market, the SP-24 phase change material [20] was used for the construction of the walls. This material is an inorganic material, characterized by its latent heat of fusion equal to 220 kJ/kg. The enthalpy distribution in the phase change of this material demonstrates its greatest distribution at −24 °C at its solidification point and −23 °C and −22 °C at its melting point, respectively. The distribution of enthalpy is presented in Figure 2. The remaining material data are summarized in Table 1.

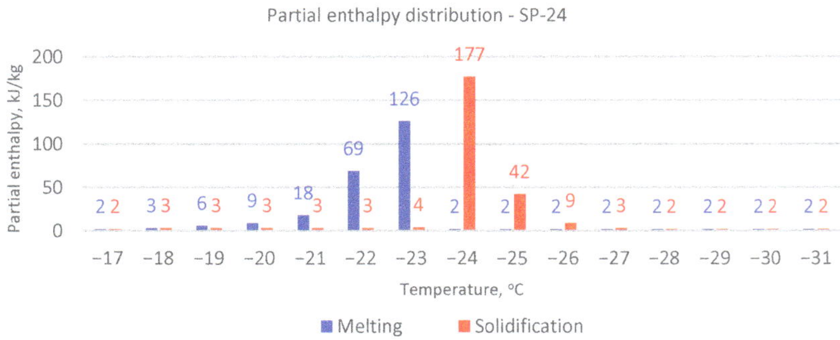

Figure 2. Distribution of the partial enthalpy of the SP-24 phase change material applied. Adapted from [20].

Table 1. Physical properties of the applied Rubiterm SP-24 phase change material. Adapted from [20].

Parameter	Value
Melting point [°C]	−23 to −24
Latent heat of fusion [J/kg]	220
Specific heat J/(kgK)	2
Thermal conductivity [$Wm^{-1}K^{-1}$]	0.6
Density [kg/m^3]	1200 (solid)/1300 (liq)

2.3. Cold Room Wall Model

In order to demonstrate the benefits of using the proposed wall structure in refrigerated semi-trailers, a computer model was developed on which a numerical simulation was carried out under conditions consistent with the conditions of the conducted experimental tests. Simulation covered both: proposed, novel wall and test stand. Numerical tests were carried out with the use of COMSOL 5.3 software [21]. Similar to the experimental tests, the modeling was executed in three variants—without the PCM layer, with a 6 mm layer, and a 10 mm layer. The results obtained in numerical tests were validated based on the results of the experiment.

The model mapped the geometry of the test stand, which consisted of a test chamber with dimensions of 1.8 m × 1.8 m × 1.6 m, containing five walls of mineral wool 40 cm thick each and one wall, the structure of which was consistent with the proposed innovative wall (polyurethane and PCM layer). There was an unventilated volume of air in the center of the test chamber. Table 2 presents the thermophysical properties of the materials used in the simulation.

Table 2. Thermophysical properties of materials used in the simulation.

Material	Melting Point [°C]	Thermal Conductivity [$Wm^{-1}K^{-1}$]	Specific Heat [J/kgK]	Latent Heat of Fusion [J/kg]	Density [kg/m^3]
Steel [22]	-	50	475	-	2700
Mineral wool [22]	-	0.035	1300	-	40
Polyurethane foam [22]	-	0.025	1460	-	40
Polycarbonate [22]	-	0.20	1200	-	1200
PCM–SP-24 [20]	−24	0.6	2000	2200	1200 (solid)/1300 (liq)

The geometry was discretized using the built-in meshing tool. Tetrahedral mesh was used in all domains and the mesh in the PCM domain was additionally densified. In addition, in the air and PCM domains, the boundary layer was densified with 8 layers and a growth rate of 1.2. The thickness of the first layer was adjusted automatically. The Figure 3 below shows the mesh view, and the mesh statistic values, with a skewness quality measure with minimum skewness of 0.038.

The heat transfer in the fluids module was used in the simulation. The simulation was performed in two steps: the first, in stationary conditions, with the objective of bringing the system to the initial conditions, i.e., full PCM freezing and stabilization of the temperature gradient in the polyurethane foam. In this simulation step, we established the temperature conditions both inside (−27 °C) and outside (22 °C) the refrigerating chamber. The second step was a time-dependent simulation with time step 900 s in which a constant temperature was set at the outer walls of the chamber (22 °C) and the tested wall.

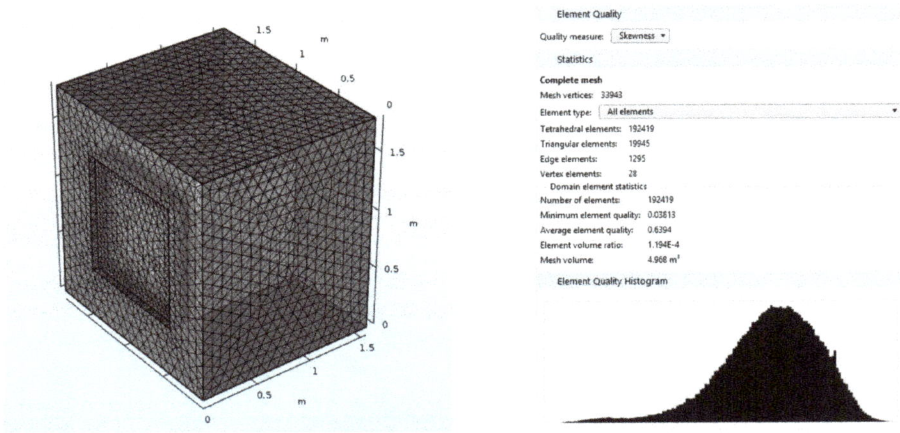

(a) (b)

Figure 3. (**a**) Mesh during numerical calculations; (**b**) Mesh statistics.

The following boundary conditions, presented in Table 3, were used in the stationary step:

Table 3. Boundary conditions employed during calculation in stationary step.

Domain/Material	Location	Boundary	Unit	Value
Climatic test chamber walls/mineral wool	External wall	Wall-temperature	°C	22
	Internal wall	Interface	-	Coupled wall
PCM layer/SP-24	External wall	Thin layer	$Wm^{-1}K^{-1}$	0.035
			m	0.0005
	Internal wall	Thin layer	$Wm^{-1}K^{-1}$	0.035
			m	0.0005
Air/air	All walls	Interface	-	Coupled wall
PU layer/PU	Internal wall	Wall-temperature	°C	−27
	Side walls	Interface	-	Coupled wall
	External wall	Wall-temperature	°C	22

The following boundary conditions, presented in Table 4, were used in the time-dependent step:

Table 4. Boundary conditions employed during calculation in time-dependent step.

Domain/Material	Location	Boundary	Unit	Value
Climatic test chamber walls/mineral wool	External wall	Wall-temperature	°C	22
	Internal wall	Wall-temperature	°C	−27
PCM layer/SP-24	External wall	Thin layer	$Wm^{-1}K^{-1}$	0.035
			m	0.0005
	Internal wall	Thin layer	$Wm^{-1}K^{-1}$	0.035
			m	0.0005
Air/air	All walls	Interface	-	Coupled wall
PU layer/PU	Internal wall	Interface	-	Coupled wall
	Side walls	Wall-temperature	°C	7
	External wall	Wall-temperature	°C	22

The boundary condition used in the time-dependent step related to the temperature of the PU layer side walls was set at 7 °C on the basis of the measured temperature on the aluminum part of the chamber, in order to take into account, the presence of thermal bridges.

2.4. Analysis of the Thermal Capacity of the Proposed Wall Structure

As part of the theoretical analysis, we can calculate the theoretical heat capacity of the wall of the refrigerated semi-trailer, which can then be compared with the results of experimental tests. In the case of a wall made in the proposed structure, its total heat capacity consists of the heat capacity of the PCM layer, taking into account the latent heat and specific heat in the entire temperature range in which the measurements were carried out, and the heat capacity of the polyurethane foam layer related to its specific heat [23]. In the case of the heat capacity of polyurethane foam, it was necessary to take the temperature gradient from −24 °C to ambient temperature in its volume into account. For calculation purposes, we adopted a linear temperature change, which is related to the homogeneous structure of the foam (the temperature gradient is presented in Figure 1b). The heat capacity values are given by the following formulas [24]. Results of theoretical calculations are presented in Table 5:

$$Q_{PCM} = \Delta h \times m_{PCM} + c_{PCM} \times m_{PCM} \times \Delta T \quad (1)$$

$$Q_W = c_{PU} \times m_{PU} \times \frac{\Delta T}{2} \quad (2)$$

$$Q = Q_{PCM} + Q_W \quad (3)$$

where:

Δh—latent heat of fusion—Δh = 220 kJkg^{-1};
m_{PCM}—mass of PCM;
c_{PCM}—specific heat of PCM—c_{PCM} = 2 kJkg^{-1}K^{-1};
ΔT—temperature difference between phase change temperature and temperature of material at the end of experiment (15 °C)—ΔT = 39 K;
c_{PU}—specific heat of polyurethane foam—c_{PU} = 1.45 kJkg^{-1}K^{-1};
m_{PU}—wall weight—m_{PU} = approximately 10 kg.

Table 5. Results of theoretical calculations of the amount of heat energy consumed.

Thermal Capacity [kJ]		
Reference Wall	**6 mm PCM Layer**	**10 mm PCM Layer**
282.75	2178.03	3632.27

The above results clearly show that the use of a PCM layer significantly increases the thermal capacity of the wall, which determines the length of time the trailer will be able to maintain the internal temperature at the required level. The performed theoretical calculations, after verification with numerical tests, allow to determine the optimal layer thickness of the phase change material that should be used in a full-scale solution.

3. Experimental Tests of Designed Wall Structure

3.1. Climatic Test Chamber

The tests of potential wall structures were carried out with the use of a specially designed climatic test chamber, which is presented in Figure 4. It consists of five adiabatic sandwich walls, each 40 cm thick, filled with mineral wool with a density of about 40 kg/m^3 and thermal conductivity of λ = 0.035 Wm^{-1}K^{-1}. The thick layer of mineral wool is intended to ensure isothermal properties of the climatic test chamber's interior. Inside the climatic test chamber, there is a space with a volume of about 1 m^3, which can be cooled to

−30 °C with the use of the integrated 0.8 kW refrigeration unit controlled by a thermostat. The climatic test chamber has no front wall, and the function of the front wall is performed by the tested wall. The climatic test chamber is placed on supports that distance it from the ground by 68 mm.

Figure 4. The climatic test chamber used to conduct experimental research.

The climatic test chamber was equipped with sensors. The temperature was measured at eight measuring points and the heat flux at three measuring points throughout the experiment. Data were collected at five-second intervals. The measurements were performed by using the following measuring equipment:

- 8 K type thermocouples:
  - T1—center of external wall;
  - T2—center of internal wall;
  - T3—center of internal sidewall;
  - T4—center of external sidewall;
  - T5—top-center of internal wall;
  - T6—top-corner of internal wall;
  - T7—air temperature outside climatic test chamber;
  - T8—air temperature at the center of climatic test chamber;
- 3 heat flux sensors–Hukseflux HFP01-05;
- 3 data loggers–TandD MCR-4V.

The T3, T4, and Q3 sensors were installed in the center of sidewall of the climatic test chamber (the adiabatic wall). T3 and Q3 sensors on its inside, and T4 sensor on its outside. The arrangement of the sensors in the entire test chamber was selected on the basis of the [25], discussing the measurements of wall insulation.

The locations of the sensors are presented in Figure 5.

3.2. The Experimental Procedure

The experiments for the purposes of this paper were carried out in three variants—for two variants of the cold store walls (with a layer of the phase change material with a thickness of 6 mm and 10 mm) and for the reference wall—without the phase change material. For each variant there have been two experiments held. At the beginning of each of the experiments, the tested wall was placed in the cavity of the climatic test chamber, and the spaces between the tested wall mounted in the cavity and the climatic test chamber walls were additionally sealed with polyurethane foam in order to limit the penetration of heat into the climatic test chamber in a way other than through the tested wall.

Each of the conducted experiments was carried out in an identical manner and was divided into two phases. The first, during which the interior of the test chamber was cooled to a temperature of −27 °C, and the temperature was kept at this level for a time sufficient for the layer of the phase change material to freeze. During the first phase of each test, the automatic control of the refrigeration union caused it to switch off after reaching the preset temperature. At that time, the space inside the climatic test chamber was heated—due to the heat transfer through the side walls absorption of energy from the PCM. After the temperature rises to approximately −24 °C, the unit started again. This mode of operation of the refrigerating unit resulted in fluctuations in the heat flux recorded by sensors placed inside the climatic test chamber, but it did not affect the main objective of this phase, i.e., bringing it to the initial state, which would take place in the case of the refrigerated transport, when the interior of the semi-trailer is cooled to the required temperature, and the phase change material layer is frozen in its entire volume.

After a sufficient period of time, 24 h for the 6 mm layer and 40 h for the 10 mm layer, the refrigerating unit was turned off and the second test phase began, i.e., free heating of the climatic test chamber's interior. The tests were terminated after the inside of the climatic test chamber reached the temperature of 15 °C—measured on T8.

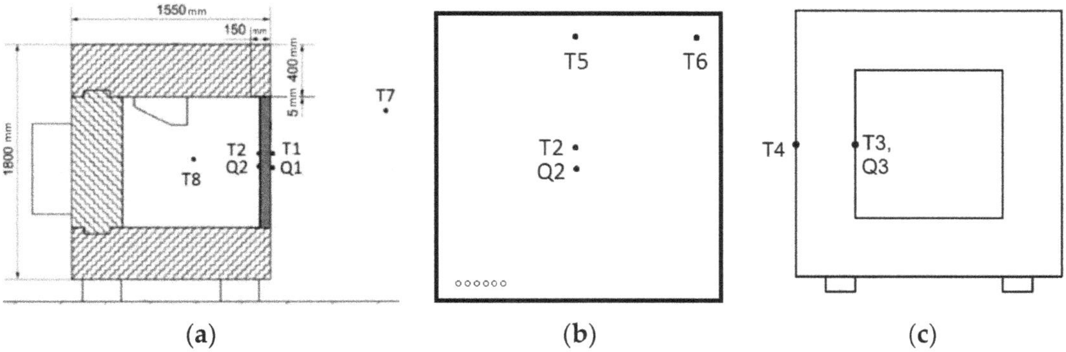

Figure 5. Location of the sensors during the tests: (**a**) cross-section view; (**b**) view from the inside of the tested wall; (**c**) front view.

3.3. Experimental Test Results

From the point of view of the purpose of the present research, the second phase of each of the experiments was more important, because the modernization of the wall structure was to keep the temperature inside the chamber low, and this was the focus of the analysis of the research results. Figure 6a–c present the temperature change curves at all eight measurement points, successively for the reference walls, the wall with a 6 mm PCM layer and the wall with a 10 mm PCM layer. The t_0 time for the presented curves is the moment of switching off the refrigeration unit inside the chamber.

The presented graphs indicate that all the tests were conducted under similar ambient temperature conditions—T7 in each case was in the range of 21 to 22 °C. The temperature measurements at the T1 and T4 measurement points located on the outer walls of the tested wall and climatic test chamber were similar.

The differences were observed in the plots for the remaining thermocouples. In the reference case, there is an immediate increase in temperature at each measurement point. In the case of a wall with a 6 mm and 10 mm PCM layer, there is a noticeable period in which the T2 and T5 indicated a constant temperature around −22 °C, which was the result of the phase change in the PCM layer. In the temperature plots, there is also noticeable the constant temperature area around 0 °C. This phenomenon may result from the properties of the phase change material SP-24, which is an aqueous salt solution, which also shows an increased heat capacity at 0 °C (phase transition temperature of water).

From the point of view of the applicability of the proposed wall structure in refrigerated transport, important is its ability to maintain the required temperature inside the chamber. The conducted experiments indicate that in each of the cases, the indications of the T8 thermocouple, located at the central point of the climatic test chamber, increased immediately after the refrigerating unit was turned off, which would indicate a result of their application that was inconsistent with the initial assumptions. It should be noted, however, that the wall with a PCM layer was used on only one of the six walls of the climatic test chamber, so that, despite the very good insulation properties of the remaining five walls, the heat penetrated through them into the climatic test chamber, heating up the air, which was characterized by low heat capacity, and consequently heating them to temperatures close to the temperature outside the climatic test chamber.

In the entire process of heat penetration inside the climatic test chamber, the influence of thermal bridges resulting from the proposed method of wall mounting in the climatic test chamber, as well as the chamber structure itself, were also significant. Despite the use of polyurethane foam for sealing and thermal insulation of the contact between the tested wall and the climatic test chamber, this connection cannot be considered isothermal. The impact of this aspect was also increased by the fact that this part of the climatic test chamber was made of aluminum, which is a good thermal conductor, enabling the heat to penetrate the chamber much faster, and contacting the PCM layer caused it to melt at the edges of the wall.

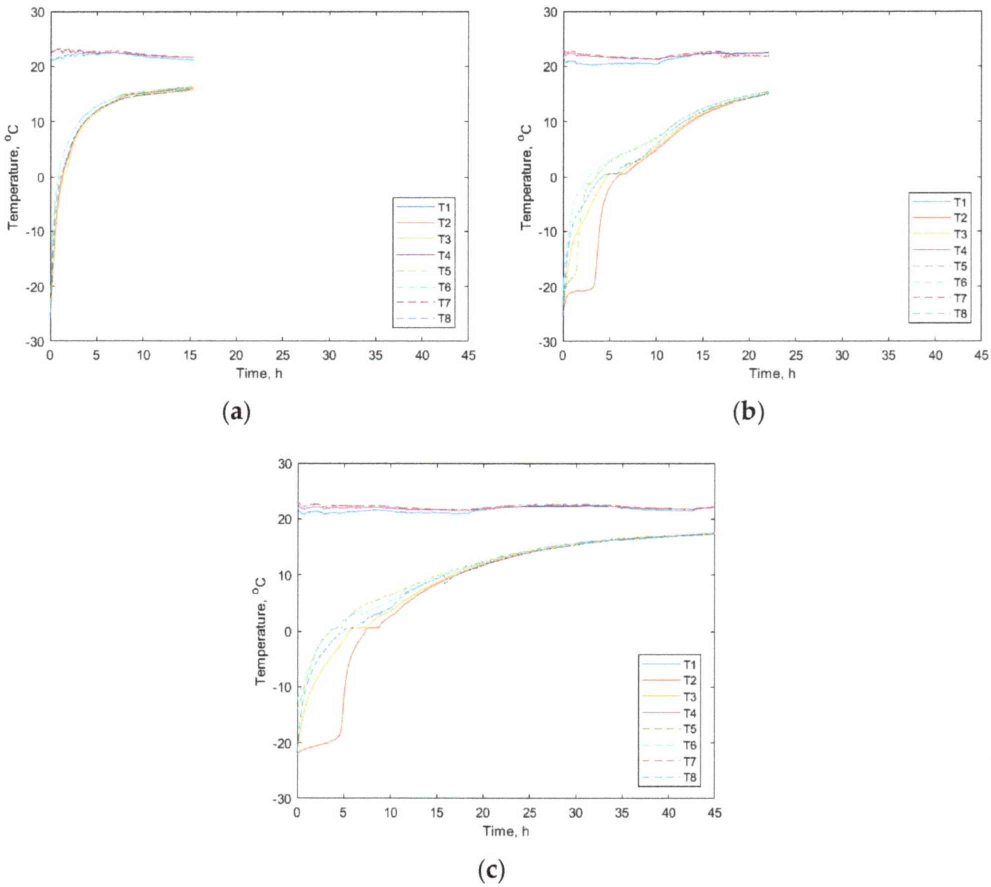

Figure 6. The plot of temperature changes at all measurement points during the test of: (**a**) the reference wall without the layer of phase change material; (**b**) the wall with a 6 mm layer of phase change material; (**c**) the wall with a 10 mm layer of phase change material.

For the reasons described above, the effect of the layer of the phase change material was significantly reduced, but when looking at the temperature changes at the T8 point in the three cases, one can clearly see the extension of the time needed to reach the ambient temperature in each of them. The temperature curves at T8 and T2 measurement points, summarized for the three variants, are presented in Figure 7a,b.

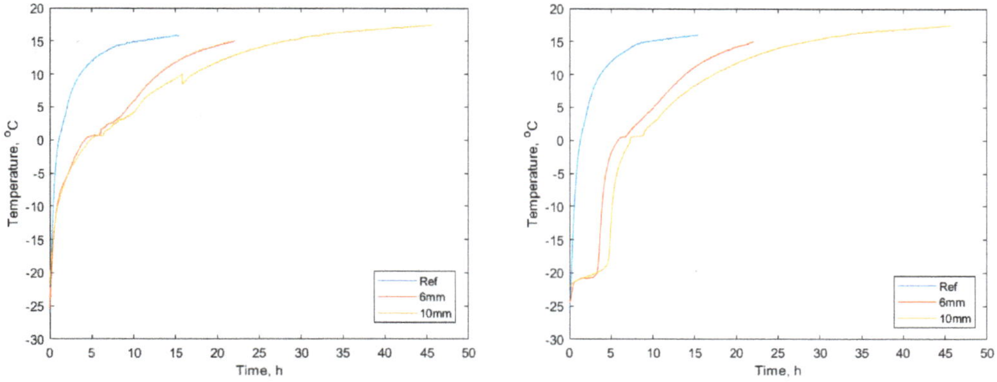

(**a**) (**b**)

Figure 7. (**a**) The plots of temperature changes at the measurement point T8 (air temperature at the center of climatic room) for the three tested variants of the wall of the refrigerated trailer; (**b**) The plots of temperature changes at the T2 (internal wall) measurement point for the three tested variants of the wall of the refrigerated trailer.

It is also worth noting that in the case of walls with a PCM layer in the first phase of heating, the temperature increases inside the chamber occurred at the same rate, regardless of the thickness of the layer. The effect of the amount of thermal energy stored in the material was noticeable only after the inside of the chamber reached a temperature of about 0 °C, when the inside of the climatic test chamber remained cooler in the case of a 10 mm PCM layer wall. This phenomenon proves that at this temperature inside the chamber, the share of the heat flux associated with heat absorption by the phase change material began to be significant compared to the heat flux penetrating through the sidewalls of the chamber and through thermal bridges.

What is of particular interest in the case of heat flux measurements, at least when we consider the purpose of our research, are the readings of the Q2 sensor located on the inner surface of the tested walls. They show the direction and intensity of thermal energy emitted or absorbed by the layer of the phase change material. In an ideal case, i.e., when the temperature inside the chamber would remain below the phase transition temperature (there would be no heat penetration through other partitions), the Q2 sensor readings would oscillate around 0 $Wm^{-2}K^{-1}$, and the PCM would be heated only through heat penetrating from the outside, through the layer of polyurethane, the amount of which would be indicated by the Q1 sensor. The graphs of changes in the value of the heat flux at the Q1 and Q2 measurement points obtained during the experimental tests are presented as Figure 8a,b.

The plots of changes in the value of the heat flux at the measurement point Q1 (the outer surface of the tested wall) is as predicted, i.e., the measured values are in the constant range of 12 to 17 $Wm^{-2}K^{-1}$, with an average of about 14 $Wm^{-2}K^{-1}$ until the phase change material melts. At the moment of melting, the value of the heat flux gradually decreases as the temperature of the PCM layer increases to values close to 0 $Wm^{-2}K^{-1}$.

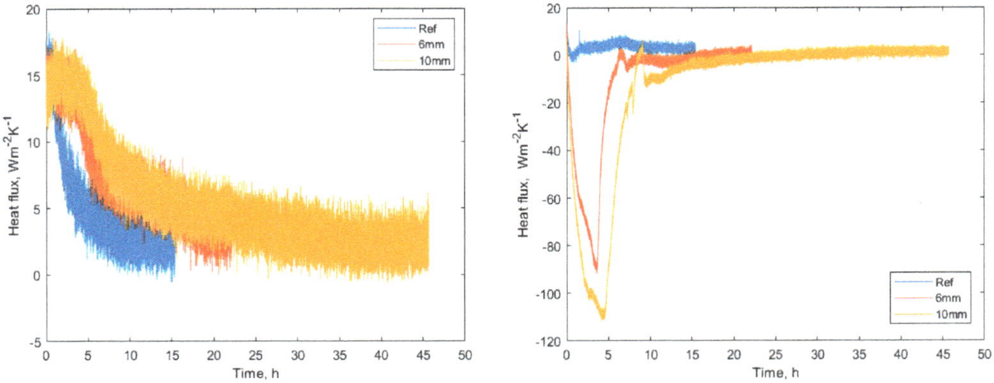

(a) (b)

Figure 8. (**a**) The plots of changes in the heat flux at the Q1 measurement point for the three tested variants of the wall of the refrigerated trailer; (**b**) The plots of changes in the heat flux at the Q2 measurement point for the three tested variants of the wall of the refrigerated trailer.

In the case of the heat flux values recorded at the Q2 measurement point, in the tests of the walls containing the phase change material layer, clearly negative values of this flow are visible, which means that the heat flux occurred in a direction opposite to that expected. This situation was caused by an increase in temperature inside the climatic test chamber. The point at which the graphs reach a minimum corresponds to the moment when the entire volume of the PCM has melted and it begins to heat up, which resulted in a decrease in ΔT between the PCM and the air inside the chamber and a decrease in the heat flux.

Heat flux data enable us to determine the amount of thermal energy that was consumed by the phase change material in each case and compare it with the theoretically calculated values. This allows us to verify the fact of the phase change occurring in the entire volume of the layer of the phase change material.

The obtained experimental results indicate that the application of a layer of phase change material in the wall of the refrigerated semi-trailer allows to maintain the temperature inside the semi-trailer at the required level, but the material must be placed on all walls, and it is particularly important to eliminate thermal bridges from the structure, the presence of which strongly affects for the time during which the phase transition takes place. In addition, the obtained results show that the differences in the heating time of the phase change material is proportional to its volume (layer thickness), therefore the selection of the layer thickness depends primarily on the temperature purpose of the semi-trailer and the required time for which the temperature inside the semi-trailer is to be maintained.

3.4. Validation of the Numerical Model

In this section, the results of experiments were compared with the results of numerical tests acquired by means of computer modeling in order to verify the correctness of these simulations. In particular, the data on the temperature inside the chamber at T2 and T8 measurement points and the heat flux at Q1 and Q2 measurement points were analyzed. The results of numerical modeling are presented in Figure 9a–d. The graphs of each of the measured parameters were compared with the plots of the corresponding values from the experiments.

In both cases—i.e., that of the 6 mm and of the 10 mm PCM layer—the temperature change and heat flux graphs from numerical modeling coincide with the experimental graphs. Achieving compliance required us to account for the influence of aluminum elements of the chamber, which in turn indicates that their presence had a significant impact on the course of experiments and distorted the heat capacity measurements of the PCM layer.

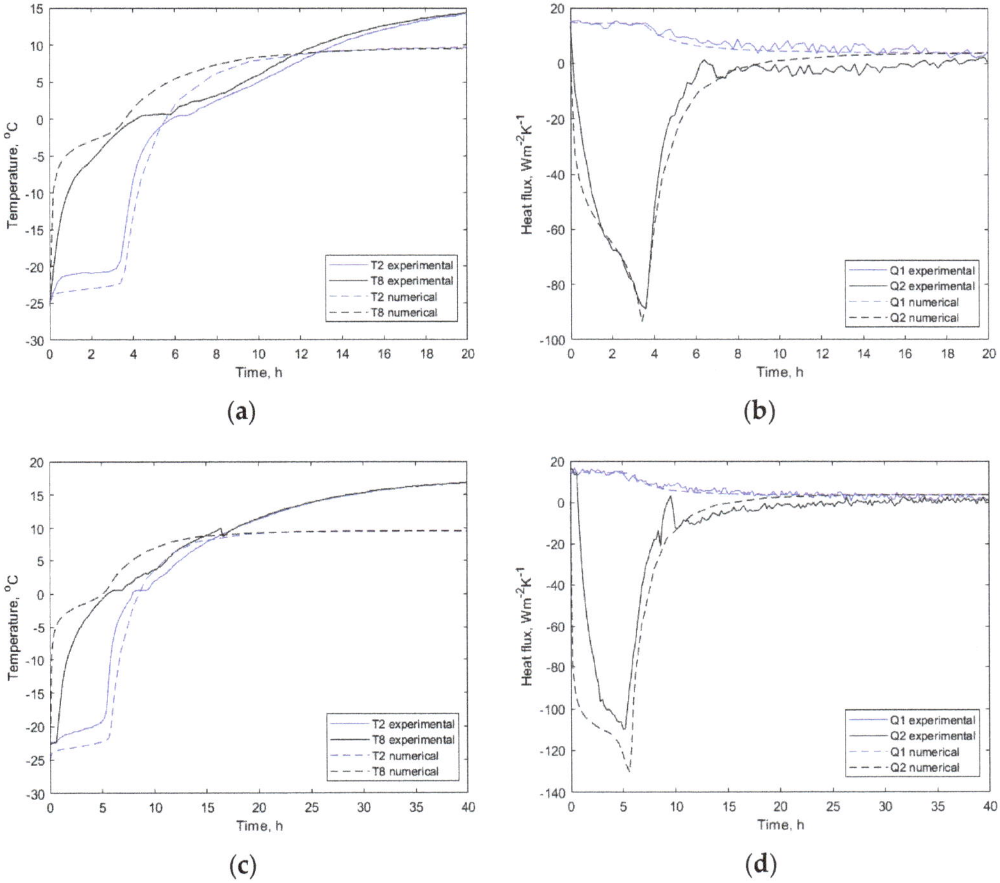

Figure 9. (**a**) Comparison of the temperature changes at the T2 and T8 measurement points in the numerical modeling with the results of experiments for a wall with a 6 mm PCM layer; (**b**) Comparison of the heat flow flux changes at Q1 and Q2 measurement points in the numerical modeling with the results of experiments for a wall with a 6 mm PCM layer; (**c**) Comparison of the temperature changes at the T2 and T8 measurement points in the numerical modeling with the results of experiments for a wall with a 10 mm PCM layer; (**d**) Comparison of the heat flow flux changes at Q1 and Q2 measurement points in the numerical modeling with the results of experiments for a wall with a 10 mm PCM layer.

4. Numerical Model of a Full-Scale Semi-Trailer

In order to verify the applicability of the proposed wall structure to a semi-trailer, as an alternative for diesel-driven refrigerator units, we performed numerical tests for a trailer with typical sizes of 13.2 m × 2.42 m × 2.45 m (inside measurements). As part of the numerical simulation, a geometric model of the semi-trailer was created, made of polyurethane foam plates, with a 6 mm layer of phase change material on the inside, and an unventilated air volume inside the semi-trailer. The geometry was discretized using the built-in meshing tool. Parameters and visualization of the created mesh are shown in Figure 10.

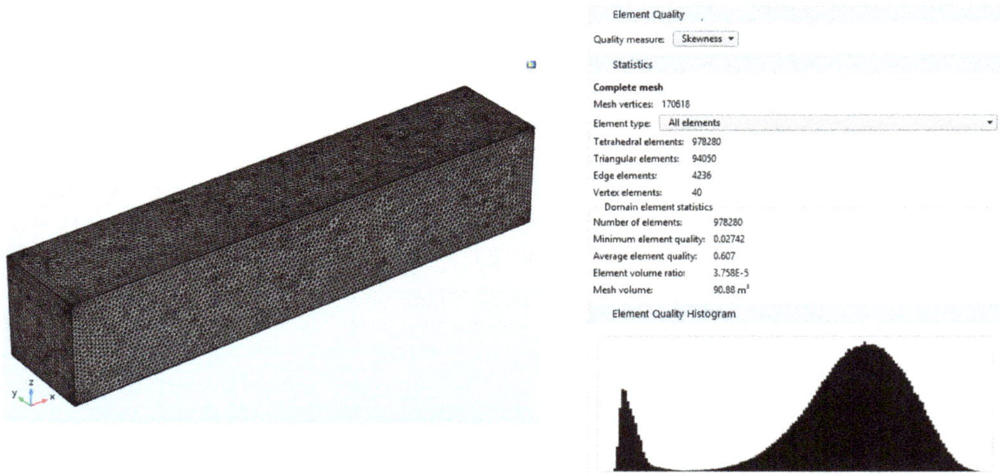

(a) (b)

Figure 10. (**a**) Mesh during numerical calculations; (**b**) Mesh statistics.

The simulation used the parameters specified in the numerical model of the climatic test chamber, but due to the inclusion of the phase change material layer in all six walls of the trailer, we were able to eliminate the influence of thermal bridges related to the aluminum elements of the test chamber. As required by the ATP [19] we assumed the external temperature of +30 °C.

In the case of a real-sized refrigerated trailer, the solution proposed was based on the use of walls with a 6 mm thick PCM layer, due to the weight of the entire solution. The total area of the internal surface of the semi-trailer that must be lined with the phase change material is 140.426 m^2. When using the 6 mm layer of Rubitherm SP-24 phase change material in the design proposed in the present publication, the weight of the phase change material would be approximately 900 kg, which is equivalent to the weight of diesel-driven refrigeration units. Would a 10 mm layer be used instead, the mass of PCM would be close to 1600 kg, which would increase the fuel consumption of the vehicle and lower its load capacity.

At the stage of numerical testing, we tested the time of maintaining the required temperature inside the trailer. The measurement was made at two measurement points—T1 located 1 m from the front wall of the trailer and T2 located in the center of the trailer. Figure 11 presents the semi-trailer model with marked measurement points.

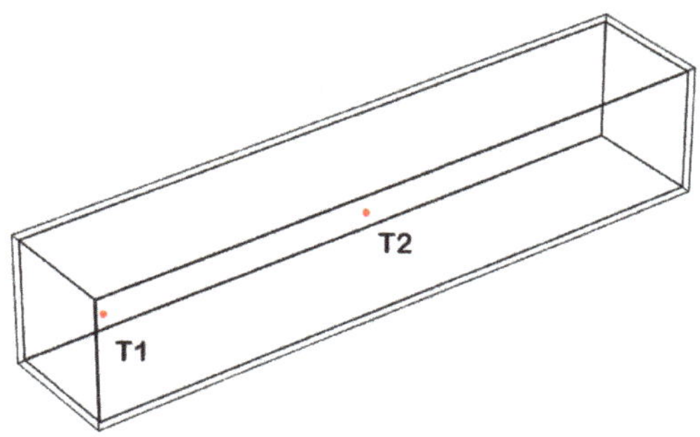

Figure 11. The model of a full-scale refrigerated semi-trailer with marked measurement points for plotting the graphs of internal temperature.

Numerical simulations, similar to the tests of the research chamber with a wall, were performed in two steps: stationary, which allowed us to reach the initial state, and time-dependent, which indicated a plot of temperature changes inside the trailer. Figure 12a–d presents the temperature distribution in the longitudinal symmetry plan of the semi-trailer in successive time steps, while Figure 13 presents a plot of temperature changes at T1 and T2 measurement points.

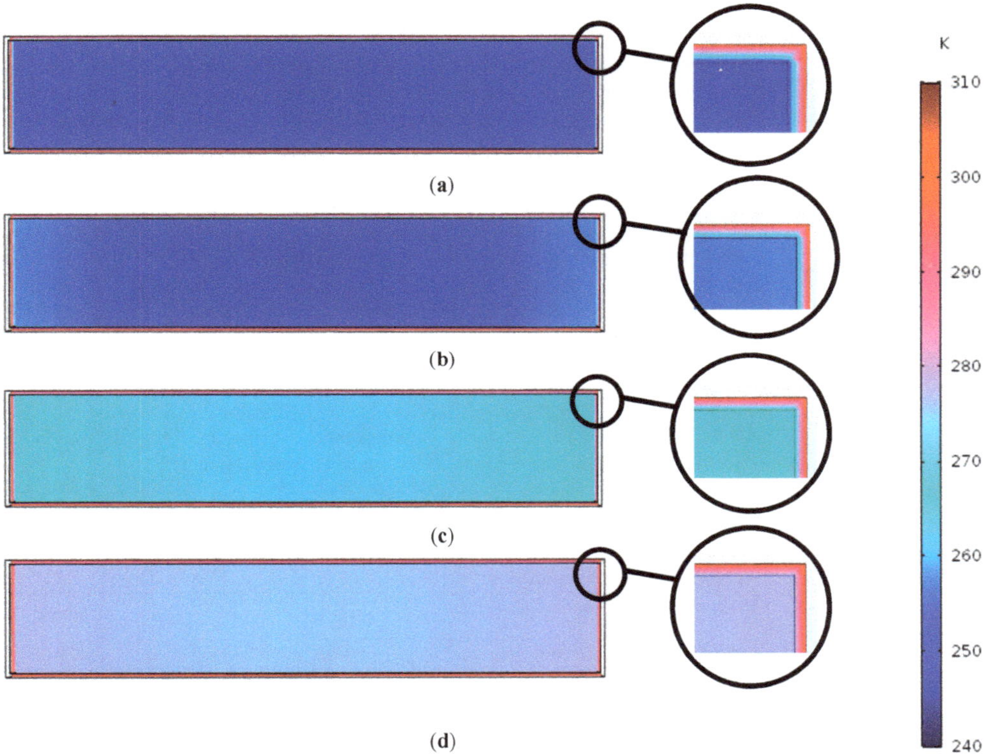

Figure 12. Distribution of temperature inside the refrigerated trailer over time (**a**) 0 h, (**b**) 24 h, (**c**) 25 h, and (**d**) 26 h.

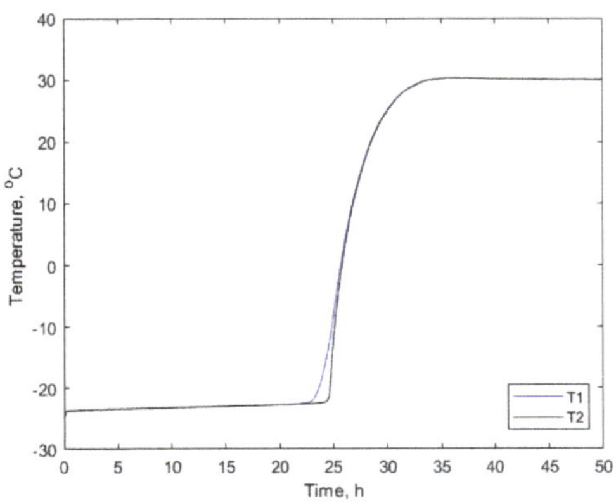

Figure 13. Graph of numerical results of temperature changes inside the refrigerated trailer at T1 and T2.

The results of numerical tests indicate that the application of the proposed wall structure of a refrigerated semi-trailer with a 6 mm layer of the phase change material allowed for maintaining the internal temperature at the level of −20 °C, as required by the ATP agreement for over 20 h. At this point, it should be stressed that the simulation was carried out for stationary mode, which does not take trailer movement and solar radiation into account, both of which could reduce this time.

Potential Energy Savings

Using the method presented by Liu et al. [3] and the data on the operating parameters of diesel-powered refrigerating units [26], and energy-efficient electrically-powered refrigerators [27], it was possible to calculate the potential energy savings resulting from the use of the wall structure proposed by the authors.

Energy savings were calculated for one transport lasting about 24 h, i.e., one that ensures full use of the capacity of the proposed wall structure. Based on the Formulas (1)–(3), the total heat capacity of the refrigerated container walls (phase change material and polyurethane foam) was calculated, which came to 305.85 MJ. Then, we went on to calculate the amount of energy required to produce that amount of cold. The calculations assume the efficiency of cooling production from diesel fuel at the level of 0.2 [26], while the COP value for electrically powered chillers is typically 1.5, with the required output temperature at the level of about −30 °C [27]. The results of the calculations are presented in Table 6.

Table 6. The results of calculations of potential energy savings resulting from application of the proposed wall structure with 6 mm PCM layer in a refrigerated trailer.

Parameter	Refrigerator Unit Driven by a Diesel Engine	Electrically Driven Stationary Refrigeration Unit
The amount of cold produced [MJ]	305.85	305.85
Cold generation efficiency (COP)	0.2	1.5
The amount of energy consumed to produce cold [MJ]	1529.10	203.88
Energy consumption reduction level	0%	86.66%

In Table 6, the results of calculations indicate that the energy savings in the case of using the proposed wall structure can amount to 86.66%. When analyzing these results, however, we should note that the calculations did not take into account the method of electricity production, which determines the result of primary energy savings and the reduction in CO_2 emissions. In addition, the calculations did not take into account the efficiency of the cooling system of the phase change material, which is the subject of further research.

5. Conclusions and Directions for Future Research

As part of this study, research has been carried out to verify the feasibility of replacing refrigeration units powered by a diesel engine in deep-frozen transport over short distances by a passive cooling system based on a layer of phase change material with a phase change temperature of about −24 °C located on the inner side of the trailer wall. The obtained test results indicate that the application of such a solution with a 6 mm layer of PCM would allow the maintenance of the inside temperature of the trailer at the level of −20 °C for up to 24 h, at an external temperature of +30 °C, and the energy savings related with carrying out such transport can be as high as 86% if RES used. According to the ATP agreement, these types of vehicles, with passive cooling, do not have a specific minimum time for which the desired temperature inside the vehicle is to be maintained, so its maximum extension is most desirable, as it would allow for the use of a semi-trailer on increasingly longer routes. The use of a 6 mm layer of PCM, in addition to ensuring nearly 24 h of transport time under the required conditions, does not increase the weight of the semi-trailer, because its weight is comparable to the refrigeration units it is intended to replace.

There should be two areas considered for further research of the subject matter discussed therein. The first should focus on the analysis of additional phenomena occurring during transport and their impact on the ability to maintain the required temperature. These phenomena include: movement of the vehicle, affecting the intensity of heat transfer at the outer surface of the trailer walls, the impact of opening the trailer door during stops along the transport route, and the solar radiation, which at selected latitudes and on certain days can significantly raise the temperature of the outer surface of the semi-trailer walls.

The second direction of research should address the design of an effective system for cooling the phase change material during the loading of goods, which, in turn, could make the proposed solution commercially implementable. Such a system must ensure fast complete freezing of the phase change material with a lower energy input than with the use of diesel refrigeration units currently fitted on semi-trailers.

Author Contributions: Conceptualization, T.U.; methodology, T.U.; software, K.Z.; validation, K.Z.; formal analysis, K.Z.; resources, T.U.; data curation, K.Z.; writing—original draft preparation, K.Z.; writing—review and editing, T.U.; visualization, K.Z.; supervision, T.U.; project administration, T.U.; funding acquisition, T.U. All authors have read and agreed to the published version of the manuscript.

Funding: This research was funded by National Centre of Research and Development grant number POIR.01.02.00-00-0216/16-00 titled; Development of design and manufacturing technology for refrigerated box semi-trailer for transport of goods in refrigerated conditions.

Institutional Review Board Statement: Not applicable.

Informed Consent Statement: Not applicable.

Acknowledgments: Thanks to National Centre of Research and Development for financial support of presented research under grant no. POIR.01.02.00-00-0216/16-00 titled; Development of design and manufacturing technology for refrigerated box semi-trailer for transport of goods in refrigerated conditions. Thanks to Wielton SA, for supporting of building of experimental setup.

Conflicts of Interest: The authors declare no conflict of interest.

Nomenclature

PCM	Phase change material
CTES	Cold thermal energy storage
CO_2	Carbon dioxide
K	Wall heat transfer coefficient ($W \cdot m^{-2}K^{-1}$)
C_p	Specific heat ($J \cdot kg^{-1}K^{-1}$)
Δh	Latent heat ($J \cdot kg^{-1}$)
T	Temperature (°C)
m	Mass (kg)
λ	Thermal conductivity ($W \cdot m^{-1}K^{-1}$)
COP	Coefficient of performance
RES	Renewable energy source

References

1. Dekker, R.; Bloemhof, J.; Mallidis, I. Operations Research for green logistics—An overview of aspects, issues, contributions and challenges. *Eur. J. Oper. Res.* **2012**, *219*, 671–679. [CrossRef]
2. Tassou, S.A.; De-Lille, G.; Ge, Y.T. Food transport refrigeration—Approaches to reduce energy consumption and environmental impacts of road transport. *Appl. Therm. Eng.* **2009**, *29*, 1467–1477. [CrossRef]
3. Liu, M.; Saman, W.; Bruno, F. Development of a novel refrigeration system for refrigerated trucks. *Appl. Energy* **2012**, *92*, 336–342. [CrossRef]
4. Michel, B.; Glouannec, P.; Fuentes, A.; Chauvelon, P. Experimental and numerical study of insulation walls containing a composite layer of PU-PCM and dedicated to refrigerated vehicle. *Appl. Therm. Eng.* **2017**, *116*, 382–391. [CrossRef]
5. Fioretti, R.; Principi, P.; Copertaro, B. A refrigerated container envelope with a PCM (Phase Change Material). *Energy Convers. Manag.* **2016**, *122*, 131–141. [CrossRef]
6. Huang, L.; Piontek, U. Improving Performance of Cold-Chain Insulated. *Appl. Sci.* **2017**, *7*, 1288. [CrossRef]

7. Stellingwerf, H.M.; Kanellopoulos, A.; van der Vorst, J.G.; Bloemhof, J.M. Reducing CO_2 emissions in temperature-controlled road. *Transp. Res. Part D* **2018**, *58*, 80–93. [CrossRef]
8. James, S.J.; James, C.J.F.R.I. The food cold-chain and climate change. *Food Res. Int.* **2010**, *43*, 1944–1956. [CrossRef]
9. Zou, R.; Huang, X. Mitigating Global Warming by Thermal Energy Storage. In *Energy Soultions to Combat Global Warming*; Springer International Publishing: Cham, Switzerland, 2017; pp. 573–594.
10. Selvnes, H.; Allouche, Y.; Manescu, R.I.; Hafner, A. Review on cold thermal energy storage applied to refrigeration systems. *Therm. Sci. Eng. Prog.* **2021**, *22*, 100807. [CrossRef]
11. Zarajabad, O.G.; Ahmadi, R. Numerical investigation of different PCM volume on cold thermal energy. *J. Energy Storage* **2018**, *17*, 515–524. [CrossRef]
12. Oró, E.; Miró, L.; Farid, M.M.; Martin, V.; Cabeza, L.F. Energy management and CO2 mitigation using phase change materials (PCM) for thermal energy storage (TES) in cold storage and transport. *Int. J. Refrig.* **2014**, *42*, 26–35. [CrossRef]
13. Du, K.; Calautit, J.; Wang, Z.; Wu, Y.; Liu, H. A review of the applications of phase change materials in cooling, heating and power generation in different temperature ranges. *Appl. Energy* **2018**, *220*, 242–273. [CrossRef]
14. Pielichowska, K.; Pielichowski, K. Phase change materials for thermal energy storage. *Prog. Mater. Sci.* **2014**, *65*, 67–123. [CrossRef]
15. Cecchel, S.; Chindamo, D.; Turrini, E.; Carnevale, C.; Cornacchia, G.; Gadola, M.; Panvini, A.; Volta, M.; Ferrario, D.; Golimbioschi, R. Impact of reduced mass of light commercial vehicles on fuel. *Sci. Total. Environ.* **2018**, *613*, 409–417. [CrossRef] [PubMed]
16. Mousazade, A.; Rafee, R.; Valipour, M.S. Thermal performance of cold panels with phase change materials in a refrigerated truck. *Int. J. Refrig.* **2020**, *120*, 119–126. [CrossRef]
17. Tong, S.; Nie, B.; Li, Z.; Li, C.; Zou, B.; Jiang, L.; Jin, Y.; Ding, Y. A phase change material (PCM) based passively cooled container for integrated road-rail cold chain transportation—An experimental study. *Appl. Therm. Eng.* **2021**, *195*, 117204. [CrossRef]
18. Glicksman, L.R. Heat transfer in Foams. In *Low Density Cellular Plastics: Physical Basis of Behaviour*; Springer: Berlin/Heidelberg, Germany, 1994; Chapter 5.
19. Agreement, A. Agreement on the International Carriage of Perishable Foodstuffs and on the Special Equipment to be Used for Such Carriage (ATP). 1970. Available online: https://unece.org/DAM/trans/main/wp11/ATP_publication/2020/ECE_TRANS_290_En_web_protected.pdf (accessed on 20 December 2021).
20. Rubitherm, Sp-24 Data Sheet. Available online: www.rubitherm.eu (accessed on 2 May 2021).
21. COMSOL Inc. *COMSOL Multiphysics Reference Manual*; COMSOL, Inc.: Burlington, VT, USA, 2017.
22. Zestawienie Parametrów Fizycznych Materiałów/Wyrobów Budowlanych wg PN-EN ISO 12524:2003, PN-EN ISO 6946:1999 i PN-91/B-0202. Available online: http://kurtz.zut.edu.pl/fileadmin/BE/Tablice_materialowe.pdf (accessed on 11 June 2021).
23. Stephan, P. Fundamentals of Heat Transfer. In *VDI Heat Atlas*; Springer: Berlin/Heidelberg, Germany, 2010; pp. 17–33.
24. Mehling, H.; Cabeza, L.F. *Heat and Cold Storage*; Springer: Berlin/Heidelberg, Germany, 2008.
25. Rasooli, A.; Itard, L. In-situ characterization of walls' thermal resistance: An extension to the ISO 9869 standard method. *Energy Build.* **2018**, *179*, 374–383. [CrossRef]
26. Chatzidakis, S.K.; Chatzidakis, K.S. Refrigerated transport and environment. *Int. J. Energy Res.* **2004**, *28*, 887–897. [CrossRef]
27. Ellis, M. *Regulatory Impact Statement: Minimum Energy Performance Standards and Alternative Strategies for Commercial Refrigeration Cabinets in Australia and New Zealand*; Mark Ellis & Associates: Wagstaffe, Australia, 2004.

energies

Article

Evaluating LNG Supply Chain Resilience Using SWOT Analysis: The Case of Qatar

Sara Al-Haidous, Mohammed Al-Breiki, Yusuf Bicer and Tareq Al-Ansari *

College of Science and Engineering, Hamad Bin Khalifa University, Qatar Foundation, Doha P.O. Box 34110, Qatar; salhaidous@hbku.edu.qa (S.A.-H.); malbreiki@hbku.edu.qa (M.A.-B.); ybicer@hbku.edu.qa (Y.B.)
* Correspondence: talansari@hbku.edu.qa

Abstract: The demand for liquefied natural gas (LNG) as an energy commodity is increasing, although its respective supply chain is subjected to risks, uncertainties, and disturbances. An analysis of experiences from the global LNG supply chain highlights many of these risks. As such, there is an incumbent need to develop resilient LNG supply chains. In this study, the risks associated with the LNG supply chain are categorized into four dimensions: Political and regulatory, safety and security, environmental effects, and reliability of new technologies. A SWOT method is then implemented to identify strengths, weaknesses, opportunities, and threats within the LNG supply chain, where the LNG supply chain of Qatar is considered as a case study. Relevant strategies are then recommended using a SWOT matrix to maximize strengths and opportunities, while avoiding or minimizing weaknesses and threats within the LNG supply chain. Finally, major parameters to be considered to develop a resilient LNG management model are listed based on the level of priority from LNG producer and receiver perspectives. Thus, as part of creating a robust LNG supply chain, decision-makers and stakeholders are urged to use the learnings from the SWOT analysis and experiences from LNG supply chain management.

Keywords: energy carrier; energy transport; liquefied natural gas; policy; strategy

Citation: Al-Haidous, S.; Al-Breiki, M.; Bicer, Y.; Al-Ansari, T. Evaluating LNG Supply Chain Resilience Using SWOT Analysis: The Case of Qatar. *Energies* **2022**, *15*, 79. https://doi.org/10.3390/en15010079

Academic Editors: Michael Gerard Pecht, Moonyong Lee and Audrius Banaitis

Received: 22 October 2021
Accepted: 14 December 2021
Published: 23 December 2021

1. Introduction

Natural gas is an energy source that offers higher environmental benefits compared to other fossil fuels. It has been promoted as a transition fuel to bridge the fossil fuel era with that of renewable fuels. Use of natural gas reduces carbon dioxide (CO_2) emissions by 30% when compared to oil and coal, with a twofold reduction in nitrogen oxide (NO_x) emissions and virtually negligible sulphur dioxide (SO_2) emissions [1]. Incidentally, the demand from the international market is the main driver influencing the production volume of liquefied natural gas (LNG). In this regard, the LNG production capacity of the state of Qatar, one of the largest producers of LNG, will increase from 77 million tons per year (MTA) to 110 MTA by 2025, with a customer portfolio spanning across the globe [2]. This expansion is based on certain assumptions related to the further globalization of the LNG market, the rationale of which are as follows: (i) oil market balance maintains high energy prices; (ii) developing nations, particularly India and China, are accelerating the minimization of greenhouse gas emissions (GHG) by increasing the use of natural gas rather than oil and coal; (iii) and natural gas starts to account for an increasing percentage in power generation compared to other forms of fuel [3]. Notably, a significant shift is underway in the LNG global market as it becomes more flexible, and LNG contracts are changing simultaneously in four dimensions: the term and volume of contracts are reduced, the prohibition on the resale of LNG is lifted, and mixed prices are transitioning from oil-based price to gas-to-gas price. The former paradigm, where purchasers of LNG are entirely dependent on producers and have to engage in long-term contracts to recover the manufacturer's investment costs and sell its natural gas for 20–30 years, is no longer preferred [4]. In addition, significant developments are occurring in the LNG market. For instance, the LNG market has become liberalized with the increase in the volume and variety of LNG traders. In 2018, the global LNG trade reached record levels of 316.5 MT [5]. High price levels in Japan have also declined, following the decline in oil-indexed LNG costs, from approximately US$ 13/MMBtu in 2015 to US$ 9/MMBtu [5]. In September 2021, the Asian LNG spot price reached a higher record of US$ 34.47/MMBtu from US$ 4.78/MMBtu in September 2020 [6]. These shifts have introduced challenges and uncertainties within the complex LNG market and supply chain encompassing production, gas processing, liquefaction, transportation, storage, regasification, and utilization. Failure in one segment might influence other segments and eventually compromise LNG supply resilience. In the drive towards competitiveness and effectiveness, supply chain management is essential in the current economy. The supply chain is composed of all the links necessary to convert consumer demands into usable products and then return those products to customers. Due to this, it is critical to ensure seamless functioning across the whole supply chain. Considering the above, this research analyzes the main issues in the supply chain considering the changing dynamics of the LNG market and provides recommendations on how to address and mitigate these challenges in the drive towards a greener and more resilient future.

LNG is natural gas that has been cooled to a liquid form for transportation and storage purposes. Natural gas has been liquefied for transportation purposes and reaches unreachable gas market by pipelines or a specialized carrier in a safe, reliable, and efficient manner [7]. The LNG supply chain comprises five key entities/processes: exploration, liquefaction, transportation, regasification, or transport to the consumer. The supply chain begins with natural gas extraction and continues with pipeline transportation to a liquefaction facility. The LNG is then transported to regasification facilities located around the globe, where it is re-gasified and supplied to consumers as natural gas. In most parts of the globe, natural gas is the fastest-growing energy source, driven by the low GHG and the ability to generate electricity with high conversion efficiencies. LNG shipping may demonstrate society's reliance on marine transport infrastructure. The high cost of LNG supply chain operations and infrastructure creates a strong incentive to develop lean and closely linked systems, which increases resource utilization and reduces costs [8]. Uncertainties in the LNG supply chain may occur in one of the five supply chain processes/entities: Production, storage, transportation, loading/unloading, and utilization. The performance of the supply chain may be affected by delays and bottlenecks caused by uncertainly in any of the connections' functions. As a result, it is essential to exert control over the numerous variables that affect the supply chain. Appropriate planning and management of uncertainties will result in improved scheduling, optimal production levels, on-time delivery, and increased customer satisfaction. Uncertainties tend to prompt decision-makers to establish safety buffers in terms of time, capability, or inventory to avoid a chain/cascade failure. With these buffers, operational performance may be limited, and competitive advantage reduced.

Resilience is a frequently used term in many areas, including engineering, environmental science, and organizational and operations research. It has encouraged scholars and practitioners to further their studies into supply chain resilience and develop conceptual frameworks. A system's capacity to recover from adversity and return to its original or better state is known as resilience [9]. Organizations consequently need to examine the resilience of their supply chains as part of continuity. In terms of energy resilience, it refers to the need to ensure that a company's energy supply is consistent and stable, in addition to factoring in the necessary back up in the case of a power outage. Whereas, energy vulnerability is defined as an energy system's susceptibility to unfavorable events and change [10]. In terms of risk management in the marine supply chain, the characteristics of a transport system that may degrade or restrict its capacity to survive are referred to as vulnerability, managing and withstanding attacks and disrupting events that occur both within and beyond the system's perimeters [11]. Resilience involves addressing the effects of a disturbance rather than only avoiding disruption. Prior to disruption, however, efforts are undertaken to build a robust system [11]. Identifying the risks within the supply chain will support a resilient supply chain, where risks can be classified into three categories: (i) internal to the firm (e.g., process and control); (ii) external to the firm but internal to the supply chain network (e.g., demand and supply); and (iii) external to the network (e.g., environment) [12].

The SWOT method has been implemented to analyze various objectives, including processes and energy transition strategies, energy resilience, energy vulnerability, environmental innovation and preservation, and business–institutions–NGOs partnership. For instance, Goers et al. (2021) adopted the SWOT method to identify strengths, weaknesses, opportunities, and threats of renewable energy deployment in energy transition for several regions. The study demonstrates strengths, identifies weaknesses, main opportunities, associated threats, and recommended strategies for guiding regional energy transitions [13]. Vasudevan (2021) examined the impediments and limits imposed by EU policies and frameworks on natural gas decarbonization based on SWOT, which includes a PESTEL (Political, Economic, Social, Technological, Environmental and Legal) macroeconomic factor assessment [14]. Moreover, strengths, weaknesses, and opportunities of the energy sector in the Arab Region was demonstrated in the report highlighting the "Energy vulnerability in the Arab Region" implemented by United Nations Economic and Social Commission for Western Asia [15]. The SWOT method has been also used for environmental innovation and preservation. For instance, the four most potential LNG bunkering systems (Truck-to-Ship, Ship-to-Ship, Port-to-Ship, and Terminal-to-Ship, and Mobile Fuel Tanks) using a SWOT analysis method in the context of the Mediterranean port was examined [16]. In addition, the SWOT method is extensively used in business–institutions–NGOs partnership as governments and international agencies seek to identify strengths, weaknesses, opportunities, and threats [17]. Moreover, existing literature has attempted to determine the optimal means of transporting LNG from one point to another. For instance, a thorough analysis of LNG supply chain resilience places a premium on system stability, reducing emissions, and delivery uptime in the case of interruptions resulting in unexpected shutoffs [18]. Bouwmeester and Osterhaven (2017) used non-linear programming to forecast the broader interregional and inter-industry consequences of natural gas supply interruptions. The model educates policymakers on the scale of the broader economic consequences of natural gas supply interruptions and identifies key gas supplier relationships [19]. Moreover, Sesini et al. (2020) presented a novel modeling framework in emergencies defined by very low temperatures and a rise in natural gas requirements [20]. A global LNG supply chain model was developed to assess vulnerabilities in identifying measures that support a resilient LNG supply chain [21]. However, there is limited work in the literature that have implemented SWOT analysis to assess the LNG supply chain and recommended strategies to enhance the resilience of the LNG supply chain, and no study has reported exogenous security threats and disruption risks across the entire LNG supply chain. As such, this study will address the aforementioned gaps using a Qatar case study described below.

Case Study and Objectives

The LNG business is becoming more competitive with over 20 countries now providing consumers throughout the globe, of which the State of Qatar, United States, Australia, Malaysia, and others are currently major LNG suppliers. As the second-largest LNG exporter in the world (at 77 MTA) in 2021, Qatar is investing heavily to retain its position as the world's most dominant player [22], by expanding LNG production capacity by approximately 64% by 2027 to achieve 126 million metric tons of LNG per year [23]. Considering that production, liquefaction, and transportation are the three key components within LNG supply chain, expanding LNG transportation capabilities can further enhance Qatar's LNG supply chain, considering that production and liquefaction are well established. The State of Qatar has gained a substantial competitive advantage by constructing one of the world's most efficient and reliable LNG end to end supply chains, thus owning and managing the whole LNG supply chain, putting it ahead of competitors in the LNG market. As such,

Qatar can be more equipped to react to future unforeseen risk occurrences and instantly recover from any potential interruptions if it boasts independent shipping capabilities, in addition to well-established production and liquefaction facilities. Therefore, the reliance of international purchasers on Qatar as a dependable LNG supplier can grow as the country's LNG supply systems become more resilient. To further elaborate on the above, this study addresses the strength, weaknesses, opportunities, and threats of the Qatari LNG supply chain and then draw strategies to obtain a resilient LNG supply chain. As such, the primary aim of this study is to strengthen the LNG supply chain's resilience via the identification of strengths, weaknesses, opportunities, and threats. The importance of the study is to mitigate all the risks associated with technical, environmental, economic, geographic, and political aspects. In this study, the LNG supply chain of Qatar is considered as a case study as part of fulfilling the following objectives:

- Examine recent global accidents that can occur for the LNG supply chain.
- Demonstrate lessons learnt in LNG supply chain experiences from around the globe.
- Address strengths, weaknesses, opportunities, and threats that can affect Qatar's LNG supply chain.
- Recommend strategies to overcome weaknesses, utilize opportunities, and avoid threats.
- Illustrate major parameters and provide a priority plan to enhance the resilience of Qatar's LNG supply chain.
- Recommends certain strategies to overcome the risks associated with the LNG supply chain.

The remainder of this report is structured as follows. Section 2 details the method implemented in this study. In Section 3, major accidents and experiences that have occurred in LNG supply chain are described. In Section 4, findings of the SWOT method are demonstrated where strengths, weaknesses, opportunities, and threats are mentioned, and recommended strategies are listed based on the SWOT matrix. Finally, Section 5 summarizes the conclusion and limitations of this study.

2. Methodology

A SWOT method is applied in this study to determine Qatar's LNG supply chain's strengths, weaknesses, opportunities, and threats. Therefore, databases and directories are widely utilized in this research to locate and gather information important to the LNG supply chain in Qatar from a technical, industrial, and market-oriented perspective. Statistical data, press announcements, and company reports have been used as resources for this research, as well as official publications and databases from a range of various organizations and associations. Most of the information gathered from these sources has been used to provide a comprehensive description of the LNG supply chain, financial system, global network of main players, market classification, downstream sectoral segmentation, regional markets, and major market and technological advancements. With these resources, researchers identified, classified, described, and forecasted changes in Qatar's LNG supply chain. The collected data are then used to recommend strategies on how to enhance the robustness of the LNG supply chain.

The SWOT method is often used for strategy creation to learn about the present status of an examined item and create future solutions to existing problems. The SWOT method may be used to identify an item's strengths (elements to leverage and develop), weaknesses (related areas of guidance and assistance), opportunities (related fields of benefit utilization), and threats (components that prevent the object from developing). Internal factors describe strengths and weaknesses, while external factors define opportunities and dangers [24]. The SWOT method was successfully used in energy discipline, for instance, in the study of sustainable energy development [25,26] and transition to renewable energy [27,28]. The main rationale for selecting a SWOT analysis is to address strengths, weaknesses, opportunities, and threats to the Qatari LNG supply chain and then draw strategies to enhance its resilience. Such analysis enables LNG suppliers to discover all elements that may influence the overall LNG market, and how that may then impact operations, allowing them to gather and compare data of various forms and objectives, which together can provide decision makers with all the necessary information to undertake critical decisions, especially with regard to future possibilities and analyzes the company's capacity to pursue them [29].

When applying the SWOT method to Qatar's LNG supply chain, internal and external factors affecting the LNG supply chain in Qatar will first be gathered and evaluated utilizing reports, literature, papers, legislation, and data. The SWOT method used in this study consists of three steps: (i) material collection; (ii) SWOT analysis; and (iii) strategy recommendations. The material collection step aims to amass all pertinent data and materials related to the research subject, including includes rules, reports, books, papers, statutes, national statistics, and legislation. The SWOT analysis phase is used to debate, analyze, and define all aspects of strengths, weaknesses, opportunities, and threats based on the data gathered. Relevant strategies are identified throughout the brainstorming process based on strengths, weaknesses, opportunities, and threats to maximize the use of strengths and opportunities while avoiding or minimizing weaknesses and threats, as illustrated in Figure 1. The SWOT matrix is then applied in four distinct strategies: strengths–opportunities (SO), weaknesses–opportunities (WO), strengths–threats (ST), and weaknesses–threats (WT). SO strategies are achieved by combining internal strengths with external opportunities and employing strengths to benefit from opportunities; internal weaknesses and external opportunities, as well as the possibility to address these weaknesses, which is the foundation of WO strategies; internal strengths and external risks are combined in ST strategies to prevent threats; internal weaknesses are combined with external threats in WT strategies, and weaknesses are reduced to avoid threats. Moreover, identifying strengths, weaknesses, opportunities, and threats can support the development of a resilient LNG supply chain. The recommended strategies can also mitigate possible risks associated, whilst utilizing available opportunities and strengths.

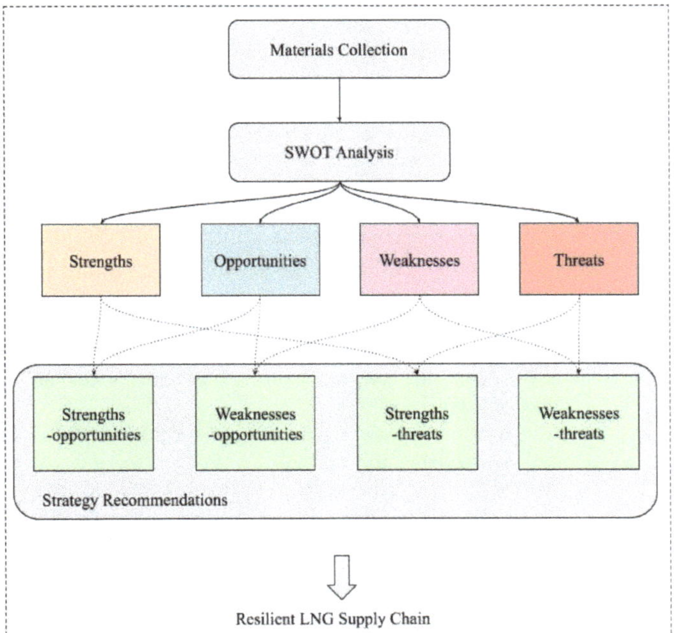

Figure 1. A flowchart of the applied methodology for the LNG supply chain via a SWOT analysis (Source: Compiled by the authors from [30]).

3. Results and Discussion

In the following sections, major accidents that could occur within the LNG supply chain are described, followed by mentioned experiences of LNG supply chain from around the globe. These experiences are characterized in four categories: Political and regulatory problems, safety and security issues, environmental effects, and reliability of new technologies. Then, the SWOT method findings are demonstrated where strengths, weaknesses, opportunities, and threats are mentioned. Finally, recommended strategies are listed based on the SWOT matrix.

3.1. Experiences in the LNG Supply Chain from around the World

Several significant accidents have occurred on the LNG supply chain. One of the substantial accidents occurred in 2013, when a 10,114 20-foot equivalent unit box ship operated by Hanjin Shipping, Hanjin Italy, collided with the 216,224 m^3 LNG carrier in the Malacca Strait. The LNG carrier carrying full LNG cargo for the Futtsu LNG terminal in Japan sailed from Ras Laffan in Qatar in December 2013. Fortunately, the cargo was not damaged, or else it could have been a full-scale disaster. The damage was limited to the forepeak. Both ships remained at the anchorage off Singapore after the accident. Futtsu LNG terminal refused to accept the cargo by damaged operating vessel. Thus, the LNG operating company decided to send another ship and transfer the cargo using a ship-to-ship operation in order to deliver the cargo on time. The damaged vessel was out of service for a particular time for maintenance [31].

Moreover, the LNG transport sector faced a new Black Swan on March 2021 when the containership Evergreen ran aground in the Suez Canal due to high-speed wind. This halted canal traffic, which averaged 50 transits a day, and caused high insecurity in global markets. The Suez Canal plays a key role in the global transit of LNG between the Atlantic basin markets and the Asia Pacific and the Middle East markets. A total of 686 LNG carriers transited the Suez Canal in 2020, comprising laden and ballast carriers. This includes 388 laden LNG vessels that crossed the canal, representing 7% of global LNG transportations. The Suez Canal blockade caused 370 ships to wait to pass through, of which 16 LNG tankers were involved. As there was no knowledge of how long it would take for the interruption to be eliminated, several LNG ships adopted alternative routes, notably to the Cape of Good Hope in Africa. This resulted in increased shipping prices as the shipping periods for LNG from the Middle East to Europe increased to 27 days, or 10 more days, instead of shipping via the Suez Canal. Meanwhile, some LNG ships chose to wait for a transit crisis solution. The blocking of the Suez Canal lasted one week, and the containership and the ship traffic began once again. In this respect, a major interruption to the market for LNG shipments was averted. Each day, about 12% of world commerce, approximately 1 million barrels of oil, and approximately 8% of LNG flow through the canal, and the canal's earnings was expected to decline by US$ 14M–15M for each day of the blockade. Prior to the epidemic, commerce through the Suez Canal accounted for 2% of Egypt's Gross Domestic Product [32].

Positive, negative, or mixed risk lessons from across the world may assist businesses and governments in responding to LNG's strong development and globalization. These lessons apply to all aspects of business, including the safe and secure passage of LNG vessels through congested waterways, the evaluation of technological innovations using the appropriate risk criteria, the management of uninformed public perceptions of unjustified LNG risks, and the assessment of political and regulatory risks. These lessons are important on their own and in combination when it comes to third-party access and multi-user implications. While the risk bank operates on all continents, experience in each nation is unique due to the nature of the government and regulatory structure. For instance, the United States and its territories serve as a testing ground for public views, security and safety, and environmental issues. Additionally, lessons about managing multiple-user terminals in a competitive setting may be learnt from US terminals. Numerous European receiving terminal projects may benefit from the growing US experience, particularly legal and multi-user issues [33]. As a result, the global LNG supply chain's acquired lessons are classified into four categories: (i) regulatory and political problems; (ii) issues of security and safety; (iii) consequences for the environment; and (iv) reliability of new technology.

3.1.1. Regulatory and Political Problems

Political and regulatory concerns exist throughout the sanctioning process and include the apparent threat of not obtaining official approval and the potential of government intervention and delays, which may decrease the competitiveness of a proposition. Regulation changes or governmental requirements may have an effect on operations and, consequently, project profitability during the post-sanction period. Political and regulatory risks may be successfully managed in some LNG-dependent countries, via a well-defined national strategy and successful government–business partnerships [34]. However, there may be challenges to expanding LNG production due to governance, legal issues, regulatory structure, and layers of authorization [35].

3.1.2. Issues of Security and Safety

For the LNG industry, safety has always been a top concern. Onshore, this has been handled via LNG-specific design standards, vessel type classification, operating procedures, and stringent operational protocols. Diverse fields use various safety regimes to avoid unwanted events, the most important difference occupying a space between consequence-based and risk-based management. In the United States and Asia, incidents are investigated using a consequence-based method based on National Fire Protection Association (NFPA) 59A. The NFPA approach requires less research than the risk-based strategy and may be simpler for the general public to understand [36]. However, this method precludes the evaluation of the efficacy of project-specific measures targeted at decreasing the probability of occurrences by concentrating only on worst-case scenario accident scenarios and disregarding the frequency of events. For instance, laws of Canada outline a risk-based approach, although it does not include specific approval standards, and Canadians rely on the NFPA for plant architecture [37]. A risk-based approach requires the acquisition of knowledge and tools for conducting cutting-edge risk assessments. Rabaska effectively implemented a risk-based approach for constructing an underground unloading line, thus minimizing public hazards [38]. If only the implications were considered without regard for probability, this design improvement would not have been suggested. As security has grown more inextricably linked to risk management by governments, businesses have realized that they would face greater examination for security and safety, particularly downstream. As public opinion and project approval are affected by safety and security, they are critical for the profitability of a project and its survival. The risk analysis of terrorist attacks on the port's LNG storage tanks was conducted and verified through case studies. The defense capabilities of the police and different prevention strategies were studied [39]. The results show significant differences in accident consequences between different defensive and emergency response forces. Different scenario analyses have been conducted, providing a theoretical basis and method support for public security and urban risk management departments' security prevention decisions [39].

3.1.3. Consequences for the Environment

Natural gas has developed a reputation for being the most environmentally friendly fossil fuel available. Nonetheless, the LNG business must continuously preserve that image by limiting environmental effects and reducing the likelihood of a disaster. The Paris Agreement was initiated and aimed to substantially reduce global GHGs to limit the global temperature increase to 2 °C above preindustrial levels while pursuing the means to limit the increase to 1.5 °C [40]. Moreover, International Maritime Organization (IMO) 2020 mandates a maximum sulphur content of 0.5% in marine fuels globally [41]. In 2018, CO_2 emissions from shipping represented 740 MT. In this regard, the IMO has been actively engaged in a global approach to enhance ships' energy efficiency further and develop measures to reduce GHG emissions from ships and provide technical cooperation and capacity-building activities [41]. Although environmental protection is a global concern, LNG companies anticipate using cutting-edge technology and operating in more active environmental areas. Offshore projects have unique difficulties associated with air pollution and the potential for marine life disruption during development and operation. There are many instances of environmental conflicts, most notably in the United States. To avoid harming the maritime environment, McMoRan's Louisiana Main Pass Energy Hub had to switch to a closed-loop system from its original opened-loop design [42]. Additionally, concerns regarding the impact of massive, permanent development in Long Island Sound on marine life are being actively addressed by Broadwater LNG [43]. Similarly, governments and businesses in Europe, Canada, and Australia have cooperated to improve the design of the built environment. For instance, only after strict measures to preserve endangered marine species were put in place by Western Australia's State Environmental Protection Program was the Gorgon project given the green light [44]. Australia's North West Shelf is now undergoing a comprehensive Joint Environmental Management Study, which presents both difficulties and opportunities for development [45]. Recent experience demonstrates that players in the LNG industry have successfully managed environmental problems via rigorous planning and risk assessment. This is a significant step forward for expanding the LNG supply chain expansion, whilst also reducing environmental effects.

3.1.4. Reliability of New Technology

As the LNG market grows more competitive, businesses are moving away from tried-and-true conventional technologies in favor of cutting-edge procedures and equipment to save money and improve flexibility, safety, and security. Alternatively, technological advances may result in increased training expenses and unanticipated problems, increasing the level of uncertainty regarding reliability to a point where many operators will not tolerate it. The most important advances have occurred in the building of offshore receiving terminals, where many companies believe that new technology would mitigate the real or perceived risks associated with a land-based posture. The Energy Bridge project aims to design a scalable method for LNG discharge. The Energy Bridge project entails equipping LNG tankers with an onboard regasification plant, a swivel mooring turret and manifold, as well as an offshore buoy and pipeline system to discharge the LNG into the land-based gas grid, all without the requirement for a complete LNG terminal. Additionally, the project allows boats to discharge LNG directly onshore at any port equipped with a gas receiving line, eliminating the requirement for a dedicated LNG single buoy mooring [46]. Nevertheless, there are dangers that should be considered; for example, a leak during berthing may cause catastrophic damage because of cryogenic loads on the terminal's top level. Auxiliary inventions inspired by FRSUs include the Submerged Turret Loading (STL) buoy [47]. As the industry grows, it faces the Research and Development (R&D) risk that high-tech firms are used to prioritizing investment in a promising concept that may or may not become economically successful in the future. For instance, The TORP HiLoad from Remora Technology, which uses suction cups in conjunction with the vessel's hull, has gained Det Norske Veritas (DNV) and the American Bureau of Shipping (ABS) Approval in Principle [48]. These credentials are needed for LNG developers to work with risk-averse businesses, particularly financial institutions. To improve efficiency, new technologies are always being created, as shown by the much bigger vessel sizes in the new fleets, such as the Q-Flex and Q-Max, each of which has a maximum capacity of 210,000 and 266,000 m^3, respectively [49]. New technology cannot be used to bypass existing quality standards and protections as the industry recognizes. Indeed, the new technology requires additional tests and measures to operate correctly straight out of the box. New technology can only help reduce overall risk if it is regarded as an integral component of the supply chain that must be handled as part of a complete risk management plan.

4. SWOT Analysis

This section presents strengths, weaknesses, opportunities, and threats of the LNG supply chain in Qatar. The obtained internal and external factors of the analysis are demonstrated in Figure 2.

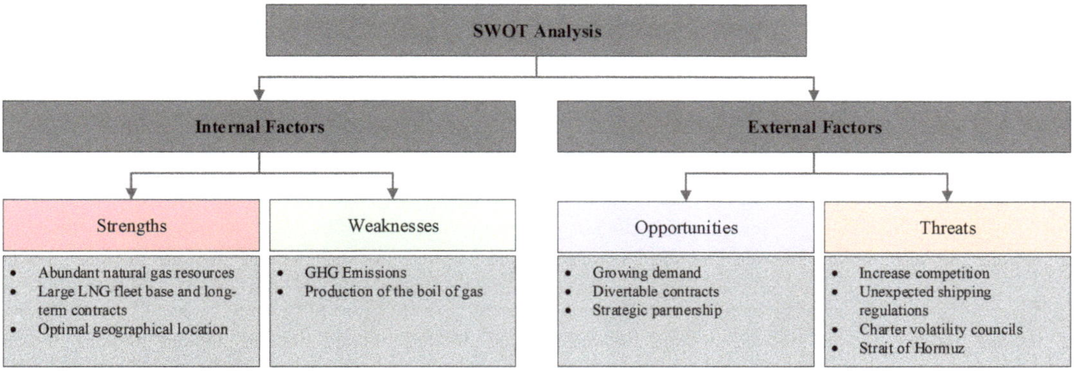

Figure 2. A summary of obtained findings (internal and external factors) from SWOT analysis for the State of Qatar.

4.1. Strengths

The term "strengths" refers to the areas where a company succeeds and differs from its competitors. Strengths include abundant natural gas resources, a large LNG fleet base, long-term contracts, and optimal geographical location.

4.1.1. Abundant Natural Gas Resources

There is 872 tcm of natural gas reserves in Qatar, making it the third-largest single non-associated gas resource in the world after Russia and Iran. This is equal to 12% of global natural gas production. These reserves are mostly confined to Qatar's vast offshore North Field, which covers half the country's land area [50]. The use of these vast reserves is a national priority to maintain national growth and prosperity. The North Gas Field is a stimulant for development in gas production and uses. This is due to the quantity of natural gas in one ideal geographical location, and meteorological conditions that ease production and lower costs in comparison to other locations across the globe. Considering the advantageous characteristics of the North Gas Field in terms of enormous reserves, in addition to its reputation as a source of cleaner energy with a long lifespan, the State of Qatar has embarked on an integrated strategy to develop the field and optimize its resources by securing new investment for the exportation of liquefied gas, the establishment of new gas-based industry segments, and the construction of the necessary facilities.

4.1.2. Large LNG Fleet Base and Long and Short-Term Contracts

Qatar is the world's biggest LNG exporter, with a fleet of 69 LNG ships (24 conventional, 31 Q-Flex, 14 Q-Max) using the most recent technology to guarantee that gas is transported safely, in ecologically friendly manner, as well as cost-effectively, across the world's oceans [51]. A major development project in the North Field is now underway, and it will increase Qatar's present LNG productivity of 77 MTA by 2025 to 110 MTA. To meet the demands of the North Field Expansion Project and Qatar's future LNG ship fleet obligations, Qatar intends to preserve LNG shipbuilding capacity up to 100 new LNG ships. This will be the world's biggest LNG shipbuilding project [52]. Due to the high number of long-term contracts Qatar has signed, there will be less fluctuation in cash flow, allowing big, long-term capital expenditures to be financed with more debt. As a result of increased leverage, borrowing costs are reduced. On the downside, long-term contracts prevent the parties to them from benefiting from lucrative short-term trading scenarios [53]. Table 1 shows some of the signed long-term contracts. However, spot and short-term contracts have increased after a drop in spot prices, which keeps long-term deals less attractive [54].

Table 1. Some examples of recent long-term contracts between Qatar and other countries.

Exporting Country	Importing Country	Type of Contract	Starts and Ends	Duration	Reference
Qatar	South Korea	Long term contract	2025–2045	20	[55]
Qatar	Singapore	Long term contract	2023–2033	10	[56]
Qatar	Pakistan	Long term contract	2018–2038	20	[57]

4.1.3. Optimal Geographical Location

Qatar is a Middle Eastern peninsula bordered by Saudi Arabia by land, and extends 160 km into the Arabian Gulf. Due to its central location between the West and the East, Qatar is an attractive LNG export destination because of the reduced transportation costs. Near the northeastern point of Qatar, is Ras Laffan Port, which is just 67 km from North Field. It is a deep-water port with one of the largest artificial harbors in the world, and comprises of six LNG terminals, six liquid product terminals, one sulphur-bearing terminal, and six dry cargo terminals presently in operation [58]. Moreover, the Ras Laffan Port is one of the largest energy export facilities, able to accommodate a broad variety of vessels, including the latest Q-Flex and Q-Max LNG carriers.

4.2. Weaknesses

Weaknesses prevent an organization from operating at maximum efficiency. These are areas in which the company may improve to stay competitive. Weaknesses include GHG emissions and production of Boil-Off Gas (BOG).

4.2.1. GHG Emissions

Globally, many organizations and governments are increasingly concerned about the impact of fossil fuels on the environment and are thus are exploring alternative energy sources [59]. Natural gas is one of those fossil fuels, although it is the cleanest, where GHG is produced throughout its supply chain phases, including production, transportation, storage, loading and unloading, and utilization. In the LNG production phase, natural gas is cooled to its liquid state to a temperature below −160 °C to produce LNG. The source of energy required to drive the liquefaction process and produce LNG is commonly sourced from fossil fuels, where the production of 1 MJ of LNG, can result in approximately 10.29 g of GHG emissions [60]. In the LNG transportation phase, transportation of LNG occurs in vessels that are fueled by fossil fuels, where the generated GHG emissions while transporting LNG by heavy fuel oil (HFO) and by an LNG fueled ocean tanker for a distance of 10,000 nautical miles are approximately 0.65 kg of CO_2 eq for 1 kg of HFO and 0.6 kg of CO_2 eq for 1 kg of LNG, respectively [60]. Moreover, GHG emissions are produced during the utilization of LNG, where approximately 57.5 g/MJ of GHG emissions are produced when LNG is burned [60]. This indicates that GHG emissions are produced in all the phases of the LNG supply chain.

4.2.2. Production of Boil Off-Gas

Boil off-gas (BOG) is an unavoidable phenomenon occurring in different phases of the LNG supply chain, namely land storage and transportation. In the transportation phase, BOG occurs in conventional vessels, whereas the vessels (Q-Flex and Q-Max) have a re-liquefaction unit that can capture BOG and convert it back into a liquid or utilize it directly as gas for propulsion purposes [61]. BOG production occurs in tanks due to temperature differences between the ambient temperature and LNG temperature. The generated BOG must be removed from LNG tanks to maintain the tank pressure. Removal of produced BOG in an LNG tank reduces LNG quantity, affecting LNG value from technical, economic, and environmental perspectives. For instance, the daily BOG production rate of storing LNG in a 160,000 m^3 land-storage tank is approximately 0.07%, and the daily BOG production rate from an LNG conventional tanker is 0.12% [62]. Production of BOG in the LNG supply chain also reduces the economic benefits of selling LNG. While accounting for produced BOG as a cost, both production and transportation costs increase, consequently reducing the economic benefits of selling LNG [63].

4.3. Opportunities

Opportunities relate to external circumstances that benefit an organization and may provide it with a competitive edge. Opportunities include growing demand, strategic partnership, and divertible contracts.

4.3.1. Growing Demand

Natural gas will be the fastest-growing fossil fuel, rising by 0.9% between 2020 and 2035 [53]. Due to the region's economic and demographic development, LNG consumption in Asia is rising much quicker than the rest of the world. Asia's LNG consumption has grown steadily throughout the first five months of 2021, with China accounting for the majority of the increase. China's LNG consumption is projected to rise by 11 MT this year, constituting more than half of the 18 MT increase in the worldwide LNG market over the forecast in 2021, and around one-third of global LNG demand growth in 2022. China's top position seems secure for years to come with its LNG demand backed by governmental backing and solid-gas market fundamentals [64]. The goal of China being carbon-neutral by 2060 is anticipated to drive an increase in the country's demand for LNG. Consequently, a supply–demand imbalance is likely to develop by the middle of this decade, with less new output entering the market than originally anticipated [65].

4.3.2. Strategic Partnerships

Various businesses are involved in energy transportation, ownership, operation, and the promotion of the LNG natural gas market, and are thus positioned as strategic partners in the LNG sector. As part of its strategy to develop new gas markets, Qatar, the world's largest LNG exporter, continues to expand its partnership with the largest energy and infrastructure conglomerate in Brazil to contribute to the development of the Brazil natural gas market through a third party [66].

4.3.3. Divertible Contracts

In order to provide the required LNG to the client at a competitive rate, pricing mechanisms can be used. In light of the growing purchasing power of buyers in the presently over-supplied market, this mechanism may provide buyers with a choice or a combination of price indices to consider. When combined with the short-term capacity and route flexibility, the power to turn back LNG cargoes at prices that the buyer deems uncompetitive is a potent combination for a successful LNG export strategy. Divertible contracts mostly applied to European contracts which the cargo diverted from one contract to another with higher price [67].

4.4. Threats

Threats are defined as events or circumstances that have the potential to cause damage to an organization. Threats include increased competition, unexpected shipping regulations, character volatility councils, and potential disturbances at the Strait of Hormoz.

4.4.1. Increase Competition

LNG Production

Unlike other major energy-producing countries, Qatar mainly focuses on LNG transportation, whereas other countries carry crude oil, processed products, and other items with LNG. The demand for vessels is enormous, and the employment of vessels that have their charters expired, and vessels presently under construction, have considerable competition. The long-term competition for LNG charters is based on the price, availability of vessels, size, age and condition of the ship, relationships with the LNG operator and quality, LNG experience, and the operator's reputation. There has been increased competition in the industry, with shipping and oil majors' firms indicating an interest in LNG transportation, as illustrated in Table 2. If the firm cannot compete with success, its revenues and market share would decrease in the long term.

Table 2. Potential competitors/companies to Qatar's LNG transportation capability (Source: Compiled by the authors from [68]).

Competitors	Location	Capacity 20-Foot Equivalent Unit (TEU)	Fleet Size
A.P. Moller—Maersk AS	Denmark	4.1 million	708
Mediterranean Shipping Company	Italy	3.8 million	560
Hapag-Lloyd AG	Germany	1.7 million	248
Evergreen Line	Taiwan	1.2 million	333

Shifting to Hydrogen/Renewable Energy

In the coming years, LNG is expected to be substituted by other renewable energy sources to limit a global temperature rise to 1.5 °C. Meanwhile, comprehensive action is needed to reduce CO_2 and methane emissions across all industries, including oil and gas. The world is increasingly transitioning to renewable energy sources to meet approximately up to 80% of global demand by 2050. As of today, clean-energy investment has been relatively flat to around 30–37% of global energy supply investment [69]. Hydrogen and other carbon-free energy carriers (i.e., ammonia) emerge as potential candidates for overseas energy transport.

4.4.2. Unexpected Shipping Regulations

LNG operations are affected by detailed rules, regulations, agreements, and standards in international waters. Strict legislation on disposal, cleaning of oil and other pollution, air emissions, water discharges, and ballast water control. LNG firms are regulated by many international bodies such as the International Maritime Organization, the European Union, the Pollution Control Authority in Norway, and many more organizations. Marine firms must comply with the amended requirements to keep their licenses for businesses. This requires higher investments and resources that influence LNG businesses' profitability.

4.4.3. Charter Volatility Councils

The market for tankers has been cyclical. The ability of LNG operators to make profitable use of their vessels would depend on influences by economic conditions in the tanker market, demand for petroleum and petroleum byproducts, global economic and financial conditions, changes in maritime transport, and other patterns. The charter prices rely on the quantity, the degradation rate of older ships, the transformation of tankers to other purposes, the number of out-of-service ships, and environmental concerns and laws. For example, charter prices of a medium-term charter business are changeable and closely linked to the current market rates. If the ships are not profitable or chartered, the financial flows of the firm may be affected.

4.4.4. Strait of Hormuz

The Strait of Hormuz is a waterway that connects the Arabian Gulf with the Gulf of Oman, and it is the sole maritime route that connects the Arabian Gulf and the Indian Ocean. In addition to the enormous quantities of oil products, condensate, and crude oil that flow through the strait, which are equal to 25% of worldwide oil consumption, the Strait of Hormuz is also one of the world's most significant chokepoints for transportation. The Strait also facilitates one-third of the world's LNG trade [70].

4.5. SWOT Matrix

SWOT matrix for the LNG supply chain of Qatar is presented in Table 3.

Table 3. SWOT matrix for the LNG supply chain of Qatar.

	Strengths (S)	Weaknesses (W)
Opportunities (O)	Strengths–Opportunities (SO) Strategies (S_1O_1) LNG expansion plan (S_2O_2) Adaptable and profitable plan (S_3O_3) Reliance by third-party	Weaknesses–Opportunities (WO) Strategies (W_1O_1) Environmental regulations and programs (W_2O_3) Enhance strategic partnerships
Threats (T)	Strengths–Threats (ST) Strategies (S_1T_1) LNG price reduction (S_3T_2) Flexible clause in the new LNG agreement (S_3T_4) Find alternative routes for LNG transportation	Weaknesses–Threats (WT) Strategies (W_1T_1) Clean LNG production (W_2T_2) Improve BOG capturing technologies and utilize BOG production effectively

4.6. Recommended Strategies

A portfolio of strategies has been created by combining internal factors encompassing "Strengths" and "Weaknesses" with external factors covering "Opportunities" and "Threats" in a SWOT analysis of Qatar's LNG supply chain. As a result, 10 strategies have been developed, as given in Figure 3.

4.6.1. Strengths–Opportunities (SO) Strategies

S_1O_1: Develop an LNG expansion plan, which can increase LNG production capacity to meet global growing demand. Noting that Qatar has implemented such a strategy, increasing LNG capacity production from 77 MTPA to 110 MTPA by implementing an expansion plan in the North Field [71].

S_2O_2: Prepare adaptable and profitable plans by applying pricing mechanisms, which describe a system in which the dynamics of supply and demand determine the cost of products and the changes in those prices. Therefore, selling LNG cargoes to higher-priced markets result in higher profits.

S_3O_3: Reliance on LNG producer third parties. Situated in an optimal geographical location can support strategic partnerships as an LNG producer third-party. The third-party owns and operates LNG cargos requested by one of the strategic partnerships to meet their requests.

4.6.2. Weaknesses–Opportunities (WO) Strategies

W_1O_1: Setting environmental regulations and supporting environmental programs, requesting emissions reduction by implementing carbon capture and storage technologies. In parallel, supporting innovative solutions that reduce produced emissions during the LNG supply chain.

W_2O_3: Enhance strategic partnerships to reduce LNG shipping distances; consequently, reduce BOG production during transportation of LNG.

4.6.3. Strengths–Threats (ST) Strategies

S_1T_1: Flexibility in varying LNG price. Since natural gas is abundant in Qatar and the number of LNG suppliers is increasing, with potentially lower production costs relative to other producers, reducing the selling price of LNG can attract more buyers.

S_3T_2: Develop flexible clauses in agreements in the case of new regulations, for instance unexpected shipping regulations.

S_3T_4: Identify alternative routes for transportation of natural gas. Since the Strait of Hormuz is the only path for LNG cargoes from Qatar, any political, geographical, environmental supply chain disruption can cause closure of this path and will disrupt LNG trading by sea. Therefore, identifying alternative routes and innovative solutions to transport natural gas/LNG from Qatar is necessary. An alternative solution can be transporting natural gas via a pipeline, however, transporting natural via pipeline is not feasible for distances over 2000 km. Thus, such a solution is optimal for shorter distances [72].

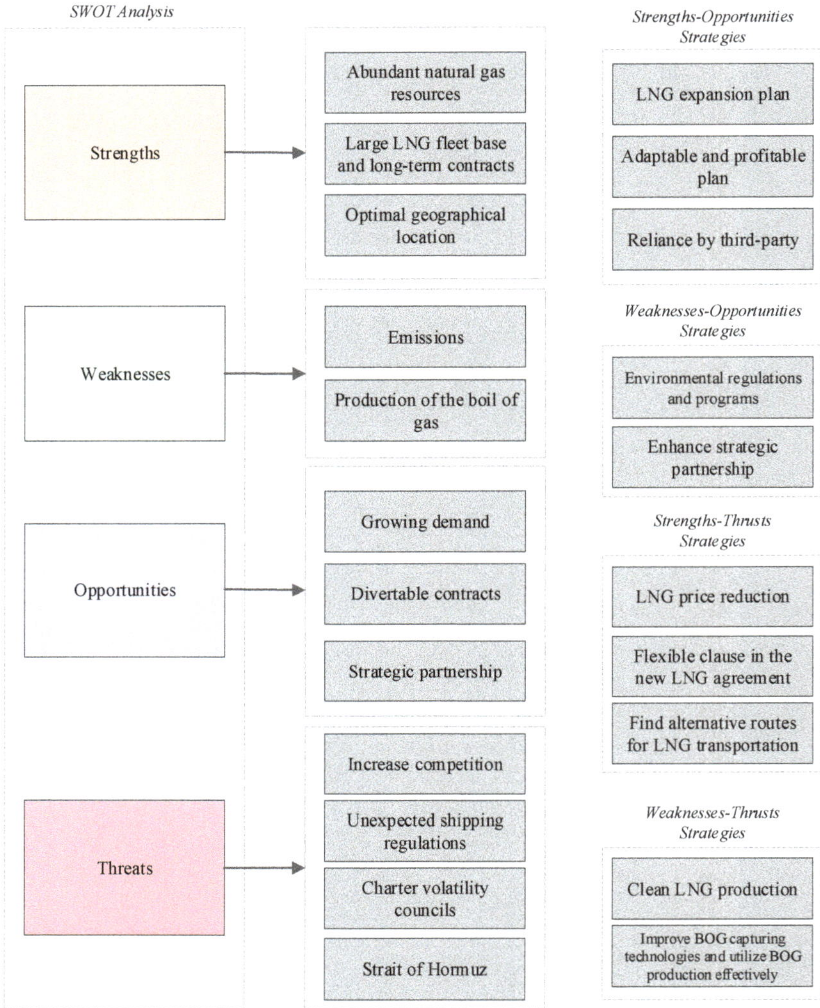

Figure 3. A summary of strategies obtained from SWOT analysis findings to be implemented for a resilient LNG supply chain.

4.6.4. Weaknesses–Threats (WT) Strategies

W_1T_1: Support production of clean LNG. As countries are aiming to reduce environmental emissions, the production of clean LNG will be favored. Therefore, LNG providers, who capture and utilize carbon emissions from the LNG supply chain are preferred.

W_2T_2: Improve BOG capturing technologies and utilize BOG production effectively. Since BOG production rates increase as the transportation distance of LNG increases, developing reliable and effective BOG capturing technologies is essential as these technologies can maintain LNG quantity while shipping trips [61].

Recommended strategies from SWOT analysis can support developing resilience in LNG supply chains. The findings from SWOT analysis, including abundant natural gas resources, large LNG fleet base and long-term contracts, and optimal geographical location are strengths that can enhance the resilience of the LNG supply chain. Furthermore, natural gas will be more resilient to accommodate the energy transition driven by climate change more than other fossil fuels since it is the cleanest fossil fuel [65]. Whereas, increased competition, unexpected shipping regulations, charter volatility are the threats that can exacerbate LNG supply chain vulnerability. In addition, opportunities obtained from SWOT analysis, namely growing demand, divertible contracts, and strategic partnership are factors that can enhance LNG supply chain resilience. Moreover, the following represents resilience LNG supply chain parameters from the LNG producer perspective based on Qatar's LNG company. Such parameters could support more informed decision-making in an efficient and reliable manner to accommodate customers' needs.

- Manage producer inventory: The inventory of LNG producers should be well planned to avoid tank top or tank bottom situations, which have an impact on vessel availability and deliver the cargo to the customer on time, respectively.
- Deliver cargo on time: The LNG producer is obligated to deliver the cargoes on the agreed time to avoid any customer end user shortage.
- Flexibility in cargo delivery: LNG producers or consumers may request to change either receiving the terminal or delivery date, and the flexibility to accommodate the raised request will add value to the requester. Moreover, the ability to deliver the cargo in an FSRU will maintain the resilience of the LNG supply chain.
- Meet customer needs: With the increasing LNG market demand, the LNG production capacity is to be increased to meet customer requirements.
- Hydrocarbon emissions: As per IMO, the carbon emissions have to be reduced to 50% by 2050 [29], which could be achieved by controlling the emissions produced during loading and transportation processes.
- Reach markets safely, efficiently, and reliably: LNG is a colorless, invisible, and non-toxic chilled methane gas with a temperature of -160 °C, which is inflammable until it comes into contact with air, at which point it quickly changes into vapors that are readily ignited. LNG fire has the potential to grow to 150 m in height and burn for an extended time. As a result, the most serious risks associated with LNG ships are explosions, fumigation, and spills.
- Cost optimization should be considered through the whole LNG supply chain, beginning from production to delivery, to recover the investment and operating costs of the LNG production and delivery.

5. Conclusions

As the demand for LNG continues to increase, the levels of risk and uncertainty within the sector also increase, which is the impetus for the development of resilient LNG supply chains and further developing resilience for the sector as a whole. This study begins by presenting major accidents that have affected the LNG supply chain. Demonstrating such accidents can provide insights into financial, technical, and environmental losses, which can be studied to eliminate such losses in similar future accidents. Moreover, examining experiences or lessons learned in global LNG supply chains can mitigate any potential risks. Any model for LNG resilience management can include the recommendations described within this article as part of its design. While resilient management techniques will not eliminate unexpected shocks, LNG businesses that incorporate them can be more robust. Models that include all significant best practices and lessons, both technical and economic, are more resilient and even less likely to leave policymakers perplexed about accounting for the so-called "unexpected" occurrences that occur in virtually every significant endeavor. Therefore, each policymaker must consider these historical lessons while evaluating LNG projects. In this study, risk drivers from the LNG supply chain are gathered and classified into four categories: Political and regulatory, safety and security, environmental effects, and reliability of new technologies. These experiences can support decision-makers in creating a resilient LNG supply chain. The limitation of the proposed method is that it can be subjective at times. Moreover, SWOT analysis is a one-dimensional assessment procedure in the categorization of attributes. However, some factors may be assigned for multi-attributes. Future research should focus on quantitative methods that can support analysis considering all aspects that may affect the resilience of the LNG supply chain. Furthermore, future can work validate some of the findings with experienced practitioners in the field.

The SWOT method broadly assesses the status of Qatar's LNG supply chain. It identifies strengths, weaknesses, opportunities, and threats within Qatar's supply chain. The strengths include abundant natural gas resources, a large LNG fleet base and long-term contracts, and optimal geographical location. The major weaknesses include the production of emissions and BOG (which impacts the carbon cost of the LNG cargo), while the major opportunities are the growing demand for LNG, issuing divertible contracts, and obtaining strategic partnerships. The major threats are increased competition, unexpected shipping regulations, charter volatility councils, and the fact there is only one path for LNG transportation, which is the Strait of Hormuz. Then, a portfolio of strategies is created by combining internal factors covering strengths and weaknesses with external factors covering opportunities and threats in a SWOT analysis of Qatar's LNG supply chain. Moreover, recommendations to enhance the resilience of the LNG supply chain are listed below:

- Develop a strategy for LNG development that allows the industry to boost its production capacity to satisfy the world's rising demand.
- Implement pricing mechanisms to create adaptive and lucrative business strategies.
- Support production of clean LNG and improve carbon capture methods.
- Improve strategic alliances to minimize LNG shipping distances, which can in turn reduce BOG generation throughout the LNG transportation process.
- Continue to minimize emissions throughout the supply chain to decrease the carbon content of the LNG cargo in line with growing demand from consumers for low carbon LNG.
- Create a provision in the agreement that allows for flexibility. In the event, that a new shipping law is implemented, a provision like this may address any unanticipated shipping restrictions.
- Support clean transportation options when delivering LNG cargoes. This can be achieved by utilization of cleaner fuels during combustion, for instance, using LNG as bunker fuel instead of using heavy fuel oil.

Author Contributions: Conceptualization, T.A.-A., Y.B., S.A.-H. and M.A.-B.; methodology, S.A.-H. and M.A.-B.; formal analysis, S.A.-H. and M.A.-B.; investigation, S.A.-H. and M.A.-B.; writing—original draft preparation, S.A.-H. and M.A.-B.; writing—review and editing, T.A.-A. and Y.B.; supervision, T.A.-A. and Y.B.; funding acquisition, T.A.-A. and Y.B. All authors have read and agreed to the published version of the manuscript.

Funding: The authors acknowledge the support provided by the Hamad Bin Khalifa University, Qatar Foundation, Qatar. The APC was funded by Hamad Bin Khalifa University, Qatar Foundation, Qatar.

Conflicts of Interest: The authors declare no conflict of interest.

Nomenclature

BOG	Boil-Off Gas
CO_2	Carbon Dioxide
Eq	Equivalent
FSRU	Floating Storage Regasification Unit
GHG	Greenhouse gas
IMO	International Maritime Organization
km	kilometers
LNG	Liquified Natural Gas

MT	Million Ton
MMBtu	Million British Thermal Unit
m^3	Cubic meter
m	meter
MJ	Megajoule
NO_x	Nitrogen Oxide
O	Opportunities
SWOT	Strengths, Weaknesses, Opportunities, Threats
S	Strengths
SO	Strengths–Opportunities
ST	Strengths–Threats
SO_2	Sulfur Dioxide
T	Threats
tcm	trillion cubic feet
TEU	20-foot equivalent unit
W	Weaknesses
WO	Weaknesses–Opportunities
WT	Weaknesses–Threats

References

1. Elengy. LNG: An Energy of the Future. 2020. Available online: https://www.elengy.com/en/lng/lng-an-energy-of-the-future.html (accessed on 10 October 2021).
2. IGU. Natural Gas: The Energy for Today and the Future. 2019. Available online: https://www.igu.org/ (accessed on 10 October 2021).
3. Razmanova, S.; Steblyanskaya, A. Arctic LNG cluster: New opportunities or new treats? In Proceedings of the IOP Conference Series: Earth and Environmental Science, Saint Petersburg, Russia, 18–19 March 2020; Volume 539. [CrossRef]
4. LNG Market Trends and Their Implications. Available online: https://doi.org/10.1787/90c2a82d-en (accessed on 10 November 2021).
5. IGU. 2019 World LNG Report. 2019. Available online: https://www.igu.org/wp-content/uploads/2019/06/IGU-Annual-Report-2019_23.pdf (accessed on 15 October 2021).
6. Jaganathan, J. *Asian LNG Spot Price Reaches Record High of $34.47/mmbtu-Platts Data*; Reuters: New York, NY, USA, 2021.
7. Mokhatab, S.; Mak, J.Y.; Valappil, J.; Wood, D.A. LNG Fundamentals. In *Handbook of Liquefied Natural Gas*; Mokhatab, S., Mak, J.Y., Valappil, J.V., Wood, D.A., Eds.; Gulf Professional Publishing: Houston, TX, USA, 2014; pp. 1–106. [CrossRef]
8. Berle, Ø.; Norstad, I.; Asbjørnslett, B.E. Optimization, risk assessment and resilience in LNG transportation systems. *Supply Chain Manag. Int. J.* **2013**, *18*, 253–264. [CrossRef]
9. Southwick, S.M.; Bonanno, G.A.; Masten, A.; Panter-Brick, C.; Yehuda, R. Resilience definitions, theory, and challenges: Interdisciplinary perspectives. *Eur. J. Psychotraumatol.* **2014**, *5*, 25338. [CrossRef]
10. Gatto, A.; Busato, F. Energy vulnerability around the world: The global energy vulnerability index (GEVI). *J. Clean. Prod.* **2020**, *253*, 118691. [CrossRef]
11. Martorell, S. Formal vulnerability assessment: A methodology for assessing and mitigating strategic vulnerabilities in maritime supply chains. In *Reliability, Risk, and Safety, Three Volume Set*; CRC Press: Boca Raton, FL, USA, 2009; pp. 1073–1080.
12. Christopher, M.; Peck, H. Building the Resilient Supply Chain. *Int. J. Logist. Manag.* **2004**, *15*, 1–14. [CrossRef]
13. Goers, S.; Rumohr, F.; Fendt, S.; Gosselin, L.; Jannuzzi, G.; Gomes, R.; Sousa, S.; Wolvers, R. The Role of Renewable Energy in Regional Energy Transitions: An Aggregate Qualitative Analysis for the Partner Regions Bavaria, Georgia, Québec, São Paulo, Shandong, Upper Austria, and Western Cape. *Sustainability* **2020**, *13*, 76. [CrossRef]
14. Vasudevan, R.A. *SWOT-PESTEL Study of Constraints to Decarbonization of the Natural Gas System in the EU: Tech-no-Economic Analysis of Hydrogen Production in Portugal*; KTH Industrial Engineering and Management: Stockholm, Sweden, 2021.
15. Economic and Social Commission for Western Asia. *Energy Vulnerability in the Arab Region*; ESCWA: Beirut, Lebanon, 2019.
16. Satta, G.; Parola, F.; Vitellaro, F.; Morchio, G. LNG Bunkering Technologies in Ports: An Empirical Application of the SWOT Analysis. *KMI Int. J. Marit. Aff. Fish.* **2021**, *13*, 1–21. [CrossRef]
17. Domeisen, N.; Sousa, P. SWOT analysis: NGOs as partners. In *International Trade Forum*; International Trade Centre: Geneva, Switzerland, 2006; p. 7.
18. Emenike, S.N.; Falcone, G. A review on energy supply chain resilience through optimization. *Renew. Sustain. Energy Rev.* **2020**, *134*, 88–110. [CrossRef]
19. Bouwmeester, M.C.; Oosterhaven, J. Economic impacts of natural gas flow disruptions between Russia and the EU. *Energy Policy* **2017**, *106*, 288–297. [CrossRef]
20. Sesini, M.; Giarola, S.; Hawkes, A.D. The impact of liquefied natural gas and storage on the EU natural gas infrastructure resilience. *Energy* **2020**, *209*, 118367. [CrossRef]

21. Libby, B.; Christiansen, A. Building resilience to climate-change-related disruptions in the global LNG supply chain. In Proceedings of the 33rd International Conference of the System Dynamics Society, Cambridge, MA, USA, 19–23 July 2015.
22. Sönnichsen, N. Global Operational LNG Export Capacity by Country. 2021. Available online: https://www.statista.com/statistics/1262074/global-lng-export-capacity-by-country/ (accessed on 15 November 2021).
23. Temizer, M. Qatar to Raise LNG Production by 64% to 126 mtpa by 2027. 2019. Available online: https://www.aa.com.tr/en/energy/general/qatar-aims-to-raise-lng-production-by-64-by-2027/27501 (accessed on 10 October 2021).
24. Speth, C. *The SWOT Analysis: A Key Tool for Developing Your Business Strategy*; 2015; pp. 1–27. Available online: https://www.50minutes.com/title/the-swot-analysis/ (accessed on 10 September 2021).
25. D'Adamo, I.; Falcone, P.M.; Gastaldi, M.; Morone, P. RES-T trajectories and an integrated SWOT-AHP analysis for biomethane. Policy implications to support a green revolution in European transport. *Energy Policy* **2020**, *138*, 111220. [CrossRef]
26. Chen, W.-M.; Kim, H.; Yamaguchi, H. Renewable energy in eastern Asia: Renewable energy policy review and comparative SWOT analysis for promoting renewable energy in Japan, South Korea, and Taiwan. *Energy Policy* **2014**, *74*, 319–329. [CrossRef]
27. Uhunamure, S.; Shale, K. A SWOT Analysis Approach for a Sustainable Transition to Renewable Energy in South Africa. *Sustainability* **2021**, *13*, 3933. [CrossRef]
28. Fertel, C.; Bahn, O.; Vaillancourt, K.; Waaub, J.-P. Canadian energy and climate policies: A SWOT analysis in search of federal/provincial coherence. *Energy Policy* **2013**, *63*, 1139–1150. [CrossRef]
29. PESTLEanalysis Contributor. 10 Reasons to Use SWOT Analysis for Your Company. 2017. Available online: https://pestleanalysis.com/10-reasons-to-use-swot-analysis-for-your-company/?nowprocket=1 (accessed on 12 September 2021).
30. Deshmukh, S. Materials Management: An Integrated Systems Approach. *J. Adv. Manag. Res.* **2015**, *12*, 226–228. [CrossRef]
31. Writer, S. Qatar LNG Vessel Involved in Collision. 2014. Available online: https://www.oilandgasmiddleeast.com/products-services/article-11636-qatar-lng-vessel-involved-in-collision (accessed on 10 October 2021).
32. Russon, M.-A. The Cost of the Suez Canal Blockage. *BBC News*, 29 March 2021. Available online: https://www.bbc.com/news/business-56559073(accessed on 25 September 2021).
33. Nafday, A.M. Public Safety Appraisal for Siting Marine Liquefied Natural Gas Terminals in California. *J. Energy Eng.* **2015**, *141*, 3. [CrossRef]
34. Ministery of Development Planning and Statistics. Qatar Second National Development Strategy 2018–2022. 2018. Available online: https://www.psa.gov.qa/en/knowledge/Documents/NDS2Final.pdf (accessed on 15 October 2021).
35. Pascal, L. Developments in the Venezuelan Hydrocarbon Sector. *Law Bus. Rev. Am.* **2009**, *15*, 531.
36. NFPA 59A. *Standard for the Production, Storage, and Handling of Liquefied Natural Gas (LNG)*; National Fire Protection Association: Quincy, MA, USA, 2009; p. 63.
37. Canadian Food Inspection Agency. Enhancing Risk Analysis: A More Systematic and Consistent Approach. 2014. Available online: https://inspection.canada.ca/DAM/DAM-aboutcfia-sujetacia/STAGING/text-texte/cfia_acco_risk_analysis_consult_1401881213845_eng.pdf (accessed on 12 November 2021).
38. MIACC. Project Rabaska-Implementation of an LNG Terminal and Related Infrastructure. 2007. Available online: https://files.pca-cpa.org/pcadocs/bi-c/1.%20Investors/3.%20Exhibits/C0530.pdf (accessed on 10 September 2021).
39. Zhu, R.; Hu, X.; Bai, Y.; Li, X. Risk analysis of terrorist attacks on LNG storage tanks at ports. *Saf. Sci.* **2021**, *137*, 105192. [CrossRef]
40. Melissa, D. Paris Climate Agreement: Everything You Need to Know. 2018. Available online: https://www.nrdc.org/stories/paris-climate-agreement-everything-you-need-know (accessed on 10 September 2021).
41. IMO. IMO 2020–Cutting Sulphur Oxide Emissions. 2020. Available online: https://www.imo.org/en/MediaCentre/HotTopics/Pages/Sulphur-2020.aspx (accessed on 15 October 2021).
42. Verdict Media Ltd. n.d. McMoRan Exploration Submits LNG Terminal Plan for Main Pass Energy Hub. Available online: https://www.hydrocarbons-technology.com/projects/main-pass/ (accessed on 15 October 2021).
43. Byer, D.; Saffert, H. Liquefied Natural Gas: An Expensive, Dirty, Foreign Fossil Fuel That Threatens Our Natural Gas Energy Independence. 2008. Available online: https://www.cleanoceanaction.org/fileadmin/editor_group1/Issues/LNG/Final_Executive_Summary_LNG_Report.pdf (accessed on 12 October 2021).
44. Review, P.E. Gorgon Gas Development Revised and Expanded Proposal. 2008. Available online: https://www.epa.wa.gov.au/sites/default/files/EPA_Report/2937_Rep1323GorgonRevPer30409.pdf (accessed on 10 September 2021).
45. CSIRO. North West Shelf Joint Environmental Management Study. In Proceedings of the West Australian Marine Science Conference, Perth, Australia, 25–26 October 2007; p. 40.
46. Songhurst, B. Floating LNG Update–Liquefaction and Import Terminals. 2019. Available online: https://www.oxfordenergy.org/wpcms/wp-content/uploads/2019/09/Floating-LNG-Update-Liquesfaction-and-Import-Terminals-NG149.pdf (accessed on 10 September 2021).
47. NOV Inc. Submerged Turret Loading. 2021. Available online: https://www.nov.com/products/submerged-turret-loading (accessed on 20 November 2021).
48. Golar LNG Invests in TORP Technology. Golar LNG. 2011. Available online: https://www.golarlng.com/investors/press-releases/archive.aspx (accessed on 10 October 2021).
49. Qatargas Oper Co. Ltd. Qatargas; Qatargas' Chartered Fleet. Available online: http://www.qatargas.com/english/operations/qatargas-chartered-fleet (accessed on 15 September 2021).

50. Fawthrop, A. Profiling the Top Five Countries with the Biggest Natural Gas Reserves. 2021. Available online: https://www.nsenergybusiness.com/features/biggest-natural-gas-reserves-countries/ (accessed on 10 October 2021).
51. Nakilat. Our Fleet. 2021. Available online: https://www.nakilat.com/our-fleet/ (accessed on 10 September 2021).
52. Bajic, A. Qatar Petroleum Issues Tender to Charter LNG Fleet. 2021. Available online: https://www.offshore-energy.biz/qatar-petroleum-issues-tender-to-charter-lng-fleet/ (accessed on 17 September 2021).
53. Agosta, A.; Dediu, D.; Heringa, B. *Global Gas Outlook to 2050*; McKinsey & Company: Brussels, Belgium, 2021.
54. NDTV. Share of Short-Term LNG Deals to Rise in Domestic Consumption: GAIL. 2020. Available online: https://www.ndtv.com/business/natural-gas-market-news-share-of-short-term-lng-deals-to-rise-in-domestic-consumption-says-gail-2316805 (accessed on 10 October 2021).
55. Verdict Media Ltd. Qatar Petroleum Finalises 20-Year LNG Supply Deal with South Korea. 2021. Available online: https://www.offshore-energy.biz/qatar-petroleum-inks-20-year-lng-agreement-with-kogas/ (accessed on 10 September 2021).
56. Enrdata. Qatar Petroleum Signs a New Long-Term LNG Contract with Singapore. 2020. Available online: https://www.enerdata.net/publications/daily-energy-news/qatar-petroleum-signs-new-long-term-lng-contract-singapore.html (accessed on 13 November 2021).
57. Qatar Pet. Qatargas and Pakistan's GEIL Sign 20-Year LNG Sales Agreement. 2018. Available online: https://www.qp.com.qa/en/Pages/BannerAdvertisement.aspx?imgname=Banner+English+01.jpg (accessed on 14 October 2021).
58. Qatar Pet. Ports Information. 2021. Available online: https://qp.com.qa/en/marketing/Pages/PortsInformation.aspx (accessed on 12 October 2021).
59. Jawerth, N. What is the clean energy transition and how does nuclear power fit in? *IAEA Bull.* **2020**, *61*, 4–5.
60. Al-Breiki, M.; Bicer, Y. Comparative life cycle assessment of sustainable energy carriers including production, storage, overseas transport and utilization. *J. Clean. Prod.* **2021**, *279*, 123481. [CrossRef]
61. Al-Kubaisi, A. An Insight into the World's Largest LNG Ships. In Proceedings of the IPTC 2008: International Petroleum Technology Conference European Association of Geoscientists & Engineers, Kuala Lumpur, Malaysia, 3–5 December 2008. [CrossRef]
62. Al-Breiki, M.; Bicer, Y. Investigating the technical feasibility of various energy carriers for alternative and sustainable overseas energy transport scenarios. *Energy Convers. Manag.* **2020**, *209*, 112652. [CrossRef]
63. Al-Breiki, M.; Bicer, Y. Comparative cost assessment of sustainable energy carriers produced from natural gas accounting for boil-off gas and social cost of carbon. *Energy Rep.* **2020**, *6*, 1897–1909. [CrossRef]
64. Thompson, G. China Becomes the World's Largest LNG Market. 2021. Available online: https://www.compressortech2.com/news/china-to-become-world-s-largest-lng-market-in-2021/8013108.article (accessed on 10 September 2021).
65. Chestney, N. Global LNG Demand Expected to Almost Double by 2040: Shell. 2021. Available online: https://www.reuters.com/article/us-shell-lng-idUSKBN2AP15S (accessed on 14 September 2021).
66. Qatar Pet. Qatar Petroleum Affiliate Ocean LNG Limited Signs a Long Term SPA with Brazil CELSE. 2021. Available online: https://www.qp.com.qa/en/Pages/BannerAdvertisement.aspx?imgname=QP+OLNG+-+CELSE+LNG+SPA+-+English.jpg (accessed on 17 September 2021).
67. Power Africa. Understanding Natural Gas and LNG Options. 2016. Available online: https://www.energy.gov/sites/prod/files/2016/12/f34/Understanding%20Natural%20Gas%20and%20LNG%20Options.pdf (accessed on 10 September 2021).
68. The World's Biggest Shipping Companies in 2020. 2020. Available online: https://www.ship-technology.com/features/the-ten-biggest-shipping-companies-in-2020/ (accessed on 14 October 2021).
69. McKinsey & Company. The Impact of Decarbonization on the Gas and LNG Industry. 2021. Available online: https://www.mckinsey.com/industries/oil-and-gas/our-insights/the-impact-of-decarbonization-on-the-gas-and-lng-industry (accessed on 14 September 2021).
70. U.S. Energy Information Administration. Strait of Hormuz Is Chokepoint for 20% of World's Oil-Today in Energy-U.S. 2012. Available online: https://www.eia.gov/todayinenergy/detail.php?id=39932 (accessed on 8 October 2021).
71. Qatar Pet. Qatar Petroleum Constructs the World's Largest LNG Project Ever, including Substantial CO2 Capture & Sequestration. 2021. Available online: https://www.euro-petrole.com/qatar-petroleum-constructs-the-worlds-largest-lng-project-ever-including-substantial-co2-capture-sequestration-n-i-21773 (accessed on 10 April 2021).
72. Bittante, A.; Jokinen, R.; Pettersson, F.; Saxén, H. Optimization of LNG Supply Chain. *Comput. Aided Chem. Eng.* **2015**, *37*, 779–784. [CrossRef]

energies
MDPI
Article

Economic Feasibility of Green Hydrogen Production by Water Electrolysis Using Wind and Geothermal Energy Resources in Asal-Ghoubbet Rift (Republic of Djibouti): A Comparative Evaluation

Mohamed Osman Awaleh [1,*], Abdi-Basid Adan [1,*], Omar Assowe Dabar [1,*], Mohamed Jalludin [1], Moussa Mahdi Ahmed [1] and Ismael Abdillahi Guirreh [2]

[1] Centre d'Etudes et de Recherches de Djibouti (CERD), Institut des Sciences de la Terre, Route de l'aeroport, Djibouti City B.P. 486, Djibouti; mohamed.jalludin@gmail.com (M.J.); moussa.mahdi@chemist.com (M.M.A.)
[2] Département Logistique & Transport, Institut Universitaire de Technologie, Université de Djibouti, Croisement RN2-RN5, Balbala B.P. 1904, Djibouti; ismael.abdillahi@gmail.com
* Correspondence: awaleh@gmail.com (M.O.A.); Abdi-Basid@outlook.com (A.-B.A.); assowe440@gmail.com (O.A.D.)

Abstract: The Republic of Djibouti has untapped potential in terms of renewable energy resources, such as geothermal, wind, and solar energy. This study examines the economic feasibility of green hydrogen production by water electrolysis using wind and geothermal energy resources in the Asal–Ghoubbet Rift (AG Rift), Republic of Djibouti. It is the first study in Africa that compares the cost per kg of green hydrogen produced by wind and geothermal energy from a single site. The unit cost of electricity produced by the wind turbine (0.042 $/kWh) is more competitive than that of a dry steam geothermal plant (0.086 $/kWh). The cost of producing hydrogen with a suitable electrolyzer powered by wind energy ranges from $0.672/kg H_2 to $1.063/kg H_2, while that produced by the high-temperature electrolyzer (HTE) powered by geothermal energy ranges from $3.31/kg H_2 to $4.78/kg H_2. Thus, the AG Rift area can produce electricity and green hydrogen at low-cost using wind energy compared to geothermal energy. The amount of carbon dioxide (CO_2) emissions reduced by using a "Yinhe GX113-2.5MW" wind turbine and a single flash geothermal power plant instead of fuel-oil generators is 2061.6 tons CO_2/MW/year and 2184.8 tons CO_2/MW/year, respectively.

Keywords: Djibouti; wind energy; geothermal energy; hydrogen; Asal-Ghoubbet Rift; cost analysis

Citation: Osman Awaleh, M.; Adan, A.-B.; Assowe Dabar, O.; Jalludin, M.; Mahdi Ahmed, M.; Abdillahi Guirreh, I. Economic Feasibility of Green Hydrogen Production by Water Electrolysis Using Wind and Geothermal Energy Resources in Asal-Ghoubbet Rift (Republic of Djibouti): A Comparative Evaluation. *Energies* **2022**, *15*, 138. https://doi.org/10.3390/en15010138

Academic Editor: Adel Merabet

Received: 30 November 2021
Accepted: 20 December 2021
Published: 26 December 2021

1. Introduction

The Republic of Djibouti is located in the Horn of Africa, at the intersection of the Red Sea and the Indian Ocean (Figure 1). Djibouti benefits from a geostrategic position over the world's main shipping road, and it is the natural outlet for the landlocked countries in the region. On the other hand, the Republic of Djibouti is one of several African countries located on the East African Rift System (EARS) (Figure 1). As in other rifting zones, the EARS activity corresponds to large seismic, tectonic, and volcanic activities [1]. This unique geodynamical environment puts the Republic of Djibouti in an excellent position to develop geothermal energy. Indeed, the Republic of Djibouti, one of the few countries with geothermal potential in Africa, is endowed with a substantial amount of this energy, which is found scattered along the Gulf of Tadjourah ridge and the western part of the country (Supplementary Figure S1) [2,3]. However, the most geologically active area in Djibouti is the Lake Asal area, and the AG Rift is one of two emergent oceanic ridges globally, the other being in Iceland [1]. Therefore, several studies have been carried out on the Republic of Djibouti geothermal areas to either select the most favorable geothermal prospects for exploration by deep drilling or estimate their geothermal potential [2–8].

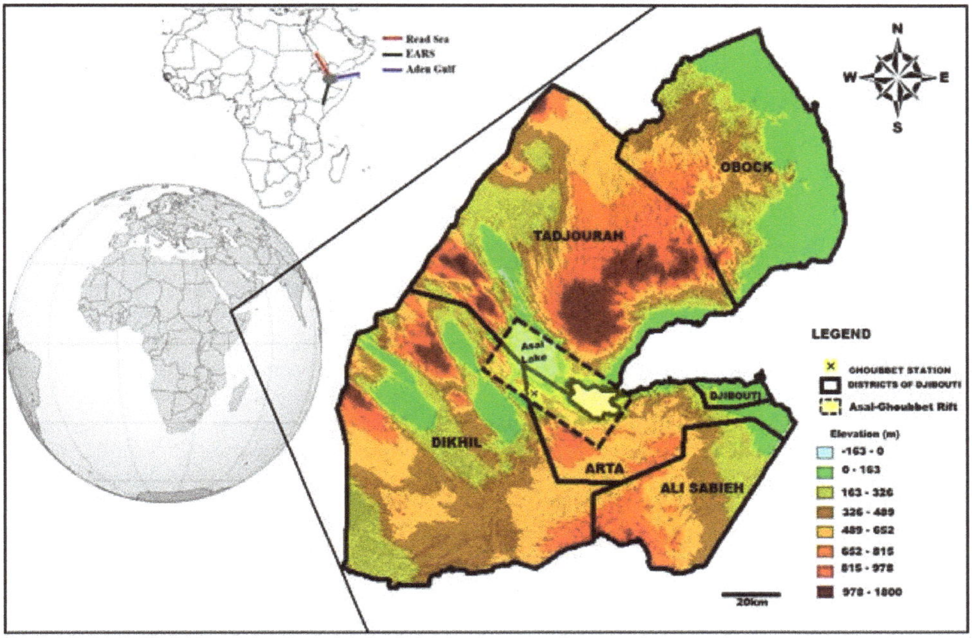

Figure 1. Locations of the wind measurement towers with topographical elevation (The dashed lines show the Asal-Ghoubbet Rift).

The Government of Djibouti has established an ambitious long-term development plan to improve energy access and energy security as a primary strategic focus by using alternative energy sources. Indeed, the Republic of Djibouti is endowed with strong potential in terms of renewable energy resources in addition to geothermal energy, such as wind and solar [9–12]. The plan aims at diversifying energy resources and focusing on renewable energy, which the country is endowed with, such as wind and solar and geothermal energy [9–12]. Furthermore, the Republic of Djibouti is among the fifteen African countries having the best wind resource potential [13], Wind energy is the world's fastest growing renewable energy technology [14,15]. For example, Zhang et al. [16] used two years of wind data measured at four heights above sea level to determine the wind potential of the southern coast of Pakistan. Soltani et al. [17] investigated the environmental, economic, and social impacts of geothermal energy systems. They provided a detailed review of barriers and highlighted options for increasing the global geothermal energy capacity and production needed to meet "net-zero" carbon emission goals [17].

In addition to electricity production, wind and geothermal energies can also generate green hydrogen via the water electrolysis process, which consists of splitting the water molecule under the effect of an electric current [18,19]. Despite the growth of the hydrogen market as a promising energy source, conventional routes for hydrogen production, i.e., from fossil fuels, have negative environmental impacts [20]. However, the power-to-gas concept, based on water electrolysis using electricity from renewable energy sources (wind, solar, geothermal), is the most environmentally friendly approach. Water electrolysis, a mature technique with relatively good efficiency [21,22], currently accounts for 4% of hydrogen production, but this is expected to expand significantly in the coming years, with a 22% market share projected for 2050 [23].

The primary consumers of hydrogen are the oil refining, metallurgy, chemical, and pharmaceutical industries. The growth of the above-mentioned industries leads to an increase in hydrogen demand. Numerous studies have investigated the integration of renewable energy for green hydrogen production. Ishaq and Dincer [24] performed a comparative assessment of renewable energy-based hydrogen production methods (i.e., Biomass, solar and geothermal). Hydrogen production using the biomass gasification technique offers higher energetic and exergetic efficiencies than the hydrogen production system using geothermal or solar energy [24]. Al-Sharafi et al. [25] investigated the potential of power generation and hydrogen production via solar and wind energy resources at different locations in Saudi Arabia. The optimization results showed that the minimum levelized cost of energy is 0.609 USD/kWh, and the cost of hydrogen production (COH) is 43.1 USD/kg. Mostafaeipour et al. [26] investigated the suitable areas in Afghanistan for harvesting wind energy for hydrogen production. On the other hand, few studies have been conducted on electricity and hydrogen production using wind energy in Africa countries [22,27]. Ayodele and Munda assessed the potential and cost-efficiency of green hydrogen production from South Africa's wind energy resources [22]. They reported a potential annual production of 6.51–226.82 metric tons of hydrogen and a minimum production cost of \$1.4–39.55/kg, depending on the wind turbine model used. On the other hand, the geothermal-assisted hydrogen production cost based on water electrolysis can compete with other renewable energy resources such as wind power [28]. Indeed, Rahmouni et al. performed an environmental investigation on a geothermal-based hydrogen production process, where the cost of the produced hydrogen was estimated at about \$8.24/kg H_2 [29]. Yilmaz et al. realized the thermoeconomic evaluation of hydrogen production by the binary geothermal power plant and estimated the cost at \$2.366/kg H_2 [30]. In another study, Yilmaz et al. [31] performed a thermodynamic and economic analysis of a geothermal energy-assisted hydrogen production system using real-time artificial neural networks on a field programmable gate array. The overall exergy efficiency of the system, the unit cost of the produced hydrogen, and the simple payback period of the system were calculated as 7978 kW, 38.37%, 1.088 \$/kg H_2, and 4.074 years, respectively.

The main contribution of this study is to compare the economic feasibility of hydrogen production by water electrolysis from wind and geothermal energy resources for a given site in Africa. Indeed, to the best of our knowledge, no study of green hydrogen production using geothermal energy has been conducted so far in Africa. In this regard, this paper analyzes the cost of green hydrogen production from wind and geothermal energy in the Asal–Ghoubbet region, located in the southwestern part of the Republic of Djibouti, thus helping potential investors and developers of hydrogen production units in the Republic of Djibouti. In sum, the contributions of this work are:

1. The analysis of the wind energy potential of the Ghoubbet region
2. The selection of the most appropriate wind turbine and electrolyzer for electricity and hydrogen production in the study area
3. The performance of thermodynamic analysis to select the appropriate geothermal energy production processes for the Rift AG region
4. Estimating the cost per Kg of hydrogen produced by combining a geothermal power plant in the Rift AG with an appropriate electrolyzer
5. Evaluating the CO_2 emission reduction of wind and geothermal energy in the study area.

2. Site Description and Data Collection

2.1. Site Description

The Republic of Djibouti is located in the Horn of Africa, bounded by Eritrea to the north, Ethiopia to the west and southwest, and Somalia to the southeast (Figure 1). The Köppen–Geiger climate classification of the area ranges from "hot desert" to "semi-arid" with a low precipitation regime and annual mean rainfall of 150 mm (climate type codes BWh and BSh, respectively) [32,33]. Two seasons predominate: a cool season (winter) from October to April with a monthly mean temperature of 20–30 °C and a hot season (summer) from May to September with a monthly mean temperature of 30–45 °C. In summer, an equatorial westerly wind zone dominates, and the mean temperature increases to between 30 and 45 °C with a high rate of evapotranspiration amounting to 2000 mm per year [7,8].

As shown in Figure 1, the AG Rift is located in an area that extends from the Arta region to the Tadjourah region. The Asal—Ghoubbet region, a land barrier of 12 km long and 10 km wide, of volcano-tectonic type, is a very recent active rift fewer than one million years old [34]. It encompasses the Asal Lake and the Ghoubbet al Kharab Gulf.

2.2. Wind Data Source

The present study uses wind data measured at meteorological stations installed by the Djibouti Centre for Research and Studies (CERD in French) at Ghoubbet (Latitude 11.5382°; Longitude 42.4119°, Figure 1). The wind speed data were collected and measured at three heights (i.e., 20 m, 40 m and 60 m) for a 10-min time interval throughout 2014–2015. The wind speeds are recorded using NRG#40C anemometers with an accuracy of 0.1 m/s in the range 1–96 m/s [35]. All data were checked thoroughly to identify any values that are outside the range of the sensors. The wind directions are recorded using an NRG 200P wind vane. A barometric pressure sensor (BP200) and temperature sensor (NRG 100S) were enclosed in a circular six-plate radiation shield to ensure accurate measurements and mounted on the tower. Further details regarding the tower, the instruments, and the support structure can be found on the website of NRG systems [35].

3. Methodology

The feasibility of green hydrogen production from wind and geothermal energy in the Republic of Djibouti was assessed using the following approach (Figure 2): (1) assessment of the wind and geothermal energy in the study area, (2) selection of the most suitable wind turbine and geothermal power plant, (3) assessment of the levelized cost of electricity ($/kWh) produced from wind and geothermal energy, (4) selection of the most appropriate electrolyzer, (5) evaluation of the levelized cost of hydrogen ($/kg), (6) evaluation of the CO_2 savings for wind and geothermal energy, (7) comparison of the levelized cost of hydrogen produced by an electrolyzer powered by wind and geothermal energy. The proposed research framework is based on the flowchart given in Figure 2.

3.1. Wind Resource Assessment

3.1.1. Weibull Density Function

The Weibull distribution model is applied for describing and analyzing wind data. This function is used to predict the characteristics of prevailing wind profile precisely [36]. To use the Weibull probability distribution, it is necessary to calculate two parameters, the scale parameter (k) and the dimensionless shape factor (c). Indeed, the two parameter Weibull distribution $f(v)$ of measured wind speed v (m/s) and the cumulative distribution of Weibull are defined by the following equation [37,38]:

$$f(v) = \left(\frac{k}{c}\right)\left(\frac{v}{c}\right)^{k-1} exp\left[-\frac{v}{c}\right]^k \quad (1)$$

$$F(v) = 1 - exp\left[-\frac{v}{c}\right]^k \quad (2)$$

In the literature, several methods to estimate the scale and shape parameters of the Weibull distribution function are reported [39]. In this study, four methods available are selected to estimate the Weibull parameters, namely the maximum likelihood method, WAsP method, moment method, and empirical method of Jestus. These methods have been detailed in the Supplementary Material. Root mean square error, determination of coefficient, and mean bias error are computed to validate and compare the computed results.

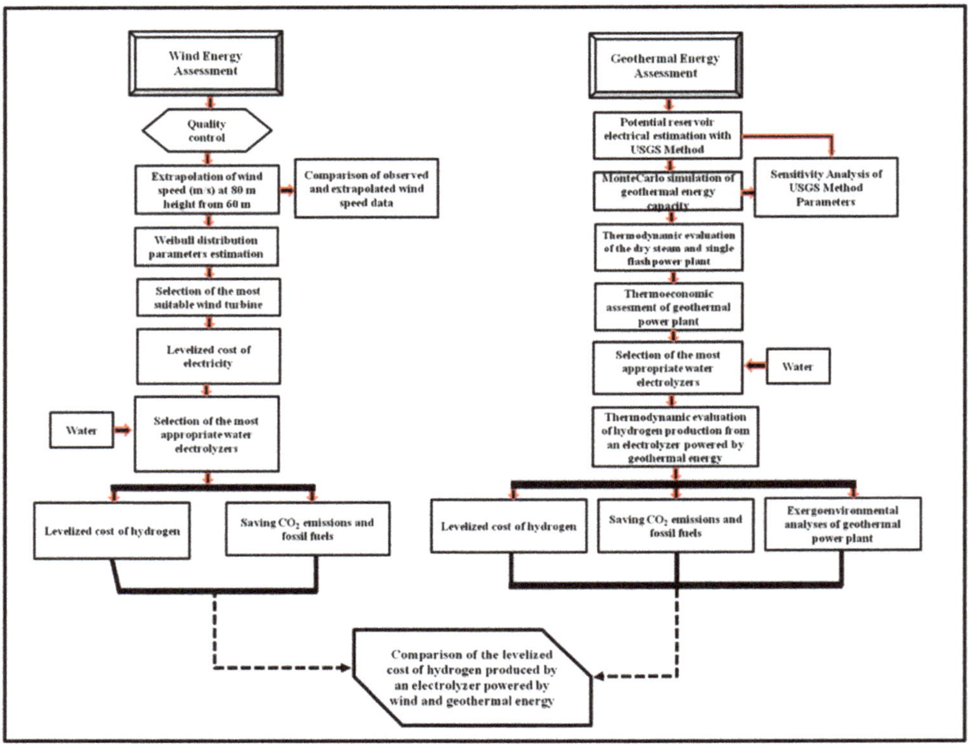

Figure 2. Methodology schematic diagram.

3.1.2. Wind Shear Coefficient

Estimating the wind power generated by any wind turbine, the wind speed should be observed at the turbine hub height. However, in most cases, wind speeds are observed at a height different from the turbine hub height. In this case, the power-law method is the most commonly used to adjust the wind speed to the turbine hub-height as follows [40]:

$$V_2 = V_1 \times \left(\frac{Z_2}{Z_1}\right)^{\alpha} \quad (3)$$

where V_2 and V_1 are the wind velocities at heights Z_2 and Z_1, respectively, and α is the wind shear coefficient or power exponent.

3.1.3. Wind Power Density

The wind power density (PD) is an indicator used to evaluate the available wind energy. The PD represents the flux of kinetic energy available in the wind per unit area, according to the wind speed V and air density ρ. It can be assessed using the following equation [41]:

$$PD = \frac{1}{N}\sum_{i=1}^{N} \rho v_i^3 \quad (4)$$

where v_i is the measured wind speed for every 10 min period expressed in m/s and N is the total number of sample data for each year. ρ is the air density in kg/m^3 (ρ = 1.225 kg/m^3 at the sea level).

3.1.4. Annual Energy Production

The total amount of energy produced and the capacity factor of a turbine are essential indicators used to assess the performance and economic sustainability of wind turbines at a specific location. The wind turbine annual energy production (E_{out}) over a desired period can be calculated as follows [42]:

$$E_{out} = 8760 \times P_{out} = 8760 \times P_r \times \left[\frac{e^{-(v_c/c)^k} - e^{-(v_r/c)^k}}{(v_r/c)^k - (v_c/c)^k} \right] - e^{-(v_f/c)^k} \quad (5)$$

Furthermore, the capacity factor (C_f), the ratio of energy produced annually from a wind turbine (E_{out}) and annual rated power (Er), can be defined as fellows [43]:

$$Cf = \frac{E_{out}}{Er} \quad (6)$$

3.2. Geothermal Resources Assessment

The geothermal resources in the Asal–Ghoubbet area were evaluated using the volumetric method according to the United States Geological Survey (USGS) [44]. This method assesses the total geothermal energy in the fluids and the rock masses of the reservoir. Moreover, the volumetric method is combined with the Monte Carlo method to minimize parameter uncertainty. A brief description of the Monte Carlo simulation method parameters is presented in the Supplementary Material. The electrical potential of the geothermal reservoir can be evaluated using the following equation [45]:

$$G_{PP} = \frac{E_{GT} \times R_f \times C_e}{L_f \times L_t} \quad (7)$$

In this study, the single flash and dry steam systems are considered for geothermal power generation. The thermodynamic performance of these two systems is evaluated in terms of energy and exergy efficiency. The details of the thermodynamic evaluation are reported in the Supplementary Material. The electrolyzer uses the energy produced by the geothermal plant to produce hydrogen from water.

3.3. Hydrogen Production from Wind and Geothermal Energy

3.3.1. Hydrogen Production from Wind Energy

Hydrogen can be generated through water electrolysis by using renewable energy, such as wind and geothermal energy. The proposed system for converting wind and geothermal energy to produce hydrogen is shown in Figure 3. An AC/DC converter delivering the quantity of energy required by the electrolyzer for its operation under real conditions is also employed. The efficiency of the converter is assumed to be 90% [46]. There are several types of electrolyzer in the literature. In this study, the polymer electrolyte membrane electrolyzer (PEME), high temperature electrolyzer (HTE), and alkaline water electrolyzer (AWE) were used because of their good ability for integration with renewable energy systems. For PEME and AWE electrolyzers, the water temperature is between 60 and 90 °C [47], while for the high temperature electrolyzer (HTE), the water temperature can vary from 800 to 1000 °C [48].

Considering the amount of hydrogen that could be produced yearly by each electrolyzer, the annual wind turbine energy has a direct relationship with production and can be formulated with the following mathematical equation [46]:

$$M_{H_2} = \frac{E_{out}}{E_{electrolyzer}} x \ \eta_{conv} \quad (8)$$

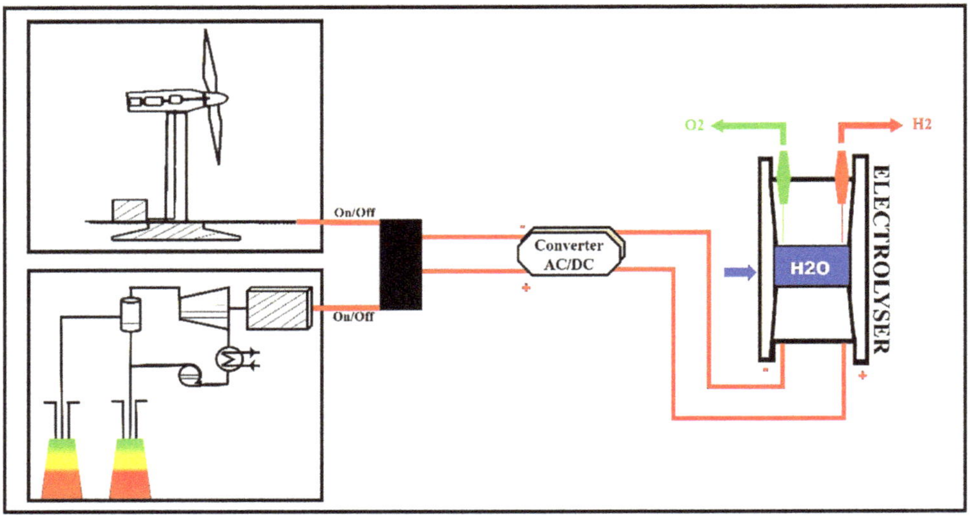

Figure 3. Simplified diagram of the hydrogen production from wind and geothermal energy.

3.3.2. Hydrogen Production from Geothermal Energy

The thermodynamic evaluation of geothermal energy is a prerequisite for using this renewable energy in green hydrogen production. Therefore, both reversible and irreversible system operations were considered to evaluate the cost of hydrogen production. The maximum specific work that can be provided by a geothermal power plant, using a resource of a reference temperature T_s in an environment whose temperature is T_0 is given by the following mathematical expression [49]:

$$W_{rev,\ geo} = c(T_s - T_o) - T_o c ln\left(\frac{T_s}{T_o}\right) \tag{9}$$

Indeed, the geofluid in the reservoir is assumed to be an incompressible liquid. The minimum work required in (kJ/kg) for the electrolyzer in ideal (Equation (10)) and non-ideal (Equation (11)) operation can be expressed as follows [49]:

$$W_{rev,\ electrolysis} = \frac{\Delta G_{electrolysis,\ H_20}}{M'_{H_2}} \tag{10}$$

$$W_{act,\ electrolysis} = \frac{W_{rev,electrolysis}}{\eta_{electrolyzer}} \tag{11}$$

The amount of hydrogen produced per unit of geofluid is defined as the ratio of the work output from the geothermal plant to that of the electrolyzer. For reversible operations, it can be evaluated with the following equation [49]:

$$y_{prod,H_2} = \frac{mass\ of\ H_2 produced}{mass\ of\ geothermal\ water\ used} = \frac{W_{rev,geo}}{W_{rev,\ electrolysis}} \tag{12}$$

3.4. Economic Assessment of Wind Energy

To estimate and compare the viability of electricity generation based on wind technology, the most widely used economic indicator is the levelized costs of energy (LCOE) [50]. It is defined as the total investment cost required for one unit of electricity produced during a period. Thus, in the case of wind power, the following mathematical expression can be used to evaluate the cost assessment [51]:

$$LCOE = \frac{PVC}{E_{out}} = \frac{C_I\left(1 + C_{O\&M}\left[\frac{(1+I_d)^t - 1}{I_d(1+I_d)^t}\right]\right)}{8760 \times P_r \times t \times C_f} \tag{13}$$

$$C_I = \text{rated power (kw)} \times C_{Aspec}(\$/kW) \times (1 + \text{variable capital cost as a fraction}) \tag{14}$$

3.5. Economic Evaluation of Geothermal Energy

The cost of geothermal electricity takes into account several factors, such as the cost of drilling and construction; annual operation and maintenance expenses; financial rates; the type of geothermal resource (steam or hot water); the productivity of the reservoir; the size and type of geothermal plant considered, etc. [52]. The levelized cost of electricity generated by geothermal energy can be calculated using the following formula [53,54]:

$$LCOE = \frac{\sum_{t=0}^{t=lifetime} C_{SU,t}\,(1+i)^{-t} + C_{O\&M,t}(1+i)^{-t} + C_{fuel}(1+i)^{-t}}{\sum_{t=1}^{t=lifetime} \dot{W}_{net}\,N\,Lf\,(1+i)^{-t}} \tag{15}$$

$$C_{surf} = 3300 \times exp^{(-0.0031(\dot{W}_{net}-5))} \tag{16}$$

$$C_{o\&m} = 2.6 \times exp^{(-0.002536(\dot{W}_{net}-5))} \tag{17}$$

However, in this study, the additional cost of capital investment for drilling and completion is taken into account and can be evaluated with the following equations [55,56]:

$$W_{DC} = a \times n \times \log(d) + b \times n \times d^2 + c \tag{18}$$

$$C_{WC} = 1.72.10e^{-8} \times d^2 + 2.3.10e^4 \times d - 0.62 \tag{19}$$

3.6. Hydrogen Cost from Wind and Geothermal Energy

In order to compare and evaluate the economic viability of hydrogen production powered by wind or geothermal energy, LCOH is used. The LCOH of green hydrogen produced from wind energy can be estimated using the following mathematical equations [46]:

$$LCOH = \frac{C_{Electrolyzer} + C_{Electricity}}{M_{H_2} \cdot T} \tag{20}$$

$$C_{Electrolyzer} = C_u \cdot \frac{M_{H_2} \text{x } E_{electrolyzer}}{8760 \text{ x } Cf \text{ x } \eta_{electrolyzer}} \tag{21}$$

$$C_{Electricity} = LCOE \text{ x } \frac{\sum_{i=1}^{t} E_{out}}{t} \tag{22}$$

The economic evaluation of hydrogen production from geothermal energy can be calculated as follows [57]:

$$c_{H_2} = E_c + M_{OM} + I_c \tag{23}$$

$$E_c = E_{electrolyzer} . C_{Electricity} \tag{24}$$

$$M_{OM} = 0.41H^{-0.23} \tag{25}$$

$$I_c = 0.5H^{-0.025} \tag{26}$$

Two models of electrolyzer were considered for green hydrogen production from geothermal energy (i.e., PEME and HTE) and wind energy (i.e., PEME, AWE). The performance characteristics of the electrolyzers are given in Supplementary Table S1.

3.7. Energy and Exergy Analysis

Energy and exergy analyses are performed on the dry steam and single flash configuration systems (Figures 4 and 5), considering the operating conditions presented in Supplementary Tables S2 and S3. Further, an average geothermal reservoir temperature and mass flow rate of 306 °C and 40 kg/s respectively were used in the thermodynamic modelling [58,59].

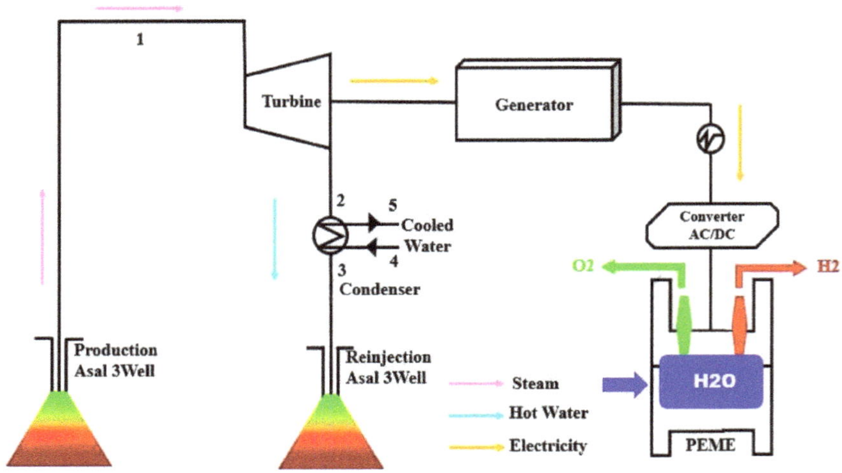

Figure 4. Schematic of the dry steam cycle with electrolyzer for Asal-Ghoubbet hydrogen production.

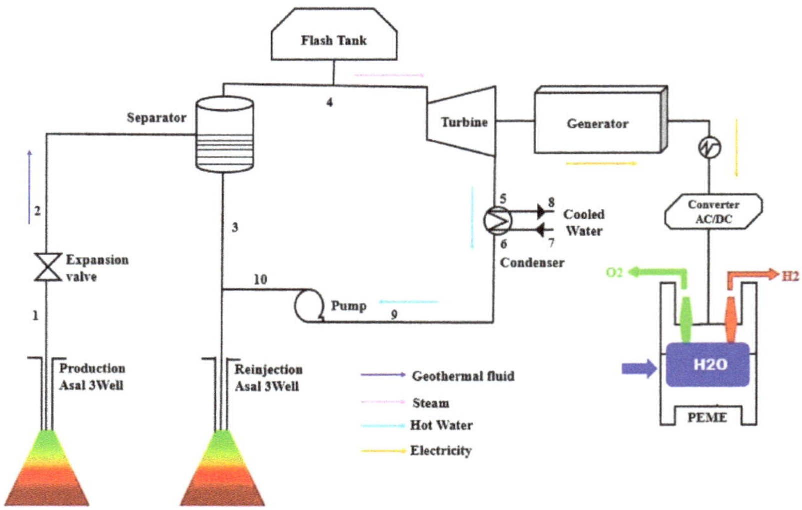

Figure 5. Schematic of the single flash cycle with electrolyzer for Asal-Ghoubbet hydrogen production.

4. Results

4.1. Wind Energy

4.1.1. Wind Resource Analysis

We analyzed measured wind data collected at the Ghoubbet site. Temporal wind speed data were recorded for two years with 10-min intervals and preprocessed by converting them into hourly and monthly mean wind speeds at 80 m. A statistical analysis of the wind data was first performed, which predicted the energy output of a typical wind energy conversion. The wind potential characterization is based on the Weibull statistical distribution. Average wind speeds and wind characteristics are inferred from the temporal distribution of the study site. Four Weibull methods (i.e., moment method (MM), empirical methods of Jestus (EMJ), WAsP, and maximum likelihood method (MLM)) were used to predict the existing wind potential at the height of 80 m. The Weibull distribution and cumulative density function at the AG Rift are shown in Figure 6a,b.

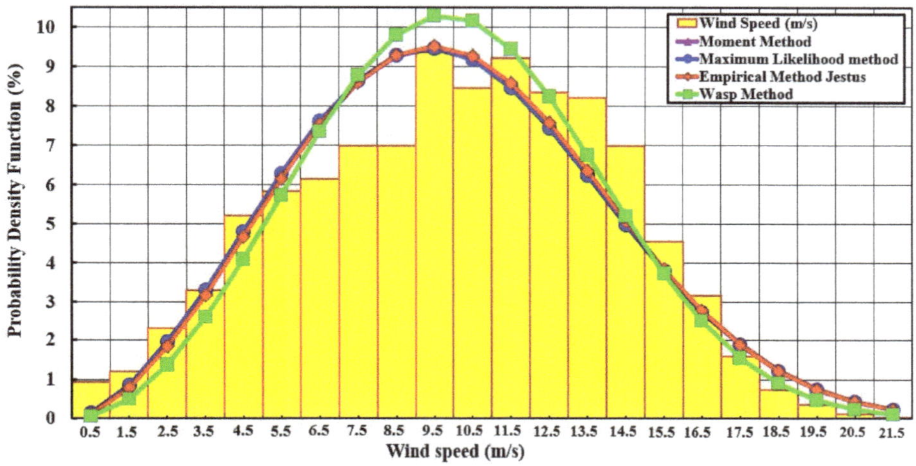

(a)

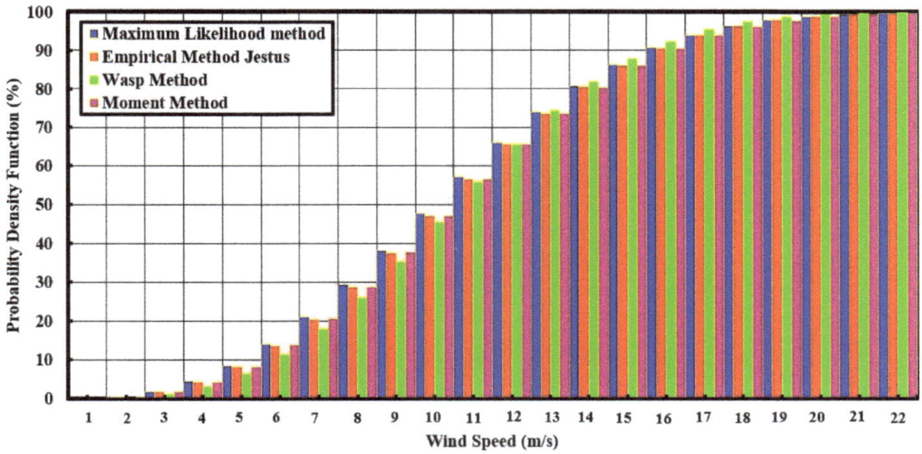

(b)

Figure 6. (**a**) Annual 80 m wind speed Weibull distribution (PDF) and (**b**) cumulative distribution function (CDF) at level of 80 m for Ghoubbet.

Analysis of the distribution diagram reveals that the moment method (MM), empirical method of Jestus (EMJ), and maximum likelihood method (MLM) have the best accuracy in terms of the Weibull distribution compared to the WAsP Method (WM), which underestimates the wind distribution (Figure 6a). This difference is also noticed in the cumulative density function (Figure 6b), where the three methods are homogeneous compared to the WM values. Statistical tests of R^2, RMSE, and MAE were also performed to strengthen this evaluation in respect of the quantitative approach. In our case, the values of the Weibull shape parameter (k) range from 2.65 to 2.96 at Asal–Ghoubbet at the height 80 m (Table 1). In our case, the Weibull shape parameter (k) values range from 2.65 to 2.96 at Asal–Ghoubbet at the height of 80 m (Table 1). The Weibull scale parameter (c) varies between 11.19 m/s and 11.25 m/s (Table 1). Indeed, the most accurate method is used to estimate the capacity factor of the Ghoubbet wind speed regime. The performance of the Weibull methods in Table 1 reveals that the moment method has the highest value of coefficient of determination (R^2) and the lowest value of mean square error (RMSE) and mean absolute error (MAE). On this basis, the moment method is the most appropriate for estimating the Weibull parameter in Asal–Ghoubbet. The EMJ method presents satisfactory results to fit the wind speed observations at a level of 80 m at Asal–Ghoubbet (Table 1). On the other hand, the average wind speeds predicted by the four methods are about 10 m/s, indicating that the match to the measured data is very high (Supplementary Table S2).

Table 1. Result of Weibull parameters at level of 80 m of Ghoubbet.

Weibull Distribution Modeling	Parameters		Performance Weibull Model			Wind Speed (m/s)
	k	c (m/s)	R^2	RMSE	MAE	Wavg (m/s)
Moment Method (MM)	2.68	11.25	0.90525	0.00984	0.00767	10.00
WAsP Method (WM)	2.96	11.25	0.88131	0.01102	0.00836	10.04
Empirical Method of Jestus (EMJ)	2.69	11.25	0.90500	0.00986	0.00769	10.00
Maximum Likelihood Method (MLM)	2.65	11.19	0.89932	0.01015	0.00780	9.947

The wind speed data at the height of 80 m were extrapolated from the wind data measured at the height of 60 m using the lowest power coefficient. The average annual power density is estimated to be about 904.45 W/m^2 for an altitude of 80 m at the Ghoubbet location.

According to the power density classification established as [11]:

- Fair (PD < 100 W/m^2)
- Fairly good (100 W/m^2 ⩽ PD < 300 W/m^2)
- Good (300 W/m^2 ⩽ PD < 700 W/m^2)
- Very good (PD ⩾ 700 W/m^2)

Wind direction analysis is essential for the planning of wind turbine installations. The frequencies of wind directions at 60 m during 2015 are presented in Supplementary Figure S2. The distribution of polar diagrams indicates that the East (90°) and South East (135°) sectors are the most reactive and have the highest wind frequencies. The average annual wind speed at Ghoubbet is estimated at 10 m/s (Supplementary Table S4).

4.1.2. Performance of Wind Turbine

Supplementary Table S5 shows the characteristics of the different wind turbine models considered in our study, ranging in capacity from 1 to 5 MW [60]. This allows for the selection of the most suitable turbine for the wind regime at the given site.

The performance of the wind turbine is evaluated in terms of capacity factors. Indeed, a turbine with a capacity factor (C_f) below the limit of 0.25 cannot be used for wind power generation. While a C_f value higher than 0.5 ensures a significant conversion of wind energy into electricity [61]. Therefore, the most suitable turbine has the highest possible C_f value. The annual energy production (E_{out}) and the C_f at 80 m hub height are reported in Supplementary Table S5. The annual wind energy production for the 42 wind turbines varies from 1677.45 MWh/year to 30,335.40 MWh/year, which depends on the wind turbine technology.

4.1.3. Cost Analysis for Wind Energy Generation

The economic evaluation is based on the selected wind turbine "Yinhe GX113-2.5MW" of 80 m height that best matches the wind regime for the Ghoubbet area. This turbine showed the most interesting capacity factor (0.8496) and annual electricity production (Table S5). Table 2 shows the economical parameters of a "Yinhe GX113-2.5MW" turbine with a capacity of 2.5 MW. The initial investment for such a turbine is estimated to be about $4,000,000 (i.e., an average specific cost of $1600 USD/kW is assumed). Installation costs, the cost of logistical transport of the equipment to the site, and other variable costs are estimated at 30% of the initial investment. The annual operation and maintenance cost (CO&M) is 25% [11]. The present value cost (PVC) of power generation from the "Yinhe GX113-2.5MW" turbine over 20 years is therefore estimated to be about US$15,640,958.47. The annual energy produced by the "Yinhe GX113-2.5MW" turbine is about 18,606.64 MWh/year. The cost of wind energy produced by the selected turbine in the Ghoubbet area is estimated at 0.042 USD per KWh (Table 2). The annual economic analysis of wind power generation is provided in Supplementary Table S6.

Table 2. Unit cost of electricity generated by Turbine "Yinhe GX113-2.5MW".

Parameters	Units	Values of Turbine T25
Specific cost of wind turbine	US$/KW	1600 [a]
Life time	Year	20 [b]
Initial investment cost	US$	4,000,000
Variable capital cost (30%)	%	1,200,000 [c]
Total investment cost	US$	5,200,000
Interest rate of Djibouti	%	0.1087 [d]
Discount rate of Djibouti	%	0.125 [e]
Operation & Maint. Cost (25%)	%	1,300,000 [f]
Capacity factor of Turbine	%	84.96
Energy output of Turbine	KWh/yr	18,606,641.39
Present Value Cost	US$	15,640,958.47
LCOE	**US$/KWh**	**0.04203**

[a,b,c,f] [11], [d,e] [62].

4.1.4. Hydrogen Production from Wind Energy

Wind energy can be easily coupled with an electrolyzer to produce green hydrogen from water splitting processes. Both the alkaline water electrolyzer (AWE) and the polymer membrane exchange electrolyzer (PEME) are commercially available in different sizes. Supplementary Table S1 shows three size classes (i.e., small, medium, and large) of these two types of electrolyzers. Furthermore, a converter with 90% efficiency was considered for the present study. The most suitable production capacity observed for the PEME electrolyzer is the medium sized one with a production of about 310.1 tons H_2/year (Supplementary Table S7). Regarding the AWE electrolyzers, the large size was the most suitable with a production capacity of about 398.7 tons H_2/year (Supplementary Table S7).

The total investment cost is related to the amount of hydrogen produced, which varies according to the available energy. The energy produced by the "Yinhe GX113-2.5MW" wind turbine could run three AWE_l or 14 $PEME_m$ electrolyzers. The capital costs for the three AWE_l and 14 $PEME_m$ electrolyzers are estimated to be $4,629,860 and $3,463,096, respectively (Table 3).

Table 3. Unit cost of hydrogen produced from wind energy.

Parameter	Unit	Value of AWE_L	Value of $PEME_m$
Rated power	KW	1000 [1]	185 [2]
Specific cost of electrolyzer	US$/KW	1547 [1]	900 [2]
Unit cost of electrolyser	US$	1,547,000	166,500
Capital investment cost	US$	4,629,860	3,463,096
Installation cost of electrolyzer	US$	185,640	19,980
Stack replacement cost	US$	618,800	66,600
Operation & Maintenance cost	US$	61,880	6660
Specific cost of converter	US$/KW	155 [1]	155 [1]
Investment of converter	US$	154,700	28,619.5
Operation & Maintenance cost of converter	US$	6188	1145
Total investment cost electrolyzer	US$	5,817,956.49	3,615,864.87
Cost of electricity	US$	782,047.92	782,047.92
Interest rate	%	10.87 [3]	10.87 [3]
Unit cost of hydrogen	US$/kg	1.045	0.672

[1] [63] [2] [22] [3] [62].

4.2. Geothermal Energy

4.2.1. Electrical Power Analysis

Four geothermal wells were drilled in the AG Rift in the late 1980s [58]. However, some parameters were not reported for these geothermal wells to use the USGS volumetric method to estimate the geothermal energy of the AG Rift system. Therefore, the missing parameters were supplemented with data from geothermal areas with similarities to the AG Rift (Supplementary Table S8). The result of the USGS volumetric method show that the geothermal potential of Asal–Goubbet could contain potential energy of about 67.18 MWe.

Table 4 shows the result of the thermodynamic assessment of dry steam and a single flash cycle. Since the highest value of thermal, operating, and isentropic efficiencies is demonstrated by dry steam compared to single flash, the former system is therefore thermodynamically interesting compared to the performance of single flash (Table 4). Further, the thermodynamics calculation predicted 22.22 MW for dry steam and 4.9 MW for single flash, considering the geothermal reservoir's preliminary mass flow and temperature conditions in the study area.

Table 4. Thermodynamic result of dry steam and single flash cycle.

Geothermal Power Plant	Dry Steam Power Plant	Single Flash Power Plant
Output power (MW)	22.22	4.91
Exergy destruction (MW)	18.83	10.81
First law efficiency (%)	21.57	9.68
Second law efficiency (%)	54.13	31.24
Isentropic efficiency (%)	76.38	67.91

4.2.2. Thermoeconomic Analysis

Based on depth (e.g., 2500 m), the average cost of geothermal drilling at AG Rift is estimated to be approximately US$12,964,527 (Supplementary Tables S10 and S11). Indeed, this cost is similar to that of the geothermal drilling carried out within the framework of the Fiale project (Asal Region, Djibouti). The results of the economic analysis of the dry steam plant and the single flash plant are shown in Supplementary Tables S9 and S10. The average cost of a 22.22 MW dry steam geothermal power plant is estimated to be about US$3157.49/KW (Table S10), while the average cost of a 4.91 MW single flash geothermal power plant is estimated to be US$3294.20/KW (Supplementary Table S11). The electrical production cost of the dry steam and single flash geothermal power plant is estimated to be 8.66 $cents/KWh and 12.53 $cents/KWh, respectively (Supplementary Tables S9 and S10).

4.2.3. Hydrogen Production from Geothermal Energy

The economic analysis of two electrolyzers (i.e., $PEME_m$ and HTE) coupled with dry steam and a single flash geothermal plant is performed. The thermodynamic analysis results show that with a temperature of 306 °C in the AG Rift geothermal reservoir and a dead state temperature of 25 °C, the maximum specific work is estimated to be about 362.05 kJ/kg of geothermal water (Table 5).

Table 5. Result of green hydrogen production supply with the geothermal energy analysis.

Thermodynamic Operation	Electrolyser Coupled with Geothermal Power Plant	Maximum Work Requirement (kJ/kg)	Minimum Work Requirement (kJ/kg)	Minimum Work Input (kWh/kg)	H_2 Produced (Tons/Year)	LCOH ($/kg)
Reversible case	Single flash-$PEME_m$	362.05	117,651	32.68	731.44	5.51
	Dry steam-$PEME_m$	362.05	117,651	32.68	3308.83	3.97
	Single flash-HTE	362.05	91,858	25.52	1426.66	4.56
	Dry steam-HTE	362.05	91,858	25.52	6453.78	3.31
Irreversible case	Single flash-$PEME_m$	362.05	190,602	52.95	451.51	8.16
	Dry steam-$PEME_m$	362.05	190,602	52.95	2042.49	5.80
	Single flash-HTE	362.05	97,721	27.14	1341.06	4.78
	Dry steam-HTE	362.05	97,721	27.14	6066.56	3.46

5. Discussion

5.1. Wind Energy

The analysis of the distribution diagram reveals that the moment method (MM), empirical method of Jestus (EMJ), and maximum likelihood method (MLM) have the best accuracy in terms of the Weibull distribution compared to the WAsP Method (WM), which underestimates the wind distribution (Figure 6a). This difference is also noticed in terms of the cumulative density function (Figure 6b), where the three methods are homogeneous compared to the WM values.

The performance of the Weibull methods in Table 1 reveals that the moment method has the highest value of the coefficient of determination (R^2) and the lowest value of mean square error (RMSE) and mean absolute error (MAE). On this basis, the moment method is the most appropriate for estimating the Weibull parameter in Asal–Ghoubbet. The EMJ method presents satisfactory results to fit the wind speed observations at a level of 80 m at Asal–Ghoubbet (Table 1). On the other hand, the average wind speeds predicted by the four methods are about 10 m/s, indicating that the match to the measured data is very high (Supplementary Table S4).

In addition, the wind potential of Ghoubbet reveals a promising site for the installation of a large-scale wind farm for electricity and hydrogen production. Moreover, the power densities predicted by the four Weibull fitting methods are estimated to be 921.3745 W/m^2, 912.1199 W/m^2, 919.1252 W/m^2, and 878.3873 W/m^2, respectively, for the MM, the MLM, the EMJ, and the WM. Nevertheless, the average monthly wind power density depends on the season. It can be observed that for the season June–September, the power density is at its lowest (192 to 194 W/m^2) for the year 2015 (Figure 7a) and can reach a wind speed below 5 m/s in several hours (Figure 7b). The performances of 42 wind turbines were analyzed in terms of energy production and capacity factors to allow for the selection of the turbine that best matches the wind regime of the given site. The calculated C_f of the 42 turbines ranges from 0.114 to 0.849 (Supplementary Table S5). The Yinhe GX113-2.5MW turbine with a height of 80 m was found to be the best turbine model with a C_f of 84.9% for Asal–Ghoubbet (Supplementary Table S5). The annual energy produced by the "Yinhe GX113-2.5MW" turbine is about 18,606.64 MWh/year. According to the result, the capacity factor is high when the values of cut-in speed, rated speed, and cut-out speed are low.

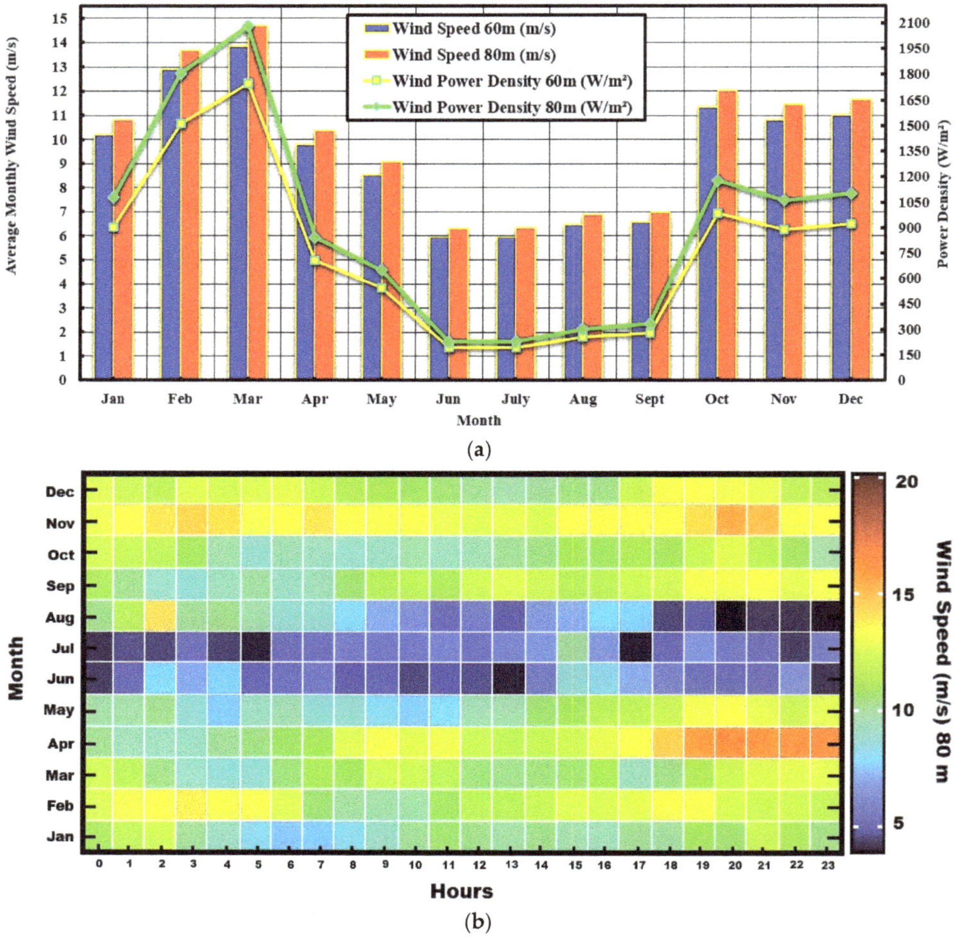

Figure 7. (**a**) Monthly average wind speed and wind power density at Ghoubbet and (**b**) Hourly pattern available wind speed at 80 m.

Therefore, the selected wind turbine was used to evaluate the economic feasibility of electricity generation and hydrogen production in the Asal–Ghoubbet area. The levelized cost of the electricity produced by this wind turbine in the Ghoubbet area is estimated at 0.042 USD per KWh (Table 2). The annual economic analysis of wind power generation is provided in Supplementary Table S6. An annual analysis of environmental issues related to wind power performance, including a degradation rate, is presented in Supplementary Table S14.

In order to compare the cost hydrogen produced using wind energy, two electrolyzer models (PEME and AWE) with different sizes were analyzed. The cost of hydrogen production from wind turbines ranges between 0.680 $/kg H_2 and 7.187 $/kg H_2, depending on the electrolyzer models and capacities (Supplementary Table S7). Furthermore, between the two suitable electrolyzers selected in this study, the $PEME_m$ has a more attractive cost than the AWE_l (Table 3, and Supplementary Table S7). The cost of hydrogen production was estimated to be 1.063 $/kg and 0.68 $/kg for AWE_l and $PEME_m$ electrolyzers, respectively (Table 3). Since the cost of green hydrogen can vary from 1.4 to 7.9 $/kg according to Ayodele and Munda, 2019 [20], our results would indicate the competitiveness of wind-generated green hydrogen at Asal–Ghoubbet Rift.

5.2. Geothermal Energy

The quantification of uncertainties in the probability distribution parameters can be dealt with quite well using the Monte Carlo simulation method (Figure 8, Supplementary Figures S4–S10). Thus, the low, best, and high estimates of AG Rift megawatts can be represented by the probability of P90, P50, and P10, which correspond to 28.385 MWe, 57.729 MWe, and 103.868 MWe, respectively (Supplementary Table S11). This result implies a preliminary assessment of the energy potential of 8.73 MWe/km^2 for a maximum duration of 25 years. However, the prediction of the geothermal potential should be studied further with a more accurate assessment (Supplementary Table S11).

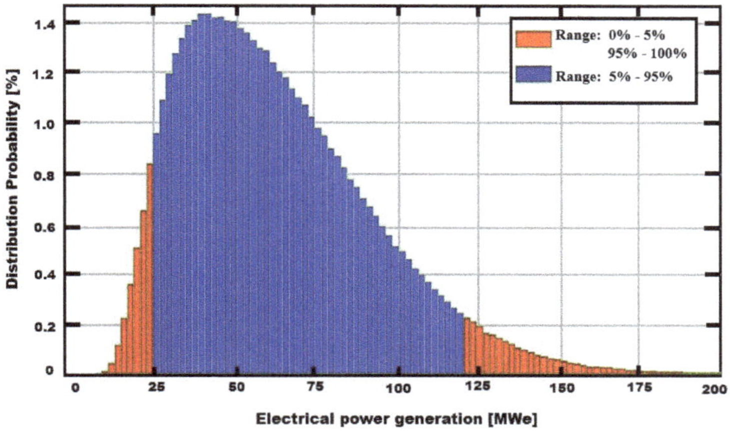

(a)

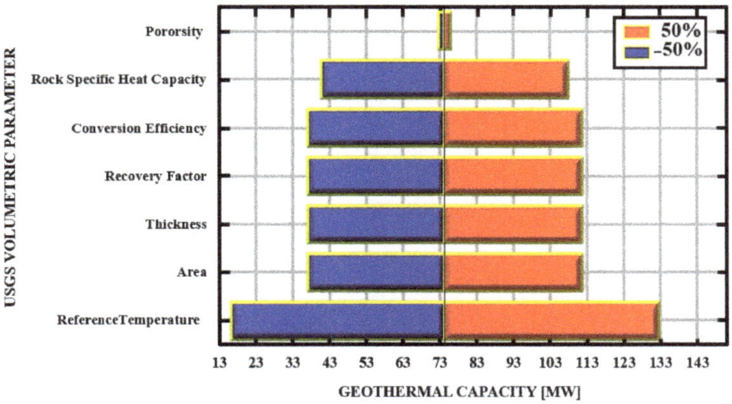

(b)

Figure 8. (**a**) Probability distribution of Monte Carlo simulation and (**b**) Tornado diagram for sensitivity result of volumetric parameters.

In order to assess the uncertainty of the volumetric parameters in the potential geothermal estimate of the Rift AG, it is useful to decipher the degree of influence of these parameters using the sensitivity analysis approach. Figure 8b shows the sensitivity of the AG Rift electric potential estimation for to the main parameters over a range of variation of ±50%. It can be noted that the degree of influence of these parameters on the geothermal potential is not the same. For example, when the area, thickness, recovery factor, conversion efficiency, and specific heat capacity parameters of the AG Rift rock separately increase by 50%, the total geothermal energy estimate increases from 74.33 MWe to 111.49 MWe. However, the porosity parameter does not directly influence the AG Rift geothermal energy estimate, whether it increases or decreases by 50% (Figure 8b). Further, an overall sensitivity analysis was performed using a $\pm 10/60\%$ increase and decrease in the mean value of the key parameters (Supplementary Figure S11).

In the Monte Carlo simulation, the most significant and positive correlation (0.73) with the geothermal energy potential of the AG rift is found to be with the recovery factor parameter (Figure S12). In other words, the higher the value of the recovery factor, the greater the geothermal potential of the AG Rift.

Thus, our result is very consistent with the LCOE proposed by Abdallah et al. [64]. Furthermore, assuming an average capacity of 2.5 MW per geothermal well at AG Rift, it should be noted that to maintain power generation at a constant level with a single flash plant, two additional 2.5 MW wells may be required for approximately eight years of service. While for a 22.22 MW dry steam plant, ten additional wells would have to be operational for three years of regular electricity production. Therefore, the total cost of 3–11 geothermal wells required to generate electricity for a single flash and a dry steam could be $38,893,581 and $142,609,797, respectively. It should also be noted that in our economic evaluation we have included the cost of drilling one well in the unit cost of electricity supplied by AG Rift's dry steam and single expansion plant (Supplementary Tables S9 and S10).

Assuming a liquid inlet state of one atm and a saturated liquid, the reversible specific work provided by the $PEME_m$ electrolyzer shows that 324.97 kg of geothermal water is required to produce 1 kg of hydrogen, whereas under non-ideal operation this could be as much as 526.45 kg of AG rift geothermal water. Moreover, we found that 253.72 kg of geothermal water using the HTE electrolyzer could produce 1 kg of hydrogen in the reversible case and can reach 296.61 kg for the non-ideal operation. Furthermore, the annual hydrogen production capacity varies from 451.51 tons H_2/year to 6453.78 tons H_2/year (Table 5). The annual O&M and yearly electrical cost for hydrogen production in different conditions (reversible and irreversible) are given in Supplementary Table S12.

The unit cost of hydrogen production ranges from \$3.31/kg H_2 to \$8.16/kg H_2 depending on the electrolyzer models/sizes (Table 5). Despite the high green hydrogen production capacity of the HTE electrolyzer relative to the PEME electrolyzer with the same energy input, it should be noted that the PEME is economically reliable, due to its 20-year life span [22,65]. At the same time, the operation life of the HTE electrolyzer is 10 years (Supplementary Table S1).

5.3. Overall Comparison

A comparative analysis is performed to evaluate the technical and economic aspects of hydrogen production with wind and geothermal energy (Table 6). In this study, wind power would produce 124.04 tons/MW/yr, while dry steam and single flash geothermal plants would produce 91.92 and 91.96 tons/MW/yr, respectively (Table 6). In addition, the fossil oil barrels saved by dry steam and single flash geothermal power plants would be about 4811.216 bbl/MW/yr and 4809.354 bbl/MW/yr, respectively (Table 6). On the other hand, wind power could save 4540.02 bbl/MW/year. An annual analysis of environmental issues related to wind power performance, including a degradation rate, is presented in Supplementary Table S14. In the present study, the cost of wind-generated electricity is $0.042/kWh. However, the cost of geothermal power in the study area is higher (i.e., $0.086/kWh for dry steam and $0.125/kWh for single flash), probably due to the higher capital cost. This is likely because the study area has a high available wind potential relative to geothermal potential. In other words, the limited data available for geothermal energy in the study area show the difficulties of exploiting this renewable energy, especially due to the very high salinity of the geothermal fluid, which leads to the clogging of geothermal wells [58].

Table 6. Comparative analysis of hydrogen production from wind and geothermal energy using PEME electrolyzer.

	Unit	Wind Turbine	Dry Steam * Power Plant	Single Flash * Power Plant
Rated power	MW	2.5	22.22	4.91
Energy output	MWh/year	18,606.6	175,186.6	38726.3
Hydrogen produced	Tons/MW/year	124.04	91.92	91.96
CO_2 avoided	Tons/MW/year	2061.6	2183.9	2184.8
Fuel oil saved	bbl/MW/year	4540.02	4809.354	4811.216
Energy cost	$/kWh	0.042	0.086	0.125
Hydrogen cost	$/kg	0.672	5.80	8.16

* Irreversible operation case.

The result of exergoenvironmental modelling showed that the impact index factors of a single flash and dry steam geothermal power plant are about 0.69 and 0.46, respectively (Supplementary Table S15). This indicates a better exergo-environmental performance of the system concerning its unusable waste exergy output and exergy destruction [66]. Furthermore, the improvement in the previous index is estimated to be 1.18 for the dry steam geothermal power plant, while it is 0.45 for a single flash geothermal power plant. This indicates that the dry steam geothermal power plant is more beneficial for the environment (Supplementary Table S15).

The cost of hydrogen production depends on the amount of hydrogen that renewable energy sources can produce. In this study, it is clear that the use of wind energy results in the lowest cost of hydrogen production ($0.672/kg) compared to the use of geothermal energy. Indeed, the greater amount of hydrogen produced by wind energy compared to geothermal energy, in this case, would probably explain the low cost of hydrogen produced by wind energy. On the other hand, it has been observed for the study area that the investment cost of wind energy seems to be lower than that of geothermal energy. The cost of hydrogen production is estimated to be about $5.80/kg H_2 and $8.16/kg H_2 for dry steam and simple flash, respectively.

Hydrogen production via electrolysis using renewable energy resources provides a sustainable and environmentally friendly energy solution. The result obtained in the present study is promising and shows that by using renewable energies, such as wind and geothermal energy, it is technically and economically feasible to produce green hydrogen at a low cost at the Asal–Ghoubbet site. However, further research is needed to simulate other green hydrogen production scenarios and to study the interface with other renewable energy sectors.

6. Conclusions

The wind speed potential in the Asal–Ghoubbet rift zone was evaluated with the Weibull distribution. A comparison of 42 wind turbines in the power range of 1–5 MW was conducted to assess their performance in adapting to the wind speed regime of Ghoubbet at 80 m height, as well as for the evaluation of unit costs of electricity produced. The Yinhe GX113-2.5MW wind turbine was found to be the most efficient in terms of energy production (18,606.6 MWh/year) and to have the lowest unit electricity cost ($0.042/KWh) for this site. $PEME_m$ and AWE_l electrolyzers could produce approximately 124.04 tons H_2/MW/year and 159.48 tons H_2/MW/year, respectively. The cost of producing hydrogen from a Yinhe GX113-2.5MW wind turbine was estimated to be $1.045/kg H_2 and $0.672/kg H_2 for the AWE_l and $PEME_m$ electrolyzers, respectively. In this case, wind energy can reduce emissions by 2061.6 tons CO_2/MW/year, assuming no degradation of wind capacity.

The potential of the Asal geothermal reservoir was evaluated using the USGS volumetric method at 67.18 MWe. A dry steam plant and a single flash power plant were selected for electricity production from a geothermal resource. Thermodynamic and thermo-economic analyses were carried out to compare the performance of these two geothermal plants in order to select the most appropriate for the Asal geothermal site. The cost of electricity from geothermal energy is evaluated at \$0.1253/KWh and \$0.0867/KWh for a single flash and dry steam geothermal power plant, respectively. The combination of a single flash power plant with HTE and $PEME_m$ electrolyzers resulted in a hydrogen production cost of \$4.78/kg H_2 and \$8.16/kg H_2, respectively. However, with the combination of dry steam with HTE and $PEME_m$ electrolyzers, the hydrogen production cost is evaluated at approximately \$3.46/kg H_2 and \$5.80/kg H_2, respectively. In addition, dry steam and single flash power plants can save 2183.9 tons CO_2/MW/year and 2184.8 tons CO_2/MW/year, respectively.

The overall results show that the Asal–Ghoubbet Rift area can produce energy and green hydrogen at a low cost using wind energy compared to geothermal energy.

Supplementary Materials: The following are available online at https://www.mdpi.com/article/10.3390/en15010138/s1: Figure S1. A schematic geological map of the Republic of Djibouti (SE Afar Rift) and hydrothermal activity of the Republic of Djibouti. In the inset: schematic map of the Afar Depression with the location of Djibouti (black rectangle); Figure S2. Annual wind direction for Ghoubbet at level of 60 m of year 2015; Figure S3. Different equipment of RO, Guangzhou Kai Yuan Water Treatment Equipment Co. Ltd; Figure S4. Probability distribution of Monte Carlo simulation of Area of Asal-Ghoubbet Rift in m^2; Figure S5. Probability distribution of Monte Carlo simulation of Thickness of Asal-Ghoubbet Rift; Figure S6. Probability distribution of Monte Carlo simulation of Recovery Factor of Asal-Ghoubbet Rift; Figure S7. Probability distribution of Monte Carlo simulation of Heat Specific of Rock of Asal-Ghoubbet Rift in J/kg°C; Figure S8. Probability distribution of Monte Carlo simulation of Reference Temperature of Asal-Ghoubbet Rift in °C; Figure S9. Probability distribution of Monte Carlo simulation of Load Factor of Asal-Ghoubbet Rift; Figure S10. Probability distribution of Monte Carlo simulation of Conversion efficiency of Asal-Ghoubbet Rift; Figure S11. Sensitivity diagram for volumetric parameters variation; Figure S12. Chart Correlation in Monte Carlo simulation; Table S1. Different size of water electrolysis model (0.4 to 800 Nm^3/h); Table S2. Important parameters for the major stages of a single flash Power Plant; Table S3. Important parameters for the major stages of a dry steam power plant; Table S4. Statistics summary of wind speed at level of 80 m of Ghoubbet; Table S5. Characteristics of the different wind turbine models considered in this study and the calculated Annual energy production and the capacity factor; Table S6. Cost prospect of the wind-powered hydrogen production system in Ghoubbet without and with degradation rate; Table S7. Electrolyzer performance analysis using wind energy; Table S8. Most probable values and distributions for the parameters of USGS volumetric method for Asal-Ghoubbet Rift; Table S9. Result of thermoeconomics analysis of dry steam Asal power plant; Table S10. Result of thermoeconomics analysis of single flash Asal power plant; Table S11. Result of Monte Carlo simulation for potential capacity of Asal Well; Table S12. Result of economic analyses of green hydrogen production supply with the geothermal energy; Table S13. Energy efficiency of the wind-powered hydrogen production system in Ghoubbet without and with degradation rate; Table S14. CO_2 and fuel oil avoided with Wind energy development; Table S15. Exergoenvironnment of dry steam and single flash power plant.

Author Contributions: Methodology, writing—original draft, review & editing, M.O.A.; software, validation, writing—original draft preparation, A.-B.A.; model formulations and software simulation, writing—original draft, review & editing, O.A.D.; methodology and results, supervision, and resources, M.J., M.M.A. and I.A.G. All authors have read and agreed to the published version of the manuscript.

Funding: The Djibouti Centre for Research and Studies (CERD in French) was funded this study.

Institutional Review Board Statement: Not applicable.

Informed Consent Statement: Not applicable.

Data Availability Statement: Not applicable.

Acknowledgments: The authors would like to thank the Djibouti Centre for Research and Studies (CERD in French) for financial support. We would like to thank Ali Ahmed from the University of Djibouti for proofreading the manuscript. We would also like to thank the three anonymous reviewers for their constructive comments that improved the manuscript.

Conflicts of Interest: The authors declare no conflict of interest.

Nomenclature

Variables

c	Specific heat of liquid water [kJ/kg K]
$C_{Electrolyzer}$	Capital cost of the electrolyzer wind system [$]
$C_{Electricity}$	Cost of wind electricity [$]
C_f	Capacity Factor [%]
C_{fuel}	Cost of fuel [$]
C_I	Total investment cost of Wind Energy [$]
CH_2	Cost of hydrogen from geothermal energy [$]
C_{Aspec}	Average specific cost of wind turbine [$/KW]
$C_{O\&M}$	Cost operating and maintenance of Wind Turbine [$]
$C_{SU,t}$	Capital cost of geothermal power plant [$]
C_u	Unit cost of electrolyzer [$/kW]
C_{surf}	Cost of construction [$]
C_{WC}	Completion well cost [$]
d	Well Depth [m]
Ec	Yearly cost of geothermal electricity [$]
$E_{electrolyzer}$	Electricity required for 1 kg H_2 [kWh/kgH_2]
Eout	Wind electricity production [kWh]
G_{PP}	Geothermal power plant [MWe]
H	Hydrogen rate production [kg/s]
i	Discount rate [%]
Ic	Capital cost of electrolyser-Geothermal system [$]
I_d	Interest rate [%]
LCOE	Levelized cost of electricity [$/kWh]
LCOH	Levelized cost of hydrogen [$/kWh]
LF	Load Factor [%]
M'_{H_2}	Molar mass of hydrogen [kg / kmol]
M_{H_2}	Amount of hydrogen produced [kg]
M_{OM}	Operation and maintenance cost of electrolyser-Geothermal power plant system [$]
N	Period of study [year]
PD	Power Density [W/m^2]
PVC	Present value cost [$]
Pr	Rated power of wind turbine [KW]
t	Life span of the commercial wind turbine [years]
T	Life time of the commercial electrolyser [years]
T_s	Reference temperature [°C]
T_0	Ambiant temperature [°C]
$W_{act,\ electrolysis}$	Minimum work required for an electrolyzer in non-ideal operation [kj/kg]
W_{DC}	Well drilling cost [$]
$\dot{W}_{net}$	Steam Turbine net power output [kW]
$W_{rev,\ geo}$	Maximum specific work [kj/kg]
$W_{rev,\ electrolysis}$	Minimum work required for an electrolyzer in ideal operation [kj/kg]

Greek symbols

α	Shear coefficient
$\eta_{electrolyzer}$	Electrolyzer efficiency [%]
η_{conv}	Efficiency of rectifier [%]
$\Delta G_{electrolysis,\ H_20}$	Change in the Gibbs function [kJ/kmol]

Subscripts

0	Dead state
1,2,3	State numbers, wind speed level
a,b,c	Drilling Coefficients
geo	Geothermal fluid
H_2O	Fresh Water
H_2	Hydrogen gas

CO_2	Carbone Dioxide
l	Large
m	Medium
n	Number of well
out	Output
rev	Reversible
s	Small
Abbreviations	
AG	Asal-Ghoubbet
AWE	Alkaline water electrolyzer
AWE_l	Large-size Alkaline water electrolyzer
AWE_m	Medium-size Alkaline water electrolyzer
AWEs	small-size Alkaline water electrolyzer
BSh	Semiarid tropical steppe climate
BWh	Hot desert climate
CDF	Cumulative probability distribution function
EARS	East African Rift System
Eq.	Equation
EMJ	Empirical Method of Jestus
HTE	High temperature electrolyzer
MAE	Mean Absolute Error
MLM	Maximum Likelihood Method
MM	Moment Method
O&M	Operation and Maintenance
PDF	Probability distribution function
PEME	Polymer electrolyte membrane electrolyzer
$PEME_l$	Large-size Polymer electrolyte membrane electrolyzer
$PEME_m$	Medium-size Polymer electrolyte membrane electrolyzer
$PEME_s$	Small-size Polymer electrolyte membrane electrolyzer
RMSE	Root Mean Square Error
USGS	United States Geological Survey
WM	Wasp Method

References

1. Mlynarski, M.; Zlotnicki, J. Fluid circulation in the active emerged Asal-Ghoubbet Rif (east Africa, Djibouti) inferred from self-Potential and Telluric—Telluric Prospecting. *Tectonophysics* **2001**, *339*, 455–472. [CrossRef]
2. Fouillac, A.M.; Fouillac, C.; Cesbron, F.; Pillard, F.; Legendre, O. Water-rock interaction between basalt and high-salinity fluids in the Asal-Ghoubbet Rif, Republic of Djibouti. *Chem. Geol.* **1989**, *76*, 271–289. [CrossRef]
3. Aden, A.H.; Raymond, J.; Giroux, B.; Sanjuan, B. New Insights into Hydrothermal Fluid Circulation Affected by Regional Groundwater Flow in the Asal Rift, Republic of Djibouti. *Energies* **2021**, *14*, 1166. [CrossRef]
4. D'Amore, F.; Giusti, D.; Abdallah, A. Geochemistry of the high-salinity geothermal field of Asal, Republic of Djibouti, Africa. *Geothermics* **1998**, *27*, 197–210. [CrossRef]
5. Awaleh, M.O.; Hoch, F.B.; Boschetti, T.; Soubaneh, Y.D.; Egueh, N.M.; Elmi, S.A.; Jalludin, M.; Khaireh, M.A. The geothermal resources of the Republic of Djibouti—II: Geochemical study of the Lake Abhe geothermal field. *J. Geochem. Explor.* **2015**, *159*, 129–147. [CrossRef]
6. Awaleh, M.O.; Boschetti, T.; Soubaneh, Y.D.; Baudron, P.; Kawalieh, A.D.; Dabar, O.A.; Ahmed, M.M.; Ahmed, S.I.; Daoud, M.A.; Egueh, N.M.; et al. Geochemical study of the Sakalol—Harralol geothermal field (Republic of Djibouti): Evidences of a low enthalpy aquifer between Manda-Inakir and Asal-Ghoubbet Rif settings. *J. Volcanol. Geotherm. Res.* **2017**, *331*, 26–52. [CrossRef]
7. Awaleh, M.O.; Boschetti, T.; Soubaneh, Y.D.; Kim, Y.; Baudron, P.; Kawalieh, A.D.; Ahmed, M.M.; Daoud, M.A.; Dabar, O.A.; Kadieh, I.H.; et al. Geochemical, multi-isotopic studies and geothermal potential evaluation of the complex Djibouti volcanic aquifer (republic of Djibouti). *Appl. Geochem.* **2018**, *97*, 301–321. [CrossRef]
8. Awaleh, M.O.; Boschetti, T.; Adaneh, A.E.; Daoud, M.A.; Ahmed, M.M.; Dabar, O.A.; Soubaneh, Y.D.; Kawalieh, A.D.; Kadieh, I.H. Hydrochemistry and multi isotope study of the waters from Hanlé-Gaggadé grabens (Republic of Djibouti, East African Rift System): A low-enthalpy geothermal resource from a transboundary aquifer. *Geothermics* **2020**, *86*, 101805. [CrossRef]
9. Pillot, B.; Muselli, M.; Philippe Poggi, P.; Haurant, P.; Hared, I. Solar energy potential atlas for planning energy system off-grid electrification in the Republic of Djibouti. *Energy Convers. Manag.* **2013**, *69*, 131–147. [CrossRef]

10. Daher, D.H.; Gaillard, L.; Amara, M.; Ménézo, C. Impact of tropical desert maritime climate on the performance of a PV grid-connected power plant. *Renew. Energy* **2018**, *125*, 729–737. [CrossRef]
11. Assowe Dabar, O.; Awaleh, M.O.; Kirk-Davidoff, D.; Olauson, J.; Söder, L.; Awaleh, S.I. Wind resource assessment and economic analysis for electricity generation in three locations of the Republic of Djibouti. *Energy* **2019**, *185*, 884–894. [CrossRef]
12. Assowe Dabar, O.; Camberlin, P.; Pohl, B.; Waberi, M.M.; Awaleh, M.O.; Silah-Eddine, S. Spatial and temporal variability of rainfall over the Republic of Djibouti from 1946 to 2017. *Int. J. Climatol.* **2021**, *41*, 2729–2748. [CrossRef]
13. Elimax, H. *Etude Strategique de Déploiement de L'énergie Eolienne en Afrique. Rapport Final*; Canadian International Development Agency: Ottawa, ON, Canada, 2004.
14. Aliyu, A.K.; Modu, B.; Tan, C.W. A review of renewable energy development in Africa: A focus in South Africa, Egypt and Nigeria. *Renew. Sustain. Energy Rev.* **2018**, *8*, 2502–2518. [CrossRef]
15. Saidur, R.; Islam, M.R.; Rahim, N.A.; Solangi, K.H. A review on global wind energy policy. *Renew. Sustain. Energy Rev.* **2010**, *14*, 1744–1762. [CrossRef]
16. Saeed, M.A.; Ahmed, Z.; Zhang, W. Wind energy potential and economic analysis with a comparison of different methods for determining the optimal distribution parameters. *Renew. Energy* **2020**, *161*, 1092–1109. [CrossRef]
17. Soltani, M.; Mohammad Jabarifar, M.; Kashkooli, F.M.; Souri, M.; Rafiei, B.; Gharali, K.; Nathwani, J.S. Environmental, economic, and social impacts of geothermal energy systems. *Renew. Sustain. Energy Rev.* **2021**, *140*, 110750. [CrossRef]
18. Mahmoud, M.; Ramadan, M.; Naher, S.; Pullen, K.; Abdelkareem, M.A.; Olabi, A.G. A review of geothermal energy-driven hydrogen production systems. *Therm. Sci. Eng. Prog.* **2021**, *22*, 100854. [CrossRef]
19. Genç, M.S.; Çelik, M.; Karasu, I. A review on wind energy and wind–hydrogen production in Turkey: A case study of hydrogen production via electrolyzer system supplied by wind energy conversion system in Central Anatolian Turkey. *Renew. Sustain. Energy Rev.* **2012**, *16*, 6631–6646. [CrossRef]
20. Short, W.; Blair, N.; Heimiller, D. *Modeling the Market Potential of Hydrogen from Wind and Competing Sources*; National Renewable Energy Lab: Golden, CO, USA, 2005.
21. Kato, T.; Kubota, M.; Kobayashi, N.; Suzuoki, Y. Effective utilization of by-product oxygen from electrolyzer hydrogen production. *Energy* **2005**, *30*, 2580–2595. [CrossRef]
22. Ayodele, T.R.; Munda, J.L. Potential and economic viability of green hydrogen production by water electrolyzer using wind energy resources in South Africa. *Int. J. Hydrogen Energy* **2019**, *44*, 17669–17687. [CrossRef]
23. Cornell, A. Hydrogen production by electrolyzer. In Proceedings of the 1st International Conference on Electrolyzer, Copenhagen, Denmark, 13–15 June 2017; pp. 12–15.
24. Ishaq, H.; Dincer, I. Comparative assessment of renewable energy-based hydrogen production methods. *Renew. Sustain. Energy Rev.* **2021**, *135*, 110192. [CrossRef]
25. Al-Sharafi, A.; Sahin, A.Z.; Ayar, T.; Yilbas, B.S. Techno-economic analysis and optimization of solar and wind energy systems for power generation and hydrogen production in Saudi Arabia. *Renew. Sustain. Energy Rev.* **2017**, *69*, 33–49. [CrossRef]
26. Mostafaeipour, A.; Dehshiri, S.J.H.; Dehshiri, S.S.H.; Jahangiri, M. Prioritization of potential locations for harnessing wind energy to produce hydrogen in Afghanistan. *Int. J. Hydrogen Energy* **2020**, *45*, 33169–33184. [CrossRef]
27. Rahmouni, S.; Negrou, B.; Settou, N.; Dominguez, J.; Gouareh, A. Prospects of hydrogen production potential from renewable resources in Algeria. *Int. J. Hydrogen Energy* **2017**, *42*, 1383–1395. [CrossRef]
28. Ghazvini, M.; Sadeghzadeh, M.; Ahmadi, M.H.; Moosavi, S.; Pourfayaz, F. Geothermal energy use in hydrogen production: A review. *Int. J. Energy Res.* **2019**, *43*, 7823–7851. [CrossRef]
29. Rahmouni, S.; Settou, N.; Chennouf, N.; Negrou, B.; Houari, M. A technical, economic and environmental analysis of combining geothermal energy with carbon sequestration for hydrogen production. *Energy Procedia* **2014**, *50*, 263–269. [CrossRef]
30. Yilmaz, C.; Kanoglu, M.; Abusoglu, A. Exergetic cost evaluation of hydrogen production powered by combined flash-binary geothermal power plant. *Int. J. Hydrogen Energy* **2015**, *40*, 14021–14030. [CrossRef]
31. Yilmaz, C.; Koyuncu, I.; Alcin, M.; Tuna, M. Artificial Neural Networks based thermodynamic and economic analysis of a hydrogen production system assisted by geothermal energy on Field Programmable Gate Array. *Int. J. Hydrogen Energy* **2019**, *44*, 17443–17459. [CrossRef]
32. Beck, H.E.; Zimmermann, N.E.; McVicar, T.R.; Vergopolan, N.; Berg, A.; Wood, E.F. Present and future Köppen-Geiger climate classification maps at 1-km resolution. *Sci. Data* **2018**, *5*, 180214. [CrossRef]
33. Geiger, R. Klassifikation der klimate nachW. Köppen. Landolt-Börnstein– Zahlenwerte Und Funkt. *Aust. Phys. Chem. Astron. Geophys. Technol.* **1954**, *3*, 603–607.
34. Manighetti, I.; Tapponnier, P.; Gillot, P.Y.; Jacques, E.; Courtillot, V.; Armijo, R.; Ruegg, J.C.; King, G. Propagation of rifting along the Arabia-Somalia plate boundary: Into Afar. *J. Geophys. Res.* **1998**, *103*, 4947–4974. [CrossRef]
35. NRG Systems. 60m XHD NOW System. 2021. Available online: https://www.nrgsystems.com/products/complete-met-systems/wind-resource-assessment-systems/detail/34m-xhd-now-system/ (accessed on 27 January 2021).
36. Mohammadi, K.; Mostafaeipour, A. Using different methods for comprehensive study of wind turbine utilization in Zarrineh, Iran. *Energy Convers. Manag.* **2013**, *65*, 463–470. [CrossRef]
37. Akdag, S.A.; Dinler, A. A new method to estimate Weibull parameters for wind energy applications. *Energy Convers. Manag.* **2009**, *50*, 1761–1766. [CrossRef]
38. Kwon, S.D. Uncertainty analysis of wind energy potential assessment. *Appl. Energy* **2010**, *87*, 856–865. [CrossRef]

39. Ombeni, J.M. Performance evaluation of Weibull analytical methods using several empirical methods for predicting wind speed distribution. *Energy Sources Part A Recovery Util. Environ. Eff.* **2020**. [CrossRef]
40. Bahrami, A.; Teimourian, A.; Okoye, C.O.; Hiri, H. Technical and economic analysis of wind energy potential in Uzbekistan. *J. Clean. Prod.* **2019**, *20*, 801–814. [CrossRef]
41. Mentis, D.; Hermann, S.; Howells, M.; Welsch, M.; Siyal, S.H. Assessing the technical wind energy potential in Africa a GIS-based approach. *Renew. Energy* **2015**, *83*, 110–125. [CrossRef]
42. Abam, F.I.; Ohunakin, O.S. Economics of wind energy utilisation for water pumping and CO_2 mitigation potential in Niger Delta, Nigeria. *Int. J. Ambient Energy* **2015**, *38*, 229–239. [CrossRef]
43. Ahmed, A.S.; Hanitsch, R. Electricity generation and wind potential assessment at Hurghada, Egypt. *Renew. Energy* **2008**, *33*, 141–148. [CrossRef]
44. Muffler, P.; Cataldi, R. Methods for regional assessment of geothermal resources. *Geothermics* **1978**, *7*, 53–89. [CrossRef]
45. Ciriaco, A.E.; Zarrouk, S.J.; Zakeri, G. Geothermal resource and reserve assessment methodology: Overview, analysis and future directions. *Renew. Sustain. Energy Rev.* **2020**, *119*, 109515. [CrossRef]
46. Rezaei, M.; Khozani, N.N.; Jafari, N. Wind energy utilization for hydrogen production in an underdeveloped country: An economic investigation. *Renew. Energy* **2019**, *147*, 1044–1057. [CrossRef]
47. D' Amore-Domenech, R.; Santiago, Ó.; Leo, T.J. Multicriteria analysis of seawater electrolyzer technologies for green hydrogen production at sea. *Renew. Sustain. Energy Rev.* **2020**, *133*, 110166. [CrossRef]
48. Jónsson, V.K.; Gunnarsson, R.L.; Árnason, B.; Sigfússon, T.I. The feasibility of using geothermal energy in hydrogen production. *Geothermics* **1992**, *21*, 673–681. [CrossRef]
49. Kanoglu, M.; Bolatturk, A.; Yilmaz, C. Thermodynamic analysis of models used in hydrogen production by geothermal energy. *Int. J. Hydrogen Energy* **2010**, *35*, 8783–8791. [CrossRef]
50. Ueckerdt, F.; Hirth, L.; Luderer, G.; Edenhofer, O. System LCOE: What are the costs of variable renewables? *Energy* **2013**, *63*, 61–75. [CrossRef]
51. Li, Y.; Wu, X.P.; Li, Q.S.; Tee, K.F. Assessment of onshore wind energy potential under different geographical climate conditions in China. *Energy* **2018**, *152*, 498–511. [CrossRef]
52. Coskun, A.; Bolatturk, A.; Kanoglu, M. Thermodynamic and economic analysis and optimization of power cycles for a medium temperature geothermal resource. *Energy Convers. Manag.* **2014**, *78*, 39–49. [CrossRef]
53. Sun, J.; Liu, Q.; Duan, Y. Effects of evaporator pinch point temperature difference on thermo-economic performance of geothermal organic Rankine cycle systems. *Geothermics* **2018**, *75*, 249–258. [CrossRef]
54. Fallah, M.; Ghiasi, R.A.; Mokarram, N.H. A comprehensive comparison among different types of geothermal plants from exergy and thermoeconomic points of view. *Therm. Sci. Eng. Prog.* **2018**, *5*, 15–24. [CrossRef]
55. Shamoushaki, M.; Fiaschi, D.; Manfrida, G.; Niknam, P.H.; Talluri, L. Feasibility study and economic analysis of geothermal well drilling. *Int. J. Environ. Stud.* **2021**, *78*, 1022–1036. [CrossRef]
56. Lukawski, M.Z.; Anderson, B.J.; Augustine, C.; Capuano, L.E.; Beckers, K.F.; Livesay, B.; Tester, J.W. Cost analysis of oil, gas, and geothermal well drilling. *J. Pet. Sci. Eng.* **2014**, *118*, 1–14. [CrossRef]
57. Yilmaz, C.; Kanoglu, M.; Bolatturk, A.; Gadalla, M. Economics of hydrogen production and liquefaction by geothermal energy. *Int. J. Hydrogen Energy* **2012**, *37*, 2058–2069. [CrossRef]
58. Aquater. *Djibouti Geothermal Exploration Project Republic of Djibouti: Final Report*; Aquater: Djibouti, Republic of Djibouti, 1989; p. 159.
59. Virkir-Orkint Consulting Group Ltd. *Geothermal Scaling And corrosion Study, Final Report*; Virkir-Orkint Consulting Group Ltd.: Reykjanik, Iceland, 1990.
60. Lucas, B.; Silvio, M. The Big Portal for Wind Energy. Available online: https://en.wind-turbine-models.com/models (accessed on 25 July 2021).
61. Ayodele, T.R.; Ogunjuyigbe, A.S.O.; Amusan, T.O. Wind power utilization assessment and economic analysis of wind turbines across fifteen locations in the six geographical zones of Nigeria. *J. Clean. Prod.* **2016**, *129*, 341–349. [CrossRef]
62. Banque Centrale de Djibouti (BCD). Rapport Annuel. 2019. Available online: https://banque-centrale.dj/wp-content/uploads/2020/09/BCD-Rapport-Annuel-2019.pdf (accessed on 22 December 2021).
63. Greiner, C.; Korpas, M.; Holen, A. A Norwegian case study on the production of hydrogen from wind power. *Int. J. Hydrogen Energy* **2007**, *32*, 1500–1507. [CrossRef]
64. Abdallah, A.; Gandino, A.; Sommaruga, C. Technical-economic studies of geothermal projects: The Djibouti case. *Geothermics* **1985**, *14*, 327–334. [CrossRef]
65. Motazedi, K.; Salkuyeh, Y.K.; Laurenzi, I.J.; MacLean, H.L.; Bergerson, J.A. Economic and environmental competitiveness of high temperature electrolyzer for hydrogen production. *Int. J. Hydrogen Energy* **2021**, *46*, 21274–21288. [CrossRef]
66. Adebayo, V.; Abid, M.; Adedeji, M.; Hussain Ratlamwala, T.A. Energy, exergy and exergo-environmental impact assessment of a solid oxide fuel cell coupled with absorption chiller & cascaded closed loop ORC for multi-generation. *Int. J. Hydrogen Energy* 2021, *in press*.

energies

MDPI

Article

Energy-Saving Potential of Thermal Diode Tank Assisted Refrigeration and Air-Conditioning Systems

Mingzhen Wang, Eric Hu * and Lei Chen

School of Mechanical Engineering, The University of Adelaide, Adelaide, SA 5005, Australia; mingzhen.wang@adelaide.edu.au (M.W.); lei.chen@adelaide.edu.au (L.C.)
* Correspondence: eric.hu@adelaide.edu.au

Abstract: Lowering the condensing temperature of the Refrigeration and Air-conditioning (RAC) system has been proven to effectively increase the system's Coefficient of Performance (COP). This paper revolves around evaluating the energy-saving generated by applying a Thermal Diode Tank (TDT) in the RAC systems. The TDT is a novel invention, which is an insulated water tank equipped with gravity heat pipes. If the TDT was placed outdoors overnight, its inside water would theoretically be at the minimum ambient temperature of the previous night. When the TDT water is used to cool the condenser of RAC systems that operate during the daytime, a higher COP of this TDT assisted RAC (TDT-RAC) system could be achieved compared with the baseline system. In this study, a steady-state performance simulation model for TDT-RAC cycles has been developed. The model reveals that the COP of the TDT-RAC cycle can be improved by 10~59% over the baseline cycle depending on the compressor types. The TDT-RAC cycle with a variable speed compressor can save more energy than that with a fixed speed compressor. In addition, TDT-RAC cycles can save more energy with a higher day/night ambient temperature difference. There is a threshold tank size for a given TDT-RAC cycle to save energy, and the energy-saving can be improved by enlarging the tank size. A desk-top case study based on real weather data for Adelaide in January 2021 shows that 9~40% energy could be saved by TDT-RAC systems every summer day on average.

Keywords: refrigeration; heat pipe; thermal diode tank; COP improvement; vapor compression cycle; potential energy savings

Citation: Wang, M.; Hu, E.; Chen, L. Energy-Saving Potential of Thermal Diode Tank Assisted Refrigeration and Air-Conditioning Systems. *Energies* **2022**, *15*, 206. https://doi.org/10.3390/en15010206

Academic Editors: Wei-Hsin Chen, Aristotle T. Ubando, Chih-Che Chueh and Liwen Jin

Received: 30 November 2021
Accepted: 28 December 2021
Published: 29 December 2021

1. Introduction

Conserving energy is a major global challenge, and one of the most effective methods is to find solutions from energy efficiency improvements. Due to the rapid development of the world economy, ever more public buildings have been built, leading to an increasing requirement of heating, ventilation, and air-conditioning (HVAC) [1]. According to the projections of Australian Government's Department of Climate Change and Energy Efficiency in 2012 [2], electricity is the main source of energy for commercial buildings, accounting for about 50% of the above-mentioned total energy consumption, of which HVAC are usually the largest end-users of electricity. In addition, about 20~30% of energy consumption is yielded by the refrigeration and air-conditioning (RAC) system, producing over 34.7 Mt of total carbon emissions [3].

The report 'State of the Climate 2020: Bureau of Meteorology' [4] predicted that Australia is facing a trend of continued climate warming. Australia straddles a mainly temperate climate in densely populated coastal areas, which is characterised by large day and night ambient temperature differences in summer [5]. Since the average annual temperature in Australia is relatively high, and the hot period is much longer than the cold period, the demand for cooling is much greater than for heating. Moreover, traditional cooling demand industries such as food and pharmaceutical factories, datacentres, and wine industries, all make growing demands for more refrigeration and air-conditioning [1].

Consequently, improving the energy efficiency of cooling systems has greater significance than improving heating systems.

Lowering the condensing temperature of the RAC system has been validated to effectively increase the system's Coefficient of Performance (COP). Walsh [6] found that the COP for a RAC system normally ranges from 0 to 7 if no condenser cooling measures are applied. A 1 °C decrease in the condensing temperature would result in a 3.23% augmentation in the COP [7]. One of the successful examples of condenser cooling measure is the application of Ground-sourced RAC (GSRAC) systems, as the average underground soil temperature is lower than the ambient temperature during the day time in summer. A typical GSRAC system can reduce 30~40% energy compared with a normal RAC system (baseline system). However, GSRAC system would occupy considerable land resources to drill and install boreholes, which could be up to 200 m deep, and their payback periods are too long, basically over 7 years [8–10].

In addition to GSRAC systems, many other measures have been developed to lower the condensing temperature and improve the COP of the RAC system. Zhang et al. [11] examined an integrated RAC system with evaporative air-coolers that cool the air before it enters the condenser. The COP of this integrated system shows a 39% increase in the COP under the hot-arid area of China. Wang et al. [12] also utilised an evaporative cooling condenser in an existing air-conditioning system, which yielded a COP augmentation by 12%. Another experiment conducted by Yu and Chan [13] found that the evaporative coolers could increase the COP of air-cooled chillers by up to 20%. However, the energy-saving performance of these evaporative coolers would be reduced with the higher humidity, so their applications in most humid regions are limited. Waly et al. [14] also experimented with an evaporative cooling method, in which the inlet air was cooled by the spraying water before entering the condenser, and their results indicated an increase in COP by 36%. In this experiment, a problem was faced that the condenser should be placed away from obstructions because the shading of condenser would reduce the COP.

Chen et al. [15] assessed the ability of liquid–vapor separation condensers (LSC) and found that the COP of the LSC system was improved by 11.1% compared with the baseline system. To further improve the LSC system, Zhong et al. [16] studied the impact of a double-row liquid–vapor separation condenser (DLSC) on the RAC system cooling performance. This novel condenser achieved an improvement in the COP by about 30%. In both researches, the cooling performance of separation condensers is constrained by the ambient temperature; if the ambient temperature was over 40 °C, these systems would even have lower COPs.

While utilising the mist generator to supply cold mist for the RAC system's condenser, Yang et al. [17] achieved a considerable COP improvement of 21.3~51.5%, which is higher than those of most evaporative cooling methods. In application, the mist generator requires an open-loop water tower that consumes water to produce mist, which would be a waste of water resource.

Above literatures revealed that the performance enhancement of RAC systems was achieved using different condenser cooling methods, with significant COP improvements. However, there is little investigation on how to calculate the theoretical value of COP in vapor compression refrigeration cycles when the condenser cooling methods are applied. In addition to the defects and deficiencies summarized above, extra energy is also required to run the cooling units, such as air-coolers and mist generators.

In this paper, the Thermal Diode Tank (TDT) is a potential cost-effective alternative of other condenser cooling measures to supply water-cooling for RAC systems. Unlike above-mentioned studies, the TDT works automatically and passively that does not consume additional energy or resources. Basically, a TDT is an insulated water tank equipped with gravity heat pipes, which can passively release heat to the ambient air if the ambient temperature is lower than the water temperature inside TDT, but would not allow the heat to enter the tank even when the ambient temperature is higher. The RAC system equipped with a TDT is expected to generate a reduction of up to 40% in its energy consumption compared with a baseline system in the presence of day and night ambient temperature differences. However, since TDT is a newly invented technology, there has been little systematic research into it. Correlations between the design and operation parameters of TDT and its improvement are unknown. The aim of this study is to understand the potential of TDT when it is used to supply cooling water to a baseline system, by conducting desk-top case studies.

2. Thermal Diode Tank (TDT)

The TDT was proposed by Hu and Kimenkowski in 2019, which is constructed as a well-insulated water tank equipped with gravity heat pipes. The Gravity Heat Pipe (GHP, shown in Figure 1) is an isolated pipe containing a working fluid that coexists in liquid and vapor states [18]. When the temperature at the evaporator section of the GHP is higher, the working fluid (water, R134a, ammonia, etc.) will evaporate from liquid to vapor phase. The vapor rises upwards where the temperature is lower than that at the evaporator section, and condenses to release heat at the condenser section. After the vapor condenses, the condensate (liquid working fluid) flows back to the evaporator section due to the gravity [19]. A continuous circulation inside the GHP can be established if the evaporator temperature is higher than the condenser temperature. Oppositely, the heat will not be transferred downwards even though the condenser temperature is higher than the evaporator temperature. Therefore, the GHP functions as a thermal diode and automatically transfers heat only in one direction, from evaporator to condenser [20]. The thermal resistance inside a GHP is almost nil, so this device is very efficient in heat transfer. In practice, if the GHP evaporator temperature was higher than its condenser temperature by 1 °C or more, it would start operating.

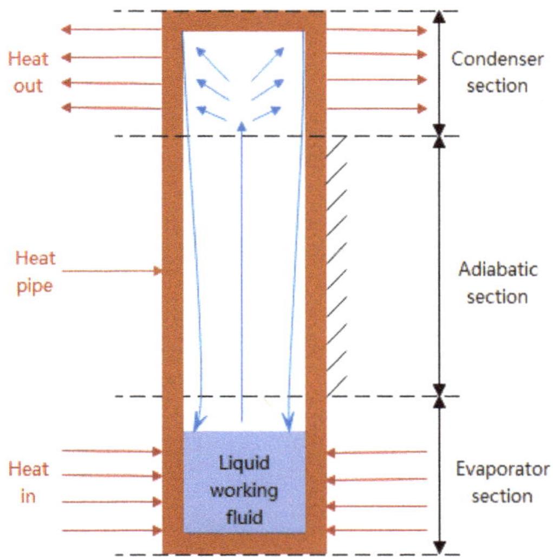

Figure 1. Gravity heat pipe working diagram.

The conceptual working diagram of the TDT is illustrated in Figure 2. With the one-direction heat transfer capability provided by GHPs, the TDT can release heat to the ambient air only if its inside water temperature is higher than the ambient temperature, but the heat from outside cannot enter the TDT. The adiabatic section of the GHPs prevents the heat from being transferred from the condenser section to the evaporator section by conduction even with scorching ambient temperatures. If such a device was placed outdoors overnight, the TDT water would theoretically be at the minimum ambient temperature of the last night. Thus, the TDT can passively harvest cold energy from the day/night ambient temperature differences to produce cooling water to lower the condensing temperature of the RAC system.

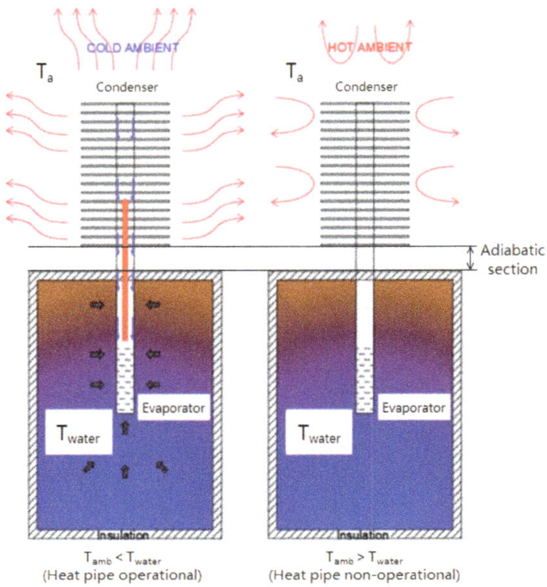

Figure 2. TDT working diagram.

Figure 3 shows the schematic diagram of the proposed TDT-RAC system. When the water at the minimum ambient temperature of the previous night is used to cool the RAC system's condenser, the COP of this integrated system would increase significantly. Namely, during the day time, the TDT-RAC system virtually works at a minimum ambient temperature of the last night. It is predicted that a lower night ambient temperature and a larger tank size are more beneficial to the TDT-RAC system, since colder and more cooling water can be produced for the RAC system's condenser.

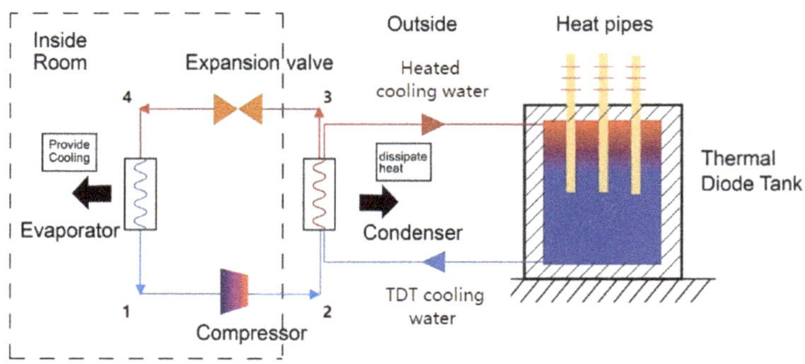

Figure 3. Schematic diagram of a TDT-RAC system.

3. Modelling of TDT-RAC Cycles

To evaluate the energy-saving performance of TDT-RAC systems compared with the baseline system, steady-state performance simulation models for TDT-RAC cycles were developed.

3.1. Baseline Cycle

The cycle *COP* of a vapour compression *RAC* system in Figure 4 can be expressed in terms of enthalpies [21]:

$$COP_{cycle\ 1} = \frac{\dot{Q}_{RAC}}{\dot{W}_C} = \frac{h_1 - h_4}{h_2 - h_1} \quad (1)$$

where

$$\dot{Q}_{RAC} = \dot{m}_1(h_1 - h_4) \quad (2)$$

$$\dot{W}_C = \dot{m}_1(h_2 - h_1) \quad (3)$$

and $\dot{m}_1$ (kg/s) is the refrigerant mass flow rate of the baseline cycle (Cycle 1), and h (kJ/kg) represents the specific enthalpy.

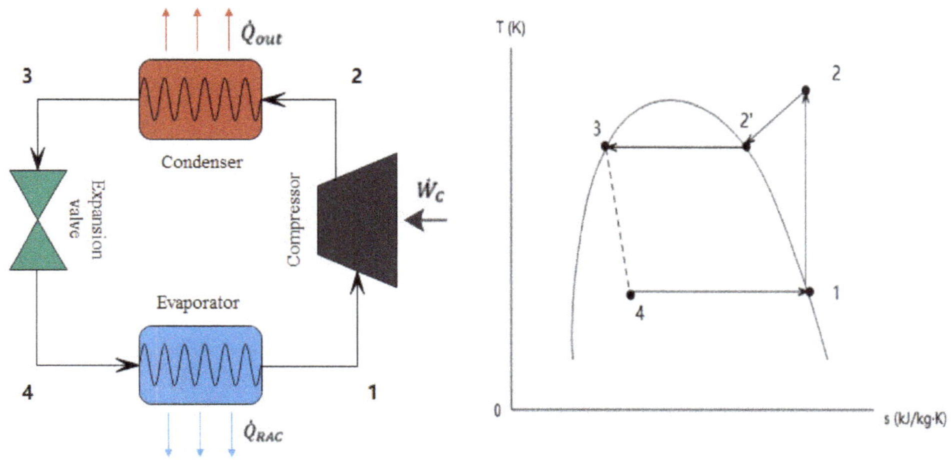

(a) (b)

Figure 4. (**a**) Vapor compression RAC system; (**b**) Baseline cycle (Cycle 1) in T-s diagram (1→2→3→4→1).

From Equation (1), it is observed that the COP can be improved theoretically by either increasing the enthalpy gap between h_1 and h_4, and/or decreasing the enthalpy gap between h_2 and h_1. It is noted that h_1 depends on the required evaporating temperature (T_1), which it is unable to vary, otherwise the cooling demand cannot be met. Therefore, in engineering practice, h_4 and/or h_2 are lowered to improve the COP of the RAC system. From state 3 to 4, the refrigerant passes the expansion valve where there is no energy transfer, so the enthalpy during this process is constant ($h_3 = h_4$). Therefore, h_4 and h_2 both depend on the condensing temperature (T_3), and an improvement in COP can be achieved by lowering the condensing temperature of the RAC system.

3.2. TDT-RAC Cycles

Depending on the compressor type (i.e., fixed speed or variable speed) of RAC systems, the impacts of the TDT on the RAC cycle are different. A fixed speed compressor always runs at a full capacity, while a variable speed compressor virtually runs at 25% to 100% capacity [22]. Figure 5 illustrates the differences in cycles in T-s diagrams with different compressors when the condensing temperature decreases. Figure 5a shows the TDT-RAC cycle with a fixed speed compressor (Cycle 2), in which the cycle 1→2→3→4→1 represents the baseline cycle, and Figure 5b shows the TDT-RAC cycle with a variable speed compressor (Cycle 3):

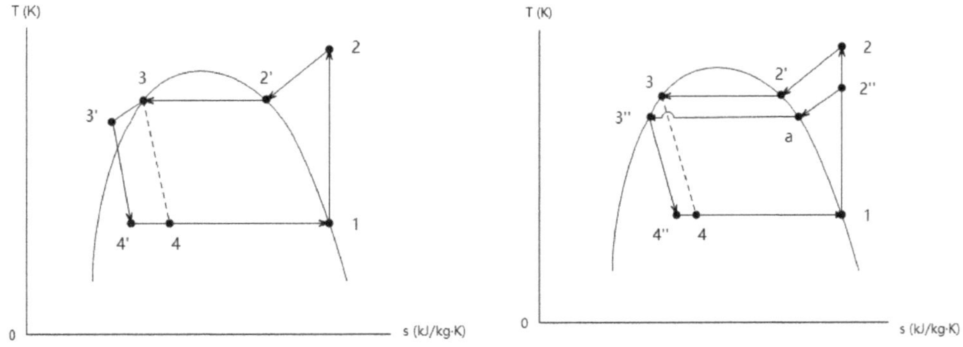

(a) (b)

Figure 5. TDT-RAC cycles (T-s diagram) with: (**a**) fixed speed compressors (1→2→3′→4′→1); (**b**) variable speed compressors (1→2″→3″→4″→1).

For the TDT-RAC Cycle 2 (1→2→3′→4′→1) in Figure 5a or Figure 6a, the refrigerant state at the condenser would be state 3′ (undercooled liquid) rather than state 3 (saturated liquid) because the condensing temperature ($T_{3'}$) is lowered by the TDT water, but the pressure at state 3′ is still equal to that at state 3. Hence, $h_{4'}$ is decreased from h_4, while the compressor work remains unaffected. Consequently, the refrigeration effect (h_1–$h_{4'}$) in the TDT-RAC cycle is greater than that in the baseline cycle, (h_1–h_4).

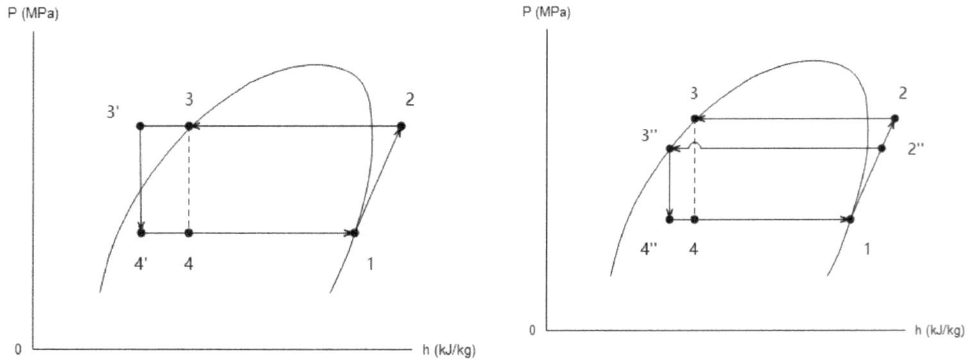

(a) (b)

Figure 6. TDT-RAC cycles (P-h diagram) with: (**a**) fixed speed compressors (1→2→3′→4′→1); (**b**) variable speed compressors (1→2″→3″→4″→1).

For Cycle 3 (1→2″→3″→4″→1) in Figure 5b or Figure 6b, the refrigerant state at the condenser would be state 3″ (saturated liquid) instead of state 3 due to the reduced condensing temperature ($T_{3''}$) generated by the TDT. The saturated refrigerant is cooled, causing a pressure drop, and $h_{4''}$ becomes smaller than h_4. Meanwhile, the variable speed compressor works at a lower speed to produce less heat, so that the refrigerant is cooled from state 2 to state 2″. Therefore, the refrigeration effect (h_1–$h_{4''}$) in the TDT-RAC cycle is greater than that in the normal RAC cycle (h_1–h_4), while the compressor work in the TDT-RAC cycle ($h_{2''}$–h_1) is lower than that in the normal RAC cycle (h_2–h_1).

3.3. Mathematical Model

The mathematical model developed in this study is a steady-state performance simulation model and its inputs and outputs are presented in Table 1. Following assumptions are made in the model:

1. The TDT is well insulated and has heat transfer with surroundings only through heat pipes;
2. The temperature of water inside TDT is uniform at any times;
3. The process from 1 to 2 in the RAC system compressor is isentropic;
4. The heat released by the RAC system's condenser is completely absorbed by the TDT water;
5. There is no energy loss in the expansion valve, so $h_3 = h_4$, $h_{3'} = h_{4'}$ and $h_{3''} = h_{4''}$;
6. The convection heat transfer coefficient of water is higher than that of air, so the condensing temperature is 10 °C over the day time ambient temperature for the baseline cycle, and 5 °C over the TDT water temperature for the TDT-RAC cycles;
7. The evaporating temperature is 10 °C lower than the room temperature set;
8. There is sufficient heat transfer capacity of heat pipes installed with the TDT, so the TDT water temperature in the early morning is assumed to be 3 °C higher than the minimum ambient temperature of last night.

Table 1. Inputs and outputs of performance simulation (mathematical) model.

	Inputs
$\dot{Q}_{RAC}$	RAC system cooling capacity (kW)
T_{room}	Room temperature set (°C)
T_a	Day time ambient temperature (°C)
T_{night}	Minimum ambient temperature (last night) (°C)
T_{water}	TDT water temperature (early morning) (°C)
Δt	Time interval (hour)
t_{RAC}	Total operation time (hour)
V	Tank size (m^3)
C_P	Specific heat capacity of water = 4.18 (kJ/kg °C)
ρ	Density of TDT water = 1000 (kg/m^3)
	Outputs
COP	COP for each time interval
$COP_{average}$	Average COP over t_{RAC}

Accordingly, three different cycles were modelled mathematically:

Cycle 1. A baseline cycle (Cycle 1→2→3→4→1 in Figure 5);

Cycle 2. A TDT-RAC cycle with a fixed speed compressor (Cycle 1→2→3′→4′→1 in Figure 5a);

Cycle 3. A TDT-RAC cycle with a variable speed compressor (Cycle 1→2″→3″→4″→1 in Figure 5b);

3.3.1. Cycle 1: A Baseline Cycle

Refer to Figure 4b and Equation (1), the COP for Cycle 1 is constant at any time when the baseline cycle is running, which is:

$$COP_{cycle\ 1} = \frac{h_1 - h_4}{h_2 - h_1} \tag{4}$$

3.3.2. Cycle 2: A TDT-RAC Cycle with a Fixed Speed Compressor

During the day when the TDT-RAC is running and the ambient temperature is higher than that inside the TDT, the heat pipe is non-operational. The change in TDT water temperature (ΔT_{water}) over the time interval (Δt) is expressed as:

$$\Delta T_{water} = \frac{Q_{water}}{C_P \rho V} \tag{5}$$

where C_P (kJ/kg °C) is the specific heat capacity of water, ρ (kg/m^3) is the density of water, V (m^3) is the water volume/tank size, and Q_{water} (kJ) is the heat absorbed by TDT water over the time interval (Δt), which is the condensing heat discharged from the condenser. Therefore, Q_{water} for the Cycle 2 is:

$$Q_{water} = \dot{m}_2 (h_2 - h_{3'}) \Delta t \tag{6}$$

Whilst the heat pipe is non-operational during the day time, the history of TDT water temperature can be found by Equation (7).

$$T_{water}{}^{i+1} = T_{water}{}^{i} + \frac{\dot{m}_2 (h_2 - h_{3'}{}^{i}) \Delta t}{C_P \rho V}, \; i \geq 1 \tag{7}$$

where i represents the order of time interval (e.g., i = 1 means the first time interval)

For the fixed speed compressor, h_2 is constant at every time interval when the TDT-RAC system runs, whereas $h_{3'}$ and $h_{4'}$ depend on $T_{3'}$, which are assumed as 5 °C higher than T_{water} in this model. Consequently, the COP at the corresponding time interval i would be different, which is:

$$COP_{cycle\ 2}{}^{i} = \frac{h_1 - h_{4'}{}^{i}}{h_2 - h_1} \tag{8}$$

3.3.3. Cycle 3. A TDT-RAC Cycle with a Variable Speed Compressor

For Cycle 3, the TDT water temperature corresponding to each time interval is:

$$T_{water}{}^{i+1} = T_{water}{}^{i} + \frac{\dot{m}_3 (h_{2''}{}^{i} - h_{3''}{}^{i}) \Delta t}{C_P \rho V}, \; i \geq 1 \tag{9}$$

where for the variable speed compressor, $h_{2''}$, $h_{3''}$ and $h_{4''}$ all depend on $T_{3''}$, which are variable at different time intervals. According to Figure 5b, the COP at the corresponding time interval is:

$$COP_{cycle\ 3}{}^{i} = \frac{h_1 - h_{4''}{}^{i}}{h_{2''}{}^{i} - h_1} \tag{10}$$

For the Cycle 1, the average COP is constant at any time. However, the hourly COPs (Δt = 1 hour) for the Cycles 2 and 3 would change over time. Therefore, the average COP over the total TDT-RAC system operation time (t_{RAC}) is the average of the COPs for all h.

$$t_{RAC} = n \times \Delta t \tag{11}$$

$$COP_{average,cycle\ j} = \frac{1}{n} \sum_{i=1}^{n} COP_{cycle\ j}{}^{i}, \; j = 2 \; or \; 3 \tag{12}$$

where i represents the order of hour, n is the total number of hour in which the Cycle j runs. For example, if t_{RAC} is 6 hours and Δt is 1 hour, then n equals 6, and the hourly COPs could be found with Equations (8) and (10), as well as the average COP over t_{RAC} with Equation (12).

The mathematical model of the TDT-RAC cycle was established by applying Equations (4) to (12). According to the theoretical cycles presented in Figure 5, this model can calculate the COP at each corresponding time interval and the average COP of the

TDT-RAC cycle if all required inputs listed in Table 1 are known. As the aim of this study is to compare the energy consumption and/or COP values of three cycles, and the equations used in this model are all well-established traditional energy conservation equations, it is considered that the performance simulation model can be verified although it is not experimentally validated again in this study.

4. Energy-Saving of TDT-RAC Systems

4.1. Energy-Saving Indicator

To compare the energy-saving performance of TDT-RAC systems, the energy-saving percentage (ESP) is defined for both Cycles 2 and 3. For TDT-RAC cycles with a given cooling capacity, the hourly ESP at each time interval (Δt = 1 hour) is calculated as the ratio of saved power consumption of the given TDT-RAC cycle to the power consumption of the Cycle 1:

$$ESP_{hourly,\ cycle\ j}{}^{i} = \frac{\dot{W}_{C,cycle\ 1} - \dot{W}_{C,cycle\ j}{}^{i}}{\dot{W}_{C,cycle\ j}{}^{i}} \times 100\%,\ j = 2\ or\ 3 \quad (13)$$

where

$$\dot{W}_{C,cycle\ j}{}^{i} = \frac{\dot{Q}_{RAC}}{COP_{cycle\ j}{}^{i}} \quad (14)$$

It is noted that the cooling capacity ($\dot{Q}_{RAC}$) of three cycles are the same. From definition, the hourly ESP can be expressed in terms of enthalpies:

$$ESP_{hourly,\ cycle\ 2}{}^{i} = 1 - \frac{h_1 - h_4}{h_1 - h_{4'}{}^{i}} \quad (15)$$

$$ESP_{hourly,\ cycle\ 3}{}^{i} = 1 - \frac{(h_1 - h_4)(h_{2''}{}^{i} - h_1)}{(h_1 - h_{4''}{}^{i})(h_2 - h_1)} \quad (16)$$

The average ESP over the total operation time (t_{RAC} in Equation (11)) is the sum of each hourly ESP divided by the total operation time in h.

$$ESP_{average,cycle\ j} = \frac{1}{n}\sum_{i=1}^{n} ESP_{hourly,\ cycle\ j}{}^{i},\ j = 2\ or\ 3 \quad (17)$$

4.2. The Reference Case

A reference case was set in this study, and the details of its inputs and outputs are given in Table 2:

Table 2. Reference case.

Inputs for the Model	
RAC system refrigerant type	R134a
Day time ambient temperature, T_a (°C)	35
Room temperature set, T_{room} (°C)	20
TDT water temperature (early morning), T_{water} (°C)	25
Minimum ambient temperature (last night), T_{night} (°C)	22
Tank size, V (m^3)	4
RAC system cooling capacity, Q_{RAC} (kW)	7
RAC system daily operating time, t_{RAC} (hour)	6

Table 2. *Cont.*

Outputs for the Reference Case	
Cycle 1: Baseline Cycle:	
$COP_{average}$	6.8
$ESP_{average}$ (%)	0
Cycle 2: TDT-RAC Cycle with a Fixed Speed Compressor:	
$COP_{average}$	7.5
$ESP_{average}$ (%)	9.9
Cycle 3: TDT-RAC Cycle with a Variable Speed Compressor:	
$COP_{average}$	10.8
$ESP_{average}$ (%)	35.5

It can be ascertained from Table 2 that the proposed TDT can indeed save energy for the RAC system. The greatest COP and ESP of each TDT-RAC system are found at the beginning of the system operation, but the overall TDT water temperature would rise over time, causing the COP and ESP to decrease. The average COPs of Cycles 2 and 3 are 7.5 and 10.8, respectively. In addition, about 9.9% and 35.5% energy can be saved with Cycles 2 and 3, respectively, compared with the baseline cycle.

5. Sensitivity Analysis

The energy-saving performance of the TDT-RAC cycle depends on some key design and operation parameters. To understand the correlations between the crucial parameters and the average ESPs of both TDT-RAC cycles, sensitivities upon the following factors were studied:

- Day time and night ambient temperatures;
- Tank size;
- RAC system cooling capacity;
- Coupled effect of tank size and cooling capacity.

5.1. Day and Night Ambient Temperatures

In Figure 7, as the day time ambient temperature increases while the night ambient temperature is fixed at 22 °C, the curve of the Cycle 3 grows more rapidly than the curve of the Cycle 2 does; the difference in the average ESPs between Cycles 2 and 3 becomes greater.

Figure 8 shows that a higher night ambient temperature decreases the average ESPs of Cycles 2 and 3, while the day time ambient temperature is fixed at 35 °C. This is because a higher night ambient temperature would increase the TDT water temperature, alongside the condensing temperatures of both TDT-RAC cycles. However, the change in night ambient temperature has no effect on the condensing temperature of the baseline cycle. Therefore, the ESPs of TDT-RAC cycles would decline as the night ambient temperature increases.

It can be concluded from Figures 7 and 8 that a greater day and night ambient temperature difference can increase the average ESPs of both TDT-RAC cycles.

5.2. Tank Size/Water Volume

From Figure 9, the average ESPs of Cycles 2 and 3 would rise for an increasing tank size, but both increasing curves would flatten out eventually. With a larger tank size but the same amount of heat absorbed by the TDT water, the condenser of the TDT-RAC cycle would be less heated; hence, their average ESPs would increase. However, this trend would flatten out if the TDT size was over a certain value. The maximum energy-saving percentages for Cycles 2 and 3 are found as 13.4% and 48.99%, respectively. This observation implies that the cost-effectiveness should be considered when determining the TDT size for a given TDT-RAC system.

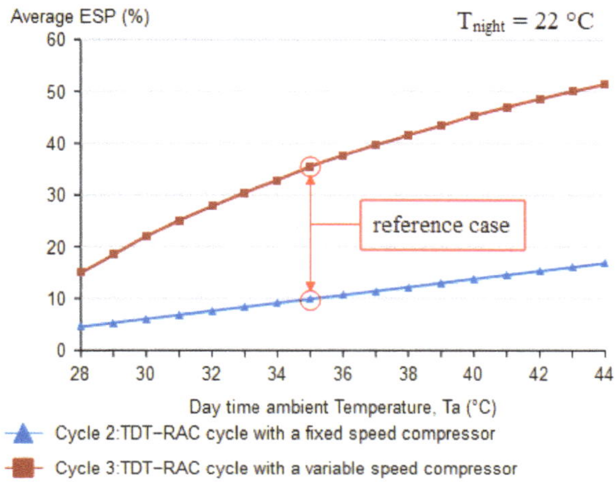

Figure 7. Average ESP versus day time ambient temperature (the minimum night ambient temperature is set as 22 °C).

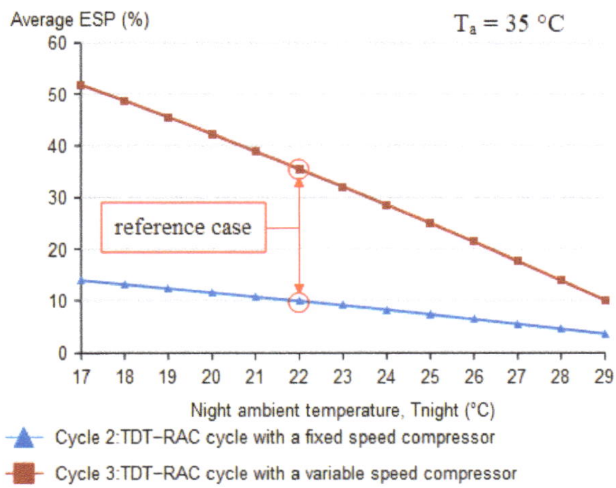

Figure 8. Average ESP versus night ambient temperature (the day time ambient temperature is set as 35 °C).

In addition, the tank size has a greater impact on the average ESP of the Cycle 3 compared with the Cycle 2. This is because a lower condensing temperature can decrease the input energy of the variable speed compressor, but not for the fixed speed compressor. Therefore, the ESP of the Cycle 3 would be more sensitive to the tank size compared with the Cycle 2.

It is also ascertained that there is a threshold tank size existing for each TDT-RAC cycle to achieve energy-saving. If the tank size is below this threshold, the ESP would be negative, which indicates that the TDT-RAC cycle would consume more energy compared with the baseline cycle. In this study case, the threshold tank size to achieve energy-saving for Cycles 2 and 3 is around 1.22 m^3. Finally, the threshold tank sizes for Cycles 2 and 3 are equal. Without other parameter changes, only when the condensing temperatures of both

TDT-RAC cycles are equal to that of a normal RAC cycle, their ESPs should be zero. In this case, their TDT sizes would be the same.

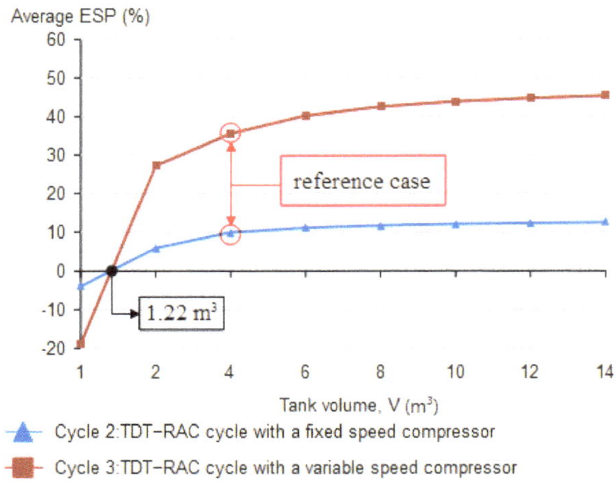

Figure 9. Average ESP versus day time ambient temperature (the threshold of both TDT-RAC curves is the same, which is about 1.22 m^3).

5.3. RAC System Cooling Capacity

The cooling load is defined as the power required to cool the room at a desirable temperature, which equals the RAC system cooling capacity. In Figure 10, when the cooling capacity (load) increases, the average ESPs of TDT-RAC cycles would decrease. The heat released by the condenser would increase as the cooling capacity increases, alongside the condensing temperatures of Cycles 2 and 3. Therefore, if the tank size remains unchanged while the condensing temperatures increase, Cycles 2 and 3 would have lower ESPs. This finding implies that for a RAC system with a greater cooling demand would require a larger TDT size, in order to satisfy the required energy-saving percentages.

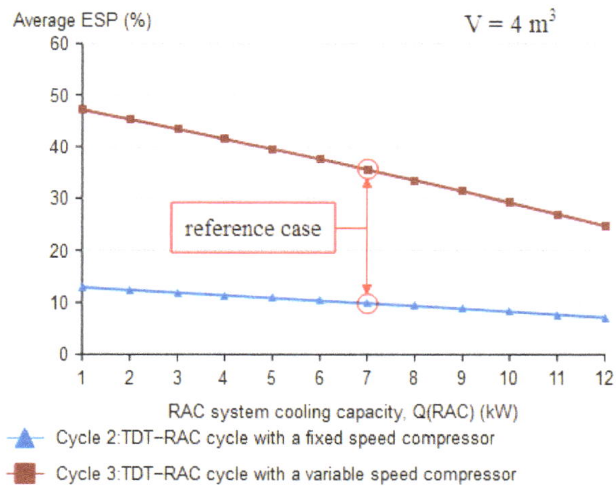

Figure 10. Average ESP versus RAC system cooling capacity (the tank size is set as 4 m^3).

5.4. Coupled Effect of Tank Size and Cooling Capacity

To understand the coupled effect of tank size and the RAC system cooling capacity on the average ESP, the simulations that generate the results in Figures 9 and 10 are combined.

The combined results are presented in terms of "Tank Size on Cooling Capacity (TS/CC)" versus the average ESPs in Figure 11. It is observed that the ESPs of Cycles 2 and 3 increase with a larger TS/CC value, but the trendlines would flatten out when the TS/CC value is over 1 m^3/kW. In addition, there is a threshold for the TS/CC value that complies with the results presented in Figure 11. Namely, only if the TS/CC value is above the threshold of 0.174 m^3/kW, the TDT-RAC cycles would have higher COPs and reduced energy consumptions compared with the baseline cycle, under the conditions of this study.

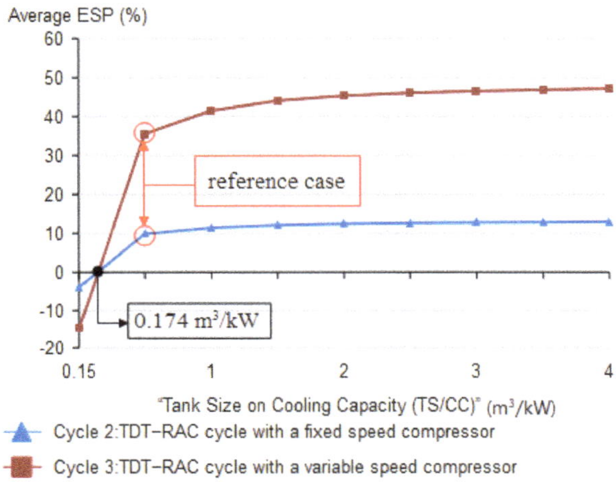

Figure 11. Average ESP versus "Tank Size on Cooling Capacity (TS/CC)" (the threshold of both TDT-RAC cycles is the same, which is about 0.174 m^3/kW).

6. Case Study

The study case is that a house requires a cooling load of 7 kW over 6 h per day, which is met by a TDT-RAC system whose characteristics are listed in Table 3. The weather data of Adelaide (Kent Town) in January 2021 are used and shown as the bar chart in Figure 12. The results of the case study, i.e., the average ESP of each day for both TDT-RAC systems with fixed speed and variable speed compressors, are presented in Figure 12.

Table 3. Characteristics of the TDT-RAC system used in the case study.

Parameter	Value/Comment
RAC system refrigerant	R134a
RAC system cooling capacity, Q_{RAC} (kW)	7
RAC system daily operating time, t_{RAC} (h)	6
Room temperature set, T_{room} (°C)	20
Tank size, V (m^3)	4
Day time ambient temperature, T_a (°C) Minimum ambient temperature (last night), T_{night} (°C) TDT water temperature (early morning), T_{water} (°C) *	Refer to the real weather data in Figure 11

* The TDT water temperature in the early morning is assumed to be 3 °C higher than the minimum ambient temperature of the last night.

In Figure 12, the dates of the maximum and the minimum day/night ambient temperature differences (ΔT_{daily}) are highlighted, which also correspond to the highest and the lowest average ESPs of both Cycles 2 and 3. The results agree with the findings in Section 5.1 (Figures 7 and 8) that a greater day/night ambient temperature difference can increase the average ESPs of TDT-RAC cycles. The results also reveal that the Cycle 2 achieved an average ESP ranging from 2.4 to 18.1%, whereas the average ESP of the Cycle 3 is from 14.9 to 61.4%. Over the whole month, Cycles 2 and 3 would save 9.1% and 40.4% energy on average, compared with the case where a baseline cycle was used.

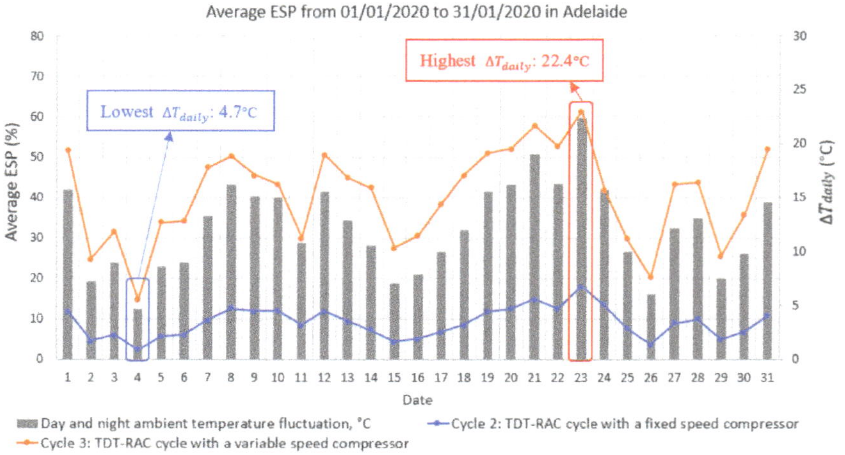

Figure 12. Results of the case study: the average ESP over 6 h of each day in January 2021, Adelaide (Kent Town).

7. Conclusions

A novel TDT-RAC system and its corresponding cycles were proposed and modelled for demonstrating its energy-saving potential. The cycle-based steady state model shows that the COPs of TDT-RAC cycles could theoretically increase by 10.3~58.8% over the baseline cycle in the reference case. The sensitivities of key design and operation parameters which could affect the ESP of the TDT-RAC cycle were analysed, including day and night ambient temperatures, tank size, RAC system cooling capacity, and coupled effect of tank size and cooling capacity. Moreover, a desk-top case study with typical Adelaide summer weather data was conducted. From the study, the following conclusions can be drawn:

1. The Thermal Diode Tank (TDT) is proved that it can theoretically reduce the energy consumption of the RAC system if they are properly equipped;
2. An increasing day/night ambient temperature difference can improve the energy-saving percentages of TDT-RAC cycles;
3. There is a threshold tank size existing for a given TDT-RAC cycle to save energy. If the tank size was below the threshold value, the TDT-RAC cycle would not save energy. Increasing the tank sizes above the threshold can increase the energy-saving percentages;
4. With the tank size above the threshold while other parameters remain the same, the TDT-RAC cycle with a variable speed compressor always achieves a higher energy-saving percentage compared to that with a fixed speed compressor;
5. The threshold and optimum value for the "Tank Size on Cooling Capacity (TS/CC)" in Figure 11 are found as 0.174 m^3/kW and 1 m^3/kW, respectively, under the conditions of this study. For future experiment, the range of tank size can be determined based on these two values with a given cooling capacity.
6. The case study shows that, compared with the baseline cycle, the TDT-RAC cycle with a fixed speed compressor could, on average, save 2.4~18.1% energy, while the TDT-RAC cycle with a variable speed compressor could save 14.9~61.4% energy. Over the whole month, 9.1~40.4% energy can be saved by TDT-RAC cycles every day on average.

Based on this initial study on TDT-RAC systems/cycles, future investigations on this topic may include developing validated dynamic models for TDT-RAC systems rather than current cycle-based models in steady state. Detailed studies on the TDT itself may include understanding the water temperature profile and thermal stratification inside the TDT under different conditions.

8. Patents

Eric Jing Hu has patent "A HEAT TRANSFER ARRANGEMENT FOR IMPROVED ENERGY EFFICIENCY OF AN AIR CONDITIONING SYSTEM–A thermal 'Diode Tank'" issued to AU 2014202998 Al.

Author Contributions: Conceptualization, E.H.; methodology, M.W.; software, M.W.; formal analysis, M.W.; investigation, M.W.; resources, M.W.; data curation, M.W. and E.H.; writing—original draft preparation, M.W.; writing—review and editing, M.W., E.H. and L.C.; supervision, E.H. and L.C.; project administration, E.H. and L.C. All authors have read and agreed to the published version of the manuscript.

Funding: This research received no external funding.

Institutional Review Board Statement: Not applicable.

Informed Consent Statement: Not applicable.

Data Availability Statement: The data presented in this study are available on request from the corresponding author.

Acknowledgments: This research received no external funding. The authors would like to thank the University of Adelaide for providing the University of Adelaide Research Scholarship for this study.

Conflicts of Interest: The authors declare no conflict of interest.

Nomenclature

C_P	Specific heat capacity (kJ/kg °C)
E_{saved}	Saved energy (kJ)
h	Specific enthalpy (kJ/kg)
$\dot{m}$	Mass flow rate (kg/s)
Q_{out}	Heat through to condenser(kJ)
$\dot{Q}_{out}$	The rate of heat through to condenser (kW)
Q_{RAC}	Cooling energy (kJ)
$\dot{Q}_{RAC}$	Cooling capacity (kW)
T_a	Day time ambient temperature (°C)
T_{night}	Night time ambient temperature (°C)
T_{room}	Room temperature (°C)
T_{water}	Water temperature (°C)
ΔT_{daily}	Daily day and night ambient temperature difference (°C)
t_{RAC}	Refrigeration and air-conditioning system daily operating time (h or s)
Δt	Refrigeration and air-conditioning system operating time interval (h or s)
V	Tank size (m^3)
$\dot{W}_C$	Input power by compressor (kW)
W_C	Input energy by compressor (kJ)
Greek Symbols	
ρ	Density (kg/m^3)
Abbreviations	
CHP	Convective Heat Pipe
COP	Coefficient of Performance
EER	Energy Efficiency Ratio
ESP	Energy-Saving Percentage
GHP	Gravity Heat Pipe
GSHP	Ground-source Heat Pump
HVAC	Heating, Ventilation, and Air-conditioning
RAC	Refrigeration and Air-conditioning

TDT	Thermal Diode Tank
TDT-RAC	Thermal Diode Tank-assisted Refrigeration and Air-conditioning

References

1. Sugarman, S.C. *HVAC Fundamentals*; River Publishers: Gistrup, Denmark, 2020; p. 509.
2. Department of Climate Change and Energy Efficiency (DCCEE). *Australia's Emissions Projections, 2012*; DCCEE: Canberra, Australia, 2012.
3. Department of Industry, Science, Energy and Resources (DISER). *Energy Update: Australian Energy Consumption and Production*; DISER: Canberra, Australia, 2020.
4. State of the Climate 2020: Bureau of Meteorology. Available online: http://www.bom.gov.au/state-of-the-climate/ (accessed on 23 September 2021).
5. Head, L.; Adams, M.; McGregor, H.V.; Toole, S. Climate change and Australia. *Wiley Interdiscip. Rev. Clim. Chang.* **2014**, *5*, 175–197. [CrossRef]
6. Walsh, J.A. Obesity & the First Law of Thermodynamics. *Am. Biol. Teach.* **2013**, *75*, 413–415. [CrossRef]
7. Kumbhar, A.; Gulhane, N.; Pandure, S. Effect of Various Parameters on Working Condition of Chiller. *Energy Procedia* **2017**, *109*, 479–486. [CrossRef]
8. Qiao, Z.; Long, T.; Li, W.; Zeng, L.; Li, Y.; Lu, J.; Cheng, Y.; Xie, L.; Yang, L. Performance assessment of ground-source heat pumps (GSHPs) in the Southwestern and Northwestern China: In situ measurement. *Renew. Energy* **2020**, *153*, 214–227. [CrossRef]
9. Girard, A.; Gago, E.J.; Muneer, T.; Caceres, G. Higher ground source heat pump COP in a residential building through the use of solar thermal collectors. *Renew. Energy* **2015**, *80*, 26–39. [CrossRef]
10. Christodoulides, P.; Aresti, L.; Florides, G. Air-conditioning of a typical house in moderate climates with Ground Source Heat Pumps and cost comparison with Air Source Heat Pumps. *Appl. Therm. Eng.* **2019**, *158*, 113772. [CrossRef]
11. HZhang You, S.; Yang, H.; Niu, J. Enhanced performance of air-cooled chillers using evaporative cooling. *Build. Serv. Eng. Res. Technol.* **2016**, *21*, 213–217. [CrossRef]
12. Wang, T.; Sheng, C.; Nnanna, A.A. Experimental investigation of air conditioning system using evaporative cooling condenser. *Energy Build.* **2014**, *81*, 435–443. [CrossRef]
13. Yu, F.W.; Chan, K.T. Application of Direct Evaporative Coolers for Improving the Energy Efficiency of Air-Cooled Chillers. *J. Sol. Energy Eng.* **2005**, *127*, 430–433. [CrossRef]
14. Waly, M.; Chakroun, W.; Al-Mutawa, N.K.; Al-Mutawa, N.K. Effect of pre-cooling of inlet air to condensers of air-conditioning units. *Int. J. Energy Res.* **2005**, *29*, 781–794. [CrossRef]
15. Chen, X.; Chen, Y.; Deng, L.; Mo, S.; Zhang, H. Experimental verification of a condenser with liquid–vapor separation in an air conditioning system. *Appl. Therm. Eng.* **2013**, *51*, 48–54. [CrossRef]
16. Zhong, T.; Ding, L.; Chen, S.; Chen, Y.; Yang, Q.; Luo, Y. Effect of a double-row liquid–vapor separation condenser on an air-conditioning unit performance. *Appl. Therm. Eng.* **2018**, *142*, 476–482. [CrossRef]
17. Yang, J.; Chan, K.; Wu, X.; Yu, F.; Yang, X. An analysis on the energy efficiency of air-cooled chillers with water mist system. *Energy Build.* **2012**, *55*, 273–284. [CrossRef]
18. Sözen, A.; Filiz, Ç.; Aytaç, I.; Martin, K.; Ali, H.M.; Boran, K.; Yetişken, Y. Upgrading of the Performance of an Air-to-Air Heat Exchanger Using Graphene/Water Nanofluid. *Int. J. Thermophys.* **2021**, *42*, 35. [CrossRef]
19. Dobre, T.; Pârvulescu, O.C.; Stoica, A.; Iavorschi, G. Characterization of cooling systems based on heat pipe principle to control operation temperature of high-tech electronic components. *Appl. Therm. Eng.* **2010**, *30*, 2435–2441. [CrossRef]
20. Zohuri, B. Basic Principles of Heat Pipes and History. *Heat Pipe Des. Technol.* **2016**, *3*, 1–41. [CrossRef]
21. Moran, M.J.; Shapiro, H.N. (Eds.) *A Review of: Fundamentals of Engineering Thermodynamics*, 2nd ed.; John Wiley & Sons: Chichester, UK, 1993; Volume 18, p. 215. ISBN 0471592757. [CrossRef]
22. Sound Energy: Compressor Speeds and What They Mean. Available online: https://soundenergycorp.com/about-us/blog/compressor-speeds-and-what-they-mean/ (accessed on 24 November 2021).

wind

MDPI

Article

Blade-Resolved CFD Simulations of a Periodic Array of NREL 5 MW Rotors with and without Towers

Lun Ma [1,*], Pierre-Luc Delafin [2], Panagiotis Tsoutsanis [1], Antonis Antoniadis [1] and Takafumi Nishino [3]

1 School of Aerospace, Transport and Manufacturing, Cranfield University, Bedfordshire MK43 0AL, UK; panagiotis.tsoutsanis@cranfield.ac.uk (P.T.); a.f.antoniadis@cranfield.ac.uk (A.A.)
2 Laboratoire des Ecoulements Géophysiques et Industriels, Université Grenoble Alpes, CNRS, Grenoble INP, 38000 Grenoble, France; pierre-luc.delafin@univ-grenoble-alpes.fr
3 Department of Engineering Science, University of Oxford, Oxford OX1 3PJ, UK; takafumi.nishino@eng.ox.ac.uk
* Correspondence: l.ma@cranfield.ac.uk

Abstract: A fully resolved (FR) NREL 5 MW turbine model is employed in two unsteady Reynolds-averaged Navier–Stokes (URANS) simulations (one with and one without the turbine tower) of a periodic atmospheric boundary layer (ABL) to study the performance of an infinitely large wind farm. The results show that the power reduction due to the tower drag is about 5% under the assumption that the driving force of the ABL is unchanged. Two additional simulations using an actuator disc (AD) model are also conducted. The AD and FR results show nearly identical tower-induced reductions of the wind speed above the wind farm, supporting the argument that the AD model is sufficient to predict the wind farm blockage effect. We also investigate the feasibility of performing delayed-detached-eddy simulations (DDES) using the same FR turbine model and periodic domain setup. The results show complex turbulent flow characteristics within the farm, such as the interaction of large-scale hairpin-like vortices with smaller-scale blade-tip vortices. The computational cost of the DDES required for a given number of rotor revolutions is found to be similar to the corresponding URANS simulation, but the sampling period required to obtain meaningful time-averaged results seems much longer due to the existence of long-timescale fluctuations.

Keywords: actuator disc theory; two-scale momentum theory; wind farm modelling; wind farm blockage; URANS; DDES

Citation: Ma, L.; Delafin, P.-L.; Tsoutsanis, P.; Antoniadis, A.; Nishino, T. Blade-Resolved CFD Simulations of a Periodic Array of NREL 5 MW Rotors with and without Towers. *Wind* **2022**, 2, 51–67. https://doi.org/10.3390/wind2010004

Academic Editors: Andrés Elías Feijóo Lorenzo, Wei-Hsin Chen, Aristotle T. Ubando, Chih-Che Chueh and Liwen Jin

Received: 8 November 2021
Accepted: 7 January 2022
Published: 14 January 2022

Publisher's Note: MDPI stays neutral with regard to jurisdictional claims in published maps and institutional affiliations.

1. Introduction

A variety of approaches exist to model wind turbines of different sizes and types, for different purposes of use. Each type of model has its specific strengths and weaknesses. The actuator disc (AD) model has been widely adopted in wind energy studies [1–4] as it is relatively simple to implement and has low computational cost; however, it can represent only limited details of the turbine. The actuator line (AL) model considers the rotational motion of turbine blades and uses predetermined aerofoil data to represent the loading on each blade [5–7] with a medium computational cost. The most comprehensive and computationally intensive method is the fully resolved (FR) model, where the exact 3D geometry of the wind turbine is resolved during simulations; therefore, it models turbine motions and flow behaviours in a much more realistic manner [8–10].

Due to the complex and computationally expensive nature of the FR turbine model, it has been primarily used for simulations of an isolated turbine or a very small number (commonly two) of turbines [11–14], not for large wind farm cases. In addition, although the turbine tower is commonly present in FR turbine models, the tower effects on the efficiency of turbines or a wind farm are rarely discussed in detail.

In 2009, Zahle et al. [15] conducted investigations on wind turbine rotor–tower interaction using an FR NREL Phase VI turbine model [16]. Two simulations (using EllipSys3D)

were presented in their study: one for an isolated rotor and the other for a downwind configuration of a full turbine (with tower) under tunnel flow conditions. The results of the simulations showed good agreement with existing experimental data. The CFD data showed a clear interaction between the tower wake and rotor blades, which caused noticeable velocity deficit in the wake, and the blades had a strong effect on the tower shedding frequency. However, there was no discussion on how the tower affects the turbine efficiency in this study.

In a more recent study, Rodrigues and Lengsfeld [9,17] developed a twin FR turbine model (parallel configuration) based on the MEXICO (Model Experiments in Controlled Conditions) experiment [18] to investigate the performance of a turbine in a wind farm. Their approach to mimic wind farm flow conditions was to perform sequential simulations, where the outlet of one simulation (for the first row of turbines in a wind farm) was used as the inlet of the other simulation (for the second row of turbines). The results showed clear tower wakes in both near- and far-wake flow fields. Noticeable velocity deficit and increased turbulence intensity caused by the turbine towers were also reported. Their approach of modelling wind farm flow has the advantage of not having to deploy multiple rows of turbines in the domain, dramatically reducing the mesh size and, hence, simulation cost. However, the wind farm layout investigated was limited to fully aligned and 10D spacing between each row of turbines, which are less commonly seen in modern large wind farms.

As a continuation of a recent study by Delafin and Nishino [19], in this paper, we employ an FR turbine model in a horizontally periodic computational domain to investigate the performance of an infinitely large wind farm with a fully staggered layout, with and without the turbine towers. The FR turbine model is based on the NREL 5 MW horizontal-axis turbine design [20,21], and the wind flow is simulated by solving the unsteady Reynolds-averaged Navier–Stokes (URANS) equations. A fully developed wind farm boundary layer is achieved by applying periodic boundary conditions, which enable the simulation of a single turbine to represent a turbine operating inside an infinitely large wind farm. The FR-URANS results are compared with corresponding AD-RANS results, as well as a theoretical model based on the two-scale momentum theory [22,23]. In addition, we also perform a delayed-detached-eddy simulation (DDES) using the same FR turbine model, to demonstrate the possibility of eddy-resolving FR turbine simulations for wind farm modelling.

2. Methodology

2.1. Turbine and Domain Geometries

The turbine design used in this study was the NREL 5 MW three-blade model [20], which has a diameter (D) of 126 m. We also considered the alterations suggested by Sandia National Laboratories [21]: an additional smoothing process was applied to blade thickness distribution, as well as root–blade transition. Furthermore, as the tip shape was not specified in the reference designs, a 'rounded' tip design approach was used in this study (Figure 1), in a similar way to Bazilev et al. [24]. In addition, to simplify the construction process of a fully structured mesh, the hub diameter was increased to 5.4 m (3 m in the original design). As a consequence, a small part of the root of each blade was hidden by the enlarged hub, which was believed to not significantly affect the aerodynamics of the whole rotor. The hub centre was 90 m above the ground, equivalent to $0.21D$ ground clearance. There was 5° (respect to the horizontal plane) tilt angle applied to the rotational axis and a 2.5° precone, as defined in [20]. The original report states that the turbine tower base diameter is 6 m and the top diameter (at 87.6 m where the tower is connected to the nacelle) is 3.87 m [20]. To simplify the 3D modelling and meshing process, we used a slightly narrower design for the tower in this study. The base diameter was still 6 m, and the tower diameter at the nacelle and tower interaction was 3.14 m.

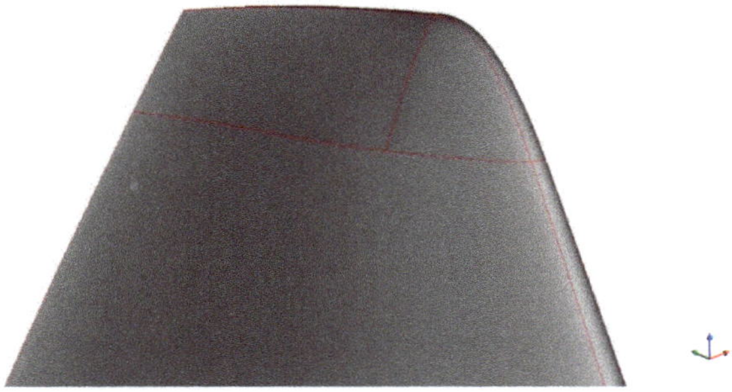

Figure 1. Rounded tip shape of the blade.

All simulations in this study considered a fully developed boundary layer flow over an infinitely large wind farm, which was modelled as a fully periodic array of turbines (Figure 2a). The outer domain size was $L_x = L_y = 6D$ and $L_z = 8D$, representing reasonable turbine spacing within a wind farm. The turbine rotor was placed inside a rotating cylindrical subdomain (diameter = $1.1D$ and thickness = $0.079D$ (10 m)), as shown in Figure 2b. As for the AD model, the configuration was the same as in the previous study [19], using a 126 m diameter disc whose centre was located at $x = y = 0$ m and $z = 90$ m, to allow a direct comparison with the FR model.

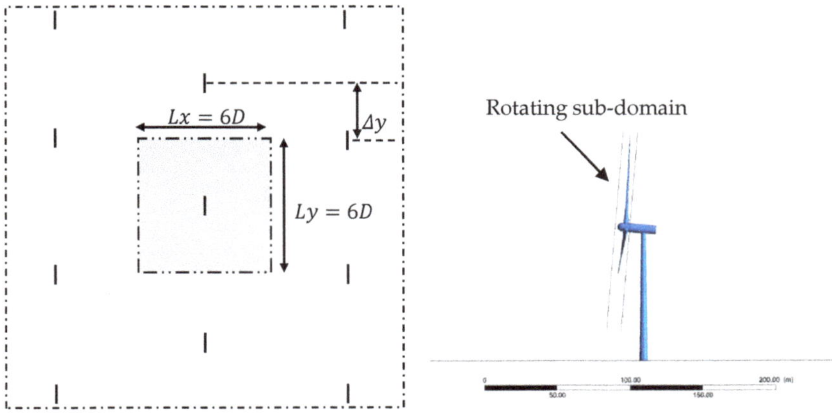

(a) (b)

Figure 2. (**a**) Staggered turbine array; (**b**) turbine geometry and rotating subdomain.

2.2. Meshing

For the purpose of this study, the boundary layers on the blades and tower were resolved (with $y^+_{max} \sim 1$), but wall functions were applied on the hub and nacelle surfaces in order to reduce the total number of cells. In addition, wall boundary conditions with a specific roughness height were applied on the bottom surface (representing a hypothetical ground or sea surface conditions) [19,22]. Therefore, the bottom surface was treated as a solid wall during this meshing stage.

The surface mesh of the blades was created on one blade with O-grid topology and then copied to the other two blades. There were 111 nodes in the blade spanwise direction. The spanwise cell sizes at each end of the blade were 0.001 m (next to the hub) and 0.0015 m (at the tip), and the largest size was 1.8 m (in the middle section of the blade). In the chordwise direction, 190 nodes were distributed. Although it was infeasible to conduct a mesh sensitivity study for the entire rotor, we carefully designed the blade surface mesh according to the results of existing mesh sensitivity studies in the literature. For example, a recent study showed that, for an NACA 64-618 aerofoil profile, under similar flow conditions, grid independence could be achieved when the total node number was above 180 around the aerofoil [25]. Figure 3 shows part of the complete surface mesh of the rotor. The streamwise cell sizes were 0.5 m and 0.1 m immediately upstream and downstream of the rotating subdomain, respectively. Away from the rotating subdomain, the largest streamwise spacing was 6 m and largest lateral spacing was 7 m. The first cell above the ground was 1 m high across the entire domain for the 'without tower' mesh. For the 'with tower' case, the tower was added to the existing outer domain mesh, with moderate modifications. For the node distribution on the tower surface, there were 115 nodes distributed in the circumferential direction of the tower (Figure 4), and 87 nodes were used along the tower height. Near to the tower bottom, the height of the first cell above the ground was reduced slightly, for better consistency and keeping reasonable cell aspect ratios. Overall, this mesh for the FR with tower contained 17.1×10^6 cells.

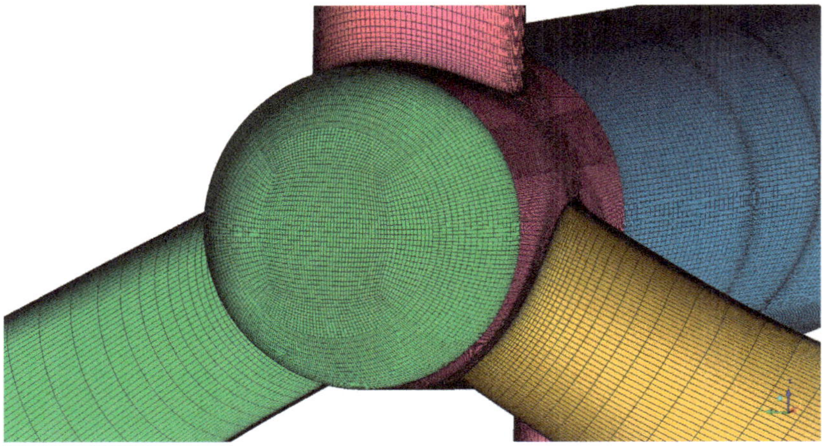

Figure 3. Surface mesh of the turbine.

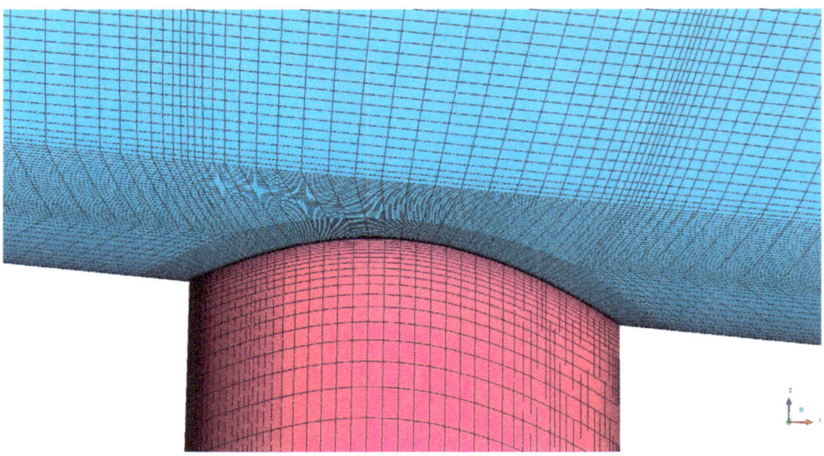

Figure 4. Nacelle and tower intersection mesh details.

We used slightly different meshing approaches for the AD with and without tower models. The AD model had 104 nodes along the disc edge, and an 'O-grid' mesh topology was used inside and around the disc (Figure 5). For the AD only case, the rest of the mesh was built around the disc without creating a subdomain. The streamwise node spacing started from 0.5 m and expanded to 11 m at the inlet and outlet boundaries, and the total number of cells for the AD only model was 8×10^5. For the AD + tower case, we used the same outer domain mesh as the FR + tower case (which contained the nacelle and tower). Then, the blade-resolved rotor model was replaced by the AD model inside the rotor subdomain (Figure 6). The mesh construction was slightly modified inside the subdomain in order to accommodate the AD model. The 'AD + tower' mesh contained 12.2×10^6 cells. A summary of the four different meshes created is shown in Table 1.

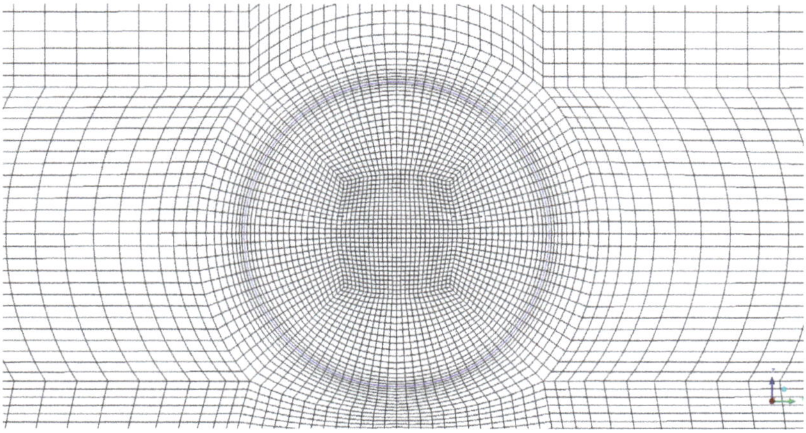

Figure 5. Cross-sectional view around the disc (for AD only case).

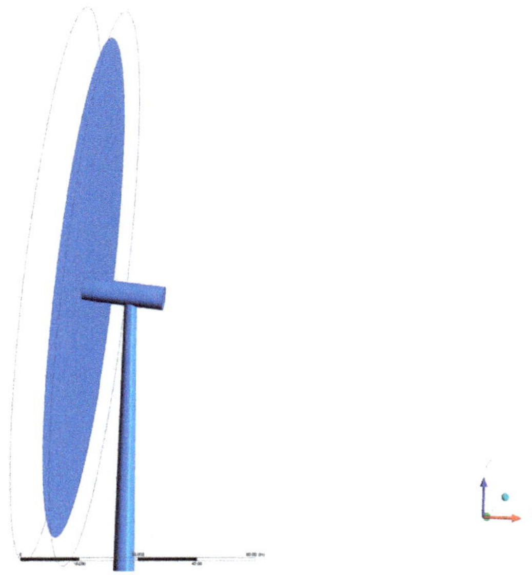

Figure 6. AD placement inside the subdomain.

Table 1. Summary of the number of cells for each case.

	FR + Tower	FR	AD + Tower	AD
Rotor subdomain	5.8×10^6	5.8×10^6	0.9×10^6	N/A
Outer domain	11.3×10^6	6.8×10^6	11.3×10^6	N/A
Total	17.1×10^6	12.6×10^6	12.2×10^6	0.8×10^6

The FR and AD models were previously tested in an isolated (not periodic) flow condition [19], and the results were in good agreement with [20].

2.3. *RANS Setup*

For the majority of the CFD simulations in this study, the Reynolds-averaged Navier–Stokes (RANS) equations were solved using the commercial CFD solver 'ANSYS FLUENT 17.2' [26] with the k–ω SST turbulence model [27]. The density and dynamic viscosity of the working fluid (air) were $\rho = 1.225\ \mathrm{kg/m^3}$ and $\mu = 1.7894 \times 10^{-5}\ \mathrm{kg/(ms)}$, respectively. The bottom boundary condition for all simulations was set as 'wall' [26], with a nominal roughness height $K_s = 1$ m and the empirical constant E and roughness constant C_s values were kept as 9.793 and 0.5, respectively (this corresponds to an aerodynamic roughness length of $z_0 = 0.051$ m since $K_s = (E/C_S)z_0$ [28]). The top boundary was set as 'symmetry'. Periodic conditions were applied to both streamwise and lateral ends of the domain. However, the streamwise ends (or the inlet and outlet interface) were split vertically in the middle into two parts, where the right inlet was connected to the left outlet and vice versa. This allowed us to simulate an infinitely large fully staggered turbine array, as previously shown in Figure 2a.

In order to decide the background pressure gradient inside the periodic domain, an 'empty box' simulation [19,22] with a fixed mass flow rate was conducted first. This simulation (steady RANS) was carried out using the 'AD only' mesh, with the disc boundary condition set as 'interior'. This fixed mass flow rate was calculated using a power-law velocity profile (with an exponent of 0.1) with reference wind speed $U_{ref} = 15\mathrm{m/s}$ at $z = 90$ m (hub level). This resulted in a farm-averaged wind speed below the rated wind speed for the NREL 5 MW turbine (11.4 m/s) in this study. The pressure gradient obtained from this empty box simulation was -7.49×10^{-4} Pa/m, which was used as the driving force of the wind flow for all wind farm simulations in this study. This was based on the assumption that, for an infinitely large wind farm, the driving force of the wind is not affected by the existence of wind turbines [1,22]. In reality, the driving force of wind over a large (finite-size) wind farm should depend on the characteristics of mesoscale atmospheric response [29,30]. To obtain the correct driving force and, thus, the correct wind speed for a real wind farm, we would need to assess the 'extractability' of wind at a given wind farm site using a mesoscale weather model [30], but this is outside the scope of the present study.

2.3.1. AD Simulations

In the AD simulations (using steady RANS), the effect of the rotor was modelled as streamwise momentum losses, i.e., as a stationary porous surface of zero thickness with a momentum loss factor K [22]. To obtain the most appropriate K value, a few isolated actuator disc simulations were conducted by Delafin et al. [19] prior to the current study. Different K values were tested to obtain a C_T (thrust coefficient) value as close as possible to an isolated FR turbine simulation (with TSR = 7.0) in [19]. The most appropriate value was found to be $K = 1.2825$, yielding an area-averaged streamwise velocity over the AD, $U_{T\ single} = 8.666$ m/s. The SIMPLE algorithm was used for pressure–velocity coupling, and the second-order upwind scheme was used for the discretisation of the momentum and turbulence model equations. Both AD simulations (with and without tower) were run for 10^6 iterations to obtain fully converged results.

2.3.2. FR Simulations

The FR simulations were carried out using unsteady RANS. All turbine surfaces were considered as 'wall'. The rotating motion of the rotor (with a fixed rotating speed of 7.703 rpm) was achieved by applying the 'sliding mesh' method to the subdomain containing the turbine rotor. Since the velocity profile inside the fully developed farm boundary layer was not known a priori, the rotating speed was determined using AD simulation results, to match the tip speed ratio (TSR) with the 'optimal' value (TSR = 7.0) reported in [20]. Specifically, we assumed a proportional relationship between the rotor rpm and U_T values (obtained from AD simulations for the single turbine case and for the periodic farm case, respectively) to determine the rotational speed for the FR farm simulations.

$$\frac{RPM_{single}}{RPM_{farm}} = \frac{U_{T\ single}}{U_{T\ farm}} \rightarrow \frac{12.1\ \text{rpm}}{RPM_{farm}} = \frac{8.666\text{m/s}}{5.517\text{m/s}} \rightarrow RPM_{farm} = 7.703\ \text{rpm} \quad (1)$$

where $U_{T\ farm}$ was obtained from the 'AD only' (without tower) farm simulation. Once the rotational speed was determined, the timestep size was set as $\Delta t = 0.0108$ s (which led to a rotor azimuthal angle of $\Delta\theta = 0.5°$ per timestep), and 15 iterations were conducted per timestep. It should be noted that this rotational speed was for the 'FR rotor only' (without tower) simulation. As shown later, adding the tower to the model reduced the average wind speed across the domain. This means that the rotational speed would need to be slightly altered after adding the tower, in order to achieve the same TSR as the rotor only case. However, in this study, we used the same rotational speed (7.703 rpm) for both 'tower' and 'no tower' cases for simplicity. After completing the FR simulations, the difference in TSR between the 'tower' and 'no tower' cases was approximately 2.1%, which can be considered negligible. According to a recent study [31], a small TSR variation would not cause noticeable changes in turbine performance (approximately 0.53% difference in power output).

To quickly generate a fully developed atmospheric boundary layer flow for the FR simulations, the flow field obtained from the AD simulations was applied as the initial flow conditions. The 'Coupled' algorithm [26] was employed for pressure–velocity coupling. A second-order implicit scheme was used for the temporal discretisation, and the second-order upwind scheme was used for spatial discretisation of the momentum and turbulence model equations. The FR simulations (with and without tower) were run for 100 revolutions first, and then additional 18 revolutions were carried out to calculate time-averaged results to be compared with the theoretical model [22]. For the FR with tower case, it took about 13 h to run one revolution simulation on 128 CPUs.

2.4. *DDES Setup*

In addition to the URANS simulations, we also attempted to run DDES for the FR with tower case. The k–ω SST turbulence model was switched from its URANS mode to DDES mode available in FLUENT [26] after the first 100 revolutions of URANS simulations. All boundary conditions were unchanged. We also used the same fixed RPM as we used in URANS, although this meant that the TSR fluctuated more in DDES (about 20.2%) than in URANS (about 3.5%). A bounded central differencing scheme was used for the spatial discretisation of the momentum equations, and a bounded second-order implicit scheme was employed for the temporal discretisation. The DDES was run only for 54 revolutions due to the limited availability of computational resources.

2.5. *Postprocessing of URANS Results*

The rotor moment coefficient, C_m, is defined as

$$C_m = \frac{\tau}{\frac{1}{2}\rho A r U_{F0}^2} \quad (2)$$

where τ is the torque of the rotor, A is the swept area of the rotor ($A = \pi r^2$), r is the radius of the rotor, and U_{F0} is the natural farm-layer wind speed obtained from the 'empty box' simulation. The power output of the rotor is calculated as,

$$P = \tau \cdot \omega \tag{3}$$

where ω is the rotational speed of the turbine (0.8067 rad/s for the farm simulations and 1.267 rad/s for the isolated turbine simulation).

3. Results and Discussion

3.1. URANS

Figure 7 shows the time histories of the turbine power output observed in the FR-URANS simulations with and without the tower. It can be seen that, for both cases, a large number of rotor revolutions were required for the power to reach a quasi-stationary state. A consistent difference of about 0.06 MW (5%) between the two cases was observed after around 100 revolutions. This clearly shows that the predicted power was reduced by adding the turbine tower to the model.

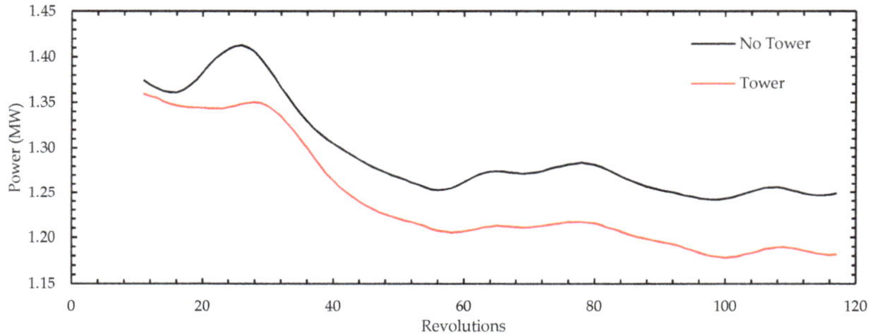

Figure 7. Convergence history of revolution-averaged power output (FR-URANS model with and without tower).

The flow field comparison between 'tower' and 'no tower' cases is shown in Figures 8 and 9; the former is plotted on a vertical plane that goes through the centre of the turbine, and the latter is plotted on the horizontal plane at the hub centre level (90 m above the ground). The tower generated its own wake and, thus, enlarged the velocity deficit behind the lower part of the rotor. This can also be observed in the velocity profiles plotted in Figure 10, especially at $x/D = 0.5$, where there was a noticeable speed reduction close to the ground due to the turbine tower. However, a more important effect of the tower in this study is that it reduced the wind speed above the turbine (Figure 8). This is because the additional flow resistance (or drag) created by the tower contributed to a stronger development of the internal boundary layer (IBL). In other words, the slower wind speed observed above a given turbine was not because of the tower of that turbine, but because of the towers of other turbines located upstream (note that there are an infinite number of turbines located upstream of the turbine shown in Figure 8, due to the periodicity). It is clear from Figure 10 that this reduction in wind speed above the turbine was observed at any streamwise positions, confirming that this is a farm-scale (rather than a local/turbine-scale) wind speed reduction. On the other hand, the wind speed difference observed behind the tower diminished quickly towards downstream due to strong turbine-scale wake mixing, and the flow profiles further downstream appeared almost identical between 'no tower' and 'tower' cases (Figure 10d).

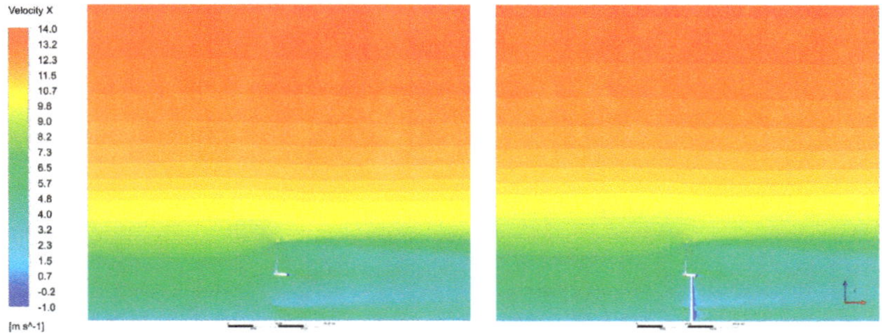

(**a**) (**b**)

Figure 8. Contours of instantaneous streamwise velocity (m/s) at the vertical centre plane (taken at 100th revolution from FR-URANS): (**a**) no tower; (**b**) with tower.

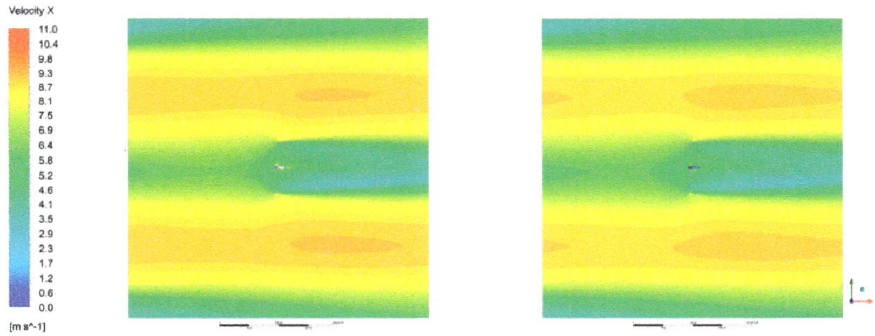

(**a**) (**b**)

Figure 9. Contours of instantaneous streamwise velocity (m/s) at the rotor hub height (taken at 100th revolution from FR-URANS): (**a**) no tower; (**b**) with tower.

The AD (with and without tower) RANS results are also plotted in Figure 10 for comparison. Between the AD and FR cases, the differences in flow profiles were mostly within the rotor region (approximately $0.21 < z/D < 1.21$). In addition, the AD results also showed that the tower reduced the wind velocity at higher positions. It is remarkable that the wind profiles above the top of the turbine ($z/D > 1.21$) were almost identical between the AD with/without tower and the FR with/without tower cases. This highlights a key finding that the choice of the rotor model (AD or FR) does not affect the prediction of the tower effect on the farm-scale wind speed reduction. This finding supports the argument that the AD model (with tower) would be sufficient to predict the wind farm blockage effect. This is discussed further in Section 3.3.

3.2. DDES

As shown in Figure 11a, the turbine power output started to deviate from the URANS results about 13 revolutions after switching to DDES, and then large fluctuations started to occur. Since the average wind speed at the hub height was about 8 m/s, these 13 revolutions approximately correspond to the time required for the hub height wind to travel through the periodic domain once. Although not shown here for brevity, the turbine thrust also fluctuated in a very similar manner to the power output. Such fluctuations of the blade loadings were also reported in a previous LES study of a single rotating blade of the NREL 5 MW turbine interacting with realistic atmospheric turbulence [32]. The time history (recorded every four timesteps) of the streamwise velocity measured at $2.5D$ upstream of the turbine (at the hub height) is plotted in Figure 11b. By comparing the two figures, it can be seen that the power output fluctuations were closely correlated with the upstream velocity fluctuations.

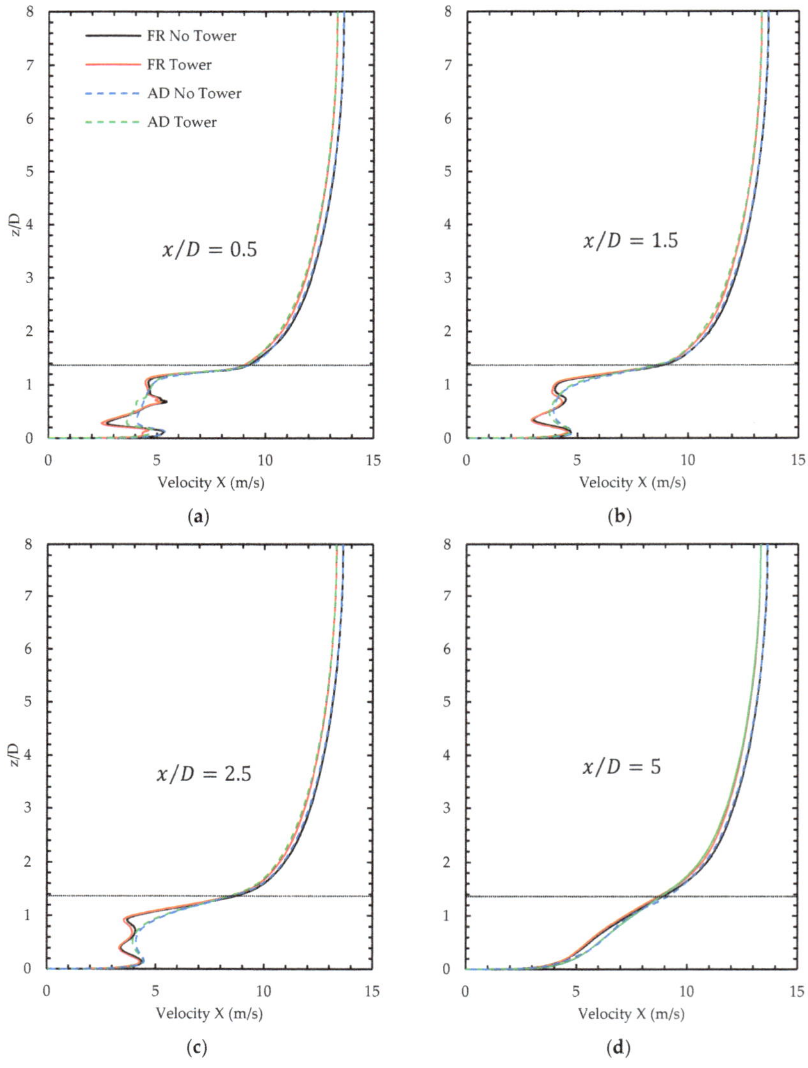

Figure 10. Streamwise velocity profiles at different locations behind the turbine: (**a**) $x/D = 0.5$, (**b**) $x/D = 1.5$, (**c**) $x/D = 2.5$, (**d**) $x/D = 5$. The dotted horizontal line indicates the top position of the turbine.

Figure 12 shows an example of instantaneous velocity contours on vertical and horizontal planes, visualising the turbulent flow field around the turbine. The flow in the top half of the domain did not contain any turbulent eddies; this is presumably because the simulation was not run long enough, but this may also be because of the relatively small horizontal size of the domain used in this study. Nevertheless, the flow at the turbine level appeared to be already fully turbulent (Figures 13 and 14).

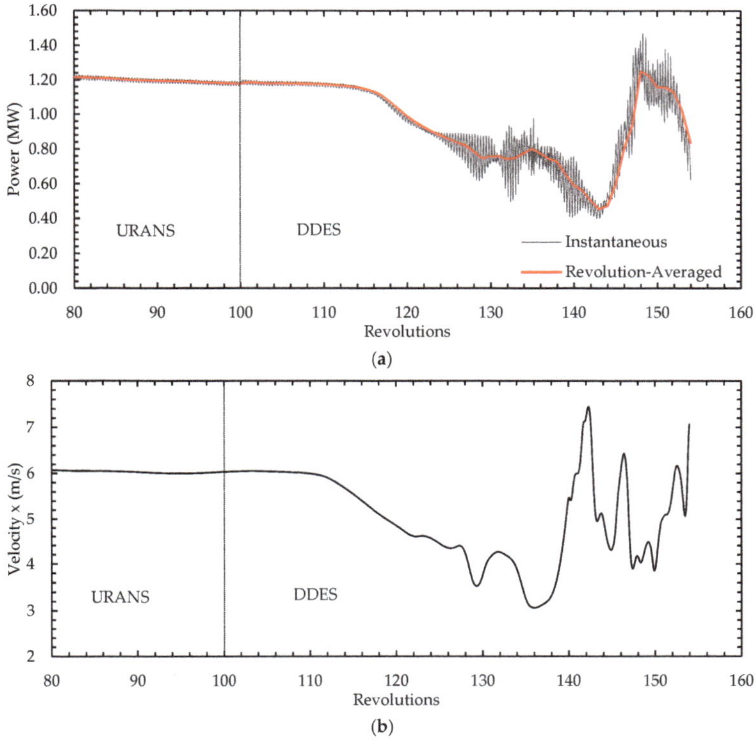

Figure 11. (**a**) Time history of instantaneous and revolution-averaged power outputs; (**b**) time history of streamwise velocity measured at hub height ($z = 90$ m) and 2.5D upstream of the turbine ($x = -315$ m).

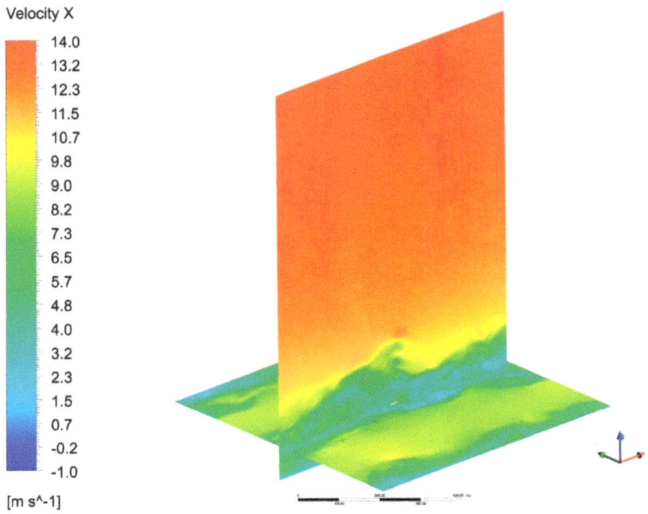

Figure 12. Instantaneous streamwise velocity contours (m/s) on the streamwise vertical plane across the centre of a turbine and the horizontal plane at the turbine hub-height (taken at the 154th revolution).

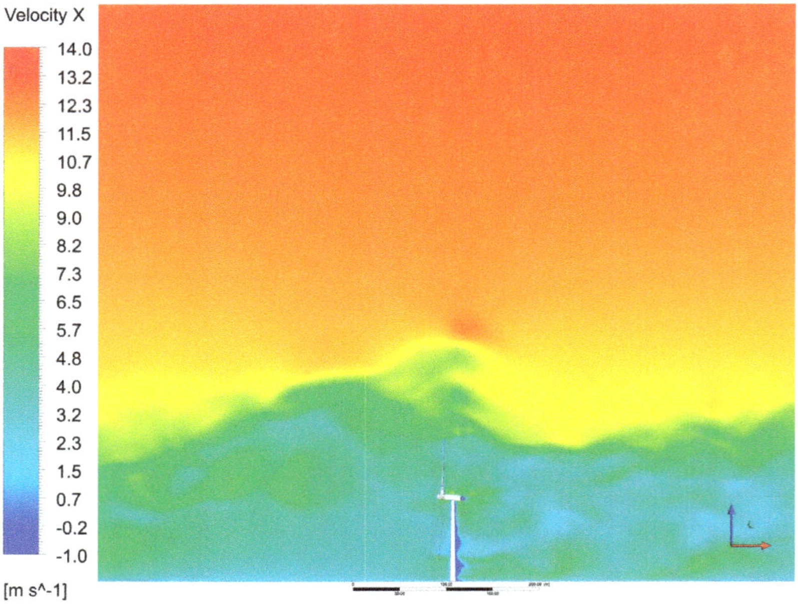

Figure 13. Contours of instantaneous streamwise velocity (m/s) at vertical centre plane (taken at the 154th revolution).

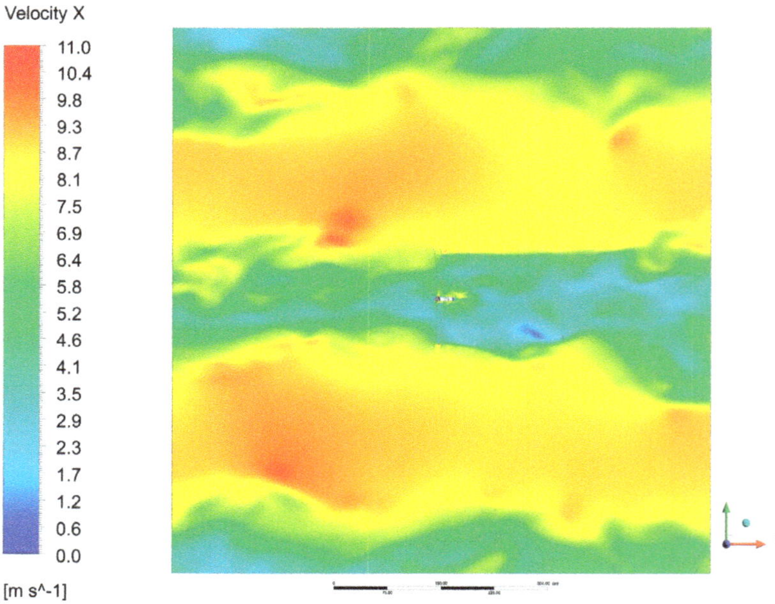

Figure 14. Contours of instantaneous streamwise velocity (m/s) at rotor hub height (taken at the 154th revolution).

Figures 15 and 16 show the instantaneous vortical flow structures obtained from URANS and DDES results. These vortex structures were visualised using iso-surfaces of the Q-Criterion, and the colour scheme is based on the instantaneous streamwise velocity.

Seven identical flow fields obtained from the original single-turbine domain are plotted together in these figures, in order to better visualise the flow structures over the staggered array of turbines (see Figure 2). As expected, the URANS results (Figure 15) show only some strong turbine-generated vortices, such as tip and hub vortices. On the other hand, the DDES results (Figure 16, using the same Q-Criterion values as in Figure 15) show much more complex turbulent flow structures created in the wake region and how they interacted with the strong turbine-generated vortices. It is remarkable that some large hairpin-like vortices were also visible in front of the turbine, similarly to those commonly appearing near the top of a turbulent boundary layer [33].

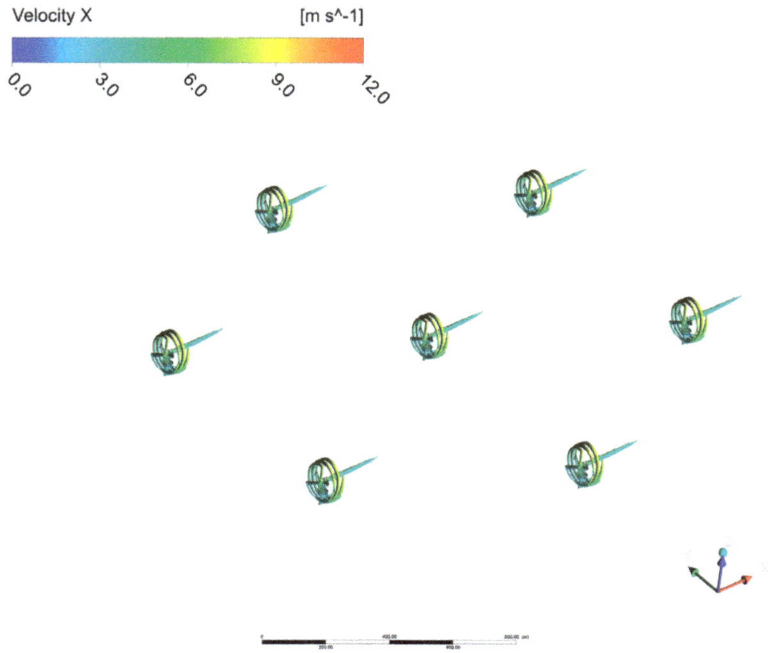

Figure 15. Vortical flow structures over a staggered array of turbines, obtained from the FR-URANS simulation with tower (visualised using iso-surfaces of Q-Criterion).

3.3. *Comparison with Theoretical Model*

In this section, we compare the AD-RANS and FR-URANS simulation results with the theoretical model of an infinitely large wind farm reported earlier in [22]. For the calculation of the tower resistance in the theoretical model, we employed a typical drag coefficient of a circular cylinder (at a corresponding Reynolds number) of 0.6. A summary of the comparison is shown in Table 2. The wind farm power and thrust coefficients (C_P and C_T) were calculated as $C_P = P/\frac{1}{2}\rho A U_{F0}^3$ and $C_T = T/\frac{1}{2}\rho A U_{F0}^2$, respectively, whereas the 'local' power and thrust coefficients (C_P^* and C_T^*) were calculated using U_F (farm-layer-average wind speed, see [22] for details) instead of U_{F0}. Note that the C_P and C_T values were very small in this study as we considered an infinitely large wind farm with a fixed driving force of the wind. The C_P values calculated from the AD-RANS simulations were very close to the theoretical model predictions, and the C_T values were also closer to the theoretical results compared to the FR-URANS results. This agrees with the findings in the previous study (for the same NREL 5 MW rotor without a tower) [19].

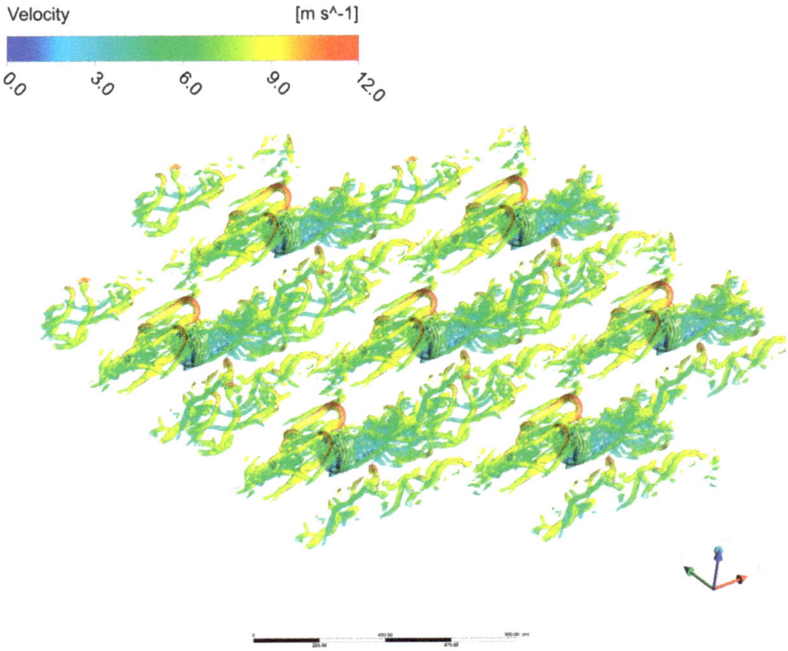

Figure 16. Vortical flow structures over a staggered array of turbines, obtained from FR-DDES with tower (visualised using iso-surfaces of Q-Criterion, taken at the 154th revolution).

Table 2. Comparison of AD-RANS, FR-URANS, and theoretical model predictions.

		C_P	C_P^*	C_T	C_T^*
AD-RANS	No Tower	0.0651	0.414	0.176	0.603
	Tower	0.0641	0.412	0.174	0.601
	Diff.%	1.55%	0.52%	1.04%	0.35%
FR-URANS	No Tower	0.0495	0.303	0.167	0.561
	Tower	0.0469	0.306	0.163	0.570
	Diff.%	5.21%	−0.90%	2.48%	−1.67%
Theoretical model [22]	No Tower	0.0622	0.380	0.198	0.570
	Tower	0.0607	0.380	0.195	0.570
	Diff.%	2.50%	0.00%	1.68%	0.00%

As for the effect of the tower, both AD-RANS and FR-URANS simulations agreed fairly well with the theoretical model predictions as summarised in Table 2. The C_P and C_T values were slightly (about 1 to 5%) lower due to the additional farm-scale wind speed reduction caused by the tower, whereas its effects on the C_P^* and C_T^* values seemed negligibly small. The effect of the tower on C_P was a little underestimated by the AD-RANS (1.55%) and the theoretical model (2.50%) compared to the FR-URANS (5.21%), showing the limitation of the methods based on the classical actuator disc theory. A possible solution to this problem would be to replace the actuator disc part of the theoretical model with the blade element momentum (BEM) theory. This modification (called the 'BEM-farm-momentum' method) has already been reported in [34] but only for the case without the tower. Therefore, it would be useful to also apply this method to the case with the tower in a future study.

4. Conclusions

In this paper, we presented blade-resolved CFD simulations of the NREL 5 MW wind turbine placed in a horizontally periodic domain, representing an infinitely large wind farm. Three main wind farm simulations were conducted in this study, namely, FR-URANS 'without tower', FR-URANS 'with tower', and FR-DDES 'with tower'. Two additional wind farm simulations using steady RANS with a simple AD model were also conducted for comparison. The FR-URANS results showed substantial effects of the tower on the overall wind farm efficiency, with a decrease of about 5% in the total power due to the additional tower drag (under the assumption that the driving force of wind over the wind farm was unchanged). This further confirms the importance of including turbine support structures in the modelling of large wind farms, which agrees with the findings reported earlier in [22,35].

We also showed that the AD-RANS and FR-URANS simulations predicted nearly identical velocity profiles above the wind farm (for each of the cases with and without the tower). This suggests that the simple AD model can capture the farm-scale wind speed reduction as accurately as the FR model does. In addition, both AD and FR simulations agreed fairly well with the theoretical model reported earlier in [22] in terms of the tower effect on the overall thrust (or drag) of the wind farm. However, the FR-URANS results showed a larger power reduction due to the tower (5.21%) compared to the AD-RANS (1.55%) and the theoretical model (2.50%).

The discrepancy between the FR-URANS and the theoretical model seemed largely due to the limitation of the actuator disc theory employed in the theoretical model. It may, therefore, be beneficial to replace the actuator disc part of the current theoretical model with BEM (as proposed for the case without the tower in [34]) in future studies. The FR-URANS results could be used to validate and/or improve such an extended theoretical wind farm model in future studies.

The FR-DDES results presented in this paper also provide useful information on 3D vortical flow structures over a periodic array of wind turbines, e.g., the interaction of large hairpin-like vortices with small but strong blade-tip vortices. However, due to limited computational resources, it was not possible in this study to obtain meaningful time-averaged results from the DDES for the wind farm performance. We found that the computational cost of the FR-DDES required for a given number of turbine revolutions is very similar to the corresponding FR-URANS simulation, but the sampling period required is much longer as the timescale for the fluctuation of the flow field is much longer than that for the turbine revolution.

Author Contributions: Conceptualisation, L.M. and T.N.; methodology, L.M. and T.N.; software, L.M. and P.-L.D.; investigation, L.M.; writing—original draft preparation, L.M.; writing—review and editing, T.N., P.-L.D., P.T. and A.A.; supervision, P.T., A.A. and T.N. All authors read and agreed to the published version of the manuscript.

Funding: This research received no external funding.

Institutional Review Board Statement: Not applicable.

Informed Consent Statement: Not applicable.

Data Availability Statement: The data presented in this study are available on reasonable request from the corresponding author.

Conflicts of Interest: The authors declare no conflict of interest.

References

1. Nishino, T. Two-scale momentum theory for very large wind farms. *J. Phys. Conf. Ser.* **2016**, *753*, 32054. [CrossRef]
2. Wu, Y.T.; Porté-Agel, F. Large-Eddy Simulation of Wind-Turbine Wakes: Evaluation of Turbine Parametrisations. *Bound.-Layer Meteorol.* **2011**, *138*, 345–366. [CrossRef]

3. Naderi, S.; Parvanehmasiha, S.; Torabi, F. Modeling of horizontal axis wind turbine wakes in Horns Rev offshore wind farm using an improved actuator disc model coupled with computational fluid dynamic. *Energy Convers. Manag.* **2018**, *171*, 953–968. [CrossRef]
4. Richmond, M.; Antoniadis, A.; Wang, L.; Kolios, A.; Al-Sanad, S.; Parol, J. Evaluation of an offshore wind farm computational fluid dynamics model against operational site data. *Ocean Eng.* **2019**, *193*, 106579. [CrossRef]
5. Stevens, R.J.; Martínez-Tossas, L.A.; Meneveau, C. Comparison of wind farm large eddy simulations using actuator disk and actuator line models with wind tunnel experiments. *Renew. Energy* **2017**, *116*, 470–478. [CrossRef]
6. Sørensen, J.N.; Mikkelsen, R.F.; Henningson, D.S.; Ivanell, S.; Sarmast, S.; Andersen, S.J. Simulation of wind turbine wakes using the actuator line technique. *Phil. Trans. R. Soc.* **2015**, *373*, 20140071. [CrossRef]
7. Kalvig, S.; Manger, E.; Hjertager, B. Comparing different CFD wind turbine modelling approaches with wind tunnel measurements. *J. Phys. Conf. Ser.* **2014**, *555*, 012056. [CrossRef]
8. Madsen, M.H.A.; Zahle, F.; Sørensen, N.N.; Martins, J.R.R.A. Multipoint high-fidelity CFD- based aerodynamic shape optimization of a 10 MW wind turbine. *Wind Energ. Sci.* **2019**, *4*, 163–192. [CrossRef]
9. Rodrigues, R.V.; Lengsfeld, C. Development of a Computational System to Improve Wind Farm Layout, Part I: Model Validation and Near Wake Analysis. *Energies* **2019**, *12*, 940. [CrossRef]
10. Stergiannis, N.; Lacor, C.; Beeck, J.V.; Donnelly, R. CFD modelling approaches against single wind turbine wake measurements using RANS. *J. Phys. Conf. Ser.* **2016**, *753*, 032062. [CrossRef]
11. Uchida, T.; Taniyama, Y.; Fukatani, Y.; Nakano, M.; Bai, Z.; Yoshida, T.; Inui, M. A New Wind Turbine CFD Modeling Method Based on a Porous Disk Approach for Practical Wind Farm Design. *Energies* **2020**, *13*, 3197. [CrossRef]
12. Weihing, P.; Schulz, C.; Lutz, T.; Krämer, E. Comparison of the Actuator Line Model with Fully Resolved Simulations in Complex Environmental Conditions. *J. Phys. Conf. Ser.* **2017**, *854*, 012049. [CrossRef]
13. Miao, W.; Li, C.; Pavesi, G.; Yang, J.; Xie, X. Investigation of wake characteristics of a yawed HAWT and its impactson the inline downstream wind turbine using unsteady CFD. *J. Wind. Eng. Ind. Aerodyn.* **2017**, *168*, 60–71. [CrossRef]
14. Wilson, J.M.; Davis, C.J.; Venayagamoorthy, S.K.; Heyliger, P.R. Comparisons of Horizontal-Axis Wind Turbine Wake Interaction Models. *J. Sol. Energ. Eng.* **2015**, *137*, 031001. [CrossRef]
15. Zahle, F.; Sørensen, N.N.; Johansen, J. Wind Turbine Rotor-Tower Interaction Using an Incompressible Overset Grid Method. *Wind Energy* **2009**, *12*, 594–619. [CrossRef]
16. Hand, M.M.; Simms, D.A.; Fingersh, L.J.; Jager, D.W.; Cotrell, J.R.; Schreck, S.; Larwood, S.M. Unsteady Aerodynamics Experiment Phase VI: Wind Tunnel Test Configurations and Available Data Campaigns. *Tech. Rep.* **2001**, 15000240. [CrossRef]
17. Rodrigues, R.V.; Lengsfeld, C. Development of a Computational System to Improve Wind Farm Layout, Part II: Wind Turbine Wakes Interaction. *Energies* **2019**, *12*, 1328. [CrossRef]
18. Schepers, J.G.; Boorsma, K.; Cho, T.; Gomez-Iradi, S.; Schaffarczyk, P.; Jeromin, A.; Shen, W.Z.; Lutz, T.; Meister, K.; Stoevesandt, B. *Final Report of IEA Task 29, Mexnet (Phase 1): Analysis of Mexico Wind Tunnel Measurements*; Technical University of Denmark: Lyngby, Denmark, 2012.
19. Delafin, P.-L.; Nishino, T. Momentum balance in a fully developed boundary layer over a staggered array of NREL 5MW rotors. *J. Phys. Conf. Ser.* **2017**, *854*, 12009. [CrossRef]
20. Jonkman, J.; Butterfield, S.; Musial, W.; Scoot, G. Definition of a 5-MW Reference Wind Turbine for Offshore System Development. *Tech. Rep.* **2009**, 947422. [CrossRef]
21. Resor, B.R. Definition of a 5MW/61.5m wind turbine blade reference model. *Tech. Rep.* **2013**, 1095962. [CrossRef]
22. Ma, L.; Nishino, T.; Antoniadis, A. Prediction of the impact of support structures on the aerodynamic performance of large wind farms. *J. Renew. Sustain. Energy* **2019**, *11*, 063306. [CrossRef]
23. Nishino, T.; Dunstan, T.D. Two-scale momentum theory for time-dependent modelling of large wind farms. *J. Fluid Mech.* **2020**, *894*, A2. [CrossRef]
24. Bazilevs, Y.; Hsu, M.-C.; Akkerman, I.; Wright, S.; Takizawa, K.; Henicke, B.; Spielman, T.; Tezduyar, T.E. 3D simulation of wind turbine rotors at full scale. Part I: Geometry modeling and aerodynamics. *Int. J. Numer. Meth. Fluids* **2011**, *65*, 207–235. [CrossRef]
25. Major, D.; Palacios, J.; Maufhumer, M.; Schmitz, S. A Numerical Model for the Analysis of Leading-Edge Protection Tapes for Wind Turbine Blades. *J. Phys. Conf. Ser.* **2020**, *1452*, 12058. [CrossRef]
26. ANSYS Inc. *ANSYS Fluent User's Guide Release 17.2*; ANSYS Inc.: Canonsburg, PA, USA, 2016.
27. Menter, F.R. Two-Equation Eddy-Viscosity Turbulence Models for Engineering Applications. *AIAA J.* **1994**, *32*, 1598–1605. [CrossRef]
28. Blocken, B.; Stathopoulos, T.; Carmeliet, J. CFD simulation of the atmospheric boundary layer: Wall function problems. *Atmos. Environ.* **2007**, *41*, 238–252. [CrossRef]
29. Bleeg, J.; Purcell, M.; Ruisi, R.; Traiger, E. Wind farm blockage and the consequences of neglecting its impact on energy production. *Energies* **2018**, *11*, 1609. [CrossRef]
30. Patel, K.; Dunstan, T.D.; Nishino, T. Time-dependent upper limits to the performance of large wind farms due to mesoscale atmospheric response. *Energies* **2021**, *14*, 6437. [CrossRef]
31. Pinto, M.L.; Franzini, G.R.; Simos, A.S. A CFD analysis of NREL's 5MW wind turbine in full and model scales. *J. Ocean Eng. Mar. Energy* **2020**, *6*, 211–220. [CrossRef]

32. Vijayakumar, G.; Brasseur, J.G.; Lavely, A.; Jayaraman, B. Interaction of Atmospheric Turbulence with Blade Boundary Layer Dynamics on a 5MW Wind Turbine using Blade-boundary-layer-resolved CFD with hybrid URANS-LES. In Proceedings of the 34th Wind Energy Symposium, AIAA SciTech Forum, San Diego, CA, USA, 4–8 January 2016. [CrossRef]
33. Eitel-Amor, G.; Flores, O.; Schlatter, P. Hairpin vortices in turbulent boundary layers. *Phys. Fluids* **2015**, *27*, 025108. [CrossRef]
34. Nishino, T.; Hunter, W. Tuning turbine rotor design for very large wind farms. *Proc. R. Soc. A* **2018**, *474*, 237. [CrossRef]
35. Ma, L.; Nishino, T. Preliminary estimate of the impact of support structures on the aerodynamic performance of very large wind farms. *J. Phys. Conf. Ser.* **2018**, *1037*, 72036. [CrossRef]

Journal of
Marine Science and Engineering

MDPI

Article

Equivalent Aerodynamic Design of Blade for Offshore Floating Wind Turbine Model

Jiahuan Lin, Huawei Duan, Baoming Xu, Yangwei Wang and Jun Zhang *

School of Mechanical Engineering and Automation, Fuzhou University, Fuzhou 350108, China; n190220050@fzu.edu.cn (J.L.); 200227065@fzu.edu.cn (H.D.); n190220016@fzu.edu.cn (B.X.); m170210011@fzu.edu.cn (Y.W.)
* Correspondence: zhang_jun@fzu.edu.cn

Abstract: The Froude-scaled offshore floating wind turbine model is inevitably affected by the Reynolds number effect, making the model unable to correctly reproduce the thrust performance of the reference wind turbine (RWT). To solve this problem, an Xfoil-AirfoilPrep-Matlab (XAM) system and a wide tip speed ratio search method (WTSM) are proposed to design a wide tip speed ratio (TSR) thrust-match model blade. The XAM system is utilized to select the best airfoil for WTSM by calculating the lift and drag coefficients of several airfoils. The WTSM is utilized to optimize the blade chord and twist. It formalizes the blade chord and twist by polynomials and then optimizes the polynomial coefficients. The thrust coefficients construct the optimization object at different TSRs. For validating the effect of the redesigned blade, the thrust performance is compared to that of the RWT blade. In addition, the thrust performance of redesigned blade at different pitch angles is also calculated and compared to those of the RWT blade. Results show that the thrust performance of redesigned blade matches well with that of the RWT blade at 0 pitch angle, and it can also match the variations of that of the RWT blade at the other pitch angles well.

Keywords: offshore floating wind turbine; redesigned blade; Reynolds number effect; grasshopper optimization algorithm

Citation: Lin, J.; Duan, H.; Xu, B.; Wang, Y.; Zhang, J. Equivalent Aerodynamic Design of Blade for Offshore Floating Wind Turbine Model. *J. Mar. Sci. Eng.* **2022**, *10*, 132. https://doi.org/10.3390/jmse10020132

Academic Editor: Barbara Zanuttigh

Received: 7 December 2021
Accepted: 12 January 2022
Published: 20 January 2022

1. Introduction

With the development of the offshore floating wind turbine (OFWT) technique, the capacity and geometry of OFWTs are rapidly increasing [1–5]. The increases promote the need for the OFWT design cost reduction approach [6]. In this scenario, the scaled model method is the most commonly used one [7,8]. It works by constructing a model to validate the static and dynamic characteristics of OFWT. However, due to the Reynolds number effect (RNE), the Froude-scaled OFWT model cannot correctly reproduce the aerodynamic performance of the reference wind turbine (RWT) [9–12], especially the thrust performance.

To solve this problem, the redesigning blade method is proposed and widely studied [13]. It reproduces the thrust performance of the RWT blade by designing a dedicated blade. The design includes two steps. The first step is choosing a thin airfoil to replace the original airfoils used in the RWT blade. The second step is adjusting the blade geometry, which is mainly described by the blade chord and twist, by manual or optimization algorithm to obtain a redesigned blade matching the RWT blade in thrust performance.

In the first step, most researchers have just selected an airfoil with an advantage performance over the original airfoils at a low Reynolds number [14–18]. Only a few studies are focusing on how to get an airfoil with better performance for redesigning the blade. For example, Timmer et al. [19] focused on the Delft University wind turbine dedicated airfoils. They revealed the influence of Gurney flaps, trailing edge wedges, vortex generators and tripwires on the airfoil performance. Zhang et al. [20] proposed a method to design the new airfoils using characteristics. A new airfoil based on E387 airfoil working at $Re = 2.0 \times 10^5$ (Reynolds number is 2.0×10^5) and another new airfoil based on

PSU 94-097 airfoil working at $Re = 4.0 \times 10^5$ were designed with enhanced aerodynamic properties. Obviously, these studies cannot fully guide the airfoil selection.

Compared to the first step, the second step of redesigning the blade has received more attention. Martin et al. [21] designed a dedicated blade based on AG 04 airfoil by manually adjusting the blade chord and twist. This design aimed to create a blade matching the thrust coefficient C_T, which is usually used to assess the thrust performance, at the rated tip speed ratio (TSR, approximately 7) and maximized the peak value of the power coefficient, which is usually used to assess the power performance. After that, some verifications were carried out by simulation tools and model tests. Bayati et al. [22] designed a model blade based on SD7032 airfoil by analytical method. This blade was designed to match the C_T–TSR curve with the RWT blade at a special TSR. Then, it was applied for some simulations and model tests. Chen et al. [23] proposed a high-order redesign method for model blade optimization, describing the blade chord and twist by a quaternary polynomial and a quadratic polynomial. The polynomial coefficients were taken as the variables and optimized for the C_T at rated TSR. The redesigned blade can meet the C_T of the RWT blade and make up for the discrepancy of the rotor thrust by the pitch angle adjustment at other TSRs. Wen et al. [24] proposed a Maximum Lift Tracking approach and Load Distribution Match approach to design a blade with the desired thrust at rated TSR. Both approaches had been verified by simulation tools. Du et al. [25] designed a model blade based on NACA 4412 airfoil by Pattern Search (PS) method. The design object guaranteed the C_T of the blade at rated TSR and considered the weight of the model blade. It can be found that there has been a great deal of redesigned blades in the past decades. However, the optimization of the geometrical parameters of the redesigned blade is almost limited to the objective of several special TSRs. In actual operations, the TSR of OFWTs is time-varying under the unsteady inflow wind, wake and control strategy [26–29]. Thus, it is required that a kind of redesigned blade has the thrust performance matching that of RWT blade in a wide range of TSR.

To solve these problems, this study first proposes an Xfoil-AirfoilPrep-Matlab (XAM) system to reference alternative airfoil selection and a wide TSR search method (WTSM) to optimize the chord and twist of the redesigned blade. The XAM system is operated based on Matlab. It runs the Xfoil and the AirfoilPrep to calculate the aerodynamic parameters of airfoils. Then, the parameters of different airfoils will be compared to select the best alternative airfoil. The WTSM takes thrust coefficients at multi-TSR as the design objective to optimize the blade chord and twist. Then, the grasshopper optimization algorithm (GOA) is utilized to search for the best solution in this study. Thus, the thrust performance of the redesigned blade can match the RWT blade at a wide range of TSR. The result is more consistent with the actual working condition.

The following contents can be organized as follows. In Section 2, the scaling laws of the OFWT model and RNE are introduced. In Section 3, the XAM system is introduced in detail, and a comparison among 767 airfoils is discussed. In Section 4, the WTSM is constructed and carried out. In Section 5, some conclusions are summarized.

2. Scaling Laws and RNE

To design a wide TSR thrust matched blade, the reason for RNE and its effect mechanism should be deeply studied. Thus, the scaling laws of the Froude-scaled OFWT model are first introduced in this section. Then, the inevitability of RNE is analyzed based on the scaling laws, and the effect mechanism of RNE is constructed. Finally, based on the effect mechanism of RNE, a preliminary airfoil selection method is proposed to guidance redesigning the blade.

2.1. Scaling Laws

The OFWT model is commonly designed to satisfy the geometry, kinematics and dynamics similarity [30]. To reach these goals, the scaling factors of different properties should be fully considered. First of all, the geometric scaling factor should be defined as follows:

$$\lambda = \frac{L_f}{L_m} \quad (1)$$

where λ is the geometric scaling factor (λ = 80 in this study), L is the characteristic length, the subscript f and m are the RWT (NREL 5MW wind turbine [31] in this study) and OFWT model, respectively.

Besides, as marine machinery, the OFWT model should be designed for hydrodynamics similarity. Thus, it should satisfy the Froude scaling law. The Froude scaling can be described as follows:

$$Fr = \frac{U_f}{\sqrt{g_f L_f}} = \frac{U_m}{\sqrt{g_m L_m}} \quad (2)$$

where Fr is the Froude number, U is the velocity of fluid and g is the gravitational acceleration.

In addition, the kinematic characteristics of wind turbines are indicated by the TSR. The equation of TSR is shown as follows:

$$\mathrm{TSR} = \frac{\omega R}{U} \quad (3)$$

where ω is the angular velocity of the rotor and R is the radius of the rotor.

Due to the $g_f = g_m$, the scaling factors of different properties can be obtained through the dimensional analysis based on the definitions of geometric scaling factors and the Froude scaling law. According to the related scaling factors, it can be found that the kinematics similarity of the OFWT model is also guaranteed well.

2.2. RNE

The Reynolds number is a dimensionless parameter used to characterize the fluid flow (i.e., wind), and it can be described as follows:

$$Re = \frac{LU}{\nu} \quad (4)$$

where Re is the Reynolds number ($Re = 11.5 \times 10^6$ for the RWT [21]), ν is the kinematic viscosity of fluid flow.

According to the related scaling factors obtained through the dimensional analysis, the mapping from the Reynolds number of RWT to the OFWT model can be concluded as follows:

$$Re_m = \lambda^{-1.5} Re_f \quad (5)$$

Equation (5) points out that the Reynolds number of the OFWT model is inevitable smaller than that of the RWT. This difference will lead to a significant decrease in the aerodynamic performance of the OFWT model, which is named RNE. To accurately study the influence of RNE on the performance of airfoils, the DU40 airfoil is taken as the example to compare the lift and drag coefficients (C_L and C_D) of airfoils at the reference and model Reynolds numbers. The results are shown in Figure 1, and α is the angle of attack. It shows that there are several significant changes in airfoil performance resulting from RNE. The C_L significantly decreases, and the C_D slightly increases. It means the peak C_L/C_D also decreases significantly. As the Blade Element Momentum (BEM) points that, the C_T of blade sections is mainly influenced by the peak C_L of airfoil, the valley C_D and the peak C_L/C_D of airfoil [32,33]. Thus, to design a thrust match blade, airfoils with larger peak C_L, smaller valley C_D and larger C_L/C_D are needed first.

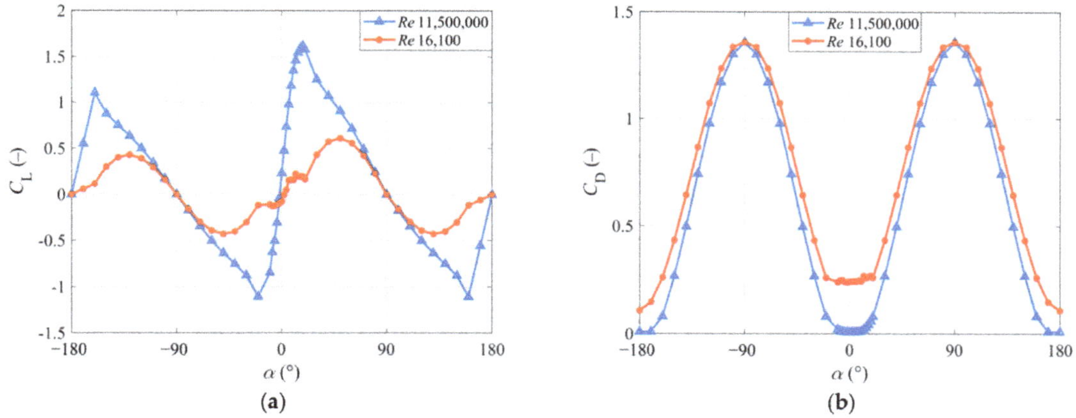

Figure 1. The aerodynamic performance of DU40 airfoil at the reference and model Reynolds numbers. (**a**) The lift coefficient. (**b**) The drag coefficient.

3. Airfoil Selection

3.1. XAM System

To find the required airfoils, the C_L and C_D of several airfoils should be calculated and compared with each other. Based on this goal, a XAM system, which integrates Xfoil [34], AirfoilPrep [35] and Matlab software, is proposed in this study. This system can fulfill the calculation of the C_L and C_D of several airfoils at the full 360-degree angle of attack, the extraction of the critical data and the comparison of the aerodynamic performance of airfoils.

The flowchart of the XAM system is shown in Figure 2. Firstly, the studied airfoil number, airfoil geometries and *Re* are read. Secondly, a bat file is written based on airfoil geometries and used to drive the Xfoil. Thirdly, the C_L and C_D at a small range of α are processed to the full 360-degree C_L and C_D data by AirfoilPrep. Fourthly, the preliminary screening conditions are determined according to the maximum peak C_L, the minimum valley C_D and the maximum peak C_L/C_D of airfoils. Then, several airfoils are selected based on the preliminary screening conditions. Afterward, the scores of the selected airfoils are calculated by a dedicated score method. Finally, the C_L and C_D of the airfoil with the highest score are output.

3.2. Result and Discussion of Airfoil Performance

In this study, the preliminary screening conditions of the XAM system are chosen through some tests. The mentioned preliminary screening conditions are set as follows: (1) the peak C_L should be greater than 75% of the maximum, (2) the valley C_D should be less than the 125% of minimum, (3) the peak C_L/C_D should be greater than the 75% of the maximum.

In addition, the dedicated score method is defined as follows. Firstly, the peak C_L, the valley C_D and the peak C_L/C_D were normalized and scored. Secondly, in this study, the weights of the peak C_L, the valley C_D and the peak C_L/C_D is set as 1/3 for the complicated relationship between the mentioned parameters and the aerodynamic performance of airfoils. The optimal study of the weights will be carried out in our future researches. Finally, the score of an airfoil is formularized as follows:

$$Q = \frac{Q_L + Q_D + Q_{L/D}}{3} \tag{6}$$

$$Q_L = \frac{C_{\mathrm{Lp}} - C_{\mathrm{Lp,min}}}{C_{\mathrm{Lp,max}} - C_{\mathrm{Lp,min}}} \tag{7}$$

$$Q_D = 1 - \frac{C_{Dv} - C_{Dv,min}}{C_{Dv,max} - C_{Dv,min}} \quad (8)$$

$$Q_{L/D} = \frac{(C_L/C_D)_p - (C_L/C_D)_{min}}{(C_L/C_D)_{max} - (C_L/C_D)_{min}} \quad (9)$$

where Q is the final score of an airfoil, Q_L is the score of an airfoil in the peak C_L, Q_D is the score of an airfoil in the valley C_D, $Q_{L/D}$ is the score of an airfoil in the peak C_L/C_D, C_{Lp} is the peak C_L of the airfoil, $C_{Lp,min}$ and $C_{Lp,max}$ are the boundary values of peak C_L of the selected airfoils, C_{Dv} is the valley C_D of the airfoil, $C_{Dv,min}$ and $C_{Dv,max}$ are the boundary values of valley C_D of the selected airfoils, $(C_L/C_D)_p$ is the peak C_L/C_D of the airfoil, $(C_L/C_D)_{p,min}$ and $(C_L/C_D)_{p,max}$ are the boundary values of peak C_L/C_D of the selected airfoils.

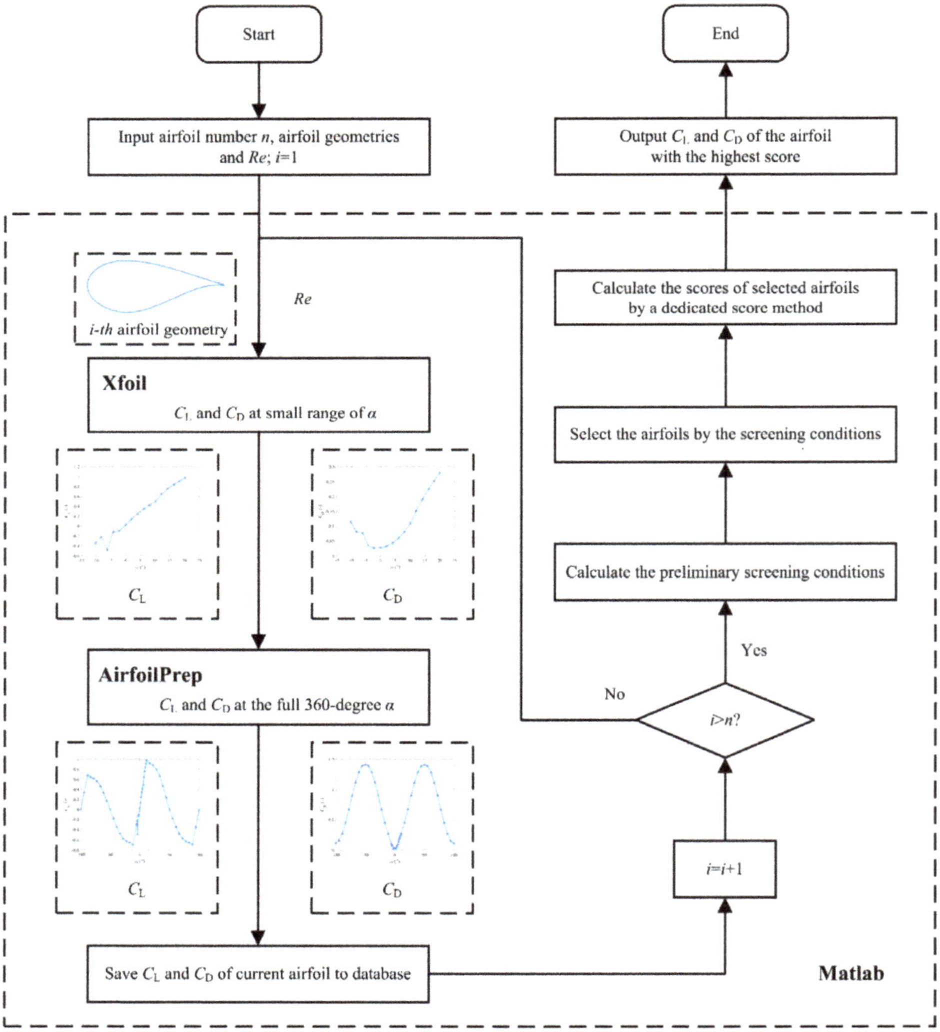

Figure 2. The flow-chart of XAM system.

Seven hundred and sixty-seven airfoils with a thickness between 1% and 8% are served for our study. Moreover, 28 airfoils are selected out by the preliminary screening conditions. These airfoils have a peak C_L greater than 1.0286, a valley C_D less than 0.0298 and a peak C_L/C_D greater than 10.1181, simultaneously. Then, the scores of these airfoils are calculated based on the dedicated score method. The peak C_L, the valley C_D, the peak C_L/C_D and the final scores of 28 selected airfoils are shown in Table 1.

Table 1. The peak C_L, the valley C_D, the peak C_L/C_D and the final scores of the 28 selected airfoils.

Airfoil	Peak C_L	Valley C_D	Peak C_L/C_D	Score
AG 04	1.0498	0.0250	10.3591	0.3347
AG 08	1.0606	0.0248	11.1918	0.4629
AG 09	1.0675	0.0244	11.8643	0.5784
AG 10	1.0717	0.0244	11.9592	0.5983
AG 11	1.0588	0.0256	10.4947	0.3291
AG 12	1.0808	0.0246	11.2672	0.5307
AG 13	1.0922	0.0245	11.8873	0.6305
AG 14	1.1047	0.0245	12.5317	0.7286
AG 16	1.0731	0.0252	10.4601	0.3852
AG 17	1.0862	0.0249	11.2889	0.5250
AG 18	1.0992	0.0247	12.1169	0.6576
AG 26	1.1010	0.0256	10.9973	0.4796
AG 27	1.1182	0.0260	11.9633	0.5960
BE6453B	1.1774	0.0292	12.1544	0.5340
GOE492	1.0931	0.0293	12.5829	0.3811
GRANTX16	1.0468	0.0255	11.0466	0.3681
HN-1070	1.0482	0.0253	11.0657	0.3870
HN-998	1.1063	0.0282	10.7751	0.2909
HN-999	1.1124	0.0286	10.6772	0.2670
MA409(original)	1.1514	0.0279	11.6869	0.5126
MA409(smoothed)	1.1154	0.0288	11.8652	0.3885
Ritz2-30-5	1.0567	0.0248	12.0982	0.5518
Ritz2-30-6	1.0313	0.0252	10.4219	0.2857
Ritz3-30-5	1.0947	0.0257	11.5023	0.5130
SIMPLEX5	1.0529	0.0258	12.2410	0.4906
SIMPLEX6	1.0497	0.0278	10.4016	0.1486
STCYR-53	1.0315	0.0283	13.4464	0.4018
USA46	1.0677	0.0259	10.4566	0.3249

As shown in Table 1, the AG 14 airfoil has the highest score, 0.7286. Thus, it is the best choice among the 767 airfoils. The geometry and aerodynamic performance of AG 14 airfoil at the model Reynolds number are shown in Figures 3 and 4, respectively. It is worth noting that a different result may be obtained using other criteria, and it will be studied in our future research.

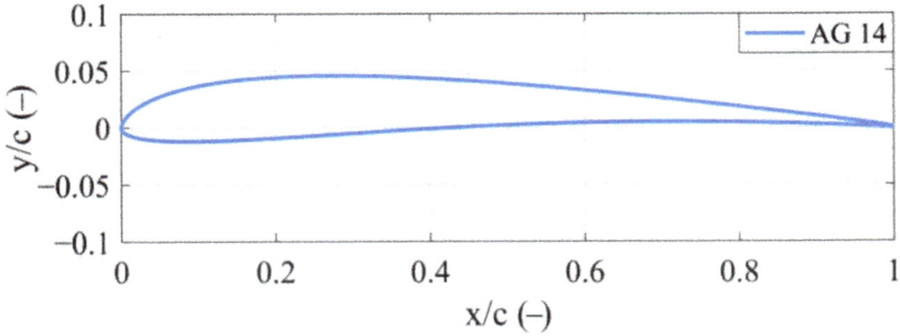

Figure 3. The geometry of AG 14 airfoil.

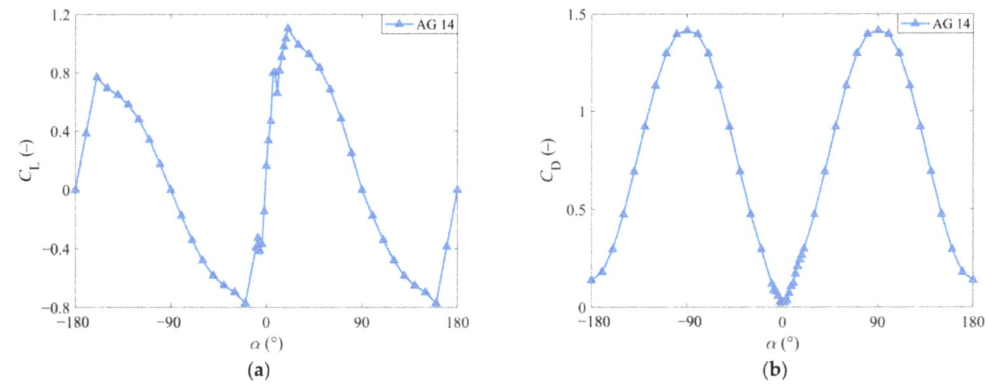

Figure 4. The aerodynamic performance of AG 14 airfoil at the model Reynolds number. (**a**) The lift coefficient. (**b**) The drag coefficient.

4. Redesigning Blade Solution

4.1. WTSM

The rated thrust coefficient is usually taken as the optimization object of redesigning the blade in the previous studies. However, it cannot promise the thrust performance of the OFWT model matching the RWT blade under other working conditions. This will lead to deviations in research related to unsteady inflow wind, wake and control strategy. To solve this problem, a WTSM is proposed to design a wide TSR thrust-match blade in this study. The WTSM reaches this goal by taking the thrust coefficients at multi-TSR as the optimization object in the design process. The flowchart of WTSM is shown in Figure 5. First is fitting the blade chord and twist of the RWT blade to a quaternary polynomial and a quadratic polynomial, respectively. Second is clarifying the constraints of blade chord and twist. Then, selecting an appropriate alternative airfoil through the XAM system (AG 14 airfoil in this study). Afterward, the blade chord and twist polynomial coefficients are taken as the design variables to search for the result with a minimum target value. Finally, the blade chord and twist are output.

In the WTSM, the constraints are mainly caused by the blade geometry. Firstly, the chord and twist of every blade section should be positive. Secondly, the blade tip chord should not be a large value, and the blade root twist should not be a large value. These requirements on blade chord and twist can be described as follows:

$$\begin{cases} c_{\text{tip}}/R < \varepsilon_1 \\ c_r/R > 0 \\ \beta_{\text{root}} < \varepsilon_2 \\ \beta_r > 0 \end{cases} \tag{10}$$

where c is the chord of blade section, β is the twist of blade section, the ε_1 and ε_2 are constants (refer to the RWT, ε_1 = 0.04, ε_2 = 14 in this study), the subscript tip, r and root are the tip of the blade, the position with distance $_r$ from the hub center and the root of the blade, respectively.

In terms of the optimization object, due to the WTSM taking the thrust coefficients at multi-TSR as the optimization object in the design process, it should be defined as follows:

$$F = \min(\sum_{1}^{n}(C_{\text{Tr}i}(\delta) - C_{\text{Tf}i})) \tag{11}$$

where n is the number of involved TSRs, δ is an 8-dimensional vector that represents all design variables (the coefficients of blade chord polynomial and blade twist polynomial),

C_{Tri} is the thrust coefficient of the redesigned blade in *i-th* TSR point, C_{Tfi} is the thrust coefficient of the RWT blade in *i-th* TSR point

It is obvious that the number of considered TSRs is one of the critical factors affecting the result. In this study, 22 TSR points are taken into consideration, and the corresponding C_T of the RWT blade can be obtained through OpenFAST.

Finally, the lower boundary (*lb*) and upper boundary (*ub*) of the search range are shown in Table 2. [*lb*, *ub*] is the limited range of δ and is determined by experience.

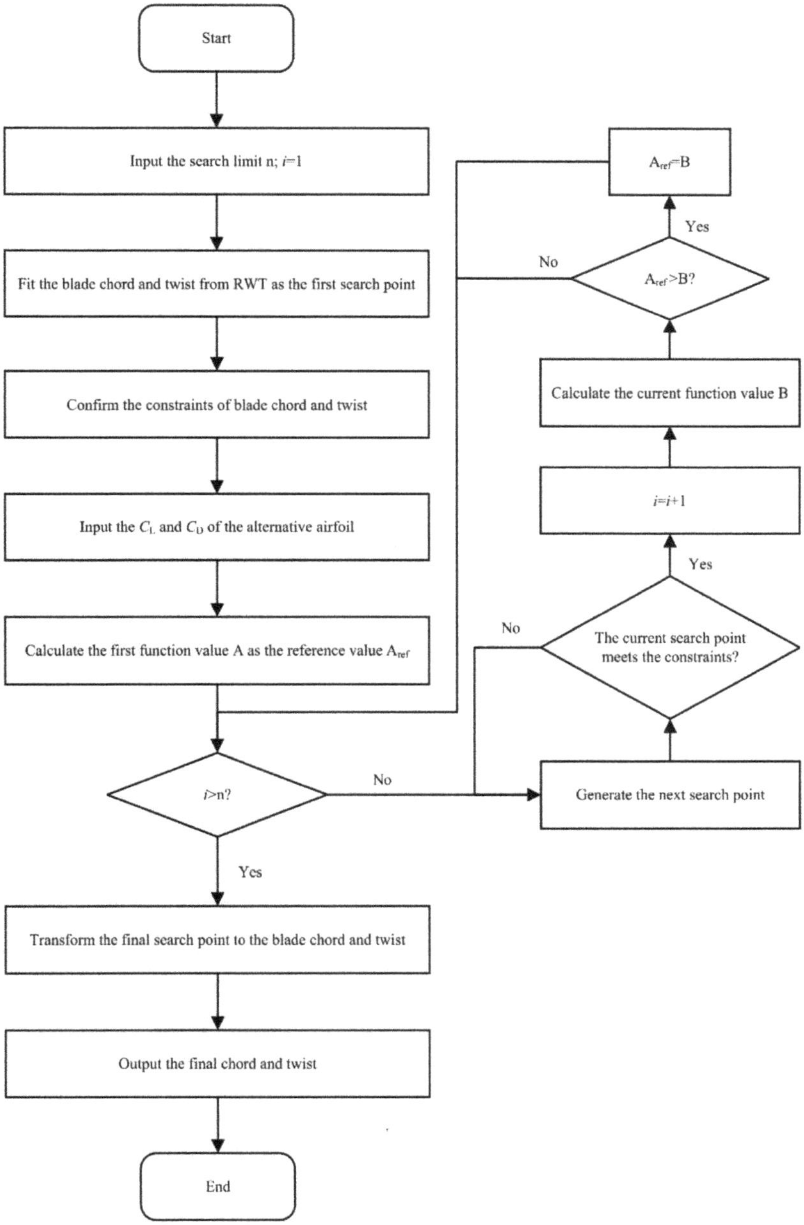

Figure 5. The flow-chart of WTSM.

Table 2. The optimization range.

lb	−1.5	0	−2.5	−0.1	0	6	−30
ub	1	3	−0.5	1	0.5	15	−16

4.2. Some Modifications in GOA

GOA is chosen to complete the search process of WTSM for its superior global optimization capacity [36]. It randomly generates the initial positions (the initial value) in the given optimization range and then calculates their corresponding objective values. Then, the current position (the current value) X_i in the *d-th* dimension of the variable are updated for the next iteration by the following equation [37]:

$$X_i^d = C\left(\sum_{\substack{j=1\\ j\neq i}}^{N} C\frac{ub_d - lb_d}{2} s\left(\left|x_j^d - x_i^d\right|\right)\frac{x_j - x_i}{d_{ij}}\right) + T_d \tag{12}$$

where N is the population, $[lb_d, ub_d]$ is the limited range of the *d-th* variable, d_{ij} is the distance from individual x_i to x_j, and T_d is the target position (the best position so far). C is a decreasing factor to balance the global search ability of GOA, and it can be defined as follows:

$$C = C_{\max} - l\frac{C_{\max} - C_{\min}}{T_{\max}} \tag{13}$$

where $C_{\max}$ and $C_{\min}$ are the boundaries of the variation interval of C, l is the present iteration number, and $T_{\max}$ is the iterative upper bound.

While s is a defined nonlinear action function used to adjust the social force, which is formulated as follows:

$$s(d) = fe^{-\frac{d}{l}} - e^{-d} \tag{14}$$

where f and d are two constants affecting the intensities of attraction and attractive length scale, respectively. Commonly, the value can be fixed as $f = 1.5$ and $d = 0.5$, respectively [36].

It is worth noting that there are some constraints in the redesigning blade process. Thus, some modifications have been done to the generation rules of the initial points. When a new initial value is generated, it should be verified to the constraints, if the constraints are satisfied, the initial value is recorded, otherwise, it should be regenerated.

4.3. Results and Discussion

In this study, the population of GOA is set as 4000, and the iteration is set as 10. The blade chord and twist results are compared with the RWT blade in Figure 6. It shows that the redesigned blade has an increase in chord and a decrease in twist. The increased blade chord can catch more wind power and slightly increase the Reynolds number. The decreased twist will lead the airfoil, which has poor performance due to the RNE, to operate at a better angle of attack. Similar findings can be found in other studies [21,24]. In addition, the smoothness of the chord and twist curve also prove the advantage of using the polynomial coefficients as the design variable.

The comparison of the thrust coefficient between the redesigned blade and the RWT blade is shown in Figure 7a. The relative error is shown in Figure 7b. To show the effectiveness of WTSM, a blade redesigned by Pattern Search (PS) in 1/50th with Du et al.'s study is also present in Figure 7a and denoted as PS [25].

As shown in Figure 7a, the C_T-TSR curve of the PS blade, which optimizes the blade chord and twist in a single TSR by PS method, can only match the RWT blade at a special TSR. In contrast, the C_T-TSR curve of the redesigned blade is matched well with the RWT blade from TSR = 3 to TSR = 11. Moreover, it can be found in Figure 7b that the relative errors of the redesigned blade to RWT blade are almost less than 5%, except for the two points TSR = 3 (8%) and TSR = 5 (6%). This proves the redesigned blade based on WTSM can reproduce the thrust performance of RWT at a wide range of TSR.

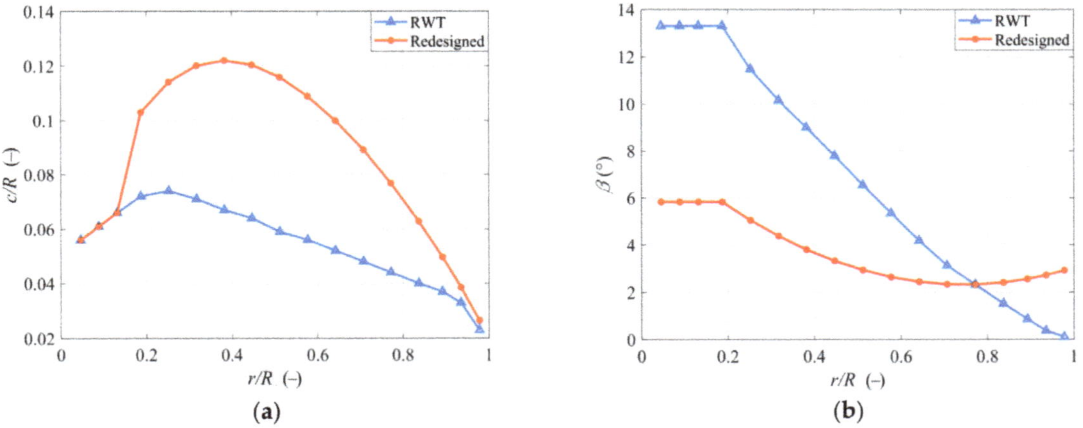

Figure 6. Comparison of the redesigned blade and the RWT blade. (**a**) Chord. (**b**) Twist.

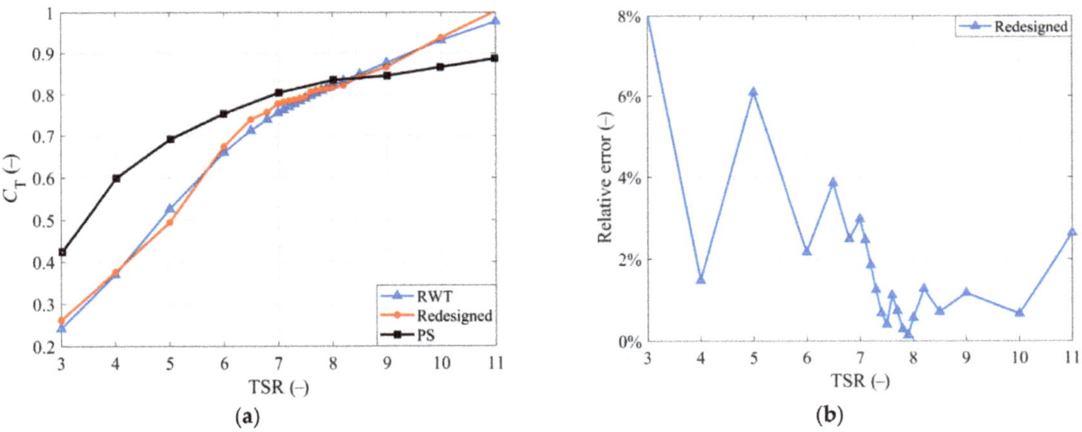

Figure 7. (**a**) The thrust coefficient comparison of redesigned blade, RWT blade and PS blade. (**b**) The relative error of redesigned blade to RWT blade.

4.4. Aerodynamic Analysis with Different Pitch Angle

To further study the capacity of the redesigned blade in researches related to control strategy, some C_T-TSR curves of the redesigned blade and RWT blade with different pitch angles are obtained from OpenFAST and presented in Figure 8. It shows that the C_T of the RWT blade decrease with the increase of pitch angle, and this decrease is more obvious with the increase of TSR. Similar trends can also be found on the C_T-TSR curves of the redesigned blade. In this scenario, though some deviations can be found between the redesigned blade and the RWT blade except the 0 pitch angle, it still has a great potential in serving for research related to control strategy.

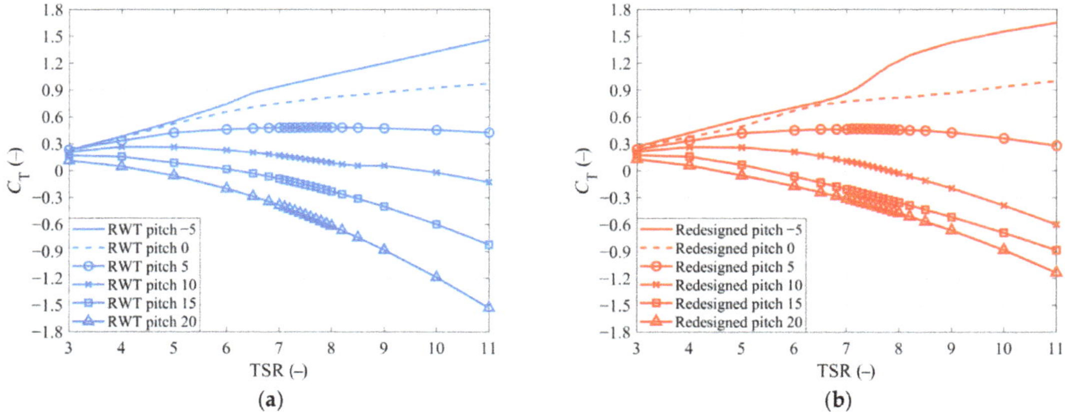

Figure 8. The C_T-TSR curves with different pitch angles. (**a**) RWT blade. (**b**) Redesigned blade.

5. Conclusions

This paper proposes a XAM system to select the best airfoil for redesigning the blade first. Then a WTSM based on GOA is put forward to optimize the blade chord and twist of the redesigned blade. The main findings are summarized as follows:

(1) A XAM system is constructed to guide the airfoil selection. Based on the XAM system, 767 airfoils are compared with each other according to the preliminary screening conditions and the dedicated score method. Finally, the AG 14 airfoil is selected out as the best airfoil.

(2) A WTSM is proposed for a wide TSR thrust-match blade. Based on the WTSM, the blade chord and twist of the redesigned blade with AG 14 airfoil is optimized for a better thrust performance. The C_T-TSR curve of the redesigned blade is compared with the RWT blade. Moreover, most of the relative errors are less than 5%, except for the working conditions that TSR = 3 (8%) and TSR = 5 (6%). It proves the thrust performance of the redesigned blade matches that of the RWT blade well at a 0 pitch angle.

(3) The thrust coefficients of the proposed redesigned blade in other pitch angles are calculated and compared to those of the RWT blade. Results show that the variations are consistent with the RWT blade. It means that the redesigned blade has great potential in the studies related to control strategy.

It is worth noting that though this paper takes as a reference the NREL 5MW wind turbine, the proposed XAM system and WTSM are also valid for other references.

Author Contributions: Conceptualization, J.L.; methodology, J.L.; software, J.L. and B.X.; validation, J.L. and H.D.; investigation, J.L., Y.W. and H.D.; writing—original draft preparation, J.L.; writing—review and editing, J.Z.; supervision, J.Z.; funding acquisition, J.Z. All authors have read and agreed to the published version of the manuscript.

Funding: This research was funded by [National Natural Science Foundation of China] grant number [51875105].

Institutional Review Board Statement: Not applicable.

Informed Consent Statement: Not applicable.

Data Availability Statement: Not applicable.

Conflicts of Interest: The authors declare no conflict of interest.

References

1. Global Wind Energy Council. *GWEC Global Wind Report*; Global Wind Energy Council (GWEC): Brussels, Belgium, 2021.
2. Ju, S.; Huang, Y.; Huang, Y. Study of optimal large-scale offshore wind turbines. *Renew. Energy* **2020**, *154*, 161–174. [CrossRef]
3. Finnegan, W.; Jiang, Y.; Dumergue, N.; Davies, P.; Goggins, J. Investigation and Validation of Numerical Models for Composite Wind Turbine Blades. *J. Mar. Sci. Eng.* **2021**, *9*, 525. [CrossRef]
4. Medina, C.; Álamo, G.M.; Quevedo-Reina, R. Evolution of the Seismic Response of Monopile-Supported Offshore Wind Turbines of Increasing Size from 5 to 15 MW including Dynamic Soil-Structure Interaction. *J. Mar. Sci. Eng.* **2021**, *9*, 1285. [CrossRef]
5. Chatterjee, J.; Dethlefs, N. Deep learning with knowledge transfer for explainable anomaly prediction in wind turbines. *Wind Energy* **2020**, *23*, 1693–1710. [CrossRef]
6. Ghigo, A.; Cottura, L.; Caradonna, R.; Bracco, G.; Mattiazzo, G. Platform optimization and cost analysis in a floating offshore wind farm. *J. Mar. Sci. Eng.* **2020**, *8*, 835. [CrossRef]
7. Dao, C.; Kazemtabrizi, B.; Crabtree, C. Wind turbine reliability data review and impacts on levelised cost of energy. *Wind Energy* **2019**, *22*, 1848–1871. [CrossRef]
8. Crabtree, C.J.; Zappalá, D.; Hogg, S.I. Wind energy: UK experiences and offshore operational challenges. *Proc. Inst. Mech. Eng. Part A J. Power Energy* **2015**, *229*, 727–746. [CrossRef]
9. Lalonde, E.; Vischschraper, B.; Bitsuamlak, G.; Dai, K. Evaluation of a neural network-based surrogate aerodynamic wind turbine model. In Proceedings of the International Conference on Advances in Wind and Structures, Seoul, Korea, 8 July 2020.
10. Martin, H.R. *Development of a Scale Model Wind Turbine for Testing of Offshore Floating Wind Turbine Systems*; University of Maine: Orono, ME, USA, 2011; pp. 1–181.
11. Lee, H.; Lee, D.J. Low Reynolds number effects on aerodynamic loads of a small scale wind turbine. *Renew. Energy* **2020**, *154*, 1283–1293. [CrossRef]
12. Gueydon, S.; Judge, F.M.; O'shea, M.; Lyden, E.; Le Boulluec, M.; Caverne, J.; Ohana, J.; Kim, S.; Bouscasse, B.; Thiebaut, F.; et al. Round robin laboratory testing of a scaled 10 mw floating horizontal axis wind turbine. *J. Mar. Sci. Eng.* **2021**, *9*, 988. [CrossRef]
13. Meng, L.; He, Y.; Zhao, Y.; Peng, T.; Yang, J. Experimental Study on Aerodynamic Characteristics of the Model Wind Rotor System and on Characterization of a Wind Generation System. *China Ocean Eng.* **2019**, *33*, 137–147. [CrossRef]
14. Fowler, M.J.; Kimball, R.W.; Thomas, D.A.; Goupee, A.J. Design and testing of scale model wind turbines for use in wind/wave basin model tests of floating offshore wind turbines. In Proceedings of the International Conference on Offshore Mechanics and Arctic Engineering, Nantes, France, 9–14 June 2013; Volume 8, pp. 1–11.
15. Duan, F.; Hu, Z.; Niedzwecki, J.M. Model test investigatiaon of a spar floating wind turbine. *Mar. Struct.* **2016**, *49*, 76–96. [CrossRef]
16. Chen, J.; Hu, Z.; Wan, D.; Xiao, Q. Comparisons of the dynamical characteristics of a semi-submersible floating offshore wind turbine based on two different blade concepts. *Ocean Eng.* **2018**, *153*, 305–318. [CrossRef]
17. Duan, F.; Hu, Z.; Liu, G.; Wang, J. Experimental comparisons of dynamic properties of floating wind turbine systems based on two different rotor concepts. *Appl. Ocean Res.* **2016**, *58*, 266–280. [CrossRef]
18. Tahir, A.; Elgabaili, M.; Rajab, Z.; Buaossa, N.; Khalil, A.; Mohamed, F. Optimization of small wind turbine blades using improved blade element momentum theory. *Wind Eng.* **2019**, *43*, 299–310. [CrossRef]
19. Timmer, W.A.; van Rooij, R.P.J.O.M. Summary of the Delft University Wind Turbine Dedicated Airfoils. *J. Sol. Energy Eng.* **2003**, *125*, 488–496. [CrossRef]
20. Zhang, S.; Li, H.; Abbasi, A.A. Design methodology using characteristic parameters control for low Reynolds number airfoils. *Aerosp. Sci. Technol.* **2019**, *86*, 143–152. [CrossRef]
21. Martin, H.R.; Kimball, R.W.; Viselli, A.M.; Goupee, A.J. Methodology for wind/wave basin testing of floating offshore wind turbines. *J. Offshore Mech. Arct. Eng.* **2014**, *136*, 020905. [CrossRef]
22. Bayati, I.; Belloli, M.; Bernini, L.; Zasso, A. Aerodynamic design methodology for wind tunnel tests of wind turbine rotors. *J. Wind Eng. Ind. Aerodyn.* **2017**, *167*, 217–227. [CrossRef]
23. Chen, Z.; He, Y.; Zhao, Y.; Meng, L.; He, C.; Yang, H.; Han, Z.; Liu, Y. High-order redesign method for wind turbine blade optimization in model test considering aerodynamic similarity. *Ocean Eng.* **2020**, *202*, 107156. [CrossRef]
24. Wen, B.; Tian, X.; Dong, X.; Li, Z.; Peng, Z. Design approaches of performance-scaled rotor for wave basin model tests of floating wind turbines. *Renew. Energy* **2020**, *148*, 573–584. [CrossRef]
25. Du, W.; Zhao, Y.; He, Y.; Liu, Y. Design, analysis and test of a model turbine blade for a wave basin test of floating wind turbines. *Renew. Energy* **2016**, *97*, 414–421. [CrossRef]
26. Wang, Y.; Lin, J.; Duan, H.; Zhang, J. Investigation on Thrust Characteristics of a Downstream Offshore Floating Wind Turbine under Yawed Inflow Conditions. *J. Mar. Sci. Eng.* **2021**, *9*, 1215. [CrossRef]
27. Wang, Y.; Wang, J.; Zhang, J. Effects of Wind Rotor Tilt Angle on Aerodynamic Power of Wind Turbine under Typical Periodic Disturbances. *Mech. Mach. Sci.* **2019**, *73*, 3459–3468.
28. Wang, Y.; Lin, J.; Zhang, J. Investigation of a new analytical wake prediction method for offshore floating wind turbines considering an accurate incoming wind flow. *Renew. Energy* **2022**, *185*, 827–849. [CrossRef]
29. Li, S.; Han, Y.; Pan, W.; Liu, S.; Hou, M. Variable-Gain Higher-Order Sliding Mode Pitch Control of Floating Offshore Wind Turbine. *J. Mar. Sci. Eng.* **2021**, *9*, 1172. [CrossRef]

30. Ruzzo, C.; Muggiasca, S.; Malara, G.; Taruffi, F.; Belloli, M.; Collu, M.; Li, L.; Brizzi, G.; Arena, F. Scaling strategies for multi-purpose floating structures physical modeling: State of art and new perspectives. *Appl. Ocean Res.* **2021**, *108*, 102487. [CrossRef]
31. Jonkman, J.; Butterfield, S.; Musial, W.; Scott, G. *Definition of a 5-MW Reference Wind Turbine for Offshore System Development*; National Renewable Energy Laboratory (NREL): Golden, CO, USA, 2009.
32. Macháček, M.; Pospíšil, S.; Kozmar, H. Scaling of wind turbine aerodynamics: Wind tunnel experiments. *MATEC Web Conf.* **2020**, *313*, 00053. [CrossRef]
33. Gajardo, D.; Escauriaza, C.; Ingram, D.M. Capturing the development and interactions of wakes in tidal turbine arrays using a coupled BEM-DES model. *Ocean Eng.* **2019**, *181*, 71–88. [CrossRef]
34. Drela, M. XFOIL: An Analysis and Design System for Low Reynolds Number Airfoils. In Proceedings of the Conference on Low Reynolds Number Airfoil Aerodynamics, Notre Darre, IN, USA, 5–7 June 1989; pp. 1–12.
35. Hansen, C. AirfoilPrep: An Excel Workbook for Generating Airfoil Tables for AeroDyn and WT_Perf. Available online: https://www.nrel.gov/wind/nwtc/airfoil-prep.html (accessed on 3 December 2021).
36. Zhang, J.; Zhang, J.; Zhong, M.; Zheng, J.; Yao, L. A GOA-MSVM based strategy to achieve high fault identification accuracy for rotating machinery under different load conditions. *Meas. J. Int. Meas. Confed.* **2020**, *163*, 108067. [CrossRef]
37. Chen, F.; Tang, B.; Song, T.; Li, L. Multi-fault diagnosis study on roller bearing based on multi-kernel support vector machine with chaotic particle swarm optimization. *Meas. J. Int. Meas. Confed.* **2014**, *47*, 576–590. [CrossRef]

energies

Article

Investigation of Biomass Integrated Air Gasification Regenerative Gas Turbine Power Plants

Momin Elhadi Abdalla [1], Salah Ahmed Abdalla [2], Syed Ali Ammar Taqvi [3], Salman Raza Naqvi [4] and Wei-Hsin Chen [5,6,7,*]

1 Chemical Engineering Department, University of Khartoum, Khartoum 11111, Sudan; mominhadi@uofk.edu
2 Energy Research Centre, University of Khartoum, Khartoum 11115, Sudan; saabdalla@uofk.edu
3 Department of Chemical Engineering, NED University of Engineering & Technology, Karachi 75270, Pakistan; aliammar@neduet.edu.pk
4 School of Chemical & Materials Engineering, National University of Sciences & Technology, Islamabad 44000, Pakistan; salman.raza@scme.nust.edu.pk
5 Department of Aeronautics and Astronautics, National Cheng Kung University, Tainan 701, Taiwan
6 Research Center for Smart Sustainable Circular Economy, Tunghai University, Taichung 407, Taiwan
7 Department of Mechanical Engineering, National Chin-Yi University of Technology, Taichung 411, Taiwan
* Correspondence: chenwh@mail.ncku.edu.tw

Abstract: The results show that Wood Chips of Acacia Nilotica trees available in Sudan lands can be successfully used in the gasification process and, on the same basis, as a bio-renewable energy resource. Simulation models were used to characterize the air gasification process integrated with a Regenerative Gas Turbine Unit. The results revealed that at a moisture content of 12%, gasification temperature of 1500 K, pressure of 20 bar, and air-like gasification medium, the biomass gasifier's flow rate is higher at higher syngas rates. The results verified that there is an optimum ER for each syngas rate, in which the slow growth of the ER revealed the maximum gasifier biomass flow rate. For ER growth at lower levels, the specific fuel consumption (SFC) of the RGT Unit declines sharply from the maximum value reached at 0.27 kg/kW·h at an ER of 5% to the minimum value reached at 0.80 kg/kW·h at an ER of 25% for the lowest gasification temperature of 1000 K. Moreover, ER growths at low levels have a significant effect on the RGT plant's performance, leading to increased RGT thermal efficiency. The increase in the biomass moisture content led to a sharp decrease in the RGT thermal efficiency. The RGT thermal efficiency remains high at higher gasification pressure. The results revealed that the syngas lower heating value remains high at lower produced syngas rates. At the optimum ER, the H_2 mole fraction depicted a value of 1.25%, 0.85% of CO, and 10.50% of CH_4 for a lower heating value of 38 MJ/kg syngas. It is shown that the gasification air entered into the gasifier decreases amid the increase in the biomass moisture content. At different syngas rates (3–10 kg/s) and optimum ER, the results predicted that the Wood Chip biomass flow rates decrease when the gasifier efficiency increases. The simulation model revealed that ER growths at lower levels have a significant effect on increasing the power of the RGT plant.

Keywords: Acacia Nilotica; air gasification; syngas; equivalence ratio; regenerative; gas turbine

Citation: Abdalla, M.E.; Abdalla, S.A.; Taqvi, S.A.A.; Naqvi, S.R.; Chen, W.-H. Investigation of Biomass Integrated Air Gasification Regenerative Gas Turbine Power Plants. *Energies* **2022**, *15*, 741. https://doi.org/10.3390/en15030741

Academic Editors: Davide Astolfi and Bruno Facchini

Received: 9 December 2021
Accepted: 13 January 2022
Published: 20 January 2022

Publisher's Note: MDPI stays neutral with regard to jurisdictional claims in published maps and institutional affiliations.

1. Introduction

Agriculture is considered the driver force of income and livelihood in Sudan. This sector occupies between 60% and 80% of the population and is regarded as the engine for raising other economic zone, such as trade, industry, and transport [1,2].

In Sudan, biomass is a vital renewable energy source because the carbon in biomass is observed as part of the biological carbon cycle widely produced on the land [3]. Biomass is a biological substance that consists of forest, agricultural, and paper waste [4,5] and various organic materials [6]. Because of the absence of a power supply in the rustic population, the rural population principally relies on biomass as a wellspring of energy [1]. Just around 10% of the whole population uses fossil fuels, primarily in urban areas. Biomass powers such as wood, cow waste, and agriculture garbage are primarily gathered from the local regime and have turned into an exchanged ware as cooking fuel. Biomass (such as wood, agricultural residue, and municipal waste) [7–10], with its sustainability and overall accessibility, is relied upon by numerous individuals to assume a key part in future energy scenarios.

The gasification process is usually accomplished in three steps [11]. Firstly, the gasification of biomass particles [12,13] happens through the drying step, which is pursued by the second process of the pyrolytic reactions [14,15], in a minimum oxygen environment, which results in a DE volatilization and declining of the biomass particle [16]. The end process is the formation of tar and char. The pyrolysis step occurs progressively from the surface and moves towards the center of the biomass particles [11]. During the gasification reactions, only a limited amount of air will be supplied to the gasifier to avoid any occurrence of the combustion reactions. The air provided should not be exactly the measure of air utilized for complete oxidation of the biomass. Through this process, the biomass will be converted into valuable products, such as a combustible mixture of gases known as syngas, producer gas, or wood gas [17–20]. This producer gas is also known as synthetic gas, which is composed of carbon monoxide (CO), methane (CH_4), along with carbon dioxide (CO_2), hydrogen (H_2), and nitrogen (N_2) [17,21]. Nitrogen is not an inflammable gas; thus, it is not preferred as a component in the producer gas, as it dilutes the syngas and has no energy value. The proportion of the thick biomass stage to the total reactor volume is an important factor in classifying the biomass gasifiers. According to this procedure, the gasifier can be categorized into (a) thick-phase gasifiers and (b) lean-phase gasifiers. In the lean-phase gasifier, such as the fluidized bed, the volume occupied by the biomass is very limited to about 0.05–0.20 m^3. Most of the gasifiers used for heavy-duty utilization, particularly in the progressing countries, are the dense-phase reactors, such as the fixed bed reactors, with a dense factor of 0.30–0.08 m^3. Other factors that affect the choice gasifier are the fuel, reactor size, ash content, and moisture. The fixed bed gasifiers have reasonably limited scope power age units and industrial heating applications [22].

There are four types of biomass gasifiers: co-current or downdraft gasifiers, counter or updraft current gasifiers, fluidized bed gasifiers, and cross-draft gasifiers. Furthermore, the gasification technique is chosen based on the accessible fuel value, capacity range, and producer gas quality [23–27]. Downdraft gasifiers have a thermal capacity range of 1 kW to 1 MW, whereas updraft gasifiers have a thermal capacity range of 1.1 MW to 12 MW, fluidized-bed gasifiers have a thermal capacity range of 1100 KW to 50,000 KW, and cross-draft gasifiers have a thermal capacity range of 10,000 KW to 200,000 KW [28]. A review [11,29] demonstrated that 2.5% were updraft type, 75% of the designs were downdraft type, 20% were fluidized bed systems, and 2.5% were of different other designs according to the gasifiers' manufacturers in Canada, the United States, and Europe [29].

Due to the low initial cost and the stability of the electrical supply under different circumstances, the use of Gas Turbine Units in electricity generation has gained more opportunities. Another outstanding feature [30] of this equipment is its capability of quick starting using a wide variety of fuels [31], from natural gas, syngas to residual oil or powdered coal [32–36]. In addition, the availability of better materials for construction and the use of adequate blade cooling systems [37,38] to counter the inlet gas temperature, which can often exceed 1200 °C [39], have improved the use of GT power supply units. As a result, the overall thermal efficiency of a GT plant can be about 35%, which is almost the same as that of a conventional steam power plant [35]. It is also becoming more common in land vehicles, such as buses and trucks, trains, and ships. The GT is frequently used to power auxiliaries, such as pumps, blowers, and compressors, in the oil and gas industry [35,40]. Researchers have conducted research and utilized different methods [41] to raise the thermal efficiency of reformative GT cycles [35,42]. One of which is the reheating process used to raise the thermal efficiency of gas and steam turbine cycles. Similarly, regeneration improves both the simple GT and the steam turbine cycles' thermal efficiency. The combined cycle [43,44], which combines a gas turbine and a steam turbine cycle, is a more-essential procedure for increasing the thermal efficiency of the power plant cycle [35,45,46].

This work aims to reinforce the understanding of a regenerative RGT as a thermal process utilizing the integration of the Biomass Gasification Unit at the combustor and applying similar design parameters to Khartoum North Station (GT,187 MW) in Sudan. The study pushes for establishing a qualified operational and conceptual design procedure and reviews for the Integrated Biomass Gasification for Regenerative Gas Turbine Unit "IBGRGT". The work also presents a preliminary strategy to identify the performance and evaluation criterion of the Gasification of Biomass Process utilizing the effect of various operating conditions.

2. Materials and Methods

Acacia Nilotica Wood Chips, available in Sudan with an average size of 1 × 2 × 3 cm, were used as a feedstock and experimentally characterized according to the standard literature data [47–51]. The wood chip samples were processed for proximate analysis to identify the moisture content, ash content, volatile, and fixed carbon using the ASTM Standards mentioned in our previously published studies [19,21,26]. To identify the carbon, hydrogen, nitrogen, and oxygen content, ultimate analysis was performed using a Perkin Elemental Analyzer. The data from both analyses are shown in Table 1. As per the proximate analysis, the sample showed higher moisture content and volatiles with low ash content. The sample showed higher carbon and oxygen content. To reveal the heating value of the sample, bomb calorimetry was used. The results include the measured higher heating value of the Nilotica Biomass, which was a value of 19,094.94 kJ/kg, via a Calorimeter setup. Air was used as a gasification agent. The gasification system considered in this study is the cross-draft gasifier.

Table 1. Ultimate and proximate analysis of Acacia Nilotica Wood Chips.

Proximate Analysis (wt%)				Ultimate Analysis (wt%)					Lower Value of Heat (kJ/kg)
Water	Ash	Volatile	Fixed Carbon	C	H	O	N	S	
37.88	1.43	68.49	30.08	48	6	44	0.40	-	19,094.94

3. Modeling of Components

A schematic of a regenerative gas turbine and Biomass Integrated Unit (BIGT) [3] with a syngas producer is shown in Figure 1. The system comprises a hot air-driven gas turbine, compressor, combustor, and regenerator beside the Gassifier Unit. By using the gasification of dry biomass, the gasifier produces syngas [3]. The biomass material will be fed to the gasifier [52] at surrounding conditions described briefly in Figure 1. Since air is the gasification medium, gasification occurs in compressed air and produces syngas that enters the combustion chamber. A thermodynamic equilibrium method [7] based on a stoichiometric approach according to the method of Jarungthammachote and Dutta [53] was used for modeling the gasifier. The syngas composition is predicted using this model at the gasifier's operational temperature and pressure. The global reaction of gasification can be written as [12,22,51]:

$$CH_xO_yN_z + wH_2O + m(O_2 + 3.76N_2) \rightarrow x_1H_2 + x_2CO + x_3CO_2 + x_4H_2O + x_5CH_4 + x_6N_2 \quad (1)$$

All coefficients from x_1 to x_6 will be calculated using atomic balancing and equilibrium constant equations. The following are the steps to take [3,42]:

$$x_2 + x_3 + x_5 = 1 \quad (2)$$

$$2x_1 + 2x_4 + 4x_5 = x + 2w \quad (3)$$

$$x_2 + 2x_3 + x_4 = y + w + 2m \quad (4)$$

$$2x_6 = z + 3.76 \times 2m \quad (5)$$

The secondary gas-phase reactions can be derived by examining the equilibrium constant expression of the oxidation reactions [11] in the gasifier, which virtually approaches an equilibrium state. These are the reactions:

$$C + 2H_2 \rightarrow CH_4 \quad (6)$$

$$CO + H_2O \rightarrow CO_2 + H_2 \quad (7)$$

The reactions described above [3] are known as the methanation and the gas–water shift reaction, and their equilibrium constants are as follows:

$$K_1 = \frac{P_{CH_4}}{P_{H_2}^2} = \frac{x_5}{x_1^2} \quad (8)$$

$$K_2 = \frac{P_{H_2}P_{CO_2}}{P_{CO}P_{H_2O}} = \frac{x_3x_1}{x_2x_4} \quad (9)$$

Finally, the energy balance is applied for the calculation of the gasification temperature (T_{gasif}) as follows [54]:

$$\overline{h}^o_{f,Biomass} + w\overline{h}^o_{f,H_2O} = x_1\left(\overline{h}^o_{f,H_2} + \Delta\overline{h}\right) + x_2\left(\overline{h}^o_{f,CO} + \Delta\overline{h}\right) + x_3\left(\overline{h}^o_{f,CO_2} + \Delta\overline{h}\right) + x_4\left(\overline{h}^o_{f,H_2O} + \Delta\overline{h}\right) + x_5\left(\overline{h}^o_{f,CH_4} + \Delta\overline{h}\right) + x_6\left(\overline{h}^o_{f,N_2} + \Delta\overline{h}\right) \quad (10)$$

which is the formation enthalpy in kJ/kmol, its value in the reference state is zero for all chemical compositions, and the enthalpy difference between the given state and the reference state [3]. The RGT power plants comprise four parts: compressor, combustion chamber (CC), turbine, and generator. The integrated biomass regenerative combined cycle arrangement considered in Figure 1 demonstrates how to utilize the hot turbine exhaust gas. The fresh atmospheric air from the surroundings is filtered and drawn continuously into the circuit; the energy is added by fuel combustion in the chamber unit [35]. The combustion products are spread through the turbine [55] and consequently produce electrical work, while the rest of the exhaust gases are discharged into the Biomass Gasifier and Regenerator Units. The needed power output of the gasifier, Q (MWth), is an essential input parameter set by the client [17]. Based on this, the designer calculates the amount of fuel to feed into the gasifier and the amount of gasifying medium required. The volume flow rate of the product gas, Vg (Nm^3/s), is calculated by subtracting its lower heating value, LHVg (MJ/Nm^3), from its ideal higher heating value, LHVg (MJ/Nm^3) [17].

$$V_{gas} = \frac{Q}{LHV_{gas}}\left(\frac{Nm^3}{sec}\right) \quad (11)$$

The lower heating value (LHV) or net heating value can be determined from gas formation (y_i), according to [17]:

$$LHV_{gas} = \sum_i^N y_i \times LHV_i \quad (12)$$

The net heating value or higher heating value (HHV) can be determined from gas formation (y_i), according to [17]:

$$HHV_{Syngas} = \sum_i^N y_i \times HHV_i\left(\frac{kJ}{kmol}\right) \quad (13)$$

$$HHV_{Syngas} = \frac{HHV_{Syngas}}{Mw_{Syngas}}\left(\frac{kJ}{kg}\right) \quad (14)$$

To find the biomass feed rate, the following formula is used, in which M_f, the needed power output, is divided by the LHV of the biomass (LHV_{bm}) and by the gasifier efficiency (η_{gef}) [17]:

$$M_f = \frac{Q}{LHV_{Biomass} \times \eta_{gef}} \quad (15)$$

The following unified correlation for HHV is based on 15 existing correlations and 50 fuels, including liquid, coal, biomass, and gas, developed by Channiwala and Parikh [17].

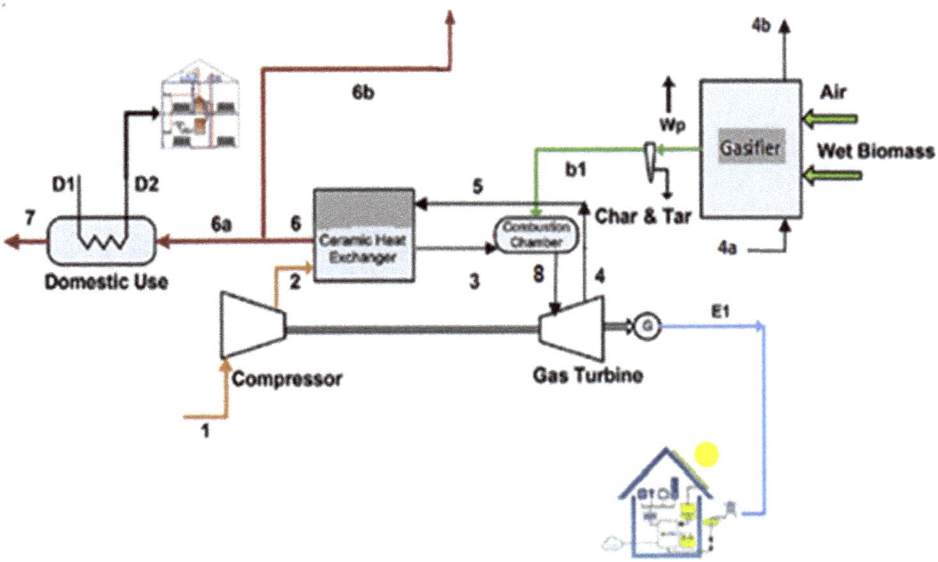

Figure 1. The regenerative gas turbine cycle integrated with the Biomass Gasifier Unit.

The percentages of carbon, hydrogen, sulfur, oxygen, nitrogen, and ash, as determined by a final analysis on a dry basis, are C, H, S, O, N, and ASH. A critical parameter is the theoretical air demand for the entire combustion of a unit mass of fuel, m_{th}, called the stoichiometric air requirement. Equation (16) [15] shows how to calculate it:

$$M_{th} = \left[0.1153C + 0.3434\left(H - \frac{O}{8}\right) + 0.0434S\right] \left(\frac{kg{\cdot}air}{kg{\cdot}\ dry\ fuel}\right) \quad (16)$$

How to calculare the amount of air needed, M_a, is shown below [17]:

$$M_a = M_{th}{\cdot}ER \quad (17)$$

The gasification of a unit mass of biomass in an air-blown gasifier is calculated by multiplying it by another parameter, ER.

The air requirement of the gasifier, M_{fa}, at a fuel feed rate of M_f is the amount of actual air [17]:

$$M_{fa} = M_{th}{\cdot}ER{\cdot}M_f \quad (18)$$

The equivalence ratio (ER) [17] is a crucial design parameter for gasifiers. The stoichiometric air–fuel ratio is the ratio of the actual air–fuel ratio to the stoichiometric air–fuel ratio. This word refers to conditions when there is a lack of oxygen, such as those seen in a gasifier [17]:

$$ER(<1.0)_{Gasification} = \frac{\text{Actual Air}}{\text{Stoichiometric Air}} = EA(>1.0)_{Combustion} \quad (19)$$

EA denotes the excess air coefficient [17]. Fuel is gasified rather than combusted when the ER value is very below 1.0, which has a big impact on the quality of gas produced by a gasifier. A gasifier's oxygen requirements, an air supply, or air-separation equipment that removes oxygen from the air can be used to fulfil this need. [17] Gasification efficiency is measured in cold-gas efficiency or hot-gas efficiency [17]. The energy intake above the potential energy output is known as cold-gas efficiency. With an LHV of Qg, the efficiency of gasifying Mf kg of solid fuel to produce Mg kg of the product gas is given as [17]:

$$\eta_{cg} = \frac{Q_g \cdot M_g}{LHV_{Biomass} \cdot M_f} \quad (20)$$

where LHV_f is the solid fuel's lower heating value (LHV). The hot-gas efficiency, abbreviated as η_{hg}, is defined as [17]:

$$\eta_{hg} = \frac{Q_g \cdot M_g + M_g . C_{Pg} \cdot (T_f - T_0)}{LHV_{Biomass} \cdot M_f} \quad (21)$$

T_0 is the fuel temperature entering the gasifier, and T_f is the gas temperature at the gasifier egress or the burner's access [17]. Accordingly, the intake pressure at the compressor inlet was modeled with the following equation [16]:

$$P_1 = P_{ATM} - \Delta P_{intake} \quad (22)$$

where the intake pressure drop (ΔP_{intake}) was taken to be 0.005 bar, and the intake temperature was modeled as the ambient temperature [56]. The process on the temperature-entropy diagram [42] is represented in Figure 1. The compressor compression ratio (r_P) can be defined as [35,41]:

$$r_P = \frac{P_2}{P_1} \quad (23)$$

where P_1 and P_2 are the compressor's inlet and outlet air pressure, respectively [18]. The isentropic outlet temperature leaving the compressor is modeled by the equation [35,57,58]:

$$\frac{T_1}{T_{2s}} = \left(\frac{P_1}{P_2}\right)^{\frac{\gamma_a - 1}{\gamma_a}} \quad (24)$$

The specific heat ratio for air γ_a was taken as 1.4 and was predicted at $\gamma_g = 1.3$ for the gas. The isentropic efficiency of the compressor and turbine was taken to be in the range of 85% to 90%. The isentropic compressor efficiency is expressed by the equation [59,60]:

$$\eta_c = \frac{T_{2s} - T_1}{T_2 - T_1} \quad (25)$$

where T_1 and T_2 are the compressor inlet and outlet air temperatures, respectively, and T_{2s} is the compressor isentropic outlet temperature. The specific work required to run the compressor work (W_C) is modeled with the following equation [60]:

$$\dot{W}_c = \dot{m}_a C_{P_a} (T_2 - T_1) = \dot{m}_a C_{P_a} T_1 \left[\frac{r_p^{\frac{\gamma_a - 1}{\gamma_a}} - 1}{\eta_c}\right] \quad (26)$$

Additionally, the specific heat of air is taken as $C_{P_{air}} = 1.005\ \frac{kJ}{kgK}$. The specific heat of flue gas (C_{pg}) is given by [60]:

$$C_{pg} = 1.8083 - 2.3127 \times 10^{-3}T + 4.045 \times 10^{-6}T^2 - 1.7363 \times 10^{-9}T^3 \quad (27)$$

From the energy balance, the combustion chamber equation is as follows [18,51]:

$$\dot{m}_a C_{P_a} T_x + \dot{m}_f LHV + \dot{m}_f C_{P_f} T_f = (\dot{m}_a + \dot{m}_f) C_{Pg} T_{it} \quad (28)$$

where $\dot{m}_f$ is the fuel mass flow rate in (kg/s), $\dot{m}_a$ is the air mass flow rate (kg/s), LHV is the fuel's low heat value, T_{it} is the turbine inlet temperature, C_{P_f} is the specific heat of fuel, and T_f is the temperature of the fuel. The specific heat of the flue gas was modeled with $C_{P_g} = 1.07$ kJ/kg·K; efficiency was set at 95%, and a pressure drop of $\Delta P_{C,C} = 0.4785$ bar was set in the combustor. Accordingly, the efficiency of the combustor is modeled as [35]:

$$\eta_{C,C} = \frac{\dot{m}_g C_{Pg} TIT - \dot{m}_a C_{P_a} T_x}{\dot{m}_f LHV_g} \quad (29)$$

The air–fuel ratio at the combustor was modeled according to the following equation [35]:

$$AFR = \frac{A}{F} = \frac{\dot{m}_a}{\dot{m}_f} \quad (30)$$

where the total mass flow rate is given by [35]:

$$\dot{m}_g = \dot{m}_a + \dot{m}_f \quad (31)$$

The discharge gas of the turbine is predicted according to the equation [35]:

$$\frac{T_8}{T_{4s}} = \left(\frac{P_8}{P_4}\right)^{\frac{\gamma_g - 1}{\gamma_g}} \quad (32)$$

where the actual outlet temperature leaving the turbine at isentropic conditions is modeled according to [35]:

$$\eta_t = \frac{T_8 - T_4}{T_8 - T_{4s}} \quad (33)$$

The regenerator effectiveness ε is modeled according to the equation [59]:

$$\varepsilon = \frac{T_3 - T_2}{T_4 - T_2} \quad (34)$$

where T_3 is the combustor inlet temperature. The shaft work produced from the turbine is calculated by the equation [35]:

$$\dot{W}_{RGT} = \dot{m}_g C_{Pg} (T_4 - TIT) = \dot{m}_g C_{Pg} TIT \eta_t \left[1 - \frac{1}{r_p^{\frac{\gamma_g - 1}{\gamma_g}}}\right] \quad (35)$$

The network from the GT Unit is expressed by the equation [35,61]:

$$\dot{W}_{RGT,Net} = \dot{W}_{RGT} - \dot{W}_c = \dot{m}_g C_{Pg} TIT \eta_t \left[1 - \frac{1}{r_p^{\frac{\gamma_g - 1}{\gamma_g}}}\right] - \dot{m}_a C_{P_a} T_1 \left[\frac{r_p^{\frac{\gamma_a - 1}{\gamma_a}} - 1}{\eta_c}\right] \quad (36)$$

The output power from the GT is expressed with the equation [35,60,61]:

$$P_{RGT} = \left[\dot{W}_{RGT} - \dot{W}_c\right] \times \eta_{Mech}\eta_{Gen} \quad (37)$$

The mechanical (η_{Mech}) and generator (η_{Gen}) efficiencies were taken to be 92% and 95%, respectively.

The heat supplied is expressed with the equation [35]:

$$\dot{Q}_{add} = \dot{m}_g C_{Pg} TIT - \dot{m}_a C_{P_a} T_3 \quad (38)$$

The heat supplied (per kg. air) to the combustor was modeled according to the equation [35]:

$$\dot{Q}_{add} = \frac{\dot{m}_f LHV_g \eta_{C,C}}{\dot{m}_{air}} = \frac{LHV_g * \eta_{C,C}}{AFR} \quad (39)$$

The GT efficiency is determined by the equation [35]:

$$\eta_{over,RGT} = \frac{\dot{W}_{RGT,Net}}{\dot{Q}_{add}} \quad (40)$$

Accordingly, the heat rate (HR), which is defined as the spent heat to generate unit energy of electricity, is determined by the equation [35,62]:

$$HR = \frac{3600 \times m_f \times LHV_g}{\eta_{over,RGT}} \quad (41)$$

The specific fuel consumption (SFC) is determined by the equation [35]:

$$SFC = \frac{3600 \times \dot{m}_f}{\dot{W}_{RGT,Net}} \quad (42)$$

4. Results and Discussion

The gasification process was simulated using Acacia Nilotica Wood Chips biomass available in Sudan. A detailed energy analysis was derived from investigating the optimal design conditions of the system that integrates the Biomass Unit with the Regenerative Gas Turbine Power Plant. The work was executed with the Thermodynamics Engineering Equation Solver (EES) codes. The work investigated air as a gasification medium to produce syngas fuel for Wood Chips (Acacia Nilotica). Figure 2 depicts the effect of the equivalence ratio (ER) on the gasifier's biomass mass flow rate at different syngas rates (5–20 kg/s). As observed, at a moisture content of 12%, gasification temperature of 1500 K, pressure of 20 bar, and air-like gasification medium, the biomass gasifier's flow rate is higher at a higher syngas rate. The higher value of the biomass flow rate reached 35 kg/s of the wood chips at a yielded syngas rate of 20 kg/s. The results verified that there is an optimum ER for each syngas rate, in which the slow growth of the ER revealed the maximum gasifier's biomass mass flow rate. After that, the further increases in the ER showed a slow downtrend due to increasing combustion products, changes of syngas constituents amid CO_2, water vapor, nitrogen, and a massive decrease in the lower heating value of the syngas. The point of inflection can be regarded as the point of the optimal design of the system to derive the required biomass rate for the necessary air gasification to produce the assigned syngas rates. The produced syngas composition varies with the amount of supplied air to the process.

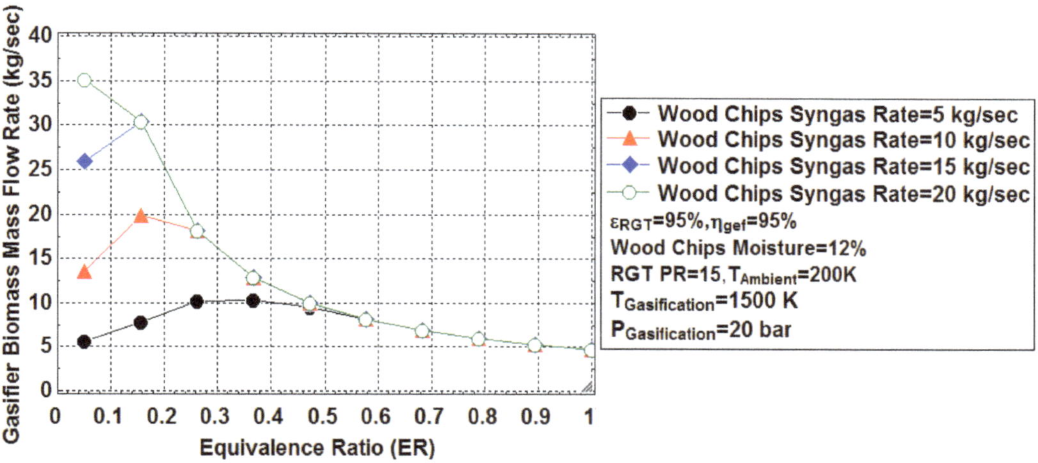

Figure 2. Variation of the equivalence ratio with gasifier's biomass mass flow rate at different Wood Chip syngas rates.

Important to note that increasing the biomass content leads to a sharp decrease in the gasifier's efficiency. The simulation results observed that an increase in the syngas rate revealed an increase in the required biomass rate for gasification. However, the quality of the produced gas depends mainly on the amount of qualified air to achieve the high energy content of the gasification products [63]. Figure 3 displays the relationship between biomass moisture content and the RGT thermal efficiency at different gasification pressures (5–20 bar). As observed, the increase in the biomass moisture content leads to a sharp decrease in the RGT thermal efficiency.

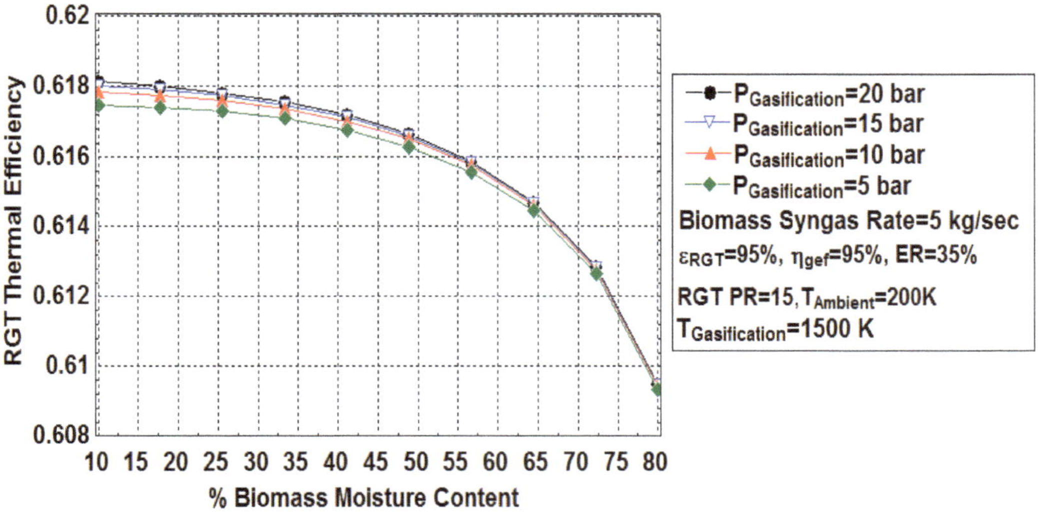

Figure 3. Variation of biomass moisture content with RGT thermal efficiency at different gasification pressures.

At 200 K and a gasification temperature of 1500 K, the RGT thermal efficiency revealed a higher value of 62% at the lowest biomass moisture content of 10% and the lowest value of 61% at higher rates of moistures. The RGT thermal efficiency remains high at higher gasification pressure. Higher values of biomass moisture content led to the low energy content of the syngas, which yielded low values of RGT thermal efficiency. As the amount of moisture content decreases, higher amounts of water have to be evaporated [64]. Since the heat needed to vaporize this water is transferred from the syngas constituents at the gasifier outlet, the temperature's products decrease, thus lowering the energy content and the syngas' physical and chemical exergy [64].

Lowering the moisture content of the biomass entering the gasifier raises the efficiencies of both the RGT and the Gasifier Unit. Further removal of the moisture will increase the chemical exergy of the product gas, and the physical exergy will be slightly increased [64]. The exergetic efficiency of the gasification process [17] decreases with the moisture content increase. In addition, with the increase in the moisture content, the CO_2 and H_2O production rate increases [65]. The increase in the gasification pressure has influenced the RGT thermal efficiency and the syngas LHV due to enhancement of the gasification reactions and the growth of the major constituents of the syngas product. Figure 4 shows the variation of the equivalence ratio (ER) with syngas lower heating value at different wood chips syngas rates (5–25 kg/s), at gasification pressure of 20 bar, the temperature of 1500 K, and moisture content of 12%. The profile exhibited an increase in the syngas LHV with the ER, till an optimum amount of the ER, subsequently the further increase in the ER has led to slow down the syngas LHV.

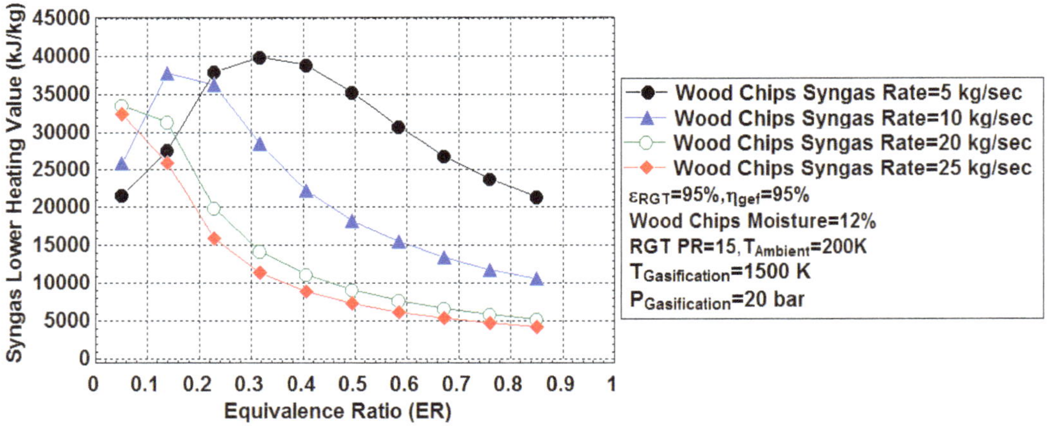

Figure 4. Equivalence ratio versus syngas lower heating value at different Wood Chip syngas rates.

The results revealed that the syngas lower heating value remains high at lower produced syngas rates. This is attributed to the increase in the raw biomass materials, which influences increasing the irreversibilities sharply inside the gasifier, which causes a reduction in the gasifier efficiency. The calorific value of the produced syngas depends on the heating value of the consumed biomass [11]. The higher the biomass heating value, the higher the calorific value of the produced syngas. Tar content in the gasification product is always higher for fuel richer with ash [66]. The results observed that the amount of consumed biomass plays a significant role in the final syngas energy content.

The gradual increase in the biomass should have a certain limit since a small amount of the required air will result in incomplete gasification of the biomass and produce less syngas energy content. The effect of the equivalence ratio (ER) on the RGT thermal efficiency at different gasification temperatures is presented in Figure 5. As observed, the simulation results revealed an optimum ER in which a higher RGT thermal efficiency occurred; thereafter, the additional amount of ER led to a decline in the thermal efficiency of the RGT Unit gradually. ER growths at low values have a significant effect on the performance of the RGT and lead to an increase in the RGT thermal efficiency. The low, moderate, and optimum ERs produce a high amount of the required syngas constituents, which revealed high syngas energy content and improved efficiency.

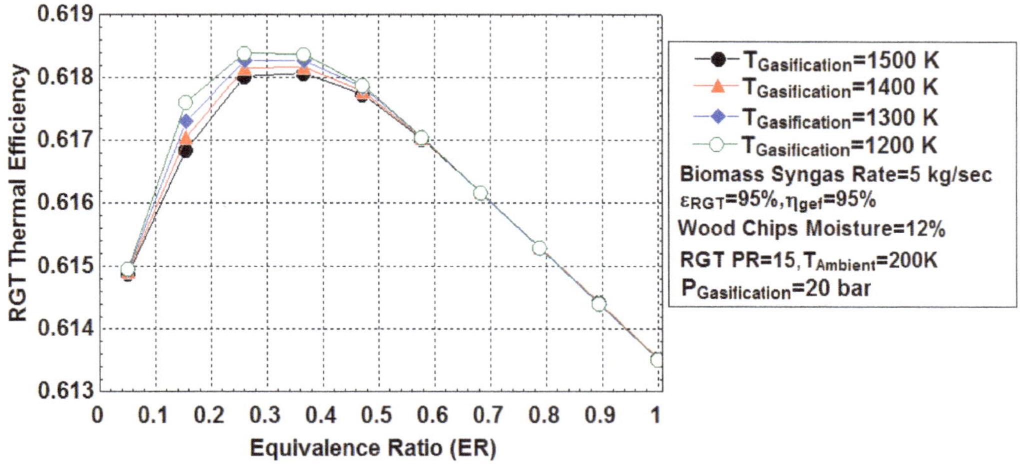

Figure 5. Effect of the equivalence ratio on the RGT thermal efficiency for different gasification temperatures.

To some extent, the reduction in the temperature promotes the growth of CO and H_2 concentrations and the decrease in the CH_4 concentration in the syngas [67]. The increase in the temperature results in primary and secondary water–gas shift reactions, secondary cracking, and reforming of heavy hydrocarbons activity. Thus this results in increasing the concentration of the H_2 in the syngas [54,68]. The activity of water gas shift reaction and Boudouard reactions will significantly increase due to temperature growth.

Consequently, the carbon reacts with CO and H_2O vapor produces a lot of CO. However, the further temperature growth promotes combustion reactions and thus decreases of CH_4 amount at the syngas [54,68] final products. Figure 6 presents the variation of the equivalence ratio (ER) and the percentage mole fractions of the syngas at a gasification temperature of 1500 K, pressure of 20 bar, regenerator effectiveness of 95%, gasifier efficiency of 95%, moisture content of 12%, and a produced amount of 5 kg/s syngas. The influence of the significant parameter, ER, is displayed and observed.

Additionally, the RGT thermal efficiency remains high at lower gasification temperatures. The optimum ER can be regarded as the point of the optimal design of the system to identify the amount of the required air to produce qualified syngas constituents and achieve a higher RGT thermal efficiency. Many reasons promote the temperature decrease for the gasification process, such as the ash, moisture, and the conditions of the gasification reactions.

The simulation results verified an increase in the syngas mole fractions amid increasing the ER to an optimum ER amount. The trend predicted a slowdown among the further increase if the ER. The moisture content of 12% is considered low and was reflected in the composition of the final syngas product. Beyond the optimal point of the ER, a massive decrease in the syngas energy content and the growth of the oxidative combustion reactions are occurring. The N_2 and H_2O vapor formation increase with the increase in the ER, while the formation of CO_2, CH_4, H_2, and CO exhibited decreases upon increasing the ER. At the optimum ER, the H_2 mole fraction depicted a value of 1.25%, 0.85% of CO, and 10.50% of CH_4, for a lower heating value of 38 MJ/kg syngas. The highest value of the syngas composition is depicted by N_2 since air is the gasification medium. Although a higher syngas energy content is observed, the amount of N_2 in the produced gas could be reduced by using steam or oxygen as oxidizing agents [11].

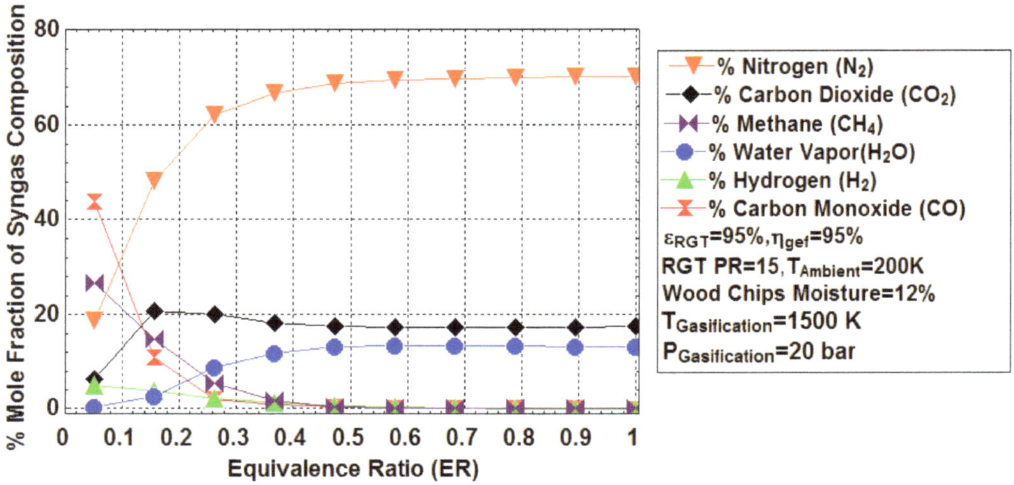

Figure 6. Influence of rquivalence ratio on the composition of the syngas.

The influence of the biomass moisture content on the RGT actual gasification's airflow rate can be observed in Figure 7. It is shown that the gasification air entered into the gasifier decreases amid the increase in the biomass moisture content. The biomass moisture content affects the quality of the consumed biomass. At a constant ER and syngas rate, increases in moisture content led to a decrease in the energy content of the syngas and an increase in the amount and the demand of the biomass flow rates, thus influencing a lower demand for the required air for the complete gasification process.

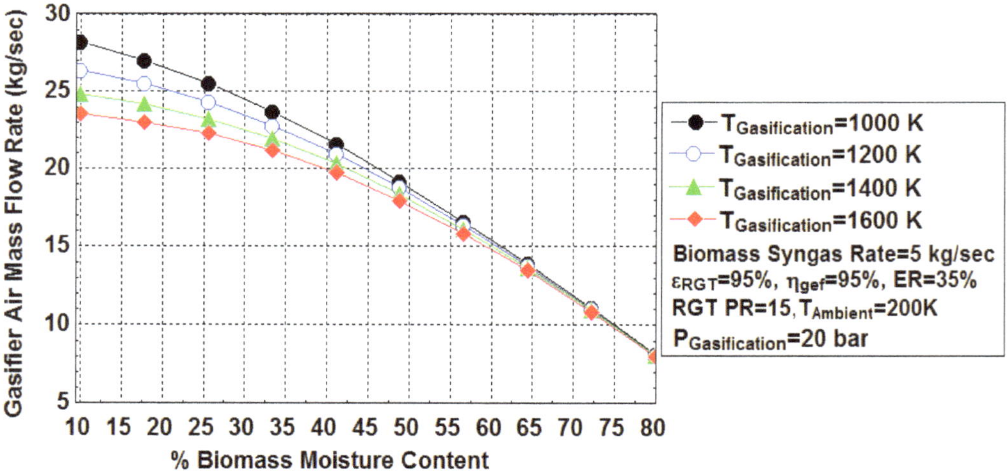

Figure 7. Effect of biomass moisture content on the gasifier's airflow rate at different gasification temperatures.

Further reduction in the gasification's air amid increasing the biomass moisture content will diminish the gasification process and produce an unqualified syngas final product. In addition, the results demonstrate that increasing the gasification temperature has slowed down the amount of the actual air required for gasification. Literature data revealed that, as the initial moisture content increases, the gasification temperature slows down due to changes in the syngas composition and the effect of the water–gas shift, Boudouard, and combustion reactions [69–71]. Moreover, the influence of the gasification pressure resulted in an increase in the amount of the actual air required for gasification (see Figure 8) since the biomass flow rate is higher at a higher gasification pressure. With air used as a gasification medium, it was observed that the increase in the gasification pressure led to an acceleration of the reaction rate and the product's conversion of the gasification reactions. Figure 9 shows the influence of the gasifier efficiency on the biomass flow rate for the gasification process. At different syngas rates (3–10 kg/s), an optimum ER of 35%, gasification temperature of 1500 K, pressure of 20 bar, and moisture content of 12%, the results predicted that the Acacia Nilotica Wood Chip biomass flow rate required for a complete gasification process decreases amid increases in gasifier efficiency. In addition, the demand for a higher biomass flow rate is attained at higher Wood Chip syngas rates. The irreversibilities of the gasifier are attributed to various factors during the gasification process, such as the quantity of the biomass flow rate, friction losses, pressure drops, releases of the high amount of gases, water corrosion, scale formation, the fusion of metals, tar and char formations, etc. In addition, results show that there is an optimum gasifier efficiency, in which the assigned biomass flow rate should be identified to minimize the possible source of irreversibilities at the gasification process.

As plotted in Figure 10, at different gasification temperatures (1000–1600 K), pressure of 20 bar, biomass moisture content of 12%, and syngas rate of 5 kg/s, the ER shows a significant impact on the specific fuel consumption (SFC) of the RGT power plant. With ER growth at lower levels, the specific fuel consumption (SFC) of the RGT Unit sharply declines from the maximum value of 0.27 kg/kW·hr at an ER of 5% to the minimum value of 0.80 kg/kW·hr at an ER of 25% for the lowest gasification temperature of 1000 K. The results show that there is an optimum ER for each gasification temperature that ensures minimizing the SFC of the RGT power plant. Beyond the optimum ER, the trends predicted an inflection point, which exhibited an increase in the SFC. This is attributed to the excess amount of air released, resulting in more combustion reactions, changing the syngas composition, and slowing down the syngas' lower heating value. The SFC of the RGT power plant reached 0.30 kg/kW·h at the higher ER of 95%. In addition, the results show that the SFC remains low at lower gasification temperatures. The influence of the ER on the RGT power plant at different syngas rates is shown in Figure 11. The simulation model revealed that ER growths at lower levels significantly affect the RGT power. The optimum ER delivered the highest RGT power, after which the syngas composition and combustion reaction will diminish the power due to a decrease in the LHV of the syngas fuel. At an ER of 5%, maximum power of 400 MW is reached at Acacia Nilotica Wood Chips' syngas rate of 20 kg/s, whereas minimum power is reached at 50 MW for a syngas rate of 5 kg/s.

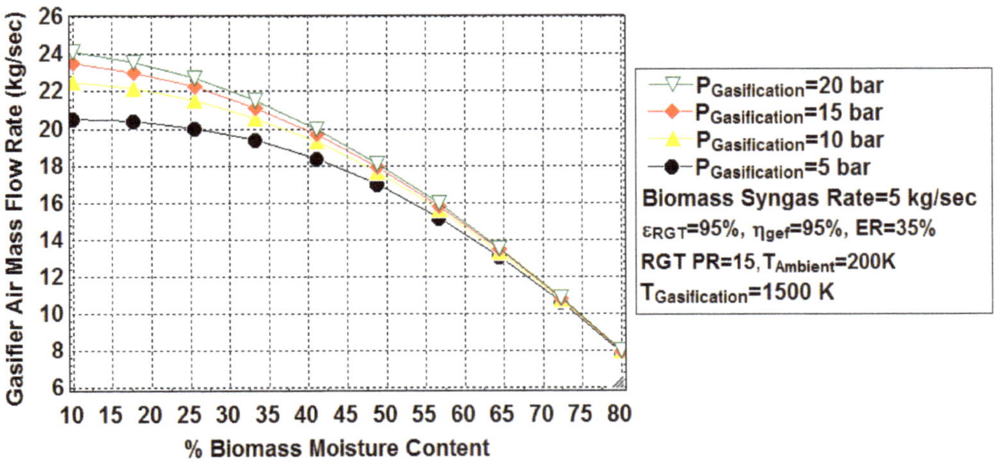

Figure 8. Variation of biomass moisture content with the gasifier's airflow rate at different gasification pressures.

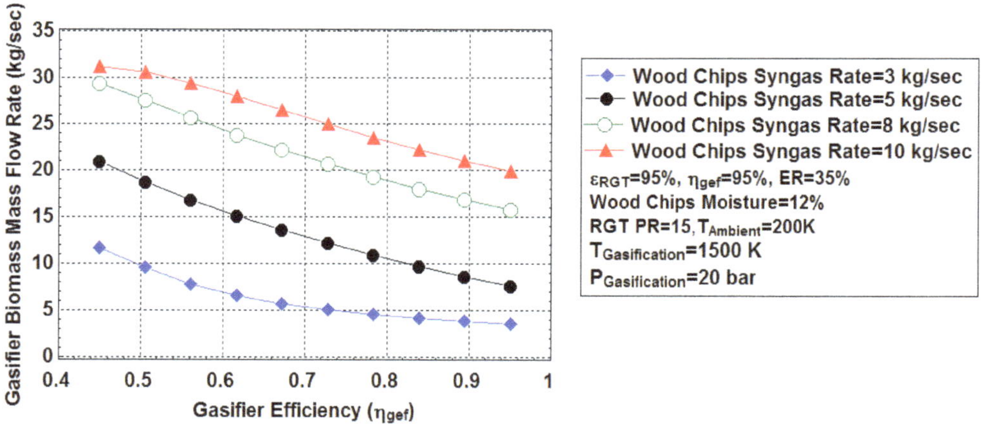

Figure 9. Variation of gasifier efficiency with biomass flow rate at different Wood Chip syngas rates.

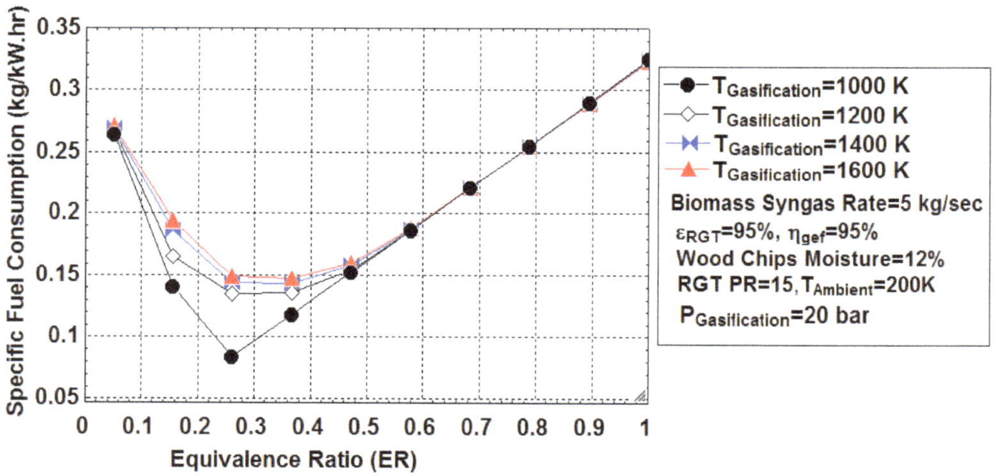

Figure 10. Variation of equivalence ratio with the specific fuel consumption at different gasification temperatures.

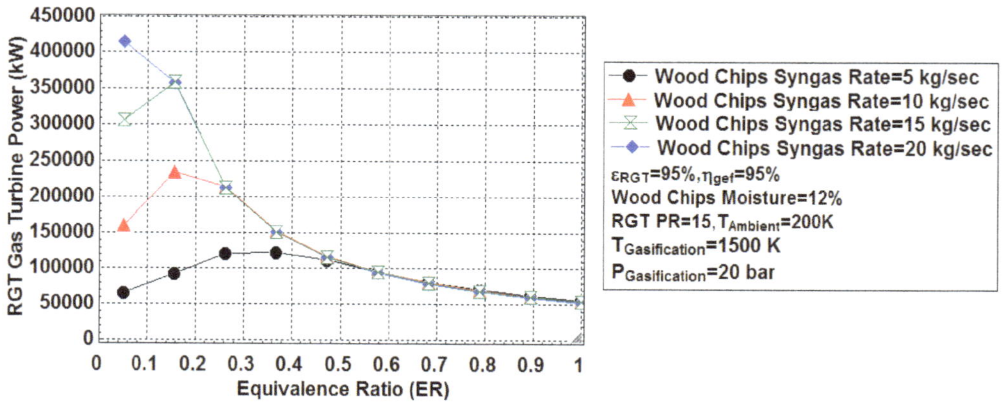

Figure 11. Effect of the rquivalence ratio on the RGT Power at different Wood Chip syngas rates.

5. Conclusions

This work reviewed and investigated the integration of the Biomass Gasifier Unit with a Regenerative Gas Turbine power plant, including the effect of various parameters. Simulation models were used to characterize the gasification process of Acacia Nilotica Wood Chip biomass in Sudan. This type of biomass (Acacia Nilotica Wood) is of great interest due to its wide domestic uses and agricultural lands. A parametric analysis of the released syngas composition, actual gasifier air, temperature, pressure, LHV, moisture content, equivalence ratio, gasifier efficiency, thermal efficiency, power, and specific fuel consumption of the regenerative gas turbine power plant was carefully investigated to identify the optimal design points of the gasifier system and the working conditions of the RGT power unit using this type of biomass. With an average syngas LHV of 30,000 MJ/kg, the results revealed that such a type of biomass (Acacia Nilotica Wood Chips) could achieve high thermal efficiency and a valuable energy-saving process with the Regenerative Gas Turbine Unit.

Author Contributions: Conceptualization, M.E.A. and S.R.N.; methodology, M.E.A.; software, S.A.A.T.; validation, S.A.A.; formal analysis, M.E.A.; investigation, S.A.A.; resources, W.-H.C.; data curation, S.A.A.; writing—original draft preparation, M.E.A.; writing—review and editing, W.-H.C.; visualization, S.A.A.T.; supervision, S.R.N.; project administration, W.-H.C.; funding acquisition, W.-H.C. All authors have read and agreed to the published version of the manuscript.

Funding: This research was funded by IsdB-TWAS, grant number "506798/2019" under the scheme of the Postdoctoral Fellowship Program.

Institutional Review Board Statement: Not applicable.

Informed Consent Statement: Not applicable.

Data Availability Statement: The data used to support the findings of this study are included within the article.

Acknowledgments: The authors greatly acknowledge the technical support of the Chemical Engineering Department and the Energy Research Centre of the University of Khartoum, Faculty of Engineering, for providing the laboratory and facilitating the work. The financial, technical, and academic supervisions of the World Academy of Science (TWAS) and the Islamic Development Bank (IsDB) under the scheme of the Postdoctoral Fellowship program at the National University of Sciences & Technology, Pakistan, are greatly appreciated. The authors also acknowledge the financial support of the Ministry of Science and Technology, Taiwan, under the grant numbers MOST 109-2221-E-006-040-MY3, MOST 110-2622-E-006-001-CC1, and MOST 110-3116-F-006-003- for this research. This research is also supported in part by the Higher Education Sprout Project, Ministry of Education at the Headquarters of the University Advancement at National Cheng Kung University (NCKU).

Conflicts of Interest: The authors proclaimed that there is no conflict of interest for this work.

Nomenclature

T	Temperature	(K)
S	Entropy	(kJ/kg·K)
P	Pressure	(kPa)
r_P	Compression Ratio	-
γ	Specific Heat Ratio	-
η_C	Isentropic Compressor Efficiency	-
η_{gef}	Gasifier Efficiency	-
T_S	Compressor Isentropic Temperature	(K)
$\dot{W}_C$	Specific Compressor Work	(MW)
$\dot{m}_a$	Air Mass	(kgair)
$\dot{m}_f$	Fuel Mass	(kg·fuel)
$\dot{m}_g$	Gas Mass	(kg·gas)

M_{th}	Gasifier Stoichiometric Air Flow Rate	(kg air/kg dry fuel)
CC	Combustion Chamber	-
ER	Equivalence Ratio	-
RGT	Regenerative Gas Turbine	-
IBG	Integrated Biomass Gasification	-
Mech	Mechanical	-
Gen	Generator	-
EES	Engineering Equation Solver	-
T_x	Combustor Inlet Temperature	(K)
$\Delta P_{C,C}$	Combustor Pressure Drop	(bar)
$\eta_{C,C}$	Combustor Efficiency	-
ε	Regenerator Effectiveness	-
$\dot{W}_{GT}$	Turbine Shaft Work	(MW)
η_T	Turbine Efficiency	-
y_i	Syngas Mole Fraction	-
P_{GT}	GT Power	(MW)
$\dot{Q}_{add}$	Heat Supplied	(kW)
Cp_a	Heat Capacity of Air	(kJ/kg·K)
Cp_f	Heat Capacity of Fuel	(kJ/kg·K)
Cp_g	Heat Capacity of Flue Gas	(kJ/kg·K)
M_{fa}	Gasifier Actual Air Flow Rate	(kg air)
ATM	Atmospheric	-
HHV	Higher Heating Value	(kJ/kg)
LHV	Lower HeatingValue	(kJ/kg)
TIT	Turbine Inlet Temperature	(K)
ASH	Ash Content	(wt%)
HR	Heat Rate	(MW)
SFC	Specific Fuel Consumption	(kg/kW·hr)

References

1. Mahgoub, F. *Current Status of Agriculture and Future Challenges in Sudan*; Nordiska Afrikainstitutet: Uppsala, Sweden, 2014.
2. Naqvi, S.R.; Naqvi, M.; Taqvi, S.A.A.; Iqbal, F.; Inayat, A.; Khoja, A.H.; Mehran, M.T.; Ayoub, M.; Shahbaz, M.; Amin, N.A.S. Agro-Industrial residue gasification feasibility in captive power plants: A South-Asian case study. *Energy* **2021**, *214*, 118952. [CrossRef]
3. Seitarides, T.; Athanasiou, C.; Zabaniotou, A. Modular biomass gasification-based solid oxide fuel cells (SOFC) for sustainable development. *Renew. Sustain. Energy Rev.* **2008**, *12*, 1251–1276. [CrossRef]
4. Sanjay, Y.; Singh, O.; Prasad, B. Energy and exergy analysis of steam cooled reheat gas–steam combined cycle. *Appl. Therm. Eng.* **2007**, *27*, 2779–2790. [CrossRef]
5. Ioannidou, O.; Zabaniotou, A. Agricultural residues as precursors for activated carbon production—A review. *Renew. Sustain. Energy Rev.* **2007**, *11*, 1966–2005. [CrossRef]
6. Chandra, H.; Arora, A.; Kaushik, S.C.; Tripathi, A.; Rai, A. Thermodynamic analysis and parametric study of an intercooled–reheat closed-cycle gas turbine on the basis of a new isentropic exponent. *Int. J. Sustain. Energy* **2011**, *30*, 82–97. [CrossRef]
7. Parvez, M.; Khalid, F. Thermodynamic investigation on sawdust and rice husk biomass integrated gasification for combined power and ejector cooling cycle. *Curr. Altern. Energy* **2017**, *2*, 19–26. [CrossRef]
8. Rashid, T.; Taqvi, S.A.A.; Sher, F.; Rubab, S.; Thanabalan, M.; Bilal, M.; Islam, B. Enhanced lignin extraction and optimisation from oil palm biomass using neural network modelling. *Fuel* **2021**, *293*, 120485. [CrossRef]
9. Beig, B.; Riaz, M.; Naqvi, S.R.; Hassan, M.; Zheng, Z.; Karimi, K.; Pugazhendhi, A.; Atabani, A.E.; Chi, N.T.L. Current challenges and innovative developments in pretreatment of lignocellulosic residues for biofuel production: A review. *Fuel* **2021**, *287*, 119670. [CrossRef]
10. Shahbaz, M.; Al-Ansari, T.; Aslam, M.; Khan, Z.; Inayat, A.; Athar, M.; Naqvi, S.R.; Ahmed, M.A.; McKay, G. A state of the art review on biomass processing and conversion technologies to produce hydrogen and its recovery via membrane separation. *Int. J. Hydrog. Energy* **2020**, *45*, 15166–15195. [CrossRef]
11. Ibrahim, T.K.; Rahman, M. Thermal impact of operating conditions on the performance of a combined cycle gas turbine. *J. Appl. Res. Technol.* **2012**, *10*, 567–577. [CrossRef]
12. Littlewood, K. Gasification: Theory and application. *Prog. Energy Combust. Sci.* **1977**, *3*, 35–71. [CrossRef]
13. Naqvi, S.R.; Taqvi, S.A.A.; Mehran, T.; Khoja, A.H.; Naqvi, M.; Bokhari, A.; Amin, N.A.S. Catalytic pyrolysis of biomass using shape-selective zeolites for bio-oil enhancement. In *Bioenergy Resources and Technologies*; Elsevier: Amsterdam, The Netherlands, 2021; pp. 39–60.
14. French, R.; Czernik, S. Catalytic pyrolysis of biomass for biofuels production. *Fuel Process. Technol.* **2010**, *91*, 25–32. [CrossRef]

15. Garcia-Perez, M.; Wang, X.S.; Shen, J.; Rhodes, M.J.; Tian, F.; Lee, W.J.; Wu, H.; Li, C.-Z. Fast pyrolysis of oil mallee woody biomass: Effect of temperature on the yield and quality of pyrolysis products. *Ind. Eng. Chem. Res.* **2008**, *47*, 1846–1854. [CrossRef]
16. Elliott, C.D.; Neuenschwander, G.G.; Hart, T.R. Hydroprocessing bio-oil and products separation for coke production. *Acs Sustain. Chem. Eng.* **2013**, *1*, 389–392. [CrossRef]
17. Basu, P. *Biomass Gasification and Pyrolysis: Practical Design and Theory*; Academic Press: Cambridge, MA, USA, 2010.
18. Kollmann, F.F.; Côté, W.A., Jr. *Principles of Wood Science and Technology. Volume I Solid Wood*; Springer: Cham, Switzerland, 1968.
19. Naqvi, S.R.; Tariq, R.; Hameed, Z.; Ali, I.; Naqvi, M.; Chen, W.-H.; Ceylan, S.; Rashid, H.; Ahmad, J.; Taqvi, S.A.; et al. Pyrolysis of high ash sewage sludge: Kinetics and thermodynamic analysis using Coats-Redfern method. *Renew. Energy* **2019**, *131*, 854–860. [CrossRef]
20. Naqvi, S.R.; Jamshaid, S.; Naqvi, M.; Farooq, W.; Niazi, M.B.K.; Aman, Z.; Zubair, M.; Ali, M.; Shahbaz, M.; Inayat, A.; et al. Potential of biomass for bioenergy in Pakistan based on present case and future perspectives. *Renew. Sustain. Energy Rev.* **2018**, *81*, 1247–1258. [CrossRef]
21. Naqvi, S.R.; Hameed, Z.; Tariq, R.; Taqvi, S.A.; Ali, I.; Niazi, M.B.K.; Noor, T.; Hussain, A.; Iqbal, N.; Shahbaz, M. Synergistic effect on co-pyrolysis of rice husk and sewage sludge by thermal behavior, kinetics, thermodynamic parameters and artificial neural network. *Waste Manag.* **2019**, *85*, 131–140. [CrossRef]
22. Khanmohammadi., S.; Atashkari, K.; Kamali, R.K. Performance assessment and multi-objective optimization of a trrigeneration system with a modified biomass gasification model. *Modares Mech. Eng.* **2015**, *15*, 209–222.
23. Hoque, M.E.; Rashid, F. Gasification Process Using Downdraft Fixed-Bed Gasifier for Different Feedstock. In *Gasification*; IntechOpen: London, UK, 2021.
24. Kumar, R.M.D.; Anand, R. Production of biofuel from biomass downdraft gasification and its applications. In *Advanced Biofuels*; Elsevier: Amsterdam, The Netherlands, 2019; pp. 129–151.
25. Kamble, A.D.; Saxena, V.K.; Chavan, P.D.; Mendhe, V.A. Co-Gasification of coal and biomass an emerging clean energy technology: Status and prospects of development in Indian context. *Int. J. Min. Sci. Technol.* **2019**, *29*, 171–186. [CrossRef]
26. Hameed, Z.; Hameed, Z.; Naqvi, S.R.; Naqvi, M.; Ali, I.; Taqvi, S.A.A.; Gao, N.; Hussain, S.A.; Hussain, S. A comprehensive review on thermal coconversion of biomass, sludge, coal, and their blends using thermogravimetric analysis. *J. Chem.* **2020**, *2020*, 5024369. [CrossRef]
27. Shahbaz, M.; Taqvi, S.A.; Loy, A.C.M.; Inayat, A.; Uddin, F.; Bokhari, A.; Naqvi, S.R. Artificial neural network approach for the steam gasification of palm oil waste using bottom ash and CaO. *Renew. Energy* **2019**, *132*, 243–254. [CrossRef]
28. Dayton, D. *Review of the Literature on Catalytic Biomass Tar Destruction: Milestone Completion Report*; National Renewable Energy Laboratory: Golden, CO, USA, 2002.
29. Knoef, H. *Inventory of Biomass Gasifier Manufacturers and Installations, Final Report to European Commission*; Contract DIS/1734/98-NL; Biomass Technology Group BV, University of Twente: Enschede, The Netherlands, 2000.
30. Saxena, A.N.; Panwar, A.; Mishra, S.; Shukla, A.K.; Sahu, M.K. Parametric Investigation of Basic and Evaporative Inlet Air-Cooled Gas Turbine Cycle. In *Advances in Engineering Design*; Springer: Singapore, 2021; pp. 751–768.
31. Mohapatra, A.K.; Sanjay, L.P. Parametric analysis of cooled gas turbine cycle with evaporative inlet air cooling. *Int. J. Sci. Eng. Res.* **2012**, *3*, h-2012.
32. Farouk, N.; Sheng, L. Effect of Fuel Types on the Performance of Gas Turbines. *Int. J. Comput. Sci. Issues (IJCSI)* **2013**, *10*, 436.
33. Mohapatra, A.K. Thermodynamic assessment of impact of inlet air cooling techniques on gas turbine and combined cycle performance. *Energy* **2014**, *68*, 191–203. [CrossRef]
34. ISO. *Natural Gas–Calculation of Calorific Values, Density, Relative Density and Wobbe Index from Composition*; ISO: Geneve, Switzerland, 2005.
35. Nag, P. *Engineering Thermodynamics*; Tata McGraw-Hill Education: New York, NY, USA, 2013.
36. Naqvi, S.R.; Tariq, R.; Hameed, Z.; Ali, I.; Taqvi, S.A.; Naqvi, M.; Niazi, M.B.K.; Noor, T.; Farooq, W. Pyrolysis of high-ash sewage sludge: Thermo-kinetic study using TGA and artificial neural networks. *Fuel* **2018**, *233*, 529–538. [CrossRef]
37. Rahman, M.; Ibrahim, T.K.; Kadirgama, K.; Mamat, R.; Bakar, R.A. Influence of operation conditions and ambient temperature on performance of gas turbine power plant. In *Advanced Materials Research*; Trans Tech Publications: Zurich, Switzerland, 2011.
38. Gill, P.W.; Smith, J.H. *Fundamentals of Internal Combustion Engines*; Oxford and IBH Publishing Company: Delhi, India, 1972.
39. Taylor, C.F. *The Internal-Combustion Engine in Theory and Practice: Combustion, Fuels, Materials, Design*; MIT Press: Cambridge, MA, USA, 1985; Volume 2.
40. Johnke, T.; Mast, M. Power boosters—Technologies to enhance gas turbine power output on demand. *Siemens Power Gener. Siemens Power J.* **2002**. Available online: https://www.cleanenergy.co.th/Article%20Boosting%20GT%20PowerN.html (accessed on 5 July 2021).
41. Omar, H.; Kamel, A.; Alsanousi, M. Performance of Regenerative Gas Turbine Power Plant. *Energy Power Eng.* **2017**, *9*, 136. [CrossRef]
42. Rh, P.; Green, D. *Perry's Chemical Engineers' Handbook*; McGraw-Hill: New York, NY, USA, 1997.
43. La Villetta, M.; Costa, M.; Massarotti, N. Modelling approaches to biomass gasification: A review with emphasis on the stoichiometric method. *Renew. Sustain. Energy Rev.* **2017**, *74*, 71–88. [CrossRef]
44. Ptasinski, K.J.; Prins, M.J.; Pierik, A. Exergetic evaluation of biomass gasification. *Energy* **2007**, *32*, 568–574. [CrossRef]
45. Abdalla, M.E.; Pannir, S.; Hussein Ahmed, A.; Mahyob, A.; Abdalla, S.A. *Regenerative Gas Turbine Power Plant: Performance & Evaluation*; Purdue University: West Lafayette, IN, USA, 2021.

46. Jones, R.; Goldmeer, J.; Monetti, B. *Addressing Gas Turbine Fuel Flexibility*; GE Energy Report GER4601 rev. B; SE Energy: Boston, MA, USA, 2011; Volume 29, p. 2012.
47. Wei, L.; Xu, S.; Zhang, L.; Liu, C.; Zhu, H.; Liu, S. Steam gasification of biomass for hydrogen-rich gas in a free-fall reactor. *Int. J. Hydrog. Energy* **2007**, *32*, 24–31. [CrossRef]
48. Erlich, C.; Fransson, T.H. Downdraft gasification of pellets made of wood, palm-oil residues respective bagasse: Experimental study. *Appl. Energy* **2011**, *88*, 899–908. [CrossRef]
49. Saleem, M.; Ali, M.; Siddiqi, Z.; Al Qahtani, A. Preparation of activated carbon from acacia (Vachellia seyal) Tree Branches and application to treat wastewater containing methylene blue dye. *Mod. Appl. Sci.* **2017**, *11*, 102–108. [CrossRef]
50. Khanmohammadi, S.; Atashkari, K.; Kouhikamali, R. Exergoeconomic multi-objective optimization of an externally fired gas turbine integrated with a biomass gasifier. *Appl. Therm. Eng.* **2015**, *91*, 848–859. [CrossRef]
51. Karimi, M.H.; Chitgar, N.; Emadi, M.A.; Ahmadi, P.; Rosend, M.A. Performance assessment and optimization of a biomass-based solid oxide fuel cell and micro gas turbine system integrated with an organic Rankine cycle. *Int. J. Hydrogen Energy* **2020**, *45*, 6262–6277. [CrossRef]
52. Javidmehr, M.; Joda, F.; Mohammadi, A. Thermodynamic and economic analyses and optimization of a multi-generation system composed by a compressed air storage, solar dish collector, micro gas turbine, organic Rankine cycle, and desalination system. *Energy Convers. Manag.* **2018**, *168*, 467–481. [CrossRef]
53. Behzadi, A.; Houshfar, E.; Gholamian, E.; Ashjaee, M.; Habibollahzade, A. Multi-Criteria optimization and comparative performance analysis of a power plant fed by municipal solid waste using a gasifier or digester. *Energy Convers. Manag.* **2018**, *171*, 863–878. [CrossRef]
54. Olgun, H.; Ozdogan, S.; Yinesor, G. Results with a bench scale downdraft biomass gasifier for agricultural and forestry residues. *Biomass Bioenergy* **2011**, *35*, 572–580. [CrossRef]
55. Rajput, R. *A Textbook of Power Plant Engineering*; Laxmi Publications: New Delhi, India, 2005.
56. Pode, R.; Diouf, B.; Pode, G. Sustainable rural electrification using rice husk biomass energy: A case study of Cambodia. *Renew. Sustain. Energy Rev.* **2015**, *44*, 530–542. [CrossRef]
57. Moran, M.J.; Shapiro, H.N.; Boettner, D.D.; Bailey, M.B. *Fundamentals of Engineering Thermodynamics*; John Wiley & Sons: Hoboke, NJ, USA, 2010.
58. Rahman, M.; Ibrahim, T.K.; Abdalla, A.N. Thermodynamic performance analysis of gas-turbine power-plant. *Int. J. Phys. Sci.* **2011**, *6*, 3539–3550.
59. Saravanamuttoo, H.I.; Rogers, G.F.C.; Cohen, H. *Gas Turbine Theory*; Pearson Education: London, UK, 2009.
60. Shi, H.; Si, W.; Li, X. The concept, design and performance of a novel rotary kiln type air-staged biomass gasifier. *Energies* **2016**, *9*, 67. [CrossRef]
61. Alauddin, Z.A.Z. Performance and Characteristics of a Biomass Gasifier System. Ph.D. Thesis, University of Wales, Cardiff, UK, 1996.
62. McKendry, P. Energy production from biomass (part 3): Gasification technologies. *Bioresour. Technol.* **2002**, *83*, 55–63. [CrossRef]
63. Bridgwater, A.V. *Progress in Thermochemical Biomass Conversion*; John Wiley & Sons: Hoboke, NJ, USA, 2008.
64. Kirsanovs, V.; Blumberga, D.; Veidenbergs, I.; Rochas, C.; Vigants, E.; Vigants, G. Experimental investigation of downdraft gasifier at various conditions. *Energy Procedia* **2017**, *128*, 332–338. [CrossRef]
65. Taba, L.E.; Irfan, M.F.; Wan Daud, W.A.M.; Chakrabarti, M.H. The effect of temperature on various parameters in coal, biomass and CO-gasification: A review. *Renew. Sustain. Energy Rev.* **2012**, *16*, 5584–5596. [CrossRef]
66. El-Emam, R.S.; Dincer, I. Thermal modeling and efficiency assessment of an integrated biomass gasification and solid oxide fuel cell system. *Int. J. Hydrogen Energy* **2015**, *40*, 7694–7706. [CrossRef]
67. Zhang, X.; Li, H.; Liu, L.; Bai, C.; Wang, S.; Zeng, J.; Liu, X.; Li, N.; Zhang, G. Thermodynamic and economic analysis of biomass partial gasification process. *Appl. Therm. Eng.* **2018**, *129*, 410–420. [CrossRef]
68. Jarungthammachote, S.; Dutta, A. Thermodynamic equilibrium model and second law analysis of a downdraft waste gasifier. *Energy* **2007**, *32*, 1660–1669. [CrossRef]
69. Das, B.K.; Hoque, S. Assessment of the potential of biomass gasification for electricity generation in Bangladesh. *J. Renew. Energy* **2014**, *2014*, 429518. [CrossRef]
70. Chen, W.-H.; Chen, C.-J.; Hung, C.-I.; Shen, C.-H.; Hsu, H.-W. A comparison of gasification phenomena among raw biomass, torrefied biomass and coal in an entrained-flow reactor. *Appl. Energy* **2013**, *112*, 421–430. [CrossRef]
71. Arpia, A.A.; Nguyen, T.-B.; Chen, W.-C.; Dong, C.-D.; Ok, Y.S. Microwave-assisted gasification of biomass for sustainable and energy-efficient biohydrogen and biosyngas production: A state-of-the-art review. *Chemosphere* **2022**, *287*, 132014. [CrossRef] [PubMed]

energies

MDPI

Review

Electric Vehicles: V2G for Rapid, Safe, and Green EV Penetration

Abdullah Dik *, Siddig Omer and Rabah Boukhanouf

Department of Architecture and Built Environment, Faculty of Engineering, The University of Nottingham, Nottingham NG7 2RD, UK; siddig.omer@nottingham.ac.uk (S.O.); rabah.boukhanouf@nottingham.ac.uk (R.B.)
* Correspondence: abdullah.dik@nottingham.ac.uk

Abstract: Low carbon and renewable energy sources (RESs) are fast becoming a key sustainable instrument in meeting the global growth of electricity demand while curbing carbon emissions. For example, the gradual displacement of fossil-fuelled vehicles with electrically driven counterparts will inevitably increase both the power grid baseload and peak demand. In many developed countries, the electrification process of the transport sector has already started in tandem with the installation of multi-GW renewable energy capacity, particularly wind and solar, huge investment in power storage technology, and end-user energy demand management. The expansion of the Electric Vehicle (EV) market presents a new opportunity to create a cleaner and transformative new energy carrier. For instance, a managed EV battery charging and discharging profile in conjunction with the national grid, known as the Vehicle-to-Grid system (V2G), is projected to be an important mechanism in reducing the impact of renewable energy intermittency. This paper presents an extensive literature review of the current status of EVs and allied interface technology with the power grid. The main findings and statistical details are drawn from up-to-date publications highlighting the latest technological advancements, limitations, and potential future market development. The authors believe that electric vehicle technology will bring huge technological innovation to the energy market where the vehicle will serve both as a means of transport and a dynamic energy vector interfacing with the grid (V2G), buildings (V2B), and others (V2X).

Keywords: electric vehicle (EV); vehicle-to-grid (V2G); renewable energy source (RES); power grid; battery electric vehicle (BEV); plug-in hybrid electric vehicle (PHEV); EV charging

Citation: Dik, A.; Omer, S.; Boukhanouf, R. Electric Vehicles: V2G for Rapid, Safe, and Green EV Penetration. *Energies* **2022**, *15*, 803. https://doi.org/10.3390/en15030803

Academic Editor: Wei-Hsin Chen

Received: 29 December 2021
Accepted: 18 January 2022
Published: 22 January 2022

1. Introduction

Reliable and affordable energy has been the main driver of global economic development and, in turn, an enhanced standard of living. However, deriving most of this energy from fossil fuel sources has severely impacted the Earth, with global warming posing an existential threat. Current estimates indicate an ascending trajectory in global energy consumption from the current level of about 583.9 EJ [1]. Similarly, according to the International Energy Agency (IEA) under the stated policy scenarios in 2019, it is expected that the electricity demand will increase to 37,000 TWh in 2040, which is nearly 3 times higher than that of 2000 [2], as illustrated in Figure 1. The data presented in Figure 1 were provided by International Energy Agency (IEA) [3].

Presently, fossil-based energy sources remain the most widely used form of energy to meet the increasing energy demand. It is widely believed that this is unsustainable as it contributes to the depletion of fossil fuel reserves and increased concentration of greenhouse gases (GHG) in the atmosphere [1,4]. This has been identified as a direct precursor of global warming and the ensuing extreme global weather conditions [5–7].

The energy challenge presented in Figure 1 is enormous as sustaining the projected growth of electricity consumption competes with the need to decarbonise the energy system as a whole. RESs, particularly wind and solar, are being implemented in large-scale schemes as a direct replacement for fossil-fuelled power generation plants that will form the backbone of future power generation infrastructure [8]. In developed countries, Renewable Energy (RE) generation capacity has been increasing rapidly over the last few years. Figure 2 shows the increasing primary energy consumption from RESs between 1990 and 2019. The data presented in Figure 2 were provided by Our World in Data [9].

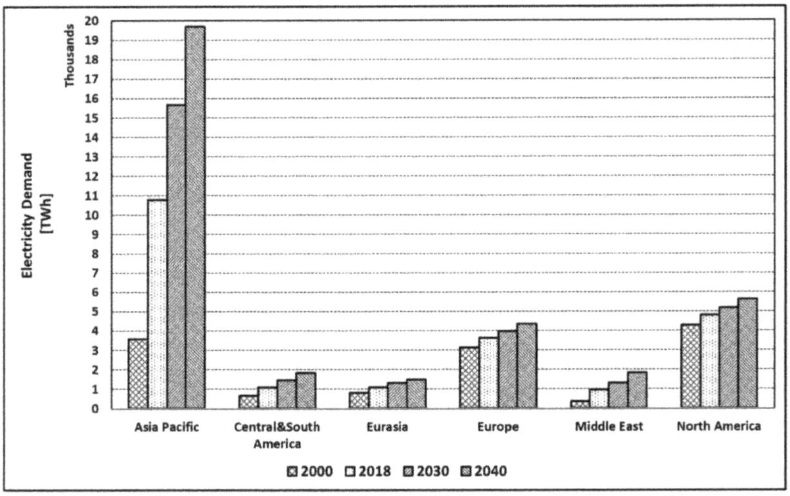

Figure 1. Projected global electricity demand between 2000 and 2040.

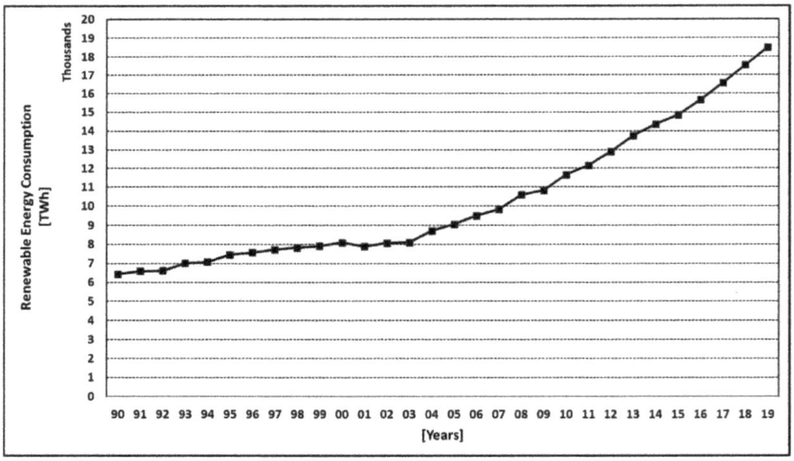

Figure 2. Global renewable energy consumption for primary energy.

For instance, in the last 5 years, the most dominant form of REs is solar and wind energy, which have grown annually by 27% and 13%, respectively [9]. Despite the progress made in many developed countries, renewables continue to account for a mere 5% of global primary energy consumption [1]. Today, the deployment of renewables at scale remains expensive and confined to countries with technical know-how and developed financial markets. In addition, the intermittency of RESs results in power quality and power flow problems, which require installation of fossil-fuelled spare power generation capacity, Demand Side Management (DSM), and Energy Storage Technologies (ESTs) [8,10–13].

Another vital sector of the economy that has presented huge challenges for decarbonise is transportation. According to IEA, in 2020, carbon emissions from transportation accounted for 24% of global emissions [14]. Currently, this is being addressed through a concerted research effort, investment in new manufacturing processes, and policy frameworks to develop EV technology to cut carbon emissions of the transportation sector as part of many developed countries commitments under successive climate change accords. The development of EVs is gathering pace in many countries, with the total number of light-duty EVs exceeding 10 million by the end of 2020 [15]. However, true decarbonisation is only possible with a fully decarbonised electricity grid, as the two sectors present a complementary synergy that can be developed into a holistic and effective decarbonisation solution [16]. V2G technology is a promising innovation providing a way to use EVs to store excess energy from RESs in the EV's battery pack and release it back to the grid during the peak hours of energy demand, maximising the efficiency of large-scale RE integration. Another future development, which is close to market commercialisation, is ancillary services (ASs) such as management systems of frequency and voltage regulation of the grid [17,18]. This is further strengthened by current data on usage of private cars, which show the daily road mileage of most vehicles in the United Kingdom (UK) is approximately 20 m and that the vehicles are in the park mode for about 90% of the daily hours [19,20].

It is in this context that this paper attempts to provide an extended overview of the current state of EVs' technological developments and market trends. The paper also examines the challenges to overcome, and the solutions being pursued by many world-leading manufacturers of EVs.

2. Overview of Electric Vehicle (EV) Technology

At the core of an EV powered by an onboard battery pack is an efficient electric motor and associated electrified powertrain. The powertrain consists of a traction inverter which optimises the motor torque, a battery management system which monitors the state of charge/discharge and health of the battery, and an onboard battery charger for recharging the battery from an AC source when the vehicles are parked.

Currently, there are four main types of EVs, namely Battery Electric Vehicles (BEVs), Plug-in Hybrid EVs (PHEVs), Hybrid EVs (HEVs), and Fuel Cell EVs (FCEVs). The configuration of the electric drivetrains of each type of these EVs is shown in Figure 3.

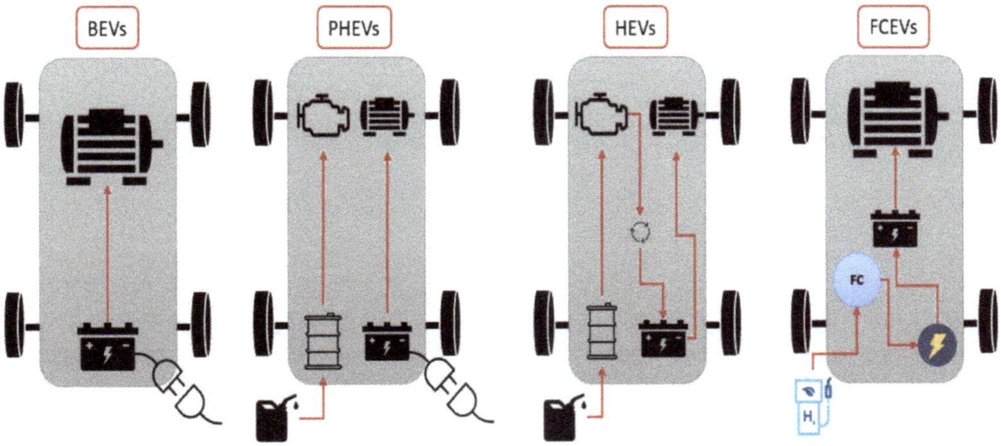

Figure 3. Types of electric vehicles and configuration of the drivetrain.

BEVs are EVs that are entirely powered by electricity supplied by the onboard battery packs. The first commercial EV, the Electrobat, was manufactured by Morris and Salom in 1894 with a maximum speed of 32 km/h and a driving range of 40 km. However, the advent of petrol and diesel engines provided superior traction power, travel range, and lower running cost, confining EV to niche markets (e.g., golf carts, warehouse forklifts, and delivery vehicles) [21]. In addition, early EVs used lead-acid batteries, which are cumbersome and have low energy density. The renaissance of EVs was practically driven by the advances of solid-state power electronics (in the 1970s) and the introduction of more powerful and lighter rechargeable lithium-ion (Li-ion) battery technology in the 1990s [21–23].

HEVs, by comparison, are powered by an internal combustion engine (ICE) and an electric motor, which is driven by the energy stored in the battery. Unlike BEVs, the battery of a HEV is charged through the mechanism of regenerative braking and by the ICE. The history of HEVs dates back to 1897 when Darracq discovered regenerative braking, which made it possible to increase the range of EVs and improve fuel efficiency. The first example of manufactured HEVs was in Belgium and France in 1899. The first HEVs consisted of an ICE, an electric motor, and a lead-acid battery. The vehicle was designed to charge the batteries by the ICE when it was coasting or idling. The first commercial success of HEVs was the introduction in 1997 of the Prius sedan car model by Toyota, followed by Honda's Insight and Civic Hybrid models [21].

Like HEVs, PHEVs have an ICE and an electric motor. PHEVs, however, are equipped with a plug-in charger for battery recharging from an external power outlet. The first invention of modern PHEVs is attributed to Prof. Dr Andy Frank in the 1990s. Since then, due to huge investments in this technology, PHEVs are one of the most preferred EVs today [24].

Finally, the fourth type of EVs is FCEVs, which use an onboard fuel cell that converts hydrogen fuel directly into electrical power and stores it in a battery to drive the electric motor and power other car ancillaries. In 1966, General Motors produced the first FCEV, called the GMC Electrovan, with a maximum speed of 112 km/h and a range of 193 km. However, this vehicle could not be put into mass production due to the requirement for hydrogen generation and fuel cell stack costs [25].

The main advantages and drawbacks of the four types of EVs are summarised in Table 1. The information presented in Table 1 is adapted from [26–29].

Table 1. Characteristics of different types of EVs.

EV Type	BEVs	PHEVs	HEVs	FCEVs
Strength	No emission or very low emission	Lower emission than HEVs	Lower emission than ICE vehicles	No emission or very low emission
	High energy efficiency	High fuel efficiency	High fuel efficiency	High efficiency
	Independent from oil/Low engine noise	Fuel diversity/No range anxiety	No charging station problem	Independent from electricity
Weakness	High purchase and battery cost	Complex technology	Complex technology/Higher cost	High fuel cell cost
	Charging station problems	Management of the energy sources	Management of the energy sources	Technical challenge and cost issue
	Battery replacement requirement	Heavier/Tail-pipe emissions	Dependence on fossil fuel/Tail-pipe emissions/Heavier	Problems of H_2 generation

Today, it can be said that the age of the EV is upon us and is in essence being imposed by the catastrophic effects of global warming. The first signs of a reduction in the carbon emissions of the transport sector have started to emerge [30] and this trend should become even clearer as the EV market continues to grow in the future.

3. The EV Market Potential

The global EV market size has been growing rapidly in the last decade, with the total number of light-duty and heavy-duty EVs reaching 10 million and 1 million, respectively by the end of 2020 [15]. According to BloombergNEF (BNEF) [31], the market outlook is for EVs to account for 58% of new car sales by 2040. It is also predicted that 8% of the total global vehicle stock, which is approximately 116 million, will be made up of PHEVs and BEVs by 2030. The UK National Grid [32] also estimates that the number of light-duty EVs in the UK will reach 36 million by 2040. Similarly, the IEA [33] estimates that there were 3 million additional EVs registered around the world in 2020, of which the BEVs share was about 66% greater than that of PHEVs [33]. The number of registered new BEVs was also estimated to increase to 4 million by the end of 2021, which is nearly double the number of registered EVs in 2020 [33,34]. Available data also support the assumption that the market for PHEVs will shrink in favour of BEVs with the advancement of battery technology and cost.

The world EV market is dominated by China, the United States (US), and Europe. The Chinese market accounted for more than 50% of global sales (1.06 million EVs) followed by Europe (561,000 EVs) and the US (327,000 EVs) in 2019 [35]. The European market, however, is developing at the fastest rate; in 2020, the number of EVs sold in Europe increased to 1.4 million, compared to China (1.2 million EVs) and the US (295,000 EVs). In Europe, the German EV market share is the biggest, with new registrations of 395,000, followed by France (185,000 EVs) and the UK (176,000 EVs) [33]. A more detailed breakdown of the world EV market is summarised in Figure 4. The data presented in Figure 4 were provided by IEA [36]. It can be seen the share of EV sales in the European Union (EU) countries is growing strongly, with countries such as Norway and the Netherlands representing 75% and 25%, respectively, of new registered cars. This is likely to accelerate further as many of these countries will be phasing out the sale of petrol and diesel cars and installing EV battery charging infrastructure in the next decade [37,38].

More detailed information related to the global EV market is summarised in Tables 2 and 3. The information presented in Tables 2 and 3 was adapted from IEA [36] and from [33,39,40], respectively.

Table 2. EV market status of some countries.

	Canada	China	France	Germany	Japan	United States
EV Fleets [units]	Total: 209,171 BEVs: 127,487 PHEVs: 81,588 FCEVs: 96	Total: 4,514,114 BEVs: 3,512,477 PHEVs: 996,191 FCEVs: 5446	Total: 416,585 BEVs: 281,603 PHEVs: 134,607 FCEVs: 375	Total: 634,236 BEVs: 330,780 PHEVs: 302,644 FCEVs: 812	Total: 297,181 BEVs: 136,700 PHEVs: 156,381 FCEVs: 4100	Total: 1,787,221 BEVs: 1,138,654 PHEVs: 639,432 FCEVs: 9135
EV Sales Share [%]	4.2% of the new cars	5.7% of the new cars	11.3% of the new cars	13.5% of the new cars	0.6% of the new cars	2.0% of the new cars

The expansion of the EV market also allows innovation in manufacturing with the introduction of new models. To date, there are some 370 different EV models from a number of traditional automotive companies and new entrants. In Europe, for example, the number of new EV models in 2020 has doubled compared to the previous year [33]. Figure 5 shows the available EV model types in the world in 2020. The data presented in Figure 5 were provided by IEA [36]. As can be seen from the figure, the Chinese market has the highest number of models (149 BEVs and 51 PHEVs), followed by the European market (42 BEVs and 59 PHEVs), and the US market (18 BEVs and 31 PHEVs). It is also worth noting that, unlike the Chinese market, the number of available PHEV models in Europe and the US is higher than that of BEVs. It can also be seen that SUV car models are more popular, with new models accounting for 55% of the new models in the world in 2020 [33].

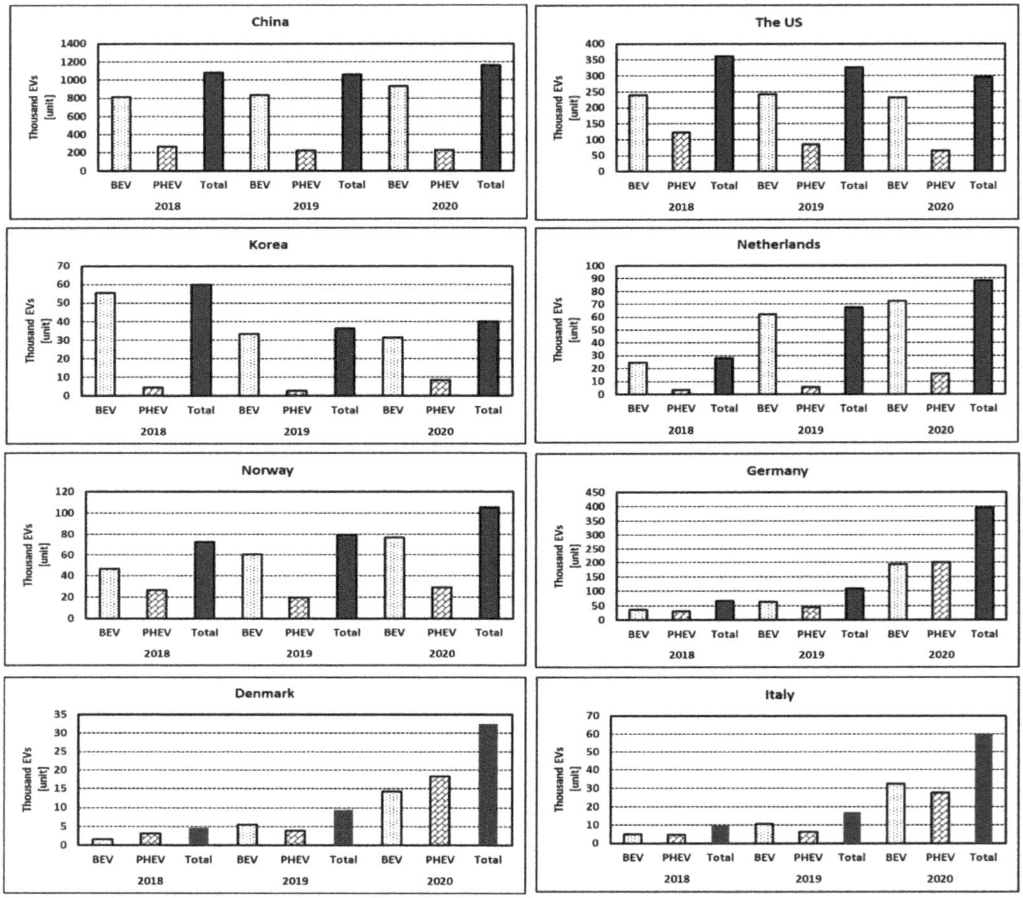

Figure 4. Global EV registration.

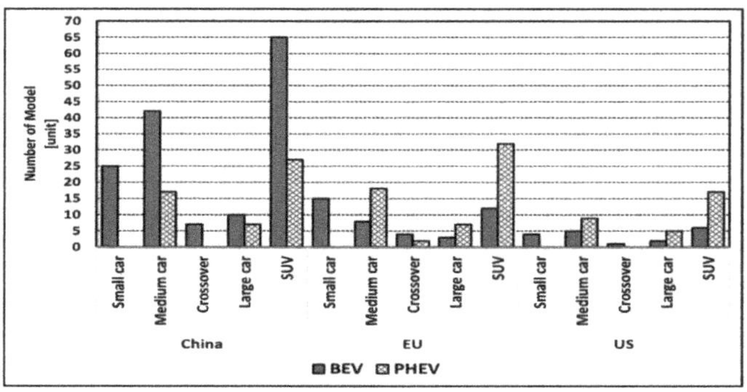

Figure 5. The number of available EV models by 2020.

Table 3. EV incentives and regulations in some countries.

	Canada	China	France	Germany	Japan	United States
Targets to Ban ICE Vehicle Sales	Not available in all states	-	✓	✓	-	Not available in all states
Targets for Low Carbon Transport	✓	✓	✓	✓	✓	✓
Vehicle Purchase Incentives	✓	✓	✓	✓	✓	✓
Charger Regulations	Not available in all regions	Not available in all regions	✓	✓	No building side regulation	Not available in all states
Charger Installation Incentives	✓	✓	✓	✓	✓	Not available in all states

Similarly, the market breakdown by EV manufacturer brands, presented in Figure 6, shows that Tesla's share of the market is the largest and is projected to maintain its current strong growth. The data presented in Figure 6 were provided by Statista [41], which is a statistic portal. The number of Tesla EVs produced in 2020 was 509,737 cars; however, by the end of 2021, the production was projected to increase to over 627,350 units, of which 228,882 units are of Tesla flagship Model 3/Y and 8941 units of Model S/X [42]. Additionally, in 2021, Tesla's Model 3 and Model Y were ranked first and third, respectively, in the number of cars sold, as shown in Figure 7. The data presented in Figure 7 were provided by INSIDEEVs [43].

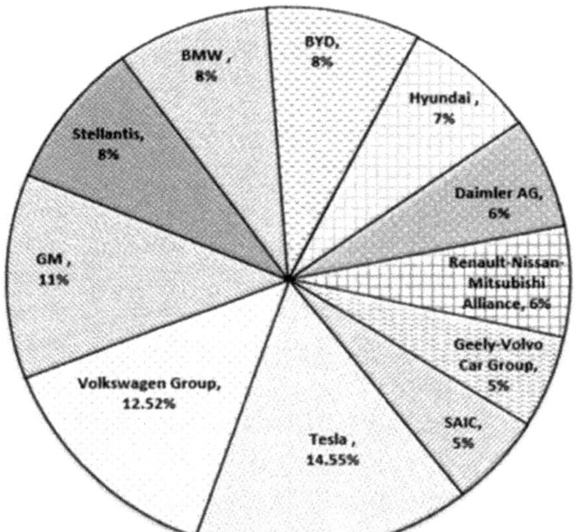

Figure 6. The global market share of the main EV brands in Q2 2021.

In Europe, Volkswagen Group, which owns Porsche, Bugatti, Skoda, Lamborghini, and SEAT, is also a strong brand with the highest growth in the EV market. The group has made a huge investment in EV manufacturing and, in 2021, increased its share of the world market, ranking second behind Tesla [41]. The group also sold a total of 293,100 units of BEV in Q3 of 2021, increasing its global BEV sales by 138% compared to 2020. The company increased its PHEV sales by 133% and achieved 246,000 unit sales in Q3 2021 [44,45]. The company also has a stated target for half of all car sales to be of BEVs by 2030 and to switch all manufacturing of cars to 100% zero-emission vehicles by 2040 [46,47].

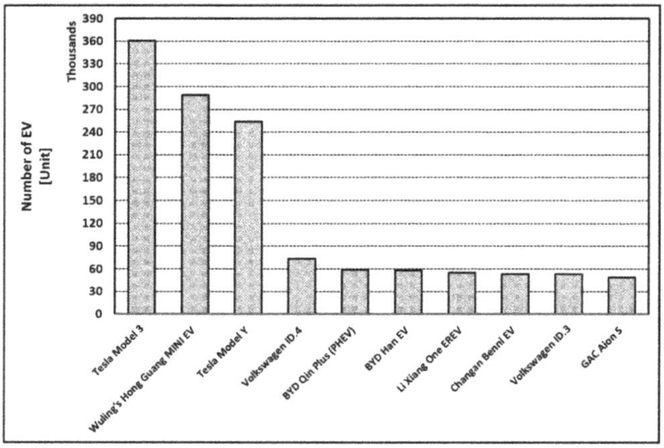

Figure 7. The numbers of best-selling EVs worldwide by model between January and September 2021.

In the US, General Motors (GM) group has a strong investment plan of USD 35 billion for e-mobility, with the aim to introduce 30 new EV models by 2025 [47]. In the first half of 2021, BEVs accounted for 221,000 units of GM's EV sales. Furthermore, mini-size EVs supplied by SGMW, a joint venture between GM, SAIC Motor Corporation Limited, and Liuzhou Wuling Motors Co Limited, gained market share since its inception in 2002, with total sales of 180,000 EVs in 2021 [34,48].

Nevertheless, the EV market is still in its infancy stage, with many new entrants each year, which will stimulate competition and drive innovation that may lead to new champions emerging in the future.

In the UK, according to the Department for Transport (DfT) [49], the number of licensed EVs increased from 8919 in 2010 to 564,694 (295,584 BEV, 246,814 PHEV, and 22,296 others) in Q2 of 2021. The main EV share of the car market is also growing, with all popular EV models being available; the most-sold models are shown in Figure 8. The data presented in Figure 8 were provided by DfT [49]. The EV market in the UK has a strong presence of Japanese brands such as Mitsubishi and Nissan.

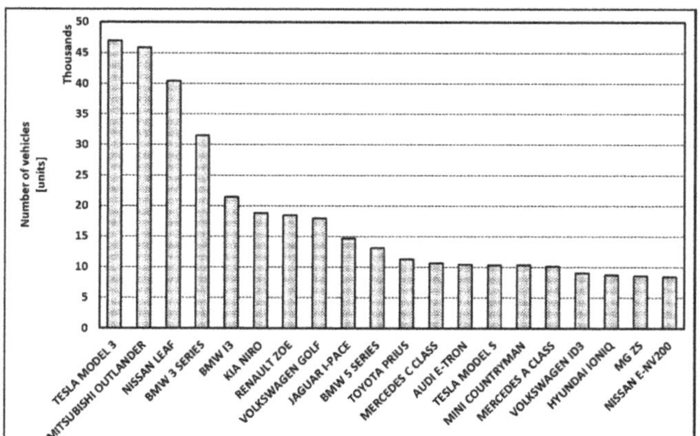

Figure 8. The most popular EVs in the UK in the second quarter of 2021.

4. Potential EV Mass Market Challenges

The high standard of driving experience of current petrol and diesel cars is the expected benchmark for new EVs, including comfort, mileage range, safety, and affordability. Deviation from this expectation may result in the EV market not reaching the planned targets. Therefore, technical development of battery technology for energy density and speed of recharging, refilling infrastructure, and affordable and reliable grid power are some of the challenges to be overcome and to enhance the confidence of the EV market.

4.1. Charging Station Infrastructure

One of the major concerns attributed to EVs is range anxiety, particularly on long journeys where recharging stations may be sparse. Therefore, the number of charging stations must be developed in phase with the increase in EV usage. It is estimated that there were 10.8 million charging stations in the world at the end of 2020, of which 9.5 million were slow chargers installed in homes and workplaces. There is currently a concerted effort in many developed countries to address this concern and make plug-in recharging stations available in public car parks, filling stations, and workplaces [33]. For example, in the UK, the number of public EV plug-in chargers has increased by almost 13 times since 2015, to reach 25,927 charging points distributed throughout the country. In addition, at the end of 2021, one-fifth of these charging installations were rapid-type chargers [50]. Figure 9 shows the increasing trend in the installation of public EV chargers and rapid chargers in the UK over the last seven years. The data presented in Figure 9 were collected from ZAP-MAP [51] by DfT [50].

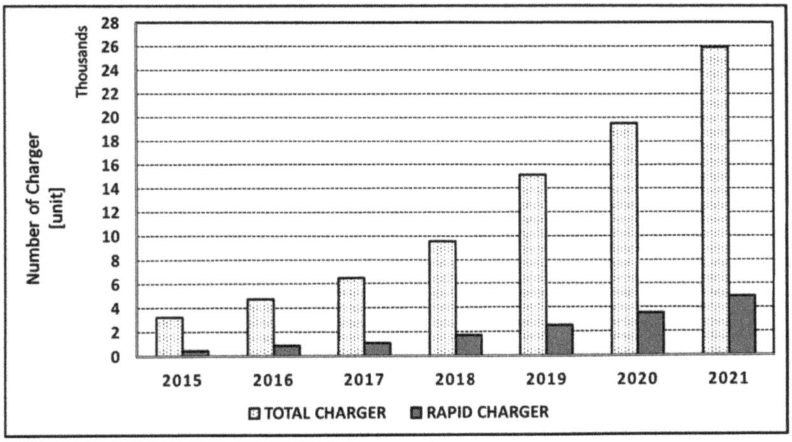

Figure 9. The number of public EV chargers in the UK.

Furthermore, the EV charger status of some countries is summarised in Table 4. The information presented in Table 4 was adapted from IEA [36].

Table 4. Number of chargers in some countries.

Canada [Unit]	China [Unit]	France [Unit]	Germany [Unit]	Japan [Unit]	United States [Unit]
Total: 13,194	Total: 807,000	Total: 46,045	Total: 44,669	Total: 29,855	Total: 103,021
[1] F & R: 2258	F & R: 309,000	F & R: 4045	F & R: 7456	F & R: 7939	F & R: 16,718
Slow: 10,936	Slow: 498,000	Slow: 42,000	Slow: 37,213	Slow: 21,916	Slow: 82,263

[1] In the table, F & R stands for fast and rapid chargers. In addition, the power of the referred F & R and slow chargers are above 22 kW and up to 22 kW, respectively.

Compared to the conventional car industry, where major manufacturers use standardised components and parts such as a universal fuel nozzle, the EV industry is still in an early stage of development, and there are currently many manufacturers vying for a share of the market using their own manufacturing standards and component specifications. Therefore, new industry policy frameworks may be sought to encourage collaboration and standardisation of manufacturing processes to minimise component supply bottlenecks. An obvious example is the many types of EV battery chargers (slow, fast, and rapid) and plug-in connectors' physical configurations that vary between manufacturers and countries.

Currently, there are four main types of plug-in chargers in the UK namely: slow, fast and rapid chargers, in addition to Tesla's supercharger. As shown in Table 5, the slow charger has a power rating of 3 to 6 kW and is suitable for home or workplace plug-in. The fast chargers are rated between 7 and 22 kW and are installed in public premises such as supermarkets, car parks, and leisure centres. The rapid chargers have a power rating higher than 43 kW and are mainly intended for use in dedicated petrol and service stations for EVs equipped with the rapid charging standard. The chargers are also classified according to the physical configuration of the connector, including a UK standard 3-pin plug, type 1 and 2, Commando, CHAdeMO, and CCS connector depending on the charger type. Although type 1 and type 2 connectors can be used for slow and fast charging, Type 2, CHAdeMO, and CCS connectors are used for rapid charging. In addition, European-based manufacturers such as Audi, BMW, Renault, Mercedes, Volkswagen, and Volvo generally prefer to use type 2 and CCS connectors for the vehicle-side connectors, whereas Asian-based manufacturers such as Nissan and Mitsubishi prefer to use type 1 and CHAdeMO connectors. A different connector configuration is used for the DC rapid charger. As EV manufacturers attempt to gain world market share by offering competitive prices, the differences in charger technologies may start to narrow in a few standard types [52].

Table 5. EV charging parameters in the UK.

Charger Type		Charge Mode	Power Rating	Connector Type	Charging Time
Slow Charger	-	Mode 2	3–6 kW	3-Pin Type 1 Type 2 Commando	8–10 h
Fast Charger	-	Mode 3	7 kW/11 kW/ 22 kW	Type 1 Type 2 Commando	3–4 h
Rapid Charger	AC Rapid	Mode 3	43 kW	Type 2	30–60 min
	DC Rapid	Mode 4	50 kW	CHAdeMO/ CCS	
	DC ultra-Rapid	Mode 4	100–350 kW	CHAdeMO CCS	
Tesla Supercharger	-	Mode 4	Up to 250 kW	Tesla CCS Type 2 CCS CHAdeMO	

It is also projected that future EV plug-in connectors will be equipped with communication capability to interface between the battery management system, the power grid, and the charging site (home, workplace, etc.) to realise the full potential of V2G, V2B, and V2X technology. In the UK market for fast EV chargers, CHAdeMO and CCS connectors are emerging as the dominant type, due to consistently increasing their market share for the last five years, as shown in Figure 10. The data presented in Figure 10 were provided by Zap-Map [53], which is the EV and charging point platform.

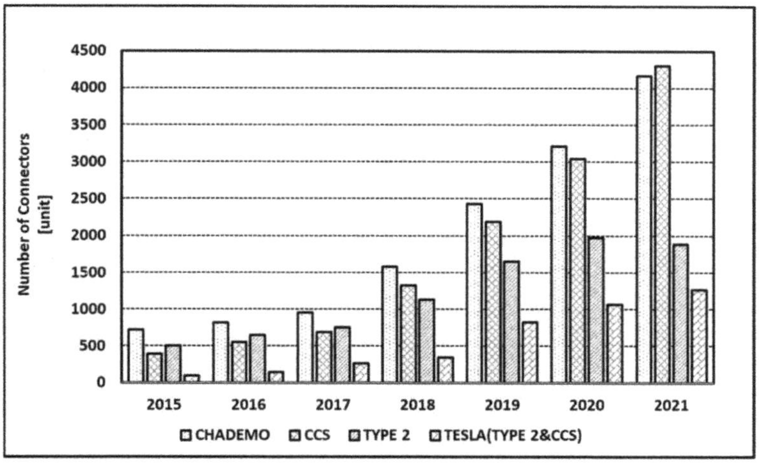

Figure 10. The number of public rapid charging connectors in the UK.

There are several international standards used to determine the power output of an EV charger. Power parameters of plug and socket systems in the UK are fixed with the BS 1363-13 standard. Additionally, BS EN 61851 and BS EN 62196 are the two basic standards used in the UK for charging electric vehicles. BS EN 62196 includes details about plugs and sockets used in EVs, whereas the BS EN 61851 standard covers the parameters associated with EV chargers. According to the BS EN 61851 standard, there are four different charging modes in the UK, Modes 1, 2, 3, and 4 [54]. Because there is no residual current device in Mode 1 charging, it is not allowed to be used in the UK [55].

The British Standards Institution (BSI) [56], the UK National Standards Body, has also published two new standards, namely, PAS 1878 and PAS 1879, in order to support low carbon emission systems in 2021. According to the BSI report, whereas PAS 1878 specifies the criteria required for any energy appliance to work and be produced, such as an energy-smart appliance, PAS 1879 defines the rules of the demand-side response service for electricity consumers and supports the operation of smart energy devices. It is likely that these two standards will be used in the application of smart EV charging stations in the future.

4.2. Recharging Time

Compared to conventional vehicles where refuelling is a quick process, the best EV battery charging technology is still slow. For example, the slow charger used in individual homes or workplaces takes up to 10 h to fully charge an EV battery. The fast and rapid chargers rated between 7 and 43 kW that are found at public sites, such as supermarkets, car parks, and leisure areas, can achieve full charging in a time scale as short as one hour, whereas rapid chargers having a power rating higher than 43 kW can fully recharge an EV's battery in less than 60 min. The rapid battery charger is becoming commonplace; however, the technology remains expensive and is not suitable for all EVs or all sites.

A research study by Zhao et al. [57] found the utilisation of rapid charging, performed by increasing the charging current and voltage, carries a higher risk of fire and the authors suggested that the battery needs to be well insulated.

4.3. EVs Battery Development and Driving Rang

Another main challenge facing mass marketing of EVs is the limited driving range of EVs, which is associated particularly with the low energy density of EV batteries. Furthermore, the battery pack requires a sophisticated electronic management system and constitutes a large part of the overall EV's cost. Today, significant research programmes are being implemented by the industry and academics to develop new materials to increase battery energy density and lower the cost.

In an effort to enhance energy storage density, three main battery technology candidates are being used namely: Li-ion, lead-acid, and nickel-metal hydride (NiMH) batteries. Today, Li-ion batteries, which entered the market in the early 1990s, have the largest market share due to their high efficiency and energy density, fast charge capability, and long service life. However, the drawback of Li-ion batteries is the high cost of Li ore, making long term commercial prospects unsustainable [58,59]. This has led many researchers to consider introducing novel designs, such as replacing graphite electrodes with a cheap silicon material to increase Li-ion batteries' energy density, decrease charging time, and reduce degradation [60–62]. For example, Wang et al. [60] observed that replacing the anode of a Li-ion battery with silicon can increase the energy density and lifespan of the battery.

In comparison, lead-acid batteries were introduced in the 1960s and achieved huge success due to their simplicity, low cost, and recyclability. Today, they are preferred for applications in HEVs and large-scale power storage schemes due to their high efficiency (80–90%) and average service life of around 1500 cycles [63].

NiMH batteries are also relatively new (first commercialised in 1991). They have higher energy density than lead-acid battery counterparts, but they are also more expensive [64]. They have been used in EVs by manufacturers such as General Motors, Honda, and Ford.

To understand the patterns of daily driving behaviour of road users, the Joint Research Centre (JRC) conducted a survey study [65] in which about 1000 vehicle owners, in 28 European countries, participated. It was found that the duration of the most frequent journeys of the participants was 39 min, whereas the average journey was 25 km (15.5 m) per day. A similar study conducted in the US in 2016 by the National Renewable Energy Laboratory (NREL) [66] found that, for an average American shopper, an EV having a range of 300 m is appropriate. Furthermore, the UK DfT [20] recommended in 2019 that the average annual mileage consumption of UK's car owners is about 7400 m per vehicle, i.e., about 20 m per day per vehicle.

These studies demonstrate that current BEVs are capable of achieving a range of over 200 m. A literature review shows that the average range of the BEV models has increased [33]. Table 6 shows some BEVs with a high range. The data presented in Table 6 were provided by EV Database [67].

Table 6. Some BEVs having a range of 200+ m.

BEVs	Battery Capacity [kWh]	Specific Energy Consumption [Wh/mi]	Range [mi]
Hyundai Kona Electric	64	0.26	245
Tesla Model 3 Long Range	70	0.245	285
Polestar 2	75	0.305	245
Tesla Model 3 Performance	76	0.265	285
Volkswagen ID.3 Pro S	77	0.275	280
Volkswagen ID.4	77	0.31	245
Mercedes EQC 400	80	0.345	230
Porsche Taycan 4S Plus	83.7	0.31	270
Porsche Taycan Turbo	83.7	0.34	245
Audi e-tron Sportback	86.5	0.375	230

Available data from current BEVs manufacturers also suggest that battery capacity is appropriate for the average daily driving range. Additionally, the performance of EVs is increasing consistently with the ever-increasing range, as can be seen in Figure 11. The data presented in Figure 11 were provided by IEA [36]. However, charging station infrastructure and rapid charging issues still need to be addressed to eliminate long-distance trip challenges.

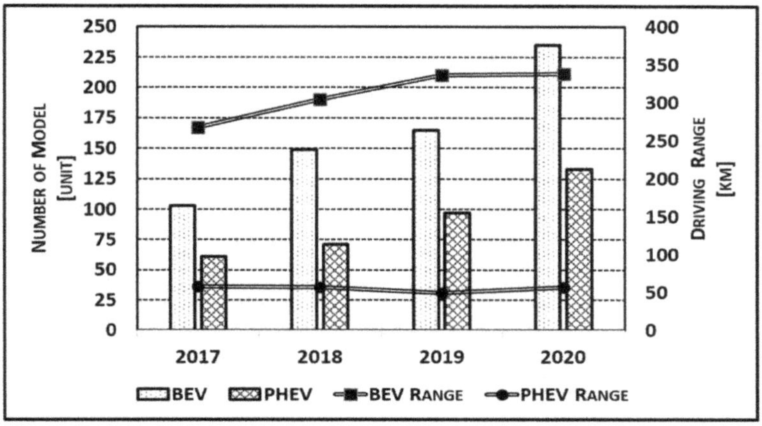

Figure 11. Available global EV models and average range.

4.4. Cost and Affordability

The current cost of EVs is higher than equivalent petrol or diesel cars but, as the market expands, the manufacturing cost will decrease through standardisation of processes and rationalisation of components supply chains. However, a significant part of the reduction in the cost of an EV will still depend on the availability of valuable metals used in battery and other components manufacturing. These critical metals include misch metal alloys used in NiMH batteries, lithium and cobalt in Li-ion batteries, platinum in the fuel cell, and neodymium and dysprosium used in permanent magnet electric motors [68]. For instance, a light vehicle with a battery capacity of 30–45 kWh requires up to 80 kg of lithium in the manufacturing of the battery [69]. As a result of the increase in the share of Li-ion batteries in the EV market, it is projected that world demand for lithium will increase from 65 kilotons (kt) in 2000 to 530 kt by 2025 [70]. It is also a fact that some of these critical rare earth metals are mined unsustainably and from limited sites in unstable regions of the world. Therefore, developing efficient recycling and recovery processes for these materials may contribute to reducing EV costs.

4.5. Manufacturing and Recycling

Although the impact of EVs in reducing operational carbon emissions cannot be ignored, it is necessary to examine the carbon emissions caused by their charging needs and the life-cycle emissions of EVs to take a broader perspective.

It can be said that the source of electric energy that EVs use as fuel can change the effect of EVs on the reduction in carbon emissions. Today, EVs generally use grid systems for their charging needs if no external source is used. Because the energy sources used in electricity generation vary, the carbon footprint of the grid system may differ according to energy policies, geographical locations, and even time of the day. For example, in Estonia, which uses a large amount of coal for electricity generation, the GHG emission rate from electricity generation is 831 gCO_2e/kWh in 2019, whereas this rate is only 19 gCO_2e/kWh in Norway because of their high-RES usage rate [71,72]. Therefore, the quantity of GHG emissions released into the atmosphere to charge an EV in Estonia is much higher than that in Norway. A study conducted by Hoehne and Chester [73] showed that late-night EV charging causes more GHG emissions because energy sources used to meet peak time demands during the day in the US contain less CO_2. Similarly, Lin and Wu [74] stated that increasing EV demand in China may increase coal-based electricity generation and therefore carbon emissions. However, the researchers added that 2348 million tons of carbon emissions could be avoided by 2040 if low carbon scenarios with a high usage share of RESs can be realized. As a result, considering the future EV predictions, it is vital to meet the charging needs with low-carbon energy sources. On the contrary, e-mobility, which increases with the aim of reducing emissions, may accidentally increase carbon rates in the atmosphere.

In order to produce a vehicle in the automobile manufacturing industry, many processes are carried out from the mining to the assembly of vehicle parts, and a large amount of energy is used for these processes. Sato and Nakata [75] detailed all the materials used to produce an ICE vehicle and calculated that 62 GJ of energy is required to produce a Honda Accord. The research also showed that 28% of the electricity in the production phase is used for alumina reduction processes. Similar to ICE vehicles, the EV manufacturing industry includes high energy-consuming processes, such as mining and assembly processes for the EVs and their components, such as batteries and motors. Although this situation varies depending on the variety of energy sources used, under the global average electricity production rates, EVs may cause higher manufacturing emissions than conventional ICE vehicles due to high energy consumption in battery production processes [76,77]. Kurland [78], for instance, noted that the use of electrical energy is between 50 and 65 kWh per kWh battery in a Li-ion battery cell production facility. However, in this study, the electrical energy required for mining processes was not taken into account. In addition, Cox et al. [77] showed that high-capacity batteries cause 40% more life cycle emissions compared to low-capacity batteries. However, a study by Hall and Lutsey [79] emphasised that this high emission in EV manufacturing can be reduced within 2 years compared to operational emissions of ICE vehicles. To compare the energy used in the production of EVs and ICE vehicles, and the GHG emissions generated, in another study, Bieker [80] conducted a life-cycle assessment. The researcher claimed that producing a BEV battery with a capacity of 45 kWh in Europe may result in 2.7 tCO2e of GHG emissions, and the production of a Li-ion battery with an NMC811-graphite cathode would cause 20% less GHG emissions by 2030. Another notable result of the study is that the life cycle emissions of BEVs cause 63–69% less GHG emissions compared to petroleum vehicles. The research would have been more interesting if a wider range of battery GHG emissions were examined by exploring the recycling and reuse of batteries for other purposes.

Along with the increasing number of EVs, used batteries may become an environmental waste problem due to hazardous chemicals within them. According to the sustainable development scenarios of the IEA [81], it is supposed that the Li-ion battery capacity spent by EVs will reach 10.1 GWh in 2025 and 1089.6 GWh in 2040. Therefore, waste management mechanisms, such as the recycling of end-of-life batteries, should be encouraged and regulated. More importantly, if these battery wastes can be recycled and reused, they can become an important opportunity due to the precious metals they contain, such as lithium (Li), nickel (Ni), and cobalt (Co). It is known that the demand for Li-ion batteries is also increasing with the increasing number of EVs. IHS Markit [82] declared that estimates show that 9300 kt of Li, 9800 kt of Co and 55,000 kt of Ni metal will be needed to meet the demand of Li-ion batteries between 2020 and 2050. Moreover, they claimed that if suitable recycling mechanisms can be established, the required 48% of Li, 47% of Ni, and 60% of Co metals may be able to be met by recycling. However, the current global raw material recycling rate is just 1301 kt/year [82]. Many techniques, such as pyrometallurgical (smelting), hydrometallurgical (leaching) and direct recycling approaches, have been introduced to recycle Li-ion batteries; however, a comprehensive technique that can recycle all battery types has not been discovered yet [83]. Although it is seen that battery recycling is a promising opportunity to recover precious metals and, therefore, a reduction in GHG emissions, their use remains limited due to some challenges. One of the biggest challenges of the recycling industry is the problem of cost. A recycling facility involves high investment costs and can only be economical with high-capacity batteries or large quantities of batteries arriving at the facility [84]. Gaines et al. [83] noted that it is not easy to remove batteries from vehicles, and discussed the difficulties of collecting the spent EV batteries from EV owners. They underlined that transportation of the collected batteries to the facility where the recycling process will be carried out can cause serious costs. The researchers also emphasized that there are possible flaming and fire risks due to the occurrence of chemical reactions during the transportation of Li-ion batteries, which have high energy potential. Interestingly, the study claimed that the legal regulations created to prevent these dangers may play a role in increasing transportation costs. The study also indicated that metals and fluorides that are harmful to health may be released during the melting of the batteries in the facility, which makes the recycling processes dangerous. In summary, the recycling technology of batteries is in an early stage and, therefore, further technical, economic, and logistical studies are required to gain additional understanding about recovering precious metals from EV batteries.

5. Impact of EVs Uptake on Power Grid Capacity

One of the key energy performance indicators of an EV is electrical energy consumption per mile driven (mile/kWh). This efficiency indicator, coupled with the growing number of EVs on the market, will have a significant impact on the power grid to satisfy the additional load from EV demand. Globally, according to BNEF's EV outlook of 2021 [31], a fully decarbonised transport industry could add about 8500 TWh of electricity consumption by 2050, which is 25% greater than the baseline case. Adding such a demand on the grid without developing adequate energy management mechanisms can lead to overloading of the grid and reduce the reliability and quality of the power supply.

To date, few studies have examined the implication of the increase in EV numbers on the grid system [85–91]. A study conducted by Papadopoulos et al. [92] examined how charging EVs may affect the grid power demand profile in the UK by 2030. The study shows that, compared to the baseline scenario of no EVs, the high uptake of EVs will double power demand from the grid at the evening peak demand, throughout the year. A peak demand of up to 100 GW is expected in winter. However, if decarbonisation of heat in domestic buildings progresses at planned using heat pumps, additional strain will be placed on the national grid, with peak demand for heat also coinciding with the evening peak demand for electricity. However, increasing the peak load is an undesirable situation for the grid because it requires relatively high energy in a short time, and may incur further costs due to loss of load and power quality [93]. Therefore, in order to deal with this challenge, a flexible grid with a large and fast-response reserve capacity may be required.

A study by Clement et al. [85] investigated the impact of controlled and uncontrolled EV battery charging on the grid power quality and losses under different market uptake levels. It was found that grid voltage supply may deviate by more than the maximum accepted norm of 10%, and Distribution Network Operators (DNOs) may resort to increasing electricity tariffs to compensate for any additional power loss. Two notable studies on the impact of EVs on the UK grid showed that grid power demand will increase. One of these studies was carried out by Mu et al. [88], who investigated large scale penetration of EVs using a spatial-temporal model to assess the load of EVs on branches in a distribution network under a 50% penetration scenario. The study found the use of an uncontrolled charge and smart charge increases the peak load by 74% and 47%, respectively. The other study was conducted by Qian et al. [89], who adopted a stochastic approach and found that 20% EV penetration would increase the grid peak load by 35.8% under the uncontrolled charging scenario.

6. Potential Solutions of EVs' Uptake | Vehicle-to-Grid (V2G) Energy Carrier

6.1. V2G Concept

The energy management of additional power demand on national grids using EVs presents another challenge in the decarbonisation agenda of the transport sector. A report published by the UK National Grid company [94] in 2021 entitled 'Future Energy Scenarios' projected that energy consumption for transportation will decrease to 139 TWh by 2050 from the 2020 level of 400 GWh. In addition, it estimates that a load of 38 GW can be managed through peak shaving with the use of EVs and smart charging. This report demonstrates that, as the number of EVs plugged into the grid increases, coupled with advancement in battery technology and smart management systems, EVs will form an integral part of future smart grids. The concept of using EVs to recharge when RE generation is high and discharge when peak demand is high was proposed by Kempton and Letendre [17] in 1997. A study by Toniato et al. [95] aimed to model the reduction in the peak load demand for charging a fleet of 138 V2G-enabled e-busses at a depot. First, the daily use of the 138 diesel buses was examined and, under the uncontrolled charging scenario, the optimization algorithm showed the average peak load was 7959 kW and occurred between 7 and 8 p.m., whereas, by using V2G technology, the peak load can be reduced by as much as 83%. The feasibility of V2G was also demonstrated by Fathabadi [96] using 15 EV chargers equipped with V2G technology and fed during off-peak hours from wind-generated power and discharged back to the grid during peak hours to balance supply and demand of electricity.

A basic schematic outline of this interaction of charging and discharging EVs with the grid, otherwise known as V2G, is shown in Figure 12.

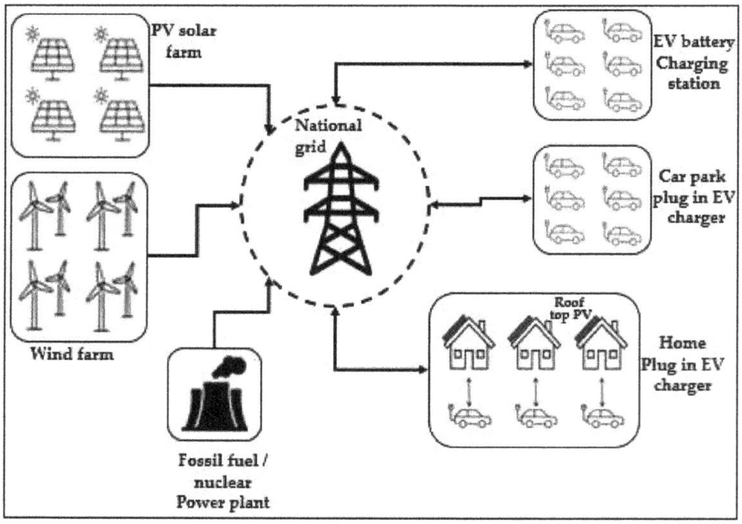

Figure 12. Schematic layout of V2G architecture.

Furthermore, V2G technology can be used for ancillary services such as energy arbitrage, peak load shaving, spinning reserve, regulation, and RE support, providing benefits to grids, EV manufacturers, EV customers, energy suppliers, network operators etc.

6.2. Financial Revenue Incentives of V2G

The prospective dual usage of EVs for transportation and electrical power storage services for the grid (V2G) makes the EV a potential source of income for the owner. Most of the studies on the feasibility of V2G were conducted in the past two decades and focused particularly on the US energy market. In assessing the potential monetary benefit of V2G, Kempton and Tomic [97] suggested that the use of EVs for transportation needs accounts for only about 4% of the daytime, whereas the remaining 96% of the time could potentially be used to earn income by providing a grid service. The authors developed a mathematical model in which they considered parameters such as peak power demand, spinning reserve, baseload power demand, and regulation services, and found that, under ideal scenario conditions, EV owners could annually earn about USD 4900 using V2G in the United States of America (USA). With a similar optimistic conclusion, Noori et al. [98] constructed a model in which payments and regulation uncertainty and battery degradation were among the considered parameters. The study found, with V2G services, annual gains of approximately USD 2600 in the USA. Similarly, Ginigeme and Wang [99] developed a charge control algorithm to integrate EVs with V2G into the grid using a real-time pricing model. It was established that, with the proposed strategy, the annual revenue per vehicle from the V2G was around USD 550. In a less optimistic study, Peterson et al. [100], found that the annual revenue of EVs in three different regions of the USA from energy arbitrage with V2G can be USD 140–250 (under best conditions) and a mere USD 6–72 under a more realistic scenario. In another study, the authors developed an algorithm for charging/discharging EVs equipped with V2G and showed that charging costs can be reduced by 13.6% for a fleet of 5000 EVs [101].

The EV users' willingness to adopt V2G under contract to provide grid services was studied by Hidrue and Parsons [102] in a survey of 3029 participants. It was found that the participants were concerned about range anxiety, freedom restrictions with V2G contract, and battery replacement cost.

6.3. *Effects of V2G on the EV Batteries' Longivity*

The number of charge/discharge cycles of a Li-ion battery ranges between 500 and 3000 cycles [103]. However, the lifespan of the batteries may not only be related to a single phenomenon, as the life cycle of batteries can be affected by factors such as depth of discharge (DoD) level, temperature, and charging power parameters [104,105]. It is often noted that V2G rapidly reduces the EV battery life span, as it charges and discharges the batteries many times during the day. Conversely, because EVs that provide a regulation service, unlike deep charging/discharging, operate within a small fraction of the battery capacity, the wear effect of the daily driving of the vehicles may be much higher than the V2G effect [98]. Similarly, a broader perspective was adopted by Peterson et al. [106], who argued that battery degradation caused by daily transportation may have double the impact compared to V2G usage for load shifting in PHEVs. However, studies in the literature also state that the degradation effect of V2G cannot be ignored. For example, Bishop et al. [107], examining the corrosive effects of V2G on PHEV and BEV batteries, observed that, even in the best scenarios, the V2G service may lead to multiple battery changes over the lifetime of the vehicle. Although the study analysed the degradation by considering different battery capacities, charge regimes, and DoD levels, and presented remarkable findings, this research would have been more convincing if it had included the cost analysis of the degradation.

Furthermore, some studies in the literature focus on the use of V2G with control algorithms. Guo et al. [108] used various charging regimes and V2G technology to assess battery degradation in EVs. In the study, which mainly focused on capacity losses in Li-ion batteries, a charge management strategy was proposed to reduce degradation. Researchers argued that using the proposed charging strategy can reduce degradation by 13.51% and increase the battery life of EVs. With a similar result, Uddin et al. [109] developed a comprehensive battery aging model using data from degradation experiments to assess the degradation caused by energy arbitrage in EV batteries with V2G. As a result, the researchers claimed that using V2G with the proposed strategy can reduce the capacity fade in Li-ion batteries by 9.1% and power fade by 12.1%. However, the ambient temperature value accepted as constant in the study will likely affect the results. This may be because Zhou et al. [110] found that there is a causal relationship between ambient temperature,

DoD, and battery degradation. They concluded that rising temperature and growing DoD values can simultaneously increase the degradation effect on the EV batteries.

6.4. Effects of V2G on the Environment

According to the 2020 IEA Report, transportation accounted for 24% of the carbon emissions created by burning fuel [14]. In the UK, transportation is responsible for 29.8% of CO_2 emissions occurring within the country [111]. Because EVs use electricity as a fuel, unlike vehicles with ICE, it is predicted that they will play an effective role in reducing CO_2 emissions from transportation [30,73]. Additionally, EVs also seem to be capable of preventing NO_x and SO_x gas emissions, which are harmful to the environment and human body because they do not lead to exhaust gas emissions [112]. As previously noted, it is possible that EVs, which will reach large numbers in the future, may cause significant environmental problems when charging with high carbon emission networks. To avoid this problem, the main principle should be that countries utilise more RESs in their energy generation policies and that they encourage EV owners to use a controlled charging strategy such as V2G.

Turton and Moura [113] examined the benefits of V2G technology until 2100 and examined its effects under some energy scenarios. Although the researchers hypothesized that V2G would have serious positive environmental impacts under a USD 135/tCO_{2-e} carbon tax scenario, they concluded that in 2100, even under existing scenarios, V2G could reduce CO_2 emissions by 6.5%. In addition, a recent study by Jiao et al. [114] found that V2G contracts offered to EV owners in the future may also have a role in reducing carbon emissions. This study effectively illustrates that V2G contracts may be important with regard to environmental issues in the future. Therefore, further studies should be carried out to establish more environmentally friendly V2G contracts.

Several attempts have been made to emphasize that the use of V2G system for various purposes, such as regulation, reserves, and demand and supply balancing, can reduce carbon emissions [115,116]. Zhao et al. [117] published a paper in which they analysed the environmental effects of using V2G-capable electric trucks for the regulation service on the grid. The researchers examined five independent system operator regions, and pointed out that an electric truck serving the grid for this purpose can prevent CO_2 emissions of between 200 and 500 tCO_2 over its lifetime (15 years). Similarly, regulation services through light-duty V2G-capable EVs formed the central focus of a study by Noori et al. [98], in which the author found that, in a region where the majority of vehicles on the roads are expected to be EVs by 2030, regional emissions savings of up to 500,000 tCO_2 per year will be achieved. Another study by Sioshansi and Denholm [118] highlighted the environmental effects of providing a spinning reserve service with PHEVs with V2G. Unlike in other studies, the researchers analysed CO_2 emissions and other harmful pollutants such as SO_2 and NO_x. The results suggest that adjusting the charging times of vehicles according to the availability of generators with high energy efficiency will result in a significant reduction in emissions from energy generation. A remarkable finding of the study is that replacing 1% of an EV fleet with EVs offering V2G service can reduce generator CO_2 emissions by 25%. Furthermore, the authors reported that reductions in CO_2 and NO_x emissions are less than those in SO_x in the process of V2G usage for the spinning reserve service because this service is usually provided with natural gas sources containing a small amount of sulphur. Hoehne and Chester [73], who investigated the environmental effects of providing a demand and supply balance in the grid by using the batteries of PHEVs with V2G, revealed that the designed smart charging strategy has the potential to reduce carbon emissions by 59% in V2G mode usage, and compared time-adjusted charging strategies (charging in only certain hours). The most striking result to emerge from the study is that the use of V2G may also increase CO_2 emissions in the scenarios. The researchers observed that carbon emissions were increased by 369 gCO_2/mile when EVs were charged for V2G use after 12 p.m. in the Midwest region of the USA. The result is linked to the electricity generation profile of the region after 12 p.m. The finding clearly indicates the necessity of considering V2G technology and RESs together so that V2G does not become a technology that increases CO_2 emissions.

6.5. Integration of RESs via V2G

Recent developments in the field of RES have led to greater interest in the usage of RESs in the grid system. Because the clean RE produced is highly valuable, it should not be wasted. However, its use is restricted by the unpredictable power output of RESs, particularly solar and wind energy. It is clear that the Duck Curve, which is shown in Figure 13, is one of the best examples used to explain this phenomenon. The Duck Curve is the pattern of California's net energy demand curve. The demand data presented in Figure 13 represent 1 January 2021, and were provided by California ISO (CAISO) [119]. The increase in the share of RESs in the California grid, which had a Camel Curve type energy demand curve until 2012, similar to that of the UK, triggered significant ramping problems in the power network. A downward ramp that occurs in energy demand due to high amounts of sunshine at noon leads to the formation of a steep upward ramp that must be covered quickly within three hours and increases the risk of overgeneration at noontime. In this case, a phenomenon known as RE curtailment occurs and the TNOs (Transmission Network Operators), who are responsible for network security, may have to ignore the wastage of produced RE [120]. Today, the problem of RE curtailment may be faced by any country that increases the share of solar PV energy in its grid system and does not limit the usage of RESs.

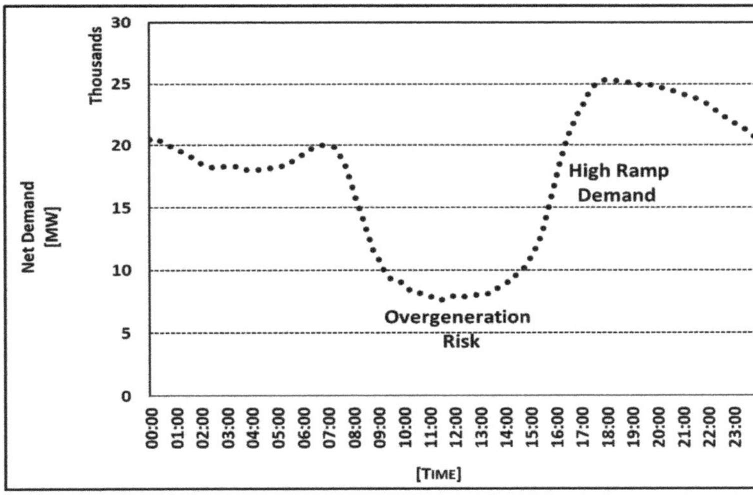

Figure 13. California's Duck Curve.

By comparison, the increasing installation of solar PV capacity will likely accelerate the formation of downward and upward ramps, and therefore increase over-generation and under-generation risks. Storage systems can play an important role in eliminating these risks. However, it is clear that V2G technology has more advantages, not only in the elimination of the risks without storage costs, but also in providing a solution to the load peak problem by sending the excess stored energy back to the system during the peak hours of energy consumption (i.e., the head of the duck).

During the last decade, the use of the link between EVs and RESs to provide a balance between demand and supply has attracted a significant amount of attention [121–125]. Borba et al. [16] examined PHEVs that were fully charged with wind energy in order to eliminate the imbalance between energy production and consumption, taking into account the future wind energy capacity increase in a selected region in Brazil. The researchers asserted that the excess energy produced by wind turbines in the region in 2030 can be completely consumed with the charge of 1.5 million PHEVs. A study by Schuller et al. [126] focused on a similar type of charging strategy and emphasized that charging EVs with RES can increase the RES usage by more than 100% compared to uncontrolled charging. However, in both studies, it was assumed that the battery capacity of the vehicles is constant, and V2G technology was not taken into account.

In the past two decades, there has been a surge of interest in studies that compare different charging strategies, different EV penetration rates, and different levels of RES integrations to maximize the interaction between EVs and RES [113,127–130]. Schetinger et al. [131], for example, presented an economic analysis of Toyota Prius PHEVs on a university campus in Brazil, and critically examined three scenarios in which there were no RES, EVs charging with RES (solar and wind), and EVs discharging with V2G. The researcher contended that the lowest Net Present Value (NPV) and Levelized Cost of Energy (LCOE) can be achieved with the combination of RES and V2G technology. In another related study, Freire et al. [132], who claimed V2G may be a supportive innovation for RE integration, concentrated on different vehicle charging schemas—uncontrolled, controlled, and V2G—by considering that all vehicles on the roads of Portugal would be EVs. Similarly, under the 100% EV penetration scenario, Colmenar-Santos et al. [19] developed a control strategy regulating the charge and discharge of EVs to increase the percentage of RESs in the grid. Additionally, taking a different approach by using heat pumps (HPs) in addition to EVs to consume the electricity produced by solar PVs, Zhang et al. [133] showed that the excess energy from installed PVs having 30 GWp can be reduced by up to 3 TWh (corresponding to 3% of the total electricity produced) with the use of one million EVs and one million HPs. The method used in the paper was to capture the excess energy from PVs using EVs and HPs, and then utilise EVs with V2G to avoid the need to increase the installed power capacity and the additional upgrading costs. The results of an article published by Claus Ekman [134] cast doubt on the researchers' method. Ekman compared the uncontrolled, night, smart charge, and V2G charging strategies of EVs, and concluded that, although 500,000 EVs have the ability to reduce the excess energy from an 8 GW wind farm by 800 MW, this EV fleet is also not enough to balance demand and supply in the grid when the RE usage rate is over 50%, and added that V2G is not effective in reducing the required installed power capacity. The difference between the results of the two papers may be caused by dissimilar parameters, such as the size of the selected EV fleets, RE generation capacities, and RES types and locations. A study by Hassan et al. [135] should be mentioned here. The researchers discussed the use of batteries of EVs as storage devices for RE, and demonstrated that wind power generation of 80 TWh in Great Britain under a 70% EV penetration scenario does not cause excess energy generation. Furthermore, when the amount of generated wind energy is increased to 180 TWh, the reference excess energy amount of 59 TWh (under a no EVs scenario) can be decreased to 40.21 TWh (32% reduction) using V2G technology.

7. Conclusions

The EV industry has made substantial progress in recent years. The introduction of new policy frameworks to address climate change and implementation of substantial financial incentives in many developed countries have provided the required confidence for the automotive industry to make sweeping decisions, such as discontinuing production of petrol and diesel engines and reshaping long-term strategies for the development of EV technology for a sustainable mass market. As a result of the current transition to full electrification of transport, some EV types, such as BEVs, PHEVs, HEVs, and FCEVs, are being developed to instil confidence in the technology. At present, significant technical development is required in the engineering of lighter and more powerful batteries, smart management systems, and manufacturing processes to reduce cost.

This paper attempts to highlight some of the technological advancements, areas of required further development, and future prospects of EVs as a means of sustainable transportation and as an interfacial energy carrier. The findings of this work are summarised as follows:

- The number of EVs sold is consistently increasing in many major economies of the world, thus displacing the process of ICE vehicles. To date, there are over 10 million EVs globally, of which BEVs and PHEVs are the best-selling types.
- China is the largest market for EVs, accounting for over 50% of global sales. However, Europe is emerging as the fastest-growing market with the most EV sales at the end of 2020.
- All major automotive companies have committed to discontinuing the production of ICE cars in the next decade or soon after. They have also embarked on large investments, often supported by government grants, for research, development and manufacturing, a sign that the EV market is gaining momentum; for example, the number of available EV models increased by 40% at the end of 2020 compared to a year earlier.
- However, the speed of market uptake will experience challenges, which could relate to both technology and affordability. For example, the number of plug-in charging installations is consistently increasing (currently at over 11 million chargers), but until this reaches a critical threshold to overcome users concerns about recharging, hesitancy will remain about switching to EVs. In addition, the development of rapid EV chargers will reduce the charging time, enabling EVs to compete with ICE vehicles.
- Concerns regarding range anxiety among potential users of EVs still needs to be addressed through the development of lighter and high energy density battery technology. However, any future technological advancement should consider the sustainability of sourcing rare earth materials, waste, and recycling issues.
- The deployment of the EV market will add significant pressure on grid infrastructure to increase both generation capacity and faster response times. However, EVs may also play a major role, in addition to their primary purpose of transportation, as an energy carrier supporting future smart power grids. V2G technology can, for example, be deployed through smart controllers to store electrical power generated from renewable sources at times of low demand, and discharge to the grid at times of peak demand. However, this still requires further research and development to address technical limitations, such as the impact on battery service life and users' behaviour and acceptability.
- The mass market uptake of EVs will constitute a major global re-industrialisation, in conjunction with the shift to renewable energy power generation. These technological and social transformations will result in immense benefits, thus improving the quality of life and the environment.

Author Contributions: Conceptualization, A.D.; Supervision, S.O. and R.B. All authors have read and agreed to the published version of the manuscript.

Funding: This research received no external funding.

Institutional Review Board Statement: Not applicable.

Informed Consent Statement: Not applicable.

Conflicts of Interest: The authors declare no conflict of interest.

References

1. Looney, B. *BP Statistical Review of World Energy 2020*, 69th ed.; British Petroleum Co.: London, UK, 2020; p. 66.
2. Cozzi, L.; Gould, T.; Bouckaert, S.; Kim, T.Y.; McNamara, K.; Wanner, B.; Mcglade, C.; Olejarnik, P.; Adam, Z.; Sarazola, L.A.; et al. *World Energy Outlook 2019*; IEA: Paris, France, 2019.
3. Cozzi, L.; Gould, T.; Bouckaert, S.; Crow, D.; Kim, T.Y.; Mcglade, C.; Olejarnik, P.; Wanner, B.; Wetzel, D.; Adam, Z.; et al. *World Energy Outlook 2020*; IEA: Paris, Frace, 2020.
4. Chen, K.; Winter, R.C.; Bergman, M.K. Carbon dioxide from fossil fuels: Adapting to uncertainty. *Energy Policy* **1980**, *8*, 318–330. [CrossRef]

5. BGS. The Greenhouse Effect. Available online: https://www.bgs.ac.uk/discovering-geology/climate-change/how-does-the-greenhouse-effect-work/ (accessed on 25 November 2021).
6. BBC. Climate Change—Impact of Human Activity. Available online: https://www.bbc.co.uk/bitesize/guides/zt6sfg8/revision/2 (accessed on 25 November 2021).
7. BBC. Climate Change: Last Decade UK's 'Second Hottest in 100 Years'. 2020. Available online: https://www.bbc.co.uk/news/science-environment-50976909 (accessed on 25 November 2021).
8. Sinsel, S.R.; Riemke, R.L.; Hoffmann, V.H. Challenges and solution technologies for the integration of variable renewable energy sources—A review. *Renew. Energy* **2020**, *145*, 2271–2285. [CrossRef]
9. Ritchie, H.; Roser, M. Energy. 2020. Available online: https://ourworldindata.org/energy (accessed on 22 December 2021).
10. Li, D.; Chiu, W.; Sun, H. Demand Side Management in Microgrid Control Systems. In *Microgrid: Advanced Control Methods and Renewable Energy System Integration*; Mahmoud, M.S., Ed.; Butterworth-Heinemann: Oxford, UK, 2017; pp. 203–230.
11. Strbac, G. Demand side management: Benefits and challenges. *Energy Policy* **2008**, *36*, 4419–4426. [CrossRef]
12. Su, H.-I.; El Gamal, A. Modeling and analysis of the role of energy storage for renewable integration: Power balancing. *IEEE Trans. Power Syst.* **2013**, *28*, 4109–4117. [CrossRef]
13. Muruganantham, B.; Gnanadass, R.; Padhy, N. Challenges with renewable energy sources and storage in practical distribution systems. *Renew. Sustain. Energy Rev.* **2017**, *73*, 125–134. [CrossRef]
14. Teter, J.; Tattini, J.; Petropoulos, A. *Tracking Transport 2020*; IEA: Paris, France, 2020.
15. McKerracher, C.; O'Donovan, A.; Albanese, N.; Soulopoulos, N.; Doherty, D.; Boers, M.; Fisher, R.; Cantor, C.; Frith, J.; Mi, S.; et al. *Electric Vehicle Outlook (EVO) 2021*; BloombergNEF (BNEF): London, UK, 2021.
16. Borba, B.S.M.; Szklo, A.; Schaeffer, R. Plug-in hybrid electric vehicles as a way to maximize the integration of variable renewable energy in power systems: The case of wind generation in northeastern Brazil. *Energy* **2012**, *37*, 469–481. [CrossRef]
17. Kempton, W.; Letendre, S.E. Electric vehicles as a new power source for electric utilities. *Transp. Res. Part D Transp. Environ.* **1997**, *2*, 157–175. [CrossRef]
18. Kempton, W.; Tomić, J. Vehicle-to-grid power implementation: From stabilizing the grid to supporting large-scale renewable energy. *J. Power Sources* **2005**, *144*, 280–294. [CrossRef]
19. Colmenar-Santos, A.; Muñoz-Gómez, A.M.; Rosales-Asensio, E.; López-Rey, Á. Electric vehicle charging strategy to support renewable energy sources in Europe 2050 low-carbon scenario. *Energy* **2019**, *183*, 61–74. [CrossRef]
20. DfT. *Annual Mileage of Cars by Ownership and Trip Purpose: England*; Department for Transport (DfT) Statistics: London, UK, 2020.
21. Ehsani, M.; Gao, Y.; Longo, S.; Ebrahimi, K.M. *Modern Electric, Hybrid Electric, And Fuel Cell Vehicles*; CRC Press: Boca Raton, FL, USA, 2018.
22. Standage, T. The Lost History of the Electric Car—And What It Tells Us about the Future of Transport. 2021. Available online: https://www.theguardian.com/technology/2021/aug/03/lost-history-electric-car-future-transport (accessed on 21 December 2021).
23. NationalGrid. The (Surprisingly Long) History of Electric Vehicles. 2021. Available online: https://www.nationalgrid.com/stories/journey-to-net-zero-stories/surprisingly-long-history-electric-vehicles (accessed on 21 December 2021).
24. Mui, S.; Shelby, M.; Chartier, D.; Ganss, D. *Plug-In Hybrids: A Scenario Analysis*; US Environmental Protection Agency: Washington, DC, USA, 2007.
25. Qin, N.; Raissi, A.; Brooker, P. *Analysis of Fuel Cell Vehicle Developments*; The Florida Solar Energy Center (FSEC): Cocoa, FL, USA, 2014.
26. Kebriaei, M.; Niasar, A.H.; Asaei, B. Hybrid electric vehicles: An overview. In Proceedings of the 2015 International Conference on Connected Vehicles and Expo (ICCVE), Shenzhen, China, 19–23 October 2015.
27. Agarwal, O.P.; Jhunjhunwala, A.; Kaur, P.; Yadav, N.; Chakrabarty, S.; Kumar, P.; Pai, M.; Bhatt, A. *A Guidance Document on Accelerating Electric Mobility in India*; WRI India: Mumbai, India, 2019.
28. Skoda. Types of Electric Vehicles—Do You Know THEM All? 2019. Available online: https://www.skoda-storyboard.com/en/skoda-world/innovation-and-technology/types-of-electric-vehicles-do-you-know-them-all/transport (accessed on 28 December 2021).
29. Un-Noor, F.; Padmanaban, S.; Mihet-Popa, L.; Mollah, M.N.; Hossain, E. A comprehensive study of key electric vehicle (EV) components, technologies, challenges, impacts, and future direction of development. *Energies* **2017**, *10*, 1217. [CrossRef]
30. Rao, H.S.; Hettige, H.; Singru, N.; Lumain, R.; Roldan, C. *Reducing Carbon Emissions from Transport Projects. Evaluation Knowledge Brief*; Dalkmann, H., Ed.; OECD: Paris, France, 2010. Available online: https://www.oecd.org/derec/adb/47170274.pdf (accessed on 10 January 2022).
31. McKerracher, C.; Izadi-Najafabadi, A.; O'Donovan, A.; Albanese, N.; Soulopolous, N.; Doherty, D.; Boers, M.; Fisher, R.; Cantor, C.; Frith, J.; et al. *Electric Vehicle Outlook (EVO) 2020*; BloombergNEF (BNEF): London, UK, 2020.
32. Hirst, D.; Winnet, J.; Hinson, S. *Electric Vehicles and Infrastructure*; House of Commons Library: London, UK, 2021; pp. 5–10.
33. Bigra, E.M.; Connelly, E.; Gorner, M.; Lowans, C.; Paoli, L.; Tattini, J.; Teter, J.; LeCroy, C.; MacDonnell, O.; Welch, D.; et al. *Global EV Outlook 2021 | Accelerating Ambitions Despite the Pandemic*; IEA: Paris, France, 2021.
34. Irle, R. Global EV Sales for 2021 H1. 2021. Available online: https://www.ev-volumes.com/country/total-world-plug-in-vehicle-volumes/ (accessed on 9 December 2021).

35. Abergel, T.; Bunsen, T.; Gorner, M.; Leduc, P.; Pal, S.; Paoli, L.; Raghavan, S.; Tattini, J.; Teter, J.; Wachche, S.; et al. *Global EV Outlook 2020 | Entering the Decade of Electric Drive?* IEA: Paris, France, 2020.
36. IEA. Global EV Data Explorer. 2021. Available online: https://www.iea.org/articles/global-ev-data-explorer (accessed on 28 December 2021).
37. DfT. Outcome and Response to Ending the Sale of New Petrol, Diesel and Hybrid Cars and Vans. 2021. Available online: https://www.gov.uk/government/consultations/consulting-on-ending-the-sale-of-new-petrol-diesel-and-hybrid-cars-and-vans/outcome/ending-the-sale-of-new-petrol-diesel-and-hybrid-cars-and-vans-government-response (accessed on 9 December 2021).
38. Pickett, L.; Winnet, J.; Carver, D.; Bolton, P. *Electric Vehicles and Infrastructure*; House of Commons Library: London, UK, 2021.
39. Wappelhorst, S. *Update on Government Targets for Phasing out New Sales of Internal Combustion Engine Passenger Cars*; International Council on Clean Transportation (ICCT): Washington, DC, USA, 2021.
40. Carey, N.; Steitz, C. EU Proposes Effective Ban for New Fossil-Fuel Cars from 2035. 2021. Available online: https://www.reuters.com/business/retail-consumer/eu-proposes-effective-ban-new-fossil-fuel-car-sales-2035-2021-07-14/response (accessed on 7 January 2022).
41. EV-Volumes. Global Plug-In Electric Vehicle Market Share in the First Half of 2021, by Main Producer. 2021. Available online: https://www.statista.com/statistics/541390/global-sales-of-plug-in-electric-vehicle-manufacturers/ (accessed on 22 December 2021).
42. TESLA. Investor Relations | Tesla Releases Third Quarter 2021 Financial Results. 2021. Available online: https://ir.tesla.com/#tab-quarterly-disclosure (accessed on 23 December 2021).
43. Kane, M. Global Plug-In Car Sales September 2021: Doubled To A New Record. 2021. Available online: https://insideevs.com/news/544743/global-plugin-car-sales-september2021/ (accessed on 23 December 2021).
44. VOLKSWAGEN. E-Mobility. 2021. Available online: https://www.volkswagenag.com/en/group/e-mobility.html (accessed on 23 December 2021).
45. VOLKSWAGEN. Volkswagen Group Doubles Deliveries of Pure E-Vehicles in Third Quarter. 2021. Available online: https://www.volkswagen-newsroom.com/en/press-releases/volkswagen-group-doubles-deliveries-of-pure-e-vehicles-in-third-quarter-7569 (accessed on 23 December 2021).
46. Taylor, C. Volkswagen Wants Half of Its Vehicle Sales to Be Electric by 2030. 2021. Available online: https://www.cnbc.com/2021/07/13/volkswagen-wants-half-of-its-vehicle-sales-to-be-electric-by-2030.html (accessed on 23 December 2021).
47. Riley, C. The Great Electric Car Race is Just Beginning. 2019. Available online: https://edition.cnn.com/interactive/2019/08/business/electric-cars-audi-volkswagen-tesla/ (accessed on 7 March 2021).
48. GM. GM Commits Full Support to SGMW's NEV Development. 2020. Available online: https://media.gm.com/media/cn/en/gm/news.detail.html/content/Pages/news/cn/en/2020/Dec/1214-GM.html (accessed on 23 December 2021).
49. DfT. *Licensed Ultra Low Emission Vehicles by Body Type and Propulsion or Fuel Type*; Department for Transport (DfT) Statistics: London, UK, 2021.
50. DfT. *Electric Vehicle Charging Device Statistics: October 2021*; Department for Transport (DfT) Statistics: London, UK, 2021.
51. ZAP-MAP. 2021. Available online: https://www.zap-map.com/ (accessed on 23 December 2021).
52. Lilly, C. EV Connector Types. 2020. Available online: https://www.zap-map.com/charge-points/connectors-speeds/ (accessed on 28 December 2021).
53. ZAP-MAP. EV Charging Stats 2021. 2021. Available online: https://www.zap-map.com/statistics/ (accessed on 24 December 2021).
54. Chen, T.; Zhan, X.; Wang, J.; Li, J.; Wu, C.; Hu, M.; Bian, H. A review on electric vehicle charging infrastructure development in the UK. *J. Mod. Power Syst. Clean Energy* **2020**, *8*, 193–205. [CrossRef]
55. DfT. *Electric Vehicle Charging in Residential and Non-Residential Buildings*; Department for Transport (DfT): London, UK, 2019.
56. BSI. PAS 1878 Energy Smart Appliances—System Functionality and Architecture. 2021. Available online: https://www.bsigroup.com/en-GB/about-bsi/uk-national-standards-body/about-standards/Innovation/energy-smart-appliances-programme/pas-1878/ (accessed on 10 December 2021).
57. Zhao, H.; Wang, L.; Chen, Z.; He, X. Challenges of fast charging for electric vehicles and the role of red phosphorous as anode material. *Energies* **2019**, *12*, 3897. [CrossRef]
58. Zhang, C.; Wei, Y.L.; Cao, P.F.; Lin, M.C. Energy storage system: Current studies on batteries and power condition system. *Renew. Sustain. Energy Rev.* **2018**, *82*, 3091–3106. [CrossRef]
59. Mirzaeian, M.; Abbas, Q.; Hunt, M.R.C.; Galeyeva, A.; Raza, R. Na-Ion Batteries. *Adv. Funct. Mater.* **2021**, *23*, 947–958.
60. Wang, Y.; Satoh, M.; Arao, M.; Matsumoto, M.; Imai, H.; Nishihara, H. High-energy, Long-cycle-life Secondary Battery with electrochemically pre-doped Silicon Anode. *Sci. Rep.* **2020**, *10*, 3208. [CrossRef]
61. Almarzooqi, A.; Mnatsakanyan, A.; Muruaga, E. Management of Used Lithium Ion Batteries of EV in Dubai. In Proceedings of the 2019 IEEE International Smart Cities Conference (ISC2), Casablanca, Morocco, 14–17 October 2019.
62. Sun, P.; Bisschop, R.; Niu, H.; Huang, X. A review of battery fires in electric vehicles. *Fire Technol.* **2020**, *56*, 1361–1410. [CrossRef]
63. Pinnangudi, B.; Kuykendal, M.; Bhadra, S. Smart Grid Energy Storage. In *The Power Grid*; D'Andrade, B.A., Ed.; Academic Press: Boston, MA, USA, 2017; pp. 93–135.

64. Kurzweil, P.; Garche, J. Overview of Batteries for Future Automobiles. In *Lead-Acid Batteries for Future Automobiles;* Elsevier: Amsterdam, The Netherlands, 2017; pp. 27–96.
65. Fiorello, D.; Zani, L. *EU Survey on Issues Related to Transport and Mobility;* JRC Science and Policy Report: Sevile, Spain, 2015.
66. Singer, M. *Consumer Views on Plug-in Electric Vehicles—National Benchmark Report;* National Renewable Energy Lab. (NREL): Golden, CO, USA, 2016.
67. Electiric Vehicle Database (EV-Database). *All Electric Vehicles.* Available online: https://ev-database.uk/#sort:path~type~order=.erange_real~number~desc | range-slider-range:prev~next=0~600 | range-slider-bijtelling:prev~next=0~600 | range-slider-acceleration:prev~next=2~23 | range-slider-fastcharge:prev~next=0~1100 | range-slider-lease:prev~next=150~2500 | range-slider-topspeed:prev~next=60~260 | paging:currentPage=0 | paging:number=9 (accessed on 10 January 2022).
68. Fishman, T.; Myers, R.J.; Rios, O.; Graedel, T.E. Implications of emerging vehicle technologies on rare earth supply and demand in the United States. *Resources* **2018**, *7*, 9. [CrossRef]
69. Schwarzer, S. *Challenges for the Growth of the Electric Vehicle Market;* The United Nations Environment Programme: Nairobi, Kenya, 2020.
70. Swiss Resource Capital AG. *Battery Metals Report 2019: Everything You Need to Know about the Battery Metals Lithium, Cobalt, Nickel and Vanadium;* Swiss Resource Capital AG: Herisau, Switzerland, 2019.
71. EEA. EEA Greenhouse Gases Data Viewer | Greenhouse Gas Emission Intensity of Electricity Generation. 2022. Available online: https://www.eea.europa.eu/data-and-maps/daviz/co2-emission-intensity-6/#tab-googlechartid_googlechartid_chart_111_filters=%7B%22rowFilters%22%3A%7B%7D%3B%22columnFilters%22%3A%7B%22pre_config_date%22%3A%5B2018%3B2019%5D%7D%3B%22sortFilter%22%3A%5B%22index_2018%22%5D%7D (accessed on 10 January 2022).
72. IEA. IEA Data Browser | Electricity. 2022. Available online: https://www.iea.org/fuels-and-technologies/electricity (accessed on 10 January 2021).
73. Hoehne, C.G.; Chester, M.V. Optimizing plug-in electric vehicle and vehicle-to-grid charge scheduling to minimize carbon emissions. *Energy* **2016**, *115*, 646–657. [CrossRef]
74. Lin, B.; Wu, W. The impact of electric vehicle penetration: A recursive dynamic CGE analysis of China. *Energy Econ.* **2021**, *94*, 105086. [CrossRef]
75. Sato, F.E.K.; Nakata, T. Energy consumption analysis for vehicle production through a material flow approach. *Energies* **2020**, *13*, 2396. [CrossRef]
76. Knobloch, F.; Hanssen, S.V.; Lam, A.; Pollitt, H.; Salas, P.; Chewpreecha, U.; Huijbregts, M.A.; Mercure, J.F. Net emission reductions from electric cars and heat pumps in 59 world regions over time. *Nat. Sustain.* **2020**, *3*, 437–447. [CrossRef] [PubMed]
77. Cox, B.; Mutel, C.L.; Bauer, C.; Mendoza Beltran, A.; Van Vuuren, D.P. Uncertain environmental footprint of current and future battery electric vehicles. *Environ. Sci. Technol.* **2018**, *52*, 4989–4995. [CrossRef] [PubMed]
78. Kurland, S.D. Energy use for GWh-scale lithium-ion battery production. *Environ. Res. Commun.* **2019**, *2*, 012001.
79. Hall, D.; Lutsey, N. *Effects of Battery Manufacturing on Electric Vehicle Life-Cycle Greenhouse Gas Emissions;* The Internal Council on Clean Transportation (ICCT): Washington, DC, USA, 2018.
80. Bieker, G. A global comparison of the life-cycle greenhouse gas emissions of combustion engine and electric passenger cars. *Communications* **2021**, *49*, 847129-102.
81. IEA. Amount of Spent Lithium-Ion Batteries from Electric Vehicles and Storage in the Sustainable Development Scenario, 2020–2040. 2021. Available online: https://www.iea.org/data-and-statistics/charts/amount-of-spent-lithium-ion-batteries-from-electric-vehicles-and-storage-in-the-sustainable-development-scenario-2020-2040 (accessed on 9 January 2022).
82. Saiyid, A. Threefold Increase in Recycling Needed to Help Meet 2030 Demand for Lithium-Ion EV Batteries. 2021. Available online: https://cleanenergynews.ihsmarkit.com/research-analysis/threefold-increase-in-recycling-needed-to-help-meet-2030-deman.html (accessed on 9 January 2022).
83. Gaines, L.; Richa, K.; Spangenberger, J. Key issues for Li-ion battery recycling. *MRS Energy Sustain.* **2018**, *5*, E14. [CrossRef]
84. Rohr, S.; Wagner, S.; Baumann, M.; Müller, S.; Lienkamp, M. A techno-economic analysis of end of life value chains for lithium-ion batteries from electric vehicles. In Proceedings of the 2017 Twelfth International Conference on Ecological Vehicles and Renewable Energies (EVER), Monte Carlo, Monaco, 11–13 April 2017.
85. Clement-Nyns, K.; Haesen, E.; Driesen, J. The impact of charging plug-in hybrid electric vehicles on a residential distribution grid. *IEEE Trans. Power Syst.* **2009**, *25*, 371–380. [CrossRef]
86. Kapustin, N.O.; Grushevenko, D.A. Long-term electric vehicles outlook and their potential impact on electric grid. *Energy Policy* **2020**, *137*, 111103. [CrossRef]
87. Kim, S.; Hur, J. A Probabilistic Modeling Based on Monte Carlo Simulation of Wind Powered EV Charging Stations for Steady-States Security Analysis. *Energies* **2020**, *13*, 5260. [CrossRef]
88. Mu, Y.; Wu, J.; Jenkins, N.; Jia, H.; Wang, C. A spatial–temporal model for grid impact analysis of plug-in electric vehicles. *Appl. Energy* **2014**, *114*, 456–465. [CrossRef]
89. Qian, K.; Zhou, C.; Allan, M.; Yuan, Y. Modeling of load demand due to EV battery charging in distribution systems. *IEEE Trans. Power Syst.* **2010**, *26*, 802–810. [CrossRef]
90. Qian, K.; Zhou, C.; Allan, M.; Yuan, Y. Load model for prediction of electric vehicle charging demand. In Proceedings of the 2010 International Conference on Power System Technology, Hangzhou, China, 24–28 October 2010.

91. Wang, S.; Zhang, N.; Li, Z.; Shahidehpour, M. Modeling and impact analysis of large scale V2G electric vehicles on the power grid. In Proceedings of the IEEE PES Innovative Smart Grid Technologies, Tianjin, China, 21–24 May 2012.
92. Papadopoulos, P.; Akizu, O.; Cipcigan, L.M.; Jenkins, N.; Zabala, E. Electricity demand with electric cars in 2030: Comparing Great Britain and Spain. *Proc. Inst. Mech. Eng. Part A J. Power Energy* **2011**, *225*, 551–566. [CrossRef]
93. Ueckerdt, F.; Kempener, R. *From Baseload to Peak: Renewables Provide a Reliable Solution*; International Renewable Energy Agency (IREA): Abu Dhabi, United Arab Emirates, 2015.
94. National Grid ESO. *Future Energy Scenarios (FES) 2021*; National Grid ESO: Warwick, UK, 2021.
95. Toniato, E.; Mehta, P.; Marinkovic, S.; Tiefenbeck, V. Peak load minimization of an e-bus depot: Impacts of user-set conditions in optimization algorithms. *Energy Inform.* **2021**, *4*, 23. [CrossRef]
96. Fathabadi, H. Novel wind powered electric vehicle charging station with vehicle-to-grid (V2G) connection capability. *Energy Convers. Manag.* **2017**, *136*, 229–239. [CrossRef]
97. Kempton, W.; Tomić, J. Vehicle-to-grid power fundamentals: Calculating capacity and net revenue. *J. Power Sources* **2005**, *144*, 268–279. [CrossRef]
98. Noori, M.; Zhao, Y.; Onat, N.C.; Gardner, S.; Tatari, O. Light-duty electric vehicles to improve the integrity of the electricity grid through Vehicle-to-Grid technology: Analysis of regional net revenue and emissions savings. *Appl. Energy* **2016**, *168*, 146–158. [CrossRef]
99. Ginigeme, K.; Wang, Z. Distributed optimal vehicle-to-grid approaches with consideration of battery degradation cost under real-time pricing. *IEEE Access* **2020**, *8*, 5225–5235. [CrossRef]
100. Peterson, S.B.; Whitacre, J.; Apt, J. The economics of using plug-in hybrid electric vehicle battery packs for grid storage. *J. Power Sources* **2010**, *195*, 2377–2384. [CrossRef]
101. Kiaee, M.; Cruden, A.; Sharkh, S. Estimation of cost savings from participation of electric vehicles in vehicle to grid (V2G) schemes. *J. Mod. Power Syst. Clean Energy* **2015**, *3*, 249–258. [CrossRef]
102. Hidrue, M.K.; Parsons, G.R. Is there a near-term market for vehicle-to-grid electric vehicles? *Appl. Energy* **2015**, *151*, 67–76. [CrossRef]
103. Cicconi, P.; Landi, D.; Morbidoni, A.; Germani, M. Feasibility analysis of second life applications for Li-Ion cells used in electric powertrain using environmental indicators. In Proceedings of the 2012 IEEE International Energy Conference and Exhibition (ENERGYCON), Florence, Italy, 9–12 September 2012.
104. Ahmadian, A.; Sedghi, M.; Mohammadi-ivatloo, B.; Elkamel, A.; Golkar, M.A.; Fowler, M. Cost-benefit analysis of V2G implementation in distribution networks considering PEVs battery degradation. *IEEE Trans. Sustain. Energy* **2017**, *9*, 961–970. [CrossRef]
105. Xiong, S. A Study of the Factors That Affect Lithium Ion Battery Degradation. Master's Thesis, University of Missouri, Columbia, MI, USA, May 2019.
106. Peterson, S.B.; Apt, J.; Whitacre, J. Lithium-ion battery cell degradation resulting from realistic vehicle and vehicle-to-grid utilization. *J. Power Sources* **2010**, *195*, 2385–2392. [CrossRef]
107. Bishop, J.D.; Axon, C.J.; Bonilla, D.; Tran, M.; Banister, D.; McCulloch, M.D. Evaluating the impact of V2G services on the degradation of batteries in PHEV and EV. *Appl. Energy* **2013**, *111*, 206–218. [CrossRef]
108. Guo, J.; Yang, J.; Cao, W.; Serrano, C. Evaluation of EV battery degradation under different charging strategies and V2G schemes. In Proceedings of the 8th Renewable Power Generation Conference (RPG 2019), Shanghai, China, 24–25 October 2019.
109. Uddin, K.; Jackson, T.; Widanage, W.D.; Chouchelamane, G.; Jennings, P.A.; Marco, J. On the possibility of extending the lifetime of lithium-ion batteries through optimal V2G facilitated by an integrated vehicle and smart-grid system. *Energy* **2017**, *133*, 710–722. [CrossRef]
110. Zhou, C.; Qian, K.; Allan, M.; Zhou, W. Modeling of the cost of EV battery wear due to V2G application in power systems. *IEEE Trans. Energy Convers.* **2011**, *26*, 1041–1050. [CrossRef]
111. BEIS. *2020 UK Greenhouse Gas Emissions, Provisional Figures*; Department for Business, Energy and Industrial Strategy (BEIS): London, UK, 2021.
112. Buekers, J.; Van Holderbeke, M.; Bierkens, J.; Panis, L.I. Health and environmental benefits related to electric vehicle introduction in EU countries. *Transp. Res. Part D Transp. Environ.* **2014**, *33*, 26–38. [CrossRef]
113. Turton, H.; Moura, F. Vehicle-to-grid systems for sustainable development: An integrated energy analysis. *Technol. Forecast. Soc. Chang.* **2008**, *75*, 1091–1108. [CrossRef]
114. Jiao, Z.; Yin, Y.; Ran, L.; Gao, Z. Integrating Vehicle-to-Grid Contract Design with Power Dispatching Optimization: Managerial Insights, and Carbon Footprints Mitigation. *Int. J. Prod. Res.* **2021**, 1–26. [CrossRef]
115. Ali, H.; Hussain, S.; Khan, H.A.; Arshad, N.; Khan, I.A. Economic and Environmental Impact of Vehicle-to-Grid (V2G) Integration in an Intermittent Utility Grid. In Proceedings of the 2020 2nd International Conference on Smart Power & Internet Energy Systems (SPIES), Bangkok, Thailand, 15–18 September 2020.
116. Bamisile, O.; Obiora, S.; Huang, Q.; Okonkwo, E.C.; Olagoke, O.; Shokanbi, A.; Kumar, R. Towards a sustainable and cleaner environment in China: Dynamic analysis of vehicle-to-grid, batteries and hydro storage for optimal RE integration. *Sustain. Energy Technol. Assess.* **2020**, *42*, 100872. [CrossRef]
117. Zhao, Y.; Noori, M.; Tatari, O. Vehicle to Grid regulation services of electric delivery trucks: Economic and environmental benefit analysis. *Appl. Energy* **2016**, *170*, 161–175. [CrossRef]

118. Sioshansi, R.; Denholm, P. Emissions impacts and benefits of plug-in hybrid electric vehicles and vehicle-to-grid services. *Environ. Sci. Technol.* **2009**, *43*, 1199–1204. [CrossRef]
119. California ISO (CAISO). Today's Outlook | Net Demand Trend. 2022. Available online: http://www.caiso.com/TodaysOutlook/Pages/default.aspx#section-net-demand-trend (accessed on 6 January 2022).
120. California ISO (CASIO). What the Duck Curve Tells Us about Managing a Green Grid. Available online: https://www.caiso.com/documents/flexibleresourceshelprenewables_fastfacts.pdf (accessed on 10 January 2022).
121. Atia, R.; Yamada, N. More accurate sizing of renewable energy sources under high levels of electric vehicle integration. *Renew. Energy* **2015**, *81*, 918–925. [CrossRef]
122. McPherson, M.; Ismail, M.; Hoornweg, D.; Metcalfe, M. Planning for variable renewable energy and electric vehicle integration under varying degrees of decentralization: A case study in Lusaka, Zambia. *Energy* **2018**, *151*, 332–346. [CrossRef]
123. Shi, R.; Li, S.; Zhang, P.; Lee, K.Y. Integration of renewable energy sources and electric vehicles in V2G network with adjustable robust optimization. *Renew. Energy* **2020**, *153*, 1067–1080. [CrossRef]
124. Sharifi, P.; Banerjee, A.; Feizollahi, M.J. Leveraging owners' flexibility in smart charge/discharge scheduling of electric vehicles to support renewable energy integration. *Comput. Ind. Eng.* **2020**, *149*, 106762. [CrossRef]
125. Child, M.; Nordling, A.; Breyer, C. Scenarios for a sustainable energy system in the Åland Islands in 2030. *Energy Convers. Manag.* **2017**, *137*, 49–60. [CrossRef]
126. Schuller, A.; Flath, C.M.; Gottwalt, S. Quantifying load flexibility of electric vehicles for renewable energy integration. *Appl. Energy* **2015**, *151*, 335–344. [CrossRef]
127. Richardson, D.B. Electric vehicles and the electric grid: A review of modeling approaches, Impacts, and renewable energy integration. *Renew. Sustain. Energy Rev.* **2013**, *19*, 247–254. [CrossRef]
128. Nguyen, H.N.; Zhang, C.; Mahmud, M.A. Optimal coordination of G2V and V2G to support power grids with high penetration of renewable energy. *IEEE Trans. Transp. Electrif.* **2015**, *1*, 188–195. [CrossRef]
129. Pfeifer, A.; Dobravec, V.; Pavlinek, L.; Krajačić, G.; Duić, N. Integration of renewable energy and demand response technologies in interconnected energy systems. *Energy* **2018**, *161*, 447–455. [CrossRef]
130. Šare, A.; Krajačić, G.; Pukšec, T.; Duić, N. The integration of renewable energy sources and electric vehicles into the power system of the Dubrovnik region. *Energy Sustain. Soc.* **2015**, *5*, 27. [CrossRef]
131. Schetinger, A.M.; Dias, D.H.N.; Borba, B.S.M.C.; Pimentel da Silva, G.D. Techno-economic feasibility study on electric vehicle and renewable energy integration: A case study. *Energy Storage* **2020**, *2*, e197. [CrossRef]
132. Freire, R.; Delgado, J.; Santos, J.M.; De Almeida, A.T. Integration of renewable energy generation with EV charging strategies to optimize grid load balancing. In Proceedings of the 13th International IEEE Conference on Intelligent Transportation Systems, Funchal, Portugal, 19–22 September 2010.
133. Zhang, Q.; Tezuka, T.; Ishihara, K.N.; Mclellan, B.C. Integration of PV power into future low-carbon smart electricity systems with EV and HP in Kansai Area, Japan. *Renew. Energy* **2012**, *44*, 99–108. [CrossRef]
134. Ekman, C.K. On the synergy between large electric vehicle fleet and high wind penetration–An analysis of the Danish case. *Renew. Energy* **2011**, *36*, 546–553. [CrossRef]
135. Hassan, A.S.; Marmaras, C.E.; Xydas, E.S.; Cipcigan, L.M.; Jenkins, N. Integration of Wind Power Using V2G as a Flexible Storage. In Proceedings of the IET Conference on Power in Unity: A Whole System Approach, London, UK, 16–17 October 2013.

wind

MDPI

Article

Savonius Wind Turbine Numerical Parametric Analysis Using Space-Filling Design and Gaussian Stochastic Process

Aristotle T. Ubando [1,2,3,*], Rathana San [4] and John Deric P. Cruz [1]

1 Department of Mechanical Engineering, De La Salle University, Manila 0922, Philippines; john_deric_cruz@dlsu.edu.ph
2 Thermomechanical Analysis Laboratory, De La Salle University, Laguna Campus, Biñan 4024, Philippines
3 Center for Engineering and Sustainable Development Research, De La Salle University, Manila 0922, Philippines
4 Industrial and Mechanical Engineering Department, Institute of Technology of Cambodia, Phnom Penh 12150, Cambodia; san.rathana@itc.edu.kh
* Correspondence: aristotle.ubando@dlsu.edu.ph

Abstract: Wind energy is an alternative source of clean energy to address the growing energy demand and provide pollution-free electricity. With the rapid development of urban areas, high wind energy resources such as high-rise building rooftops are excellent locations for urban wind turbine installation. One of the practical and simple urban wind turbines is the Savonius design. It has a simple design, easy to maintain, and is very affordable. This work focuses on the design evaluation of a Savonius wind turbine (SWT) by varying the rotor diameter, rotor height, and twist angle for urban applications. A transient computational fluid dynamics (CFD) approach is applied to assess the various design treatments using a space-filling design of experiments. To address the spaces in the hypercube statistical design, a sphere packing design method was adopted which suited the evaluation of computational simulations results such as that of the CFD. The Gaussian stochastic process model was applied to establish the trend of the parametric performance of the optimized SWT design through the model fitting. The results have shown that optimized SWT performs well with its self-starting capability compared to the traditional Savonius design. In addition, the optimized SWT has shown a better peak power coefficient compared with the results of previous works on the design of SWT.

Keywords: computational fluid dynamics (CFD); space-filling design; Savonius wind turbine; Gaussian stochastic process; urban wind turbines; sphere packing design

Citation: Ubando, A.T.; San, R.; Cruz, J.D.P. Savonius Wind Turbine Numerical Parametric Analysis Using Space-Filling Design and Gaussian Stochastic Process. *Wind* **2022**, 2, 113–128. https://doi.org/10.3390/wind2010007

Academic Editor: Andrés Elías Feijóo Lorenzo

Received: 25 January 2022
Accepted: 16 February 2022
Published: 18 February 2022

1. Introduction

The power generation industry has been one of the most important industries in the world especially as the demand of world energy consumption continues to increase. The IEA [1] projected the increase of energy demand by the year 2035 as twice that of the 1990s. Thus, the need to construct different power generation plants increases as well to meet the demand. However, most of these conventional power plants consume non-renewable raw materials such as coal, petroleum, and other fossil fuels. Accordingly, Calautit et al. [2] emphasized that the continuous consumption of fossil fuels would lead to more generation of greenhouse gas emissions. These greenhouse gas emissions play a significant role in worsening the global warming and climate change situation. To address these global concerns, the industry continues to seek renewable power generation methods that can sustainably generate the projected energy demand. Renewable energy sources for power generation have provided an alternative means of power generation for the past few decades. These vary between solar, hydroelectric, wind, and biomass, among others. Among these alternative energy sources, wind energy has gained interest from engineers and scientists to potentially generate energy through wind turbine design for the past

20 years. According to the WWEA [3], the total energy generation from wind energy sources skyrocketed as the awareness of the use of the renewable source for energy generation was widely accepted. Traditional wind turbines are large windmill-like structures found in flat-terrain areas which are the main source of wind energy [4]. However, due to the size of each of these wind turbines, the creation of wind farms has become an issue due to the requirement of a large area. Hence, a small-scale wind turbine for the urban areas significantly gained traction. A wind turbine for urban areas requires further evaluation since its performance characteristics are still under development [2]. Škvorc and Kozmar [5] outlined the significance of urban wind turbines to the progress of smart and sustainable cities due to their potential advantages. These advantages range from enhanced energy efficiency, clean energy, and economic advancements [6]. In addition, Naderipour et al. [7] highlighted that urban wind turbines aid in satisfying the power requirement of an area as this remarkably reduces the power transmission losses from the power plants. Moreover, urban wind turbines have the potential to limit greenhouse gas emissions [8]. However, Tasneem et al. [9] have noted that the minimum emission of greenhouse gases from urban wind turbines is attributed to their production process. Although urban wind turbines consist of various advantages, they still pose different disadvantages. Chrysochoidis-Antsos et al. [10] have recognized that urban wind turbines can be disturbing in terms of their visual representation and loud noise. Thus, the development of various urban wind turbines was geared towards the improvement of the turbine performance.

One of the known urban wind turbine designs is the Savonius wind turbine (SWT). Kim and Cheong [11] have revealed that the most appropriate type of small-scale wind turbine is the SWT owing to its multi-directional starting capability and its low noise. Moreover, further research and development suggested other advantages such as its good performance, simple design, and low cost in terms of production and maintenance [9]. Even though the SWT has the potential to address the problems of urban wind turbines, its low efficiency continues to pose a challenge for its use [12–14]. Various works were performed to address this problem [15]. Previous studies in the enhancement of the performance of the SWT are discussed as follows. Yahya et al. [16] investigated the effects of multi-bladed SWT on its rotor performance. Their results have shown that the number of blades significantly increased the performance of the SWT [16]. Saad et al. [17] evaluated the performance of the rotor of the SWT where the effect of a multi-stage was assessed. Their results revealed the multi-stage SWT self-starting capability and enhance torque oscillation [17]. Meanwhile, the single-stage SWT generates a higher dynamic moment. Moreover, Irabu and Roy [18] presented an SWT wind tunnel with the evaluation of the effect of a guide-box tunnel to improve its performance. The area ratio of this guide box is 0.43 compared to the SWT. In this study, it was found that the use of the guide-box tunnel increased its efficiency by 150% [18]. However, Dewan et al. [15] argued that the addition of this guide-box tunnel further complicated the SWT system and limited the capability of the turbine to adapt from the wind conditions.

Recent studies have performed parametric design improvements to enhance the performance of the SWT. Aboujaoude et al. [14] proposed to include an axisymmetric deflector to the SWT to elevate its aerodynamic performance. Tjahjana et al. [19] assessed the influence of the slotted blades in an SWT that resulted in a 34% increase in the power coefficient. Xu et al. [20] recently introduced the use of plasma excitation flow control to the SWT that enabled a significant improvement in its output torque and efficiency. Al-Ghriybah et al. [21] investigated the influence of the blade spacing in an SWT that resulted in the improvement of its power coefficient by one third. To further increase the power production of the SWT, Berhanu et al. [22] evaluated its use as an exhaust air energy recovery in a central air-conditioning system. Their result has shown an overall improvement in the system efficiency of about 87% [22]. To further enhance the aerodynamic performance of an SWT, a computational approach was adopted such as the use of the ANSYS software [23]. Using the same method, Marinić-Kragić et al. [13] had multiple arc blades that enabled the SWT to improve the performance by 50%. On the other hand, Marzec et al. [24]

further explored the expansion of the computational approach by the development of a fluid–structure interface (FSI) of an SWT. The FSI model was then validated with an experimental result and found to influence the centrifugal forces in the turbine operation. With the use of a computational fluid dynamic (CFD) approach, Marinić-Kragić et al. [25] introduced a scoop-based design of an SWT that enabled a 10% to 39% improvement in its power coefficient.

Various modelling techniques were developed to further enhance the performance of urban wind turbines. An adaptive neuro-fuzzy inference system was developed by Elsisi et al. [26] for the improved blade pitch control of urban wind turbines. A numerical approach coupled with an experimental test was performed by Zalhaf et al. [27] to understand the transient behavior of wind turbine blades that are hit by lightning. Recently, a hybrid numerical approach was developed by Sayed et al. [28] to incorporate dust loaded wire-duct precipitators in the computational domain of urban wind turbines. With the transmission of electricity from the wind turbine, Abouelatta et al. [29] developed a numerical method that integrated a full multigrid method to evaluate the fast corona discharge with varying wind speeds.

These studies are significant in enhancing the performance of the SWT. However, the design enhancement approach using the CFD method coupled with space-filling design with the sphere packing method has not been explored in previous works. CFD is a numerical approach that allows a design of an SWT to be evaluated computationally [30]. On the other hand, space-filling design is a type of statistical design of experiment tool that addresses the non-uniformity of the inside experimental space [31]. The sphere packing method is used in a hypercube which can be well represented by results of computational approaches such as the CFD method [32]. Moreover, a Gaussian stochastic process (GaSP) model was used to fit in a multivariate normal distribution the developed model results from the space-filling design [33]. Based on the available literature, no studies have applied CFD and space-filling design with the GaSP model to design an SWT. Hence, the novelty of this work lies in the methodology to redesign the SWT using a CFD approach with space-filling design of experiment and GaSP model. The designed SWT was evaluated based on the maximum power coefficient and assessed the influences of the considered design parameters.

2. Performance of the Savonius Wind Turbine

The performance of the SWT is best described by its power coefficients where the moment coefficient and the power coefficient are given as C_m and C_p, respectively. These coefficients were obtained from the ratios between the actual parameter over the theoretical parameter. Two of the important parameters are the moment and the power coefficients [34]. The moment coefficient is quantified using Equation (1) while the power coefficient is determined using Equation (2).

$$C_m = \frac{M}{M_w} = \frac{M}{(1/2)\rho r A V^2} \tag{1}$$

$$C_P = \frac{P_w}{(1/2)\rho A V^3} = \frac{M}{(1/2)\rho A V^2 r}\frac{r\omega}{V} \tag{2}$$

where A provides the swept area of the turbine in m^2, M represents the rotor moment in N-mm, P_w indicates the extracted power in W, V depicts the air velocity in m/s, r shows the radius of the rotor in m, ρ covers the density of air in kg/m^3, and λ denotes the tip-speed ratio. Barlow et al. [35] described the model frontal area of the turbine concerning the cross-sectional area of the air domain as shown in Equation (3).

$$\epsilon = \left(\frac{Model\ frontal\ area}{Cross - sectional\ area}\right) = \frac{A}{C} \tag{3}$$

where C indicates the cross-sectional area in m^2 and ϵ represents the percent blockage ratio.

To establish a deterministic model validation for the simulation results, the Gaussian Stochastic process (GaSP) model was considered. The GaSP model consists of an $n \times 1$ vector for its output response, $y(x)$, which is given as $N(\mu 1_n, \sigma^2 R(X, \theta)$. The parameter $R(X, \theta)$ in the output response is a $n \times n$ correlation matrix and can take different forms depending on the situation [36]. Equation (4) provides the correlation function of the GaSP model while Equation (5) shows the fitted Gaussian prediction function. The GaSP equations shown in both Equations (4) and (5) were adopted from Jones and Johnson [37].

$$(R_{ij}(X,\theta)) = exp\left(-\sum_{k=1}^{P}\theta_k\left(x_{ik} - x_{jk}\right)^2\right) \quad (4)$$

$$\hat{y}(x) = \hat{\mu} + r'\,(x,\,\hat{\theta})\,R^{-1}\,(x,\hat{\theta})\,(y - \hat{\mu}1_n) \quad (5)$$

where $R(X,\theta)$ is a square correlation matrix wherein it is a function of the design space and points as well as some unknown thetas, $\hat{\mu}$ provides the fitted mean, $\hat{\sigma}$ indicates the variance, and $r(x,\hat{\theta})$ represents the $n \times 1$ vector with a condition if $(\theta_k \geq 0)$.

3. Methodology

Different analytical and simulation software was used to perform the study. The discussion on the methodology is divided into three portions such as the CFD model, the design of experiments through the space-filling design of experiment, and the GaSP model.

3.1. Computational Fluid Dynamics

The CFD approach was performed using the commercial software ANSYS Fluent. The $K - \varepsilon$ turbulence model was used since this model provides accurate results with the SWT with relatively lesser computational time. These numerical simulations were conducted to obtain the various aerodynamic performance of each design treatment while determining its respective power and moment coefficients.

3.1.1. Simulation Set-Up

The geometries of the developed model were constructed using the Design Modeler of ANSYS. The geometry of the SWT rotor is shown in Figure 1 where the rotor diameter (D) was evaluated to change from 0.29 m to 0.40 m, the rotor height (H) was varied from 0.16 m to 0.35 m, and the twist angle was changed from 15° to 65°.

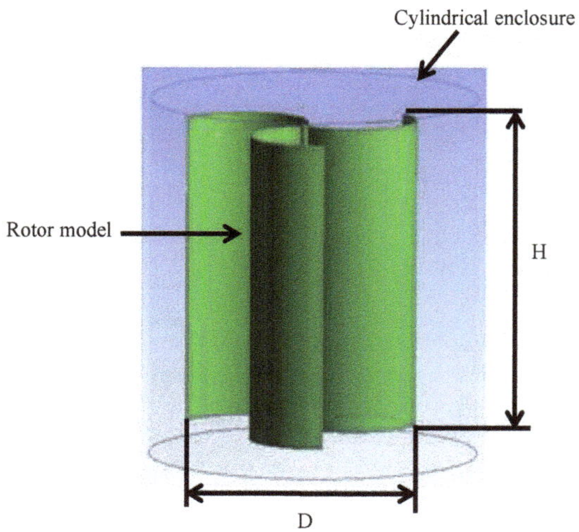

Figure 1. An example of the geometric model of the SWT.

In the development of the geometry, a 21.6% percent blockage ratio (ϵ) [38] was considered as shown in Equation (3). Meanwhile, the upwind distance of the rotor used was 0.30 m while a downstream distance of the rotor utilized was 1.20 m. In the development of the geometry of the SWT, two domains were identified, which are the stationary zone and the rotational zone as shown in Figure 2. The stationary zone is the wind tunnel domain which consists of a rectangular-shaped air domain; while the rotational zone is a cylindrical-shaped domain, as shown in the figure, that allows rotation of the SWT shaft axis.

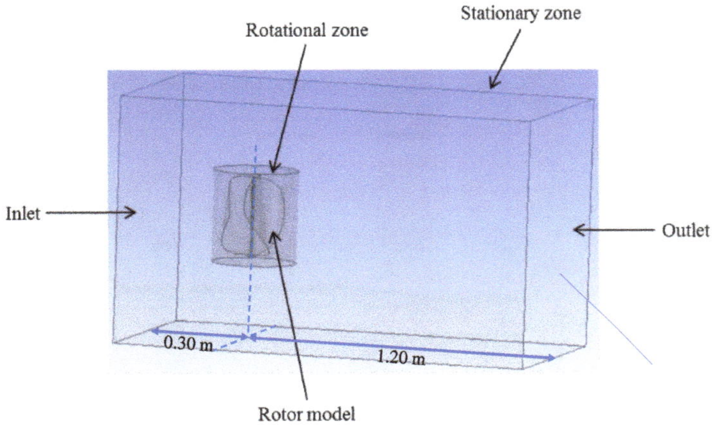

Figure 2. The geometry simulation setup.

As shown in Figure 2, the inlet and the outlet zone were identified, while all the other sides of the rectangular zone were set to a walled domain. An air inlet velocity of 6.5 m/s was used together with an outlet pressure of 1 bar. A convergence criterion of 1×10^{-6} was used with a maximum of 1000 iterations per simulation run. Figure 3 shows the separated cell zone of the simulation in a section view.

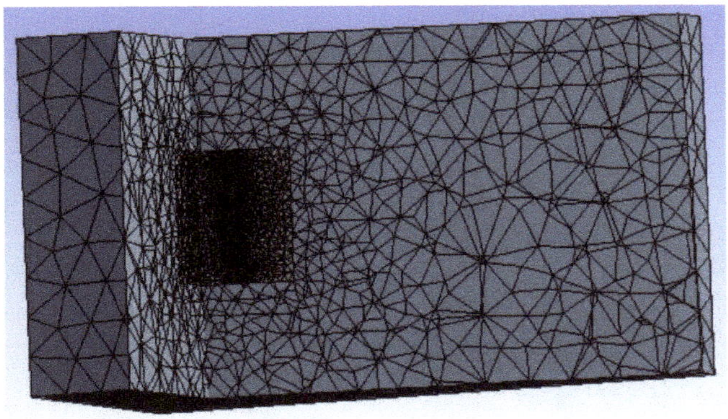

Figure 3. The separated cell zone of the simulation.

3.1.2. Evaluation of the Simulation Mesh

The quality of the mesh of the geometries for the simulation is an important step to consider to attain an accurate solution with low computational cost. The mesh setup of the developed SWT model is shown in Figure 4. The top view of the rotor that showcase the near wall mesh and the mesh surrounding the rotor is shown in Figure 5.

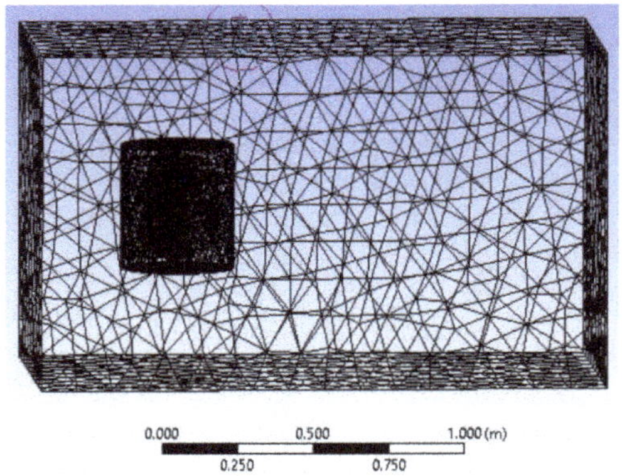

Figure 4. Mesh of the simulation setup.

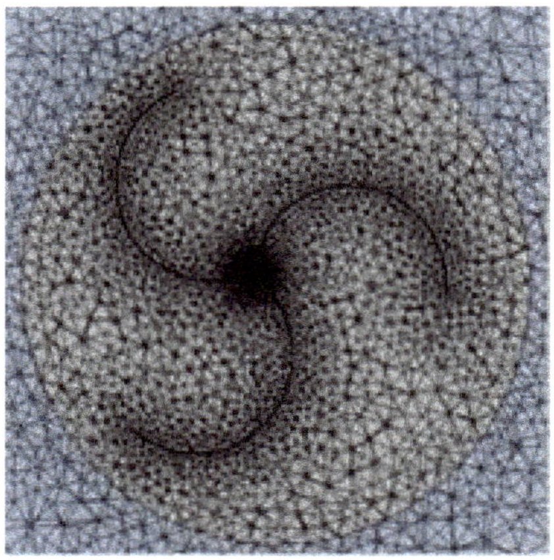

Figure 5. Top view of the rotor showing the mesh near the walls of the rotor.

The mesh of the developed SWT model was performed to evaluate the skewness value, aspect ratio, and orthogonal quality. A tetrahedral grid shape was adopted as the mesh shape as it was able to capture the complex curvature shape of the SWT. With these factors of interest, a mesh independence test was performed by changing the mesh sizes of each geometry from −100 to 100 as shown in Table 1. Figure 6 illustrates the mesh independent test of the simulation setup in terms of skewness value and orthogonal quality.

Table 1. Summary table of the values of the mesh independent testing.

Mesh Sizing (Relevance Center, Relevance)	Number of Nodes	Aspect Ratio	Skewness Value	Orthogonal Quality
Coarse (−100)	29,636	1.942	0.254	0.844
Coarse (0)	68,709	1.898	0.245	0.85
Medium (0)	77,040	1.895	0.243	0.851
Fine (0)	107,690	1.88	0.24	0.853
Fine (100)	313,822	1.836	0.221	0.862

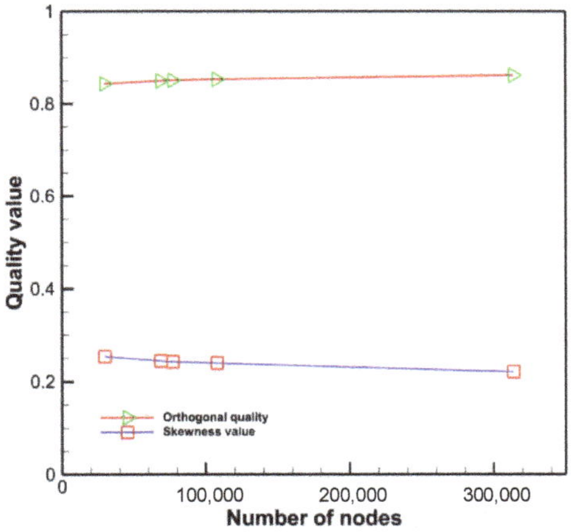

Figure 6. Mesh independence test in terms of orthogonal quality and skewness value.

The design of experiments was performed where each geometry model of the SWT was developed. The developed SWT designs were evaluated into two types of simulations, namely, the static condition and the dynamic condition. The static condition simulation was done using the moving reference frame (MRF) while the dynamic condition simulation was performed using the dynamic sliding mesh motion (SMM).

3.2. *Design of Experiments*

The JMP statistical analysis software was utilized to determine the simulation treatments for the study. The sphere-packing design method under the space-filling design is used to evaluate each design for the different parameters considered. The parametric factors considered are shown in Table 2 where the values were adopted from Lee et al. [38] which showed indicative promising results for the SWT.

Table 2. Summary of the parameters used in the sphere-packing design method adopted from Lee et al. [38].

Factors	Role	Values
Rotor Diameter (m)	Continuous	0.29 to 0.40
Rotor Height (m)	Continuous	0.16 to 0.35
Twist Angle (degree)	Continuous	15 to 65

From Table 2, the total number of simulation treatments obtained from the space-filling design of the experiment is 30. Hence, 30 unique SWT geometries were prepared and evaluated based on the moment coefficient and power coefficient performances.

3.3. *Gaussian Stochastic Process (GaSP) Model*

After the CFD simulation of the 30 treatments, the GaSP model was adopted to assess the effects of the height and the diameter of the rotor as well as its twist angle with regards to the moment coefficient and power coefficient performances. Afterward, model fitting was performed using the GaSP model to derive the model equations of the input and output factors. It was found that there were two responses to be obtained from this model, namely, the model coefficients and power coefficients. For the GaSP model validation, the model also considered the moment and the coefficient of moment as a factor on top of the considered 3 parametric factors defined in Table 2. These 5 parameters were evaluated based on the power coefficient response.

4. Results and Discussion

The results of the study are elaborated in this section along with its discussion.

4.1. Gaussian Stochastic Process Results

The results of the model fitting of the 30 simulation treatment results to the GaSP model are shown in Figure 7, which highlights the power coefficient prediction plot for the design treatments.

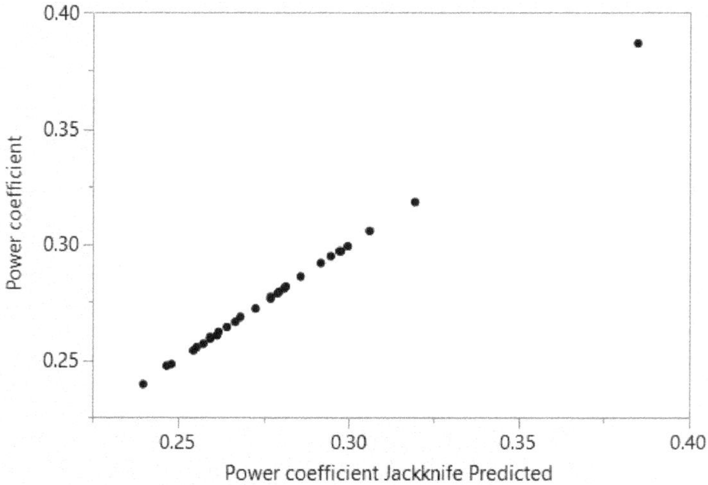

Figure 7. Model fitting of the power coefficient using the GaSP model.

Based on this graph, the actual power coefficient is directly proportional to the predicted power coefficient. The 45° plot line shown in the figure signifies that the model fitted the predicted model. Aside from this, the θ-value of each performance parameter is presented in Table 3.

Table 3. Summary table of the θ-values of each parameter.

Parameters	θ-Values
Rotor Diameter	0.93
Rotor Height	7.80×10^{-7}
Twist Angle	0
Moment	0
Moment Coefficient (C_m)	0.63

The scale of the θ-values ranges from 0 to 1 which indicates the relative correlation of each of the parameters. This means that the rotor height, the twist angle, and the moment have a relatively higher correlation with the respect to the power coefficient, while the rotor diameter and the moment coefficient have a relatively lower correlation to the power coefficient. Moreover, Figure 8 shows the interaction plot of each of the considered parameters concerning the power coefficient. It was found that there is no interaction effect between the response and each performance parameters since the lines in the graphs are all parallel. However, it is interesting to note how the range of values for each of the performance parameters varies in terms of the change in the power coefficient. The gaps between the red line and the blue line represent the difference of the power coefficient at changing parameters. The larger the gap between the red and the blue lines indicates that the power coefficient is sensitive to the changing values of the specific parameter. Hence, no gap between the red and the blue line signifies the insensitivity of the power coefficient on the specified parameter.

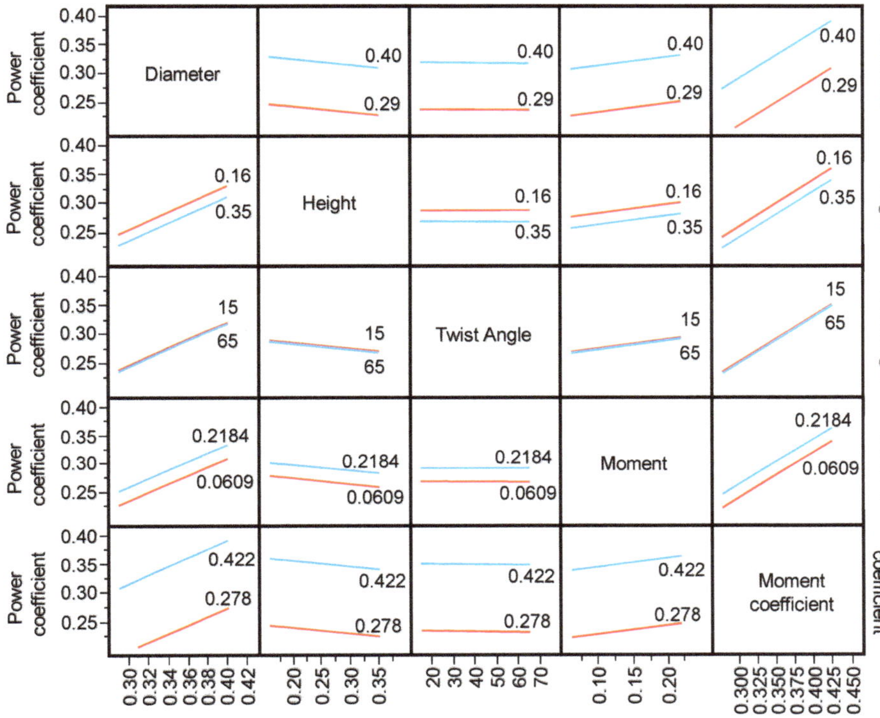

Figure 8. The interaction plot of the power coefficient versus the performance parameters.

Subsequently, the profiler plots of each performance parameter in terms of the power coefficient are shown in Figure 9a–e. It can be seen in these graphs that the height and the diameter of the rotor are directly proportional to the power coefficient. It is to be noted that the power coefficient increases as the vacuum pressure and the moment length of the blade increases. Similarly, the power coefficient was found to be directly proportional to the moment coefficient as well as the moment. This can be attributed to the direct relationship of the moment or torque from the power calculation of the blades of the wind turbine. Among the performance parameters considered, it is only the twist angle that causes the power coefficient to decrease when the parameter increases. This is due to the decrease of suction pressure and moment arm as the twist angle is increased.

The resulting fitted model for the performance parameters with respect to the power coefficient is shown in Equation (6). The establishment of the coefficient is shown in the fitted equation as shown below.

$$y = 0.296 + \sum_{i=1}^{n} r_i' e^{-(0.93\ (x_1-D_i)^2+0.78\times10^{-6}\ (x_2-H_i)^2+0.63\ (x_3-C_{m_i})^2)} \quad (6)$$

where y is the projected power coefficient, D_i indicates the diameter, H_i shows the height, C_{mi} denotes the moment coefficient, r' is the stochastic process in a finite set of i. The values of the fitted mean and the variance for the GaSP model fitting are shown in Table 4 while the factor r' values are shown in Table 5.

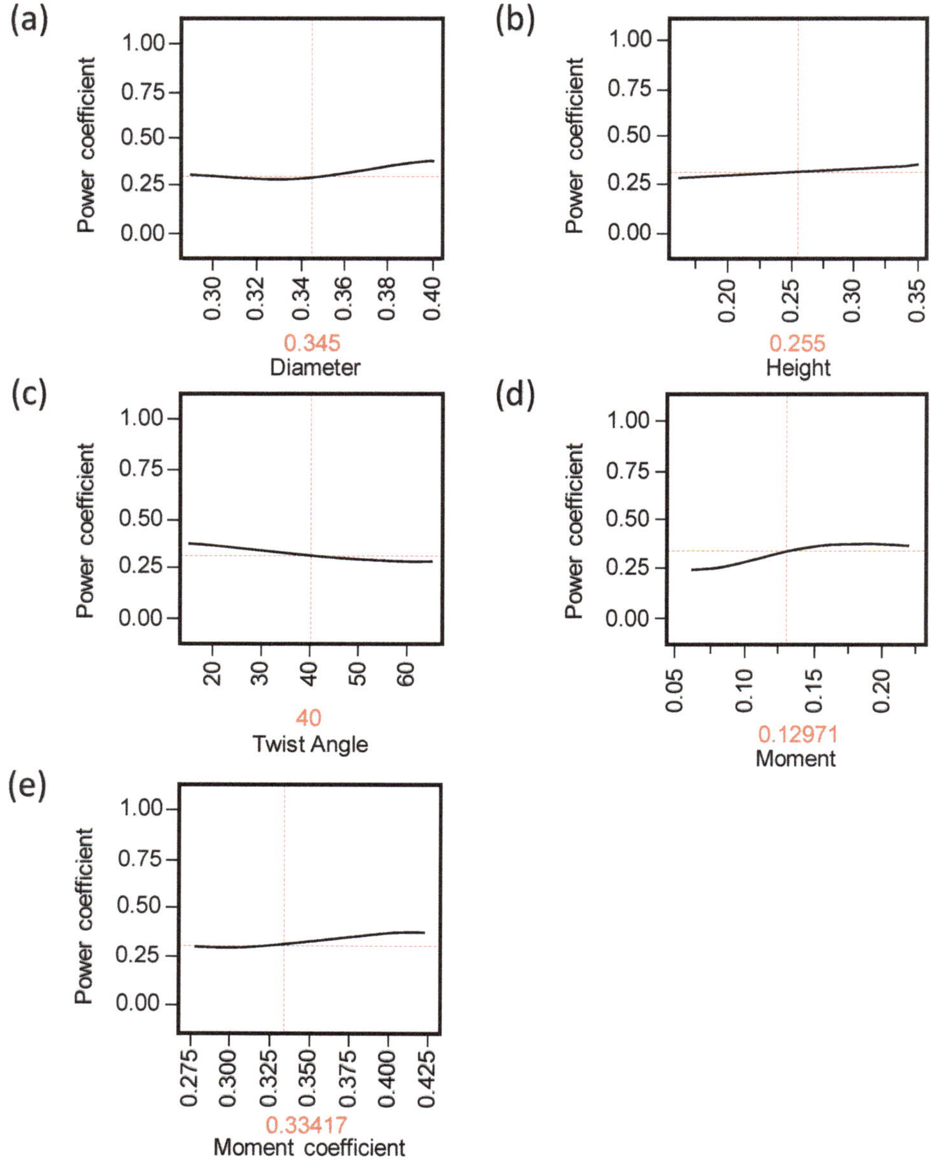

Figure 9. The profiler plot regarding the power coefficient versus the (**a**) diameter, (**b**) height, (**c**) twist angle, (**d**) moment, and (**e**) moment coefficient.

Table 4. The results of the GaSP model fit.

Fitted Mean (μ)	Variance (σ^2)	Nugget
0.296	0.564	8.11×10^{-8}

Table 5. The r' values.

No.	R'
1	1866.17
2	−2635.16
3	−2803.35
4	1206.91
5	4019.12
6	−2056.81
7	3222.55
8	86.46
9	2043.26
10	1946.31
11	2394.82
12	−4339.64
13	−1982.76
14	2257.21
15	2195.78
16	183.21
17	945.57
18	1150.75
19	−1056.43
20	−1425.85
21	3564.55
22	−1442.13
23	−420.78
24	−2082.96
25	−64.14
26	−5626.42
27	−826.45
28	1014.12
29	−540.61
30	−793.3

4.2. Static Condition Analysis

The static condition of the simulation represents the starting position of the wind turbine. The moment coefficients of the new SWT were analyzed and compared with the traditional SWT with a semi-cylindrical shape SWT. The moment coefficients in the static condition of both the new and the old are provided in Figures 10 and 11. It can be seen in Figure 10 that the static moment coefficient of the new design remains positive. It is also shown here that the maximum static moment coefficient can be observed at the angle of attack values of 90° to 160° and 450° to 540°. Meanwhile, Figure 11 shows that the static moment coefficient of the traditional design crossed through the negative values. Additionally, it can be recognized that the minimum values of static moment coefficients are from 20° to 80° and 330° to 380°, respectively. With this said, it was found that the optimized design can start by itself since its moment coefficients remain positive while the SWT with a semi-cylindrical shape has negative moment coefficients which hinder its ability to self-start [39]. Therefore, the new SWT has a relatively superior self-starting capability compared to the traditional SWT.

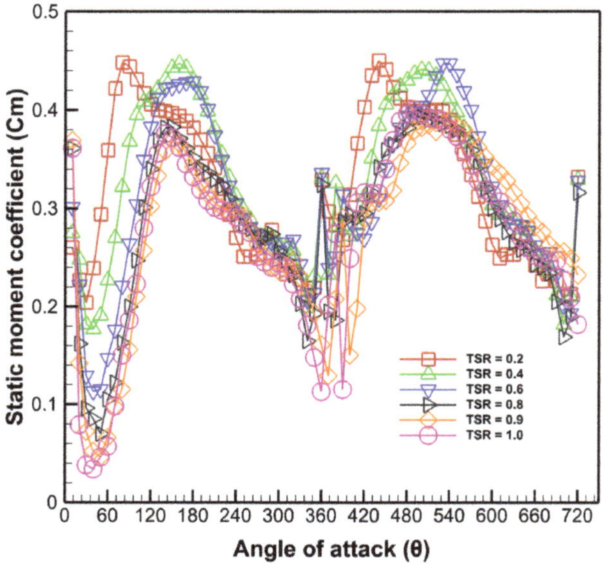

Figure 10. The relationship between the static moment coefficient of the new SWT versus its angle of attack.

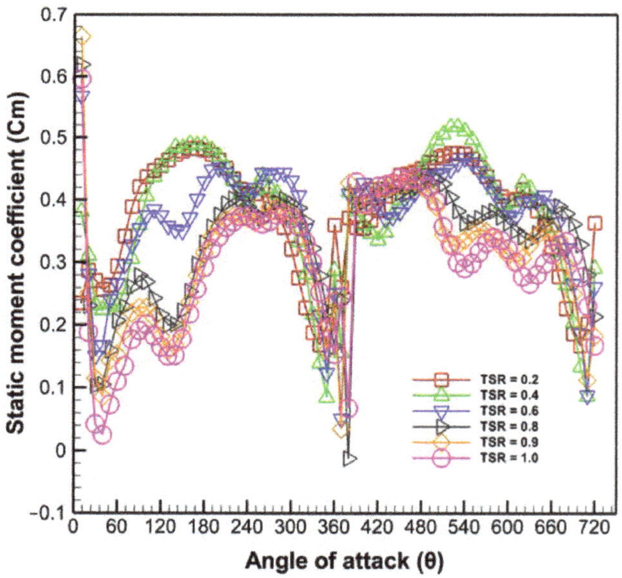

Figure 11. The relationship between the static moment coefficient of the traditional SWT wind turbine with semi-cylindrical shape versus its angle of attack.

4.3. Dynamic Condition Analysis

The parameters that were tested for the dynamic condition include the power coefficient and the dynamic moment coefficient of the wind turbines. Figure 12 showed the comparative study of the different SWT configurations with respect to the dynamic moment coefficient, while Figure 13 provided the comparative study of the different SWT configurations in terms of their power coefficients. It was recognized that the tip-speed ratio

(TSR) also played a role in the performance of the SWT. The dynamic moment coefficient of each wind turbine decreases as the TSR is increased. However, the decrease of the new SWT is not as drastic compared to the other SWT configurations. A possible explanation for this is that the new design consists of smaller curvatures that help in maintaining the solution pressure and limit the wake-flow present at the rear of the rotor of the turbine. On the contrary, the power coefficient of each wind turbine followed a bell-curve relative to the TSR where the values increased then decreased until it reached a certain point that indicated its corresponding maximum power coefficient. The maximum power coefficient of the new design was observed at 0.2525 at a TSR of 0.80.

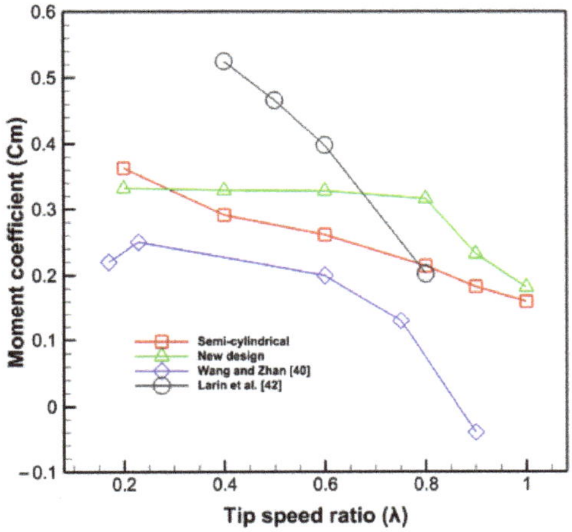

Figure 12. The behavior of the dynamic moment coefficient versus the TSR of the different SWT.

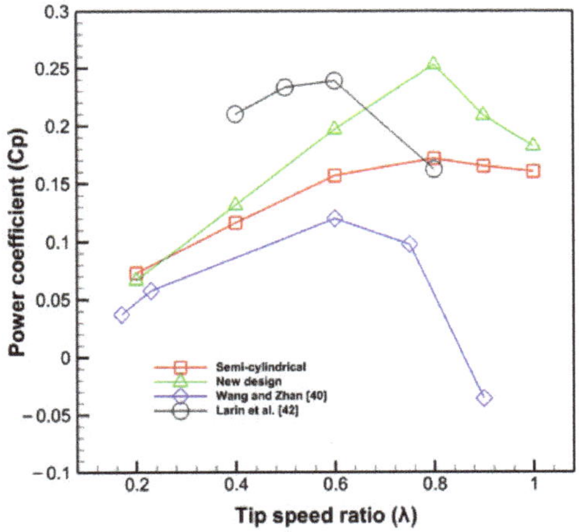

Figure 13. The behavior of the power coefficient versus the TSR of the various SWT.

4.4. Discussion on the Comparison of the Results with Other Studies

In the dynamic condition analysis of the new SWT, the performance of the new design to the traditional SWT and other designs from past literature were compared. The maximum power coefficient of this new design is relatively higher than the previous designs presented by Wang and Zhan [40], the traditional semi-cylindrical SWT [40], the model of Wekesa et al. [41], and the prototype of Larin et al. [42] by 13.28%, 8.16%, 2.32%, and 1.38%, respectively. Hence, this work highlights the improvement of the maximum power coefficient compared with other studies.

5. Conclusions

A computational fluid dynamics (CFD) model was developed for a Savonius wind turbine (SWT) that considered various geometric parameters based on a space-filling design of experiments. The space-filling design considered the sphere packing method which worked well with the results of computational simulations such as CFD. A Gaussian stochastic process (GaSP) model was used to model fit the space-filling design results for the dependent variables and the response. With the application of the space-filling design, the study considered 30 distinct designs based on the combination of the parametric factors such as the diameter and height of the rotor as well as its angle of twist. The results have shown that the model fitted the GaSP model. The results revealed that the power coefficient is directly proportional to the performance parameters of the moment, moment coefficient, rotor height, and rotor diameter, while it was found that the power coefficient has shown a slight decrease while the twist angle increases. It was found that the maximum power coefficient of the optimized SWT design is 0.2525 at a tip speed ratio of 0.80. The maximum power coefficient was found to be superior with the SWT design of previous works. In addition, the results revealed that the optimized SWT design is capable of self-starting with relatively better performance compared to the other designs of previous works. Future work involves the simultaneous effects of the overlap ratio of the SWT and the influence of the inlet air velocity.

Author Contributions: Conceptualization, A.T.U. and R.S.; methodology, R.S.; validation, A.T.U., R.S. and J.D.P.C.; formal analysis, R.S.; investigation, R.S.; resources, A.T.U.; data curation, J.D.P.C.; writing—original draft preparation, R.S.; writing—review and editing, A.T.U. and J.D.P.C.; visualization, A.T.U. and J.D.P.C.; supervision, A.T.U.; project administration, A.T.U.; funding acquisition, A.T.U. and R.S. All authors have read and agreed to the published version of the manuscript.

Funding: This research received no external funding.

Institutional Review Board Statement: Not applicable.

Informed Consent Statement: Not applicable.

Data Availability Statement: Data available in a publicly accessible repository. The data presented in this study are openly available in [40,42].

Acknowledgments: The second author gratefully acknowledges the graduate scholarship support of the Government of Cambodia and the AUN/SEED-net.

Conflicts of Interest: The authors declare no conflict of interest.

References

1. IEA. *World Energy Outlook 2021*; International Energy Agency: Paris, France, 2021; pp. 1–379.
2. Calautit, K.; Aquino, A.; Calautit, J.K.; Nejat, P.; Jomehzadeh, F.; Hughes, B.R. A Review of Numerical Modelling of Multi-Scale Wind Turbines and Their Environment. *Computation* **2018**, *6*, 24. [CrossRef]
3. WWEA. *New Study Proves: Community Power Is Increasingly Being Marginalised*; The World Wind Energy Association: Bonn, Germany, 2019.
4. Micallef, D.; Van Bussel, G. A Review of Urban Wind Energy Research: Aerodynamics and Other Challenges. *Energies* **2018**, *11*, 2204. [CrossRef]
5. Škvorc, P.; Kozmar, H. Wind energy harnessing on tall buildings in urban environments. *Renew. Sustain. Energy Rev.* **2021**, *152*, 111662. [CrossRef]
6. Ramsebner, J.; Haas, R.; Auer, H.; Ajanovic, A.; Gawlik, W.; Maier, C.; Nemec-Begluk, S.; Nacht, T.; Puchegger, M. From single to multi-energy and hybrid grids: Historic growth and future vision. *Renew. Sustain. Energy Rev.* **2021**, *151*, 111520. [CrossRef]
7. Naderipour, A.; Abdul-Malek, Z.; Mustafa, M.W.B.; Guerrero, J.M. A multi-objective artificial electric field optimization algorithm for allocation of wind turbines in distribution systems. *Appl. Soft Comput.* **2021**, *105*, 107278. [CrossRef]
8. Hand, B.; Kelly, G.; Cashman, A. Aerodynamic design and performance parameters of a lift-type vertical axis wind turbine: A comprehensive review. *Renew. Sustain. Energy Rev.* **2021**, *139*, 110699. [CrossRef]

9. Tasneem, Z.; Al Noman, A.; Das, S.K.; Saha, D.K.; Islam, M.R.; Ali, M.F.; Badal, R.M.F.; Ahamed, M.H.; Moyeen, S.I.; Alam, F. An analytical review on the evaluation of wind resource and wind turbine for urban application: Prospect and challenges. *Dev. Built Environ.* **2020**, *4*, 100033. [CrossRef]
10. Chrysochoidis-Antsos, N.; Amoros, A.V.; van Bussel, G.J.W.; Mertens, S.M.; van Wijk, A.J.M. Wind resource characteristics and energy yield for micro wind turbines integrated on noise barriers—An experimental study. *J. Wind. Eng. Ind. Aerodyn.* **2020**, *203*, 104206. [CrossRef]
11. Kim, S.; Cheong, C. Development of low-noise drag-type vertical wind turbines. *Renew. Energy* **2015**, *79*, 199–208. [CrossRef]
12. Mohamed, M.H.; Alqurashi, F.; Thévenin, D. Performance enhancement of a Savonius turbine under effect of frontal guiding plates. *Energy Rep.* **2021**, *7*, 6069–6076. [CrossRef]
13. Marinić-Kragić, I.; Vučina, D.; Milas, Z. Global optimization of Savonius-type vertical axis wind turbine with multiple circular-arc blades using validated 3D CFD model. *Energy* **2022**, *241*, 122841. [CrossRef]
14. Aboujaoude, H.; Beaumont, F.; Murer, S.; Polidori, G.; Bogard, F. Aerodynamic performance enhancement of a Savonius wind turbine using an axisymmetric deflector. *J. Wind. Eng. Ind. Aerodyn.* **2022**, *220*, 104882. [CrossRef]
15. Dewan, A.; Gautam, A.; Goyal, R. Savonius wind turbines: A review of recent advances in design and performance enhancements. *Mater. Today Proc.* **2021**, *47*, 2976–2983. [CrossRef]
16. Yahya, W.; Ziming, K.; Juan, W.; Qurashi, M.S.; Al-Nehari, M.; Salim, E. Influence of tilt angle and the number of guide vane blades towards the Savonius rotor performance. *Energy Rep.* **2021**, *7*, 3317–3327. [CrossRef]
17. Saad, A.S.; Elwardany, A.; El-Sharkawy, I.I.; Ookawara, S.; Ahmed, M. Performance evaluation of a novel vertical axis wind turbine using twisted blades in multi-stage Savonius rotors. *Energy Convers. Manag.* **2021**, *235*, 114013. [CrossRef]
18. Irabu, K.; Roy, J.N. Characteristics of wind power on Savonius rotor using a guide-box tunnel. *Exp. Therm. Fluid Sci.* **2007**, *32*, 580–586. [CrossRef]
19. Tjahjana, D.D.D.P.; Arifin, Z.; Suyitno, S.; Juwana, W.E.; Prabowo, A.R.; Harsito, C. Experimental study of the effect of slotted blades on the Savonius wind turbine performance. *Theor. Appl. Mech. Lett.* **2021**, *11*, 100249. [CrossRef]
20. Xu, W.; Li, C.-C.; Huang, S.-X.; Wang, Y. Aerodynamic performance improvement analysis of Savonius Vertical Axis Wind Turbine utilizing plasma excitation flow control. *Energy* **2022**, *239*, 122133. [CrossRef]
21. Al-Ghriybah, M.; Fadhli Zulkafli, M.; Hissein Didane, D.; Mohd, S. The effect of spacing between inner blades on the performance of the Savonius wind turbine. *Sustain. Energy Technol. Assess.* **2021**, *43*, 100988. [CrossRef]
22. Berhanu, H.; Gudeta, D.; Haiter Lenin, A.; Karthikeyan, B. Numerical and experimental investigation of an exhaust air energy recovery Savonius wind turbine for power production. *Mater. Today: Proc.* **2021**, *46*, 4142–4152. [CrossRef]
23. Pranta, M.H.; Rabbi, M.S.; Roshid, M.M. A computational study on the aerodynamic performance of modified savonius wind turbine. *Results Eng.* **2021**, *10*, 100237. [CrossRef]
24. Marzec, Ł.; Buliński, Z.; Krysiński, T. Fluid structure interaction analysis of the operating Savonius wind turbine. *Renew. Energy* **2021**, *164*, 272–284. [CrossRef]
25. Marinić-Kragić, I.; Vučina, D.; Milas, Z. Computational analysis of Savonius wind turbine modifications including novel scooplet-based design attained via smart numerical optimization. *J. Clean. Prod.* **2020**, *262*, 121310. [CrossRef]
26. Elsisi, M.; Tran, M.Q.; Mahmoud, K.; Lehtonen, M.; Darwish, M.M.F. Robust Design of ANFIS-Based Blade Pitch Controller for Wind Energy Conversion Systems Against Wind Speed Fluctuations. *IEEE Access* **2021**, *9*, 37894–37904. [CrossRef]
27. Zalhaf, A.S.; Mansour, D.E.A.; Han, Y.; Yang, P.; Yang, P.; Darwish, M.M.F. Numerical and Experimental Analysis of the Transient Behavior of Wind Turbines when Two Blades are Simultaneously Struck by Lightning. *IEEE Trans. Instrum. Meas.* **2021**, 1. [CrossRef]
28. Sayed, A.M.; Abouelatta, M.A.; Badawi, M.; Mahmoud, K.; Lehtonen, M.; Darwish, M.M.F. Novel accurate modeling of dust loaded wire-duct precipitators using FDM-FMG method on one fine computational domains. *Electr. Power Syst. Res.* **2022**, *203*, 107634. [CrossRef]
29. Abouelatta, M.A.; Ward, S.A.; Sayed, A.M.; Mahmoud, K.; Lehtonen, M.; Darwish, M.M.F. Fast Corona Discharge Assessment Using FDM integrated With Full Multigrid Method in HVDC Transmission Lines Considering Wind Impact. *IEEE Access* **2020**, *8*, 225872–225883. [CrossRef]
30. Sonawane, C.R.; Sasar, Y.; Shaikh, M.; Kokande, Y.; Mustafa, M.; Pandey, A. Numerical simulation of Savonius rotors used for low wind speed application. *Mater. Today Proc.* **2022**, *49*, 1610–1616. [CrossRef]
31. Xiong, Z.; Liu, L.; Ning, J.; Qin, H. Sphere packing design for experiments with mixtures. *Stat. Probab. Lett.* **2020**, *164*, 108807. [CrossRef]
32. Azadani, L.N.; Gharouni, N. Multi objective optimization of cylindrical shape roughness parameters in a solar air heater. *Renew. Energy* **2021**, *179*, 1156–1168. [CrossRef]
33. Glyn-Davies, A.; Girolami, M. Anomaly detection in streaming data with gaussian process based stochastic differential equations. *Pattern Recognit. Lett.* **2022**, *153*, 254–260. [CrossRef]
34. Mathew, S. *Wind Energy: Fundamentals, Resource Analysis and Economics*; Springer: Berlin/Heidelberg, Germany, 2006.
35. Barlow, J.B.; Rae, W.H.; Pope, A. *Low-Speed Wind Tunnel Testing*; Wiley: Hoboken, NJ, USA, 1999.
36. Fang, K.-T.; Lin, Y.; Peng, H. A new type of robust designs for chemometrics and computer experiments. *Chemom. Intell. Lab. Syst.* **2022**, *221*, 104474. [CrossRef]
37. Jones, B.; Johnson, R.T. Design and analysis for the Gaussian process model. *Qual. Reliab. Eng. Int.* **2009**, *25*, 515–524. [CrossRef]

38. Lee, J.-H.; Lee, Y.-T.; Lim, H.-C. Effect of twist angle on the performance of Savonius wind turbine. *Renew. Energy* **2016**, *89*, 231–244. [CrossRef]
39. Saad, A.S.; El-Sharkawy, I.I.; Ookawara, S.; Ahmed, M. Performance enhancement of twisted-bladed Savonius vertical axis wind turbines. *Energy Convers. Manag.* **2020**, *209*, 112673. [CrossRef]
40. Wang, Y.-F.; Zhan, M.-S. 3-Dimensional CFD simulation and analysis on performance of a micro-wind turbine resembling lotus in shape. *Energy Build.* **2013**, *65*, 66–74. [CrossRef]
41. Wekesa, D.; Wang, C.; Wei, Y.-J.; Zhu, W. Experimental and numerical study of turbulence effect on aerodynamic performance of a small-scale vertical axis wind turbine. *J. Wind. Eng. Ind. Aerodyn.* **2016**, *157*, 1–14. [CrossRef]
42. Larin, P.; Paraschivoiu, M.; Aygun, C. CFD based synergistic analysis of wind turbines for roof mounted integration. *J. Wind. Eng. Ind. Aerodyn.* **2016**, *156*, 1–13. [CrossRef]

energies

MDPI

Article

Numerical and Experimental Investigation of the Effect of Design Parameters on Savonius-Type Hydrokinetic Turbine Performance

Kuo-Tsai Wu [1], Kuo-Hao Lo [1], Ruey-Chy Kao [1] and Sheng-Jye Hwang [2,*]

[1] Tainan Hydraulics Laboratory, National Cheng Kung University, Tainan 709, Taiwan; adiwu@thl.ncku.edu.tw (K.-T.W.); oscar77841245@gmail.com (K.-H.L.); rckao@thl.ncku.edu.tw (R.-C.K.)
[2] Department of Mechanical Engineering, National Cheng Kung University, Tainan 701, Taiwan
* Correspondence: jimppl@mail.ncku.edu.tw; Tel.: +886-6-275-7575 (ext. 62148)

Abstract: To meet the increased demand of hydroelectric power generation, a novel drag-based Savonius turbine with the characteristics of a simpler fabrication process and good starting characteristics is designed, fabricated, and analyzed. The newly designed turbine is suitable to be installed in rivers, irrigation channels, ocean currents, etc., for small-scale hydroelectric power generation. In the present study, experiments are carried out to investigate the influence of the design parameters of this turbine on its power performance in order to improve its efficiency, including blade arc angles (180°, 135°), blade placement angles (0°, ±22.5°), and the number of blades (2, 3, 6, and 8). Further, three-dimensional CFD simulations are performed with $Re = 6.72 \times 10^5$, matching the experimental conditions, in order to study the changes in the flow field and the rotation characteristics of the turbine. The research results indicate that a six-bladed turbine with a blade arc angle of 135° and a blade placement angle of 0° has higher torque and better power performance, which makes it the most suitable design when also considering cost. Furthermore, it was found that an increase in the number of turbine blades contributes to improving the performance of the turbine. The maximum power coefficient is 0.099 at a tip speed ratio of 0.34.

Keywords: Savonius hydrokinetic turbine (SHT); power coefficient (C_p); moment coefficient (C_m); towing tank; tip speed ratio (*TSR*); computational fluid dynamics (CFD)

Citation: Wu, K.-T.; Lo, K.-H.; Kao, R.-C.; Hwang, S.-J. Numerical and Experimental Investigation of the Effect of Design Parameters on Savonius-Type Hydrokinetic Turbine Performance. *Energies* **2022**, *15*, 1856. https://doi.org/10.3390/en15051856

Academic Editor: Eduardo Álvarez Álvarez

Received: 30 December 2021
Accepted: 25 February 2022
Published: 2 March 2022

1. Introduction

Due to global warming [1], the global environment has changed in various ways since the industrial revolution because of damage to the environment and limitations in the amount of available fossil fuels [2,3]. The dependence of humans on electricity has become increasingly more serious with the growth of the population and economic results, so countries have begun to pay attention to the development and utilization of green energy, including biomass energy, geothermal energy, solar energy, hydropower, and wind energy [4,5]. Because water accounts for more than 75% of the earth's surface, compared with other energy sources, hydropower is a cleaner and more predictable source, and generates lower CO_2 emissions [6,7]. Therefore, water resources that harvest hydrokinetic energy from flowing rivers, tidal currents, ocean water currents, irrigation canals, or other artificial waterways are considered to be viable energy sources.

There are two categories of turbines used to harvest energy from water flow: non-turbine systems [8] and turbine systems. A turbine system includes the use of hydrostatic energy (potential energy) and hydrokinetic energy turbines to generate electricity. Among them, a hydrokinetic energy turbine uses the kinetic energy of natural water resources to replace potential energy to generate electricity, without using a water storage structure which leads to environmental damage.

According to the alignment of the axis of rotation relative to the direction of the water flow, there are mainly two technologies applied to hydrokinetic turbines: horizontal-axis turbines [9] and vertical-axis turbines [10]. Horizontal-axis turbines have an advantage in terms of tidal and wind energy, but for small-scale energy conversion they are not cost-effective [11,12]. Vertical-axis turbines are mainly divided into two types [13]: drag-based Savonius and lift-based Darrieus. The Savonius turbine was invented by a Finnish architect, Sigurd J. Savonius, for wind applications in 1920 [14]. It is an energy conversion device that uses the difference in resistance between concave blades and convex blades to convert energy.

Although Savonius turbines were originally used for wind power (SWT), water is 856 times denser than air, which has led researchers to study the potential of Savonius hydrokinetic turbines (SHTs) [15].

Furthermore, the fluid dynamic characteristics in the flow field around turbines are considerably different because of the fundamental difference between wind and water flow. Wind flow is mainly controlled by pressure differences, while water flow is mainly affected by gravity [16].

Although the power performance of the Savonius turbine 20% less that that of the lift-based turbine [17], it has many advantages, including being suitable for low velocity streams, a simple turbine structure, better self-starting characteristics, the ability to harness water flow from different directions, and being less noisy, which reduces the vibration and damage to the turbine structure [18]. In addition, maintenance costs are low because the power equipment can be installed above the water.

In recent years, many researchers have proposed various numerical and experimental methods to improve Savonius hydrokinetic turbine (SHT) technology. The aspect ratio, the overlap ratio, the blade profile, the number of blades, and deflectors are the main factors that directly affect the efficiency of SHTs.

Reviews on the early stages of research on vertical axis Savonius turbines all focused on wind energy. Blackwell et al. [19] proposed that increasing the Reynolds number and aspect ratio would improve performance. Sivasegaram et al. [20] investigated the performance of different blade numbers, shapes, sizes, and positions. According to the literature, the Reynolds number, the wake aspect ratio, the size of the end plate and the shaft diameter all influence turbine performance. Kahn [21], Ushiyama, and Nagai et al. [22] studied the effects of the aspect ratio of the rotor, the shape of the rotor, the overlap, and separation gap between the rotor blades, the profile of the blade cross section, and the number of blades on the performance characteristics. Fujisawa and Gotoh [23] tested the pressure distribution on the blade surface under various rotor angles and tip speed ratios to research the aerodynamic performance of the Savonius turbine. Sheldahl et al. [24] experimentally tested the performance of two- and three-bladed SWTs. Their tests revealed that, compared with the three-bladed turbine, the C_p value for the two-bladed turbine was approximately 1.5 times higher than that for the three-bladed turbine. Emmanuel et al. [25] demonstrated that the performance of the six-bladed SWT is better than that of the two-bladed SWT. According to the results of a CFD simulation, the power coefficient is expected to increase from 0.25 to 0.3. (If a deflector was installed in the six-bladed SWT, this could be increased to 0.5.) Mahmoud et al. [26] developed a two-bladed turbine that is more efficient than both a three-bladed turbine and four-bladed turbine, where a rotor with end plates was found to be more efficient than a rotor without end plates. Compared with a single-stage rotor, a two-stage rotor had better performance. A rotor without an overlap ratio performed better than a rotor with an overlap ratio in terms of operation. The results also showed that the power factor increased with the aspect ratio. Wenehenubun et al. [27] found that, compared with two-bladed and four-bladed wind turbines, three-bladed wind turbines have higher rotational speeds and tip speed ratios. Compared with two-bladed or three-bladed wind rotors, four-bladed wind turbine rotors had higher torque; four-bladed wind turbines had good performance at lower tip speed ratios, but three-bladed wind turbines had the best performance at higher tip speed ratios. Banerjee [28] and Alom [29] used experiments and simulations to study elliptical blades. The results indicated that the flow characteristics of the elliptical blade turbine were better than the traditional semicircular design. Khaled

R.Abdelaziz et al. [30] studied the effect of adding curved and straight auxiliary blades at the turbine blade on the performance of the Savonius turbine numerically.

In the past decade, research on Savonius hydrokinetic turbines (SHT) has made great progress.

Faizal et al. [31] designed the five-bladed Savonius and studied how orbital motion in waves affected the energy extracted by SHT. Increasing both wave height and frequency amplifies the speed of the turbine, according to their results. As a result of this, the performance of SHTs is determined by their design parameters and the surrounding fluid behavior. Yaakob et al. [32] analyzed the performance of a two-bladed semicircular SHT in two- and three-dimensional simulations with a constant current velocity of 0.56 m/s and reported a C_{pmax} of 0.275 at a TSR = 0.7, which validated the experimental results. In 2011, Kailash Golecha et al. [12] tested two stage and three stage modified Savonius rotors for a hydrokinetic turbine to study the influence of the deflector plate at the optimal position. Deflection plates for two-stage deflection plates for two-stage 0° phase shifts, 90° phase shifts, and three-stage modified Savonius rotors have each seen significant increase in power coefficient of 42%, 31%, and 17% respectively. In addition, this study investigated whether by placing the deflector appropriately upstream of the river, the performance of the Savonius rotor could be significantly improved. Sarma et al. [15] examined a three-bladed turbine in an open channel at a water velocity of 0.3–0.9 m/s. The test revealed that a C_{pmax} of 0.39 at TSR = 0.77. Compared with the conventional SWT under the same input value, this was 61.32% higher. Khan et al. [33] studied the performance of single-stage, two-stage and three-stage semicircular blade SHTs. The C_{pmax} value of the two-stage rotor was approximately 0.5, which was approximately 4% better performance than that of single-stage and three-stage rotors. However, these differences were small, which indicates that the rotor should be selected based on the simplicity of the structure rather than the power output. Golecha et al. [34] investigated whether a straight arc shape leads to a higher power coefficient. The study also showed that the use of deflection plates on both the advancing blade side and the returning blade side improved the performance of the Savonius rotor.The impact of overlap ratios and aspect ratios on performance of Savonius turbines in an open channel was studied by Patel et al. [35]. The results showed that the C_p value for an overlap ratio was approximately 0.11 for Savonius turbines, where the maximum aspect ratios less were than 0.6. Kumar and Saini [36] investigated the influence of twist angle and Reynolds numbers on the hydrokinetic performance of a Savonius turbine. A numerical simulation was carried out using ANSYS Fluent in their study. A C_{pmax} of 0.39 at a TSR = 0.9 was obtained for a given water velocity of 2 m/s. Parag K et al. [37] studied the performance of SHTs through experiments and CFD. The performance of two- and three-bladed SHT, as well as elliptical blade and semi-circular blades, is compared. The results showed that the performance of two-bladed and semi-circular blade SHT was better than that of other alternatives. Mabrouk M et al. [38] investigated the performance of the Helical Savonius in irrigation channels. An improved deflector system design was proposed in order to enhance the performance of the studied water turbine. Fen G et al. [39] researched the influence of a rear deflector on a Savonius turbine with a towing box experiment and a two-dimensional simulation. Although deflector devices can improve the performance of an SHT [12,34,39], the additional equipment also make the entire turbine system more complicated.

1.1. Motivation

Wind and water are currently the two most common energy sources for Savonius turbines; however the flow field performance of open channel water flow differs considerably from wind. Water flow is controlled by gravity, while wind is mainly influenced by pressure differences. In addition, researchers are interested in the higher power density of water to study the potential of SHT. Therefore, this study is aimed toward modifying the design parameters of SHTs to improve their performance. In previous literature on SWTs, there have been different opinions on the optimal number of blades. Some studies [24,26,37,40] have shown that two blades provide better performance efficiency than three blades. However, some studies have suggested that more than two blades will lead to better performance [25,27,29]. The above results were based on the findings related to SWTs, but there are still few studies examining the effects of the number of blades on SHT performance, thus, an evaluation of this parameter is necessary.

1.2. Present Objective

The literature on SWTs has discussed many different types of blades, but SHT studies have mostly discussed only semi-circular blades. In this study, an attempt is made to change the shape and angle of the blades to observe the efficiency of the turbine and verify the results through experiments and simulations. In brief, the novelty of this research lies in the experimental testing of the effects of the number of blades and blade shape on SHTs. First, six-bladed SHTs with arc angles of 135° and 180° (semi-circular) are tested to find the best blade shape based on the power coefficient. Subsequently, the six-bladed SHT with a 135° arc angle is tested with different placement angles of 0° and ±22.5° in the experiments. Further, SHTs with two, three, six, and eight blades were tested to find the optimal blade shape and number. In order to obtain a deeper understanding of the hydrodynamic performance and flow field changes of the SHT, CFD simulations using ANSYS-CFX were carried out.

2. Turbine Design and Performance Parameters

The parameters that influence the performance of a vertical drag hydrokinetic turbine include the aspect ratio, blade shapes, number of blades, and end plates. The aspect ratio (AR) has an important influence on efficiency performance, application location, structural capacity and power demand of the turbine, and it is typically expressed by the following equation:

$$AR = \frac{H}{D} \quad (1)$$

where H and D represent the turbine height and diameter, respectively. A lower aspect ratio ($AR < 1.5$) cause the turbine to have a more stable structure and greater stability [37]. In this study, the turbine is design with an $AR = 1$, for which the design parameters are shown in Figure 1. A hydrokinetic turbine with blades having two different profiles is considered for the analysis using blades with arc angles (ψ) of 180° and 135°, as shown in Figure 2. Each blade includes a concave and convex profile and is divided into an advancing blade and a returning blade according to the rotation. The pressure difference between the concave and convex surfaces of the blades is used to drive the turbine to rotate. In order to reduce the effect of the vortex inside the turbine and allow water to flow smoothly through the turbine, the blade arc angle (ψ) is designed to be 135°. The turbine blades are 3.75 mm thick. The height (H) and diameter (D) of the turbine are both 600 mm. In this study, the blade placement angles were varied to investigate their influence on the performance of the turbine. The dimensions of the blade placement angles (α) are −22.5°, 0°, and +22.5°, as shown in Figure 3. In addition, several blades were developed with the same specifications (ψ = 135°, α = 0°) to study the effect of the number of blades on the efficiency of a turbine, as shown in Figure 4.

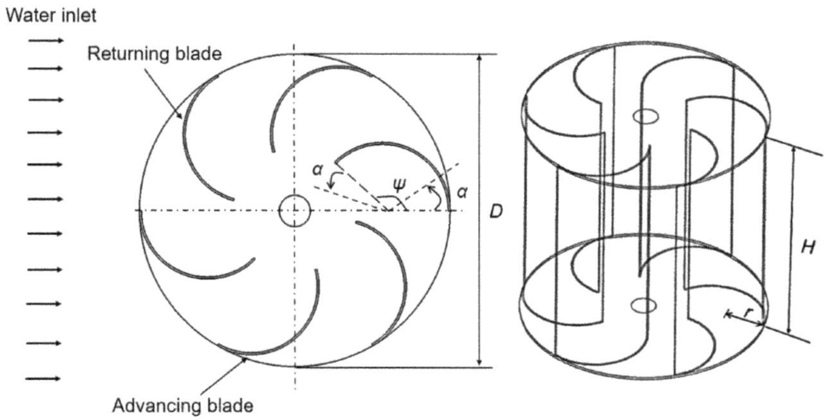

Figure 1. Schematic diagram of a 6-bladed turbine.

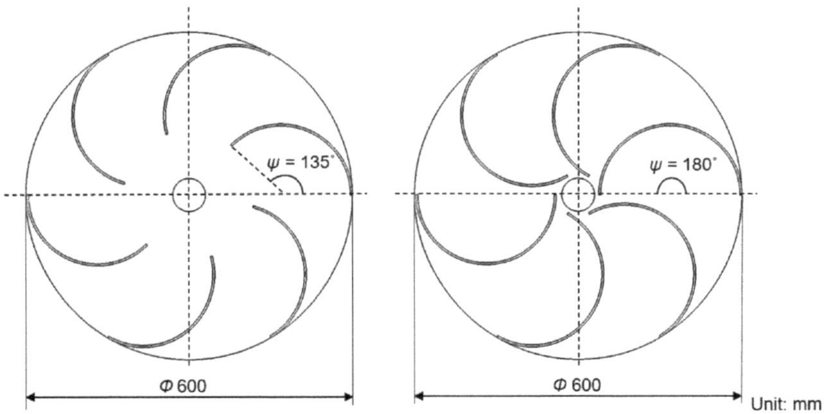

Figure 2. The geometry of the turbine blades with different arc angles (ψ = 135° and 180°).

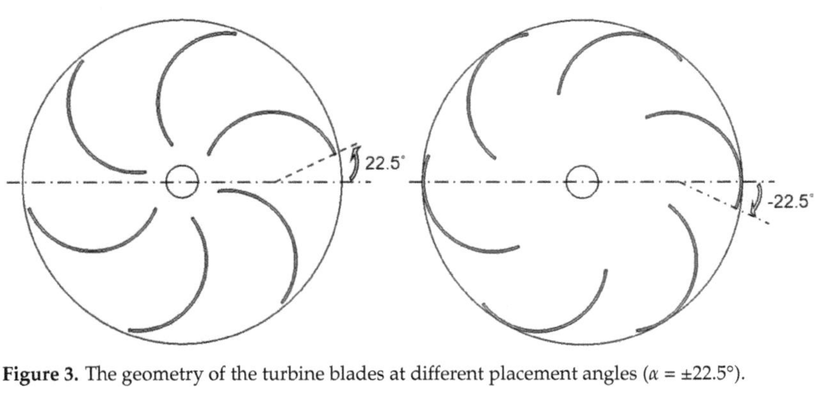

Figure 3. The geometry of the turbine blades at different placement angles (α = ±22.5°).

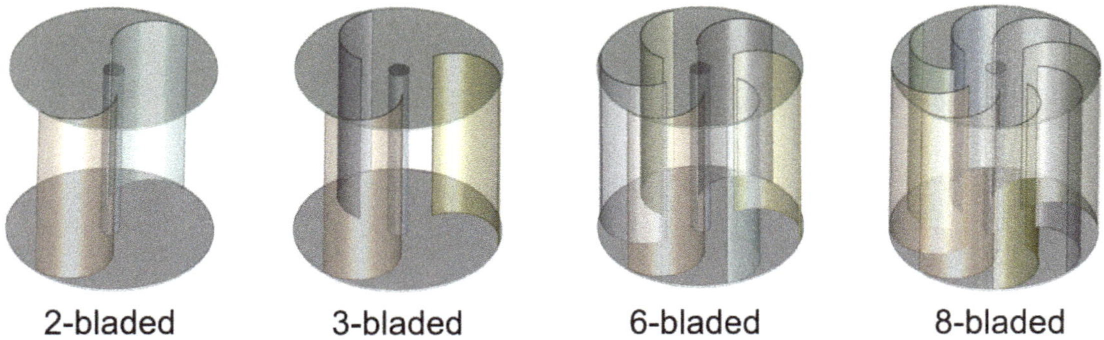

Figure 4. The geometry of the turbine for different numbers of blades (α = 0°, ψ = 135°).

Data Reduction

To evaluate the performance of water turbines, dimensionless coefficients are typically used. The most commonly used evaluation characteristic coefficients include the following: For open channels, the Reynolds number based on the rotor diameter is given by:

$$Re = \frac{\rho VD}{\mu} \quad (2)$$

where ρ is the density of water; V is the free stream velocity; D is the rotor diameter; and μ is the viscosity of water.

The power coefficient of the turbine represents the ratio of the mechanical power produced by the turbine to the actual water kinetic energy available in the same area covered by the turbine [41], which is defined as follows:

$$C_p = \frac{T\omega}{\frac{1}{2}\rho AV^3} = \frac{P_{output}}{P_{available}} \quad (3)$$

where T is the torque produced by the turbine; ω is the angular velocity; and A is the projected turbine area.

The tip speed ratio is given by the following expression:

$$TSR = \frac{R\omega}{V} \quad (4)$$

where R is the turbine radius.

The turbine moment coefficient is defined by the following equation:

$$C_m = \frac{T}{\frac{1}{2}\rho ARV^3} \quad (5)$$

The blockage ratio is given by the expression below. When the blockage ratio is less than 30%, the performance of the turbine is less affected by the boundary effect [42]. The blockage rate in this study is 8%.

$$BR = \frac{frontal\ area\ of\ the\ object}{cross\ Section\ of\ the\ tunnel} \quad (6)$$

3. Experimental Methodology

The turbine material settings and experiment model were all made of stainless steel. The turbine proposed in this study was considered to be able to actually operate and generate electricity, so a large turbine (60 cm × 60 cm) was used in the scale. However, due to the limitation of the site, it was difficult to conduct experiments. A towing tank test platform [33,39,43] was set up to conduct the experiments. The width of the towing tank was approximately 5 m, and the turbine was fully immersed. The depth of water in the towing tank was 0.9 m, and the distance from the water surface to the top of the turbine was approximately 15 cm. In order to measure the torque and rotational speed generated by the turbine, a torque sensor with an accuracy of ±1% and a tachometer with an operating range of 10–29,999 RPM and accuracy of ±0.05% were mounted on the towing tank experimental platform, as shown in Figures 5 and 6. The setup consisted of trailer, a sliding rail, a braking pulley, and an acoustic Doppler velocimeter (ADV), which was used to measure the velocity of the water flow. The trailer drove the turbine to move forward and rotate at a fixed speed of 1 m/s by the slide rail and also drove the relative movement of the turbine and the water flow, so it successfully simulated the impact of the water flow on the turbine. Because the position of the turbine was driven by the trailer moving forward, the turbine was not affected by the size of the tank or the backflow caused by the water hitting the wall. The trailer was set to 1 m/s, corresponding to Re = 6.72×10^5. The flow states were turbulent. Figure 7 shows the rotation of the turbine in the towing tank. This paper is to study the performance of the Savonius-type hydrokinetic turbine. The amount of hydrokinetic energy contained in flowing water current depends on the water velocity. Therefore, it can be known from the definition of the power coefficient, the actual water kinetic energy captured by the turbine is dominated by the free stream velocity and the free flow cross-sectional area, not by the height of the effective head, as shown in Equation (3). Moreover, the cross-sectional area of water passing through the turbine is affected by the height of the upstream water surface rather than the downstream water surface. Therefore, the water depth is determined when the turbine is stationary in this study In the experimental setup, the depth of the water is ensured so that the entire turbine is submerged in water during operation. The main purpose is to study and discuss the operation and flow field changes of the turbine as a whole immersed in the water.

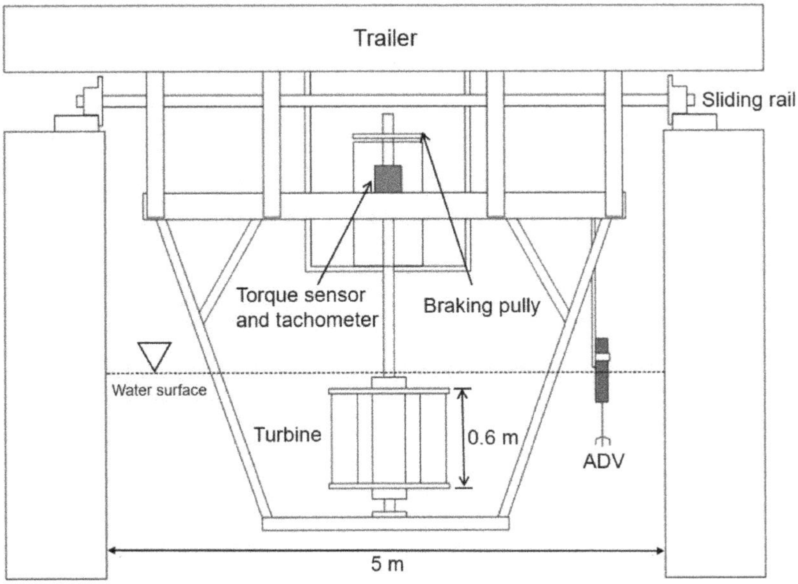

Figure 5. Schematic diagram of the towing tank.

Figure 6. Towing tank experimental platform.

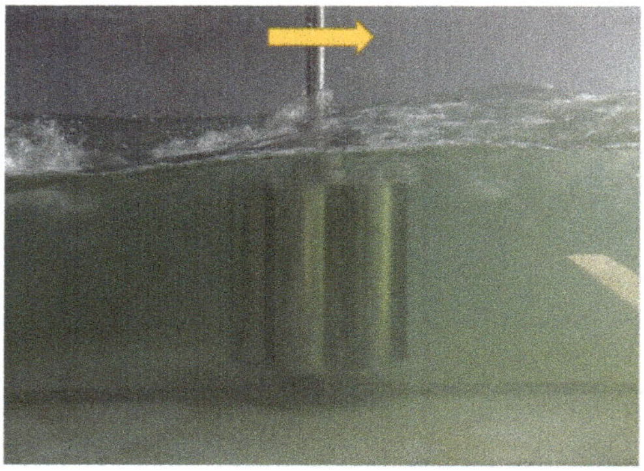

Figure 7. Rotation of the 6-bladed turbine in the tank.

4. Computational Methodology

Turbines will cause a vortex and wake region disturbance phenomenon when rotating, so it is impossible to clearly observe the changes in the water flow experimentally. In order to solve this problem, CFD is a very suitable tool. For an axisymmetric flow field, two-dimensional modeling can be used to calculate the flow field changes simply and quickly [43–46], or three-dimensional modeling can be used for more complex and real calculations to improve accuracy. Many previous studies only discussed dynamic simulations by giving the turbine a fixed angular velocity to facilitate rapid calculation [25,47], but there was a lack of important data for the turbine startup phase. Therefore, in this study, the passive rotation of the turbine for a three-dimensional CFD model under the impact of water flow is examined in order to capture continuous changes in the flow field.

4.1. Governing Equations

In the present study, the CFX flow solver is used to solve the Navier–Stokes equations and turbulence model with finite volume method discretization in order to deal with viscous incompressible turbulent flow field problems, where the body force is considered. Semi-Implicit Methods for Pressure-Linked Equation (SIMPLE) is used for the pressure-velocity coupling with second order upwind scheme. Convergence criteria are set at 10^{-5} for RMS residuals equations of momentum, continuity and turbulence characteristics.The details of the computational settings are shown in Table 1. The continuous equations and momentum equations are as follows [48,49]:

$$\frac{\partial(\rho)}{\partial t} + \frac{\partial(\rho u_j)}{\partial x_j} = 0 \tag{7}$$

$$\frac{\partial(\overline{\rho u_i})}{\partial t} + \frac{\partial(\overline{\rho u_i u_j})}{\partial x_j} = -\frac{\partial(\overline{p})}{\partial x_i} + \frac{\partial}{\partial x_i}[\mu(\frac{\partial \overline{u_i}}{\partial x_j} + \frac{\partial \overline{u_j}}{\partial x_i} - \frac{2}{3}\delta_{ij}\frac{\partial \overline{u}_m}{\partial \overline{x}_m})] + \frac{\partial}{\partial x_j}(-\overline{\rho u_i' u_j'}) \quad (8)$$

The Boussinesq's gradient transport hypothesis is used to convert the Reynolds stress term into the relationships among the average flow velocity gradient.

$$-\overline{\rho u_i' u_j'} = \mu_t(\frac{\partial \overline{u_i}}{\partial x_j} + \frac{\partial \overline{u_j}}{\partial x_i}) - \frac{2}{3}(\rho k + \mu_t \frac{\partial \overline{u}_m}{\partial \overline{x}_m}) \quad (9)$$

where μ_t is the turbulent viscosity, which is defined as:

$$\mu_t = \rho \frac{C_\mu k^2}{\epsilon} \quad (10)$$

where ρ is the density; k is the turbulent kinetic energy; ϵ is the Turbulent dissipation rate; and C_μ is the constant of the standard k-ϵ turbulence model.

Table 1. Details of the computational settings.

Characteristic	Value
Spatial discretization method	Finite Volume Method (FVM)
Convergence criteria for residuals	10^{-5}
Turbulence model	Standard k-ϵ
Skewness	0.58

4.2. Turbulence Modelling

The flow field around the rotation of the turbine creates high levels of turbulence in nature. In order to simulate actual flow field performance, the choice of turbulence is very important, so a turbulent flow with a Reynolds number approximately 6.72×10^5 was used in this study. Many previous studies have used the standard k-ϵ turbulence model for simulations [15,50–52], so the standard k-ϵ, which contains two semi-empirical equations, is used in this study. This two-equation semi-empirical model can be expressed as:

- Turbulent kinetic energy, k

$$\frac{\partial(\rho k)}{\partial t} + \frac{\partial(\rho k \mu_i)}{\partial x_i} = \frac{\partial}{\partial x_j}[(\mu + \frac{\mu_t}{\sigma_k})\frac{\partial k}{\partial x_j}] + G_k - G_b - \rho\epsilon - Y_M + Sk \quad (11)$$

- Turbulent dissipation rate, ϵ

$$\frac{\partial(\rho \epsilon)}{\partial t} + \frac{\partial(\rho \epsilon \mu_i)}{\partial x_i} = \frac{\partial}{\partial x_j}[(\mu + \frac{\mu_t}{\sigma_\epsilon})\frac{\partial k}{\partial x_j}] + C_{1\epsilon}\frac{\epsilon}{k}(G_k + G_{2\epsilon}G_b) - C_{2\epsilon}\rho\frac{\epsilon^2}{k} + S\epsilon \quad (12)$$

where σ_ϵ is the turbulent Prandtl number for ϵ; σ_k is the turbulent Prandtl number for k; G_k is the turbulent kinetic energy generation due to the mean velocity gradient; Y_M is the effect of changes in dilatation of the compressible turbulence on the overall dissipation rate; and G_b is the generation of turbulence kinetic energy due to buoyancy. In the above equation, C_μ, $C_{1\epsilon}$, $C_{2\epsilon}$, σ_ϵ, and σ_k are empirical coefficients, for which the values are as follows: C_μ = 0.09, $C_{1\epsilon}$ = 1.44, $C_{2\epsilon}$ = 1.92, σ_ϵ = 1.3, and σ_k = 1.0.

4.3. Computational Domains and Boundary Settings

The overall three-dimensional computational domains were mainly divided into rotating domain and stationary domain. The rotating domain contains turbines, and the other domains were stationary domains. Since in the experiment, the trailer setting method is used, the turbine will not be affected by the disturbed water flow generated after hitting the wall and will thus not affect the efficiency performance. Therefore, in terms of the simulation, for the basin, no-slip wall surfaces have been used. This design ensures that the water flow hitting the wall surface will not bounce back to the turbine and affect its rotation. After the water flow hits the wall surface, the velocity on the wall surface is 0, and the water flow moves along the wall surface and away from the turbine, which is the same as concept for the experimental setup. The computational domains are set to be the same as those used in the experiment, for which the total length is 7.5 m, which is 12.5 times the diameter of the turbine, and the domain width is 5 m, which is 8.3 times the diameter of the turbine. A velocity of $V = 1$ m/s is set at the inlet, and a pressure outlet is set at the downstream outlet, which ensures that the upstream and downstream pressures are the same. The rest of the walls are all set to the boundary conditions of the stationary wall, which are adiabatic boundaries and do not affect any condition of the fluid in the analysis, as shown in Figure 8.

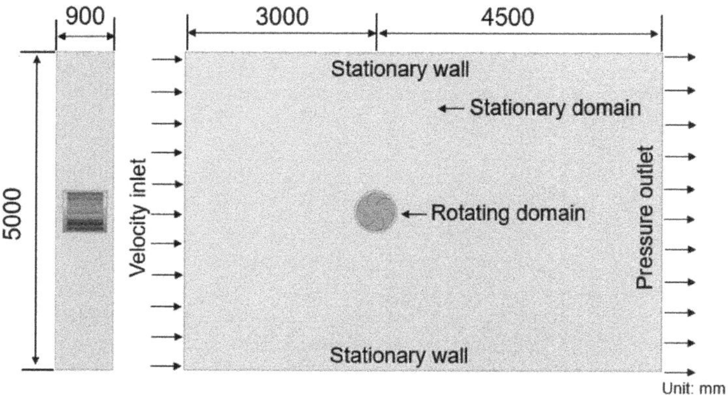

Figure 8. Dimensions of the flow field.

4.4. Grid Independence Test

The ideal grid design should consider the characteristics of the flow field and the computer performance to select the appropriate grid size and perform partial mesh refinement processing in the rotating area. Because of the complex geometry of the turbine, an unstructured tetrahedron mesh was used to create the mesh for the entire domain. However, considering that the viscosity of the fluid near the solid boundary layer changes significantly, for the surface of the turbine blade, a structured prism mesh was adopted. The inflation layer of the blade surface boundary layer consisted of the prism mesh with a first layer thickness of 0.5 mm and a maximum of 10 layers with a growth rate of 1.2, as shown in Figure 9. In the CFD simulation, the independence of the mesh had to be analyzed first. As shown in Table 2, a total of 7 sets of grid numbers ranging from 256,726 to 2,547,070 with average skewness of 0.58 which were generated to analyze the effect of the mesh number on the performance of the turbine and the hydrodynamic characteristics around the turbine. As the number of mesh elements increased from 256,726 to 1,093,773, the value of C_p continued increased. However, when the number of mesh elements increased from 1,093,773 to 2,547,070, the C_p value did not change to any significant degree. Therefore, 1,093,773 has indicated the convergence had been reached. In order to reduce the calculation costs, the total mesh number was set at 1,093,773 for the simulation.

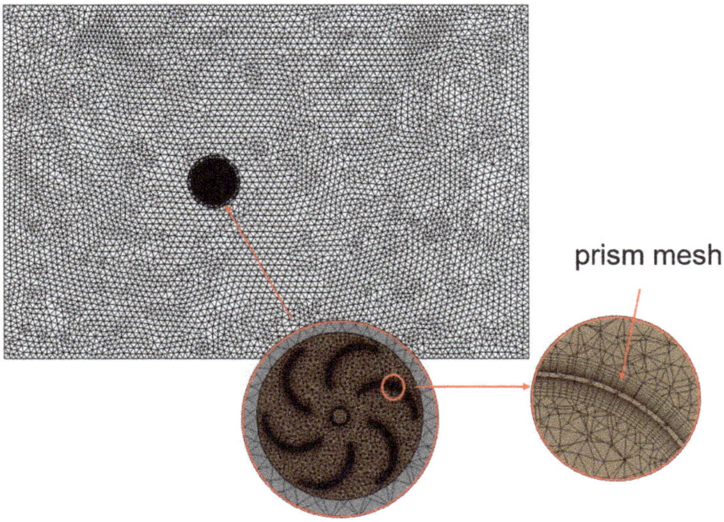

Figure 9. Computational domain mesh details (prism mesh).

Table 2. Variations in C_{pmax} relative to different amounts of elements.

Number of Elements in the Rotating Zone	Number of Elements in the Fixed Zone	Total Elements	C_{pmax}
240,088	16,638	256,726	0.0731
364,272	23,106	387,387	0.0812
627,904	32,654	660,558	0.0872
836,707	40,905	877,612	0.0924
1,010,098	83,675	1,093,934	0.0979
1,465,038	160,895	1,625,934	0.0984
2,261,104	285,966	2,547,070	0.0978

5. Results and Discussion

5.1. *Effect of Blade Arc Angles* (ψ)

Many studies have pointed out that the shape of the blade has a great influence on the output power of a turbine [53,54]. First, the difference in the measured efficiency of the 6-bladed turbine (α = 0°) is discussed when the blade arc angles were 180° and 135°. As shown in Figure 10, the moment coefficient and the power coefficient were calculated to correspond to the *TSR* to evaluate the efficiency of the turbine. Without applying load, the *TSR* for a blade arc angle of 135° was approximately 0.75, which was 0.4 greater than that obtained for a blade arc angle of 180°, as shown in Figure 10a. Additionally, as the mechanical load increased, the rotational speed of the turbine decreased and at the same time, the torque generated by the turbine increased, so the C_m of the 135° blade arc angle was greater than that for a blade arc angle of 180°. Furthermore, it can be seen that the C_p value and *TSR* exhibited a quadratic curve. Figure 10b shows a C_{pmax} of 0.099 at a *TSR* of 0.34 when the blade arc angle was 135°, and a C_{pmax} of 0.0249 at a *TSR* of 0.244 when the blade arc angle was 180°. It is obvious that the blade arc angle of 135° had better efficiency performance because the increase in the vorticities decreased the hydrodynamic performance of the turbine.

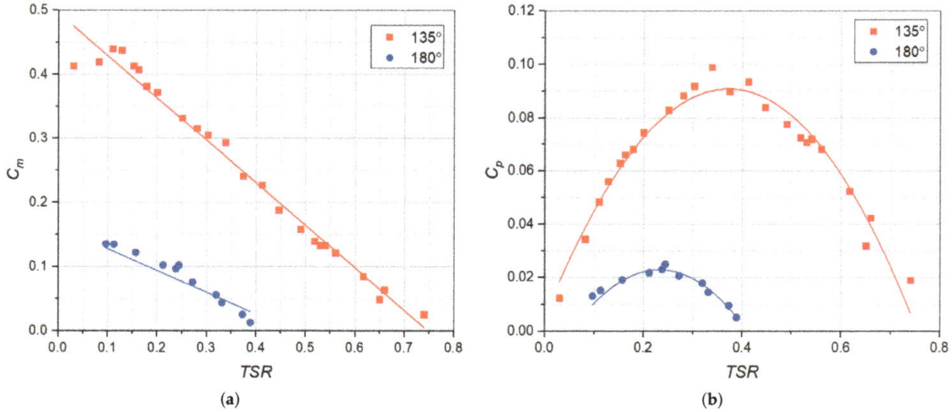

Figure 10. Variation in C_m and C_p with the *TSR* for the different blade arc angles (ψ). (**a**) C_m vs. *TSR*, (**b**) C_p vs. *TSR*.

5.2. Effect of the Blade Placement Angles (α)

From the previous section, it is known that when the blade arc angle is 135°, the efficiency of the turbine is better, so we further measured the performance differences for different placement angles (α = −22.5°, 0°, and 22.5°) in the experiments. The experiments were conducted on six-bladed turbine with a blade arc angle of 135°. Figure 11a shows the variations in C_m with the *TSR* for the different blade placement angles. Obviously, the torque and angular velocity generated by the forward and reverse placement design are less than the 0° placement design. Furthermore, it can be seen in Figure 11b that the efficiency of the forward rotation of 22.5° (C_p = 0.043, *TSR* = 0.309) is better than that of the reversed rotation of 22.5° (C_p = 0.0294, *TSR* = 0.255). However, the efficiency is still less than half that of the blade placement angles of 0°. The main reason for this phenomenon is that the reverse placement of the blade makes the water flow unable to effectively impact the concave surface of the forward blade to generate positive torque, and the turbine thus cannot smoothly capture the energy provided by the water flow. On the other hand, the design of the forward placement of the blades causes the distance between the blades and the shaft to be shortened, so when the turbine is rotating, the chance of generating vortexes on the concave surface of the blade increases.

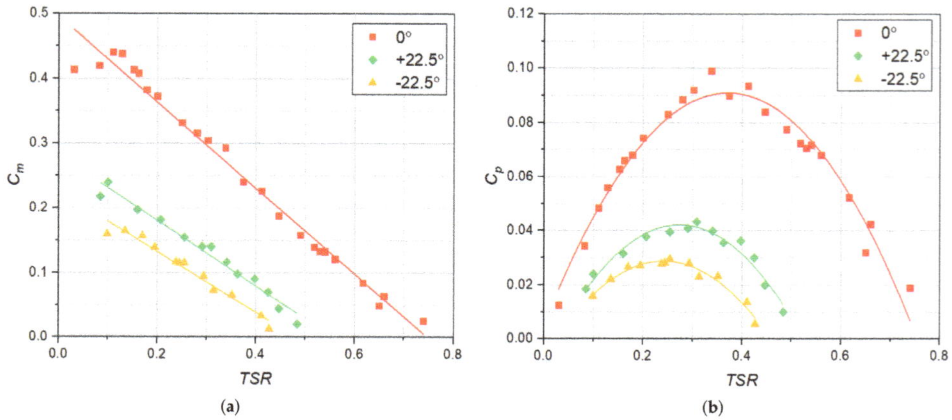

Figure 11. Variation in C_m and C_p with the *TSR* for the different blade placement angles (α). (**a**) C_m vs. *TSR*, (**b**) C_p vs. *TSR*.

5.3. Validation and Effect of the Number of Blades

To understand the dynamics of water flow, CFD simulations were carried out to capture the variations in the water flow and rotation of the turbine. Because the number of turbine blades affects the rotational speed and torque of the turbine, simulations were carried out to investigate the effect of the number of blades. According to the results discussed in the previous chapters, the power differences among two, three, six, and eight bladed turbines with $\alpha = 0°$ and $\psi = 135°$ were simulated in the study when the inlet velocity was 1 m/s, and the results were compared with the experimental values, as shown in Figure 12. It can be observed that the simulation results had the same trend as that of the experimental results but with minor deviation and the overall characteristic curve obtain from simulations is a bit to the right of the experimental results. In the experiments, frictional drag is generated when the turbine rotates, so the angular velocity of the turbine is slower than that in the simulations. Both the experimental and simulation results showed that a greater number of blades led to a wider range of turbine speeds and greater power generation.

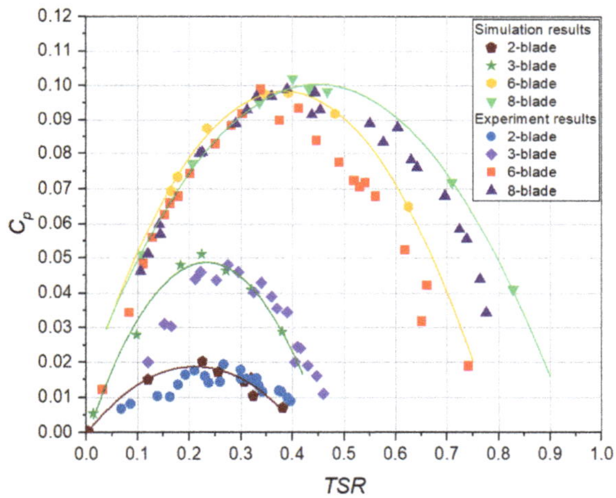

Figure 12. Effect of the number of blades on the power coefficient for different *TSR* values (experimental and simulation results).

The performance of the six- and eight-bladed turbine was much better than that for the two- and three-bladed turbines. The maximum C_p for the six-bladed turbine was 0.0979 at a *TSR* = 0.39, and for eight-bladed turbine, it was 0.102 at a *TSR* = 0.4. When the number of blades was increased to six, the efficiency produced by the turbine appeared to reach its limit.The greater the number of blades, the greater the torque load that the overall turbine can withstand, and the turbine can also reach higher speeds. The C_m range for the six-bladed turbine was below 0.55; the *TSR* range was within 0.75; the C_m range for the eight-bladed was slightly less than 0.5; and the *TSR* range was within 0.9. As shown in Figure 13, when the turbine speed was faster, the torque generated by the turbine was small. The C_m and the *TSR* exhibited a linearly decreasing relationship. It can be seen that the *TSR* and C_m values of the six-bladed turbine were very close to those for the eight-bladed distribution, which means that the two can produce similar speed ranges and torques.

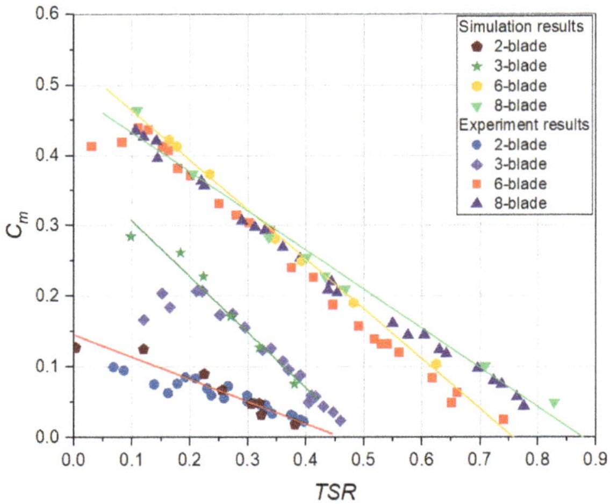

Figure 13. Effect of the number of blades on the moment coefficient for different *TSR* values (experimental and simulation results).

5.4. Pressure Contours

Figure 14 shows the distribution of the pressure contours among the two, three, six, and eight-bladed turbines at *TSR* values corresponding to the maximum power coefficient. It was observed that a pressure drop through the turbine occurred from upstream to downstream. The blades were driven to rotate because the concave surface pressure of the forward blade was greater than that of the convex surface. The eight-bladed turbine produced more positive torque due to a greater number of advancing blades, but at the same time, there were also more returning blades that provided negative torque, so the output power compared to the six-bladed reached convergence. However, its output power was still much higher than that of the two- and three-bladed turbines, which was determined by comparing the pressure difference between the upstream and downstream of the turbine. It was also observed that the high pressure zone on the convex surface of the returning blade decreased with increases in the number of blades, which reduced the amount of negative torque generated, and also indicated the consistency between the results for the simulation and the experiment. Based on the CFD simulation results, it was observed that increasing the pressure area on the concave surface of the advancing blade and reduced the pressure distribution on the convex surface of the returning blade affected by the impact of the water flow, which was the key to improve the performance of the turbine.

5.5. Velocity Contours

Figure 15 presents the distribution of the velocity contours among the two, three, six, and eight-bladed turbines at a *TSR* corresponding to the maximum power coefficient. The velocity of the water flow was constant before entering the turbine rotating domain. When the turbine rotated, it was observed that there was a high-velocity area at the tip of the advancing blade. On the other hand, a wake zone was generated downstream of the turbine. It is worth noting that the advancing blade tip of the two-bladed turbine had a more intense velocity flowing from the tip of the blade than was the case for the other turbines with more blades. This phenomenon led to a negative impact on the blade–fluid interactions, which deteriorated the efficiency of the turbine [15]. A higher velocity zone was observed at the top and bottom side, as shown in side view near the end plate. It can be observed that as the number of blades increased, this phenomenon decreased significantly, in turn improving the efficiency performance. Moreover, the overlapping flow, which can enhance the torque of the turbine, was observed to be strong on the concave surfaces of the returning blades in the case of the six- and eight-bladed turbines. However, for the eight-bladed turbine, the overlapping flow increased the hydraulic resistance of the turbine without sufficient hydrokinetic energy to impact the blades, so the performance of the six- and eight-bladed turbines was about the same. On the other hand, it was observed that the overlapping flow in the two- and three-bladed turbine left the turbine without impacting the other blades and passed through the turbine downstream. This produced vortices on the concave surfaces of the returning blade and caused a decline in the performance of the turbine.

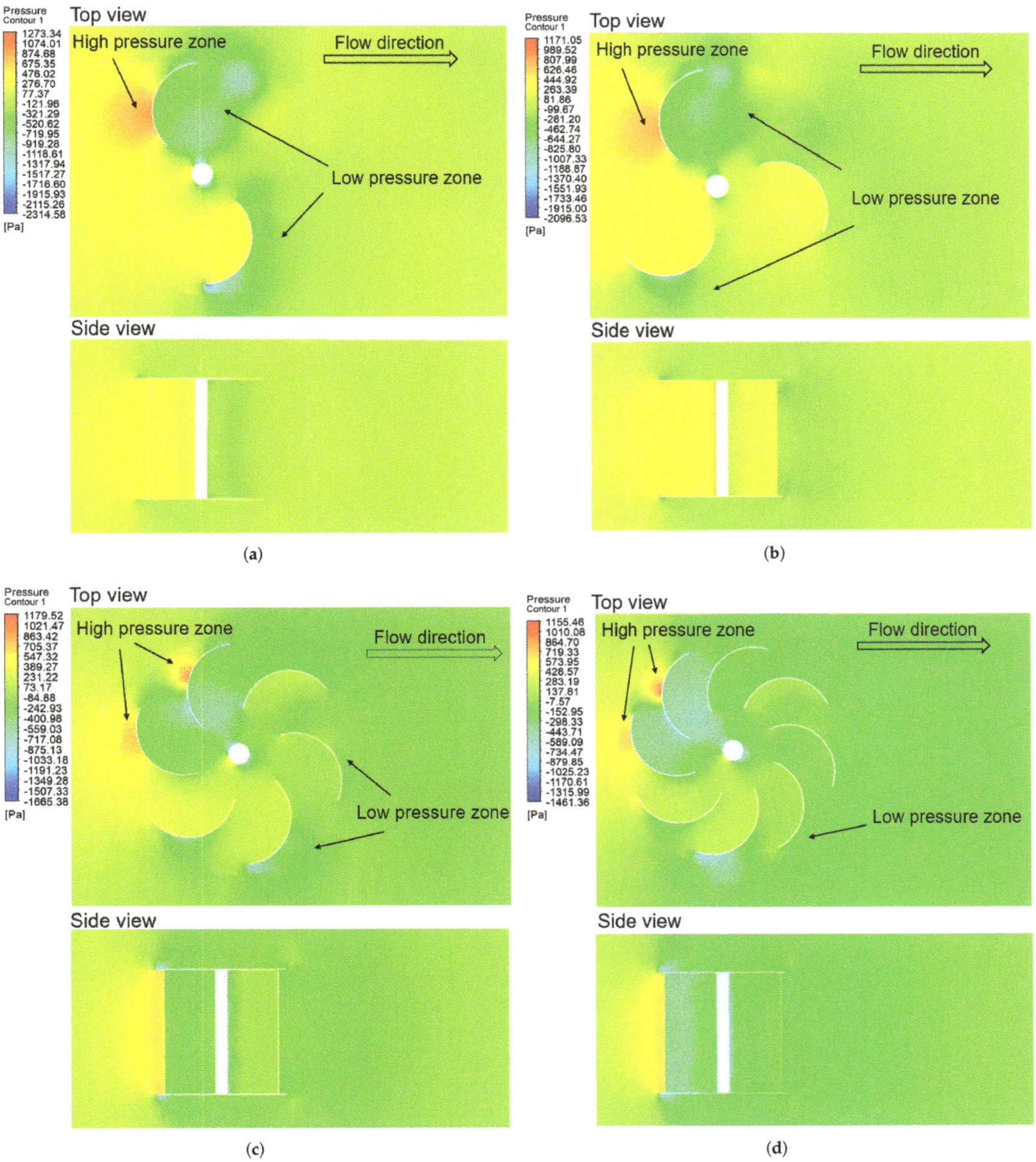

Figure 14. Comparison of the pressure distribution around the blades for (**a**) 2-bladed turbine, (**b**) 3-bladed turbine, (**c**) 6-bladed turbine, and (**d**) 8-bladed turbine.

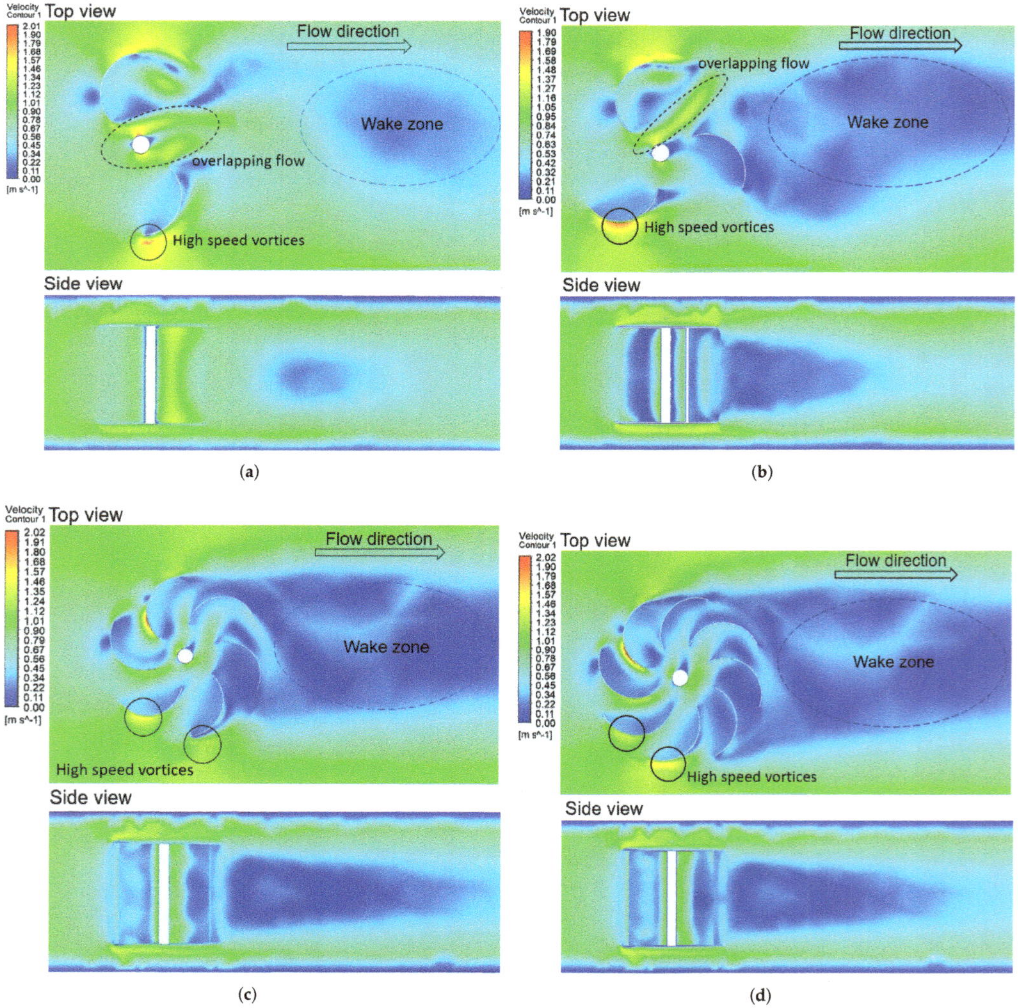

Figure 15. Comparison of the velocity distribution around the blades for (**a**) 2-bladed turbine, (**b**) 3-bladed turbine, (**c**) 6-bladed turbine, and (**d**) 8-bladed turbine.

6. Conclusions

The purpose of the current study was to investigate the influence of the blade arc angles, blade placement angle, and the number of blades on the Savonius-type hydrokinetic performance of a turbine. Considering various design parameters, the turbines were investigated and tested in an open towing tank at $V = 1$ m/s. Further, three-dimensional CFD simulations were performed to capture the variations in the flow field and the rotation of the turbine, which were in good agreement with the experimental results. The measured data was expressed as a dimensionless coefficient to evaluate the performance of the turbine. Moreover, the study is considered as the calibration of the CFD model for further tests and providing experiments that are difficult to be carried out. The main conclusions drawn from this study are summarized below:

1. The six-bladed SHT with $\psi = 135°$ and $\alpha = 0°$ produced the highest C_{pmax} of 0.099 at a TSR of 0.34.

2. The performance of the blade angle of 135° was approximately 2.96 times better than the turbine of the blade angle of 180°.
3. Compared with the design of the blade placement position for the reverse rotation, the forward rotation of the blade placement had less influence on the efficiency of the turbine.
4. The C_p and the *TSR* showed a quadratic curve relationship, and the C_m and the *TSR* showed a linearly decreasing relationship.
5. Based on the simulation results and the experimental results, it was found that the range of the rotational speed of the turbine became wider by increasing the number of blades due to the fact that they could harvest more hydrokinetic energy. However, when the number of blades was increased to eight, the mechanical power of the turbine reached its limit due to hydraulic resistance.
6. Compared with the other turbines with varying numbers of blades, the six-bladed turbine had fewer high-speed vortices on advancing blades and more overlapping flow collisions with the returning blades. Thus, the six-bladed turbine converted more available hydro energy to the mechanical power.

In the present study, a turbine was tested at a fixed *AR* and water velocity in both experiments and simulations. In order to improve the performance of a turbine, it is necessary to investigate a variety of *AR* values, Reynolds numbers, and other design parameters to find the optimum design. There are divergent opinions on the performance of the number of SWTs in the previous literature, but in the case of the SHT, this study clearly points out that the hydrokinetic performance of the SHT increases with an increase in the number of blades, while the performance of the eight-bladed turbine reached convergence.

Author Contributions: Conceptualization, K.-T.W.; methodology, K.-T.W.; software, K.-T.W.; validation, K.-T.W.; formal analysis, K.-H.L.; investigation, K.-T.W.; resources, S.-J.H.; data curation, K.-H.L.; writing—original draft preparation, K.-H.L.; writing—review and editing, S.-J.H.; supervision, R.-C.K.; project administration, R.-C.K. All authors have read and agreed to the published version of the manuscript.

Funding: This work was funded by the Energy Bureau, Ministry of Economic Affairs under grant Project No. 106-D0102.

Conflicts of Interest: The authors declare no conflict of interest.

References

1. VanZwieten, J.; McAnally, W.; Ahmad, J.; Davis, T.; Martin, J.; Bevelhimer, M.; Cribbs, A.; Lippert, R.; Hudon, T.; Trudeau, M. In-stream hydrokinetic power: Review and appraisal. *J. Energy Eng.* **2015**, *141*, 04014024. [CrossRef]
2. Kosai, S. Dynamic vulnerability in standalone hybrid renewable energy system. *Energy Convers. Manag.* **2019**, *180*, 258–268. [CrossRef]
3. Hdom, H.A. Examining carbon dioxide emissions, fossil & renewable electricity generation and economic growth: Evidence from a panel of South American countries. *Renew. Energy* **2019**, *139*, 186–197.
4. Peng, Z.; Guo, W. Saturation characteristics for stability of hydro-turbine governing system with surge tank. *Renew. Energy* **2019**, *131*, 318–332. [CrossRef]
5. Rostami, R.; Khoshnava, S.M.; Lamit, H.; Streimikiene, D.; Mardani, A. An overview of Afghanistan's trends toward renewable and sustainable energies. *Renew. Sustain. Energy Rev.* **2017**, *76*, 1440–1464. [CrossRef]
6. Kougias, I.; Aggidis, G.; Avellan, F.; Deniz, S.; Lundin, U.; Moro, A.; Muntean, S.; Novara, D.; Pérez-Díaz, J.I.; Quaranta, E.; et al. Analysis of emerging technologies in the hydropower sector. *Renew. Sustain. Energy Rev.* **2019**, *113*, 109257. [CrossRef]
7. Amponsah, N.Y.; Troldborg, M.; Kington, B.; Aalders, I.; Hough, R.L. Greenhouse gas emissions from renewable energy sources: A review of lifecycle considerations. *Renew. Sustain. Energy Rev.* **2014**, *39*, 461–475. [CrossRef]
8. Rostami, A.B.; Armandei, M. Renewable energy harvesting by vortex-induced motions: Review and benchmarking of technologies. *Renew. Sustain. Energy Rev.* **2017**, *70*, 193–214. [CrossRef]
9. Kumar, D.; Sarkar, S. A review on the technology, performance, design optimization, reliability, techno-economics and environmental impacts of hydrokinetic energy conversion systems. *Renew. Sustain. Energy Rev.* **2016**, *58*, 796–813. [CrossRef]
10. Khan, M.; Bhuyan, G.; Iqbal, M.; Quaicoe, J. Hydrokinetic energy conversion systems and assessment of horizontal and vertical axis turbines for river and tidal applications: A technology status review. *Appl. Energy* **2009**, *86*, 1823–1835. [CrossRef]
11. Rourke, F.O.; Boyle, F.; Reynolds, A. Marine current energy devices: Current status and possible future applications in Ireland. *Renew. Sustain. Energy Rev.* **2010**, *14*, 1026–1036. [CrossRef]

12. Golecha, K.; Eldho, T.; Prabhu, S. Influence of the deflector plate on the performance of modified Savonius water turbine. *Appl. Energy* **2011**, *88*, 3207–3217. [CrossRef]
13. Chen, L.; Chen, J.; Zhang, Z. Review of the Savonius rotor's blade profile and its performance. *J. Renew. Sustain. Energy* **2018**, *10*, 013306. [CrossRef]
14. Fleming, P.; Probert, S.; Tanton, D. Designs and performances of flexible and taut sail Savonius-type wind-turbines. *Appl. Energy* **1985**, *19*, 97–110. [CrossRef]
15. Sarma, N.; Biswas, A.; Misra, R. Experimental and computational evaluation of Savonius hydrokinetic turbine for low velocity condition with comparison to Savonius wind turbine at the same input power. *Energy Convers. Manag.* **2014**, *83*, 88–98. [CrossRef]
16. Chen, B.; Nagata, S.; Murakami, T.; Ning, D. Improvement of sinusoidal pitch for vertical-axis hydrokinetic turbines and influence of rotational inertia. *Ocean. Eng.* **2019**, *179*, 273–284. [CrossRef]
17. Gorelov, D.; Krivospitsky, V. Prospects for development of wind turbines with orthogonal rotor. *Thermophys. Aeromech.* **2008**, *15*, 153–157. [CrossRef]
18. Patel, V.; Eldho, T.; Prabhu, S. Theoretical study on the prediction of the hydrodynamic performance of a Savonius turbine based on stagnation pressure and impulse momentum principle. *Energy Convers. Manag.* **2018**, *168*, 545–563. [CrossRef]
19. Blackwell, B.F.; Feltz, L.V.; Sheldahl, R.E. *Wind Tunnel Performance Data for Two- and Three-Bucket Savonius Rotors*; Sandia Laboratories Albuquerque: Albuquerque, NM, USA, 1977.
20. Sivasegaram, S. Secondary parameters affecting the performance of resistance-type vertical-axis wind rotors. *Wind Eng.* **1978**, *2* , 49–58.
21. Khan, M.H. Model and prototype performance characteristics of Savonius rotor windmill. *Wind Eng.* **1978**, *2*, 75–85.
22. Ushiyama, I.; Nagai, H. Optimum design configurations and performance of Savonius rotors. *Wind Eng.* **1988**, *12*, 59–75.
23. Fujisawa, N.; Gotoh, F. Experimental study on the aerodynamic performance of a Savonius rotor. *J. Sol. Energy Eng.* **1994**, *116*, 148–152. [CrossRef]
24. Sheldahl, R.E.; Blackwell, B.F.; Feltz, L.V. Wind tunnel performance data for two-and three-bucket Savonius rotors. *J. Energy* **1978**, *2*, 160–164. [CrossRef]
25. Emmanuel, B.; Jun, W. Numerical study of a six-bladed Savonius wind turbine. *J. Sol. Energy Eng.* **2011**, *133*, 044503. [CrossRef]
26. Mahmoud, N.; El-Haroun, A.; Wahba, E.; Nasef, M. An experimental study on improvement of Savonius rotor performance. *Alex. Eng. J.* **2012**, *51*, 19–25. [CrossRef]
27. Wenehenubun, F.; Saputra, A.; Sutanto, H. An experimental study on the performance of Savonius wind turbines related with the number of blades. *Energy Procedia* **2015**, *68*, 297–304. [CrossRef]
28. Banerjee, A.; Roy, S.; Mukherjee, P.; Saha, U.K. Unsteady flow analysis around an elliptic-bladed Savonius-style wind turbine. In Proceedings of the Gas Turbine India Conference, New Delhi, India, 15–17 December 2014; American Society of Mechanical Engineers: New York, NY, USA, 2014; Volume 49644, p. V001T05A001.
29. Alom, N.; Kolaparthi, S.C.; Gadde, S.C.; Saha, U.K. Aerodynamic design optimization of elliptical-bladed Savonius-Style wind turbine by numerical simulations. In Proceedings of the International Conference on Offshore Mechanics and Arctic Engineering, Busan, Korea, 19–24 June 2016; American Society of Mechanical Engineers: New York, NY, USA, 2016; Volume 49972, p. V006T09A009.
30. Abdelaziz, K.R.; Nawar, M.A.; Ramadan, A.; Attai, Y.A.; Mohamed, M.H. Performance improvement of a Savonius turbine by using auxiliary blades. *Energy* **2021**, *244*, 122575. [CrossRef]
31. Faizal, M.; Ahmed, M.R.; Lee, Y.H. On utilizing the orbital motion in water waves to drive a Savonius rotor. *Renew. Energy* **2010**, *35*, 164–169. [CrossRef]
32. Yaakob, O.; Ahmed, Y.M.; Ismail, M.A. Validation study for Savonius vertical axis marine current turbine using CFD simulation. In Proceedings of the 6th Asia-Pacific Workshop on Marine Hydrodynamics-APHydro2012, Johor Bahru, Malaysia, 3–4 September 2012; pp. 3–4.
33. Khan, M.; Islam, N.; Iqbal, T.; Hinchey, M.; Masek, V. Performance of Savonius rotor as a water current turbine. *J. Ocean. Technol.* **2009**, *4*, 71–83.
34. Kailash, G.; Eldho, T.; Prabhu, S. Performance study of modified Savonius water turbine with two deflector plates. *Int. J. Rotating Mach.* **2012**, *2012*, 679247. [CrossRef]
35. Patel, V.; Bhat, G.; Eldho, T.; Prabhu, S. Influence of overlap ratio and aspect ratio on the performance of Savonius hydrokinetic turbine. *Int. J. Energy Res.* **2017**, *41*, 829–844. [CrossRef]
36. Kumar, A.; Saini, R. Performance analysis of a Savonius hydrokinetic turbine having twisted blades. *Renew. Energy* **2017**, *108*, 502–522. [CrossRef]
37. Talukdar, P.K.; Sardar, A.; Kulkarni, V.; Saha, U.K. Parametric analysis of model Savonius hydrokinetic turbines through experimental and computational investigations. *Energy Convers. Manag.* **2018**, *158*, 36–49. [CrossRef]
38. Mosbahi, M.; Ayadi, A.; Chouaibi, Y.; Driss, Z.; Tucciarelli, T. Performance study of a Helical Savonius hydrokinetic turbine with a new deflector system design. *Energy Convers. Manag.* **2019**, *194*, 55–74. [CrossRef]
39. Guo, F.; Song, B.; Mao, Z.; Tian, W. Experimental and numerical validation of the influence on Savonius turbine caused by rear deflector. *Energy* **2020**, *196*, 117132. [CrossRef]

40. Zhao, Z.; Zheng, Y.; Xu, X.; Liu, W.; Hu, G. Research on the improvement of the performance of Savonius rotor based on numerical study. In Proceedings of the 2009 International Conference on Sustainable Power Generation and Supply, Nanjing, China, 6–7 April 2009; pp. 1–6.
41. Ragheb, M.; Ragheb, A.M. Wind turbines theory-the betz equation and optimal rotor tip speed ratio. *Fundam. Adv. Top. Wind. Power* **2011**, *1*, 19–38.
42. Alexander, A.; Holownia, B. Wind tunnel tests on a Savonius rotor. *J. Wind. Eng. Ind. Aerodyn.* **1978**, *3*, 343–351. [CrossRef]
43. Yao, J.; Li, F.; Chen, J.; Yuan, Z.; Mai, W. Parameter analysis of Savonius hydraulic turbine considering the effect of reducing flow velocity. *Energies* **2019**, *13*, 24. [CrossRef]
44. Tian, W.; Mao, Z.; Zhang, B.; Li, Y. Shape optimization of a Savonius wind rotor with different convex and concave sides. *Renew. Energy* **2018**, *117*, 287–299. [CrossRef]
45. Kacprzak, K.; Liskiewicz, G.; Sobczak, K. Numerical investigation of conventional and modified Savonius wind turbines. *Renew. Energy* **2013**, *60*, 578–585. [CrossRef]
46. Mohamed, M.; Janiga, G.; Pap, E.; Thévenin, D. Optimal blade shape of a modified Savonius turbine using an obstacle shielding the returning blade. *Energy Convers. Manag.* **2011**, *52*, 236–242. [CrossRef]
47. Mari, M.; Venturini, M.; Beyene, A. A novel geometry for vertical axis wind turbines based on the Savonius concept. *J. Energy Resour. Technol.* **2017**, *139*, 061202. [CrossRef]
48. Versteeg, H.K.; Malalasekera, W. *An Introduction to Computational Fluid Dynamics: The Finite Volume Method*; Pearson Education : London, UK, 2007.
49. Succi, S.; Pappetti, F.; Succi, S. *An Introduction to Parallel Computational Fluid Dynamics*; Nova Publishers: Hauppauge, NY, USA, 1996.
50. Wahyudi, B.; Soeparman, S.; Wahyudi, S.; Denny, W. A simulation study of Flow and Pressure distribution patterns in and around of Tandem Blade Rotor of Savonius (TBS) Hydrokinetic turbine model. *J. Clean Energy Technol.* **2013**, *1*, 286–291. [CrossRef]
51. Driss, Z.; Jemni, M.; Chelly, A.; Abid, M. Computational study of a vertical axis water turbine placed in a hydrodynamic test bench. *Int. J. Mech. Appl.* **2013**, *3*, 98–104.
52. Roy, S.; Saha, U.K. Computational study to assess the influence of overlap ratio on static torque characteristics of a vertical axis wind turbine. *Procedia Eng.* **2013**, *51*, 694–702. [CrossRef]
53. Roy, S.; Saha, U.K. Wind tunnel experiments of a newly developed two-bladed Savonius-style wind turbine. *Appl. Energy* **2015**, *137*, 117–125. [CrossRef]
54. Mao, Z.; Tian, W. Effect of the blade arc angle on the performance of a Savonius wind turbine. *Adv. Mech. Eng.* **2015**, *7*, 1687814015584247. [CrossRef]

energies

MDPI

Article

Scenario-Based Comparative Analysis for Coupling Electricity and Hydrogen Storage in Clean Oilfield Energy Supply System

Fengyuan Yan [1], Xiaolong Han [1], Qianwei Cheng [2,*], Yamin Yan [1,*], Qi Liao [1] and Yongtu Liang [1]

1 Beijing Key Laboratory of Urban Oil and Gas Distribution Technology, China University of Petroleum-Beijing, Beijing 102249, China; yfy@cup.edu.cn (F.Y.); cuphxl@163.com (X.H.); qliao@cup.edu.cn (Q.L.); yongtuliang@126.com (Y.L.)
2 Center for Spatial Information Science, The University of Tokyo, Chiba 277-8568, Japan
* Correspondence: cheng@csis.u-tokyo.ac.jp (Q.C.); yanym0910@163.com (Y.Y.)

Abstract: In response to the objective of fully attaining carbon neutrality by 2060, people from all walks of life are pursuing low-carbon transformation. Due to the high water cut in the middle and late phases of development, the oilfield's energy consumption will be quite high, and the rise in energy consumption will lead to an increase in carbon emission at the same time. As a result, the traditional energy model is incapable of meeting the energy consumption requirement of high water cut oilfields in their middle and later phases of development. The present wind hydrogen coupling energy system was researched and coupled with the classic dispersed oilfield energy system to produce energy for the oilfields in this study. This study compares four future energy system models to existing ones, computes the energy cost and net present value of an oilfield in Northwest China, and proposes a set of economic evaluation tools for oilfield energy systems. The study's findings indicate that scenario four provides the most economic and environmental benefits. This scenario effectively addresses the issue of high energy consumption associated with aging oilfields at this point, significantly reduces carbon emissions, absorbs renewable energy locally, and reduces the burden on the power grid system. Finally, sensitivity analysis is utilized to determine the effect of wind speed, electricity cost, and oilfield gas output on the system's economic performance. The results indicate that the system developed in this study can be applied to other oilfields.

Keywords: scenario-based comparative analysis; electricity and hydrogen storage; energy supply system; techno-economic assessment

Citation: Yan, F.; Han, X.; Cheng, Q.; Yan, Y.; Liao, Q.; Liang, Y. Scenario-Based Comparative Analysis for Coupling Electricity and Hydrogen Storage in Clean Oilfield Energy Supply System. *Energies* **2022**, *15*, 1957. https://doi.org/10.3390/en15061957

Academic Editors: Wei-Hsin Chen, Aristotle T. Ubando, Chih-Che Chueh, Liwen Jin and Ricardo J. Bessa

Received: 6 January 2022
Accepted: 1 March 2022
Published: 8 March 2022

1. Introduction

1.1. Background

Currently, widespread usage of fossil fuels results in energy shortages, climate system degradation, and global warming, all of which pose a threat to the earth's biosphere [1]. Since the development of some Chinese oilfields has progressed to the point of high water cut, oilfield energy consumption has increased, resulting in a huge quantity of carbon emissions [2]. It is critical to improve the existing oilfield process, reduce carbon emissions, and improve economic benefits and energy utilization to achieve the goal of carbon peak and carbon neutralization [3] (i.e., strive to achieve the peak of domestic carbon dioxide emissions by 2030 and achieve carbon neutrality by 2060 [4]).

Wind energy is gaining popularity as a result of its clean, carbon-free, and long-term characteristics [1]. Wind energy resources are abundant in China, with huge and consistent wind speeds ranging from 300 to 500 W/m^2 [5]. They are mostly spread in the three northeastern provinces, Xinjiang, and other fairly vast areas. Wind energy deployment on a wide scale, however, is hampered by supply unpredictability and poor utilization. Because wind is intrinsically variable, it is not appropriate for grid power delivery, resulting in energy waste, as the stringent operating regulations of electricity markets have a big impact on wind turbines. Wind energy storage is a viable option for improving energy supply stability. Traditional battery energy storage has a poor storage density and a high cost of ownership, making it unsuitable for large-scale power storage in oilfields. Since the last century, hydrogen energy has been proposed by some experts and scholars at home and abroad as a medium for energy storage. Aziz. M [6] and A.T. Wijayanta [7] et al. proposed a new hydrogen energy storage method by combining hydrogen energy with NH_3. Converting renewable energy to hydrogen storage is an efficient and clean energy storage means, and developing hydrogen production from renewable energy is an ideal way to convert wind energy into electricity and store it on a large scale [8].

Hydrogen is an excellent energy storage medium due to its low toxicity and high density. The hydrogen fuel cell technology used in vehicles produces no pollutants or carbon dioxide, contributing to the mitigation of global energy scarcity and climate change [9]. At present, the major technical techniques for hydrogen preparation are thermochemical hydrogen synthesis and water electrolysis hydrogen generation [10]. Two of the most prevalent thermochemical hydrogen generation technologies are hydrogen synthesis from fossil fuels and hydrogen production from chemical raw ingredients [11]. Due to the high carbon emissions associated with hydrogen generation using fossil fuels, this research focuses only on hydrogen production using renewable energy electrolytic water.

The transformation and optimization of energy systems for high water cut oilfields is currently receiving a lot of attention. According to the foregoing study, there is an urgent need for innovative energy systems suitable for high energy consumption oilfields in the course of oilfield development and aging. To reduce carbon emissions, it must produce enough electricity for the central processing facility and make full use of renewable energy. Motivated by this aim, we offer a novel oilfield energy system with wind hydrogen connection and attempt to tackle the three questions below:

(1) How can renewable energy be combined with oilfield energy systems to provide a new set of viable, safe, and low-carbon oilfield distributed energy systems?
(2) How can we fully utilize and store renewable energy, enhance the energy efficiency, and provide adequate and steady energy for oilfields?
(3) How can we establish an economic evaluation method suitable for most oilfield energy systems and verify the economic and environmental benefits of the system described in this paper?

1.2. Related Works

At present, China's oilfield development has fallen into the stage of excessive energy consumption and reduced economic benefits. Renewable energy, as a sustainable energy source that is different from fossil energy, can largely solve the problems of oilfield development at this stage due to its cleanness and renewability. The use of renewable energy in combination with the oilfield energy system is the main direction of the next stage. However, in the development and utilization of renewable energy, how to rationally use renewable energy is a big problem. Experts and scholars at home and abroad have also carried out extensive research on this problem. For example, Li, Xu et al. [9] based their research on the associated geothermal energy in the oilfield and proposed a polygeneration energy system driven by the associated geothermal water during the high water cut period of the oilfield. It solves the problem of high energy consumption in high water cut oilfields. Wang, Yuan et al. [10] proposed a risk–opportunity assessment management framework based on existing assets, data, technology, and geothermal energy obtained from oil wells, and optimized the utilization of geothermal energy in different oilfields in practice. Li, Liu et al. [12,13] studied the utilization of geothermal energy in high water cut oilfields, developed a multi-power generation energy system driven by the relevant mixture of petroleum and geothermal water, evaluated the technical and economic performance, and used geothermal energy to solve the problem of high energy consumption of high water cut oilfields. In addition to geothermal energy being applied to oilfield systems, other forms of renewable energy have also been developed. For example, Spielman, Brey et al. [14]

proposed a sustainable offshore wind farm energy system based on the abundant wind energy resources in offshore oil and gas fields. It fills the blank of the energy system for offshore oil and gas field development. Zou, Qiu et al. [15] proposed adding hydrogen energy as storage energy based on the sustainable development plan of offshore oil and gas fields, realizing the efficient utilization and storage of renewable energy.

Although there are many kinds of energy systems proposed by domestic and foreign experts and scholars, given the current situation of high energy consumption, high water content, and low income in the development of oilfields in China, there is a need for a system that can not only realize large-scale energy storage but also meet the needs of various oilfield energy sources. Oilfield energy systems that consume high amounts of energy demand renewable solutions. At present, the means of storing renewable energy mainly include electricity storage and thermal storage. For example, Huang, Gu et al. [16] took Daqing and Shengli oilfields as examples to comprehensively evaluate oil and gas resources, and converted part of the fossil energy in the oilfields into electricity for storage, solving the problem of energy storage. The excessive consumption problem was addressed by Hu Sun et al. [17], who proposed an energy system for coal mines that converts coal energy into electrical energy storage. F. Monfaredi et al. [18] developed a hybrid power and natural gas coupled energy system, converting natural gas into electrical energy storage and applying it in oilfields. Amir et al. proposed a thermoelectric generator energy system for recycling waste energy, which converts waste energy into thermal energy for storage and utilization. Huang, Yu et al. [19] proposed a two-stage energy management system of thermoelectric integration, which can store electrical energy and thermal energy at the same time and convert the two forms of energy into each other. However, thermal storage cannot meet the various forms of energy demand in oilfields, the electric storage model cannot store energy on a large scale, and livestock batteries are expensive to manufacture. Therefore, some experts and scholars have proposed hydrogen energy storage as a new energy storage method. For example, L. Bartolucci et al. [20] proposed an energy system based on hydrogen, which uses hydrogen as storage energy for the energy consumption of residential buildings. Dong Wu et al. [21] proposed a hydrogen-based energy integration system combining hydrogen and water storage. Using hydrogen as storage energy can effectively improve storage efficiency. At the same time, it can also realize the mutual conversion of hydrogen energy and other forms of energy through fuel cells and other methods. Under the dual-carbon goal, hydrogen energy, as a clean energy source, can effectively reduce carbon emissions and promote carbon neutrality. Therefore, drawing on this energy system that uses hydrogen as a renewable energy storage method and introducing it into oilfields, this paper proposes a new wind–hydrogen coupled oilfield energy system and conducts economic and reliability analysis.

The system's primary flow is depicted in Figure 1. The related gas and crude oil generated by the oilfield can be used to provide heat and electricity for the oilfield system. Simultaneously, when utilized in conjunction with a wind turbine, it can provide electric energy for the oilfield, with the power grid serving as a backup system for the oilfield system. Simultaneously, wind energy's residual electricity is utilized to produce hydrogen, which is used to store energy and can also be used to power the central processing facility via a fuel cell or combustion [22,23]. This system has three major advantages.

1.3. Paper Organization

The structure for the remainder of this paper is organized as follows. Section 2 presents the hybrid energy system that utilizes the wind farm's electric power to electrolyze water to produce hydrogen, which is then stored in a hydrogen storage tank. The oilfield's associated gas is burned to generate heat energy for the oilfield's system. In Section 3, the mathematical model of the system is given. In Section 4, a detailed description of the examples and the results are provided. Section 5 gives the conclusions of the complete manuscript.

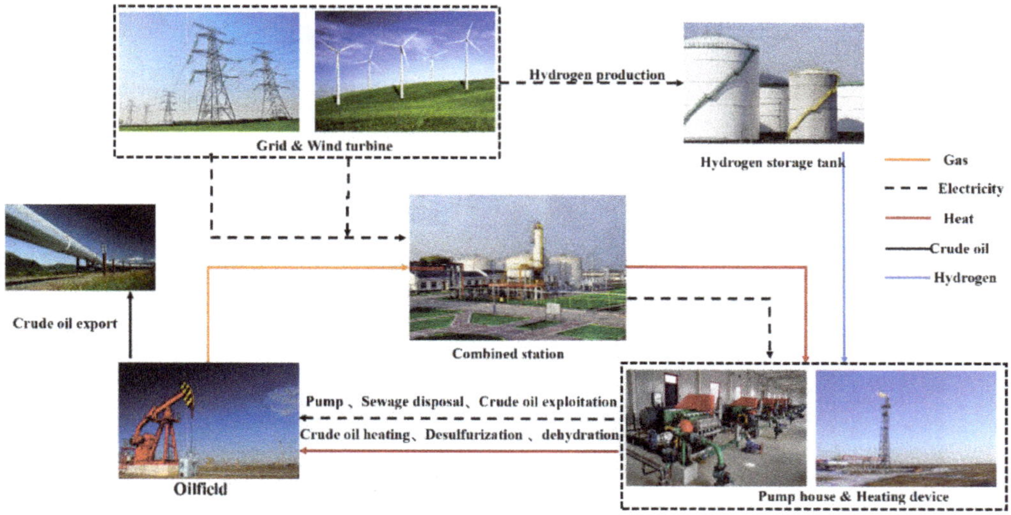

Figure 1. Oilfield hybrid energy system.

(1) The system combines renewable energy with gas turbine power generation to meet the electricity and heat requirements of the oilfields while assisting to achieve economic benefits and dual-carbon targets.
(2) The system considers converting electric energy to hydrogen, which eliminates the instabilities associated with wind power generation and the original system's high battery cost, as well as makes energy storage and peak shaving more feasible.
(3) The Energy Internet is established to improve system resilience and avoid production and operation problems caused by energy shortages since the new fuel cell and hydrogen storage tank can be used in place of the battery, and the stored hydrogen can be sold to increase the economic benefits.

2. Problem Statement

As an oilfield enters the middle and late stages of development, its energy consumption increases with the decline of production and the increase of water content. The classic oilfield distributed energy system relies heavily on gas turbines and boilers to provide electricity and heat energy. However, the system has poor energy efficiency and produces significant amounts of carbon dioxide. Therefore, this study presents and compares five novel energy systems based on improved distributed energy systems, as shown in Table 1. Specifically, scenario one is a standard distributed energy system that is powered by a gas turbine and a crude oil heater. On this basis, wind turbines are erected in scenario two, and the generated electric energy is immediately linked to the grid to provide electricity for the oilfield. In scenario three, the wind energy is not connected to the power grid but is directly connected to the storage battery with peak shaving and valley filling function to meet the electricity demand of the oilfield. Compared to scenario two, scenario three avoids the grid connection of renewable energy, which can reduce the pressure on the power grid system and eliminate other adverse effects. In scenario four, hydrogen generation from residual power is an indirect energy storage method, and it can be powered by fuel cells in case of need, which may bring greater economic benefits. Based on scenario four, scenario five uses crude oil instead of natural gas to provide heat energy for the oilfield joint station. These five schemes cover a variety of practical and feasible hydrogen storage and electricity storage situations and are of practical significance.

Table 1. Literature summary of the oilfield energy system.

Oilfield Energy System	Author	Input Energy	Output Energy	Document Number
Associated geothermal drive	Li, Xu et al.	Geothermal energy, electric energy	Thermal energy, electric energy	6
Oilfield geothermal energy system	Wang, Yuan et al.	Geothermal energy, electric energy	Thermal energy, electric energy	7
The energy system of abandoned oilfields in North China	Guo, T. et al.	Geothermal energy	Thermal energy, electric energy	9
Multipower generation energy system in high water cut oilfield	Li, Liu et al.	Fossil energy, geothermal energy, electric energy	Thermal energy, electric energy	10
Sustainable offshore wind farm energy system	Spielman, Brey et al.	Wind energy and electric energy	Electric energy, hydrogen energy	11
Sustainable development of oilfield energy system	Zou, Qiu et al.	Renewable energy	Thermal energy, electric energy	12
Daqing Oilfield sustainable energy system	Huang gu et al.	Electric energy	Thermal energy, electric energy	13
Sustainable offshore oil and gas field development system	Hu Sun et al.	Wind energy and electric energy	Thermal energy, electric energy, hydrogen energy	14

The energy flow of the five scenarios is introduced as follows:

(1) As the original power supply system of the oilfield, only the electricity purchased from the grid can meet the electricity demand of the oilfield. This scenario is used as a control group to compare the economics of the other four schemes. Figure 2 shows the energy system framework of scenario one;

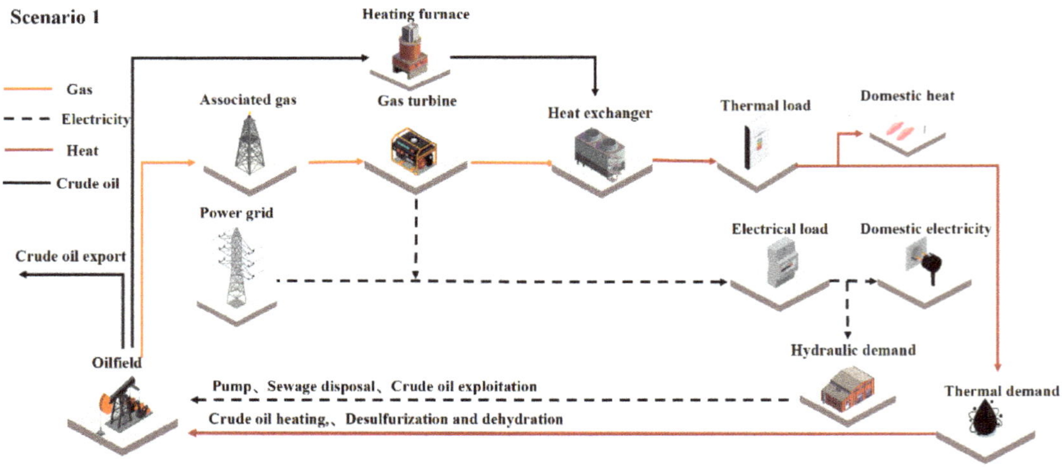

Figure 2. The oilfield energy system framework of scenario 1.

(2) In the case of grid power supply, wind turbines can be used to provide electricity demand for oilfields. However, it is worth noting that wind power generation is unstable and the grid needs to be used as a backup solution. This solution can compare the economics and feasibility of directly using renewable energy as a means of power supply. Figure 3 shows the energy system framework of scenario two;

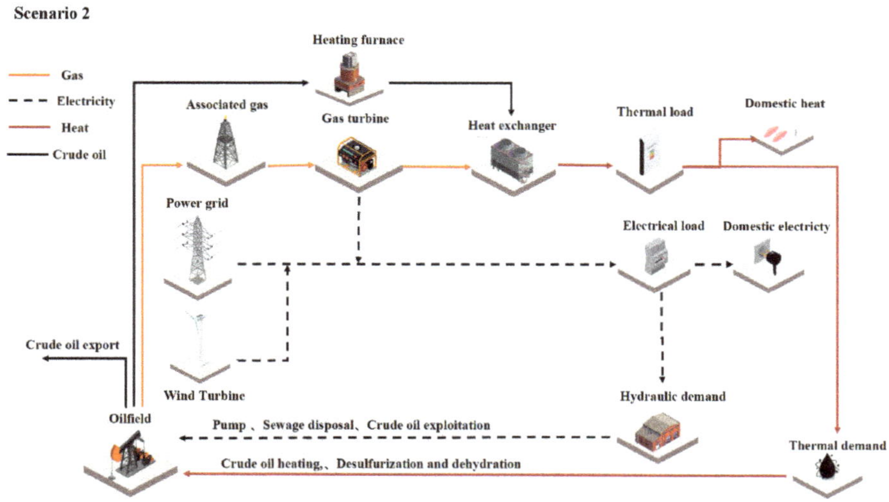

Figure 3. The oilfield energy system framework of scenario 2.

(3) When wind turbines are used to provide energy consumption for oilfields, the remaining electricity is used for hydrogen production. The obtained hydrogen can be stored in a hydrogen storage tank. During the peak period of electricity consumption, fuel cell combustion is used to generate electricity to achieve the purpose of energy storage and peak shaving. The program introduces hydrogen based on renewable energy power generation to explore the possibility of hydrogen used in oilfield energy systems. Figure 4 shows the energy system framework of scenario three;

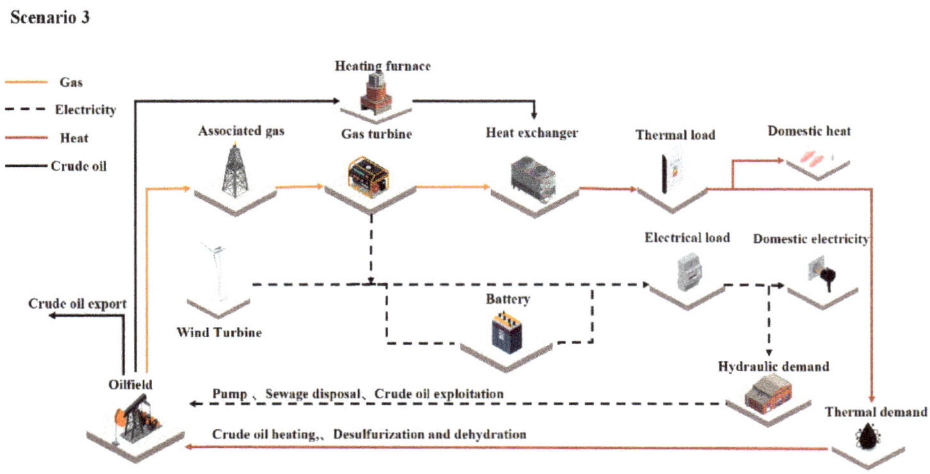

Figure 4. The oilfield energy system framework of scenario 3.

(4) In scenario four, the gas turbine also burns associated gas from the oilfield to generate electricity to meet part of the power demand of the oilfield system. At the same time, the electricity generated by the wind turbines is used in two parts: the first part is used for part of the power needs of the oilfield system, and the second part is used to supply the electrolyzer, which electrolyzes water to produce hydrogen. The produced hydrogen will be stored in a high-pressure liquid hydrogen storage tank. When the power supply of other systems in the oilfield is insufficient, the hydrogen in the output

tank will be converted into hydrogen energy and electric energy through a hydrogen fuel cell, thereby reaching the hydrogen energy level.

Secondly, the heat energy consumption of the oilfield comes from the high-temperature steam generated by the gas turbine, and the high-temperature steam generates heat through the heat exchanger (boiler, etc.) to provide heat consumption for the oilfield. Figure 5 shows the energy system framework for scenario four;

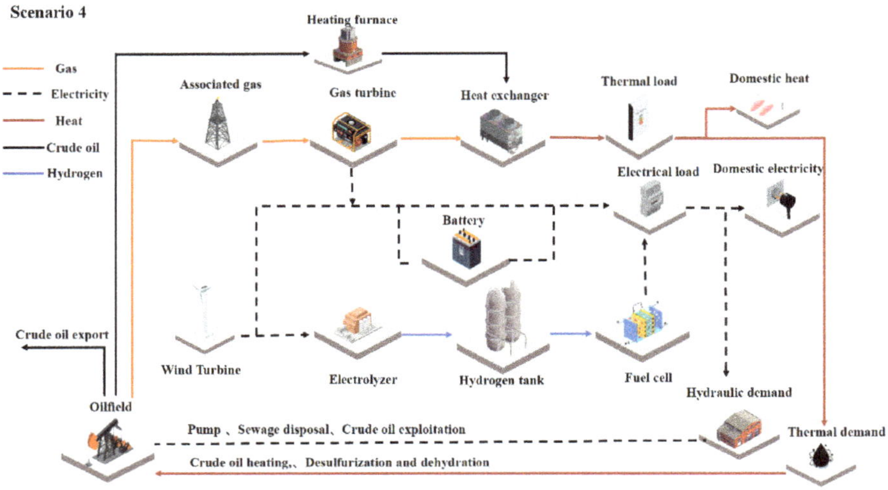

Figure 5. The oilfield energy system framework of scenario 4.

(5) Scenario five (Figure 6) and scenario four are control groups, and the energy system of the oilfield is the same, but considering the non-environmental nature of the crude oil heating furnace, we replaced it with a more environmentally friendly hydrogen combustion furnace as the heating equipment in the oilfield. At the same time, all the hydrogen produced in this scenario is only used for combustion to provide heat for the oilfield, not for electricity generation.

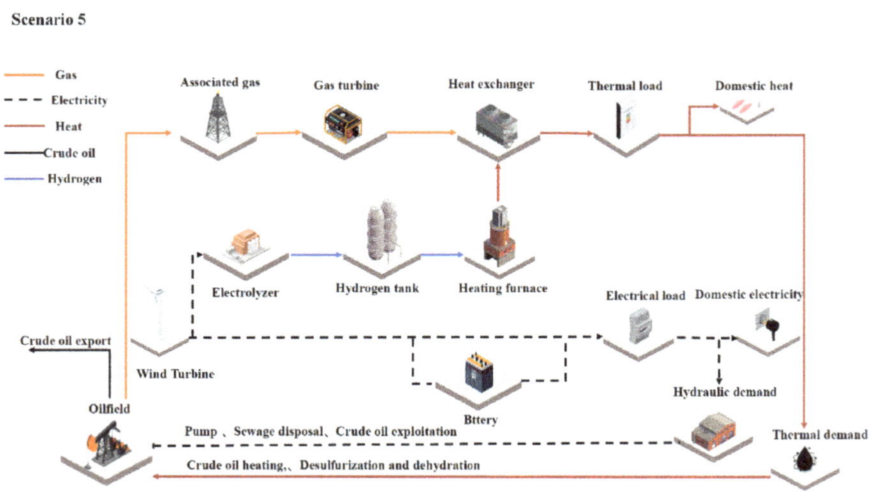

Figure 6. The oilfield energy system framework of scenario 5.

The source of heat energy is not only the high-temperature gas produced by the gas turbine but also the high-quality heat energy provided directly by natural gas combustion. However, because the heat energy consumed to heat crude oil or household heat is insignificant in an oilfield system, the heat generated by traditional gas turbines and heat exchangers is sufficient to compensate for the oilfield's heat loss.

Apart from being a green, clean, and carbon-free renewable energy source, hydrogen may also help save energy, which is a significant advantage of the hybrid energy system. Hydrogen can be transferred as a gas via pipelines or as a liquid via oil tankers, as well as stored in hydrogen tanks. In terms of peak control, wind energy is favored to supply power to the oilfield system in the event of excess wind energy. The residual energy can be transformed to hydrogen in the electrolyzer via an electrochemical reaction. The hydrogen produced can be retained in the hydrogen storage tank in case of an emergency, or it can be sold commercially via other means such as pipelines. On the other hand, if wind energy is insufficient to power the oilfield, hydrogen in the hydrogen storage tank will be used, as would the proton exchange membrane fuel cell, and hydrogen will be used to generate electricity to meet the oilfield's power needs.

To determine the system's practicality, the following work is required:

(1) Existing data indicate that energy, heat energy, economic benefits, and carbon emissions can be reduced;
(2) A comprehensive evaluation model of the hybrid energy system's technology and economics is established, including the quantification of the project's net present value and investment recovery period;
(3) A sensitivity analysis was conducted to determine the effect of variables such as hydrogen, power price, and wind speed on the economic performance of all energy systems.

3. Methodology

This section introduces each component of the system to evaluate its technical and economic performance. Section 3.1 discusses the operation costs, as well as the technical details of electricity and hydrogen production technology. Section 3.2 describes the design and operation costs, as well as the technical details of the gas turbine heating and power supply system. Finally, Section 3.3 proposes a method for computing the NPV and LCOE of the entire energy system based on its composition.

3.1. Electricity and Hydrogen Production Technology

3.1.1. Wind Turbine

The mechanical system, the electrical system, the control system, and the auxiliary system all comprise the wind turbine system. Wind turbines are usually divided according to their spin orientation into horizontal axis and vertical axis wind turbines. Wind energy generation is a technique that converts the kinetic energy of the wind to mechanical energy, which is subsequently turned into electric energy. Wind energy is generated by utilizing the wind to drive the blades of a windmill to rotate, and then increasing the speed of the blades via a speed-boosting mechanism to encourage the generator to create electricity. To generate power, windmill technology requires a wind speed (degree of breeze) of around three meters per second [24]. In this paper, wind turbines are mainly used in scenarios 2, 3, 4, and 5 to convert the wind resources of the oilfield into electrical energy. The formula for calculating the annual wind energy of the oilfield and the cost of wind turbines is as follows.

(1) Annual electricity production (AEP)

Wind resources are seasonal and unevenly distributed, which adds complexity to the computation. Wind speed, on the other hand, follows a specific random distribution law. According to statistical analysis, the probability density of the wind speed distribution (wind profile distribution) can be defined by the Weibull distribution function, which provides a general mathematical description for the wind speed distribution [25]:

$$f(\nu) = \frac{k}{c}\left(\frac{\nu}{c}\right)exp\left[-\left(\frac{\nu}{c}\right)^k\right], \quad k > 0, \nu > 0, c > 0 \tag{1}$$

where, c is the proportional parameter (M/s); k is the shape parameter; v is the wind speed.

Weibull parameters k and c can be calculated according to the mean and variance of local wind speed.

$$k = (\sigma/\nu)^{-1.086}, \quad 1 \leq k \leq 10 \tag{2}$$

$$c = \nu/\Gamma\left(1 + \frac{1}{k}\right) \tag{3}$$

In this way, the theoretical power of wind power generation can be obtained. Then, the correction coefficient CP is introduced. The maximum theoretical power of the turbine is:

$$P_{WT} = \frac{1}{2}c_p(\lambda, \beta)\rho A v^3 \tag{4}$$

where, ρ is the air density; A is the swept area of the blade; v is the wind speed; C_P is the limit of Betz (59.3%).

The annual power generation (AEP) is calculated by combining the product of the wind profile described by function with the power curve of the selected turbine [26]:

$$AEP = \sum_{T=1}^{Tmax} P_{WT} \times (1 - \mu)/1000 \tag{5}$$

where, μ is the total generation loss.

(2) Cost of wind turbines

Among various cost models, the cost model developed by NREL is a detailed wind turbine cost model widely used in the research community [27]. The total wind turbine cost in the NREL wind turbine cost model is:

$$Cost = FCR \times ICC + AOC \tag{6}$$

where, *Cost* is the total cost of the wind turbine; *ICC* is the initial capital cost of the wind turbine; *FCR* is the fixed charge rate; *AOC* is the annual operating cost of the wind turbine.

It should be clear that the wind turbine cost calculated here is the average cost over the designed wind turbine's life.

$$ICC = C_{MS}^{WT} + C_{ES}^{WT} + C_{Control}^{WT} + C_{AS}^{WT} + C_{In}^{WT} \tag{7}$$

where, *ICC* is the sum of the balance of wind turbine system cost and plant cost; the wind turbine system is a complex system, including multiple subsystems, including mechanical (C_{MS}^{WT}), electrical (C_{ES}^{WT}) and electronic control systems ($C_{Control}^{WT}$) and some auxiliary systems (C_{AS}^{WT}), as well as the cost of infrastructures, such as foundation, road, license, electrical connection, installation, and transportation [28].

The initial capital cost of the steam turbine is shown in Table 2. The cost of each component or infrastructure is a function of turbine rotor radius, turbine power rating, and hub height.

$$AOE = OC_t^{LIC} + OC_t^{LRC} + OC_t^{O\&M} \tag{8}$$

Table 2. Scenarios for production, transportation, and storage of energy.

Equipment	Scenario I	Scenario II	Scenario III	Scenario IV	Scenario V
	Grid-GT	Grid-GT-WT	WT-GT	WT-GT-H_2	WT-GT-Fur
Wind Turbine	-	√	√	√	√
Electrolyzer	-	-	-	√	√
Furnace (Crude oil)	√	√	√	√	-
France (H_2)	-	-	-	-	√
Heat exchanger	√	√	√	√	√
Gas Turbine	√	√	√	√	√
H_2 tank	-	-	-	√	√
Fuel cell	-	-	-	√	√
Battery	-	-	√	√	√
Power grid	√	√	-	-	-

The annual operation cost (AOE) of a steam turbine includes land lease fee (OC_t^{LIC}), (USD); leveling operation and maintenance fee (OC_t^{LRC}) and leveling and resetting fee ($OC_t^{O\&M}$), (USD). These costs are incurred when the rated power of the turbine or the annual energy of the turbine is constant. See Table 3 for details of each cost.

Table 3. The initial capital cost of the wind turbine [18,19].

Subsystem	Components	Cost Model (USD)
Mechanical system	Blade	$(0.4019R^3 + 2.7445R^{2.5025} - 955.24)/0.72$
	Gearbox	$16.45 \times (0.001P_r)^{1.249}$
	Low-speed shaft	$0.1 \times (2R)^{2.887}$
	Main bearings	$(0.64768R/75 - 0.01068672) \times (2R)^{2.5}$
	Mechanical brake	$1.9894 \times 10^{-3}P_r - 0.1141$
Electrical system	Generator	$0.065P_r$
	Power converter	$0.079P_r$
	Electrical connection	$0.04P_r$
Control system	Pitch system	$0.480168 \times (2R)^{2.6578}$
	Yaw system	$0.0678 \times (2R)^{2.964}$
	Control safety system	35,000
Auxiliary system	Hydraulic, cooling system	$0.012P_r$
	Hub	$2.0061666R^{2.53} + 24,141.275$
	Nose cone	$206.69R - 2899.185$
	Mainframe	$11.9173875 \times (2R)^{1.953}$
	Nacelle cover	$1.1537 \times 10^{-2}P_r + 3849.7$
	Tower	$0.59595\pi R^2 H - 2121$
Infrastructures	Foundation	$303.24 \times (\pi R^2 H)^{0.4037}$
	Roads civil work	$2.17 \times 10^{-15}P_r^3 - 1.45 \times 10^{-8}P_r^2 + 0.06954P_r$
	Interface connections	$3.49 \times 10^{-15}P_r^3 - 2.21 \times 10^{-8}P_r^2 + 0.1097P_r$
	Engineering permits	$9.94 \times 10^{-10}P_r^2 + 0.02031P_r$
	Transportation	$1.581 \times 10^{-14}P_r^3 - 3.75 \times 10^{-8}P_r^2 + 0.0547P_r$
	Installation	$1.965 \times (2HR)^{1.1736}$

It is worth noting that the original capital cost is divided into five components: the mechanical system, the electrical system, the control system, the auxiliary system, and the infrastructures, as shown in Table 2. Here, is symbolized the rotor radius, (m), the wind turbine's rated power, (kW), and the hub height, (m). It is discovered that the cost model for each component of the subsystem is characterized as a function of the wind turbine's rotor radius, rated power, and hub height. This implies that altering any physical parameter has an effect on the system's cost, and the effect can be quantified. The yearly running costs are shown in Table 4, including the Levelized replacement cost, the land leasing cost, and the

Levelized operations and maintenance cost. Table 4 also reflects the wind turbine's yearly energy output, which can be computed using Equation (4). Similarly, the cost model is expressed as a function of the rotor radius of the turbine and its yearly energy output. The overall cost of the wind turbine may be determined by multiplying the original capital cost by the yearly running cost.

Table 4. Annual operating cost of a wind turbine [18,19].

Type	Cost Model (USD)
Levelized replacement cost	$0.00107P_r$
Land lease costs	$1.08 \times 10^{-6}AEP$
O&M	$7 \times 10^{-6}AEP$

3.1.2. Electrolytic Water Technology

The electrolytic water hydrogen production technique has a long history and can be classified into alkaline water electrolysis, proton exchange membrane water electrolysis, and solid oxide water electrolysis hydrogen production. Among them, the most established and commonly used technology is alkaline water electrolysis hydrogen generation [29,30]. Table 5 compares the three electrolytic water technologies. Electrolytic cells can be classified into three types based on their electrolytes: alkaline electrolytic cells (AE) [31], proton exchange membrane (PEM) electrolytic cells [32], and solid oxide (SOEC) electrolytic cells [33]. The characteristics of several electrolytic cells are listed in Table 6. The electrolyzer is used in scenarios four and five in the energy system, in which the surplus electricity generated by the wind turbine is used to supply the electrolyzer to electrolyze water to produce hydrogen, and the produced hydrogen is stored in the hydrogen storage tank for storage.

Table 5. Comparison of three typical hydrogen production technologies from electrolytic water [34,35].

Project	Hydrogen Production from Alkaline Electrolytic Water	Hydrogen Production by Electrolysis of Water with Proton Exchange Membrane	Hydrogen Production by Electrolysis of Water with Solid Oxide
Energy efficiency/%	60~75	70~90	85~95
Operating temperature/°C	70~90	70~80	700~1000
Current density/(A·cm^2)	0.2~0.4	1.0~2.0	1.0~10.0
Power consumption/(kWh·m^3)	4.5~5.5	3.8~5.0	2.6~3.6
Response speed	Faster	Fast	Low
Power quality requirements	Steady point power supply	Steady point or fluctuation	Steady point power supply
System operation and maintenance	Corrosive liquid, later stage complex operation, and maintenance and high cost	Complex operation and maintenance and high cost	No corrosive liquid, simple operation, and maintenance, low cost, the project is still in the laboratory stage, and the operation and maintenance demand is not clear for the time being
Cell life/(h)	12,000	10,000	-
Equipment cost/(USD·kWh)	300	1200	-
Characteristic	The technology is mature and has realized large-scale industrial application with low cost	It has good renewable energy adaptability, no pollution, high cost at this stage, has not been commercialized in China, high PEM replacement cost, and the catalyst is a precious metal	Some electric energy can be replaced by thermal energy, with high conversion efficiency, difficult material selection under high-temperature conditions, and has not been commercialized

Table 6. Characteristic parameters of electrolyzers [36,37].

Electrolyte Type		Current Density/A·cm^{-2}	Voltage Range/V	Oxygen Production/m^3·h^{-1}	Power Consumption/(kW·h)·m^{-3}	Operating Temperature/°C	Operating Pressure/Mpa	Efficiency/%
Basylous	Single-stage	0.1~0.2	2.0~2.4	≤8 × 104	5.0	60~80	<3	75~90
	Multistage	0.2~0.4	1.8~2.0	≤2 × 105	4.3~4.6	60~80	<3	75~90
PME		0.6~2.0	1.8~2.2	≤2.5 × 104	4.5	50~80	<20	80~90
High-temperature solid-state oxidation		0.3~2.0	0.7~1.5	≤2.5 × 104	3.9~4.0	650~1000	<2.5	80~90

A typical water electrolysis hydrogen generation process consists primarily of water electrolytic cells, a gas–liquid separator, a gas scrubber, an electrolyte circulation pump, an electrolyte filter, a pressure regulator, measuring and control equipment, and power supply equipment. The electrolytic cell is the central component that completes the hydrogen generation process by water electrolysis; the generated gas is then sent through a gas–liquid separator and a gas scrubber to achieve pure hydrogen. Hydrogen synthesis via hydro hydrolysis requires tight control of process parameters such as electrolytic current, voltage, and electrolyte density.

The alkaline electrolyzer technique is the most developed and least expensive. It is currently the most extensively used electrolyzer. It is well suited for large-scale hydrogen generation systems based on wind energy. At the moment, the solid oxide electrolyzer technology is in its infancy and has not been widely marketed. Due to its high current density, battery efficiency, hydrogen purity, high working pressure, and adaptability, the PEM electrolyzer has become the most widely used hydrogen generation technique via water electrolysis. However, the cost of a PEM electrolyzer is prohibitively high, and wind power variability reduces the electrolyzer's service life [34]. As a result, this work does not use a PEM electrolyzer but rather a more mature alkaline electrolyzer.

Alkaline electrolyzers operate at low temperatures (60–80 °C) and employ aqueous potassium hydroxide or sodium hydroxide solutions as electrolytes with a concentration of approximately 20–30% [38]. The septum of alkaline electrolyzers is made of asbestos, while the electrodes are made of nickel materials. The created hydrogen is around 99 percent pure, alkaline electrolyte's maximum operating current density is less than 400 mA/cm^2, and the power consumption for H_2 generation is approximately 4.5–5.5 k [39].

The efficiency is approximately 60%. To estimate the cost of an electrolytic cell, we frequently utilize a cost per kilowatt-hour of hydrogen output based on a lower heating value (LHV). The cost of the electrolytic cell is calculated as a function of the hydrogen generation capacity and current density. The following equation provides an estimate of the cost of an electrolytic cell [40]:

$$C_{WE} = 29,450 \times \left(V_{hy}\right)^{\alpha} \times \left(A_{hy}\right)^{\beta} \tag{9}$$

where, α and β all are coefficients, $\alpha = 0.79$, $\beta = -0.32$; C_{WE} is the cost of water electrolier, (USD); V_{hy} hydrogen generation capacity, (m^3); A_{hy} is rating current density, ($A{\cdot}m^3$).

3.1.3. Hydrogen Storage Technology

There are numerous methods for storing hydrogen, which can be classified into physical and chemical hydrogen storage based on the storage principle. Physical hydrogen storage mostly consists of liquefaction, high-pressure storage, and low-temperature compression. Chemical hydrogen storage methods include metal hydride storage, activated carbon adsorption storage, carbon fiber and carbon nanotube storage, organic liquid hydride storage, and inorganic hydrogen storage, among others. Due to the technical constraints and current state of development, high-pressure hydrogen storage, liquefaction storage, and transportation, as well as metal hydride hydrogen storage, are the most suitable modes of hydrogen storage for oilfield energy storage. Table 7 provides an overview of the various hydrogen storage modes.

Table 7. Comparison of several hydrogen storage methods [41].

Hydrogen Storage Technologies	Hydrogen Storage Capacity	Cost	Ease of Operation	Security	Ease of Transport	Technology Maturity	Ooperating Pressure/Mpa	Domestic Applications
High-pressure gaseous hydrogen storage	Small	Low	Easy	Poor	Convenient	Maturity	15.2~70.9	Extensive
Low-temperature liquid hydrogen storage	large	Very high	Difficult	Poor	Normal	Immaturity	5~10	For aerospace only
Metal hydride hydrogen storage	large	High	Easy	Security	Convenient	General	1~6	Very little
Organic liquid hydrogen storage	large	High	Easy	Security	Convenient	Immaturity	10~14.2	Very little

Liquefied hydrogen storage is a rather mature form of hydrogen storage. At normal temperature and pressure, the density of liquid hydrogen is 845 times that of gaseous hydrogen, and the volume energy density is several times that of compressed storage [42]. Compressing hydrogen and cooling it to temperatures below 21 k liquefies it into liquid hydrogen, which is then stored in a special insulated vacuum container. The term "high-pressure gaseous hydrogen storage" refers to the process of storing gaseous hydrogen at a pressure greater than the hydrogen's critical temperature. At the moment, it is a frequently utilized method of hydrogen storage.

In general, the gas tank is employed as a container, which is convenient and straightforward. Its advantages include low energy consumption during storage, low cost (when the pressure is not too high), rapid charging and venting, the ability to release hydrogen at room temperature, and the ability to operate normally at temperatures tens of degrees below zero. Additionally, the hydrogen release can be managed via the pressure-reducing valve. Due to the advantages outlined above, high-pressure gaseous hydrogen storage has developed into a more mature hydrogen storage technology.

Metal hydride hydrogen storage is a novel type of hydrogen storage that stores and releases hydrogen using metal hydride hydrogen storage materials. It makes use of certain metals or alloys to absorb hydrogen in the form of metal hydride after interacting with hydrogen to create metal hydride. Following that, hydrogen is released. Metal hydride hydrogen storage tanks have a hydrogen storage density of 1000 times that of hydrogen in its standard state, which is comparable to or even greater than that of liquid hydrogen [43]. They can be used not only to collect and store hydrogen but also to provide hydrogen to devices that require hydrogen. Metal hydride hydrogen storage tanks have a high hydrogen storage mass density ratio, a high hydrogen storage volume ratio (more than high-pressure compression and liquid hydrogen storage), a high level of safety, and a high level of hydrogen purity [44]. In comparison to liquid hydrogen storage and gaseous high-pressure hydrogen storage, hydrogen in metal hydride is a safer fuel. However, this technology is still in its infancy and can be used exclusively in new cars and laboratory research.

High-pressure gaseous hydrogen storage is used in this system. The high-pressure gaseous hydrogen storage tank in this paper is used in conjunction with the electrolyzer and is applied in scenarios four and five. The capacity of the hydrogen storage tank is 5000 kg, and the two hydrogen storage tanks can store 10 t of hydrogen from the electrolyzer. The investment cost of a hydrogen storage tank generally depends on the volume of the hydrogen tank and the quality of stored hydrogen. At the same time, another part of the cost comes from the operation and maintenance cost of the storage tank:

$$C_{hyst} = 45,890 \times {M_{storetank}}^{\alpha} \quad (10)$$

$$O_{hyst} = UOM^{store} \times {M_{storetank}}^{\alpha} \quad (11)$$

where, C_{hyst} represents the cost of a hydrogen storage tank, (USD); $M_{storetank}$ represents the weight of storage tank, (kg); O_{hyst} represents operation and maintenance cost, (USD); UOM^{store} is the unit capital and operation and maintenance cost of the tank, (USD/kg); α is constant, $\alpha = 0.55$.

3.1.4. Fuel Cell

Hydrogen produced by electrolysis of water is mixed with oxygen in the fuel cell, and the chemical energy of hydrogen and oxygen is converted into electrical energy under the action of a catalyst, and the reaction product is only water, realizing the conversion of energy. The fuel cell is mainly composed of motor, electrolyte diaphragm, and electrolyte, which is mainly divided into the high-temperature fuel cell and low-temperature fuel cell [45]. Among them, the high-temperature fuel cell is divided into solid oxide type and molten carbonate type, and the low-temperature fuel cell is alkaline type proton exchange membrane type and carbonic acid type. Due to its high cost, the fuel cell used in this paper is only used as a hydrogen-to-electricity conversion tool in scenario four to convert hydrogen energy into more useful electricity to provide electricity for oilfields. The characteristic parameters of these five fuel cells are shown in Table 8.

Table 8. The characteristic parameters of these five fuel cells.

FC Type	Current Density/ ($A\cdot cm^{-2}$)	End Voltage/V	Power Density/ ($W\cdot cm^{-2}$)	H_2 Consumption/ ($cm^3\cdot(min\cdot A)^{-1}$)	O_2 Consumption/ ($cm^3\cdot(min\cdot A)^{-1}$)	Pressure/ Mpa	Temperature/ °C	Efficiency	
								Battery	System
Proton Exchange Membrane	0.1~0.9	0.6~0.8	0.35~0.70	7.0	3.5	0.1~0.2	30~100	50~70	30~50
Alkali	0.1~0.9	0.6~0.8	0.10~0.30	7.0	3.5	0.1	80	60~70	62
Phosphoric acid	0.1~0.9	0.6~0.8	0.12~0.14	7.0	3.5	0.1	150~200	55	40
Fused carbonates	0.1~0.9	0.6~0.8	0.10~0.12	7.0	3.5	0.1~1.0	600~700	55	47
Solid Oxide	0.1~0.9	0.6~0.8	0.15~0.70	7.0	3.5	0.1	850~1100	60~65	55~60

At present, an alkaline fuel cell has the fastest development, but it is mainly used in the field of aerospace. The phosphoric acid battery has been put into commercial use, but its efficiency is low. Solid oxide fuel cell has wide adaptability and high energy conversion efficiency, so solid oxide fuel cell is used as an electric hydrogen conversion tool in this system. The cost calculation of fuel cells is related to the cell capacity [46]:

$$C_{cell} = UC^{cell} \times V_{cell} + OC^{cell} \tag{12}$$

where, C_{cell} is the cost of fuel cell, (USD); UC^{cell} is fuel cell capacity, (USD); UC^{cell} is the unit construction cost of the fuel cell, (USD/kWh); OC^{cell} is the operating cost of the fuel cell, (USD); V_{cell} is fuel cell capacity, (kWh);

3.2. *Gas Turbine Heating and Power Supply System*

3.2.1. Gas Turbine

A compressor is the primary component of a gas turbine. The combustion chamber and gas turbine components of a gas turbine cycle are generally referred to as simple cycles [47]. The majority of gas turbines operate on a basic cycle. Compressors function on the premise of sucking air from the external atmospheric environment, pressurizing it after multistage compression, and increasing the air temperature correspondingly. Compressed air is drawn into the combustion chamber and combined with the injected fuel to produce high-temperature, high-pressure gas. Following expansion, the high-temperature, high-pressure gas performs external work, completing the process of chemical energy conversion to electrical energy. As an important piece energy equipment in the traditional oilfield energy system, the gas turbine also plays an indispensable role in the energy system of this paper. Therefore, gas turbines are used in all scenarios, and the associated gas produced by burning oilfields provides energy for the combined station. Simultaneously, the waste heat gas is routed through a heat exchanger for heat exchange, and the resulting heat is sent to the central processing facility.

For the cost of gas turbine, we use the cost of gas turbine per unit power to estimate the cost of gas turbine:

$$C_{GT} = \frac{U_{GT} \times w_{GT}}{Ef_{GT}} \tag{13}$$

where, C_{GT} is represent a cost of a gas turbine, (USD); U_{GT} is represents unit capital cost, (USD/kW); w_{GT} is represent rated power, (kW); Ef_{GT} is represents the efficiency of a gas turbine, $Ef_{GT} = 0.45$.

3.2.2. Heat Exchanger

A heat exchanger is a piece of energy-efficient equipment used to transfer heat between two or more fluids of varying temperatures. Heat exchangers, as important pieces of energy conversion equipment in traditional energy systems, are usually used in conjunction with gas turbines to receive high-temperature and high-pressure steam from gas turbines to store thermal energy through heat exchange media. They are used to transfer heat from a fluid with a higher temperature to a fluid with a lower temperature for the fluid temperature to reach the process-specified indications to fulfill process requirements. Simultaneously, a heat exchanger is a critical piece of equipment for optimizing energy utilization exchange because it is able to operate with a variety of media, operating circumstances, temperatures, and pressures. Additionally, the structural kinds and exchange types of heat exchangers vary, including interval heat exchangers, regenerative heat exchangers, and direct contact heat exchangers.

The shell and tube (also known as tubular) heat exchanger is employed in this arrangement, which is primarily formed of four components: the shells, the tube bundle, the tube plate, and the head. The shell is generally circular and contains a parallel tube bundle or spiral tube, with both ends of the tube bundle secured to the tube plate. In shell and tube heat exchangers, two types of fluids are used for heat exchange. One runs within the tube, and its stroke is referred to as tube side; the other flows externally to the tube. The heat exchanger's cost estimate is identical to that of the gas turbine:

$$C_{HE} = \frac{U_{HE} \times w_{HE}}{Ef_{HE}} \tag{14}$$

where, the unit power cost of the heat exchanger is 1235 USD/kW; C_{HE} represents cost of a heat exchanger, (USD); U_{HE} represents unit capital cost; w_{HE} is represents rated power, (kW); Ef_{HE} has represented the efficiency of a heat exchanger, $Ef_{HE} = 0.33$.

3.2.3. Oil Heating Furnace

The crude oil combustion furnace burns crude oil as fuel to generate heat energy. When the gas turbine cannot provide a heat supply for the combined station system, burning crude oil provides thermal energy for the combined station, which is applied in scenarios one, two, three and four in this paper. Its main structure includes a furnace shell, heat exchanger, flue pipe, flue, and combustion device. Cost calculation of crude oil combustion furnace is as follows:

$$C_{OHF} = 24,010 + UC_{OHF} \times w_{OHF} \tag{15}$$

$$OC_{OHF} = UO_{OHF} \times w_{OHF} \tag{16}$$

where, C_{OHF} is the investment cost of crude oil combustion furnace, (USD); UC_{OHF} is the unit investment cost of crude oil combustion furnace, (USD/kW); OC_{OHF} is the operation and maintenance crude oil combustion furnace, (USD); OC_{OHF} is the operation and maintenance cost of crude oil combustion furnace; UO_{OHF} is the unit operation and maintenance cost, (USD/kW); w_{OHF} is the power of combustion furnace, (kW).

3.2.4. Gas Burner

The gas combustion furnace burns combustible gas to generate high-temperature flue gas, and the high-temperature flue gas is discharged from the furnace and out of the furnace through the smoke exhaust system. Heat is exchanged in heat exchangers to provide thermal energy for oilfield energy systems. The whole system includes a heating furnace, heating system, smoke exhaust system, and cooling system. Since the combustion products of different combustible gases are also different, to compare the environmental impacts, especially the carbon emissions, of the combustion of associated gas (mainly combustion of CH_4) and hydrogen (H_2) in oilfields, this paper uses a gas-fired combustion furnace for comparison in scenario 5. The economic and environmental benefits of burning associated gas and hydrogen are compared. The cost of a gas burner is calculated as follows:

$$C_{HHF} = 24,410 + UC_{HHF} \times w_{HHF} \tag{17}$$

$$OC_{HHF} = UO_{HHF} \times w_{HHF} \tag{18}$$

where, C_{HHF} is the investment cost of hydrogen heating furnace, (USD); UC_{HHF} is the unit investment cost of hydrogen heating furnace, (USD/kW); OC_{HHF} is the operation and maintenance cost of hydrogen heating furnace, (USD); UO_{HHF} is the unit operation and maintenance cost, (USD/kW); w_{HHF} is the power of hydrogen heating furnace, (kW).

3.2.5. Battery

The energy storage battery used in the system is a lead-acid battery. The lead-acid battery has the following characteristics: it has a wide temperature range and can operate normally under a temperature environment of −30~60 °C. The low-temperature performance of the battery is good. It can be used even in areas with relatively low temperatures. The capacity consistency is good, and the consistency is maintained in the series and parallel use of the battery. In the unstable charging environment, it has stronger charging acceptance ability, long service life, reduces repair and maintenance costs, and reduces the overall investment of the system. The battery used in this paper can store 4500 kW of power from wind turbines and fuel cells, but it is only used in scenarios three and four due to its high cost.

$$C_{BT} = UC_{BT} \times V_{BT} \tag{19}$$

$$OC_{BT} = UO_{BT} \times V_{BT} \tag{20}$$

where, C_{BT} is the cost of battery, (USD); UC_{BT} is the unit capacity cost of battery, (USD/kWh); V_{BT} is the battery capacity, (kWh); OC_{BT} is the maintenance operation cost of battery, (USD); UO_{BT} is the maintenance and operation cost per unit capacity of battery, (USD/kWh).

3.3. *NPV and LCOE Model*

After the introduction of equipment, it is necessary to evaluate the economic and environmental performance of the whole oilfield energy system. The system uses payback period, net present value (NPV) (an index indicating the difference between future cash inflow and cash flow), and the Levelized cost of energy (LCOE) (a more accurate horizontal power cost index for cost estimation) to describe the economic performance of the system. It should be noted that LCOH refers to the Levelized cost of hydrogen:

$$PP = \frac{IN}{C_{in}} \tag{21}$$

$$NPV = \sum_{i=1}^{n} \frac{NCF_i}{(1+r)^i} \tag{22}$$

$$LCOE = \frac{\sum_{i=1}^{n} \frac{NCF_i}{(1+r)^i}}{\sum_{i=1}^{n} \frac{ED_i}{(1+r)^i}} \tag{23}$$

$$LCOH = \frac{\sum_{i=1}^{n} \frac{C_i}{(1+r)^i}}{\sum_{i=1}^{n} \frac{M_{H_2,i}}{(1+r)^i}} \tag{24}$$

$$C_{in} = NI_i - NE_i \tag{25}$$

where, PP represents the payback period, [a]; IN represents initial investment, (USD); C_{in} represents cash inflows, (USD); NCF_i represents net cash flow in year i, (USD); r represents the discount rate, $r = 0.12$; ED_i is MWh of electricity deposited on the grid in the year i; $M_{H2,i}$ is the mass of hydrogen produced in year i, (kg); $LCOE$ is the energy cost of the system, (USD/MWh); C_i is the total annual cost of hydrogen, (USD); $LCOH$ is unit hydrogen cost, (USD/kg); NI_i is net income in year i; NE_i is net expenditure in year i.

In addition to measuring the economic benefits of the system, it is also necessary to calculate the environmental benefits that the system can produce, which is measured by reducing carbon dioxide emissions when fossil fuels are replaced by hydrogen with equivalent calorific value:

$$EB = \frac{CV_{H_2} \times Q_a}{CV_{ff}} \times EF_{ff} \tag{26}$$

where, EB is the environmental benefit of the system, kg carbon dioxide; CV_{H_2} and CV_{ff} are the low calorific value of hydrogen fuel and fossil fuel, respectively; Q_a is the annual hydrogen output, kg; EF_{ff} is the emission coefficient of fossil fuels.

4. Results and Discussions

4.1. Input Parameters

Taking an oilfield central processing facility located in Northwest China as an example, Figure 7 shows the specific location of the central processing facility. The central processing facility's annual gas production is 1.1×10^7 m^3. According to statistics, China's oilfield central processing facility production ranges from 0.56×10^7 m^3~2.8×10^8 m^3 [20]. As previously stated, the system's electric energy powers gas turbines and hydrogen synthesis utilizing renewable energy. In addition, the yearly electricity generation is split into two sections. The first is combustion-related gas power production using gas turbines. For example, a 50 MW gas turbine can create 420 GWh of electricity each day from combustion-related gases. Then, we need to figure out how much electricity the wind turbine generates every year. Considering that the central processing facility is situated in Northwest China, we collected wind speed data from the northwest wind tower between 1 January 2019 and 31 December 2019, as shown in Figure 8 below.

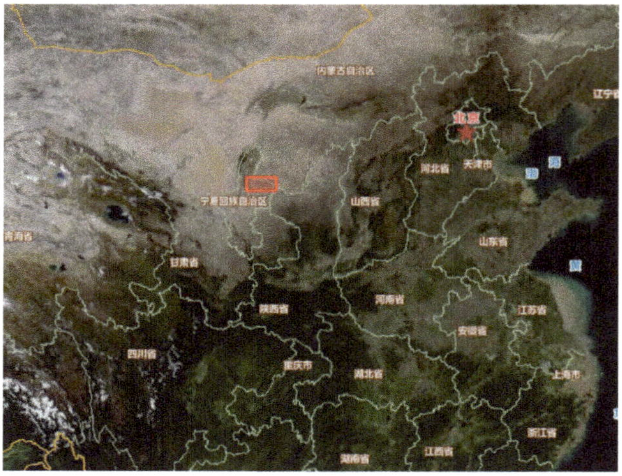

Figure 7. The geographic locations of an oilfield in China.

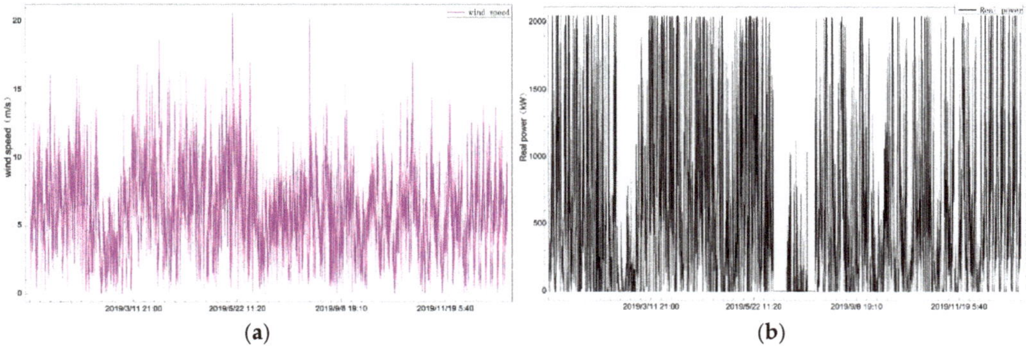

Figure 8. (**a**) Wind speed data and (**b**) output power production in one year.

4.1.1. Energy Demand

The power consumption of the central processing facility includes compressor, dehydration pump, electric dehydrator, export pump, sewage pump unit, and domestic power. The total annual power demand of the central processing facility in 2019 was 238 GWh, and the proportion of electric energy consumption of each system is shown in Figure 9a. The thermal energy consumption includes crude oil heating and dehydration, crude oil export, and domestic heat. In 2019, the total thermal demand of the central processing facility was 776 GWh, and the proportion of heat energy consumption of each system is shown in Figure 9b.

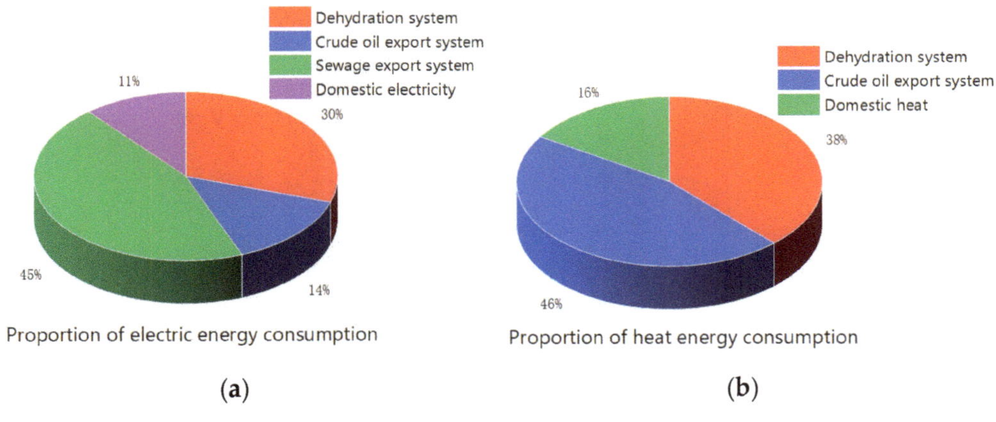

Figure 9. (**a**) Proportion of system electric energy consumption and (**b**) proportion of system heat energy consumption.

4.1.2. Technical and Cost Information on Equipment

In this case study, since there is no economic and technical information on the facilities recommended by the market or the government, the project parameters are estimated based on field surveys, industry statistics, and academic literature, as shown in Table 9. Specifically, the lifetime and the linear capacity-dependent cost and O&M cost are collected from HOMER Pro's website. The expenditure of each scheme is calculated in turn according to the formula mentioned above through the collected data, and the income mainly comes from the cost of energy produced by the system.

Table 9. Parameter of some equipment.

Facility	Linear Capacity-Dependent Cost (USD/kW, MWh, kg)	O&M (USD/op.Hour, Year)	Lifetime (Year)	Rated Power (%)
Wind Turbine	550	20	20	45
Electrolyzer	145	140	15	80
Furnace (Crude oil)	250	30	25	80
France (H_2)	250	30	25	80
Heat exchanger	350	10	25	55
Gas Turbine	300	6	25	45
H_2 tank	380	1800	25	80
Fuel cell	1000	120	18	80
Battery	210,000	1400	20	80

4.2. Economic Performance of Five Scenarios

The NPV of the five scenarios is positive, indicating that they are economically feasible and can yield advantages over the system's 20-year life, as shown in Figure 10. It is worth mentioning that the NPV for the five scenarios varies according to the equipment utilized in each scenario and the corresponding basic investment cost. We address solely the power supply in this section. To begin, consider scenario one, with the lowest profit margin, only gas turbines, and the power infrastructures employed to generate electricity. In this scenario, the gas turbine is unable to meet the central processing facility's power consumption, and thus power must be purchased from the power grid. Although scenario one requires less equipment and lower initial investment, the income generated by scenario one is low due to the expense of power grid power purchase. In contrast, the profit of scenario three with the highest profit is USD 46.21 million during its service life.

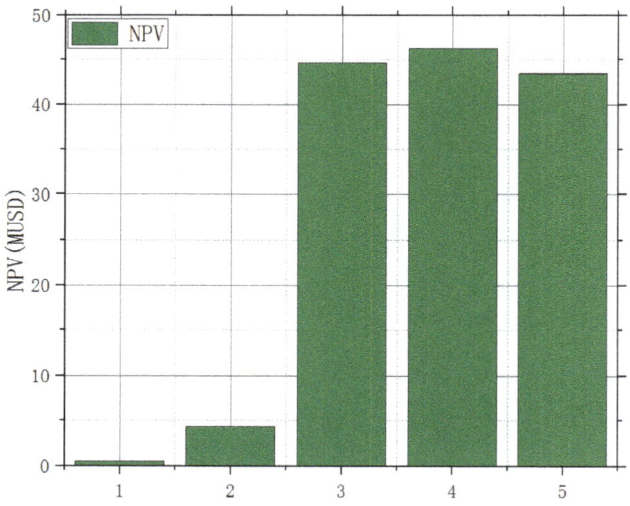

Figure 10. NPV and initial investment cost of five scenarios.

Since the design principles are the same, the advantages of the five scenarios are comparable when the central processing facility's scale is known. We find that the economics of the five scenarios are mostly determined by annual spending and equipment operation and maintenance costs. Scenario three, which employs wind turbines and gas turbines for power generation, eliminates the requirement for the central processing facility to purchase electricity from the power grid, significantly decreasing the annual cost of power in comparison to the traditional technology. Second, unlike scenarios four and five, scenario three delivers the entire amount of electricity created by the wind turbine generator to the battery.

Due to the elimination of the electrolytic cell, hydrogen storage tank, fuel cell, and other equipment in scenario three, the basic investment cost is relatively low, resulting in the highest net present value. Scenarios four and five have hydrogen production and energy conversion equipment with extremely high investment and operating costs, thus the net present value is slightly less than in scenario three. However, hydrogen has the potential to store energy on a large scale, and is more capable of shaving peaks and filling valleys than batteries.

The terms LCOE and LCOH are developed to facilitate the analysis of the unit cost of oilfield system capacity. As shown in Figure 11, the chart depicts the unit capacity costs for five different situations. In scenarios four and five, where hydrogen and wind energy are combined, LCOH is used to calculate the cost of hydrogen produced per unit mass by the system. As can be seen, the system's unit capacity cost is positively correlated with its net present value. Scenario four has the lowest unit capacity cost, and the cost of producing hydrogen per kilogram is USD 7.23, whereas it is USD 6.20 in scenario five. The reason for this is that scenario four employs large-capacity energy storage batteries to store hydrogen energy. Due to the exponential growth in the cost of energy storage batteries as capacity increases, the unit hydrogen production cost in scenario four is high.

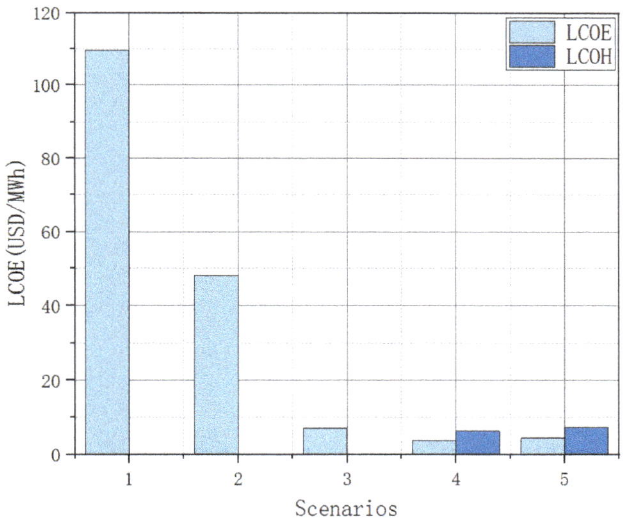

Figure 11. LCOE and LCOH of five scenarios.

4.3. Analysis of Associated Gas Production

The generation of associated gas in the oilfield has a direct effect on the heat and power generated by gas turbines, which is a critical link in the oilfield's operation. Due to the central processing facility's reliance on gas turbines, it is a significant component impacting system economics. Therefore, sensitivity analysis of natural gas production to system economic cost is carried out in this section.

As shown in Figure 12, the chart illustrates the NPV performance of five scenarios in which the oil and gas field's associated gas production ranges between 0.5×10^7 m^3 and 1.0×10^7 m^3. As can be observed, the oilfield's associated gas production has a bigger impact on scenarios one, two, and three, but has a lesser impact on scenarios four and five. Because the entire oilfield system is solely reliant on gas turbines and the power grid for power purchase in the first scenarios, the output of oilfield-associated gas has a significant impact on the system's economy. When associated gas production falls below 1.0×10^7 m^3 the oilfield energy system that is solely based on the combustion and generation of associated gas can no longer provide economic benefits to the oilfield, indicating that the investment cost of the entire system, including the cost of power purchased from the power grid, is significantly greater than the cost of the associated gas-produced power.

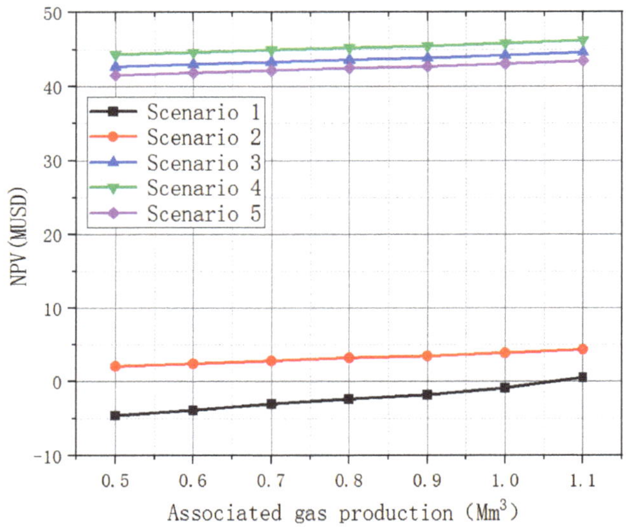

Figure 12. NPV of associated gas production under different scenarios.

In addition to depending on gas turbines for power generation, scenarios three, four, and five combine wind and solar energy, resulting in a negligible impact on the net present value of scenarios three, four, and five. As the oilfield development progresses to the later stages, the annual output of the oilfield becomes more unstable, and the risk resistance of the system to associated gas production increases, which has a beneficial effect on the application of high water cut oilfields in the later stages of development, and which is also one of the contributions of the scenarios proposed in this study to engineering practice.

4.4. Analysis of Electricity Price

Due to the project's sole income stream being the sale of electricity and hydrogen, it is vital to examine the influence of energy prices on the project's NPV. Both scenarios one and two introduce the power grid as a backup energy source for the oilfield. Due to the oilfield's lengthy water cut development phase, the energy consumption associated with oilfield development is considerable. Thus, in both scenarios one and two, electricity must be purchased from the power grid, and the cost of power purchase will climb in lockstep with the increase in the cost of electricity. The electric energy generated by the five energy systems is used to create money for the system. The influence of power prices ranging from USD 0.05 to 0.64 cents per kilowatt-hour is examined. As illustrated in the Figure 13, when electricity prices increase from USD 0.05 to 0.64 per kilowatt-hour, the NPV of the five scenarios increases, and the increase in electricity price is linear with the increase in net present value, but the increased range of net present value is different. In scenario one, for every 0.01 cent increase in the price of electricity, the NPV of the system increases by USD 3560, while in scenarios three, four, and five, the NPV of the system increases, even more, reaching USD 2 million.

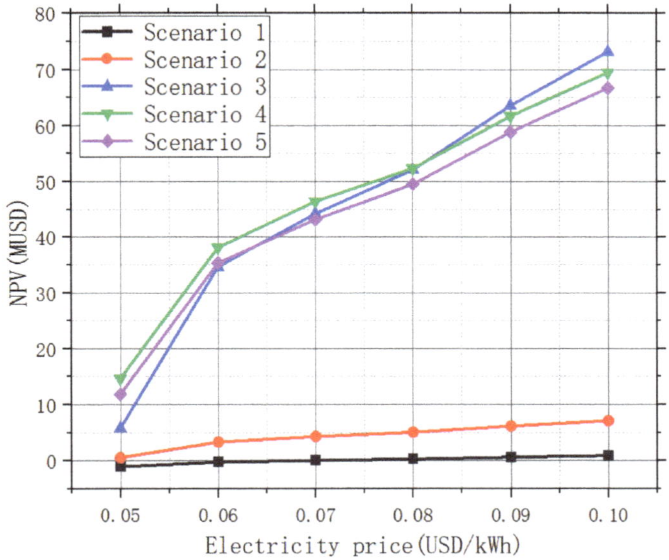

Figure 13. NPV of the model under different electricity prices.

In addition, we explore the factors that contribute to the disparate growth rates. The first reason is that in scenarios one and two, power from the power grid is utilized. This portion of power must be obtained from the power grid at market rates. While an increase in electricity prices increases the oilfield's revenue, it also raises the oilfield's cost. As a result, the low rate of rising net present value can be validated. In practice, China's electricity rates vary by region. Northwest, northeast, and other regions have cheaper electricity prices, while Southeast and South China have higher electricity prices. As a result, it is more profitable to establish scenarios four and five in Southeast China coastal cities with higher electricity rates.

4.5. Analysis of Wind Speed

In scenarios two, three, four, and five, wind energy is introduced to the standard oilfield energy system as a source of energy for the oilfield, and the electric energy generated by the wind turbine is directly proportional to the wind speed in the region. Thus, based on the NPV performance of the system at 50%, 70%, 90%, and 110% of the wind speed in this region, as shown in Figure 14, wind speed is linear with the NPV, and increasing wind speed can enhance the NPV of the system. Among these, the NPV of scenario three is the most wind-speed-dependent. The analytical reasons might be attributed to three situations. The system's primary source of revenue is the electric energy provided by the wind turbine. In comparison to scenarios four and five, wind energy is not converted to hydrogen for storage, implying that the system's peak shaving storage capacity will be relatively low.

China's oilfields are primarily dispersed in the northwest, northeast, and southwest, according to the current circumstances. The wind power density of the area where the oilfields are located is 300 W/m^2 in this article. In comparison to these data, average wind power in the northeast, such as the Daqing Oilfield, is more than 200–300 W/m^2, and average wind power in coastal and island areas, such as Zhoushan and Huangdao, is more than 500 W/m^2. The available hours range from 7000 to 8000 h. As a result, when the system proposed in this research is implemented in the oilfields in the above areas, the economic gains could exceed USD 6×10^7 in scenarios three, four, and five presented in this study.

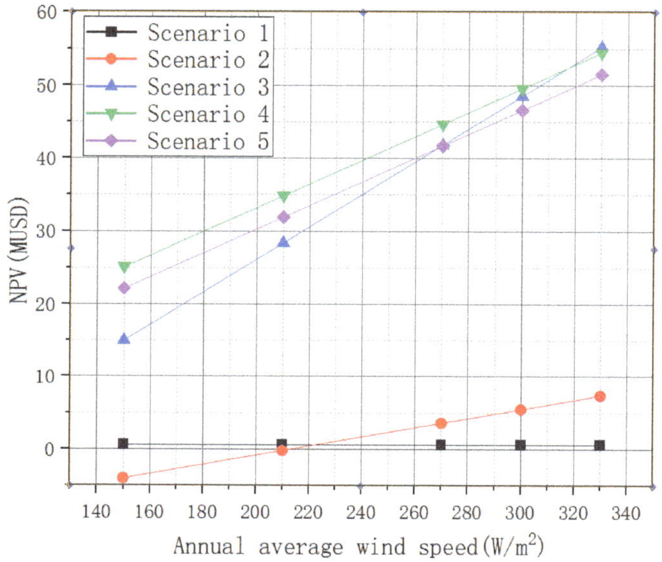

Figure 14. NPV of the model under different wind speed.

4.6. Analysis of Carbon Emissions

Along with examining the economics of five scenarios, it is critical to examine the oilfield energy system's carbon footprint. Natural gas, crude oil, and the electricity grid all have emission factors of 0.19 kg/(kWh), 0.24 kg/(kWh), and 0.89 kg/(kWh). It is worth noting that because grid power generation is treated as coal-fired generation in this article, the carbon emission coefficient is high, and so the carbon emissions of the five scenarios are calculated. As illustrated in the picture, scenario one's typical oilfield energy model generates the maximum carbon dioxide, 45,874 t/year. In the standard oilfield distributed energy system, a considerable amount of crude oil and electricity from the power grid are used to power the central processing facility, which results in a significant increase in carbon emissions due to the vast area of power purchased from the power grid. This is not consistent with national policy in this instance.

As shown in Figure 15, although wind turbines are used to generate electricity in scenario two, the oilfield's primary energy sources remain the power grid and natural gas, resulting in slightly lower annual carbon emissions than scenario one, at 39,813 tons, but still quite significant. Wind energy generation is the primary source of energy in scenarios three, four, and five, and the only source of carbon emissions is the combustion of oilfield-associated gas. Due to the clean renewable energy and the low carbon emissions associated with natural gas combustion, the carbon dioxide emissions are much lower than in scenarios one and two. Additionally, scenario four may produce 46,200 tons more carbon dioxide than scenario one during 20 years, and the system improves the environment.

4.7. Discussion

Following the preceding discussion, it is evident that we may examine the five systems' merits and disadvantages. To begin, scenario one features the shortcomings of a traditional distributed energy system, including high carbon emissions, high energy consumption, and a low rate of return. It is destined to be eliminated once oilfields are developed. While renewable energy generation is combined in scenario two, the mechanical application of renewable energy does not boost the system's benefits. Simultaneously, because of the unpredictability of wind resources, the system's carbon emission performance is suboptimal in places with limited wind resources. The third, fourth, and fifth scenarios combine wind and hydrogen to generate electricity for the oilfield, which has a low carbon footprint and economic benefits. Finally, it is determined that scenario four is the best, taking into account its relevance to the oilfield, the difficulty of construction, and the support of national policies. In comparison to the traditional method, it generates USD 46.21 million more revenue and cuts carbon emissions by 44,096 t/a. It applies to areas with abundant wind resources and low electricity rates, such as South China's coastal and northwest regions, which are defined as the WGHN system.

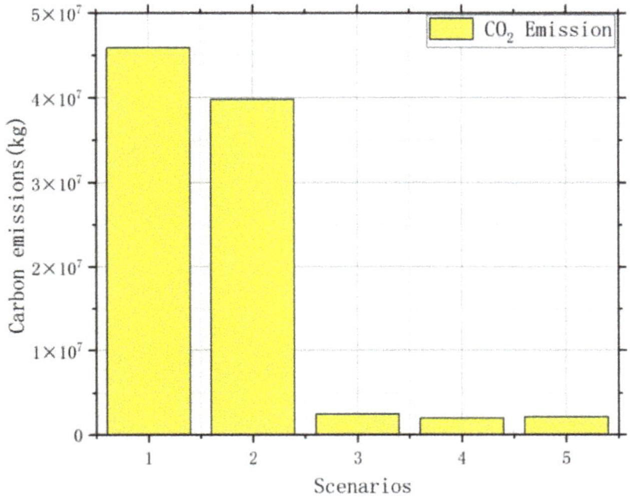

Figure 15. Carbon emissions in five scenarios.

5. Conclusions

To address the fact that oilfield development has entered a period of high water scarcity, energy consumption in the oilfield is increasing, and the oilfield system is losing money, this paper discusses four different types of wind–hydrogen coupling energy systems based on traditional distributed energy. The economics and dependability of the four scenarios are examined using data from an oilfield in Northwest China. Finally, a new form of oilfield energy system is developed that is suited for the majority of existing onshore oilfields. In comparison to a conventional energy system, this concept can increase USD 46.21 million throughout the system's 20-year life, with a payback period of less than one year. It is a low-risk, high-return investment strategy. The system is capable of utilizing renewable energy successfully and reducing carbon emissions by 44,096 t/a. It not only provides significant economic benefits but also significant environmental benefits. Wind speed, power price, and oilfield-related gas output all affect the field economics. The following conclusions are made:

(1) In comparison to other oilfield renewable energy systems, the system fully utilizes excess power, which not only provides economic and environmental benefits but also significantly alleviates the problem of wind waste;
(2) Through sensitivity analysis, it is found that the WGHN system is most sensitive to wind speed. When the wind speed is increased by 10%, the net present value of the system will increase by USD 1270,334. Secondly, the electricity price will also affect the system revenue. Each increase in the unit price of electricity is 0.017 USD/kWh, and system revenue increased by USD 877,500;
(3) Based on a comprehensive analysis of several regions in China, it is concluded that China's northeast oilfields (such as Daqing), North China coastal areas (such as Huangdao Oilfield and Bohai offshore platform), northwest oilfields (such as Karamay

and Tarim Oilfields), and southeast coastal areas (such as Zhoushan and Ningbo) can all benefit from the WGHN system due to their abundant wind resources.

Overall, this study offers a trustworthy new oilfield energy system, which has complete reference value for tackling the problems of excessive energy consumption, low energy efficiency, and poor economy caused by the development of some old oilfields. At the same time, amid today's global warming and energy scarcity, limiting carbon emissions is the key. To achieve a carbon peak in 2030 and carbon neutralization in 2060, the reform of the oilfield energy system is significantly important. The wind energy detailed in this study, as well as the connection between hydrogen and other renewable energy sources and the oilfield energy system, will become the driving trend of future development. The hydrogen produced from the residual power of wind energy suggested in this study is also a major innovation; that is, it boosts the utilization rate of energy and locally consumes the electric energy generated by renewable energy, which has substantial theoretical reference significance in the later combination of renewable energy and hydrogen energy. At the same time, scenario four of this article can examine a range of characteristics, including oilfield location, yearly wind speed distribution, oilfield-related gas output, market demand, hydrogen and power pricing. Through the technical and economic evaluation of the oilfield wind–hydrogen linked power supply system, we can better grasp the investment value of the system under various regions and scenarios.

However, the research in this article has certain flaws. First, where wind resources are limited, the economic benefits of the system are minimal. Second, because hydrogen storage technology is still in its infancy, hydrogen storage is a challenge. Finally, some infrastructure constraints need further study. Therefore, we propose the following policy recommendations.

First, an examination of cost sensitivity shows that the main cost driver is energy prices, and self-sufficiency in electricity contributes most of the revenue. Therefore, the focus should be on reducing the cost of energy infrastructure, such as fuel cells. In addition, the government should provide investors with incentives to reduce entry barriers for hydrogen projects, thereby increasing the penetration rate of renewable energy across the country, and helping the country achieve its carbon neutral goal as soon as possible. Secondly, scenario four proposed in this article is more likely to bring significant economic benefits in coastal areas rich in northwest wind resources. The main reason is that most of the project's revenue comes from wind energy. When the wind speed doubles, the output power of the wind turbine will increase eight times. Therefore, the windy location may help to further increase expected revenue. Finally, after renewable energy is connected to the grid, it is vital to ensure the stability of the power market and grid system. Currently, many wind farms are inefficiently connected to the grid. The government should formulate appropriate rules and evaluation indicators to promote the private sector to enter the renewable energy market, improve market competitiveness, stabilize energy costs, and solve the problem of mismatch between supply and demand.

Author Contributions: Conceptualization, F.Y.; data curation, Q.C.; formal analysis, F.Y., Y.Y. and Q.L.; funding acquisition, Y.L.; methodology, X.H. and Q.L.; writing—original draft, X.H.; writing—review & editing, Q.C.; investigation, X.H.; validation, Y.Y. All authors have read and agreed to the published version of the manuscript.

Funding: This work was partially supported by the National Natural Science Foundation of China (51874325) and the Grant-in-Aid for Early-Career Scientists (19K15260) from the Japan Ministry of Education, Culture, Sports, Science and Technology.

Institutional Review Board Statement: Not applicable.

Informed Consent Statement: Not applicable.

Acknowledgments: This work was partially supported by the National Natural Science Foundation of China (51874325) from the National Key Research and Development Program of China (2018YFE0196000).

Conflicts of Interest: The authors declare no conflict of interest.

Nomenclature

AEP	Annual electricity production
ICC	Initial capital cost
FCR	Fixed charge rate
AOC	Annual operating cost
WT	Wind turbine
MS	Mechanical system
ES	Electronic system
AE	Alkaline electrolytic
PEM	Proton exchange membrane
SOEC	Solid oxide electrologist cell
GT	Gas turbine
HE	Heat exchange
OHF	Oil heating furnace
HHF	Hydrogen heating furnace
BT	Battery
PP	Payback period
LCOE	Levelized cost of energy
LCOH	Levelized cost of hydrogen
FF	Fossil fuel
NPV	Net present value
IN	Initial investment
ED	Energy deposited
NI	Net income
NE	Net expenditure
WGHNl	Wind–gas turbine–hydrogen nexus

References

1. Zhang, X.; Liao, Q.; Wang, Q.; Wang, L.; Qiu, R.; Liang, Y.; Zhang, H. How to promote zero-carbon oilfield target? A technical-economic model to analyze the economic and environmental benefits of the Recycle-CCS-EOR project. *Energy* **2021**, *225*, 120297. [CrossRef]
2. Wei, X.; Qiu, R.; Liang, Y.; Liao, Q.; Klemeš, J.J.; Xue, J.; Zhang, H. Roadmap to carbon emissions neutral industrial parks: Energy, economic and environmental analysis. *Energy* **2022**, *238*, 121732. [CrossRef]
3. Yan, W.; Meng, Z.; Zou, M.; Miao, H.; Ma, F.; Yu, R.; Qiu, W.; Liu, X.Y.; Lin, N. Neutralization reaction in the synthesis of carbon materials for supercapacitors. *Chem. Eng. J.* **2020**, *381*, 122547. [CrossRef]
4. Huang, L.; Liao, Q.; Yan, J.; Liang, Y.; Zhang, H. Carbon footprint of oil products pipeline transportation. *Sci. Total Environ.* **2021**, *783*, 146906. [CrossRef] [PubMed]
5. Xing, X.; Yan, Y.; Zhang, H.; Long, Y.; Wang, Y.; Liang, Y. Optimal design of distributed energy systems for industrial parks under gas shortage based on augmented ε-constraint method. *J. Clean. Prod.* **2019**, *218*, 782–795. [CrossRef]
6. Aziz, M.; Putranto, A.; Biddinika, M.K.; Wijayanta, A.T. Energy-saving combination of N2 production, NH3 synthesis, and power generation. *Int. J. Hydrogen Energy* **2017**, *42*, 27174–27183. [CrossRef]
7. Aziz, M.; Wijayanta, A.T.; Nandiyanto, A.B.D. Ammonia as Effective Hydrogen Storage: A Review on Production, Storage and Utilization. *Energies* **2020**, *13*, 3062. [CrossRef]
8. Wijayanta, A.T.; Aziz, M. Ammonia production from algae via integrated hydrothermal gasification, chemical looping, N_2 production, and NH_3 synthesis. *Energy* **2019**, *174*, 331–338. [CrossRef]
9. Li, T.; Xu, Y.; Wang, J.; Kong, X.; Zhu, J. Poly-generation energy system driven by associated geothermal water for oilfield in high water cut stage: A theoretical study. *Geothermics* **2018**, *76*, 242–252. [CrossRef]
10. Wang, K.; Yuan, B.; Ji, G.; Wu, X. A comprehensive review of geothermal energy extraction and utilization in oilfields. *J. Pet. Sci. Eng.* **2018**, *168*, 465–477. [CrossRef]
11. Huang, C.; Gu, B.; Chen, Y.; Tan, X.; Feng, L. Energy return on energy, carbon, and water investment in oil and gas resource extraction: Methods and applications to the Daqing and Shengli oilfields. *Energy Policy* **2019**, *134*, 110979. [CrossRef]
12. Guo, T.; Zhang, Y.; He, J.; Gong, F.; Chen, M.; Liu, X. Research on geothermal development model of abandoned high temperature oil reservoir in North China oilfield. *Renew. Energy* **2021**, *177*, 1–12. [CrossRef]
13. Li, T.; Liu, Q.; Xu, Y.; Dong, Z.; Meng, N.; Jia, Y.; Qin, H. Techno-economic performance of multi-generation energy system driven by associated mixture of oil and geothermal water for oilfield in high water cut. *Geothermics* **2021**, *89*, 101991. [CrossRef]

14. Spielmann, V.; Brey, T.; Dannheim, J.; Vajhøj, J.; Ebojie, M.; Klein, J.; Eckardt, S. Integration of sustainability, stakeholder and process approaches for sustainable offshore wind farm decommissioning. *Renew. Sustain. Energy Rev.* **2021**, *147*, 111222. [CrossRef]
15. Zhang, Q.; Zhang, H.; Yan, Y.; Yan, J.; He, J.; Li, Z.; Shang, W.; Liang, Y. Sustainable and clean oilfield development: How access to wind power can make offshore platforms more sustainable with production stability. *J. Clean. Prod.* **2021**, *294*, 126225. [CrossRef]
16. Zou, X.; Qiu, R.; Yuan, M.; Liao, Q.; Yan, Y.; Liang, Y.; Zhang, H. Sustainable offshore oil and gas fields development: Techno-economic feasibility analysis of wind-hydrogen-natural gas nexus. *Energy Rep.* **2021**, *7*, 4470–4482. [CrossRef]
17. Hu, H.; Sun, X.; Zeng, B.; Gong, D.; Zhang, Y. Enhanced evolutionary multi-objective optimization-based dispatch of coal mine integrated energy system with flexible load. *Appl. Energy* **2022**, *307*, 118130. [CrossRef]
18. Monfaredi, F.; Shayeghi, H.; Bizon, N. Developing optimal energy management of integrated energy systems in the hybrid electricity and gas networks. *J. Energy Storage* **2022**, *48*, 103984. [CrossRef]
19. Huang, Y.; Wang, Y.; Liu, N. A two-stage energy management for heat-electricity integrated energy system considering dynamic pricing of Stackelberg game and operation strategy optimization. *Energy* **2021**, *244*, 122576. [CrossRef]
20. Bartolucci, L.; Cordiner, S.; Mulone, V.; Pasquale, S. Hydrogen based Multi Energy Systems: Assessment of the marginal utility of increasing hydrogen penetration on system performances. *Int. J. Hydrogen Energy* **2021**, *46*, 38588–38602. [CrossRef]
21. Dong, X.; Wu, J.; Xu, Z.; Liu, K.; Guan, X. Optimal coordination of hydrogen-based integrated energy systems with combination of hydrogen and water storage. *Appl. Energy* **2022**, *308*, 118274. [CrossRef]
22. Yan, Y.; Zhang, H.; Long, Y.; Wang, Y.; Liang, Y.; Song, X.; Yu, J.J. Multi-objective design optimization of combined cooling, heating and power system for cruise ship application. *J. Clean. Prod.* **2019**, *233*, 264–279. [CrossRef]
23. Song, D.; Liu, J.; Yang, J.; Su, M.; Yang, S.; Yang, X.; Joo, Y.H. Multi-objective energy-cost design optimization for the variable-speed wind turbine at high-altitude sites. *Energy Convers. Manag.* **2019**, *196*, 513–524. [CrossRef]
24. Maienza, C.; Avossa, A.; Ricciardelli, F.; Coiro, D.; Troise, G.; Georgakis, C. A life cycle cost model for floating offshore wind farms. *Appl. Energy* **2020**, *266*, 114716. [CrossRef]
25. Chen, L.; Liu, Y. Research on Main Constraints in Sustainable Development of China Oil-Gas Upstream Industry. *Energy Procedia* **2012**, *14*, 325–330. [CrossRef]
26. Huskey, A.; Forsyth, T. *NREL Small Wind Turbine Test Project: Mariah Power's Wind Spire Wind Turbine Test Chronology*; Technical Report NREL/TP-500-45552; National Renewable Energy Laboratory: Golden, CO, USA, 2009.
27. Chen, J.; Wang, F.; Stelson, K.A. A mathematical approach to minimizing the cost of energy for large utility wind turbines. *Appl. Energy* **2018**, *228*, 1413–1422. [CrossRef]
28. Abobkr, A.H.; El-Hawary, M.E. Evaluation of wind turbine characteristics built-in model in Matlab Simulink. In Proceedings of the 2016 IEEE Electrical Power and Energy Conference (EPEC), Ottawa, ON, Canada, 12–14 October 2016; pp. 1–4.
29. Burton, N.A.; Padilla, R.V.; Rose, A.; Habibullah, H. Increasing the efficiency of hydrogen production from solar powered water electrolysis. *Renew. Sustain. Energy Rev.* **2021**, *135*, 110255. [CrossRef]
30. Mohammadi, A.; Mehrpooya, M. A comprehensive review on coupling different types of electrolyzer to renewable energy sources. *Energy* **2018**, *158*, 632–655. [CrossRef]
31. D'Amore-Domenech, R.; Santiago, Ó.; Leo, T.J. Multicriteria analysis of seawater electrolysis technologies for green hydrogen production at sea. *Renew. Sustain. Energy Rev.* **2020**, *133*, 110166. [CrossRef]
32. Amikam, G.; Nativ, P.; Gendel, Y. Chlorine-free alkaline seawater electrolysis for hydrogen production. *Int. J. Hydrogen Energy* **2018**, *43*, 6504–6514. [CrossRef]
33. Timmerberg, S.; Kaltschmitt, M. Hydrogen from renewables: Supply from North Africa to Central Europe as blend in existing pipelines—Potentials and costs. *Appl. Energy* **2019**, *237*, 795–809. [CrossRef]
34. Nuttall, L.J.; Fickett, A.P.; Titterington, W.A. Hydrogen Generation by Solid Polymer Electrolyte Water Electrolysis. In *Hydrogen Energy: Part A*; Veziroğlu, T.N., Ed.; Springer: Boston, MA, USA, 1975; pp. 441–455.
35. Millet, P.; Ngameni, R.; Grigoriev, S.A.; Mbemba, N.; Brisset, F.; Ranjbari, A.; Etiévant, C. PEM water electrolyzers: From electrocatalysis to stack development. *Int. J. Hydrogen Energy* **2010**, *35*, 5043–5052. [CrossRef]
36. Schmidt, O.; Gambhir, A.; Staffell, I.; Hawkes, A.; Nelson, J.; Few, S. Future cost and performance of water electrolysis: An expert elicitation study. *Int. J. Hydrogen Energy* **2017**, *42*, 30470–30492. [CrossRef]
37. Grigoriev, S.A.; Fateev, V.N.; Bessarabov, D.G.; Millet, P. Current status, research trends, and challenges in water electrolysis science and technology. *Int. J. Hydrogen Energy* **2020**, *45*, 26036–26058. [CrossRef]
38. Laguna-Bercero, M.A. Recent advances in high temperature electrolysis using solid oxide fuel cells: A review. *J. Power Sources* **2012**, *203*, 4–16. [CrossRef]
39. Seyedmatin, P.; Karimian, S.; Rostamzadeh, H.; Amidpour, M. Electricity and hydrogen co-production via scramjet multi-expansion open cooling cycle coupled with a PEM electrolyzer. *Energy* **2020**, *199*, 117364. [CrossRef]
40. Oi, T.; Wada, K. Feasibility study on hydrogen refueling infrastructure for fuel cell vehicles using the off-peak power in Japan. *Int. J. Hydrogen Energy* **2004**, *29*, 347–354. [CrossRef]
41. Lümmen, N.; Karouach, A.; Tveitan, S. Thermo-economic study of waste heat recovery from condensing steam for hydrogen production by PEM electrolysis. *Energy Convers. Manag.* **2019**, *185*, 21–34. [CrossRef]
42. Miao, B.; Giordano, L.; Chan, S.H. Long-distance renewable hydrogen transmission via cables and pipelines. *Int. J. Hydrogen Energy* **2021**, *46*, 18699–18718. [CrossRef]

43. Kaiser, M.J. Offshore pipeline construction cost in the U.S. Gulf of Mexico. *Mar. Policy* **2017**, *82*, 147–166. [CrossRef]
44. Parker, N.C.; Ogden, J.M.; Fan, Y. The role of biomass in California's hydrogen economy. *Energy Policy* **2008**, *36*, 3925–3939. [CrossRef]
45. Han, J.-H.; Ryu, J.-H.; Lee, I.-B. Modeling the operation of hydrogen supply networks considering facility location. *Int. J. Hydrogen Energy* **2012**, *37*, 5328–5346. [CrossRef]
46. Holagh, S.G.; Haghghi, M.A.; Chitsaz, A. Which methane-fueled fuel cell is of superior performance in CCHP applications; solid oxide or molten carbonate? *Fuel* **2022**, *312*, 122936. [CrossRef]
47. Hosseinimaab, S.M.; Tousi, A.M. A new approach to off-design performance analysis of gas turbine engines and its application. *Energy Convers. Manag.* **2021**, *243*, 114411. [CrossRef]

energies

Article

Thermal Stability Calculation and Experimental Investigation of Common Binary Chloride Molten Salts Applied in Concentrating Solar Power Plants

Jingyu Zhong [1], Jing Ding [1,*], Jianfeng Lu [1], Xiaolan Wei [2] and Weilong Wang [1,*]

1 School of Materials Science and Engineering, Sun Yat-sen University, Guangzhou 510006, China; zhongjy35@mail2.sysu.edu.cn (J.Z.); lujfeng@mail.sysu.edu.cn (J.L.)
2 School of Chemistry and Chemical Engineering, South China University of Technology, Guangzhou 510640, China; xlwei@scut.edu.cn
* Correspondence: dingjing@mail.sysu.edu.cn (J.D.); wwlong@mail.sysu.edu.cn (W.W.)

Abstract: A computational study on thermal stability was conducted the first time, combining the modified quasi-chemical model, the Antoine equation, and the adiabatic flash evaporation calculation principle to design a method to calculate the system pressure-temperature (P-T) phase diagram of binary chloride molten salts. The evaporation temperature of the molten salt obtained by analyzing the P-T phase diagram of the eutectic molten salt clearly defined the upper limit of the optimal operating temperature of the mixed molten salt. The results indicated that the upper-temperature limits of NaCl-KCl, NaCl-$CaCl_2$, KCl-$CaCl_2$, NaCl-$MgCl_2$, and KCl-$MgCl_2$ are determined to be 1141 K, 1151 K, 1176 K, 1086 K, and 1068 K. The maximum working temperature was measured experimentally using a thermogravimetric analysis (TGA), and the relative error between the calculation and experiment was calculated. The maximum error between the calculated and experimental values of the maximum operating temperature was 6.02%, while the minimum was 1.29%, demonstrating the method's high accuracy. Combined with the lowest eutectic temperature and the upper-temperature limits of binary chloride molten salts, the stable operating temperature ranges of NaCl-KCl, NaCl-$CaCl_2$, KCl-$CaCl_2$, NaCl-$MgCl_2$, and KCl-$MgCl_2$ are 891~1141 K, 750~1151 K, 874~1176 K, 732~1086 K, and 696~1086 K.

Keywords: thermal stability calculation; upper-temperature limit; P-T phase diagram; modified quasi-chemical model; adiabatic flash calculation; binary chloride molten salts

Citation: Zhong, J.; Ding, J.; Lu, J.; Wei, X.; Wang, W. Thermal Stability Calculation and Experimental Investigation of Common Binary Chloride Molten Salts Applied in Concentrating Solar Power Plants. *Energies* **2022**, *15*, 2516. https://doi.org/10.3390/en15072516

Academic Editors: Wei-Hsin Chen and Lyes Bennamoun

Received: 8 February 2022
Accepted: 23 March 2022
Published: 29 March 2022

1. Introduction

Human beings have been struggling with global warming exacerbating in the aftermath of increasing climate change and huge energy demand issues. Renewable clean energy sources are being developed to reduce the dominance of traditional fossil energy in global energy consumption [1,2]. For example, the scientific community has proposed a cost-effective thermal energy storage (TES) strategy to be combined with concentrated solar power (CSP) plants to ensure the continuity of electricity supply [3–5]. After concentrating the sun's rays by mirrors, solar energy can be stored as thermal energy by the TES system for later use [3,6]. Among TES materials, it is possible for molten salts to meet the high-temperature use requirements of the CSP plant and the goal of TES, owing to their advantages of high heat storage density and efficiency, wide operating temperature range and low economic cost [7,8]. In commercial solar thermal power generation systems, Solar Salt ($NaNO_3$-KNO_3, 60-40 wt.%) and Hitec Salt ($NaNO_3$-KNO_3-$NaNO_2$, 7-53-40 wt.%) are two of the most extensive and mature molten salts used as heat transfer and storage media [9–11]. However, nitrates used in CSP plants operate only below 773 K, which is severely limited due to poor thermal stability, which causes nitrates to decompose easily at high temperatures [12–14]. Chloride molten salts have emerged as potential candidates for the next generation of CSP in recent years. Because of their wide working temperature range and excellent high-temperature stability, chloride molten salts have been found to be suitable for many advanced high-temperature power cycles.

Furthermore, chloride molten salts have been temporarily limited in commercial applications due to the lack of data on thermophysical properties at high temperatures even near the boiling zone, although they meet the heating temperature requirements of high-temperature thermal utilization [15]. The determination of the maximum working temperature of chloride molten salts as future work was strongly suggested in the latest review article [16]. Currently, researchers frequently employ the thermogravimetric analysis (TGA) technique to assess the thermal stability of a molten salt [17–19]. The maximum working temperature of nitrate molten salts is limited by thermal decomposition at high temperatures and is frequently determined by the temperature at which 3 wt.% mass loss occurs [17–19]. Experimental research on the maximum operating temperature of chloride molten salts has also been conducted by many researchers. A substantial mass loss (~1 wt.%) in the 100~200 °C range owing to the vaporization of a minor amount of water and other impurities was reported by conducting 10 separate TGA measurements, each with fresh salts [20]. Since all 10 tests indicated that the mass loss is small enough to be inappreciable in the 200~700 °C range for $MgCl_2$-KCl (32-68 mol.%) molten salt, there was a suggestion that the traditional TGA 3 wt.% mass loss method is inapplicable for chloride molten salts, which needs to be improved with other factors. It is demonstrated that the mass loss of chlorides is not caused by thermal decomposition as it is for nitrates, but rather by salt or water vaporization and salt impurity reactions, as investigated using the DSC equipment, TGA technique, X-ray Diffraction (XRD), and Mass Spectrometry (MS) [21,22]. Mass loss owing to the chemical decomposition of molten salt or the evaporation caused by high vapor pressure was distinguished by measuring the mass loss of chloride molten salts in three different atmospheres with DSC-TGA equipment [23]. If corrosion is not taken into account, the maximum working temperature of chloride molten salts was proposed in the above work to be determined by the salt vapor pressure and the gas atmosphere.

In summary, the maximum upper operating temperature of chloride molten salt is not limited by thermal decomposition like nitrate, but by other factors, such as salt vapor pressure or corrosivity regarding salt impurities [16]. It takes a long time to investigate the maximum operating temperatures of chloride molten salts using experiments with a large workload and a high level of uncertainty. Nonetheless, at the time of writing, the upper limit of the operation temperature of molten salts had not yet been discovered through modeling and computation research. In the past, a lot of experience has been accumulated using phase diagram calculation methods to predict the composition and lowest eutectic point of unknown multicomponent molten salt materials, which determined the lower limit of the optimal working temperature of molten salts in service conditions. Using the modified quasi-chemical model through the FactSage software, binary and ternary phase diagrams of some alkali metal and rare earth metal chlorides were optimized, which were then combined with experimental data to determine the best calculation model parameters [24–26]. The phase diagrams of the fluoride and chloride molten salt system with and without compound formation were calculated adopting the Regular Solution Model, and the influence the interaction coefficients of the binary boundary system will have on the isotherm of the ternary system was specifically analyzed [27]. It would be a significant breakthrough if we could calculate the highest working temperature of chloride molten salts using phase diagram computation, just as we can calculate the lowest eutectic point using the solid-liquid phase diagram. The computational method has the advantage of increasing the efficiency of the study on the maximum working temperature of chloride molten salts and reducing the experimental workload by reporting a predicted value in advance.

To implement molten salt materials in CSP, low vapor pressure is favorable, as vapor pressure is a thermophysical parameter required to fully characterize molten salts [17]. Taking chloride molten salt evaporation as a consideration, this work studied the evaporation temperature of five binary chloride molten salts. Based on the analysis of the vapor-liquid phase equilibrium thermodynamic model of common binary chloride molten salt systems, this paper established a computation method of the P-T phase diagram of binary molten chlorides expressed in equal molar mass proportions. The P-T phase diagrams of five binary chloride salts—NaCl-KCl, NaCl-$CaCl_2$, KCl-$CaCl_2$, NaCl-$MgCl_2$, and KCl-$MgCl_2$—were calculated using the adiabatic flash evaporation calculation principle. For the first time, the maximum working temperature of the binary chloride salts was determined through computational calculation. The calculation results were relatively accurate and verified by TGA, proving the feasibility of the method.

2. Materials and Methods

Considering the vapor pressure of chloride molten salt, the water in the salt or the salt itself may evaporate at high temperatures, causing a mass loss under service. At this point, the molten chloride system transitions from liquid to vapor, and vapor liquid equilibrium calculations can be used to investigate the evaporation of chloride molten salts. For the binary chloride molten salt with the lowest eutectic point and the corresponding optimal components ratio and the relationship between system pressure and temperature can be obtained by calculating the vapor liquid phase diagram of the salt under that ratio. By analyzing the evaporation temperature obtained from the P-T phase diagram of the eutectic molten salt, the upper optimal working temperature of molten chlorides can be clearly defined. The temperature point at one atmospheric pressure on the pressure temperature curve is the critical point of the vapor liquid two phases, where the temperature is the molten salt's evaporation temperature [28]. In the field of organics and alloys, the P-T phase diagram exploration is more common. For instance, the P-T phase diagram of the ferroelectric clathrate bisthiourea pyridine bromide was constructed through dielectric spectroscopy, neutron and X-ray diffraction and nuclear magnetic resonance, and the effects of temperature and pressure on their physical properties were studied in the meantime [29]. In another article, the P-T-x-y phase diagram of the Cd-Zn-Te system was constructed to discuss the character of the sublimation of the $Cd_{1-x}Zn_xTe$ (CZT) solid solution [30].

Before calculating, based on necessary considerations, the four following assumptions were made: (1) heat exchange could not occur with the outside world in the chloride molten salt system; (2) the system is under high temperature and normal pressure when in equilibrium; (3) ideal gas state equation is followed; and (4) the liquid phase is treated as a nonideal solution.

2.1. *Vapor Liquid Phase Equilibrium Thermodynamic Theory*

The exchange of energy and matter occurs when the two vapor liquid phases are in contact with each other. When the properties of each phase no longer change, the system is said to be in vapor liquid equilibrium. The fugacity of each component in the vapor and liquid phases is equal at this point, as shown in Equation (1) [31].

$$\hat{f}_i^V(T,P,y_i) = \hat{f}_i^L(T,P,x_i) \tag{1}$$

where $\hat{f}_i^V$ and $\hat{f}_i^L$ represent the fugacity of component i, respectively, in the vapor and liquid phases. x_i is the mole fraction of component i in the liquid phase, while y_i is the mole fraction of component i in the vapor phase; the system pressure is represented by P.

Since the system is under normal pressure without the heat exchange assumed in this work, the activity coefficient method was used for the vapor liquid phase equilibrium calculation. The fugacity coefficient and the activity coefficient are used to express the fugacity of the vapor phase component and the liquid phase component, respectively.

$$\hat{f}_i^V(T,P,y_i) = \hat{\Phi}_i^v P y_i \tag{2}$$

$$\hat{f}{}_i^L(T,P,x_i) = r_i x_i p_i^s \Phi_i^* \exp\left[V_i^l(P-p_i^s)/RT\right] \quad (3)$$

where $\overset{\wedge}{\Phi}{}_i^v$ represents the fugacity coefficient of component i in the vapor phase, and Φ_i^* represents the fugacity coefficient of component i in the liquid phase. The activity coefficient of component i in the liquid phase, the saturated vapor pressure of component i at temperature T and the molar volume of pure component i can be expressed as r_i, p_i^s, V_i^l correspondingly, while R represents the thermodynamic constant.

Based on Equations (1) and (2) [31], and (3) [31], the following Equation (4) [31] can be obtained:

$$\overset{\wedge}{\Phi}{}_i^v P y_i = \Phi_i^* p_i^s r_i x_i \exp\left[V_i^l(P-p_i^s)/RT\right] \quad (4)$$

2.2. Adiabatic Flash Calculation Theory

In this work, the adiabatic flash calculation theory was introduced to calculate the pressure temperature (P-T) phase diagram, considering that molten chloride salt with high vapor pressure will evaporate at high temperature. To obtain the P-T phase diagram of molten chlorides, the activity coefficient, obtained using the modified quasi-chemical model, and the saturated vapor pressure, calculated using the Antoine equation, were substituted into the adiabatic flash calculation. The activity coefficient and the vapor pressure were used as parameters in the adiabatic flash evaporation calculation process, and ultimately served for the calculation of the relationship of system pressure and temperature.

For chloride molten salt system, there are three basic phase characteristics in its gas and liquid two-phases: (a) bubble point state; (b) dew point state; (c) flash equilibrium state, as shown in Figure 1. The bubble point is the point at which the liquid phase begins to bubble and boil, and the dew point is the point at which the gas phase begins to condense into dew-like droplets.

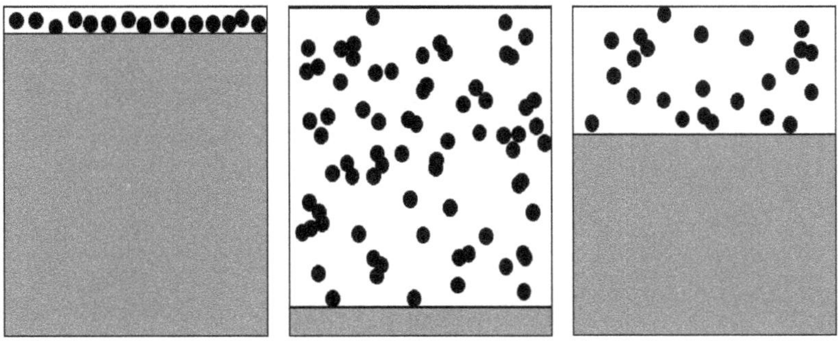

Figure 1. Basic phase characteristics of chloride molten salt system: (**a**) bubble point state; (**b**) dew point state; and (**c**) flash equilibrium state.

2.2.1. Adiabatic Flash Calculation

Flash evaporation is a single equilibrium stage distillation process. The vapor phase is enriched in volatile components, while the liquid phase is enriched in heavy components. The basic equations of flash calculation are the mass balance equation, phase balance equation, and energy balance equation, as shown below in Equations (5)–(7) [32],

$$z_i = y_i v + x_i(1-v) \quad i = 1,2,\cdots,n \quad (5)$$

$$K_i = r_i \Phi_i^s p_i^s (PF)_i \Big/ \left(\overset{\wedge}{\Phi}{}_i^v P\right) \quad i = 1,2,\cdots,n \quad (6)$$

$$H_F + Q/F = vH_v + (1-v)H_l \quad (7)$$

where z_i represents the total composition before operation of component i, while v is the vaporization rate. K_i is the vapor-liquid equilibrium ratio of component i, which is defined as the ratio of the molar fractions of the components in the two vapor liquid phases, as shown in Equation (8) [32]. Φ_i^s represents the saturated liquid fugacity coefficient of component i, respectively. The Poynting factor of component i is abbreviated as $(PF)_i$. H_F, H_v, and H_l are the molar enthalpies of the mixture, vapor, and liquid phases, respectively, while Q is the endothermic heat of the process. F is the total amount of the system, which is a known quantity, and hence, can be regarded as a constant.

$$K_i = y_i / x_i \tag{8}$$

Combining Equations (5) and (8), the two formulas for calculating the vapor phase composition and the liquid phase composition can be obtained.

$$y_i = z_i K_i / [1 + v(K_i - 1)] \quad i = 1, 2, \cdots, n \tag{9}$$

$$x_i = z_i / [1 + v(K_i - 1)] \quad i = 1, 2, \cdots, n \tag{10}$$

Since the system is under high temperature and normal pressure when in equilibrium, the vapor phase can be treated as an ideal gas mixture, $\hat{\Phi}_i^v = 1$ and $\Phi_i^s (PF)_i$ can be approximated as 1. As a result, Equation (6) can be rewritten as follows:

$$K_i = r_i p_i^s / P \quad i = 1, 2, \cdots, n \tag{11}$$

Barometric pressure 760 mmHg could be taken as the initial value of system pressure, which afterwards calculated by the following Equation (12) [32]. Due to $\sum_i y_i = 1$, combing Equations (8) and (11) and summing over all samples gives the following formation:

$$P = x_i r_i p_i^s + x_j r_j p_j^s \tag{12}$$

The activity coefficient r_i can be calculated by the modified quasi-chemical model, which will be described later in detail. As for the saturated vapor pressure p_i^s, it is an indispensable parameter in the adiabatic flash evaporation calculation process, as shown in Figures 2–5. In the calculation flow, as the loop parameter T is iteratively calculated, the corresponding p_i^s is constantly recalculated.

The saturated vapor pressure p_i^s of the pure component i can be calculated by applying the Antoine equation [32], which can provide high-accuracy data,

$$\lg p_i^s = -B_i / T + A_i \tag{13}$$

or

$$\lg p_i^s = -B_i / T - C_i T + A_i \tag{14}$$

where A_i, B_i, and C_i are Antoine constants of component i, which are published in the general physical property data manual.

Considering the adiabatic conditions of the system, $Q = 0$, and the energy balance Equation (7) can be written as follows:

$$H_F = v H_v + (1 - v) H_l \tag{15}$$

where H_v, and H_l can be calculated as Equations (16) and (17) [32]:

$$H_l = (1 - v) \sum x_i Cpl_i T \tag{16}$$

and

$$H_v = v \sum y_i Cpv_i T \tag{17}$$

where Cpl_i and Cpv_i are the liquid specific heat capacity and vapor specific heat capacity of component i, respectively.

Knowing that the temperature T is an unknown quantity, its initial value should be assumed between the precalculated bubble point and dew point temperature, and then the initial value of T should be iteratively calculated and adjusted until Equation (15) is satisfied. In addition, the temperature value T is not sensitive to the vaporization rate v for mixtures with a narrow boiling range. For calculation, in consequence, it is more appropriate to use T as the internal iteration variable and v as the external iteration variable. Accordingly, the initial value of v is calculated by interpolating the precalculated bubble point and dew point temperature as in Equation (18) [32].

$$v = (T_b - T)/(T_b - T_d) \quad (18)$$

For the narrow boiling range system, as shown in Figure 2, the adiabatic flash evaporation calculation process is a sequential algorithm, wherein the inner cycle parameter is the temperature T and the outer cycle parameter is the vaporization rate v. Both T and v are solved by the Newton iteration method, and the convergence accuracies for both are 0.001. The objective function and the Newton method iteration equation of T and v are shown below [32].

$$HT = H_F - H0 \quad (19)$$

$$T = T - HT/dHT \quad (20)$$

$$Fv = \sum z_i(K_i - 1)/[1 + v(K_i - 1)] \quad (21)$$

$$v = v - Fv/dFv \quad (22)$$

$H0$, HT, and dHT are the molar enthalpy of the system before operation, the enthalpy change, and its derivative value during the working process of the system, respectively.

$$H0 = \sum z_i Cpl_i T \quad (23)$$

$$dHT = v \cdot Cpv + (1 - v) \cdot Cpl \quad (24)$$

Since the sum of the vapor phase mole fraction and the liquid phase mole fraction are both 1, the objective function of v can be obtained as Fv, and the derivative of Fv is dFv.

$$dFv = -\sum z_i \{(K_i - 1)/[1 + v(K_i - 1)]\}^2 \quad (25)$$

The wide boiling range system in which the outer cycle parameter is T and the inner cycle parameter is v, is opposite to the narrow boiling range system (Figure 3).

2.2.2. Bubble Point and Dew Point Temperature Calculation

The iterative calculation method is commonly used to calculate the bubble point and dew point temperature, and the initial temperature value must be input at the start of the iterative algorithm. Because the system pressure is not significantly different from the saturated vapor pressure of the pure component, it can be used as a replacement in the first trial calculation. Moreover, the saturated temperature of the pure component can be inversely calculated by the Antoine equation as follows:

$$T_i^s = -B_i/(\lg P - A_i) \quad (26)$$

where T_i^s is the saturated temperature of the component i. In the case of Equation (14), however, T_i^s needs to be solved by the dichotomy numerical calculation method.

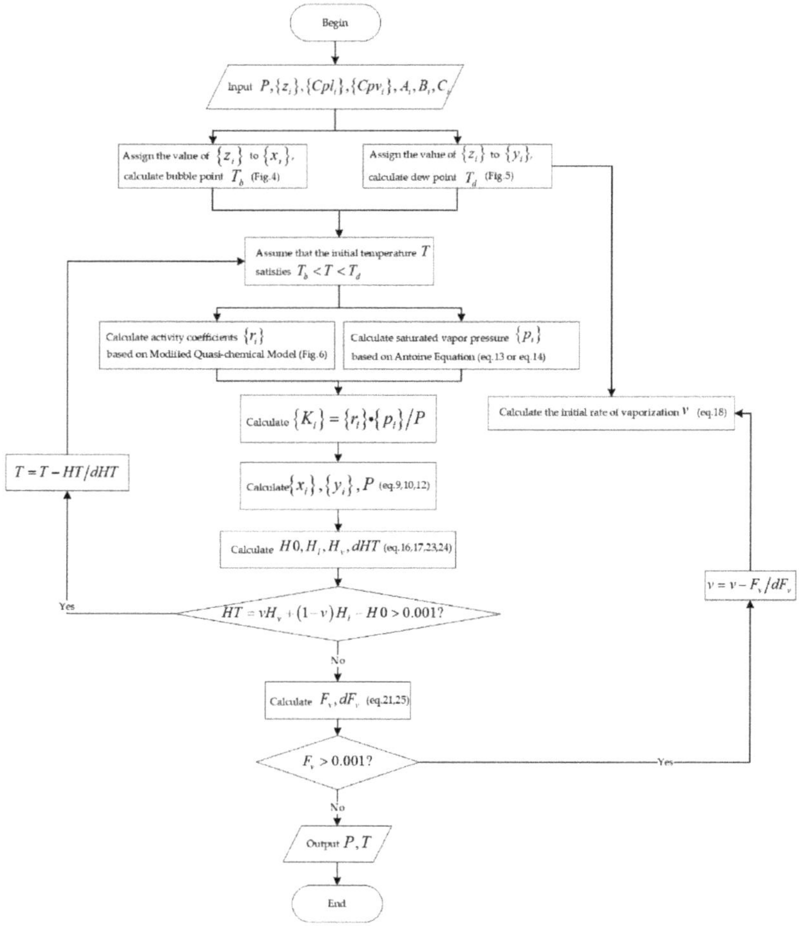

Figure 2. Flowchart for the calculation of adiabatic flash for narrow boiling range mixtures.

The initial temperature value can be calculated by linearly adding up the saturation temperature of each component. For the bubble point temperature calculation, the initial value of T is calculated as follows [32]:

$$T_0 = \sum_i x_i T_i^s \tag{27}$$

After calculating r_i using the modified quasi-chemical model and p_i using the Antoine equation, y_i is calculated by combining Equations (8), (9) and (11).

The vapor pressure ratio [32] is introduced to calculate the saturated vapor pressure of component j,

$$p_{jb} = P\sum_i (y_i/r_i)(p_j/p_i) \tag{28}$$

Once p_{jb} is determined, the Antoine equation can be used to inversely calculate the system temperature, as shown in Figure 4. After iteration, finally, the bubble point temperature T_b is calculated.

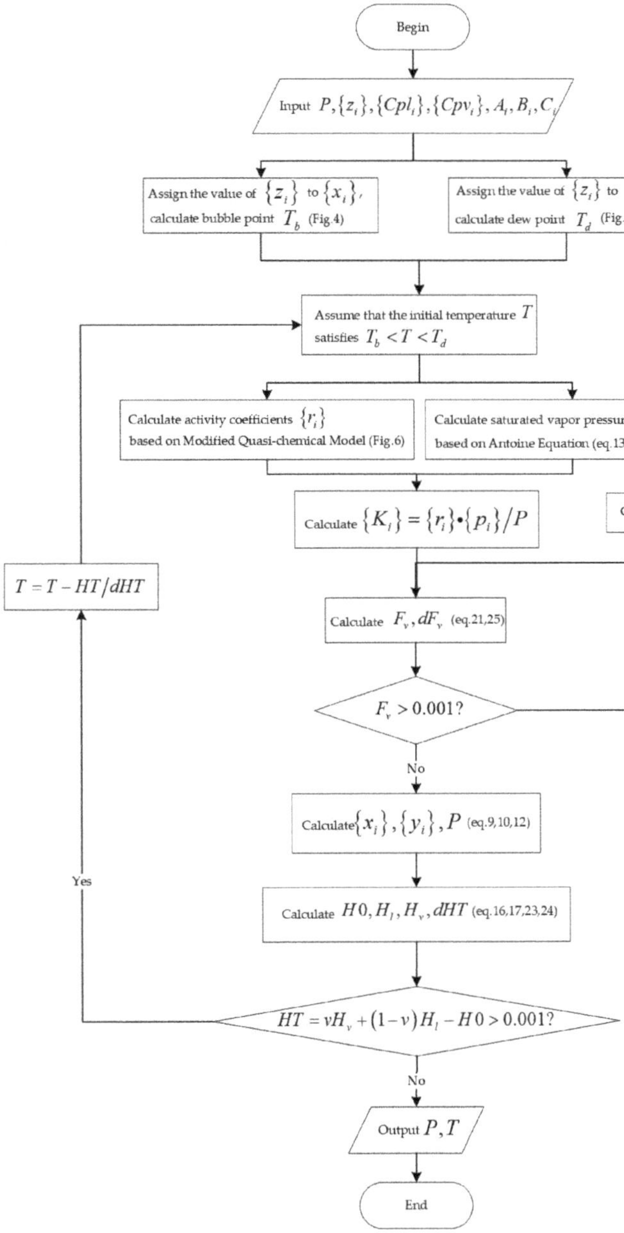

Figure 3. Flowchart for the calculation of adiabatic flash for wide boiling range mixtures.

Similarly, for the dew point temperature calculation, the initial value of T is calculated as follows [32]:

$$T_0 = \sum_i y_i T_i^s \tag{29}$$

Figure 5 illustrates the flow chart of the dew point temperature calculation. After calculating p_i using the Antoine equation, x_i is calculated by combining Equations (8), (10) and (11). T_d is then calculated according to Equation (27). Subsequently, p_{jd} is calculated according to the following equation [32] after obtaining r_{id} according to the modified quasi-chemical model.

$$p_{jd} = P \Big/ \left[\sum_i x_i r_{id} (p_i / p_j) \right]. \tag{30}$$

Further, we calculate T_d and p_{id} in turn, using the Antoine equation with p_{id}, and obtain y_{id} according to Equations (8), (9) and (11). After iteration, the dew point temperature T_d is calculated.

2.2.3. Modified Quasi-Chemical Model

Because the liquid phase of the molten salt system is treated as a nonideal solution in the assumptions made in this article, the modified quasi-chemical model [33–35] was applied to calculate the solution's activity coefficient. Taking the binary system as an example, the quasi-chemical theory only considers the interactions between the nearest neighbor bonds. It is believed that there are three coordination forms when component 1 and component 2 are mixing, as shown below in Equation (31) [33].

$$[1-1] + [2-2] \rightarrow 2[1-2] \tag{31}$$

In the equation above, the coordination form $[1-2]$ is derived from the coordination form of $[1-1]$ and $[2-2]$. Hence, the total excess Gibbs free energy (ΔG^E) is calculated as follows [34]:

$$\Delta G^E = RTZ(b_1x_1 + b_2x_2)\{Y_1 \ln(\xi - 1 + 2Y_1) - Y_1 \ln[Y_1(1+\xi)] + Y_2 \ln(\xi - 1 + 2Y_2) - Y_2 \ln[Y_2(1+\xi)]\}/2 \tag{32}$$

$$\xi = \sqrt{1 + 4x_1x_2\{\exp[2(\omega - T\eta)/(zRT)] - 1\}} \tag{33}$$

where R and z represent the thermodynamic constant and the coordination number in quasi-lattice model, respectively.

The modified quasi-chemical model pays attention to the molar enthalpy change ω and non-configurational entropy change η of the components to express the model more clearly. On this basis, the equivalent fraction Y_i is introduced [34].

$$\omega = \omega_0 + \omega_1 Y_2 + \omega_2 Y_2^2 + \omega_3 Y_2^3 + \cdots \tag{34}$$

$$\eta = \eta_0 + \eta_1 Y_2 + \eta_2 Y_2^2 + \eta_3 Y_2^3 + \cdots \tag{35}$$

$$Y_1 = b_1x_1/(b_1x_1 + b_2x_2) \tag{36}$$

$$Y_2 = b_2x_2/(b_1x_1 + b_2x_2) \tag{37}$$

The selection of appropriate b_1 and b_2 would maximize components order of the system, under which $Y_1 = Y_2 = \frac{1}{2}$ [34]. In the meanwhile, b_1, b_2 can be calculated as follows [34], where ε related to the composition of the system with maximum order.

$$b_2z = -[\ln\varepsilon + (1-\varepsilon)\ln(1-\varepsilon)/\varepsilon]/\ln 2 \tag{38}$$

$$b_1 = b_2\varepsilon/(1-\varepsilon) \tag{39}$$

$$\varepsilon = b_1/(b_1 + b_2) \tag{40}$$

The relationship between the molar excess free energy and partial molar excess free energy is as follows [33]:

$$\overline{G_i^E} = G^E + (1 - x_i)\left(\partial G^E/\partial x_i\right)_{T,P} \tag{41}$$

Incorporating Equations (32)–(37) and (41), the partial molar excess free energy of components 1 and 2 can be calculated [35].

$$\overline{G_1^E} = b_1 RTz \ln\{(\xi - 1 + 2Y_1)/[Y_1(1+\xi)]\}/2 - 2b_1 Y_1 Y_2^2[\partial(\omega - T\eta)/\partial Y_2]/(1+\xi) \quad (42)$$

$$\overline{G_2^E} = b_2 RTz \ln\{(\xi - 1 + 2Y_2)/[Y_2(1+\xi)]\}/2 + 2b_2 Y_2 Y_1^2[\partial(\omega - T\eta)/\partial Y_2]/(1+\xi) \quad (43)$$

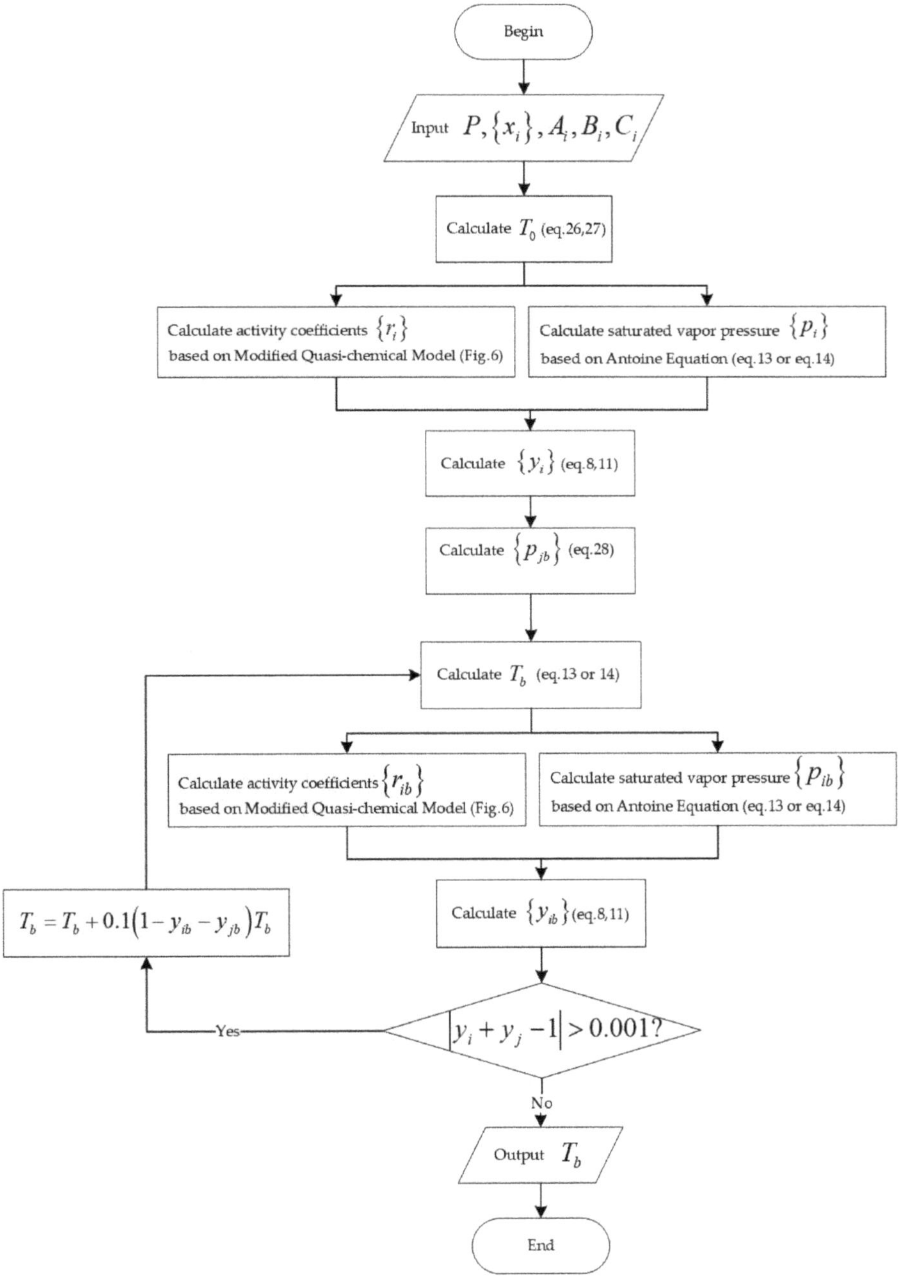

Figure 4. Flowchart for the calculation of bubble point temperature.

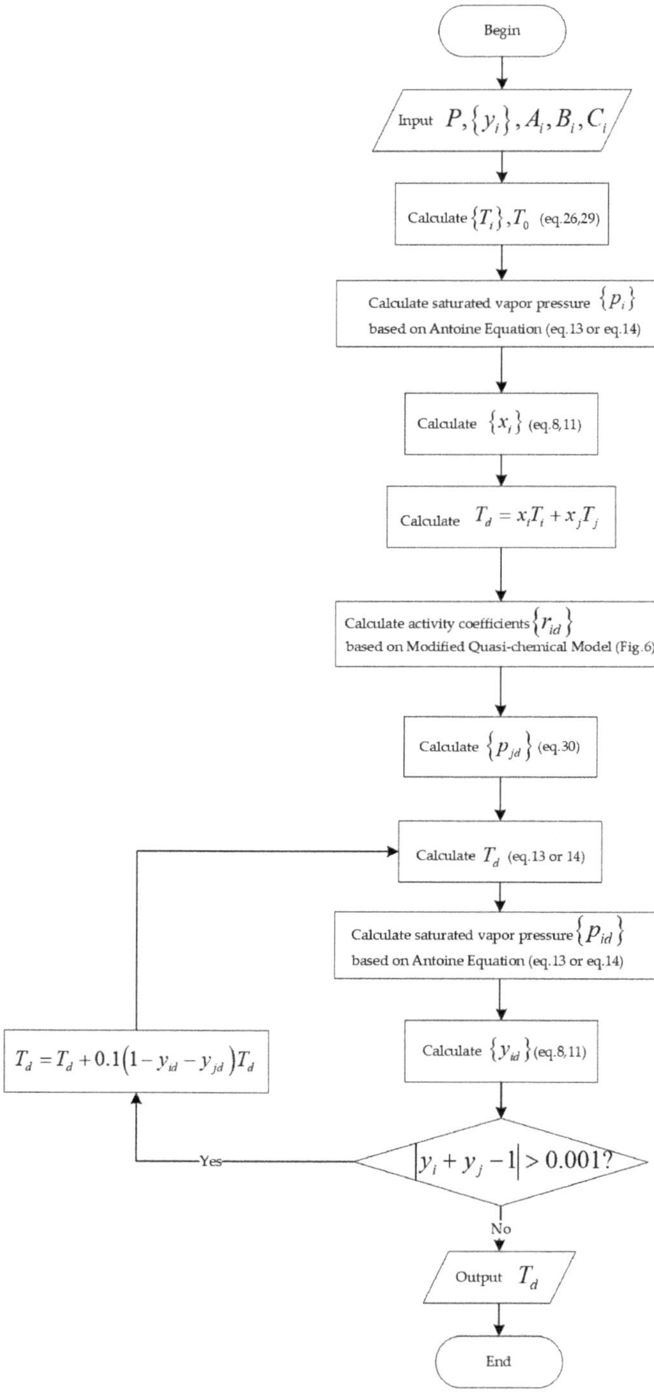

Figure 5. Flowchart for the calculation of dew point temperature.

The activity coefficient can be obtained [35] according to the following relationship between itself and partial molar excess free energy, as shown in Figure 6.

$$\overline{G_i^E} = RT \ln r_i \tag{44}$$

$$\ln r_1 = b_1 z \ln\{(\xi - 1 + 2Y_1)/[Y_1(1+\xi)]\}/2 - 2b_1 Y_1 Y_2^2 [\partial(\omega - T\eta)/\partial Y_2]/[RT(1+\xi)] \tag{45}$$

$$\ln r_2 = b_2 z \ln\{(\xi - 1 + 2Y_2)/[Y_2(1+\xi)]\}/2 + 2b_2 Y_2 Y_1^2 [\partial(\omega - T\eta)/\partial Y_2]/[RT(1+\xi)] \tag{46}$$

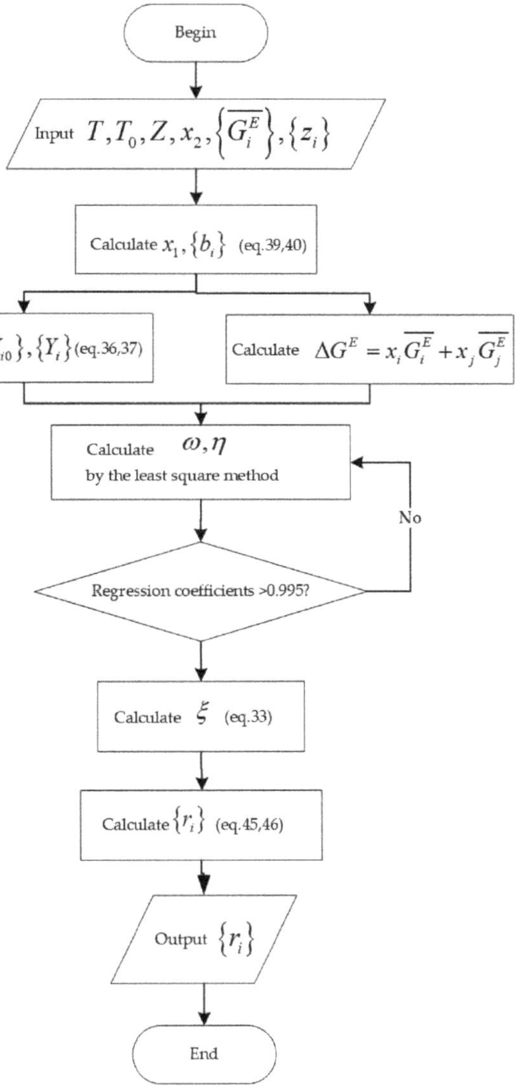

Figure 6. Flowchart for the calculation of activity coefficients.

2.3. Materials and Synthesis

After all the components were dried at 393 K in advance in an oven for 24 h, potassium chloride and magnesium chloride (A.R., from Shanghai Aladdin Biochemical Technology Co. Ltd., Shanghai, China) and sodium chloride and calcium chloride (A.R., from Guangzhou Chemical Reagent Technology Co. Ltd., Guangzhou, China) were combined in pairs to prepare five binary chloride molten salts with a corresponding mole mass ratio using the static melting method [36]. Further, five groups of two different salts were fully and evenly mixed in five corundum crucibles after being ground and crushed, and were heated using a muffle furnace from the ambient temperature (305 K) to a temperature 423 K higher than the lowest eutectic point at a heating rate of 5 K/min. The salts were fused into a homogeneous mixture, held at that temperature for 10 h, and then cooled at ambient temperature (305 K) before being ground into powder, sealed, and stored in a vacuumed desiccator.

2.4. Measurements and Procedure

At present, researchers mostly use TGA to analyze the thermal stability of molten salt. When the thermal weight loss curve is significantly reduced, the corresponding temperature can be defined as the upper limit of the working temperature of the molten salt.

In this paper, the experimental value of the maximum operating temperature of five chloride salts was obtained by a thermal gravimetric analyzer (PerkinElmer STA 8000). All five chloride molten salts were heated to 1273 K at a rate of 20 K/min with a platinum crucible in a highly purified nitrogen atmosphere at a flow rate of 100 mL/min.

3. Results and Discussion

3.1. Verification of the Modified Quasi-Chemical Model in Binary Chloride Molten Salts System

Through ω and η, the thermodynamic quantities of chloride molten salts can be calculated using the modified quasi-chemical model. For the NaCl-KCl binary chloride molten salt system, $z = 6$, $\varepsilon = \frac{1}{2}$, $b_1 = 0.3333$, and $b_2 = 0.3333$, and the multinomial expressions of ω and η at 1083 K were obtained as follows, supported by literature data [37]:

$$\omega = 0.036179 - 0.06599Y_2 + 0.027465Y_2^2 + 0.015558Y_2^3 - 0.01569Y_2^4 + 0.003203Y_2^5 - 0.00443Y_2^6 \text{ (J/mol)} \quad (47)$$

$$\eta = -39.1822 + 71.46415Y_2 - 29.7441Y_2^2 - 16.849Y_2^3 + 16.99621Y_2^4 - 3.46882Y_2^5 + 4.799909Y_2^6 \text{ (J/mol)} \quad (48)$$

The activity coefficients of the NaCl-KCl system at 1083 K were calculated by substituting Equations (47) and (48) into (45) and (46) as shown in Figure 7. The activities of the NaCl-KCl system were calculated using the relationship between activity and the activity coefficient illustrated in Equation (49), as shown in Table 1. Figure 8 depicts the comparison results of calculated values and literature data.

$$a_i = x_i \gamma_i \quad (49)$$

Table 1. Calculated activity values and literature data of the NaCl-KCl binary chloride system at 1083 K.

NaCl-KCl at 1083 K				
x_{KCl}	$a_{NaCl-lit}$	$a_{NaCl-cal}$	$a_{KCl-lit}$	$a_{KCl-cal}$
0.1	0.898	0.896	0.086	0.088
0.2	0.794	0.797	0.177	0.175
0.3	0.695	0.690	0.267	0.271
0.4	0.582	0.584	0.371	0.370
0.5	0.477	0.479	0.477	0.471
0.6	0.371	0.373	0.580	0.578
0.7	0.270	0.271	0.687	0.686
0.8	0.172	0.175	0.796	0.792
0.9	0.084	0.082	0.899	0.902

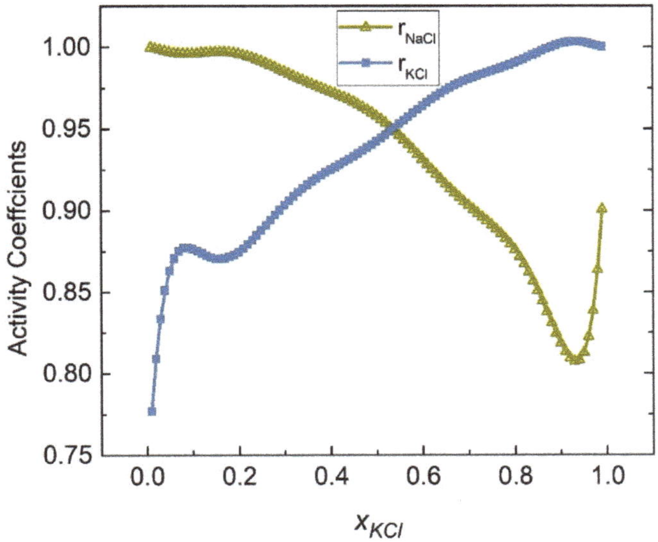

Figure 7. Calculated activity coefficients of each element of the NaCl-KCl system (1083 K).

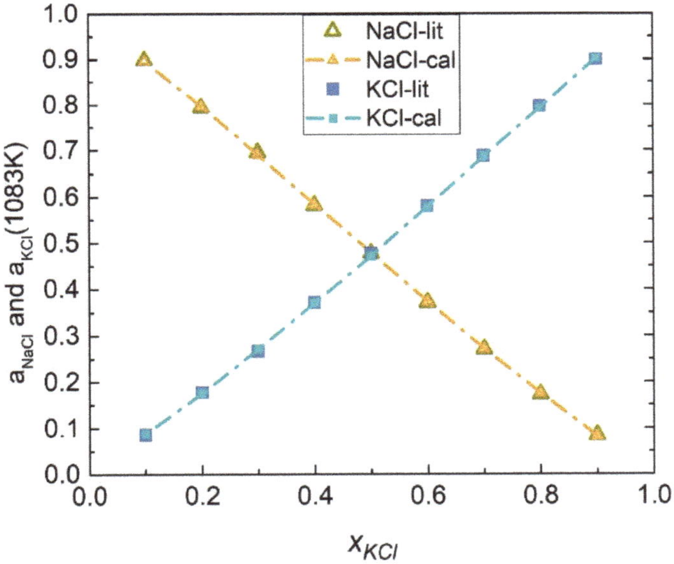

Figure 8. Comparison of the calculated activities and literature data of each element of the NaCl-KCl system (1083 K).

The average relative deviation (ARD) S_i^* and standard deviation (SD) S_i were introduced to measure the approximation between the calculated values and the literature data,

$$S_i^* = \pm(100/n)\sum_{i=1}^{n}\left|(a_{i,cal} - a_{i,lit})/a_{i,lit}\right| \quad (50)$$

$$S_i = \pm\sqrt{\left[\sum_{i=1}^{n}(a_{i.cal} - a_{i,lit})^2\right]/n} \quad (51)$$

where n is the number of literature data, $a_{i,cal}$ and $a_{i,lit}$ represent the calculated value and literature data of activities of component i, respectively. The calculation results, at 1083 K, showed that the SD and ARD of component NaCl were 0.00267 and 0.052%, while the SD and ARD of KCl were 0.00318 and 0.056%.

For the NaCl-$CaCl_2$ binary chloride molten salt system, $z = 6$, $\varepsilon = \frac{1}{2}$, $b_1 = 0.3333$, and $b_2 = 0.3333$, and the multinomial expressions of ω and η at 998 K, supported by literature data [38], were obtained as follows:

$$\omega = 27.64323 - 99.7584Y_2 + 149.3453Y_2^2 - 120.03Y_2^3 + 55.81855Y_2^4 - 15.011Y_2^5 + 2.180184Y_2^6 - 0.1692Y_2^7 \text{ (J/mol)} \quad (52)$$

$$\eta = -27587.9 + 99558.88Y_2 - 149047Y_2^2 + 119789.9Y_2^3 - 55706.9Y_2^4 + 14981.02Y_2^5 - 2175.82Y_2^6 + 168.865Y_2^7 \text{ (J/mol)} \quad (53)$$

The activity coefficients of the NaCl-$CaCl_2$ system at 998 K were calculated by substituting Equations (52) and (53) into (45) and (46), as shown in Figure 9. Table 2 shows the calculated activities of the NaCl-$CaCl_2$ system, and Figure 10 shows the comparison results of calculated values and literature data. The SD and ARD of component NaCl were 0.01282 and 5.888%, respectively, whereas the SD and ARD of $CaCl_2$ were 0.02815 and 0.71%.

Similarly, for the KCl-$CaCl_2$ binary chloride molten salt system, $z = 6$, $\varepsilon = \frac{1}{2}$, $b_1 = 0.3333$, and $b_2 = 0.3333$, and the multinomial expressions of ω and η at 1058 K, supported by literature data [39], were obtained as follows:

$$\omega = -1.09632 + 3.493849Y_2 - 4.33233Y_2^2 + 2.639964Y_2^3 - 0.87714Y_2^4 + 0.191135Y_2^5 - 0.0802Y_2^6 \text{ (J/mol)} \quad (54)$$

$$\eta = 1159.908 - 3696.49Y_2 + 4583.605Y_2^2 - 2793.08Y_2^3 + 928.0193Y_2^4 - 202.221Y_2^5 + 84.85492Y_2^6 \text{ (J/mol)} \quad (55)$$

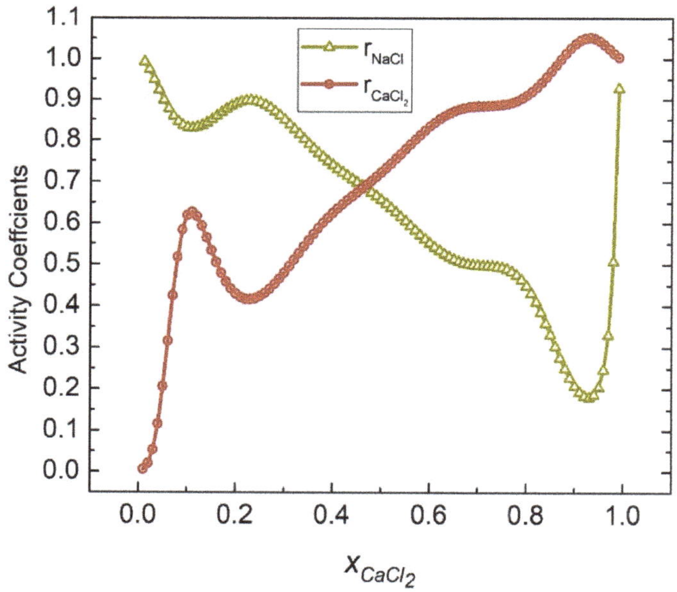

Figure 9. Calculated activity coefficients of each element of the NaCl-$CaCl_2$ system (998 K).

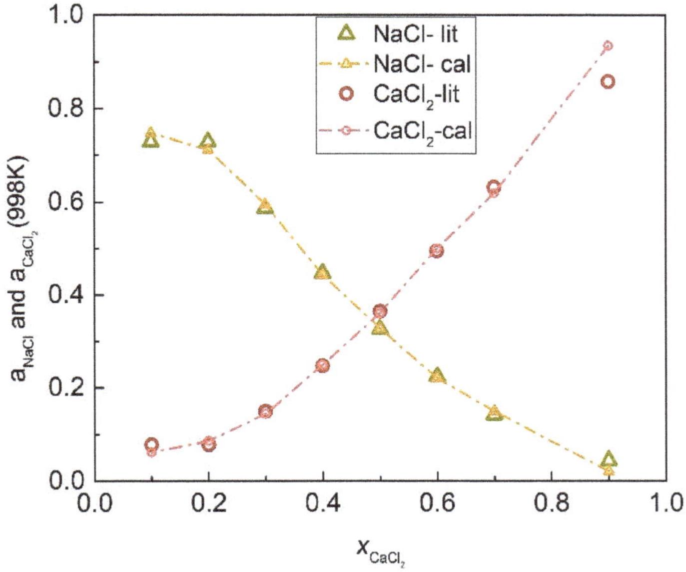

Figure 10. Comparison of the calculated activities and literature data of each element of the NaCl-$CaCl_2$ system (998 K).

Table 2. Calculated activity values and FactSage data of the NaCl-$CaCl_2$ binary chloride system at 998 K.

NaCl-$CaCl_2$ at 998 K				
x_{CaCl_2}	$a_{NaCl-fac}$	$a_{NaCl-cal}$	a_{CaCl_2-fac}	a_{CaCl_2-cal}
0.1	0.729	0.748	0.078	0.062
0.2	0.729	0.712	0.078	0.086
0.3	0.587	0.595	0.149	0.144
0.4	0.447	0.444	0.247	0.250
0.5	0.326	0.328	0.365	0.361
0.6	0.224	0.221	0.495	0.500
0.7	0.142	0.149	0.632	0.619
0.9	0.044	0.021	0.858	0.934

The activities coefficients of the KCl-$CaCl_2$ system at 1058 K were calculated by substituting Equations (54) and (55) into (45) and (46), respectively, as shown in Figure 11. Table 3 shows the calculated activities of the KCl-$CaCl_2$ system, and Figure 12 shows the comparison results of calculated values and literature data. The SD and ARD of component KCl were 0.00775 and 1.806%, respectively, while the SD and ARD of $CaCl_2$ were 0.00727 and 5.75%, respectively.

For the NaCl-$MgCl_2$ binary chloride molten salt system, $z = 6$, $\varepsilon = \frac{1}{3}$, $b_1 = 0.2294$, and $b_2 = 0.4594$, and the multinomial expressions of ω and η at 1098 K, supported by literature data [39], were obtained as follows:

$$\omega = -0.99823 + 2.031157Y_2 - 1.45713Y_2^2 + 0.405384Y_2^3 - 0.09392Y_2^4 \text{ (J/mol)} \quad (56)$$

$$\eta = 1096.053 - 2230.21Y_2 + 1599.928Y_2^2 - 445.112Y_2^3 + 103.1219Y_2^4 \text{ (J/mol)} \quad (57)$$

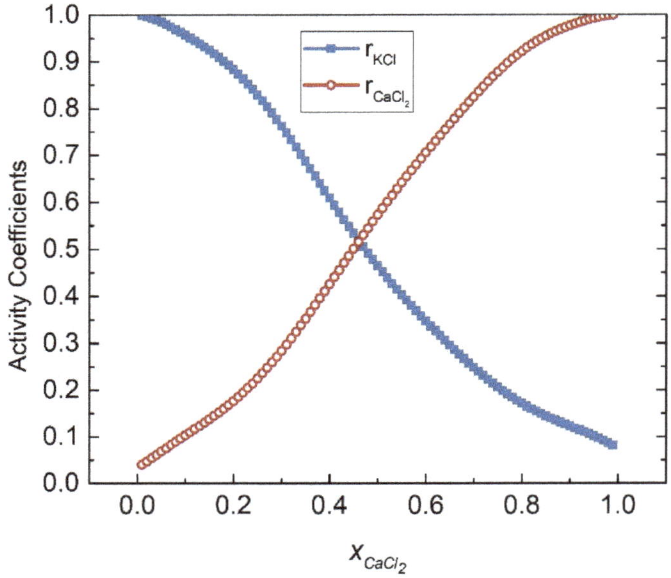

Figure 11. Calculated activity coefficients of each element of the KCl-$CaCl_2$ system (1058 K).

Table 3. Calculated activity values and literature data of the KCl-$CaCl_2$ binary chloride system at 1058 K.

KCl-$CaCl_2$ at 1058 K				
x_{CaCl_2}	$a_{KCl-lit}$	$a_{KCl-cal}$	a_{CaCl_2-lit}	a_{CaCl_2-cal}
0.1	0.867	0.862	0.009	0.010
0.2	0.704	0.707	0.031	0.035
0.3	0.524	0.534	0.074	0.085
0.4	0.353	0.366	0.155	0.170
0.5	0.218	0.232	0.280	0.287
0.6	0.133	0.139	0.418	0.423
0.7	0.073	0.075	0.579	0.577
0.8	0.035	0.034	0.735	0.738
0.9	0.012	0.012	0.884	0.879

The activities coefficients of the NaCl-$MgCl_2$ system at 1098 K were calculated by substituting Equations (56) and (57) into (45) and (46), respectively, as shown in Figure 13. Table 4 shows the calculated activities of the NaCl-$MgCl_2$ system, and Figure 14 shows the comparison results of calculated values and literature data. The SD and ARD of component NaCl were 0.01642 and 2.884%, respectively, while the SD and ARD of component $MgCl_2$ were 0.01696 and 6.604%, respectively.

For the KCl-$MgCl_2$ binary chloride molten salt system, $z = 6$, $\varepsilon = \frac{1}{3}$, $b_1 = 0.2294$, and $b_2 = 0.4594$, and the multinomial expressions of ω and η at 998 K were obtained, supported by literature data [39], as follows:

$$\omega = -0.17975 - 0.46824Y_2 + 2.864088Y_2^2 - 4.12313Y_2^3 + 2.346907Y_2^4 - 0.55023Y_2^5 - 0.07127Y_2^6 \text{ (J/mol)} \quad (58)$$

$$\eta = 179.3948 + 467.302Y_2 - 2858.36Y_2^2 + 4114.884Y_2^3 - 2342.21Y_2^4 + 549.1279Y_2^5 + 71.12457Y_2^6 \text{ (J/mol)} \quad (59)$$

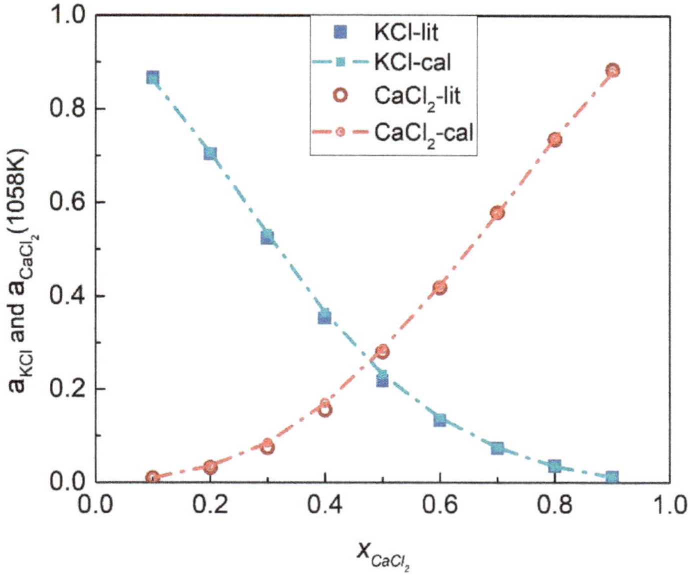

Figure 12. Comparison of the calculated activities and literature data of each element of the KCl-$CaCl_2$ system (1058 K).

The activity coefficients of the KCl-$MgCl_2$ system at 998 K were calculated by substituting Equations (58) and (59) into (45) and (46), respectively, as shown in Figure 15. Table 5 shows the calculated activities of the KCl-$MgCl_2$ system, and Figure 16 shows the comparison results of calculated values and literature data. Furthermore, the SD and ARD of component KCl and $MgCl_2$ were 0.02200 and 8.014% and 0.008407 and 12.3865%, respectively, which can be found in Table 6.

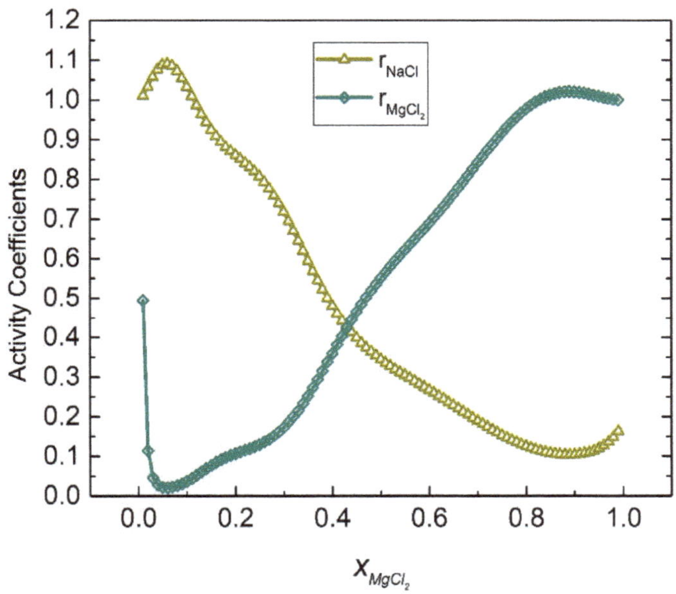

Figure 13. Calculated activity coefficients of each element of the NaCl-$MgCl_2$ system (1098 K).

Table 4. Calculated activity values and literature data of the $NaCl-MgCl_2$ binary chloride system at 1098 K.

$NaCl-MgCl_2$ at 1098 K				
x_{MgCl_2}	$a_{NaCl-lit}$	$a_{NaCl-cal}$	a_{MgCl_2-lit}	a_{MgCl_2-cal}
0.10	0.880	0.929	0.006	0.004
0.15	0.797	0.787	0.012	0.012
0.20	0.705	0.688	0.021	0.021
0.25	0.600	0.606	0.039	0.032
0.30	0.494	0.502	0.061	0.052
0.35	0.391	0.386	0.099	0.088
0.40	0.309	0.288	0.148	0.144
0.45	0.240	0.219	0.211	0.209
0.50	0.186	0.172	0.276	0.275
0.55	0.135	0.137	0.364	0.342
0.60	0.103	0.107	0.448	0.413
0.65	0.077	0.080	0.529	0.496
0.75	0.040	0.038	0.684	0.688
0.85	0.020	0.016	0.831	0.861
0.90	0.013	0.012	0.897	0.879

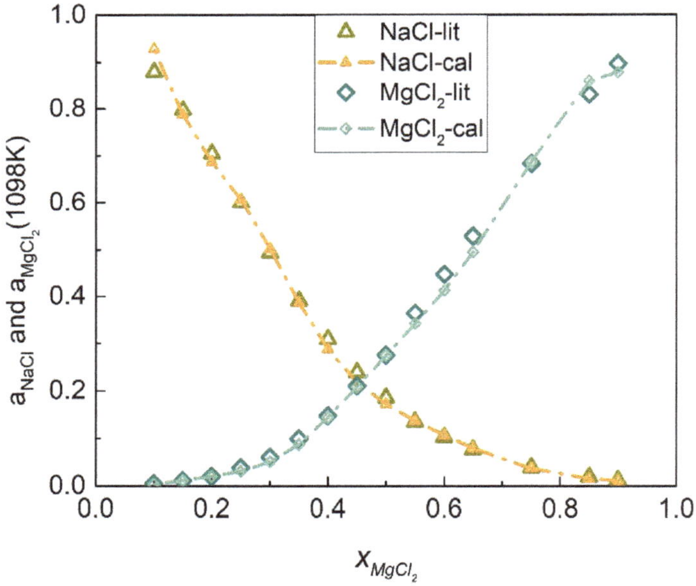

Figure 14. Comparison of the calculated activities and literature data of each element of the $NaCl-MgCl_2$ system (1098 K).

Combined with the above data, it can be shown that the modified quasi-chemical model can fit the bond energy parameters ω and η of the components of the binary chloride molten salts systems and calculate their activity coefficient. The modified quasi-chemical model had a good calculation effect on the binary chloride molten salts system, which is in good agreement with the data in the literature according to the error analysis of activity. As a result, the modified quasi-chemical model's calculated data were deemed reliable. According to the bond energy parameters ω and η calculated above, the activity coefficients of chloride salts can be obtained at any temperature.

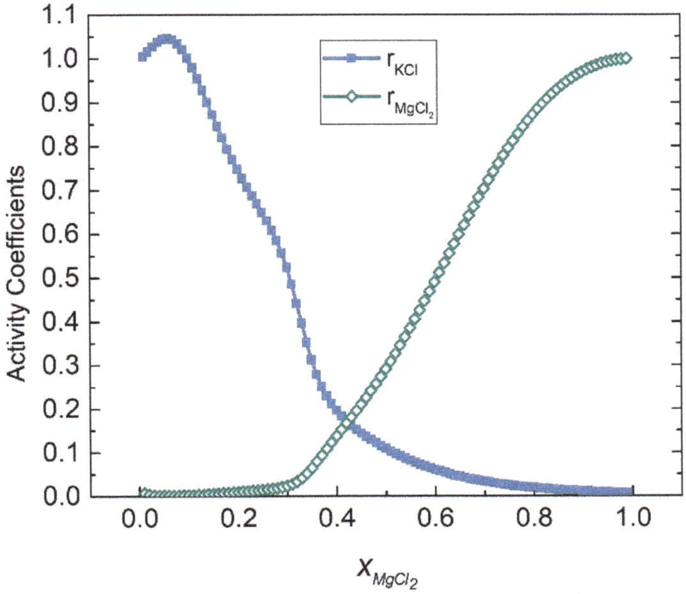

Figure 15. Calculated activity coefficients of each element of the KCl-$MgCl_2$ system (998 K).

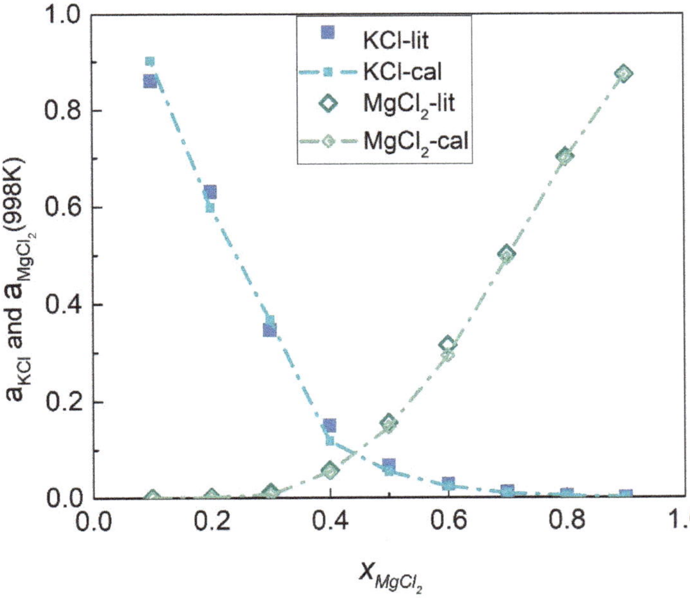

Figure 16. Comparison of calculated activities and literature data of each element of the KCl-$MgCl_2$ system (998 K).

Table 5. Calculated activity values and literature data of the KCl-$MgCl_2$ binary chloride system at 998 K.

KCl-$MgCl_2$ at 998 K				
x_{MgCl_2}	$a_{KCl-lit}$	$a_{KCl-cal}$	a_{MgCl_2-lit}	a_{MgCl_2-cal}
0.1	0.860	0.902	0.0004	0.0002
0.2	0.630	0.598	0.002	0.002
0.3	0.346	0.367	0.012	0.007
0.4	0.149	0.118	0.058	0.055
0.5	0.067	0.055	0.155	0.146
0.6	0.028	0.024	0.315	0.294
0.7	0.012	0.010	0.501	0.493
0.8	0.004	0.004	0.702	0.698
0.9	0.0012	0.0011	0.873	0.873

Table 6. Average relative deviation and standard deviation of the binary chloride systems.

$i-j$	T/K	$\pm S_{i-cal}$	$\pm S_{j-cal}$	S^*_{i-cal} (%)	S^*_{j-cal} (%)
NaCl-KCl	1083	0.00267	0.00318	0.052	0.056
NaCl-$CaCl_2$	998	0.01282	0.02815	5.888	0.710
KCl-$CaCl_2$	1058	0.00775	0.00727	1.806	5.750
NaCl-$MgCl_2$	1098	0.01642	0.01696	2.884	6.604
KCl-$MgCl_2$	998	0.02200	0.00841	8.014	12.387

3.2. Calculation of the P-T Phase Diagram

Binary chloride molten salts systems with small and big boiling point difference between the components can be regarded as narrow boiling range systems and wide boiling range systems, respectively. The boiling points and specific heat capacity of common chlorides NaCl, KCl, $CaCl_2$, and $MgCl_2$ are shown in Table 7, obtained from the FactSage software and databases. The NaCl-KCl, NaCl-$MgCl_2$, and KCl-$MgCl_2$ binary chloride molten salt systems are narrow boiling range systems due to the relatively small boiling point difference, whereas the NaCl-$CaCl_2$ and KCl-$CaCl_2$ binary chloride molten salt systems are wide boiling range systems due to the relatively large temperature range. The Antoine constants of NaCl, KCl, $CaCl_2$, and $MgCl_2$ are demonstrated in Table 8 [40,41].

Table 7. Physical parameters for the pure salts obtained from FactSage.

Component	$T_{b.p.}$ K	*Cpl* J/(mol·K)	*Cpv* J/(mol·K)
NaCl	1734	$77.763824 - 7.5312 \times 10^{-3}T$ (298–1500 K) 66.944 (1500–2000 K)	$37.333832 + 7.36384{*}10^{-4}T - 158573.6T^{-2}$
KCl	1693	73.59656	$37.145552 + 9.49768{*}10^{-4}T - 84098.4T^{-2}$
$MgCl_2$	1685	$193.40893 - 0.36201388T$ $-3788503.9T^{-2} + 3.19987074T^2$ (298–660 K) 92.048 (660–2500 K)	$62.3647 - 695753.73T^{-2} + 67611239T^{-3}$
$CaCl_2$	1873	$81.479332 - 0.00254166793T$ $-2789.1275T^{-1}$ (298–700 K) 102.533 (700–3000 K)	$62.382104 - 100.35854T^{-1} - 244513.85T^{-2}$

Table 8. Antoine constants for the pure salts.

Component	Antoine Constant A	Antoine Constant B	Antoine Constant C
NaCl	8.4459	9565	/
KCl	8.28	9032	/
$CaCl_2$	8.4073	12622	/
$MgCl_2$	23.15	11735	4.076

Combining the relevant thermodynamic data, the P-T phase diagrams of the molten salts NaCl-KCl, NaCl-$CaCl_2$, KCl-$CaCl_2$, NaCl-$MgCl_2$, and KCl-$MgCl_2$ were calculated using the adiabatic flash calculation theory and modified quasi-chemical model detailed in the previous section according to the flowchart of the wide and narrow boiling range mixtures.

For the P-T phase diagram, the system is in flash equilibrium only when the system pressure is equal to one atmosphere, and the temperature corresponding to the standard atmospheric pressure (760 mmHg) is the evaporation point of the substance. Based on this, as demonstrated in Figures 17–21, the calculated evaporation temperature of NaCl-KCl, NaCl-$CaCl_2$, KCl-$CaCl_2$, NaCl-$MgCl_2$, and KCl-$MgCl_2$, corresponding to the standard atmospheric pressure, were 1140.7 K, 1150.6 K, 1176.4 K, 1085.5 K, and 1067.9 K, at which the molten salts are in flash equilibrium. Before these temperature points, the eutectic chloride molten salts above are in liquid phase, while after these temperature points, the chloride molten salts are expected to gradually evaporate as they transition from liquid to vapor. When the evaporation process began, the weight of the chloride molten salts began to fall sharply. When the weight of the chloride molten salts dropped to a certain extent in the actual service conditions, the corresponding temperature at that time was identified as the maximum working temperature of the molten salts, which also reflects the thermal stability of the molten salts.

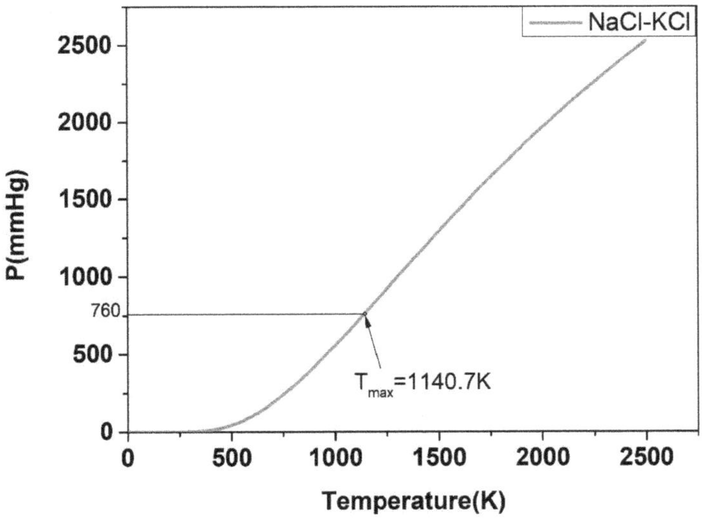

Figure 17. Calculated P-T phase diagram of the NaCl-KCl binary system.

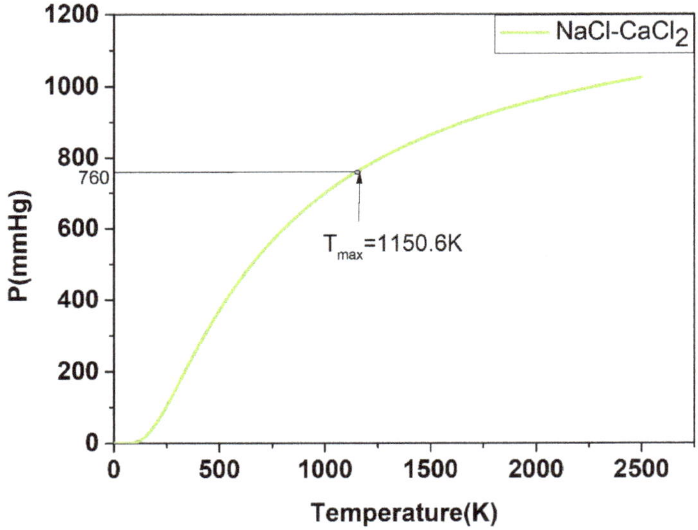

Figure 18. Calculated P-T phase diagram of the NaCl-$CaCl_2$ binary system.

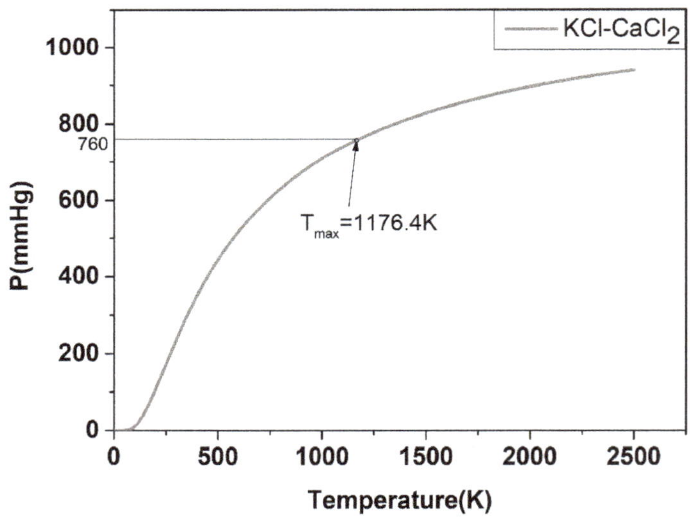

Figure 19. Calculated P-T phase diagram of the KCl-$CaCl_2$ binary system.

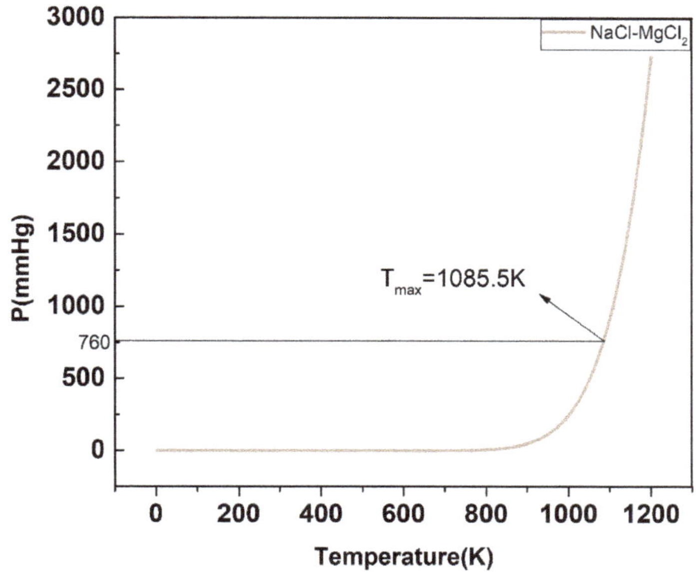

Figure 20. Calculated P-T phase diagram of the $NaCl-MgCl_2$ binary system.

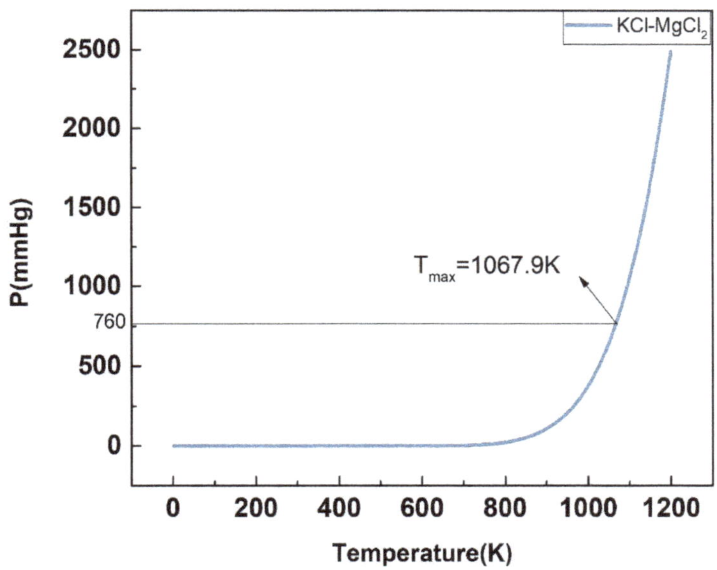

Figure 21. Calculated P-T phase diagram of the $KCl-MgCl_2$ binary system.

3.3. Thermal Gravimetric Analysis of the Binary Chloride Salts

As shown in Table 9, the molar mass ratio corresponding to the lowest eutectic temperature of NaCl-KCl, $NaCl-CaCl_2$, $KCl-CaCl_2$, $NaCl-MgCl_2$, and $KCl-MgCl_2$ chloride molten salts was obtained, referring to the FactSage software. The five salts were prepared according to the ratio and were then taken into the TGA.

Table 9. Lowest eutectic temperature and molar mass ratio of the binary chloride salts from FactSage.

Species	Lowest Eutectic Temperature K	Molar Mass Ratio
NaCl-KCl	841.14	0.47:0.53
NaCl-$CaCl_2$	749.7	0.51:0.49
KCl-$CaCl_2$	874.4	0.75:0.25
NaCl-$MgCl_2$	732.16	0.57:0.43
KCl-$MgCl_2$	695.69	0.7:0.3

The weight change of the five salts after certain thermal treatment, obtained by the thermal gravimetric analyzer, were shown below. As shown in Figures 22–26, the weight of the five types of chloride molten salts was similarly lost sharply above a certain temperature point. As a result, these temperatures are thought to be the highest working temperatures of chloride molten salts. Results show that the experimental values of the highest working temperature of NaCl-KCl, NaCl-$CaCl_2$, KCl-$CaCl_2$, NaCl-$MgCl_2$, and KCl-$MgCl_2$ were 1126.2 K, 1117.9 K, 1124.1 K, 1023.9 K, and 1017.9 K, respectively. The mass loss of the five salts at their respective maximum operating temperatures was 4.1%, 12.55%, 11.33%, 19.69%, and 8.5%. The error between the calculated values and experimental values after comparison was 1.29%, 2.93%, 4.65%, 6.02%, and 4.91%, respectively, as shown in Table 10.

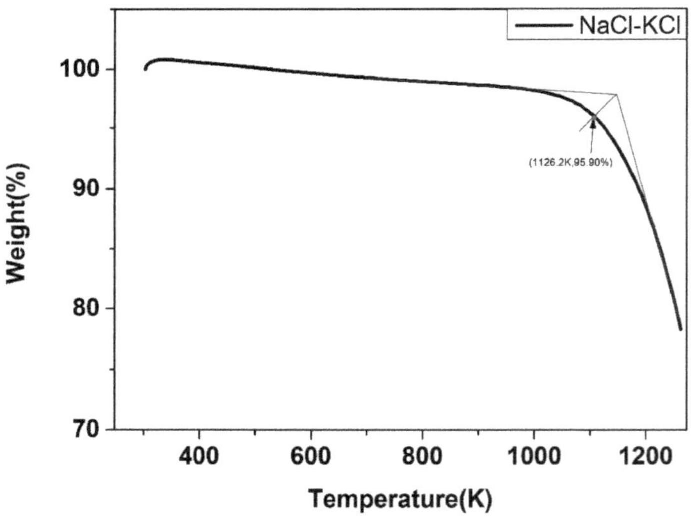

Figure 22. Weight change (wt.%) of the NaCl-KCl binary chloride salt.

Table 10. Calculated and experimental value of the upper-temperature limit of the binary chloride salts.

Species	Calculated Value of Upper-Temperature Limit K	Experimental Value of Upper-Temperature Limit K	Relative Error %
NaCl-KCl	1140.7	1126.2	1.29
NaCl-$CaCl_2$	1150.6	1117.9	2.93
KCl-$CaCl_2$	1176.4	1124.1	4.65
NaCl-$MgCl_2$	1085.5	1023.9	6.02
KCl-$MgCl_2$	1067.9	1017.9	4.91

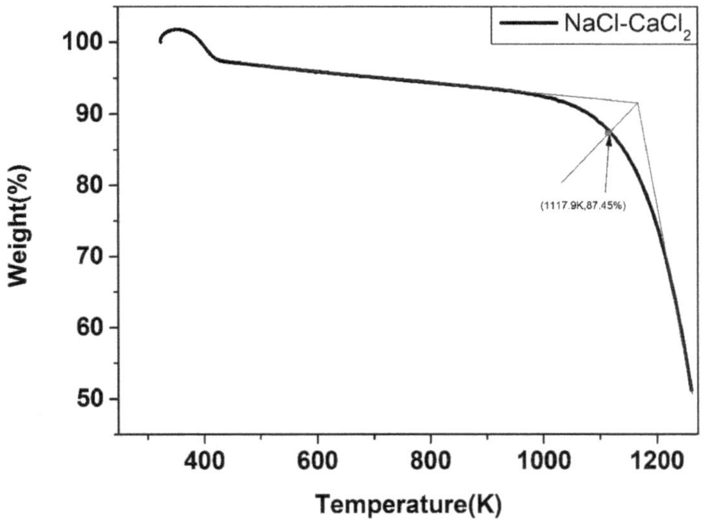

Figure 23. Weight change (wt.%) of the NaCl-$CaCl_2$ binary chloride salt.

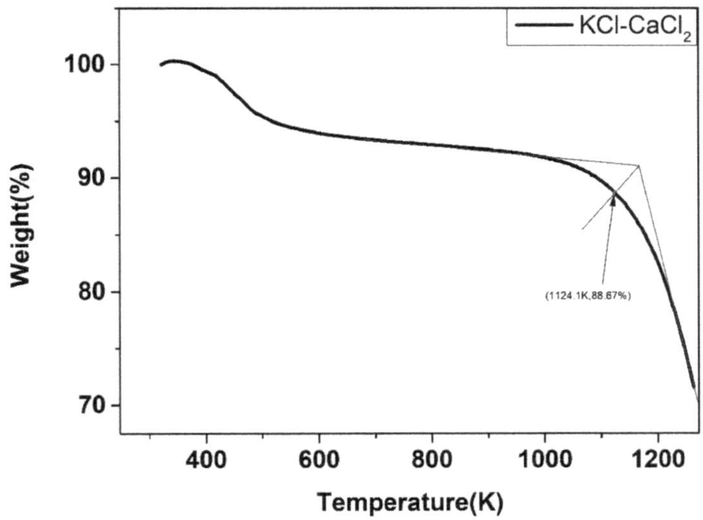

Figure 24. Weight change (wt.%) of the KCl-$CaCl_2$ binary chloride salt.

Because the maximum error between the calculated value and the experimental value of the maximum operating temperature was 6.02%, while the minimum was 1.29%, which is relatively small, we could consider that the calculation of the maximum operating temperature in this work is relatively accurate.

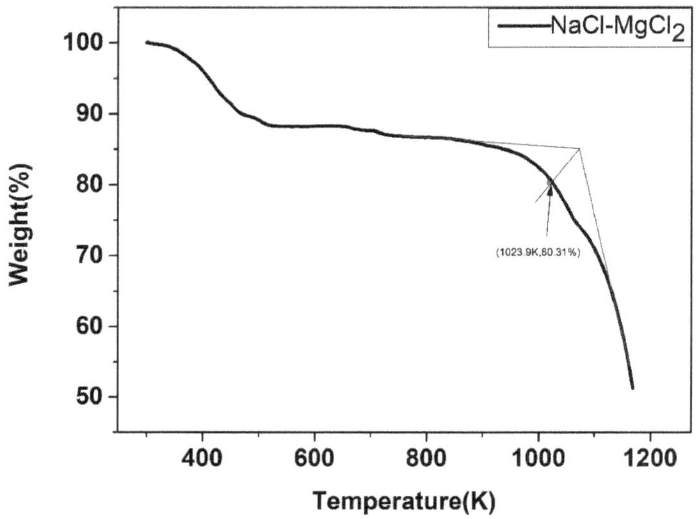

Figure 25. Weight change (wt.%) of the NaCl-$MgCl_2$ binary chloride salt.

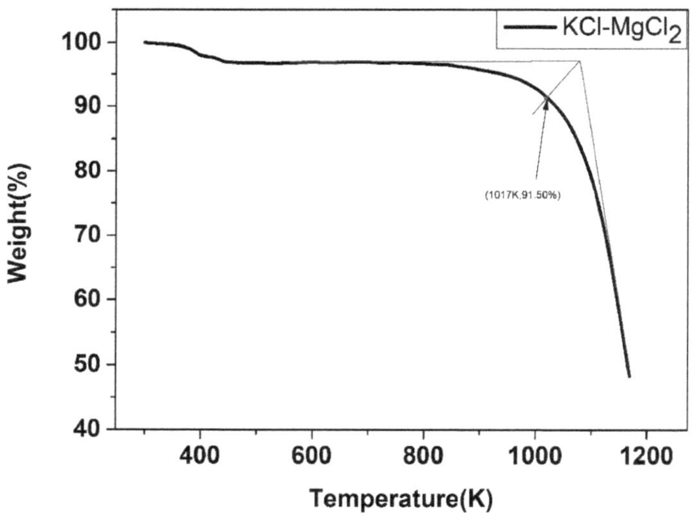

Figure 26. Weight change (wt.%) of the KCl-$MgCl_2$ binary chloride salt.

4. Conclusions

In this paper, the thermodynamic values of different binary chloride molten salt systems were calculated using the modified quasi-chemical model. The maximum SD between the calculated activity and the literature activity was 0.02815, while the minimum was 0.00267, demonstrating that the modified quasi-chemical model has a high degree of accuracy. Combining the modified quasi-chemical model, the Antoine equation, and the adiabatic flash evaporation calculation principle, the P-T phase diagram of binary chloride molten salts were obtained and the evaporation temperatures of NaCl-KCl, NaCl-$CaCl_2$, KCl-$CaCl_2$, NaCl-$MgCl_2$, and KCl-$MgCl_2$ were 1140.7 K, 1150.6 K, 1176.4 K, 1085.5 K, and 1067.9 K. The maximum operating temperatures measured experimentally were 1126.2 K, 1117.9 K, 1124.1 K, 1023.9 K, and 1017.9 K. Accordingly, the relative error between the theoretically estimated and experimental values was 1.29%, 2.93%, 4.65%, 6.02%, and 4.91%

for each system. As a result, the method is deemed extremely useful in determining the upper limit of the working temperature of binary chloride salts.

Based on these results, the upper-temperature limits of NaCl-KCl, NaCl-$CaCl_2$, KCl-$CaCl_2$, NaCl-$MgCl_2$, and KCl-$MgCl_2$ are determined to be 1141 K, 1151 K, 1176 K, 1086 K, and 1068 K. Consequently, the upper limit of the working temperature of the wide boiling range system for the alkali metal-alkaline earth metal binary chloride molten salt systems is higher than that of the narrow boiling range system. Because NaCl and KCl belong to the same main group and have similar properties, the upper limit of the operating temperature of NaCl-KCl is also higher, although it has a narrow boiling range.

The lower limit of the working temperature should be about 50 K higher than the lowest eutectic point, which was demonstrated in Table 9. Combined with the lowest eutectic temperature and the upper-temperature limits of binary chloride molten salts, the stable operating temperature ranges of NaCl-KCl, NaCl-$CaCl_2$, KCl-$CaCl_2$, NaCl-$MgCl_2$, and KCl-$MgCl_2$ are 891~1141 K, 750~1151 K, 874~1176 K, 732~1086 K, and 696~1086 K, and NaCl-$CaCl_2$, NaCl-$MgCl_2$, and KCl-$MgCl_2$ binary molten chlorides have a wider working temperature range. Therefore, in practical applications, the types of chloride molten salts used in different temperature sections can be adjusted according to the optimal working temperature range of different chloride molten salts. Furthermore, this article provides an idea for accurately predicting the maximum operating temperature using the computation method, which could hopefully promote the commercial application of chloride molten salts.

Author Contributions: Conceptualization, J.D., J.L. and X.W.; formal analysis, J.Z.; funding acquisition, J.D., J.L. and W.W.; investigation, J.Z.; methodology, J.Z.; resources, J.D. and W.W.; supervision, J.D.; validation, J.Z.; writing—original draft, J.Z.; writing—review and editing, X.W. and W.W. All authors have read and agreed to the published version of the manuscript.

Funding: This research was funded by the funding of the National Natural Science Foundation of China (52036011), Natural Science Foundation of Guangdong Province (2017B030308004), and National Natural Science Foundation of China (U1707603).

Institutional Review Board Statement: Not applicable.

Informed Consent Statement: Not applicable.

Data Availability Statement: The data are not publicly available because it is part of the current research work.

Conflicts of Interest: The authors declare no conflict of interest.

Nomenclature

A_i	Antoine constants of component i
B_i	Antoine constants of component i
C_i	Antoine constants of component i
Cpl_i	Liquid specific heat capacity of component i
Cpv_i	Vapor specific heat capacity of component i
F	Total amount of the system
G^E	Molar excess free energy
$\overline{G_i^E}$	Partial molar excess free energy of component i
H_F	Molar enthalpies of the mixture
H_l	Molar enthalpies of the liquid
H_v	Molar enthalpies of the vapor
HT	Enthalpy change
$H0$	Molar enthalpy of the system before operation
K_i	Vapor-liquid equilibrium ratio of component i
P	System pressure (mmHg)
Q	Endothermic heat of the process

R	Thermodynamic constant
S_i	Standard deviation (SD)
S_i^*	Average relative deviation (ARD)
T	Temperature (K)
T_b	Bubble point temperature
T_d	Dew point temperature
T_i^s	Saturated temperature of the component i
T_0	Initial value of T
V_i^l	Molar volume of pure component i
Y_i	Equivalent fraction
a_i	Activity of component i
$a_{i,cal}$	Calculated value of activities of component i
$a_{i,lit}$	Literature data of activities of component i
b_1	Constant of equivalent fraction
b_2	Constant of equivalent fraction
dHT	Derivative of enthalpy change
$\hat{f}_i^L$	Fugacity of component i in the liquid phase
$\hat{f}_i^V$	Fugacity of component i in the vapor phase
n	Number of literature data
p_i	Vapor pressure of component i
p_i^s	Saturated vapor pressure of component i
p_{id}	Saturated vapor pressure of component i in dew point temperature calculation
p_{jb}	Saturated vapor pressure of component j in bubble point temperature calculation
p_{jd}	Saturated vapor pressure of component j in dew point temperature calculation
r_i	Activity coefficient of component i in the liquid phase
r_{id}	Activity coefficient of component i in dew point temperature calculation
v	Vaporization rate
x_i	Mole fraction of component i in the liquid phase
y_i	Mole fraction of component i in the vapor phase
y_{id}	Mole fraction of component i in the vapor phase in dew point temperature calculation
z	Coordination number
z_i	Total composition before operation of component i
Φ_i^*	Fugacity coefficient of component i in the liquid phase
Φ_i^s	Saturated liquid fugacity coefficient of component i
$\hat{\Phi}_i^v$	Fugacity coefficient of component i in the vapor phase
ε	Composition of the system with maximum order
η	Non-configurational entropy change
ω	Molar enthalpy change
ΔG^E	Total excess Gibbs free energy
$(PF)_i$	Poynting factor of component i

References

1. Dias, P.; Vilanova, A.; Lopes, T.; Andrade, L.; Mendes, A. Extremely stable bare hematite photoanode for solar water splitting. *Nano Energy* **2016**, *23*, 70–79. [CrossRef]
2. Tian, H.; Du, L.; Wei, X.; Deng, S.; Wang, W.; Ding, J. Enhanced thermal conductivity of ternary carbonate salt phase change material with Mg particles for solar thermal energy storage. *Appl. Energy* **2017**, *204*, 525–530. [CrossRef]
3. Islam, M.T.; Huda, N.; Abdullah, A.B.; Saidur, R. A comprehensive review of state-of-the-art concentrating solar power (CSP) technologies: Current status and research trends. *Renew. Sustain. Energy Rev.* **2018**, *91*, 987–1018. [CrossRef]
4. Ferreira, A.; Kunh, S.S.; Fagnani, K.C.; De Souza, T.A.; Tonezer, C.; Dos Santos, G.R.; Coimbra-Araújo, C.H. Economic overview of the use and production of photovoltaic solar energy in brazil. *Renew. Sustain. Energy Rev.* **2018**, *81*, 181–191. [CrossRef]
5. Sun, J.; Liu, Q.; Hong, H. Numerical study of parabolic-trough direct steam generation loop in recirculation mode: Characteristics, performance and general operation strategy. *Energy Convers. Manag.* **2015**, *96*, 287–302. [CrossRef]

6. Myers, P.D., Jr.; Goswami, D.Y. Thermal energy storage using chloride salts and their eutectics. *Appl. Therm. Eng.* **2016**, *109*, 889–900. [CrossRef]
7. Aneke, M.; Wang, M. Energy storage technologies and real life applications—A state of the art review. *Appl. Energy* **2016**, *179*, 350–377. [CrossRef]
8. Liu, M.; Tay, N.S.; Bell, S.; Belusko, M.; Jacob, R.; Will, G.; Saman, W.; Bruno, F. Review on concentrating solar power plants and new developments in high temperature thermal energy storage technologies. *Renew. Sustain. Energy Rev.* **2016**, *53*, 1411–1432. [CrossRef]
9. Zavoico, A.B. *Solar Power Tower Design Basis Document*; Sandia National Laboratories: Livermore, CA, USA, 2001. [CrossRef]
10. Badger Energy Corporation. *Design, Handling, Operation and Maintenance Procedures for Hitec Molten Salt*; SAND81-8179; Sandia National Laboratories: Livermore, CA, USA, 1981.
11. Kearney, D.; Kelly, B.; Herrmann, U.; Cable, R.; Pacheco, J.; Mahoney, R.; Price, H.; Blake, D.; Nava, P.; Potrovitza, N. Engineering aspects of a molten salt heat transfer fluid in a trough solar field. *Energy* **2004**, *29*, 861–870. [CrossRef]
12. Fernández, A.; Ushak, S.; Galleguillos, H.; Perez-Trujillo, F.J. Development of new molten salts with $LiNO_3$ and $Ca(NO_3)_2$ for energy storage in CSP plants. *Appl. Energy* **2014**, *119*, 131–140. [CrossRef]
13. Nunes, V.M.B.; Queirós, C.S.; Lourenço, M.J.V.; Santos, F.J.V.; De Castro, C.N. Molten salts as engineering fluids—A review: Part I. Molten alkali nitrates. *Appl. Energy* **2016**, *11*, 603–611. [CrossRef]
14. Peng, Q.; Yang, X.; Ding, J.; Wei, X.; Yang, Y. Experimental study and mechanism analysis for high-temperature thermal stability of ternary nitrate salt. *CIESC J.* **2013**, *64*, 1507–1512. [CrossRef]
15. Ding, J.; Du, L.; Pan, G.; Lu, J.; Wei, X.; Li, J.; Wang, W.; Yan, J. Molecular dynamics simulations of the local structures and thermodynamic properties on molten alkali carbonate K_2CO_3. *Appl. Energy* **2018**, *220*, 536–544. [CrossRef]
16. Villada, C.; Ding, W.; Bonk, A.; Bauer, T. Engineering molten Mg_2–KCl–NaCl salt for high-temperature thermal energy storage: Review on salt properties and corrosion control strategies. *Sol. Energy Mater. Sol. Cells* **2021**, *232*, 111344. [CrossRef]
17. Fernández, A.G.; Gomez-Vidal, J.; Oró, E.; Kruizenga, A.; Solé, A.; Cabeza, L.F. Mainstreaming commercial CSP systems: A technology review. *Renew. Energy* **2019**, *140*, 152–176. [CrossRef]
18. Sau, S.; Tizzoni, A.; Corsaro, N.; Veca, E.; Navas, M.; Martinez-Tarifa, A. Report on protocols for standardized testing procedures and data analysis criteria (Version 1). *Zenodo* **2015**. [CrossRef]
19. Raade, J.W.; Padowitz, D. Development of Molten Salt Heat Transfer Fluid With Low Melting Point and High Thermal Stability. *J. Sol. Energy Eng.* **2011**, *133*, 031013. [CrossRef]
20. Xu, X.; Wang, X.; Li, P.; Li, Y.; Hao, Q.; Xiao, B.; Elsentriecy, H.; Gervasio, D. Experimental Test of Properties of KCl–Mg_2 Eutectic Molten Salt for Heat Transfer and Thermal Storage Fluid in Concentrated Solar Power Systems. *J. Sol. Energy Eng.* **2018**, *140*, 051011. [CrossRef]
21. Du, L.; Ding, J.; Tian, H.; Wang, W.; Wei, X.; Song, M. Thermal properties and thermal stability of the ternary eutectic salt NaCl-Ca_2-Mg_2 used in hightemperature thermal energy storage process. *Appl. Energy* **2017**, *204*, 1225–1230. [CrossRef]
22. Maksoud, L.; Bauer, T. Experimental investigation of chloride molten salts for thermal energy storage applications. In Proceedings of the 10th International Conference on Molten Salt Chemistry and Technology, Shenyang, China, 10–14 June 2015; pp. 273–280.
23. Mohan, G.; Venkataraman, M.; Gomez-Vidal, J.; Coventry, J. Assessment of a novel ternary eutectic chloride salt for next generation high-temperature sensible heat storage. *Energy Convers. Manag.* **2018**, *167*, 156–164. [CrossRef]
24. Pelton, A.D.; Chartrand, P. Thermodynamic evaluation and optimization of the LiCl-NaCl-KCl-RbCl-CsCl-Mg_2-Ca_2 system using the modified quasi-chemical model. *Metall. Mater. Trans. A* **2001**, *32*, 1361–1383. [CrossRef]
25. Robelin, C.; Chartrand, P.; Pelton, A.D. Thermodynamic evaluation and optimization of the (NaCl+KCl+Mg_2+Ca_2+Mn_2+Fe_2+Co_2 +Ni_2) System. *J. Chem. Thermodyn.* **2004**, *36*, 809–828. [CrossRef]
26. Robelin, C.; Chartrand, P. Thermodynamic evaluation and optimization of the (NaCl+KCl+Mg_2 +Ca_2+$ZnCl_2$) system. *J. Chem. Thermodyn.* **2011**, *43*, 377–391. [CrossRef]
27. Foosnaes, T.; Oestvold, T.; Oeye, H.A. Calculation of charge asymmetric additive ternary phase diagrams with and without compound formation. *Acta Chem. Scand. Ser. A Phys. Inorg. Chem.* **1978**, *32*, 973–987. [CrossRef]
28. Fried, V.; Hameka, H.F.; Blukis, U. *Physical Chemistry*; Macmillan Publishing Co., Inc.: New York, NY, USA, 1977.
29. Bilski, P.; Bobrowicz-Sarga, L.; Czarnecki, P.; Marczak, A.; Wasicki, J. The p–T phase diagram for new ferroelectric bis-thiourea pyridinium bromide. *J. Chem. Thermodyn.* **2013**, *59*, 182–187. [CrossRef]
30. Nipan, G.D. p–T–x–y phase diagram of the Cd–Zn–Te system. *J. Alloy. Compd.* **2004**, *371*, 160–163. [CrossRef]
31. Smith, J.M.; Van Ness, H.C.; Abbott, M.M. *Introduction to Chemical Engineering Thermodynamics*, 6th ed.; McGraw-Hill Book Co.: New York, NY, USA, 2001.
32. Poling, B.E.; Prausnitz, J.M.; O'Connell, J.P. *The Properties of Gases and Liquids*, 5th ed.; McGraw-Hill Book Co.: New York, NY, USA, 2001.
33. Pelton, A.D.; Blander, M. *Proceedings of the AIME Symposium on Molten Salts and Slags, Nevada, 1984*; Tahoe, L., Ed.; TMS-AIME: Warrendale, PA, USA, 1984; pp. 281–294.
34. Pelton, A.D.; Blander, M. Thermodynamic analysis of ordered liquid solutions by a modified quasichemical approach-Application to silicate salgs. *Metall. Trans. B* **1986**, *17*, 805–815. [CrossRef]
35. Blander, M.; Pelton, A.D. Thermodynamic analysis of binary liquid silicates and prediction of ternary solution properties by modified quasichemical equations. *Geochim. Cosmochim. Acta* **1987**, *51*, 85–95. [CrossRef]

36. Peng, Q.; Ding, J.; Wei, X.; Yang, J.; Yang, X. The preparation and properties of multi-component molten salts. *Appl. Energy* **2010**, *87*, 2812–2817. [CrossRef]
37. Li, R.; Gu, W. Calculation of thermodynamic properties of NaCl-KCl system. *Eng. Chem. Metall.* **1989**, *10*, 24–30.
38. Bale, C.W.; Bélisle, E.; Chartrand, P.; Decterov, S.A.; Eriksson, G.; Gheribi, A.; Hack, K.; Jung, I.-H.; Melançon, J.; Pelton, A.D.; et al. Recent Developments in factsage thermochemical software and databases. In *Celebrating the Megascale*; Springer: Cham, Switzerland, 2014; pp. 141–148. [CrossRef]
39. Egan, J.J.; Bracker, J. Thermodynamic properties of some binary fused chloride mixtures obtained from e.m.f. measuerments. *J. Chem. Thermodyn.* **1974**, *6*, 9–16. [CrossRef]
40. Janz, G.J.; Allen, C.B.; Downey, J.R., Jr.; Tomkins, R.P.T. *Physical Properties Data Compilations Relevant to Energy Storage. 1. Molten Salts: Eutectic Data*; U.S. Government Printing Office: Washington, DC, USA, 1978.
41. Janz, G.J.; Allen, C.B.; Bansal, N.P.; Murphy, R.M.; Tomkins, R.P.T. *Physical Properties Data Compilations Relevant to Energy Storage. 2. Molten Salts: Data on Single and Multi-Component Salt Systems*; U.S. Government Printing Office: Washington, DC, USA, 1981.

processes

MDPI

Review

Recent Advances in the Decontamination and Upgrading of Waste Plastic Pyrolysis Products: An Overview

Salma Belbessai [1,2], Abir Azara [1,2] and Nicolas Abatzoglou [1,*]

1 Department of Chemical & Biotechnological Engineering, Université de Sherbrooke, Sherbrooke, QC J1K 2R1, Canada; salma.belbessai@usherbrooke.ca (S.B.); abir.azara@usherbrooke.ca (A.A.)
2 Laboratoire de Valorisation des Énergies Fossiles, École National Polytechnique, 10 Avenue Hassen Badi El Harrach, P.O. Box 182, Alger 16200, Algeria
* Correspondence: nicolas.abatzoglou@usherbrooke.ca

Abstract: Extensive research on the production of energy and valuable materials from plastic waste using pyrolysis has been widely conducted during recent years. Succeeding in demonstrating the sustainability of this technology economically and technologically at an industrial scale is a great challenge. In most cases, crude pyrolysis products cannot be used directly for several reasons, including the presence of contaminants. This is confirmed by recent studies, using advanced characterization techniques such as two-dimensional gas chromatography. Thus, to overcome these limitations, post-treatment methods, such as dechlorination, distillation, catalytic upgrading and hydroprocessing, are required. Moreover, the integration of pyrolysis units into conventional refineries is only possible if the waste plastic is pre-treated, which involves sorting, washing and dehalogenation. The different studies examined in this review showed that the distillation of plastic pyrolysis oil allows the control of the carbon distribution of different fractions. The hydroprocessing of pyrolytic oil gives promising results in terms of reducing contaminants, such as chlorine, by one order of magnitude. Recent developments in plastic waste and pyrolysis product characterization methods are also reported in this review. The application of pyrolysis for energy generation or added-value material production determines the economic sustainability of the process.

Keywords: plastic pyrolysis; contamination; pre-treatment; products upgrading; pyrolysis applications

Citation: Belbessai, S.; Azara, A.; Abatzoglou, N. Recent Advances in the Decontamination and Upgrading of Waste Plastic Pyrolysis Products: An Overview. *Processes* **2022**, *10*, 733. https://doi.org/10.3390/pr10040733

Academic Editors: Wei-Hsin Chen, Aristotle T. Ubando, Chih-Che Chueh and Liwen Jin

Received: 21 March 2022
Accepted: 1 April 2022
Published: 11 April 2022

1. Introduction

Plastics have become an important part of modern life as they are ubiquitous. The demand for plastics is increasing worldwide; 367 million tons of plastics were produced in 2020 [1]. Plastic waste management is now a major concern in many countries. As an example, Canada uses 4.6 million metric tons of plastic each year, only 9% of which is recycled [2]; the rest ends up in landfills. Unfortunately, 79% of used plastic worldwide goes to landfill sites or the natural environment [3] because mechanical recycling does not tolerate mixed and contaminated plastics. Furthermore, mechanical recycling delays final disposal rather than avoiding it [3]. To this day, polyethylene terephthalate (PET) is considered the only polymer for which an efficient mechanical recycling scheme is established [4]. The degradation of plastic takes many years, creating severe risks to organisms and the environment. For instance, some animals, especially sea animals, mistake plastic for food and die from entanglement [5]. Moreover, plastic exposed to heat can decompose to greenhouse gases [6]. Plastic that ends up in oceans and rivers decomposes, releasing toxic chemical compounds that can be transferred to the human body via contaminated seafood [7].

During the COVID-19 pandemic, single-use plastic-product consumption and release have increased remarkably. For instance, personal protection equipment, such as face shields, isolation gowns, hair and shoe nets and safety glasses, comprise 72% polypropylene (PP) [8]. In addition, studies revealed that if 1% of used masks are disposed of improperly, 40 tons of masks per month are dispersed into the environment [9]. Medical devices are composed of 10% high-temperature plastics, 20% engineered plastics and 70% commodity plastics (polyethylene [PE], polystyrene [PS], polyvinylchloride [PVC] and PP) [9]. Therefore, hospital waste management is a growing concern. Still, plastic waste comes from different sources, such as municipal solid waste (MSW), hospital waste (HW), automobile shredder residue (ASR) and waste electrical and electronic equipment (WEEE), as shown in Figure 1. PE, PP, PVC, PS and PET are the main plastics present in MSW; more plastics with further additives such as acrylonitrile butadiene styrene (ABS) and poly(methyl methacrylate) (PMMA) are present in WEEE and ASR, respectively.

Thermal destruction is one possible solution that has gained attention in recent years. Because of their hydrocarbon-based nature, waste plastics can be transformed into valuable products such as fuels. Specifically, pyrolysis is a thermochemical treatment that is considered a promising alternative for the energetic and material valorization of plastic waste. Pyrolysis converts mixed and contaminated plastics into gas, liquid and solid streams using heat in the absence of oxygen. Since plastic wastes are derived from fossil-fuel sources, their composition is similar to petrochemical fuel, and they have high calorific value; thus, they can be considered a valuable source of energy [10]. A techno-economic study by Al-salam et al. [11] revealed that the calorific value of waste is amongst the most sensitive parameters that affect the economic performance of thermal treatment processes. In addition, the pyrolysis process is a flexible technology because the operating conditions can be optimized to maximize the production of the targeted stream. Compared to gasification, pyrolysis systems produce significantly more olefin products [12]. These unsaturated hydrocarbons can be re-polymerized to produce new recycled plastics, thus closing the loop of a true cyclic economy.

Nevertheless, the production of chemical materials and fuels from heterogeneous mixtures of waste is a significant challenge. The process should consider the composition of the feedstock, which is variable. For example, the heterogeneity of the feedstock was the main reason for the failure of the Rwe-ConTherm plant (Hamm) [13]. The process was affected by corrosion, resulting in the collapse of the chimney in 2009. In addition, the pyrolysis of waste plastic produces a large spectrum of aliphatic and aromatic hydrocarbons with different molecular weights, which in most cases, can not be used without further post-treatment. Moreover, plastic waste contains various contaminants, such as halogens, metals and additives, which can be present in pyrolysis products. These contaminants are responsible for many problems, including corrosion, catalyst poisoning and clogging [14]. These challenges are why the production of high-value products through plastic waste pyrolysis has not taken off industrially.

Consequently, many studies [14–16] emphasized the need to upgrade pyrolysis products to meet the specification standards of current precursors (e.g., steam cracking) or final products (e.g., fuels). Pre-treatment, such as dehalogenation, is also required to produce streams with low contaminant concentrations. Upgrades include the cracking of long-chain molecules, reforming, separation operations and decontamination.

Several reviews have examined a variety of aspects of the pyrolysis of waste plastics. Sharuddin et al. [17] reported the influence of process parameters and different plastics on oil production and quality. They also reviewed the physical and chemical characteristics of pyrolytic oils. Chen et al. [18] reported on technologies for MSW pyrolysis focusing on reactors, products and measures to mitigate the environmental impact. In another study, Miandad et al. [19] reviewed catalytic pyrolysis in terms of pyrolytic oil composition as well as the influence of operating parameters such as temperature and retention time on the pyrolysis process. Kinetic studies, along with the techno-economic evaluation, were discussed by Kunwar et al. [20]. These studies summarized the state-of-the-art approaches to waste plastic pyrolysis, especially in terms of pyrolytic oil production. However, the information on contamination and the need to upgrade pyrolysis products is rarely discussed in the open literature. Another relevant, recent review by Kusenber et al. [14] described the contaminant composition of post-consumer plastic waste pyrolysis oil and its implication for steam cracking. The study found that the contaminant level exceeded established limits by one or more orders of magnitude, concluding that intermediate upgrading steps are necessary to convert waste plastic into valuable chemicals through pyrolysis.

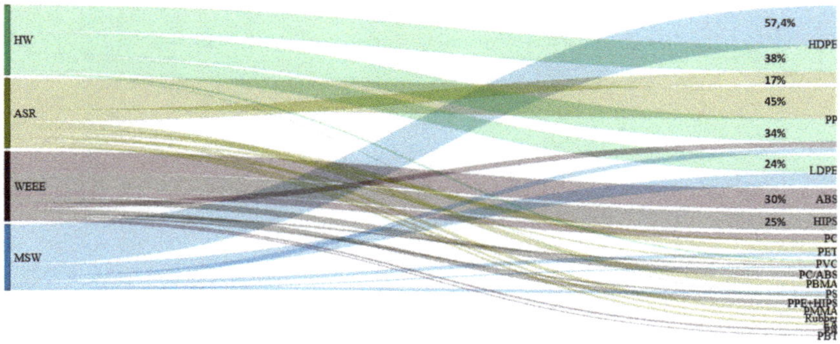

Figure 1. Sankey chart depicting the polymeric composition of different waste streams. Data were gathered from [21–24], and this distribution varies from country to country. HIPS: High impact polystyrene, PC: Polycarbonate, PBMA: Poly butyl methacrylate, EA: Ethyl acrylate, PA: Polyamide, PBT: Butylene terephthalate.

This work reviews the solutions proposed in the literature to upgrade plastic waste pyrolysis products. To the best of our knowledge, the information on these pre- and post-treatment methods has not been discussed thoroughly in previous reviews. Moreover, we discuss the effects of catalysts on pyrolysis products and contaminants. Finally, this paper shows the value of the final application of the gaseous, liquid and solid products. The intended uses of the products are key factors in determining whether the process is economically sustainable in a commercial application.

2. Pyrolysis Products and Contaminants: The Need for Pre- and Post-Treatment

2.1. Thermal Pyrolysis

Thermal pyrolysis is a thermochemical treatment that can be conducted at a wide range of temperatures (350–900 °C). Unlike combustion, which requires excess oxidizing agents or gasification, which occurs under stoichiometric conditions, pyrolysis is conducted in the absence of oxygen. Macromolecular polymers are decomposed into smaller molecules, forming various hydrocarbons, and oxygenates from plastics that contain oxygen (i.e., PET). Pyrolysis can be carried out at different temperature levels: low (<400 °C), medium (400–600 °C) and high (>600 °C). The temperature level and residence time define the type of pyrolysis and the desired products, as illustrated in Table 1.

Table 1. Pyrolysis processes based on operating conditions and targeted products adapted from [8,25].

Process	Heating Rate	Residence Time	Temperature (°C)	Major Products
Slow carbonization	Very low	days	450–600	Charcoal
Slow pyrolysis	<5 °C/s	10–60 min	450–600	Char, oil
Fast pyrolysis	10–200 °C/s	0.5–5 s	550–650	Oil
Flash pyrolysis	1000 °C/s	<1 s	450–900	Oil, gas

Table 2 shows the main results of experimental investigations on the pyrolysis products and their dependence on the plastic waste composition. The products were characterized using advanced analytical techniques, such as two-dimensional gas chromatography [15,26]. The product compositions vary significantly depending on the plastic waste material and type of pyrolysis (e.g., fast or slow). In slow pyrolysis, polyolefins produce large amounts of paraffins and olefins, whereas high concentrations of aromatics and gases are generated during fast pyrolysis.

Table 2. Effect of plastic waste composition on the oil quality and the contaminants present in the oil.

Feedstock	Reactor and Operating Conditions	Product Distribution (wt%)	PIONA (wt%)	Main Contaminants
Post-consumer plastic waste (~88% PP, ~12% PE) [15]	Continuous stirred tank reactor (CSTR) 450 °C Atmospheric pressure Feeding rate: 1 kg/h	Liquid (wax): 87 Gas: 9 Solid: 3	Liquid oil n-Parraffins:3.1 α-Olefins: 6 Diolefins: 19.5 Aromatics: 1 Isoparaffins: 4.7 Iso-olefins: 62.7 Naphthenes: 3 Gas: NR	Fe: 21 ppmw Na: 114 ppmw Pb: 6 ppmw Si: 43 ppmw Cl: 137 ppmw
Post-consumer plastic waste (~46% PP, ~53% PE and ~1% others) [15]	Continuous stirred tank reactor (CSTR) 450 °C Atmospheric pressure Feeding rate: 1 kg/h	Liquid (wax): 89 Gas: 7 Solid: 3	Liquid oil n-Parraffins: 14 α-Olefins: 12.9 Diolefins: 7.6 Aromatics: 13.6 Isoparaffins: 5.8 Iso-olefins: 39 Naphthenes: 7.1 Gas: NR	Ca: 17 ppmw Na: 82 ppmw Pb: 5 ppmw Si: 28 ppmw Cl: 474 ppmw
Post-consumer plastic waste (~1% PP, ~97% PE and ~2% others) [15]	Continuous stirred tank reactor (CSTR) 450 °C Atmospheric pressure Feeding rate: 1 kg/h	Liquid (wax): 85 Gas: 10 Solid: 5	Liquid oil n-Parraffins: 34.4 α-Olefins: 25.5 Diolefins: 4.3 Aromatics: 3.9 Isoparaffins: 6.5 Iso-olefins: 13.8 Naphthenes: 11.6 Gas: NR	Fe: 3 ppmw Na: 82 ppmw Pb: 4 ppmw Si: 47 ppmw O: 2100 ppmw Cl: 143 ppmw
Plastic solid waste (PE, PP, PS and PA) and traces of food residuals [26]	Fast pyrolysis, 430 °C, Atmospheric pressure, Vapour residence time: 1 s.	NR	Liquid oil n-Parraffins: 5 α-Olefins: 12.3 Isoparaffins: 8.2 Aromatics: 67.1 Gas n-Parraffins: 22.3 Isoparaffins: 27.6 Naphthenes: 21.0 Aromatics: 27.1	Nitrogen-containing compounds: 6.4 wt% Sulfur-containing compounds 0.6 wt% Oxygen-containing compounds 2.5 wt%

NR: not reported. PIONA: paraffinic, iso-paraffinic, olefinic, naphthenic and aromatic content.

The type of plastic waste is critical when specific products are targeted. As an example, PS decomposes at low temperatures with a high yield in oil that is rich in styrene [27]. However, PE decomposes to wax at low temperatures and gas and oil at high temperatures [17]. Compared to PE, PP produces more of its monomer. As shown in Table 2, a feedstock rich in PP produces high amounts of iso-olefins and diolefins, whereas a feedstock rich in PE produces liquids containing high concentrations of linear paraffins and olefins. This is explained by the tertiary carbon present in PP, which makes the C–C bond less stable and easy to degrade. PP and PE pyrolysis follow the random scission decomposition mechanism. Therefore, a wide range of molecules, following a Gaussian distribution [10], is produced. The fragmentation of the polymeric chains produces free radicals, which can react in different ways. The most likely reaction is β-scission, producing a new free radical and an unsaturated end. The radical can also capture a hydrogen atom, creating another radical and a saturated end, known as "intermolecular hydrogen transfer". A similar transfer can occur in the same radical, referred to as "intramolecular hydrogen transfer". This mechanism ends when two radicals meet [28]. This reaction mechanism explains the high concentration of paraffins and olefins in PE and PP pyrolysis oils. Some polymers, such as PMMA, decompose into their monomers as the polymeric chain undergoes β-scission to produce methyl methacrylate [29]; the decomposition process is known as unzipping. PS decomposes according to two mechanisms: random scission and unzipping. The most common product is styrene, accompanied by small amounts of its dimer and trimer [28]. The third decomposition mechanism is lateral-group scission, as in the case of PVC. The degradation starts with the removal of HCl from the main polymer chain. The resulting unsaturated chain turns into aromatics, such as benzene, toluene and naphthalene [30]. This divergence in the pyrolysis products requires further upgrading to make them suitable for chemical processing. For instance, when the feedstock is rich in PP, the oil produced is rich in olefins and diolefins. As it is, this oil cannot be valorized as fuel before hydrotreatment. However, if these unsaturated compounds can be removed, they become a suitable feedstock for the petrochemical industry. Table 3 shows the difference in hydrocarbon composition between plastic pyrolysis oil (PPO), vacuum gas oil (VGO), light cycle oil (LCO) and steam cracker feedstock (presented here by naphtha fraction). PPO is very rich in olefins (almost 60 wt%), which explains why it cannot be used directly as fuel or as steam cracker feedstock for monomer recovery.

Table 3. Composition and contaminants present in pyrolytic oil, VGO, LCO and stream cracker feedstock.

Elements	Plastic Pyrolysis Oil (PPO) [15]	Vacuum Gas Oil (VGO) [31]	Light Cycle Oil (LCO) [32]	Steam Cracker Feedstock
		Hydrocarbons (wt%)		
Paraffins	19.8	8.49	22.3	41.7 [1]
Olefins	59.5	-	-	-
Naphthenes	7.1	29.16	15.9	46.2 [1]
Aromatics	13.6	62.34	61.8	12.1 [1]
		Contaminants (wt%)		
S	0.0046	1.17	0.1771	0.5 [2]
N	0.1143	0.23	0.1375	Light feedstock: 0.01 [2], heavy feedstock: 0.2
O	<0.1	NR	NR	0.1 [2]
		Other contaminants (ppm)		
Cl	474	NR	NR	3 [2]
Si	28	NR	NR	1 [2]
Na	82	NR	NR	0.125 [2]

[1] Values of naphtha composition from [33]. [2] Values from [14]. NR: not reported.

Plastic wastes from different sources contain various hazardous substances that end up in the pyrolysis products. The presence of volatile chlorine and sulfur in the feedstock leads to the formation of HCl and H_2S in the gaseous stream and even the liquid products [18]. Kusenberg et al. [15] reported that pyrolysis processes yield a significant reduction in the heteroatom and metal concentration in the resulting liquid phase. The majority of the heteroatoms are found in the gaseous phase, while metals are concentrated in solid carbonaceous products. Table 4 illustrates the elemental composition of solid waste (PP ~46% PP, ~53% PE and ~1% others) and its pyrolytic oil. This indicates that most of these contaminants remain in the solid residue. However, the remaining fraction of contaminants in PPO is still problematic. A recent study by Kusenberg et al. [34] confirmed the necessity of decontaminating pyrolysis products prior to steam cracking. These researchers studied the steam cracking of PPO blended with fossil naphtha. They compared the obtained yields with those of pure naphtha steam cracking. Steam cracking of PPO/naphtha yielded ~23% of ethylene at 820 °C and ~28% at 850 °C, exceeding pure naphtha's yields at both conditions (~22 and ~27%, respectively). Nevertheless, high coke formation and heat exchanger fouling was observed with PPO/naphtha blend. This was attributed to the presence of heteroatoms and metal contaminants in plastic waste.

Table 4. Elemental composition of solid waste and pyrolysis oil [15].

Element	N	S	O	Cl	Al	Ca	Cu	Fe	K	Mg	Mn	Na	Si	Zn
Units		(wt%)							(ppmw)					
Solid waste	0.2	<0.1	0.3	3600	387.5	1599.7	22.3	120	158.1	139.0	0.3	254.5	80.8	33.3
Pyrolysis oil	0.1143	0.0046	<0.1	474	273.5	16.6	2.1	-	36.2	54.1	0.2	82.1	27.6	4.6

According to Table 3, sulfur and nitrogen are not problematic for PPO. Nevertheless, amounts of oxygen, chlorine, iron, sodium and silicon in the pyrolysis oil exceed the threshold values for industrial steam crackers [15]. These contaminants come from different sources: residual paper, biomass and additives. Additionally, Table 2 shows that the concentrations of these contaminants vary with the type of feedstock. More iron and sodium (21 and 114 ppmw, respectively) are found in a PPO of rich PP feedstock, while the highest concentration of oxygen (2100 ppmw) is in a feedstock rich in PE, which indicates PET contamination. This shows that part of these contaminants comes from the polymeric matrix and the contamination from products contained in plastic packaging (e.g., soap or food). Therefore, the current sorting and washing steps do not remove such elements completely. Toraman et al. [26] reported that the oxygenated compounds are in the form of (ketones, phenols, aldehydes and esters, while nitrogen comes in various forms, such as nitriles, pyridines, quinolines, indole and caprolactam, and sulfated compounds are in forms of thiols/sulfides, thiophenes/disulfides, benzothiophenes and dibenzothiophenes.

The origin of these contaminants (O, Cl, Fe, Na, Si) and the problems they cause are discussed in detail in a recently published review [14]. They are known to cause issues such as corrosion, clogging and downstream catalytic poisoning [14]. If the pyrolysis oils were used as fuels, these elements might trigger undesirable reactions and cause gum formation [32].

2.2. Catalytic Pyrolysis: The Effect of Catalyst on Pyrolysis Products and Contaminats

Catalytic pyrolysis has been tested at different scales with various types of plastic streams. The use of a catalyst in pyrolysis decreases the activation energy of the process, thus accelerating the reaction rate. This saves energy as the operating temperature is reduced. In catalytic pyrolysis, the C–C bonds of the polymers are broken on Brønsted acidic sites of the catalyst. Moreover, the catalyst offers better selectivity toward specific products and improves their quality [35]. Catalysts can be in contact with the plastic (in situ catalytic pyrolysis) or in a two-step process (thermal pyrolysis followed by catalytic cracking), also referred to as in-line pyrolysis or ex situ catalytic pyrolysis [36,37]. This last configuration is more advantageous as the temperature of pyrolysis and catalytic upgrading can be controlled independently [38]. In addition, the catalyst is more efficient, and its deactivation is delayed [39] as the poisoning of acid sites by the inorganic contaminants and asphaltenes/heavy waxes is reduced. Most inorganic contaminants contained in the plastic waste are expected to stay in the char inside the pyrolysis reactor, which can be removed occasionally [40]. Both homogeneous (i.e., one liquid phase) and heterogeneous (i.e., solid phase) catalysts have been used in plastic pyrolysis. The most well-known homogeneous catalysts are Lewis acids, such as $AlCl_3$ [41]. However, heterogeneous catalysts are the most commonly used for plastic pyrolysis because the catalyst can be separated from the products and recovered. The most common heterogeneous catalysts are classified as nanocrystalline zeolites; conventional solid acids, such as zeolites; fluid catalytic cracking (FCC) catalysts silica-alumina; mesostructured catalysts, such as MCM-41; and metal supported on basic oxides [17,20,42].

Zeolites are crystalline aluminosilicates, consisting of a sequence of SiO_4 and AlO_4 units; the ratio SiO_2/Al_2O_3 determines the type of zeolite and its reactivity [43]. Zeolites have been widely studied in the catalytic pyrolysis of waste polymers as one of the most effective solid catalysts for the cracking of plastic waste [44]. Generally, the use of zeolites leads to an increased yield of volatiles [45]. Both the SiO_2–Al_2O_3 ratio and the pore size of zeolites have significant influences on pyrolysis products and catalyst deactivation. Elordi et al. [46] reported that HZSM-5, having the smallest pores, was more selective to C_2–C_4 olefins with a yield of 60 wt% (29% propene, 21% butenes and 10% ethane), compared to HY-zeolite and Hβ-zeolite. Coke deposition on HZSM-5 was less than that on the other zeolite catalysts because the growth of coke precursors in zeolites with larger pores gave rise to polyaromatic structures that remain inside the pores, owing to hindered counter-diffusion. Meanwhile, HZSM-5 micropores caused a steric hindrance that limited bimolecular hydrogen transfer. Consequently, HZSM-5 deactivation was not significant when compared to deactivations by Hβ-zeolite and HY-zeolite. This phenomenon was confirmed by similar studies [47,48]. Moreover, Miskolzci et al. [49] indicated that HZSM-5 has the highest activity in double-bond isomerization in municipal plastic waste (MPW) oil, as listed in Table 5. In their study, the HZSM-5 catalyst increased the concentration of internally positioned double bonds from 17.7% to 66.9%. In addition, HZSM-5 showed excellent efficiency in oil deoxygenation and aromatic hydrocarbon formation. Oxygenated products are undesirable in PPO. They increase the oil viscosity and decrease its heating value and stability while rendering the PPO corrosive [50]. When blending PP with PC, the concentration of oxygenates in the presence of HZSM-5 was reduced from 72.3% to 2.9% [51]. By the effects of both Brønsted and Lewis acid sites, the alkenes and alkanes produced from PP trigger aromatization reactions (cyclization, Diels–Alder, dehydrogenation and hydrogen transfer reaction) [52]. These reactions provide hydrogen radicals, which are contacted with oxygenates (phenols, ethers and furans) from PC. In addition, HZSM-5 promotes the direct hydrodeoxygenation of adsorbed phenols by dehydration [51,53]. In the presence of enough light hydrocarbons, HZSM-5 can also promote Diels–Alder reactions of benzofurans into aromatic hydrocarbons [54].

Table 5. Effect of catalyst on pyrolysis products and contaminants. Concentrations are as specified in the respective reference.

Feedstock	Catalyst/Sorbent	Reactor and Operating Conditions	PIONA	Undesired Elements/Compounds in PPO (ppm)	Relevant Remarks
Municipal plastic waste (MPW) Miskolczi et al. [50,55]	No catalyst	Batch reactor, 500 °C, ratio of catalyst to MPW: 1/10	20% paraffins 23% olefins	S: 51 Cl: 618 Ca: 297 Zn: 124 Br: 253 Sb: 105	Presence of 926 ppm of Cl and 520 ppm of Br in the gas
	Y-zeolite		11.5% paraffins 18% olefins	S: 34 Cl: 457 Ca: 282 Zn: 146 Br: 194 Sb: 99	Presence of 1355 ppm of Cl and 594 ppm of Br in the gas
	β-zeolite		4.5% paraffins 9.8% olefins	S: 37 Cl: 399 Ca: 273 Zn: 128 Br: 201 Sb: 114	Presence of 1291 ppm of Cl and 601 ppm of Br in the gas
	FCC		NR	S: 44 Cl: 422 Ca: 291 Zn: 117 Br: 205 Sb: 128	Presence of 1166 ppm of Cl and 552 ppm of Br in the gas

Table 5. *Cont.*

Feedstock	Catalyst/Sorbent	Reactor and Operating Conditions	PIONA	Undesired Elements/Compounds in PPO (ppm)	Relevant Remarks
	MoO_3		22.2% paraffins 25% olefins	S: 42 Cl: 451 Ca: 299 Zn: 140 Br: 185 Sb: 91	Presence of 1352 ppm of Cl and 596 ppm of Br in the gas
	Ni-Mo-catalyst		15% paraffins 26.8% olefins	S: 39 Cl: 416 Ca: 281 Zn: 129 Br: 219 Sb: 113	Presence of 1403 ppm of Cl and 591 ppm of Br in the gas
	HZSM-5		18.5% paraffins 23.7% olefins	S: 42 Cl: 487 Ca: 304 Zn: 132 Br: 266 Sb: 104	Presence of 1210 ppm of Cl and 555 ppm of Br in the gas
	$Al(OH)_3$		10% paraffins 15% olefins	S: 29 Cl: 372 Ca: 295 Zn: 127 Br: 201 Sb: 97	Presence of 594 ppm of Cl and 407 ppm of Br in the gas
PP/PE/PS/PVC/ABS-Br (3/3/2/1/1) Brebu et al. [56]	No catalyst	Single-step fixed-bed reactor, 450 °C	High amounts of aromatics More than 50% of PPO is benzene derivatives (n-C8 n-C10)	Cl: 4972 Br: 1924 N: 1214	Bromine compounds: bromomethane, bromobutane, bromophenol and dibromophenol
	α-FeOOH			Cl: 3370 Br: 170 N: 840	More effective in Br removal
	Fe-C			Cl: 1014 Br: 170 N: 981	Faster degradation and highest amount of oil (67 wt%)
	Ca-C			Cl: 113 Br: 418 N: 1370	More effective in Cl removal
	$CaCO_3$			Cl: 355 Br: 1161 N: 1078	More effective in Cl removal
MPW Lopez-Urionabarrenechea et al. [57]	No catalyst	Semi-batch reactor, 440 °C	NR	Cl in liquid: 0.2% Cl in gas: 5.3% Cl in solid: <0.1%	
	ZSM-5	Conventional catalytic pyrolysis Semi-batch reactor, 440 °C	95.1% aromatics 2.8% olefins	Cl in liquid: 1.2% Cl in gas: 1% Cl in solid: 0.4%	81.5% of C5–C9 compounds
		Stepwise pyrolysis, 300 °C for 60 min then 440 °C	80.6% aromatic 4.8% olefins	Cl in liquid: 0.3% Cl in gas: 3% Cl in solid: 0.4%	74.4% of C5-C9 compounds More >C13 compounds Loss of catalyst activity
		Non-catalytic dechlorination + catalytic pyrolysis	94.2% aromatics 3.3% olefins	Cl in liquid: 0.3% Cl in gas: 2.2% Cl in solid: 0.4%	82.0% of C5–C9 compounds
PC/PP (1/3) Sun et al. [51]	HZSM-5	Two-staged tubular furnace, 500 °C	95.8% aromatics 4.2% oxygenates	Phenols: 4.2% Furans: 0% Ethers: 0%	The aromatics yield reached 98.1% at 700 °C The presence of PP improved the deoxygenation effect of oxygenate compounds

Table 5. *Cont.*

Feedstock	Catalyst/Sorbent	Reactor and Operating Conditions	PIONA	Undesired Elements/Compounds in PPO (ppm)	Relevant Remarks
MPW Miskolczi and Ates [45]	No catalyst	Stirred batch reactor, 500 °C	32.8% paraffins 49.5% olefins 9.7% aromatics 4.0% naphthenes 4.0% oxygenates	Cl: 1285 Br: 1533 P: 498 S: 71 Sb: 189	Oil density (at 20 °C): 0.848 g/cm^3 Oil viscosity at 40 °C: 133 mPa·s
	β-zeolite		31.8% paraffins 47.3% olefins 3.5% aromatics 4.3% naphthenes 3.1% oxygenates	Cl: 1273 Br: 1563 P: 574 S: 51 Sb: 179	Oil density (at 20 °C): 0.814 g/cm^3 Oil viscosity at 40 °C: 113 mPa·s High efficiency in increasing volatile yields
	y-zeolite		32.0% paraffins 49.1% olefins 3.1% aromatics 4.5% naphthenes 3.0% oxygenates	Cl: 1322 Br: 1407 P: 663 S: 57 Sb: 173	Oil density, g/cm^3 (at 20 °C): 0.822 Oil viscosity at 40 °C, mPas: 119
	Ni-Mo-catalysts		31.4% paraffins 49.0% olefins 2.8% aromatics 5.8% naphthenes 2.8% oxygenates	Cl: 1135 Br: 1522 P: 582 S: 65 Sb: 164	Oil density (at 20 °C): 0.828 g/cm^3 Oil viscosity at 40 °C: 126 mPa·s Increases H2 production
MPW + heavy oil (1/3) Miskolczi and Ates [45]	No catalyst	Stirred batch reactor, 500 °C	34.9% paraffins 52.1% olefins 9.1% aromatics 1.8% naphthenes 2.1% oxygenates	Cl: 173 Br: 264 P: 115 S: 16 Sb: 47	Oil density (at 20 °C): 0.832 g/cm^3 Oil viscosity at 40 °C: 216 mPa·s
	β-zeolite		27.2% paraffins 47.4% olefins 9.8% aromatics 4.1% naphthenes 2.0% oxygenates	Cl: 210 Br: 385 P: 117 S: 14 Sb: 43	Oil density, g/cm^3 (at 20 °C): 0.782 Oil viscosity at 40 °C: 168 mPa·s
	y-zeolite		30.1% paraffins 46.7% olefins 11.5% aromatics 2.7% naphthenes 2.3% oxygenates	Cl: 214 Br: 326 P: 94 S: 15 Sb: 37	Oil density (at 20 °C): 0.787 g/cm^3 Oil viscosity at 40 °C: 181 mPa·s
	Ni-Mo-catalysts		34.6% paraffins 46.2% olefins 9.0% aromatics 4.0 naphthenes 2.4% oxygenates	Cl: 195 Br: 279 P: 102 S: 9 Sb: 51	Oil density (at 20 °C): 0.792 g/cm^3 Oil viscosity at 40 °C, mPas: 202

Regarding the contaminants, Table 5 shows that most of the elements (S, Cl, Ca, Zn, Br and Sb) were found in the PPO when the gas phase had only S, Cl and Br contaminants, which was caused by the dehalogenation reactions and the formation of HCl and HBr [58]. This shift in halogens was intensified during catalytic pyrolysis; catalysts decreased the chlorine and bromine content in PPO and increased their respective amounts in the gas phase. The catalytic pyrolysis also reduced the concentration of other contaminants (Ca, Zn, Sb) compared to thermal pyrolysis, although no significant difference was observed among the different catalysts in terms of decontamination efficiency. Owing to its alkalinity, $Al(OH)_3$ was the most efficient in removing acidic contaminants in PPO. Table 5 also shows that the Ca–C composite was more effective in chlorine removal with a 97% reduction. Nonetheless, these results are from different studies with different reactors and operating conditions.

Lopez-Urionabarrenechea's study [57], described in Table 5, recommended the following configuration when the feedstock contains PVC: a low-temperature dechlorination step complemented with alkaline additives to capture HCl, followed by a catalytic step at higher temperatures to avoid the loss of catalyst activity during the dechlorination step. In another study [45], iron-based catalysts (α-FeOOH and Fe-C) tested on pyrolysis of a mix of plastic containing ABS-Br were found to be effective in removing more than 90 wt% of bromine from PPO. Nonetheless, these catalysts have small effects on the removal of organic nitrogen (20–30 wt%). Table 5 also shows that the co-pyrolysis of MPW and heavy oil with a mass ratio of 1/3 could significantly decrease the concentration of contaminants in the resulting PPO by one order of magnitude. These results show the potential of reducing contamination by diluting plastic waste or its derived PPO in petroleum-based feedstock.

Fluid catalytic cracking (FCC) catalysts are composed mostly of Y-zeolite crystals, activated alumina and kaolinite [59]. Their complex compositions make them suitable for a variety of cracking reactions. FCC catalysts are mainly used in the petroleum industry to upgrade the heavy fraction of crude oil into light fractions, such as gasoline. The FCC catalyst that is used in plastic pyrolysis is often a spent catalyst, commonly referred to as an equilibrated FCC catalyst [42]. Fortunately, this catalyst has no cost, and it is a waste material from the petroleum industry. Studies show that an FCC catalyst still has cracking ability despite the contamination from previous usage [60,61]. Due to the reduced acidity, this catalyst generates a much lower coke yield compared to a fresh FCC catalyst. The acidity is lowered because of the poisoning of active sites by metal contaminants [40]. Consequently, this catalyst is not effective in contaminant removal, as shown in Table 5.

The effect of catalytic pyrolysis on the product decomposition and distribution has been extensively studied with different configurations [62,63], reactors [38,40], catalysts [55,56] and operating conditions [64]. Nevertheless, the catalyst effect on contaminants has only been investigated minimally, to the best of our knowledge. Therefore, more investigations are required to study the behaviour of catalysts in real-world plastic waste. The effect of different contaminants on the catalyst performance and the contribution of catalysts in the decontamination process are also areas to explore.

2.3. *Advances in Characterization of Waste Plastic and Pyrolysis Products*

Plastic waste stream consists of a mixture of different polymers containing several sources of contaminates such as paper, food residue and metals. Characterizing waste stream is critical for waste management. Contaminates identification is important for identifying the recycling route for waste plastic. There are several techniques to characterize the plastic waste, including differential scanning calorimetry (DSC), Fourier transform infrared (FTIR) spectroscopy and inductively coupled plasma optical emission spectroscopy (ICP-OES) or mass spectroscopy (ICP-MS).

ICP-OES and ICP-MS are used to determine metal concentrations in the polymeric waste. This method detects ultra-trace (ppb) of metal concentration [65]. Roosen et al. [66] performed the ICP-OES analysis of different plastic packaging waste. They found that the highest concentrations in Fe (270 ppm), Zn (45.6 ppm) and Mg (186 ppm) are attributed to PP and PS packaging trays. For the determination of halogens and sulfur concentrations, combustion ion chromatography (CIC) is used [67,68]. In this method, the sample is firstly pyrolyzed in an oxidizing atmosphere; the resulting vapours are absorbed by an adequate absorbent and then introduced to the IC system for separation and quantification. This method is advantageous because it contains an automated sample preparation for both solids and liquids. The C, H, N, S and O composition of waste plastic is usually detected using an elemental analyzer [15]. Thermogravimetric analysis (TGA) is also used to investigate the thermal behaviour of the plastic waste [68]. Nowadays, the use of coupling techniques such as TG-MS, TG-FTIR [69,70] and TG-FTIR-MS [71] to analyze the degradation of waste plastic through characterization of the resulting products, is getting more popular. In this context, some researchers used Pyro-GC (Pyrolyzer-gas chromatography) to investigate the fast pyrolysis of plastic waste by analyzing the quality of the products [72]. Plastic waste is also analyzed by FTIR to identify functional groups, organic, polymeric and inorganic materials [73].

One of the most used techniques for pyrolysis products analysis is gas chromatography coupled with different detectors such as flame ionization detector (FID) for quantifying hydrocarbons; electron capture detection (ECD) for halogenated hydrocarbons; thermal conductivity detector (TCD) for CO_2, CO, H_2, O_2, Ar, N_2 analysis; MS for functional group, aromatics and double bond analysis.

Recently, a remarkable progress is achieved in identifying pyrolysis oil components due to a powerful technique, which is two-dimensional gas chromatography (GC × GC). This method provides more detailed information on the composition, compared to one-dimensional GC. The GC × GC uses two columns of different stationary phases, connected by a modulation tool. There are two types of modulation: thermal and flux modulator. The effluent passes through both columns, which creates tow retention times for each component. One detector is enough for the analysis, though several detectors can be used to take profit from their advantages [74], as it is illustrated in Table 6. The two columns are of different polarities, when the first one is nonpolar and the second is polar the arrangement is called normal phase (NP), when it is the inverse it is called reverse phase (RP).

Table 6. GC × GC coupled with different detectors, for the characterization of PPO.

2D-GC Technique	Column Arrangement	Columns Used	Molecules Detected	Reference
GC × GC-FID	NP	Two plot columns: PTMSP poly-(1-trimethylsily-1-propyne) GASPRO silica	Saturated and unsaturated hydrocarbons from C3–C8	[75]
GC × GC-FID	NP	RTX-1 PONA (Dimethyl polysiloxane) BPX-50 (50% phenyl polysilphenylene-siloxane)	Diolefins, iso-olefins, mononaphthenes, n-paraffins, iso-paraffins and monoaromatics from diesel fraction	[26,34,76,77]
	RP	Stabilwax (polyethylene glycol) Rxi-5 ms (5% diphenyl 95% dimethyl polysiloxane)	Diolefins, iso-olefins, mononaphthenes, n-paraffins, iso-paraffins and monoaromatics from diesel fraction	[77]
GC × GC-NCD (nitrogen chemiluminescence detector)	NP	RTX-1 PONA (Dimethyl polysiloxane) BPX-50 (50% phenyl polysilphenylene-siloxane)	Nitrogen compounds	[15,26]
GC × GC-SCD (sulfur chemiluminescence detector)	NP	RTX-1 PONA (Dimethyl polysiloxane) BPX-50 (50% phenyl polysilphenylene-siloxane)	Sulfur compounds	[15,26]
GC × GC-ToF-MS (time of flight MS)	RP	RTX-1 PONA (Dimethyl polysiloxane) BPX-50 (50% phenyl polysilphenylene-siloxane)	Oxygenated compounds	[26]

3. Waste Plastic Pre-Treatment

3.1. Plastic Separation

Plastics such as PVC and PET, which are present in MPW, produce dangerous substances during pyrolysis. HCl and chlorinated hydrocarbons, such as chloroform ($CHCl_3$) and dichloromethane (CH_2Cl_2), are formed from PVC [78]. These organic and inorganic chlorides can corrode the pyrolysis equipment, contaminate other products and cause air pollution without appropriate gas-emission control modules [14,78]. Moreover, if these chlorinated hydrocarbons are oxidized (burned), more harmful products, such as dioxins and furans, can result [79]. PET thermal decomposition leads to the formation of carbonic acids, such as benzoic and terephthalic acids, which are problematic to the pyrolysis facility, causing corrosion and clogging in the piping [80] (terephthalic acid is solid at room temperature). In addition, the pyrolysis of PET is less interesting than its mechanical or chemical recycling (e.g., hydrolysis, methanolysis, glycolysis, ammonolysis and aminolysis). Chemical recycling leads to the complete depolymerization of PET [81]. Consequently, the separation of mixed plastics is required before pyrolysis.

The various methods of plastic separation include manual separation, gravity separation by flotation [79,82], centrifugal separation [83], triboelectrostatic separation [84] and selective dissolution [85]. Manual separation is inefficient and labour-intensive; gravity separation is limited by the similar specific gravities of plastics, such as PVC (1.3–1.4) and PET (1.38–1.41) [84]. In triboelectrostatic separation, the tribo-charger imparts the charge on the plastic particles, for example, in a fluidized bed. The particles are charged negatively or positively depending on their work function (i.e., their relative affinity for electrons). Then, the particles are separated in an electrostatic separator where they can be deflected towards the appropriate counter electrode [84]. This method is more effective when the difference in the work functions of plastics is significantly high. The selective dissolution method consists of dissolving plastics in a solvent that targets only one polymer. The dissolved plastic is recovered by the rapid evaporation of the solvent [86] or by the addition of an appropriate "anti-solvent" to make the polymer precipitate [85]. As the solvents are toxic and expensive, this method is rather inconvenient.

For MPW, the pyrolysis is preceded by drying to reduce the moisture content before entering the reactor. The heat required by the dryer mainly comes from the combustion of part of the pyrolysis products [87]. Pre-treatment also includes size reduction by crushing and sieving the plastic, especially when working with fluidized bed reactors (FBRs).

3.2. Dehalogenation

The chlorine content in pyrolysis products is related to the presence of PVC, while bromine comes mainly from brominated flame retardants in ABS and HIPS. WEEE plastics are principally composed of HIPS, ABS, PVC and PC, as illustrated in Figure 1, which makes them rich in halogens. As an example, the pyrolysis of ABS releases different brominated products such as HBr, CH_3Br, C_2H_5Br, C_3H_5Br, C_3H_7Br and C_3H_5BrO [88]. Some researchers also reported the presence of bromophenol and dibromophenol during the pyrolysis of WEEE [89]. In order to obtain fuels or chemical products from WEEE, dehalogenation treatment is required prior, during or after pyrolysis [23]. In the literature, dehalogenation is focused on plastics rich in halogens, such as PVC and flame retardant plastics [90]. However, some researchers [14,15] concluded that the dehalogenation treatment of plastic waste, even polyolefin waste, is necessary for the PPO to meet current specifications set for steam cracker feedstock. A high level of chlorine in PPO can come from the PVC contamination of the plastic waste or from the adsorption of salt that was in the packaged product [91]. Thus, we illustrate some examples of debromination and dechlorination in the following paragraphs.

Cagnetta et al. [92] investigated the dehalogenation of PP containing the flame retardant decabromodiphenyl ethane (DecaBDE) by means of mechanochemical pre-treatment. The debromination of PP was carried out using Fe-SiO_2 or CaO-SiO_2 in a planetary ball mill at room temperature. After eight hours of dry milling, 90% of the bromide was recovered when using Fe-SiO_2, and 80% was the recovery of bromine in the case of CaO-SiO_2. Organic bromine contained in PP was mineralized into soluble inorganic bromide. With the high-energy milling and presence of SiO_2, iron particles become smaller and activated. These fine iron particles with high surface energy become electron donors [23]. The transfer of electrons to the flame retardant occurs according to the following equation [92]:

$$F + C_{12}Br_{10}O \rightarrow C_{12}Br_9O^{\cdot} + Br^- + Fe^+ \quad (1)$$

This reaction continues until the debromination and carbonization of DecaBDE [93]. The polymeric chain also captures the electrons from iron particles, which causes cleavages. This facilitates the next processing step of pyrolysis [92,93].

Grause et al. [94] studied the removal of the same flame retardant from HIPS using NaOH/ethylene glycol solution (NaOH(EG)) in both a stirred flask and a ball mill reactor between 150 and 190 °C. The debromination reached 42% in the stirred flask at 190 °C and about 98% in the ball mill reactor after 24 h. Therefore, ball milling had a more positive effect on the debromination process. Analytical methods showed that debromination was achieved by the substitution of bromine by hydroxyl groups (from NaOH) or hydrogen (from DecaBDE) [95,96]. The reaction is controlled by diffusion in both the stirred flask and ball mill reactor with an activation energy of 205 kJ mol^{-1}. The polymer matrix did not change; it was just cross-linked through the DecaBDE backbone.

As mentioned previously, when the feedstock contains PVC, a dechlorination process is needed to reduce the chlorine content in the pyrolysis products. There are several methods of dechlorination, such as stepwise pyrolysis (i.e., two-step pyrolysis), catalytic pyrolysis or the addition of adsorbents in the feedstock. In stepwise pyrolysis, the plastic is heated at a low temperature to decompose PVC and capture HCl; this step is called dehydrochlorination. In the second stage, the remainder of the plastic is heated to a high temperature. López et al. [97] performed the stepwise pyrolysis of a mixture of plastics containing PVC at different temperature and time conditions, the addition of $CaCO_3$ and the combination of both methods. They reported that 300 °C and 60 min were the optimum conditions in the dechlorination step to reduce the liquid chlorine content by 50 wt%. However, the authors noticed that stepwise pyrolysis led to the formation of heavy hydrocarbons and fewer aromatics. The addition of $CaCO_3$ was efficient in capturing HCl and reducing chlorine content in the gases significantly (to 0.9 wt%). Nevertheless, the concentration of chlorine in the liquid (0.6 wt%) was higher compared to the result of the stepwise pyrolysis (0.2 wt%). The combination of both methods led to lower HCl generation, but the liquid chlorine content was the same as in stepwise pyrolysis. The efficiency of stepwise pyrolysis for dechlorination was also reported by another study where 90% of chlorine was recovered as HCl in the dechlorination step at 350 °C for 60 min [98].

Recent studies investigated the efficiency of hydrothermal treatment for the chlorine removal of waste feedstock. The main advantage of this technology compared to other dechlorination methods is the enhancement of heat and mass transfer due to the homogeneous reaction. The supercritical or subcritical water present in the system works simultaneously as a solvent and a catalyst for acid-catalyzed reactions [99]. Li et al. [100] carried out the hydrothermal treatment of pure PVC in a batch reactor. The highest dechlorination efficiency of 94.3 wt% was obtained at 240 °C with 1% NaOH. Wang et al. [101] studied the effect of hydrothermal dechlorination pre-treatment on oil production through the fast pyrolysis of mixed plastics. Results showed that the dechlorination efficiency reached 99.9 wt%, and the total yield of oil and wax increased by 7.06 wt% after pyrolysis. Furthermore, methane selectivity increased by 17.81%, owing to the possible weakening of the C–C bond energy of the β-position during the hydrothermal pretreatment.

Nishibata et al. [102] investigated the effect of superheated steam with catalysts and adsorbents on the simultaneous dechlorination and degradation of PVC. They have found that the CaO caused more dechlorination and degradation than other metal oxides, including Fe_3O_4, SiO_2, Al_2O, $Ca(OH)_2$ and MgO, in the presence of superheated steam. The temperature is increased by the exothermic reaction of CaO with steam, which promotes PVC degradation. The newly formed HCl reacts with CaO and $Ca(OH)_2$ to form calcium chlorides such as $CaCl_2$ and CaClOH. After degradation in the presence of CaO and steam, 91 wt% of chlorine present in the sample was found in the inorganic phase.

Most industrial applications use inexpensive alkaline additives, such as calcium oxide and sodium carbonate, in the plastic feedstock to remove HCl [103]. They also employ an alkaline solution to wash the gas in a scrubber to remove all acids from the stream [104]. Agilyx [103], which uses stepwise pyrolysis, is the only current technology capable of handling plastic waste containing up to 70% PVC. In the first step, the plastic is heated under a vacuum inside a batch reactor, during which the moisture and HCl are separated from the feedstock [105]. Additionally, BASF in Germany pursues a two-step technology [42]. The dehydrochlorination step is carried out at 250–380 °C. The system handles feedstock with PVC content lower than 5%. The HCl produced is recycled for PVC production.

4. Products Upgrading

4.1. Dittillation of Pyrolytic Oil

The oil obtained from the fast pyrolysis of waste plastic is usually a dark liquid composed of various hydrocarbon compounds from C_5 to C_{30}. Although fractional distillation is frequently used in the petroleum industry, information on pyrolytic oil distillation is scarce. Some researchers used distillation to split pyrolytic oil into gasoline, diesel and heavy oil fractions for fuel recovery. Others performed distillation to recover monomers and close the loop toward new virgin plastic. For instance, Baena-González et al. [106] carried out the distillation of PPO at atmospheric pressure up to 240 °C. This operation led to a bitumen at the bottom of the column and a distilled fraction. The resulting bitumen contained aromatics (55.05 wt%) and saturates (33.41 wt%). The detailed composition of the bitumen indicated its potential to be added in asphalt or bituminous mixtures. The distilled fraction was also rich in aromatics (54.72 wt%), with styrene as the principal compound, followed by ethylbenzene and toluene. These results indicate that the feedstock contained high amounts of PS and that the pyrolysis conditions favoured aromatization reactions. The distilled fraction was subjected to a liquid–liquid extraction with sulfolane to separate aromatic compounds from other components. Another fractional distillation was carried out to separate the different aromatic compounds and recover styrene (73.26 wt%). This study demonstrated the technical feasibility of producing different materials including bitumen, olefins, toluene and styrene from the fractionalization of PPO.

Thahir et al. [107] studied the pyrolysis of waste PP in a pyrolysis reactor integrated with a distillation bubble cap plate column (Figure 2) to optimize liquid products. Experiments were conducted using 500 g of plastic waste. Vapours produced from pyrolysis of waste plastic flow through the column. Ash residue and wax stays in the reactor, whereas non-condensed vapour flows through the riser to reach the cap and eventually, forms liquid bubble (mixture of vapour and condensate). The pyrolysis temperature affected the liquid fuel characterization yielded on each tray of the column, as described in Table 7. The total liquid oil yield at 500–560 °C reached 88 wt% with the highest yield of gasoline (67 wt%). However, at 650 °C, the diesel yield reached 83 wt%. This study shows the possibility of tuning the pyrolysis temperature to optimize the desired fuel. The physicochemical characteristics of these fuels, such as density, viscosity, octane-cetane number, ash content and calorific value, are similar to those of conventional fossil fuels. However, the chemical composition was not reported.

Table 7. Distribution of fuel products along the column [107].

Temperature (°C)	Plate I	Plate II	Plate III	Plate IV
500–560	Gasoline	Gasoline	-	-
580–600	Kerosene	Gasoline	Gasoline	-
620–650	Diesel-wax	Kerosene	Gasoline	Gasoline

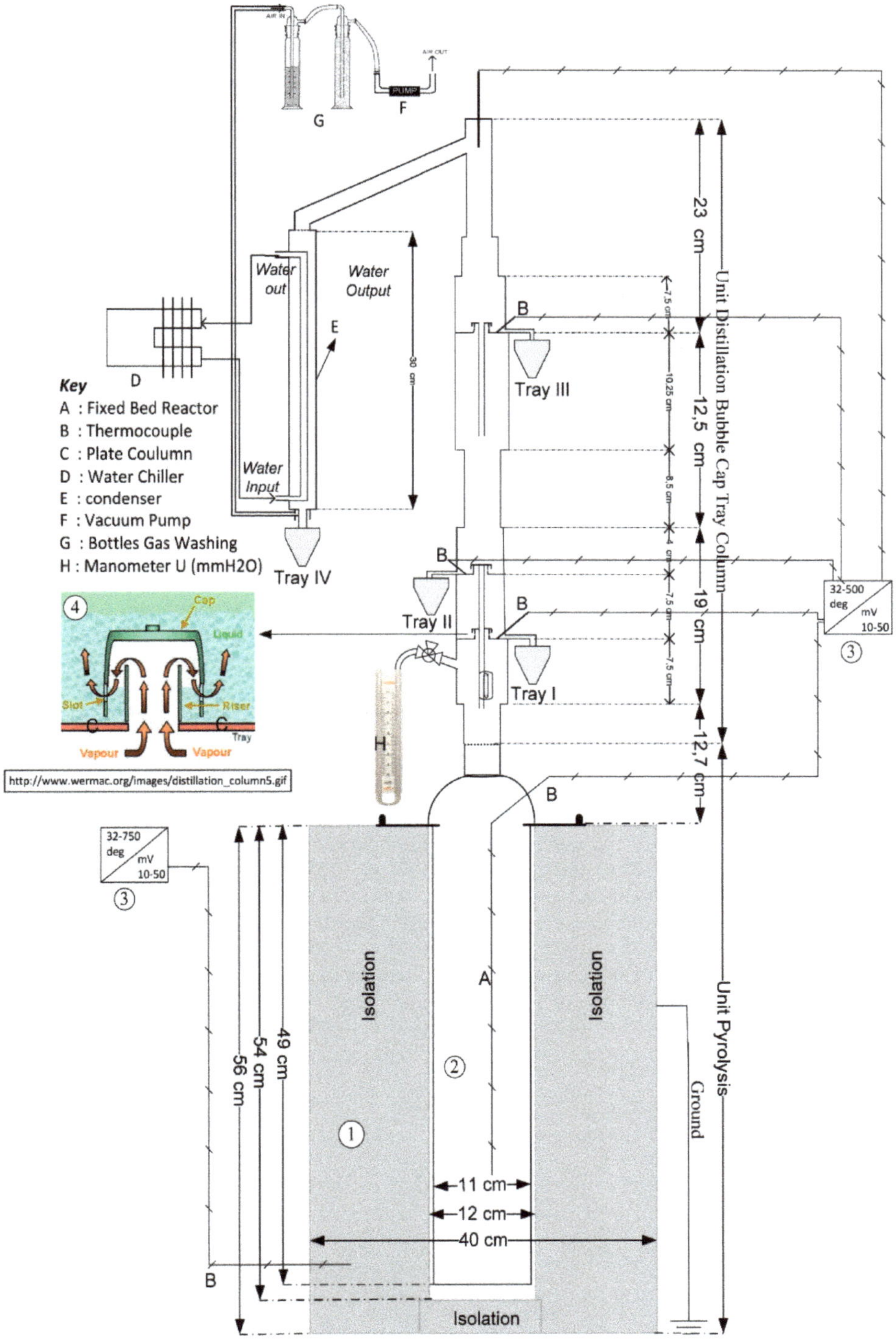

Figure 2. Schematic diagram of thermal pyrolysis integrated with a distillation bubble cap column, reproduced with permission [107].

Another study that investigated the distillation of PPO was carried out by Lee et al. [108]. The PPO came from a Korean pyrolysis kiln facility that treats ten tons/day of mixed plastic waste at approximately 450 °C. The objective was to collect pyrolysis oil fractions similar to petroleum diesel based on carbon number. First, atmospheric distillation was performed to recover the specific fractions following the boiling points of different petroleum fuels (~169 °C for gasoline, 138–278 °C for kerosene and 138–399 °C for diesel). Then, vacuum distillation was conducted to reduce the heat duty. At 100 °C lower than that for atmospheric distillation, similar carbon fractions were obtained in distillation at vacuum conditions. The main yields of the different fractions are gathered in Figure 3.

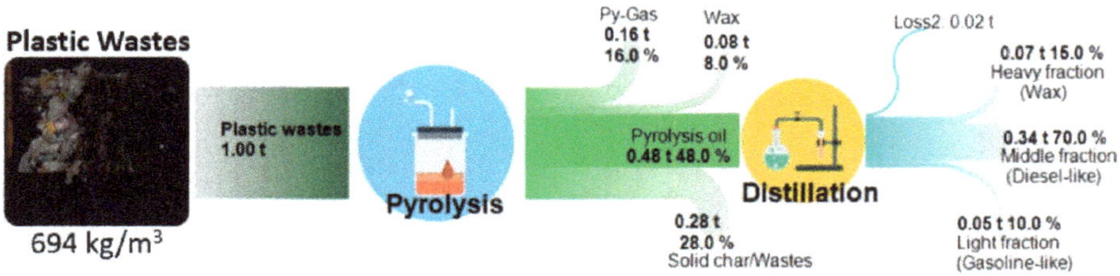

Figure 3. Plastic waste pyrolysis process mass balance from Lee et al., reproduced with permission [108].

Most commercial pyrolysis plants continuously fractionate the liquid product to control the carbon distribution of the different fractions [42].

Dao Thi et al. [77] performed a detailed group-type characterization of both naphtha (C_5–C_{11}) and diesel fractions (C_7–C_{23}) originating from the distillation of PPO by means of two-dimensional gas chromatography. Table 8 shows that both fractions were rich in olefins and diolefins, which indicated that further processing would be required, such as hydroprocessing, because, as mentioned before, high contents of unsaturated compounds negatively affect the quality of the fuels, owing to the gum formation through secondary reactions [109]. The original PPO contained high amounts of aromatics (67.1 wt%), while the naphtha and diesel fractions had an aromatic content of 9 and 2 wt%, respectively. Therefore, the aromatics present in PPO had a high carbon number, and they remained at the bottom of the distillation column. The presence of heteroatoms (S, N, O) in both fractions was reduced compared to PPO. The fractionalization led to low concentrations of heteroatom-containing compounds in both light and heavy fractions.

Table 8. PIONA and elemental composition of naphtha, diesel and PPO determined by comprehensive two-dimensional gas chromatography analysis [77].

Elements	Naphtha	Diesel	PPO [a]
		PIONA (wt%)	
Paraffins	15	28	5
Isoparaffins	2	4	8.2
α-olefins	35	36	12.3
Iso-olefins	9	9	-
Diolefins	4	4	-
Naphthenes	26	17	-
Aromatics	9	2	67.1
	Elemental composition (wt%)		
C	85.93	85.51	88
H	13.93	14.49	10.9
S	0.021	0.001	0.17
N	0.003	ND	1.06
O	0.14	0.01	0.35

ND: not detected. [a] the composition of PPO is taken from a previous study of the same research group [26].

4.2. Pyrolysis Wax Treatment in FCC Units

The cracking of polyolefinic pyrolysis waxes in an FCC unit has been studied extensively [110]. This cracking is adopted for recovering raw materials and obtaining fuels. Studies have shown that the cracking of waxes leads to higher yields of gasoline compared to the cracking of VGO [111]. Rodríguez et al. [31] investigated the FCC of HDPE pyrolysis waxes in a riser simulator reactor under industrial conditions in order to produce fuels from waxes coming from a pyrolysis plant. The waxes were obtained during a fast pyrolysis of HDPE at 500 °C in a conical spouted bed reactor. The reaction of FCC was carried out at temperatures from 500–560 °C, catalyst/oil mass ratios of C/O = 3–7 and a residence time of 6 s, which are typical values used in the industry. Conversion values of HDPE waxes varied from 36.7–5.1 wt%, increasing when the temperature and catalyst–oil ratio increased. The yield, which grouped following the distribution used in refineries, at 530 °C and C/O = 5 was the following: dry gas (4 wt%), liquefied petroleum gas (LPG) (14 wt%), naphtha (28 wt%), light cracked products (LCO) (43 wt%), heavier cracked products (HCO) (7 wt%), coke (4 wt%). Olefins were the most abundant hydrocarbons in the naphtha fraction, followed by aromatics, isoparaffins, n-paraffins and naphthenes. The temperature and C/O had substantial effects on the product distribution. High cracking temperatures increased the paraffinic fraction and reduced the aromatics.

Some authors discussed the possibility of integrating pyrolysis plants with refineries [42,112]. Liquid wax derived from the pyrolysis of waste plastic can be fed, along with oil products, into steam reforming, hydroprocessing, FCC and coking processes for fuel production, as illustrated in Figure 4. Monomers and light hydrocarbons can be directed to petrochemical plants for the production of new polymer resins [113]. This recycling configuration allows the valorization of all kinds of plastic waste and their pyrolysis products, while minimizing the landfilled fraction.

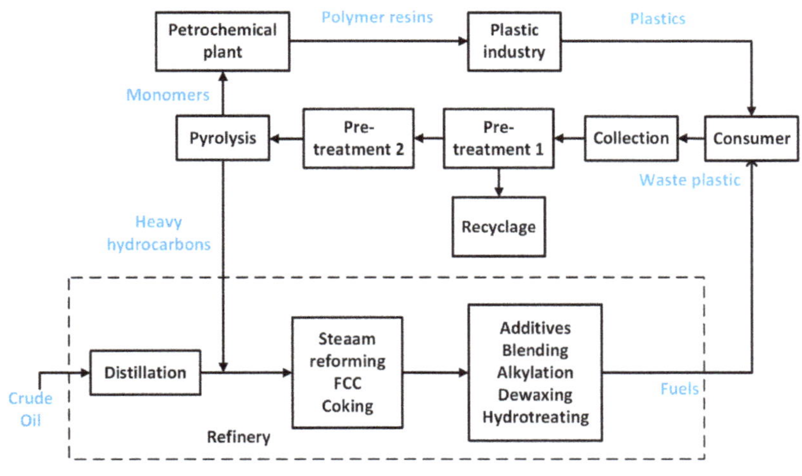

Figure 4. Integration of a plastic waste pyrolysis plant with an oil refinery.

In this context, Rodríguez et al. [114] complemented their study of HDPE pyrolysis waxes in an FCC unit, but this time they used a blend of HDPE waxes and VGO (1:4 mass ratio). The reaction was carried out in a laboratory-scale reactor mimicking the behaviour of an industrial FCC reactor. The results showed that the conversion values of the blend at 500 and 530 °C (40.6–47.6 and 49.3–55.5 wt%, respectively) were slightly lower than those of pure VGO (41.4–47.3 and 51.1–55.5 wt%, respectively). Nevertheless, at 560 °C, the blend showed a higher conversion (63.1–66.3 wt%) compared to the VGO (61.1–62.7 wt%) because the cracking of the waxes was promoted at high temperatures. The yields of naphtha and LPG increased with the blending, whereas that of dry gas decreased.

4.3. Catalytic Upgrading of Pyrolysis Liquids

The use of a catalyst can improve the pyrolysis liquids by breaking the long hydrocarbon chains and increasing the selectivity of the desired products. A catalyst can be used in the pyrolysis process, as explained in Section 2.2, or as a post-treatment for upgrading the liquid phase. Lee et al. [115] studied the effects of zeolites on catalytic upgrading of pyrolysis wax oil. This oil was obtained from the pyrolysis of MPW in a commercial rotary kiln reactor. The catalytic experiments of wax upgrading were conducted in a continuous fixed-bed reactor at 450 °C using three commercial zeolites: HZSM-5, HY-zeolite and modernite (HM). The HZSM-5 zeolite gave the highest gas yield (51.04 wt%) compared to the other zeolites, with a selectivity toward aromatic and cyclic components. HY showed medium catalytic activity with high paraffinic content, and the carbon number of these was between 5 and 6. The HM catalyst, having a one-dimensional pore structure, showed the lowest catalytic activity.

Furthermore, Wang et al. [116] designed a practical laboratory pyrolysis oil catalytic separator, which is a combination of distillation and catalytic cracking (Figure 5). The oil was from an MPW pyrolysis company, and the catalysts used were zeolite 4A and Cu-(MDC-7) and Ni-based catalysts. The temperature was kept between 320 and 380 °C, and the products were separated into three categories: F1 (gasoline-like fraction), F2 (diesel-like fraction) and F3 (wax). The results showed that the presence of catalysts decreased the mass yield of F2, owing to the loss of some gases (e.g., CO, CO_2, CH_4) through decarboxylation, decarbonylation and dehydration reactions. Compared to other catalysts, MDC-7 generated the highest mass yield of F1 (15.8 wt%), whereas the highest yield of F2 (66.3 wt%) was produced with Ni-based catalyst. The use of catalysts reduced the heavy carbon range ($>C_{23}$) from 22.1 wt% in the original oil to 0.1–1.6 wt% in F1 and 7.3–8.4 wt% in F2. Moreover, F1 and F2 fractions from catalytic separation had lower total acid number (TAN) values compared to those of thermal separation, suggesting that more deoxygenation reactions took place in the presence of catalysts. The order of deoxygenation capacity was Ni-based catalyst followed by MDC-7 and zeolite 4A. In terms of composition, MDC-7 exhibited high aromatic and naphthenic contents, while the Ni-based catalyst showed the highest content of olefins in F1 (54.49%) and F2 (36.16%). The authors suggested that the catalytic reaction mechanisms of both catalysts were as follows:

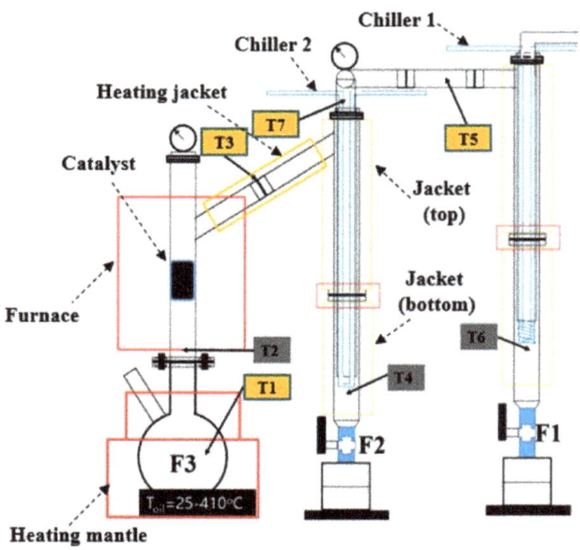

Figure 5. Schematic diagram of a laboratory pyrolysis oil separation system, reproduced with permission from [116].

For the MDC-7 catalyst, long-chain paraffin, olefins, alcohols and ester were converted into short-chain olefins and paraffins through catalytic cracking, decarboxylation, decarbonylation and dehydration. Then, aromatics and naphthenes were formed via aromatization and cyclization, respectively.

- With the Ni-based catalyst, long-chain alcohols were transferred into olefins via dehydration and catalytic cracking, while benzoic acid and phenols were transferred into aromatics through deoxygenation reactions. Some long-chain paraffins were cracked into short-chain paraffins.

4.4. Pyrolysis Oil Hydroprocessing

As previously mentioned, the composition of the pyrolytic liquid can vary depending on the feedstock and may contain undesirable compounds. Hydrotreating the liquids can help reduce the olefins and aromatics and remove heteroatoms (N, S, Cl and O). Hydrodenitrogenation occurs according to the following reaction [117]:

$$C_5H_5N + 5H_2 \rightarrow C_5H_{12} + NH_3 \quad (2)$$

The hydrodechlorination of chlorobenzene as a prevalent chlorinated hydrocarbon is shown in the reaction below [14]:

$$C_6H_5Cl + H_2 \leftrightarrow C_6H_6 + HCl \quad (3)$$

Hydrodeoxygenation of an oxygenated compound is schematically presented below [117]:

$$R - OH + H_2 \rightarrow R - H + H_2O \quad (4)$$

This operation is conducted in the presence of a hydrotreating catalyst at temperatures ranging from 190–340 °C and pressures of 20–204 atm [104]. These conditions help achieve the removal of heteroatoms while the cracking is minimized. Ding et al. [118] used bifunctional (acidic/metallic functions) catalysts (NiO/HBeta, NiO/HSAPO-11 and NiO/HMCM-41) for the hydrocracking of waxes obtained from the pyrolysis of polyolefins at 300 °C in a stirred autoclave reactor, under 20 atm of hydrogen. With a mixture of Ni/H-Beta and ZSM-5, the hydrocracking led to higher fractions of gases (30.2 wt%) and diesel (23.5 wt%). A comparison of the catalysts showed that the ratio of acid-to-metal function sites affects the mechanism of hydrocracking and hydroisomerization of waxes. When the acid strength was high and the hydrogenation power of the catalysts was weak, more isoparaffins and lighter hydrocarbons were produced, which decreased the oil pour point. On the contrary, when the catalyst had a higher metal function, the hydrogenation of olefins was the predominant reaction, which lowered the production of isoparaffins. Therefore, the authors suggested a two-stage reactor system with the use of both catalysts to have a medium composition.

Moreover, hydrocracking allows us to tailor the selectivity toward the desired fuel by adjusting the temperature, as shown in Table 9. Higher temperatures favour the end-chain cracking; thus, more light hydrocarbons are produced. The PPO nature is also an important factor; for example, when PP pyrolysis oil, which is rich in olefins, goes through complete hydrogenation, large amounts of saturated hydrocarbons are produced [119]. A comparison of the physiochemical properties of this oil, the hydrogenated oil and diesel is outlined in Table 10. Hydrogenation enhanced the density, viscosity, cetane index, flash point, fire point and pour point. The properties of the hydrogenated oil matched the EN590 standards. This hydrogenated PP oil was blended with diesel, and promising results were obtained during engine performance trials. This application will be presented in more detail in Section 5.1.

Table 9. Main results from hydrotreatment of different pyrolysis oil.

Feedstock	Reactor Type	Pressure (atm)	Catalyst	Temperature (°C)	Main Results
LDPE pyrolysis oil (~47.7 wt% gasoline (C5–C12), ~36.2 wt% light diesel (C13–C18), ~16.1 wt% heavy diesel (C_{19}–C_{40}) [120]	Stirred autoclave reactor	20	Pd/h-ZSM-5	250	Reduction of gasoline fraction through oligomerization and increasing the share of light and heavy diesel up to 41.8 wt% and 20.3 wt%, respectively
				310	Light diesel decreases to 26.2 wt% High production of isoparaffins (34.5 wt%) through hydroisomerization
				350	Cracking is dominant, producing 11.5 wt% of gases and 56.6 wt% of gasoline High production of aromatics (24.2 wt%)
HDPE pyrolysis oil (26.5 wt% naphtha, 33.1 wt% LCO, 40.4 wt% HCO) [121]	Stirred tank reactor (STR) in semi-batch regime	80	NiW/HY	400	Product distribution: LCO (~28 wt%), naphtha (~29 wt%) and gas (~10.4 wt%)
				420	Product distribution: LCO (~23.3 wt%), naphtha (~35 wt%) and gas (~30.3 wt%)
				440	Product distribution: LCO (~14.3 wt%), naphtha (~30.8 wt%) and gas (45.9 wt%) Naphta rich in isoparaffins and one-ring aromatics
PP pyrolysis oil (67 wt% alkanes, 20 wt% alkenes and traces of aromatics) [119]	Autoclave reactor	70	Ni/ZSM-5	350	Complete conversion of alkenes to alkanes, hydrogenated PP oil contained 97% alkanes. Alkanes distribution: 8.3 wt% (C1–C10), 63 wt% (C10–C20), 25 wt% (C20–C30)
LDPE pyrolysis oil (48 wt% gasoline, 35 wt% diesel and 15 wt% heavy diesel) [122]	Stirred autoclave reactor	20	Ni/h-ZSM-5, Ni/h-Beta, Ni/Al-MCM-41, Ni/Al-SBA-15	310	Complete hydrogenation of alkenes for all catalysts except Ni/h-ZSM-5, due to its high cracking activity.
LDPE pyrolysis products [123]	Stirred autoclave reactor	5–40	Ni/h-β	250–350	Higher temperatures promote aromatization reactions Higher pressures promote hydrogenation of olefins and Saturation of more than 80% of olefins
Polyolefins pyrolysis oil [124]	Stirred autoclave reactor	20	Ni/h-β	310	Saturation of more than 90% of olefins Amount of gasoline + light diesel was within 80–85%

Regarding heteroatom removal, Miller et al. [125] reported that with 1 wt% HZSM-5, the hydroprocessing reduced chlorine content from 50–70 ppm to 2–8 ppm. Similarly, Lingaiah et al. [126] studied the dehydrochlorination of MPW-derived oil using different catalysts: iron oxide, iron oxide-carbon composite, ZnO, MgO and red mud. The original oil contained almost 600 ppm of chlorine. After hydrotreatment, the concentration of chlorine was reduced to 32–140 ppm, with the iron oxide catalysts being the most effective and stable. However, a study on catalytic poisoning in the presence of different halogenic and metallic contaminants is required. Metal removal techniques such as membrane filtration [127] may be needed prior to hydroprocessing.

Table 10. Physicochemical properties of PPO, hydrogenated PPO and diesel [119].

Parameter	PPO	Hydrogenated PPO	Diesel
Density (kg/m^3)	771.4	851.5	837.5
Pour point (°C)	−30	−20	−15
Flash point (°C)	20	65	72
Fire point (°C)	30	72	82
Calculated cetane index (N/A)	60	62	52
Kinematic viscosity at 40 °C (mm^2/s)	1.78	3.5	2.31
Gross calorific value (KJ/kg)	44,957	44,915	45,593
Ash content (%)	0.01	0.01	0.01
Conradson carbon residue (%)	0.10	0.10	0.18

4.5. Fuel Properties Enhancement

If the pyrolysis liquids are saturated with paraffinic compounds, a dewaxing step is required. The presence of long-chain hydrocarbons in the fuel leads to high cloud and pour points. Acidic catalysts such as zeolites have been used for catalytic dewaxing. These catalysts have big pores, which can selectively isolate the long straight n-paraffins and crack them [104]. Dewaxing reactions are usually performed in a semi-batch system at around 450 °C and 4 atm [128].

Pyrolytic liquids are thermodynamically unstable and tend to go through polymerization and oxidation, which is mainly caused by the presence of unsaturated components. This process can lead to the formation of sediments, gums, dark colours and asphaltene agglomeration, affecting the combustion performance of the fuel. Several additives can be added to the fuel derived from waste plastic to overcome this problem and meet the required standards. Amine-based antioxidants are commonly used to prevent diesel oxidation and radical polymerization reactions [42]. The chemical compound 4-tert-butylcatechol is also used as a polymerization inhibitor in pyrolytic oil [106]. Detergents and dispersants, such as alkylphenols, are other additives that can keep oil-insoluble fractions suspended and prevent agglomeration [42].

4.6. Char Upgrading

Char is a by-product of the plastic pyrolysis process. It is a porous carbon material composed mainly of volatile matter and fixed carbon, but it can also contain mineral matter initially present in the feedstock [17]. High temperatures promote the formation of char [129]. During the pyrolysis of contaminated plastics, most of the contaminants stay in the char, as discussed in Section 2.1. Consequently, char cannot be used as raw material, and an upgrading process is necessary. This step can improve the process efficiency and sustainability and avoid the addition of char to landfill waste.

Bernardo et al. [130] tested the effect of the dichloromethane (DCM) extraction of char residue produced during the co-pyrolysis of a waste mixture composed of plastics (i.e., PE and PS), pine biomass and used tires. The analysis showed that DCM extraction removed organic contaminants of high to medium volatility, such as benzene, toluene, ethyl benzene, xylene (BTEX) and alkyl phenol compounds, effectively. In contrast, non-volatile organics and heavy metals (Cd, Pb, Zn, Cu, Hg and As) remained in the char and were not extracted. Subsequently, the treated char was classified as a hazardous and ecotoxic material. In a following study, Bernardo et al. [131] carried out a treatment to upgrade crude chars produced from the co-pyrolysis of different mixtures of plastics, biomass and tire waste. The chars were firstly treated by sequential organic solvent extractions with organic solvents (i.e., hexane, hexane acetone and acetone); then, they were subjected to an acidic demineralization with HCl. The results showed that the solvent extraction treatment allowed the recovery of 63–81% of the pyrolytic oil trapped in the crude char. The demineralization procedure was efficient in the removal of 64–86% of inorganic contaminants (Al, Mg, Pb, Cr, Na, Fe, K, Mn, Mo, Ca). The resulting chars were mainly mesoporous and macroporous materials with adsorption capacities of 3.59–22.2 mg/g for methylene blue dye.

5. Applications

Product applications indicate whether the process is profitable on an industrial scale. One pyrolysis unit can have different applications to profit from all the products and improve the process efficiency. The unit also improves the environmental sustainability of the process by minimizing the waste that would have required disposal (char) and the use of the gases to generate energy.

5.1. Liquid Oil Applications

Liquid oil from pyrolysis has exhibited great potential as a new energy resource. The experimental calorific value of polyolefin-derived oil is higher than 40 MJ/kg [132], which is considered more than enough for energy utilization. The physical properties of this oil are also similar to those of commercial diesel and gasoline. However, the crude oil needs several treatments before it can be used as fuel. When liquid oil is the desired product, the optimum pyrolysis temperature ranges from 500–550 °C [133,134]. If a catalyst is used, this temperature range is lower [17]. The use of a suitable catalyst may improve the oil yield and its quality, except for PS, which yields a high liquid fraction without catalysts [135]. A recent study [106], as mentioned in Section 4.4, demonstrated that the synthesis of PS as possible using styrene recovered from PPO with a yield of 77.64% and a molecular weight of more than 53,000 g/mol. Therefore, separating PS from other plastics is recommended to recover styrene instead of extracting it from the pyrolysis oil. In this context, Zayoud et al. [76] studied the pyrolysis of used PS in a pilot-scale reactor at different pressures (0.02, 0.5 and 1.0 bara) and temperatures (450, 500, 550 and 600 °C). The objective of the study was to maximize styrene production. Authors found that 0.02 bara and 550 °C are the optimum conditions for the production of styrene with 55.9% yield. The other compounds of the liquid pyrolyzate consist of poly-aromatics that contain styrene dimers and trimers.

Some research has been conducted on the use of PPO in diesel engines. However, a comparison of the literature results is difficult, as oils derived from different plastics do not have the same composition. In most studies, blends of PPO and conventional diesel were used to avoid modifying the engine. Mangesh et al. [136] performed an experimental investigation to identify the type of plastic that gives the most suitable oil for diesel engine fuel. HDPE, LDPE, PP and PS were pyrolyzed separately, and the oil produced for each type of plastic was analyzed and compared with diesel. PP oil was selected because its physicochemical properties (e.g., density, viscosity, cetane index) most closely matched those of diesel. Engine tests were conducted on TurbochargedEicher E483 using various blends of PP pyrolysis oil (5, 10, 15%). The blends showed an ignition delay and a decrease in engine efficiency. Moreover, CO, NOx and HC emissions were significantly higher than pure diesel. A similar study was conducted by Singh et al. [137] using different ratios of non-treated mixed PPO (10, 20, 30, 40 and 50%). The results showed comparable engine efficiencies with that of diesel fuel. However, the authors also reported that the use of different blend ratios of PPO increased the exhaust emissions, owing to the presence of oxygenated compounds.

These studies reveal that the physicochemical properties of the PPO are not the only criteria for the oil to be used in the diesel engine. In the first study, PP pyrolysis oil was rich in alkenes, which increased the combustion delay and lowered the engine efficiency. The PPO used in the second study, showed better results in terms of engine efficiency, resulted from the pyrolysis of a mixture of real waste (HDPE, LDPE, PP, PS, PET and thermoset plastics). This oil was more varied in its composition in that it contained alkenes, alkanes, aromatics and 17.54% oxygenated compounds. To enhance their results, Mangesh et al. [119] performed catalytic hydrogenation on the PP oil. Details of this hydrogenation are described in Table 8. Hydrogenated PP oil was blended with diesel in ratios of 10, 20, 30 and 40 wt%. Blends of 10 and 20 wt% showed combustion, exhaust emission and engine performance on par with pure diesel. The higher blend ratios (30 and 40 wt%) decreased the efficiency of the engine slightly and increased the CO, CO2, NOx and unburned hydrocarbon (UHC) emissions. The hydrocracking of PP pyrolysis oil yielded an oil rich in alkanes and lower in carbon number, which improved the combustion results. Nevertheless, the information regarding the economic viability of the process is lacking.

5.2. Solid Products

5.2.1. Carbon Nanotubes

Different studies have been conducted to explore the possibilities of using pyrolysis products in different applications as materials rather than energy sources. One such application is the production of nanocarbons, such as carbon nanotubes (CNTs) and nanofilaments (CNFs), with the potential for hydrogen production [138]. These materials are higher-value products that could render the pyrolysis process more efficient and techno-economically and socio-politically sustainable. This technology mixes the appropriate catalyst with the plastic waste in one reactor, or a two-stage reactor system, where the hydrocarbons produced in the first reactor interact with the catalyst in the second reactor. Of these options, the two-step approach is recommended, which allows the regeneration of the catalyst. CNTs are produced when the gases coming from the pyrolysis of waste plastic interact with a catalyst at temperatures between 600 and 1200 °C in a chemical vapour deposition (CVD) process [139]. In this process, the carbon contained in the hydrocarbons precipitates as graphitic nanofilaments at the surface of the catalyst.

CNTs are used to reinforce polymer composites because of their mechanical and electronic properties [140]. They are valuable in many applications where electrical conductivity is critical, owing to sp^2 hybridization in the carbon structure [141]. To maximize the production of CNTs, the degradation of waste polymers should be promoted into light hydrocarbons and aromatics, which are efficient precursors [142]. Azara et al. [143] comprehensively described the synthesis of filamentous carbon nanomaterial via the catalytic conversion of waste plastic pyrolysis products.

Ni-based catalysts are known to have good activity for C–C and C–H cleavage, and so they are widely used for catalytic reforming to produce CNTs. Zhang et al. [138] tested the production of CNTs from waste tires using different catalysts: Co/Al_2O_3, Cu/Al_2O_3, Fe/Al_2O_3 and Ni/Al_2O_3. The results indicated that Ni/Al_2O_3 had the highest performance for the production of multi-walled CNTs, along with a high H_2 yield. Some studies suggested using a bimetallic catalyst to gain the synergic effect of the interaction between two metals. Yao et al. [144] studied the effect of a Ni-Fe/Al_2O_3 catalyst on the production of CNTs and H_2 from waste plastic pyrolysis. The highest H_2 yield of 8.47 $g_{H2}/g_{plastic}$ and the highest yield of carbon were obtained at a high loading of Fe. In contrast, at a high Ni loading, the CNTs had narrow diameters and uniform distribution.

In another study, Yao et al. [145] investigated the synthesis of multi-walled CNTs from waste plastics, using a combination of two metals, Fe and Ni, supported on four silica-alumina materials: ZSM5, MCM41, NKF5 and H-Beta. Ni-Fe/MCM41, with the largest surface area and pore size, produced the highest carbon (55.6 wt%) and H_2 (38.1 $mmol_{H2}/g_{plastic}$) yields. The Raman spectroscopy analysis showed that the CNTs produced from Ni-Fe/MCM41 had a more graphitic nature and fewer defects than other catalysts. Hence, the formation of Fe-Ni alloys catalyzed the growth of CNTs.

This technology has yet to be scaled up because of the challenges it faces [146]. The yield and quality of CNTs depend on several parameters, such as the type of catalyst, the reforming temperature and the shapes of the metallic particles. Moreover, some of these parameters and process variables affect the production of CNTs in an interdependent way. The heterogeneity of the feedstock and the presence of contaminants also make the formation mechanism of CNTs hard to determine. Research has demonstrated that different plastics produce different yields and qualities of CNTs [147]. Moreover, the separation of

CNTs from the catalyst must be well-defined in continuous processes. Pilot-scale systems should be developed to demonstrate the efficiency of transforming waste plastic into CNTs and hydrogen.

The production of CNTs from waste plastic is a promising way to generate high-value products, reduce their cost and promote this composite-filler technology. Furthermore, a life cycle assessment study [148] has shown that integrating CNT production with the pyrolysis process benefits the environment and decreases human toxicity and terrestrial eco-toxicity potentials.

5.2.2. Char

Char can potentially be used as an adsorbent for different environmental applications. Miandad et al. [149] synthesized carbon–metal double-layered oxide (C/MnCuAl-LDOs) adsorbents to study Congo red adsorption. The char used for the preparation of this adsorbent was a by-product of PS pyrolysis. The char was crushed and thermally activated in a muffle furnace at 550 °C. Then, it was chemically activated with a solution of H2SO4 and HNO3. The final adsorbent was effective for Congo red removal, with an adsorption capacity of 345.2 mg/g at pH = 4.0. Acosta et al. [150] prepared a KOH-activated carbon from tire pyrolysis char. This adsorbent eliminated bisphenol A with a capacity of 123 mg/g.

Moreover, char can be utilized for heavy metal and metalloid adsorption. Singh et al. [151] used non-modified char derived from pyrolysis of a mixture of PVC, PET and PE for arsenic adsorption. The effect of the feedstock material on char adsorption was studied. The highest-performing char for arsenic adsorption was produced from PVC and PE, which had an efficiency of 99.4%.

Furthermore, char can be used as a filler material to produce epoxy-composite materials. Sogancioglu et al. [152] studied the behaviour of char-based epoxy-composite material using PP pyrolyzed char. Chars were obtained from pyrolysis of PP at different temperatures (300 to 700 °C). With the highest aromatic content, the pyrolyzed char at 700 °C improved the hardness of the epoxy composites. Increasing the amount of char led to more epoxy-composite electrical conductivity for all chars tested. These composite materials are used in the automobile, aircraft and microelectronics industries.

Char also has the potential to be used in energy applications. Jamradloedluk and Lertsatitthanakorn [153] reported that char manufactured from HDPE has a calorific value of 4500 cal/g. To increase its surface area, the char was crushed and thermally activated at 900 °C for three hours. Then, it was extruded to produce kilogram briquettes. One briquette was able to boil water from room temperature within 13 min.

5.3. *Gas*

Gases comprise the non-condensable fraction produced from plastic waste pyrolysis. They are mainly composed of light hydrocarbons such as H_2, CH_4, C_2H_4, C_2H_6, C_3H_8 and C_3H_6 [154]. The production of gases is favoured at high temperatures and short residence times because unsaturated gases undergo secondary reactions to form aromatics [155,156]. The presence of a catalyst promotes the formation of gaseous products [19]. The gases have high calorific values between 40 and 50 MJ/kg [129]. They can be used for energy generation or in the pyrolysis system to produce energy for endothermic decomposition. Moreover, light olefins, such as ethene and propylene, are high-value monomers that can be used in the petrochemical industry after separation from other gases.

In a recent study, Eschenbacher et al. [40] tested different steam-treated industrial FCC-type catalysts and HZSM-5 additives for the in-line catalytic upgrading of pyrolysis vapours derived from PE and real (contaminated) mixed polyolefins. The purpose of the study was to maximize the production of light olefins. The severe steaming pre-treatment of the catalyst was carried out to limit the formation of coke by reducing the acidity. The steam-treated HZSM-5 additive showed the highest selectivity toward C_2–C_4 olefins, with a yield of 69 wt% (19% C_2H_4, 22% C_3H_6, 10% 1,3-C_4H_6 and 18% other C_4 olefins), obtained at high catalyst loading and temperature (700 °C). In addition, a high yield of

C_5–C_{10} aliphatics (up to 42 wt%) was produced using the FCC catalyst. The processing of real mixed polyolefins with the HZSM-5 exhibited similar performance with even higher polypropylene production (31 wt%). The coke loads per catalyst on the steamed and unsteamed HZSM-5 were 40 and 60 $\mu g/m^2$, respectively. This study showed success with the tuning of different parameters (catalyst type, catalyst leading and temperature) to maximize the production of high value–based chemicals. Moreover, this investigation demonstrated the potential of a two-step process and a suitable catalyst to produce light monomers instead of pyrolysis oil by employing steps for upgrading (hydrotreatment and steam cracking).

To maximize the light olefin production, Santos et al. [157] designed an integrated reactor/separation system (Figure 6), where only the light hydrocarbons could leave the reactor. The pyrolysis experiments of HDPE were carried out at different temperatures (400, 450 and 500 °C) in both thermal and catalytic pyrolysis. The catalyst used was HZSM-5 with 1% (*w*/*w*) loading. Increasing the pyrolysis temperature led to an increase in the gas yields for both thermal and catalytic pyrolysis and the product distribution were in the range of C_2–C_8. At 500 °C with the catalyst, the gas yield reached almost 100%, with a product distribution in the range of C_2–C_6. Furthermore, the overall O/P ratio in catalytic pyrolysis was almost six times that in thermal pyrolysis. The gaseous yield could also be increased by increasing the coolant temperature at the reactor outlet, also increasing the average molecular weight of the products. This new simple design allows the conversion of HDPE into valuable short olefins that can be used in the petrochemical industry.

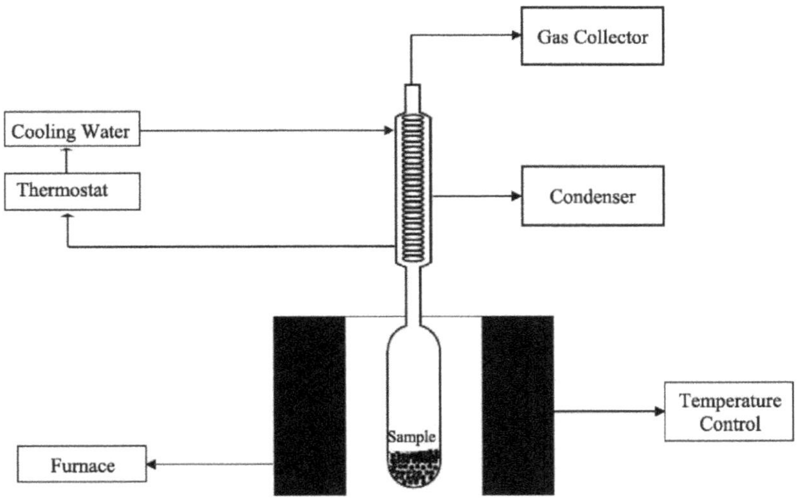

Figure 6. Reactor/separator set-up scheme, reproduced with permission [157].

6. Conclusions

Pyrolysis is a way of recovering waste plastics that cannot be mechanically recycled and will otherwise end up in the environment. The contaminants in plastic waste create challenges for the success of this technology. The use of a catalyst in pyrolysis can tailor the products for a specific application and reduce contaminants. Some alkali sorbents, such as $Al(OH)_3$, are also efficient for the removal of acidic contaminants. Pyrolysis products need further treatments either to eliminate the undesirable materials, such as HCl, or to enhance the properties of the products. Pyrolytic oil cannot be considered as a final product; therefore, the following treatments are suggested: the integration of pyrolytic plants with oil refineries, deployment of appropriate environmental safety devices and treatment of pyrolytic liquids with hydrogen-based technologies. Moreover, to achieve good-quality products, such as fuels and chemical precursors, the pretreatment of the feedstock is necessary because satisfactory products cannot be obtained from a heterogeneous mixture of waste. Moreover, the presence of contaminants such as heteroatoms and metals lead to operational problems.

The integration of pyrolytic plants with oil refineries to process pyrolytic oil in FCC, hydrocracking and steam reforming units is necessary. This integration will lead to lower contaminant levels by dilution. Moreover, the hydroprocessing of PPO gives promising results in terms of deoxygenation and decontamination.

These additional steps increase both capital and running costs, which may lead to economic challenges. To optimize efficiency, mass and energy balances should consider all the steps involved starting from the pre-treatment of the feedstock and including all the entropic heat losses. A plausible proof of self-sustainability should also be provided to evaluate the net operational efficiency. Moreover, quality standards should be formulated to match the specifications of the current refinery feedstock.

- Pyrolysis products can be used in several applications and this targeted application determines the economic sustainability of the process:
- Upgraded pyrolytic oil can be used as fuel in diesel engines or fed to steam crackers for the production of new monomers.
- CNTs with strong mechanical and electronic properties can be produced.
- Upgraded chars can be used as adsorbents.
- Gases with high calorific values can be used for energy generation or light olefin production.

Finally, reduce and re-use strategies need to take priority with the challenges facing current recycling techniques. Governments should support pyrolysis technology to reduce waste rather than make a profit.

Author Contributions: Conceptualization, S.B. and N.A.; methodology, S.B.; validation, S.B., A.A. and N.A.; investigation, S.B.; resources, N.A.; data curation, S.B.; writing—original draft preparation, S.B.; writing—review and editing, N.A. and A.A.; visualization, S.B.; supervision, N.A.; project administration, N.A.; funding acquisition, N.A. All authors have read and agreed to the published version of the manuscript.

Funding: Funding from NSERC; PRIMA Quebec, KWI Polymers Solutions Inc. and Soleno Inc. Grant No is: RDCPJ 500331-16.

Informed Consent Statement: Not applicable.

Data Availability Statement: Not applicable.

Conflicts of Interest: The authors declare no conflict of interest.

References

1. Ian Tiseo. Statista. 2020. Available online: https://www.statista.com/statistics/282732/global-production-of-plastics-since-1950/#:~:text=In2019%2Ctheglobalproduction,quarteroftheglobalproduction (accessed on 27 January 2021).
2. Buonsante, V. Drowning in Plastic: Ending Canada's Contribution to the Plastic Disaster. *Oceana* **2020**. [CrossRef]
3. Geyer, R.; Jambeck, J.R.; Law, K.L. Production, use, and fate of all plastics ever made. *Sci. Adv.* **2017**, *3*, 25–29. [CrossRef] [PubMed]
4. Ragaert, K.; Delva, L.; Van Geem, K. Mechanical and chemical recycling of solid plastic waste. *Waste Manag.* **2017**, *69*, 24–58. [CrossRef] [PubMed]
5. Mofijur, M.; Ahmed, S.F.; Rahman, S.M.A.; Yasir, S.K.; Siddiki, A.; Islam, M.S.; Shahabuddin, M.; Chyuan, H.; Mahlia, T.M.I.; Djavanroodi, F.; et al. Source, distribution and emerging threat of micro- and nanoplastics to marine organism and human health: Socio-economic impact and management strategies. *Environ. Res.* **2021**, *195*, 110857. [CrossRef]
6. Meys, R.; Bachmann, M.; Winter, B.; Zibunas, C.; Suh, S. Achieving net-zero greenhouse gas emission plastics by a circular carbon economy. *Science* **2021**, *76*, 71–76. [CrossRef]
7. Benson, N.U.; Bassey, D.E.; Palanisami, T. Heliyon COVID pollution: Impact of COVID-19 pandemic on global plastic waste footprint. *Heliyon* **2021**, *7*, e06343. [CrossRef]
8. Harussani, M.M.; Sapuan, S.M.; Rashid, U.; Khalina, A.; Ilyas, R.A. Pyrolysis of polypropylene plastic waste into carbonaceous char: Priority of plastic waste management amidst COVID-19 pandemic. *Sci. Total Environ.* **2022**, *803*, 149911. [CrossRef]
9. Qayyum, Y.; Mudasar, G.; Umer, K.; Muhammad, A.; Ijaz, W. Review of hospital plastic waste management strategies for Pakistan. *Environ. Sci. Pollut. Res.* **2022**, *29*, 9408–9421. [CrossRef]

10. Scheirs, J.; Kaminsky, W. *Feedstock Recycling and Pyrolysis of Waste Plastics: Converting Waste Plastics into Diesel and Other Fuels*; John Wiley & Sons: Hoboken, NJ, USA, 2006. [CrossRef]
11. Al-salem, S.M.; Papageorgiou, L.G.; Lettieri, P. Techno-economic assessment of thermo-chemical treatment (TCT) units in the Greater London area. *Chem. Eng. J.* **2014**, *248*, 253–263. [CrossRef]
12. Fox, J.A.; Stacey, N.T. Process targeting: An energy based comparison of waste plastic processing technologies. *Energy* **2019**, *170*, 273–283. [CrossRef]
13. Gleis, M. Gasification and Pyrolysis—Reliable Options for Waste Treatment? 2012. Available online: https://www.vivis.de/wp-content/uploads/WM3/2012_WM_403_411_Gleis.pdf (accessed on 31 March 2022).
14. Kusenberg, M.; Eschenbacher, A.; Djokic, M.R.; Zayoud, A.; Ragaert, K.; De Meester, S.; Van Geem, K.M. Opportunities and challenges for the application of post-consumer plastic waste pyrolysis oils as steam cracker feedstocks: To decontaminate or not to decontaminate? *Waste Manag.* **2022**, *138*, 83–115. [CrossRef] [PubMed]
15. Kusenberg, M.; Zayoud, A.; Roosen, M.; Dao, H.; Abbas-abadi, M.S.; Eschenbacher, A.; Kresovic, U.; De Meester, S.; Van Geem, K.M. A comprehensive experimental investigation of plastic waste pyrolysis oil quality and its dependence on the plastic waste composition. *Fuel Process. Technol.* **2022**, *227*, 107090. [CrossRef]
16. Saad, M.; Oasmaa, A.; Pihkola, H.; Deviatkin, I.; Tenhunen, A.; Mannila, J.; Minkkinen, H.; Pohjakallio, M.; Laine-ylijoki, J. Pyrolysis of plastic waste: Opportunities and challenges. *J. Anal. Appl. Pyrolysis* **2020**, *152*, 104804. [CrossRef]
17. Sharuddin, S.D.A.; Abnisa, F.; Daud, W.M.A.W.; Aroua, M.K. A review on pyrolysis of plastic wastes. *Energy Convers. Manag.* **2016**, *115*, 308–326. [CrossRef]
18. Chen, D.; Yin, L.; Wang, H.; He, P. Pyrolysis technologies for municipal solid waste: A review. *Waste Manag.* **2014**, *34*, 2466–2486. [CrossRef]
19. Miandad, R.; Barakat, M.A.; Aburiazaiza, A.S.; Rehan, M.; Nizami, A.S. Catalytic pyrolysis of plastic waste: A review. *Process Saf. Environ. Prot.* **2016**, *102*, 822–838. [CrossRef]
20. Kunwar, B.; Cheng, H.N.; Chandrashekaran, S.R.; Sharma, B.K. Plastics to fuel: A review. *Renew. Sustain. Energy Rev.* **2016**, *54*, 421–428. [CrossRef]
21. Lin, H.; Huang, M.; Luo, J.; Lin, L.; Lee, C.; Ou, K. Hydrocarbon fuels produced by catalytic pyrolysis of hospital plastic wastes in a fluidizing cracking process. *Fuel Process. Technol.* **2010**, *91*, 1355–1363. [CrossRef]
22. Srinivasa, M.; Kurose, K.; Okuda, T.; Nishijima, W.; Okada, M. Separation of polyvinyl chloride (PVC) from automobile shredder residue (ASR) by froth flotation with ozonation. *J. Hazard. Mater.* **2007**, *147*, 1051–1055. [CrossRef]
23. Yang, X.; Sun, L.; Xiang, J.; Hu, S.; Su, S. Pyrolysis and dehalogenation of plastics from waste electrical and electronic equipment (WEEE): A review. *Waste Manag.* **2013**, *33*, 462–473. [CrossRef]
24. Areeprasert, C.; Asingsamanunt, J.; Srisawat, S. Municipal Plastic Waste Composition Study at Transfer Station of Bangkok and Possibility of its Energy Recovery by Pyrolysis. *Energy Procedia* **2017**, *107*, 222–226. [CrossRef]
25. Gao, F. Pyrolysis of Waste Plastics into Fuels. Ph.D. Thesis, University of Canterbury, Christchurch, New Zealand, 2010.
26. Toraman, H.E.; Dijkmans, T.; Djokic, M.R.; Van Geem, K.M.; Marin, G.B. Detailed compositional characterization of plastic waste pyrolysis oil by comprehensive two-dimensional gas-chromatography coupled to multiple detectors. *J. Chromatogr. A* **2014**, *1359*, 237–246. [CrossRef] [PubMed]
27. Miandad, R.; Barakat, M.A.; Aburiazaiza, A.S.; Rehan, M.; Ismail, I.M.I. Effect of plastic waste types on pyrolysis liquid oil. *Int. Biodeterior. Biodegrad.* **2017**, *119*, 239–252. [CrossRef]
28. De Sousa, L.C.R. Gasification of wood, urban wastewood (Altholz) and other wastes in a fluidized bed reactor. Ph.D Thesis, Swiss Federal Institute of Technology Zurich, Zürich, Switzerland, 2001. [CrossRef]
29. Hornung, A.; Seifert, H. Rotary Kiln Pyrolysis of Polymers Containing Heteroatoms. In *Feedstock Recycling and Pyrolysis of Waste Plastics: Converting Waste Plastics into Diesel and other Fuels*; Scheirs, J., Kaminsky, W., Eds.; Jhon Wiley and Sons: New York, NY, USA, 2006; pp. 549–564.
30. Bhaskar, T.; Sakata, Y. Liquefaction of PVC Mixed Plastics. In *Feedstock Recycling and Pyrolysis of Waste Plastics*; Scheirs, J., Kaminsky, W., Eds.; Jhon Wiley and Sons: New York, NY, USA, 2006; pp. 493–525.
31. Rodríguez, E.; Gutiérrez, A.; Palos, R.; Vela, F.J.; Arandes, J.M.; Bilbao, J. Fuel production by cracking of polyolefins pyrolysis waxes under fluid catalytic cracking (FCC) operating conditions. *Waste Manag.* **2019**, *93*, 162–172. [CrossRef] [PubMed]
32. Shi, Q.; Xu, C.; Zhao, S.; Chung, K.H.; Tn, A. Characterization of Heteroatoms in Residue Fluid Catalytic Cracking (RFCC) Diesel by Gas Chromatography and Mass Spectrometry. *Energy Fuels* **2009**, *23*, 6062–6069. [CrossRef]
33. Hossein, A.; Dehaghani, S.; Badizad, M.H. Experimental Study of Iranian Heavy Crude Oil Viscosity Reduction by Diluting with Heptane, Methanol, Toluene, Gas Condensate and Naphtha. *Petroleum* **2017**, *2*, 415–424. [CrossRef]
34. Kusenberg, M.; Roosen, M.; Zayoud, A.; Djokic, M.R.; Dao, H.; De Meester, S.; Ragaert, K.; Kresovic, U.; Van Geem, K.M. Assessing the feasibility of chemical recycling via steam cracking of untreated plastic waste pyrolysis oils: Feedstock impurities, product yields and coke formation. *Waste Manag.* **2022**, *141*, 104–114. [CrossRef]
35. Al-Salem, S.M.; Antelava, A.; Constantinou, A.; Manos, G.; Dutta, A. A review on thermal and catalytic pyrolysis of plastic solid waste (PSW). *J. Environ. Manag.* **2017**, *197*, 177–198. [CrossRef]
36. Eschenbacher, A.; Saraeian, A.; Shanks, B.H.; Jensen, P.A.; Henriksen, U.B.; Ahrenfeldt, J.; Jensen, A.D. Sustainable Energy & Fuels biomass pyrolysis vapors using micro and bench-scale reactors. *Sustaiable Energy Fuels* **2020**, *4*, 3780–3796. [CrossRef]

37. Eschenbacher, A.; Arendt, P.; Birk, U.; Ahrenfeldt, J. Catalytic deoxygenation of vapors obtained from ablative fast pyrolysis of wheat straw using mesoporous HZSM-5. *Fuel Process. Technol.* **2019**, *194*, 106119. [CrossRef]
38. Akubo, K.; Nahil, M.A.; Williams, P.T. Aromatic fuel oils produced from the pyrolysis-catalysis of polyethylene plastic with metal-impregnated zeolite catalysts. *J. Energy Inst.* **2019**, *92*, 195–202. [CrossRef]
39. Artetxe, M.; Lopez, G.; Amutio, M.; Elordi, G.; Bilbao, J.; Olazar, M. Cracking of High Density Polyethylene Pyrolysis Waxes on HZSM-5 Catalysts of Different Acidity. *Ind. Eng. Chem. Res.* **2013**, *31*, 10637–10645. [CrossRef]
40. Eschenbacher, A.; Varghese, R.J.; Abbas-abadi, M.S.; Van Geem, K.M. Maximizing light olefins and aromatics as high value base chemicals via single step catalytic conversion of plastic waste. *Chem. Eng. J.* **2022**, *428*, 132087. [CrossRef]
41. Aguado, J.; Serrano, D.P.; Escola, J.M. Catalytic upgrading of plastic waste. In *Feedstock Recycling and Pyrolysis of Waste Plastics*; Scheirs, J., Kaminsky, W., Eds.; John Wiley & Sons: New York, NY, USA, 2006; pp. 73–110.
42. Butler, E.; Devlin, G.; Mcdonnell, K. Waste Polyolefins to Liquid Fuels via Pyrolysis: Review of Commercial State-of-the-Art and Recent Laboratory Research. *Waste Biomass Valor.* **2011**, *2*, 227–255. [CrossRef]
43. Manos, G. Catalytic degradation of plastic waste to fuel over microporous materials. In *Feedstock Recycling and Pyrolysis of Waste Plastics*; Scheirs, J., Kaminsky, W., Eds.; John Wiley & Sons: New York, NY, USA, 2006; pp. 193–208.
44. Syamsiro, M.; Saptoadi, H.; Norsujianto, T.; Noviasri, P. Fuel Oil Production from Municipal Plastic Wastes in Sequential Pyrolysis and Catalytic Reforming Reactors. *Energy Procedia* **2014**, *47*, 180–188. [CrossRef]
45. Miskolczi, N.; Ates, F. Thermo-catalytic co-pyrolysis of recovered heavy oil and municipal plastic wastes. *J. Anal. Appl. Pyrolysis* **2016**, *117*, 273–281. [CrossRef]
46. Elordi, G.; Olazar, M.; Lopez, G.; Castano, P.; Bilbao, J. Role of pore structure in the deactivation of zeolites (HZSM-5, H-Beta and HY) by coke in the pyrolysis of polyethylene in a conical spouted bed reactor. *Appl. Catal. B* **2011**, *102*, 224–231. [CrossRef]
47. Kissin, Y.V. Primary Products in Hydrocarbon Cracking over Solid Acidic Catalysts under Very Mild Conditions: Relation to Cracking Mechanism. *J. Catal.* **1998**, *105*, 101–105. [CrossRef]
48. Castano, P.; Elordi, G.; Olazar, M.; Aguayo, A.T.; Pawelec, B.; Bilbao, J. Insights into the coke deposited on HZSM-5, H-Beta and HY zeolites during the cracking of polyethylene. *Appl. Catal. B Environ.* **2011**, *104*, 91–100. [CrossRef]
49. Ates, F.; Miskolczi, N.; Borsodi, N. Bioresource Technology Comparision of real waste (MSW and MPW) pyrolysis in batch reactor over different catalysts. Part I: Product yields, gas and pyrolysis oil properties. *Bioresour. Technol.* **2013**, *133*, 443–454. [CrossRef]
50. Yoosuk, B.; Tumnantong, D.; Prasassarakich, P. Amorphous unsupported Ni-Mo sulfide prepared by one step hydrothermal method for phenol hydrodeoxygenation. *Fuel* **2012**, *91*, 246–252. [CrossRef]
51. Sun, K.; Wang, W.; Themelis, N.J.; Bourtsalas, A.C.T.; Huang, Q. Catalytic co-pyrolysis of polycarbonate and polyethylene/polypropylene mixtures: Promotion of oil deoxygenation and aromatic hydrocarbon formation. *Fuel* **2021**, *285*, 119143. [CrossRef]
52. Mastral, J.F.; Berrueco, C.; Gea, M.; Ceamanos, J. Catalytic degradation of high density polyethylene over nanocrystalline HZSM-5 zeolite. *Polym. Degrad. Stab.* **2006**, *91*, 3330–3338. [CrossRef]
53. Xue, Y.; Zhou, S.; Bai, X. Role of Hydrogen Transfer during Catalytic Copyrolysis of Lignin and Tetralin over HZSM-5 and HY Zeolite Catalysts. *ACS Sustain. Chem. Eng.* **2016**, *4*, 4237–4250. [CrossRef]
54. Cheng, Y.; Huber, G.W. Production of targeted aromatics by using Diels–Alder classes of reactions with furans and olefins over ZSM-5. *Green Chem.* **2012**, *14*, 3114–3125. [CrossRef]
55. Miskolczi, N.; Ates, F.; Borsodi, N. Comparison of real waste (MSW and MPW) pyrolysis in batch reactor over different catalysts. Part II: Contaminants, char and pyrolysis oil properties. *Bioresour. Technol.* **2013**, *144*, 370–379. [CrossRef]
56. Brebu, M.; Bhaskar, T.; Murai, K.; Muto, A. Removal of nitrogen, bromine, and chlorine from PP/PE/PS/PVC/ABS-Br pyrolysis liquid products using Fe- and Ca-based catalysts. *Polym. Degrad. Stab.* **2005**, *87*, 225–230. [CrossRef]
57. Lopez-Urionabarrenechea, A.; De Marco, I.; Caballero, B.M.; Laresgoiti, M.F.; Adrados, A. Catalytic stepwise pyrolysis of packaging plastic waste. *J. Anal. Appl. Pyrolysis* **2012**, *96*, 54–62. [CrossRef]
58. Nolte, M.W.; Liberatore, M.W. Viscosity of Biomass Pyrolysis Oils from Various Feedstocks. *Energy Fuels* **2010**, *24*, 6601–6608. [CrossRef]
59. Magee, J.; Mitchell, M. (Eds.) *Fluid Catalytic Cracking: Science and Technology*; Elsevier: Amsterdam, The Netherlands, 1993.
60. Lee, K.; Noh, N.; Shin, D.; Seo, Y. Comparison of plastic types for catalytic degradation of waste plastics into liquid product with spent FCC catalyst. *Polym. Degrad. Stab.* **2002**, *78*, 539–544. [CrossRef]
61. Barbarias, I.; Artetxe, M.; Arregi, A.; Alvarez, J.; Lopez, G.; Olazar, M. Catalytic Cracking of HDPE Pyrolysis Volatiles over a Spent FCC Catalyst. *Chem. Eng. Trans.* **2015**, *43*, 2029–2034. [CrossRef]
62. Wang, K.; Johnston, P.A.; Brown, R.C. Comparison of in-situ and ex-situ catalytic pyrolysis in a micro-reactor system. *Bioresour. Technol.* **2014**, *173*, 124–131. [CrossRef] [PubMed]
63. Gaurh, P.; Pramanik, H. A novel approach of solid waste management via aromatization using multiphase catalytic pyrolysis of waste polyethylene. *Waste Manag.* **2018**, *71*, 86–96. [CrossRef] [PubMed]
64. Zhang, Y.; Duan, D.; Lei, H.; Villota, E.; Ruan, R. Jet fuel production from waste plastics via catalytic pyrolysis with activated carbons. *Appl. Energy* **2019**, *251*, 113337. [CrossRef]
65. Wysocka, I. Talanta Determination of rare earth elements concentrations in natural waters – A review of ICP-MS measurement approaches. *Talanta* **2021**, *221*, 121636. [CrossRef] [PubMed]

66. Roosen, M.; Mys, N.; Kusenberg, M.; Billen, P.; Dumoulin, A.; Dewulf, J.; Van Geem, K.M.; Ragaert, K.; De Meester, S. Detailed Analysis of the Composition of Selected Plastic Packaging Waste Products and Its Implications for Mechanical and Thermochemical Recycling. *Environ. Sci. Technol.* **2020**, *54*, 13282–13293. [CrossRef]
67. Hall, W.J.; Williams, P.T. Pyrolysis of brominated feedstock plastic in a fluidised bed reactor. *J. Anal. Appl. Pyrolysis* **2006**, *77*, 75–82. [CrossRef]
68. Li, K.; Lee, S.W.; Yuan, G.; Lei, J.; Lin, S.; Weerachanchai, P.; Yang, Y.; Wang, J. Investigation into the Catalytic Activity of Microporous and Mesoporous Catalysts in the Pyrolysis of Waste Polyethylene and Polypropylene Mixture. *Energies* **2016**, *9*, 431. [CrossRef]
69. Singh, R.K.; Ruj, B.; Sadhukhan, A.K.; Gupta, P. A TG-FTIR investigation on the co-pyrolysis of the waste HDPE, PP, PS and PET under high heating conditions. *J. Energy Inst.* **2020**, *93*, 1020–1035. [CrossRef]
70. Jin, W.; Shen, D.; Liu, Q.; Xiao, R. Evaluation of the co-pyrolysis of lignin with plastic polymers by TG- FTIR and Py-GC/MS. *Polym. Degrad. Stab.* **2016**, *133*, 65–74. [CrossRef]
71. Yang, B.; Chen, M. Py–FTIR–GC/MS Analysis of Volatile Products of Automobile Shredder Residue Pyrolysis. *Polymers* **2020**, *12*, 2734. [CrossRef] [PubMed]
72. Kumar, R.; Ruj, B.; Sadhukhan, A.K.; Gupta, P. Impact of fast and slow pyrolysis on the degradation of mixed plastic waste: Product yield analysis and their characterization. *J. Energy Inst.* **2019**, *92*, 1647–1657. [CrossRef]
73. Gala, A.; Guerrero, M.; Manuel, J. Characterization of post-consumer plastic film waste from mixed MSW in Spain: A key point for the successful implementation of sustainable plastic waste management strategies. *Waste Manag.* **2020**, *111*, 22–33. [CrossRef] [PubMed]
74. Stas, M.; Auersvald, M.; Vozka, P. Two-Dimensional Gas Chromatography Characterization of Pyrolysis Bio-oils: A Review. *Energy Fuels* **2021**, *35*, 8541–8557. [CrossRef]
75. Yu, A.; Patrushev, Y.V.; Sidelnikov, V.N.; Buryak, A.K. Analysis of light components in pyrolysis products by comprehensive two-dimensional gas chromatography with PLOT columns. *Talanta* **2020**, *209*, 120448. [CrossRef]
76. Zayoud, A.; Dao, H.; Kusenberg, M.; Eschenbacher, A.; Kresovic, U.; Alderweireldt, N.; Djokic, M.; Van Geem, K.M. Pyrolysis of end-of-life polystyrene in a pilot-scale reactor: Maximizing styrene production. *Waste Manag.* **2022**, *139*, 85–95. [CrossRef]
77. Linear, C.I.; Thi, H.D.; Van Geem, K.M. Detailed Group-Type Characterization of Plastic-Waste Pyrolysis Oils: By Comprehensive Two-Dimensional Gas. *Separations* **2021**, *8*, 103. [CrossRef]
78. Arjang, S.; Motahari, K.; Saidi, M. Experimental and Modeling Study of Organic Chloride Compounds Removal from Naphtha Fraction of Contaminated Crude Oil Using Sintered γ-Al_2O_3 Nanoparticles: Equilibrium, Kinetic, and Thermodynamic Analysis. *Energy Fuels* **2018**, *32*, 4025–4039. [CrossRef]
79. Lee, K. Pyrolysis of municipal plastic wastes separated by difference of specific gravity. *J. Anal. Appl. Pyrolysis* **2007**, *79*, 362–367. [CrossRef]
80. Muhammad, C.; Onwudili, J.A.; Williams, P.T. Thermal degradation of real-world waste plastics and simulated mixed plastics in a two-stage pyrolysis-catalysis reactor for fuel production. *Energy Fuels* **2015**, *29*, 2601–2609. [CrossRef]
81. Yoshioka, T.; Handa, T.; Grause, G.; Lei, Z.; Inomata, H.; Mizoguchi, T. Effects of metal oxides on the pyrolysis of poly(ethylene terephthalate). *J. Anal. Appl. Pyrolysis* **2005**, *73*, 139–144. [CrossRef]
82. Wang, C.; Wang, H.; Fu, J.; Liu, Y. Flotation separation of waste plastics for recycling—A review. *Waste Manag.* **2015**, *41*, 28–38. [CrossRef] [PubMed]
83. Richard, G.M.; Mario, M.; Susana, T. Optimization of the recovery of plastics for recycling by density media separation cyclones. *Resour. Conserv. Recycl.* **2011**, *55*, 472–482. [CrossRef]
84. Lee, J.; Shin, J. Triboelectmstatic Separation of PVC Materials from Mixed Plastics for Waste Plastic Recycling. *Korean J. Chem. Eng.* **2002**, *19*, 267–272. [CrossRef]
85. Pappa, G.; Boukouvalas, C.; Giannaris, C.; Ntaras, N.; Zografos, V.; Magoulas, K.; Lygeros, A.; Tassios, D. The selective dissolution/precipitation technique for polymer recycling: A pilot unit application. *Resour. Conserv. Recycl.* **2001**, *34*, 33–44. [CrossRef]
86. Nauman, E.B.; Lynch, J.C. Polymer Recycling by Selective Dissolution. U.S. Patent 005278282A, 11 January 1994.
87. Dong, J.; Tang, Y.; Nzihou, A.; Chi, Y.; Weiss-Hortala, E.; Ni, M. Life cycle assessment of pyrolysis, gasification and incineration waste-to-energy technologies: Theoretical analysis and case study of commercial plants. *Sci. Total Environ.* **2018**, *626*, 744–753. [CrossRef]
88. Yu, S.; Su, W.; Wu, D.; Yao, Z.; Liu, J.; Tang, J.; Wu, W. Thermal treatment of flame retardant plastics: A case study on a waste TV plastic shell sample. *Sci. Total Environ.* **2019**, *675*, 651–657. [CrossRef]
89. Xiong, J.; Yu, S.; Wu, D.; Lü, X.; Tang, J.; Wu, W.; Yao, Z. Pyrolysis treatment of nonmetal fraction of waste printed circuit boards: Focusing on the fate of bromine. *Waste Manag. Res.* **2020**, *38*, 1251–1258. [CrossRef]
90. Shen, Y.; Zhao, R.; Wang, J.; Chen, X.; Ge, X.; Chen, M. Waste-to-energy: Dehalogenation of plastic-containing wastes. *Waste Manag.* **2016**, *49*, 287–303. [CrossRef]
91. Baumgartner, A.J.; Blaschke, M.W.; Coleman, S.T.; Kohler, R.; Paxon, T.E. Feedstock Contaminants in Ethylene Plants—An Update. In Proceedings of the AIChE Spring Meeting and Global Congress on Process Safety, New Orleans, LA, USA, 27 March 2017.
92. Cagnetta, G.; Zhang, K.; Zhang, Q.; Huang, J.; Yu, G. Mechanochemical pre-treatment for viable recycling of plastic waste containing haloorganics. *Waste Manag.* **2018**, *75*, 181–186. [CrossRef]

93. Zhang, K.; Huang, J.; Zhang, W.; Yu, Y.; Deng, S.; Yu, G. Mechanochemical degradation of tetrabromobisphenol A: Performance, products and pathway. *J. Hazard. Mater.* **2012**, *243*, 278–285. [CrossRef] [PubMed]
94. Grause, G.; Diego, J.; Tanaka, H.; Bhaskar, T.; Kameda, T.; Yoshioka, T. A novel process for the removal of bromine from styrene polymers containing brominated flame retardant. *Polym. Degrad. Stab.* **2015**, *112*, 86–93. [CrossRef]
95. Jakab, E.; Uddin, A.; Bhaskar, T.; Sakata, Y. Thermal decomposition of flame-retarded high- impact polystyrene. *J. Anal. Appl. Pyrolysis* **2003**, *69*, 83–99. [CrossRef]
96. Grause, G.; Ishibashi, J.; Kameda, T.; Bhaskar, T.; Yoshioka, T. Kinetic studies of the decomposition of flame retardant containing high-impact polystyrene. *Polym. Degrad. Stab.* **2010**, *95*, 1129–1137. [CrossRef]
97. López, A.; De Marco, I.; Caballero, B.M.; Laresgoiti, M.F.; Adrados, A. Dechlorination of fuels in pyrolysis of PVC containing plastic wastes. *Fuel Process. Technol.* **2011**, *92*, 253–260. [CrossRef]
98. Yanik, J.; Uddin, A.; Ikeuchi, K.; Sakata, Y. The catalytic effect of Red Mud on the degradation of poly (vinyl chloride) containing polymer mixture into fuel oil. *Polym. Degrad. Stab.* **2001**, *73*, 335–346. [CrossRef]
99. Kuhlmann, B.; Arnett, E.M.; Siskin, M. Classical Organic Reactions in Pure Superheated Water. *J. Org. Chem.* **1994**, *56*, 3098–3101. [CrossRef]
100. Li, T.; Zhao, P.; Lei, M.; Li, Z. Understanding Hydrothermal Dechlorination of PVC by Focusing on the Operating Conditions and Hydrochar Characteristics. *Appl. Sci.* **2017**, *7*, 256. [CrossRef]
101. Wang, Y.; Wu, K.; Liu, Q.; Zhang, H. Low chlorine oil production through fast pyrolysis of mixed plastics combined with hydrothermal dechlorination pretreatment. *Process Saf. Environ. Prot.* **2021**, *149*, 105–114. [CrossRef]
102. Nishibata, H.; Uddin, A.; Kato, Y. Simultaneous degradation and dechlorination of poly (vinyl chloride) by a combination of superheated steam and CaO catalyst/adsorbent. *Polym. Degrad. Stab.* **2020**, *179*, 109–225. [CrossRef]
103. American Chemistry Council, Conversion Technology: A Complement to Plastic Recycling. 2011. Available online: https://plastics.americanchemistry.com/Plastics-to-Oil/ (accessed on 31 March 2022).
104. Scheirs, J. Overview of commercial pyrolysis processes for waste plastics. In *Feedstock Recycling and Pyrolysis of Waste Plastics: Converting Waste Plastics into Diesel and Other Fuels*; Scheirs, J., Kaminsky, W., Eds.; John Wiley & Sons: New York, NY, USA, 2006; pp. 381–433.
105. DeWhitt, K.C. Systems and Methods for Recycling Plastic. U.S. Patent 8193403B2, 5 June 2012.
106. Baena-gonzález, J.; Santamaria-echart, A.; Luis, J.; González, S. Chemical recycling of plastic waste: Bitumen, solvents, and polystyrene from pyrolysis oil. *Waste Manag.* **2020**, *118*, 139–149. [CrossRef] [PubMed]
107. Thahir, R.; Altway, A.; Juliastuti, S.R. Production of liquid fuel from plastic waste using integrated pyrolysis method with refinery distillation bubble cap plate column. *Energy Rep.* **2019**, *5*, 70–77. [CrossRef]
108. Lee, D.; Nam, H.; Wang, S.; Kim, H.; Hwan, J.; Won, Y.; Wook, B.; Doo, Y.; Nam, H.; Lee, K.; et al. Characteristics of fractionated drop-in liquid fuel of plastic wastes from a commercial pyrolysis plant. *Waste Manag.* **2021**, *126*, 411–422. [CrossRef] [PubMed]
109. Gray, M.R.; Mccaffrey, W.C. Role of Chain Reactions and Olefin Formation in Cracking, Hydroconversion, and Coking of Petroleum and Bitumen Fractions. *Energy Fuels* **2002**, *16*, 756–766. [CrossRef]
110. Arandes, M.; Torre, I.; Azkoiti, M.J.; Castan, P.; Bilbao, J.; De Lasa, H. Effect of catalyst properties on the cracking of polypropylene pyrolysis waxes under FCC conditions. *Catal. Today* **2008**, *135*, 413–419. [CrossRef]
111. Torre, I.; Arandes, J.M.; Castano, P.; Azkoiti, M.; Bilbao, J.; De Lasa, H.I. Catalytic Cracking of Plastic Pyrolysis Waxes with Vacuum Gasoil: Effect of HZSM-5 Zeolite in the FCC Catalyst. *Int. J. Chem. React. Eng.* **2006**, *4*, A31. [CrossRef]
112. Lopez, G.; Artetxe, M.; Amutio, M.; Bilbao, J.; Olazar, M. Thermochemical routes for the valorization of waste polyolefinic plastics to produce fuels and chemicals. A review. *Renew. Sustain. Energy Rev.* **2017**, *73*, 346–368. [CrossRef]
113. Gracida-Alvarez, U.R.; Winjobi, O.; Sacramento-Rivero, J.C.; Shonnard, D.R. System Analyses of High-Value Chemicals and Fuels from a Waste High-Density Polyethylene Refinery. Part 2: Carbon Footprint Analysis and Regional Electricity Effects. *ACS Sustain. Chem. Eng.* **2019**, *7*, 18267–18278. [CrossRef]
114. Rodríguez, E.; Palos, R.; Gutiérrez, A.; Trueba, D.; Arandes, J.M.; Bilbao, J. Towards waste refinery: Co-feeding HDPE pyrolysis waxes with VGO into the catalytic cracking unit. *Energy Convers. Manag.* **2020**, *207*, 112554. [CrossRef]
115. Lee, K. Effects of the types of zeolites on catalytic upgrading of pyrolysis wax oil. *J. Anal. Appl. Pyrolysis* **2012**, *94*, 209–214. [CrossRef]
116. Wang, S.; Kim, H.; Lee, D.; Lee, Y.; Won, Y.; Wook, B.; Nam, H.; Ryu, H.; Lee, K. Drop-in fuel production with plastic waste pyrolysis oil over catalytic separation. *Fuel* **2021**, *305*, 121440. [CrossRef]
117. Prado, G.H.C.; Rao, Y.; De Klerk, A. Nitrogen Removal from Oil: A Review. *Energy Fuels* **2017**, *31*, 14–36. [CrossRef]
118. Ding, F.; Luo, C.; Zhang, H.; Xiong, L.; Chen, X.D. Hydrocracking of Polyolefin Thermal Cracking Waxes Over Ni-loaded Molecular Sieve Catalysts Hydrocracking of Polyolefin Thermal Cracking Waxes Over Ni-loaded Molecular Sieve Catalysts. *Pet. Sci. Technol.* **2015**, *33*, 1846–1852. [CrossRef]
119. Mangesh, V.L.; Padmanabhan, S.; Tamizhdurai, P.; Narayanan, S.; Ramesh, A. Combustion and emission analysis of hydrogenated waste polypropylene pyrolysis oil blended with diesel. *J. Hazard. Mater.* **2020**, *386*, 121–453. [CrossRef]
120. Serrano, D.P.; Escola, J.M.; Briones, L.; Arroyo, M. Hydroprocessing of the LDPE thermal cracking oil into transportation fuels over Pd supported on hierarchical ZSM-5 catalyst. *Fuel* **2017**, *206*, 190–198. [CrossRef]
121. Vela, F.J.; Palos, R.; Trueba, D.; Bilbao, J.; Arandes, M. Different approaches to convert waste polyolefins into automotive fuels via hydrocracking with a NiW/HY catalyst. *Fuel Process. Technol.* **2021**, *220*, 106891. [CrossRef]

122. Escola, J.M.; Aguado, J.; Serrano, D.P.; García, A.; Peral, A.; Briones, L.; Calvo, R.; Fernandez, E. Catalytic hydroreforming of the polyethylene thermal cracking oil over Ni supported hierarchical zeolites and mesostructured aluminosilicates. *Appl. Catal. B Environ.* **2011**, *106*, 405–415. [CrossRef]
123. Escola, J.M.; Aguado, J.; Serrano, D.P.; Briones, L.; De Tuesta, J.L.D.; Calvo, R.; Fernandez, E. Conversion of Polyethylene into Transportation Fuels by the Combination of Thermal Cracking and Catalytic Hydroreforming over Ni-Supported Hierarchical Beta Zeolite. *Energy Fuels* **2012**, *26*, 3187–3195. [CrossRef]
124. Escola, J.M.; Aguado, J.; Serrano, D.P.; Briones, L. Hydroreforming over Ni/H-beta of the thermal cracking products of LDPE, HDPE and PP for fuel production. *J. Mater. Cycles Waste Manag.* **2020**, *14*, 286–293. [CrossRef]
125. Miller, S.; Shah, N.; Huffmann, G. Production of Premium Oil Products from Waste Plastic by Pyrolysis and Hydroprocessing. In *Feedstock Recycling and Pyrolysis of Waste Plastics*; Scheirs, J., Kaminsky, W., Eds.; John Wiley & Sons: New York, NY, USA, 2006; pp. 345–362.
126. Lingaiah, N.; Uddin, A.; Muto, A.; Imai, T.; Sakata, Y. Removal of organic chlorine compounds by catalytic dehydrochlorination for the refinement of municipal waste plastic derived oil. *Fuel* **2001**, *80*, 1901–1905. [CrossRef]
127. Ates, N.; Uzal, N. Removal of heavy metals from aluminium anodic oxidation wastewaters by membrane filtration. *Environ. Sci. Pollut. Res.* **2018**, *25*, 22259–22272. [CrossRef]
128. Blasco, T.; Chica, A.; Corma, A.; Murphy, W.J.; Agúndez-rodríguez, J.; Pérez-pariente, J. Changing the Si distribution in SAPO-11 by synthesis with surfactants improves the hydroisomerization/dewaxing properties. *J. Catal.* **2006**, *242*, 153–161. [CrossRef]
129. Jung, S.; Cho, M.; Kang, B.; Kim, J. Pyrolysis of a fraction of waste polypropylene and polyethylene for the recovery of BTX aromatics using a fluidized bed reactor. *Fuel Process. Technol.* **2010**, *91*, 277–284. [CrossRef]
130. Bernardo, M.; Lapa, N.; Gonçalves, M.; Barbosa, R.; Mendes, B.; Pinto, F.; Gulyurtlu, I. Toxicity of char residues produced in the co-pyrolysis of different wastes. *Waste Manag.* **2010**, *30*, 628–635. [CrossRef]
131. Bernardo, M.; Lapa, N.; Gonc, M.; Mendes, B.; Pinto, F.; Fonseca, I.; Lopes, H. Physico-chemical properties of chars obtained in the co-pyrolysis of waste mixtures. *J. Haza.* **2012**, *220*, 196–202. [CrossRef]
132. Sharuddin, S.D.A.; Abnisa, F.; Daud, W.M.A.W.; Aroua, M. Pyrolysis of plastic waste for liquid fuel production as prospective energy resource. In *IOP Conference Series: Materials Science and Engineering*; IOP Publishing: Bristol, UK, 2018. [CrossRef]
133. Al-salem, S.M. Thermal pyrolysis of high density polyethylene (HDPE) in a novel fixed bed reactor system for the production of high value gasoline range hydrocarbons (HC). *Process Saf. Environ. Prot.* **2019**, *127*, 171–179. [CrossRef]
134. Panda, A.; Singh, R.K. Experimental Optimization of Process for the Thermo-catalytic Degradation of Waste Polypropylene to Liquid Fuel. *Adv. Energy Eng.* **2013**, *1*, 74–84.
135. Rehan, M.; Miandad, R.; Barakat, M.A.; Ismail, I.M.I.; Almeelbi, T.; Gardy, J.; Hassanpour, A.; Khan, M.Z.; Demirbas, A.; Nizami, A.S. Effect of zeolite catalysts on pyrolysis liquid oil. *Int. Biodeterior. Biodegrad.* **2017**, *119*, 162–175. [CrossRef]
136. Mangesh, V.L.; Padmanabhan, S.; Tamizhdurai, P.; Ramesh, A. Experimental investigation to identify the type of waste plastic pyrolysis oil suitable for conversion to diesel engine fuel. *J. Clean. Prod.* **2020**, *246*, 119066. [CrossRef]
137. Singh, R.K.; Ruj, B.; Sadhukhan, A.K.; Gupta, P.; Tigga, V.P. Waste plastic to pyrolytic oil and its utilization in CI engine: Performance analysis and combustion characteristics. *Fuel* **2020**, *262*, 116539. [CrossRef]
138. Zhang, Y.; Wu, C.; Nahil, M.A.; Williams, P. Pyrolysis—Catalytic Reforming/Gasification of Waste Tires for Production of Carbon Nanotubes and Hydrogen. *Energy Fuels* **2015**, *29*, 3328–3334. [CrossRef]
139. Zahid, M.U.; Pervaiz, E.; Hussain, A.; Shahzad, M.I.; Bilal, M.; Niazi, K. Synthesis of carbon nanomaterials from different pyrolysis techniques: A review Synthesis of carbon nanomaterials from different pyrolysis techniques: A review. *Mater. Res. Express* **2018**, *5*, 052002. [CrossRef]
140. Ansari, R.; Hasrati, E.; Shojaei, M.F.; Gholami, R.; Shahabodini, A. Forced vibration analysis of functionally graded carbon nanotube-re-inforced composite plates using a numerical strategy. *Phys. E Low-Dimens. Syst. Nanostruct.* **2015**, *69*, 294–305. [CrossRef]
141. Maiti, S.; Banerjee, P.; Purakayastha, S.; Ghosh, B. Silicon-doped carbon semiconductor from rice husk char. *Mater. Chem. Phys.* **2008**, *109*, 169–173. [CrossRef]
142. Gong, J.; Chen, X.; Tang, T. Recent progress in controlled carbonization of (waste) polymers. *Prog. Polym. Sci.* **2019**, *94*, 1–32. [CrossRef]
143. Azara, A.; Belbessai, S.; Abatzoglou, N. A review of filamentous carbon nanomaterial synthesis via catalytic conversion of waste plastic pyrolysis products. *J. Environ. Chem. Eng.* **2022**, *10*, 107049. [CrossRef]
144. Yao, D.; Wu, C.; Yang, H.; Zhang, Y.; Nahil, M.A.; Chen, Y.; Williams, P.T.; Chen, H. Co-production of hydrogen and carbon nanotubes from catalytic pyrolysis of waste plastics on Ni-Fe bimetallic catalyst. *Energy Convers. Manag.* **2017**, *148*, 692–700. [CrossRef]
145. Yao, D.; Yang, H.; Hu, Q.; Chen, Y.; Chen, H.; Williams, P.T. Carbon nanotubes from post-consumer waste plastics: Investigations into catalyst metal and support material characteristics. *Appl. Catal. B Environ.* **2020**, *280*, 119–413. [CrossRef]
146. Williams, P.T. Hydrogen and Carbon Nanotubes from Pyrolysis-Catalysis of Waste Plastics: A Review. *Waste Biomass Valoriz.* **2020**, *12*, 1–28. [CrossRef]
147. Acomb, J.C.; Wu, C.; Williams, P.T. Environmental Control of steam input to the pyrolysis-gasification of waste plastics for improved production of hydrogen or carbon nanotubes. *Appl. Catal. B Environ.* **2014**, *147*, 571–584. [CrossRef]

148. Ahamed, A.; Veksha, A.; Yin, K.; Weerachanchai, P.; Giannis, A. Environmental impact assessment of converting flexible packaging plastic waste to pyrolysis oil and multi-walled carbon nanotubes. *J. Hazard. Mater.* **2020**, *390*, 121449. [CrossRef] [PubMed]
149. Miandad, R.; Kumar, R.; Barakat, M.A.; Basheer, C.; Aburiazaiza, A.S.; Nizami, A.S.; Rehan, M. Untapped conversion of plastic waste char into carbon-metal LDOs for the adsorption of Congo red. *J. Colloid Interface Sci.* **2018**, *511*, 402–410. [CrossRef] [PubMed]
150. Acosta, R.; Nabarlatz, D.; Jagiello, J.; Gadonneix, P.; Celzard, A. Adsorption of Bisphenol A on KOH-activated tyre pyrolysis char. *J. Environ. Chem. Eng.* **2018**, *6*, 823–833. [CrossRef]
151. Singh, E.; Kumar, A.; Khapre, A.; Saikia, P.; Kumar, S. Efficient removal of arsenic using plastic waste char: Prevailing mechanism and sorption performance. *J. Water Process Eng.* **2020**, *33*, 101095. [CrossRef]
152. Sogancioglu, M.; Yel, E.; Ahmetli, G. Behaviour of waste polypropylene pyrolysis char-based epoxy composite materials. *Environ. Sci. Pollut. Res.* **2020**, *27*, 3871–3884. [CrossRef] [PubMed]
153. Jamradloedluk, J.; Lertsatitthanakorn, C. Characterization and Utilization of Char Derived from Fast Pyrolysis of Plastic Wastes. *Procedia Eng.* **2014**, *69*, 1437–1442. [CrossRef]
154. Williams, E.A.; Williams, P.T. Analysis of products derived from the fast pyrolysis of plastic waste. *J. Anal. Appl. Pyrolysis* **1997**, *41*, 347–363. [CrossRef]
155. Mastral, F.; Esperanza, E.; García, P.; Juste, M. Pyrolysis of high-density polyethylene in a fluidised bed reactor. Influence of the temperature and residence time. *J. Anal. Appl. Pyrolysis* **2002**, *63*, 1–15. [CrossRef]
156. Mastral, F.J.; Esperanza, E.; Berrueco, C.; Juste, M.; Ceamanos, J. Fluidized bed thermal degradation products of HDPE in an inert atmosphere and in air-nitrogen mixtures. *J. Anal. Appl. Pyrolysis* **2003**, *70*, 1–17. [CrossRef]
157. Santos, E.; Rijo, B.; Lemos, F.; Lemos, M.A.N.D.A. A catalytic reactive distillation approach to high density polyethylene pyrolysis—Part 1—Light olefin production. *Chem. Eng. J.* **2019**, *378*, 122077. [CrossRef]

energies

MDPI

Article

Concept for the Use of Cotton Waste Hydrolysates in Fermentation Media for Biofuel Production

Michal J. Binczarski [1], Justyna Z. Malinowska [1], Joanna Berlowska [2], Weronika Cieciura-Wloch [2], Sebastian Borowski [2], Malgorzata Cieslak [3], Dorota Puchowicz [3] and Izabela A. Witonska [1,*]

1 Institute of General and Ecological Chemistry, Lodz University of Technology, 116 Zeromskiego Street, 90-924 Lodz, Poland; michal.binczarski@p.lodz.pl (M.J.B.); justyna.malinowska@dokt.p.lodz.pl (J.Z.M.)
2 Department of Environmental Biotechnology, Lodz University of Technology, 171/173 Wolczanska Street, 90-924 Lodz, Poland; joanna.berlowska@p.lodz.pl (J.B.); weronika.cieciura-wloch@dokt.p.lodz.pl (W.C.-W.); sebastian.borowski@p.lodz.pl (S.B.)
3 Department of Chemical Textiles Technologies, ŁUKASIEWICZ Research Network-Textile Research Institute, 5/15 Brzezinska Street, 92-103 Lodz, Poland; malgorzata.cieslak@iw.lukasiewicz.gov.pl (M.C.); dorota.puchowicz@iw.lukasiewicz.gov.pl (D.P.)
* Correspondence: izabela.witonska@p.lodz.pl; Tel.: +48-42-631-30-94

Abstract: Currently, most cotton textile waste is sent to landfill. However, due to the use of synthetic additives and the chemical treatment of cotton fibers, cotton textile waste is difficult to biodegrade. Cotton textile waste can also be subjected to material recycling, or to incineration/gasification to produce energy. Here, we present the optimization of acid hydrolysis of cotton yarn fibers for glucose efficiency. The cotton yarn hydrolysates showed great potential for replacing simple sugar solutions in fermentation media. The highest glucose concentration was obtained in the hydrolysates of cotton yarn hydrolyzed in a 2% solution of sulfuric acid or phosphoric acid at 140–160 °C for 2 h. After 2 h of hydrolysis at 140 °C with 2% H_3PO_4, the concentration of glucose in the cotton yarn hydrolysate (13.19 g/L) increased fivefold compared with cotton yarn treated under the same conditions with H_2SO_4 (2.65 g/L). The structural modifications in the solid residues after acid hydrolysis were analyzed using a scanning electron microscope with energy dispersive spectroscopy (SEM-EDS), attenuated total reflectance Fourier-transform infrared spectroscopy (FTIR-ATR), and Raman spectroscopy. The SEM images, IR spectra, and Raman spectra revealed that the most significant changes in the morphology of the fibers occurred when the process was carried out at high temperatures (≥140 °C). Better growth of the yeast strains *Saccharomyces cerevisiae* Ethanol Red and *Saccharomyces cerevisiae* Tokay ŁOCK0204 was observed in the medium containing phosphoric acid hydrolysate. The maximum methane yield of 278 dm^3/kgVS and the maximum hydrogen yield of 42 dm/kgVS were reported for cotton yarn waste after pretreatment with H_3PO_4. This might have been linked to the beneficial effect of phosphorus, which is a key nutrient for anaerobic digestion. The proposed hydrolysis method does not generate fermentation inhibitors.

Keywords: cotton; acid hydrolysis of cotton; ethanol fermentation; biogas production

Citation: Binczarski, M.J.; Malinowska, J.Z.; Berlowska, J.; Cieciura-Wloch, W.; Borowski, S.; Cieslak, M.; Puchowicz, D.; Witonska, I.A. Concept for the Use of Cotton Waste Hydrolysates in Fermentation Media for Biofuel Production. *Energies* **2022**, *15*, 2856. https://doi.org/10.3390/en15082856

Academic Editors: Wei-Hsin Chen, Aristotle T. Ubando, Alok Kumar Patel, Chih-Che Chueh and Liwen Jin

Received: 10 February 2022
Accepted: 11 April 2022
Published: 13 April 2022

1. Introduction

Population growth, globalization, urbanization, and economic growth, as well as purchasing habits, are driving the development of industries throughout the world. As more and more material goods are produced, the production of waste is also increasing. By 2050, worldwide municipal solid waste generation is expected to rise by approximately 70% to 3.4 billion metric tons [1]. One of the largest and most rapidly developing industries is the textile industry. The development of the textile industry also has a negative impact on the environment, first, because textile production requires large quantities of water, energy, and chemicals, and second, because of the sewage and textile waste produced. Global production of textile materials in 2019 was estimated at approximately 100 Mt [2].

Estimated textile production in 2025 will be approximately 121 Mt [3]. In 2019, the global textile market increased to USD 1.053 trillion [4]. The global demand for textiles (clothes, decorative and utility fabrics, specialty fabrics, etc.) [5,6] is growing, and that trend looks set to continue.

Textile wastes can be divided into two main groups: pre-consumer (post-production) textile wastes, which are wastes generated during the textile production processes, and post-consumer textile waste, which is generated during use and disposal by consumers [7,8]. Post-consumer wastes cause problems related to their disposal. Currently, 63% of textiles are made from petrochemical raw materials that cannot be processed by microorganisms [9]. The other 37% are made from natural fibers (cotton, wool, linen, etc.). Large amounts of chemical substances are used to impart functional properties to natural fibers, such as water repellence, fire resistance, and protection from UV radiation. Antibacterial agents, fungicides, insecticides, and permanent pressing resins are also used [10–13]. Due to the use of synthetic additives and the chemical treatment of natural fibers, textile waste is difficult to biodegrade. Textiles, therefore, constitute a significant group of wastes. Currently, most textile waste is either sent to landfill [14,15] or incinerated [7,16,17]. New solutions to the problem of processing and disposing of textile waste are, therefore, being sought.

The main raw material used in all-natural fibers is cotton [18,19]. The annual world production of cotton in the 2018/2019 season amounted to approximately 25 million metric tons. The largest producers were India, the USA, and China [20]. In the UK, cotton accounts for 54.7% of total post-consumer clothing [21]. As a natural and biodegradable material, cotton may seem ecological; but its cultivation requires large areas of land and processing into textiles demands large amounts of water, chemicals, and energy. Therefore, there is great interest in the sustainable management of cotton waste [21–25]. Waste garments are mainly collected, sorted depending on the color/fabric type, and converted into regenerated/reclaimed fibers, which can be used in various yarns [26,27]. However, the amount of recycled fiber used for spinning is not more than 30% [28].

After it has been reused or recycled into new products, cotton finally becomes unsuitable for further textile processes. Such unusable waste includes cotton dust, post-production yarn waste, woven and knitted fabrics, scraps after processing used clothes into new clothing, and other products. In accordance with the principles of the circular economy (3R—reuse, reduce, recycle), cotton waste should be exploited as much as possible, not only by the textile industry. Various methods based on the destruction of the physical or chemical structure of cotton waste have been investigated [29,30].

Cotton is a natural polymer containing mainly cellulose (88.0–96.5%). Other components include pectins, proteins, waxes, fats, and minerals [31–33]. Cellulose is a linear polymer composed of glucopyranose residues linked by 1,4-β-glycosidic bonds. The heterocyclic rings found in cellulose contain one primary hydroxyl group ($–CH_2OH$) and two secondary hydroxyl groups (–OH) [34,35].

The chemical properties of cotton are determined by the properties of its basic component, namely, cellulose. Cellulose is involved in addition and decomposition, with a decrease in the degree of polymerization (various types of hydrolysis) and substitution or oxidation (etherification, esterification). The most important reactions of cellulose fibers are oxidative degradation, thermal and radiation modification, hydrolysis, substitution, addition, cross-linking, dyeing, and interphase polycondensation [31]. Cotton hydrolysis is the process of breaking the β-1,4-glycosidic bonds in cellulose and reducing its degree of polymerization [36]. There are several methods that can be used for cotton fiber hydrolysis: acid hydrolysis, alkaline hydrolysis, enzymatic hydrolysis, and thermal hydrolysis. Hydrolysis is influenced by various factors, such as the type of solvent, the heating source, and the duration of the process. The hydrolysis process transforms cotton into glucose and solid fiber residues. The glucose can be used for various purposes, including bioethanol production [35,37–40].

An alternative method of processing textile waste containing cotton fibers is thermally assisted acid hydrolysis. After appropriate supplementation, the hydrolysates can be used as fermentation substrates for bioenergy production (biogas, bioethanol) or other bioproducts, such as lactic acid and gluconic acid. This study investigated the influence of temperature, type of acid, and acid concentration on the hydrolysis of cotton waste. The hydrolysates after supplementation were used as fermentation media in dark fermentation processes and for ethanol fermentation.

2. Materials and Methods

2.1. Materials

Cotton yarn Z689 × 2 S 542 with a linear density of 58.73 ± 0.641 tex, fiber length 15–32 mm, specific strength 18.44 ± 1.089 cN/tex, breaking strength 1083 ± 64 cN, and relative elongation at break 5.8 ± 0.3% was first cut in an MRC RJM30D ball mill. Cotton yarn is a post-production waste. It was washed at 60 °C and rinsed three times to remove spinning preparations. Images of the cotton yarn at various magnifications are presented in Figure 1.

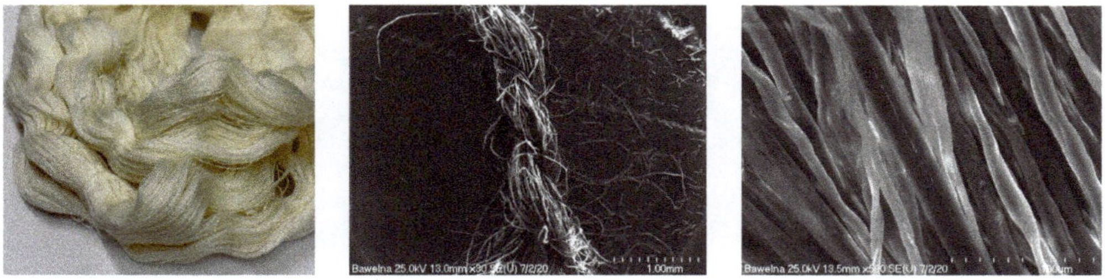

(**a**) (**b**) (**c**)

Figure 1. (**a**) Macroscopic photo of the cotton yarn used in this study; (**b**) SEM image at 30× magnification of the cotton yarn; (**c**) SEM image at 500× magnification of the cotton yarn.

2.2. Acid Hydrolysis of Cotton

Aqueous glucose solutions were obtained from the cotton yarn via thermally assisted acid hydrolysis. Cotton yarn hydrolysis was performed in pressure reactors with volumes of 50 cm^3 (Parr Instrument Company 4552 Series Mini Reactor, Moline, IL, USA), 1 L (Parr Instrument Company 4577 Series Reactor, Moline, IL, USA), and 7.99 L (Parr Instrument Company, 4551 Moveable Cart Stand Reactor with 4848 Controller, Moline, IL, USA). The hydrolysis reactions were carried out at 80–200 °C for 1–8 h in a 50 cm^3 reactor, using 1 g of comminuted cotton yarn and 20 cm^3 of an aqueous solution of sulfuric acid (H_2SO_4, 95%, Stanlab, Lublin, Poland) at concentrations of 0.5–10%. To scale up the reaction, reactors with a volume of 1 L and 7.99 L were used. The reactions were carried out under optimized conditions (140 °C for 2 h, 2% sulfuric acid/phosphoric acid (H_3PO_4, 85%, Stanlab, Lublin, Poland)), using 20 g or 150 g of crushed cotton yarn and 600 cm^3 or 4500 cm^3 of aqueous acid solution. The reaction mixtures were cooled in the reactor to room temperature, neutralized via the addition of NH_4OH (25% NH_4OH, POCh, Gliwice, Poland) to pH 7–7.5, and filtered to remove solid yarn residues on a funnel with a sintered disc lined with a quality hard filter (Filtrak, Ahlstrom-Munksjö Group, Helsinki, Finland). The glucose concentration in the hydrolysates were determined using high-performance liquid chromatography (HPLC, Sykam GmbH, Eresing, Germany, with an S1125 pump system, S 5300 autosampler, S 4115 column thermostat, and RI S 3585 detector). The sugars were separated on a SETREX IEX-H^+ column (300 × 8.0 mm ID) at 80 °C using 0.008 $mol \cdot dm^{-3}$ H_2SO_4 + 2% v/v ACN (flow 0.8 $cm^3 \cdot min^{-1}$) as the mobile phase. Quantitative analysis of glucose was performed on the basis of a calibration curve plotted for the concentration range of 0–10 $g \cdot dm^{-3}$ (the curve in the analyzed range was linear y = 0.19733 x, R^2 = 0.9998803).

2.3. SEM-EDS

A scanning electron microscope (SEM S-4700, Hitachi, Tokyo, Japan) equipped with energy dispersive spectroscopy (EDS) capability (Thermo-Noran Inc., Madison, WI, USA) was used for SEM analysis of the hydrolysates. Samples were embedded in conductive carbon pads and the excess loose powder was removed. To reduce electric charging, the samples were sputter-coated with carbon (Cressington 208 HR system). Images were acquired in back-scattered electron (BSE) mode. An accelerating voltage of 25 kV was used. For the purposes of comparison, the same samples were analyzed using an FEI Quanta 650 SEM (FEI Company, Hillsboro, OR, USA) equipped with a Bruker Energy Dispersive Spectroscopy (EDS) system (Bruker Corporation, Billerica, MA, USA). A 15 kV accelerating voltage was used with a 3.5 μA electron beam current and a 10 mm working distance. The compositions of each sample were measured at least three times at different locations that were approximately 0.25 mm^2 in size.

2.4. FTIR-ATR Spectroscopy

Infra-red (IR) analyses were made on a VERTEX 70 FTIR spectrometer (Bruker, Bremen, Germany) with an ATR Golden Gate Diamond Accessory (Specac, Orpington, UK). Spectra were obtained in absorption mode. The measuring range was from 600 cm^{-1} to 4000 cm^{-1}. The spectral resolution was one data point per 2 cm^{-1}. In most cases, 64 scans were acquired. The scans were Fourier-transformed and the values were averaged for each sample. Spectra were registered using Bruker OPUS 6.5 software (Version 6.5, Bruker, Kennewick, WA, USA), then processed with Microcal Origin 8.0 (Version 8.0, Originlab Corporation, Northampton, MA, USA) software [41].

2.5. Raman Spectroscopy

A Renishaw InVia Reflex Raman-dispersive spectrometer (Renishaw, Wotton under Edge, UK) with a Leica microscope (Leica, Wetzlar, Germany) was used. Spectra were obtained in the range of 100–3300 cm^{-1} with a spectral resolution of 1 cm^{-1}. An excitation source with λ = 785 nm at 300 mW was applied. The laser power (from 1% to 5%) that was used depended on the sample. The analysis was made in the closed microscope chamber of the spectrometer. Samples were situated in the focus of the laser light using a 50× microscope objective and CCD Camera. The results were recorded in the Renishaw WIRE 5.3 program (Version 5.3, Wotton under Edge, UK) [41] and then processed with Microcal Origin 8.0 software (Version 8.0, Originlab Corporation, Northampton, MA, USA).

2.6. Yeast Cell Multiplication Using Cotton Hydrolysates as the Cultivation Medium

To prepare the fermentation medium, 40 cm^3 of hydrolysate was collected, to which 0.4 g of yeast extract (ChemiLab, Tarnobrzeg, Poland) was added, together with 0.8 g of K-peptone (BTL sp. Z o.o., Lodz, Poland). The hydrolysate supplemented with yeast extract (1%) and K-peptone (2%) was sterilized at 121 °C for 15 min.

To describe the growth kinetics, tests in microtiter plates were carried out. The sterilized medium was transferred to a sterile microtiter 96-channel plate at 180 μL and inoculated with 20 μL of an inoculum suspension of yeast cells (containing at least 10^8 yeast cells) in MEB medium (Merck Millipore, Bedford, MA, USA). The yeast strains Saccharomyces cerevisiae Ethanol Red (Fermentis Division S.I., Lesaffre, Marcq-en-Baroeul, France) and Saccharomyces cerevisiae Tokay ŁOCK 0204 (ITFiM PŁ collection, Lodz, Poland) were used. Cultivation was carried out at 25 °C for 72 h in a Thermo Scientific Multiskan GO UV/VIS spectrophotometer equipped with a thermostat and mixing module (Thermo Fisher Scientific, Waltham, MA, USA). The absorbance of the solutions was measured at a wavelength of 620 nm. The increase in absorbance was proportional to the turbidity caused by the proliferation of microorganisms in the solution.

Probe tests were carried out to assess the level of yeast multiplication. The medium (supplemented hydrolysate) was transferred in 10 cm^3 portions to glass probes and sterilized. The medium was inoculated with 0.5 cm^3 of yeast suspensions of Saccharomyces cerevisiae Ethanol Red (Fermentis Division S.I., Lesaffre, Marcq-en-Baroeul, France) and Saccharomyces cerevisiae Tokay ŁOCK 0204 (ITFiM PŁ collection, Lodz, Poland) containing at least 10^8 yeast cells. Cultivation was carried out at 25 °C for 72 h. The number of colony-forming units was estimated using the standard plate count method.

2.7. Cultivation of a Microorganism Consortium for Methane and Hydrogen Production

In anaerobic digestion and dark fermentation tests, a consortium of anaerobic microorganisms was used as the inoculum. The inoculum represented anaerobically digested sewage sludge collected from the mesophilic anaerobic digestion tank of the Group Wastewater Treatment Plant in Lodz (Poland), with a biomass concentration of 24.59 g $DM \cdot dm^{-3}$. For the fermentation of methane and hydrogen, the cotton yarn hydrolysates were mixed with the inoculum in the ratio of 1:2 on a volatile solids basis.

For methane production (AD) tests, the anaerobic sludge was mixed with the substrate and adjusted to pH 7 with 20% NaOH (NaOH, Stanlab, Lublin, Poland), which is optimal for methanogenic archaea. To obtain the most efficient hydrogen production in dark fermentation experiments, it was necessary to eliminate hydrogen-consuming microorganisms, which were mainly methanogens. Therefore, the substrate and the inoculum were adjusted to pH 5.5 with 20% H_2SO_4 and then thermostated at 80 °C for 1.5 h. Both batch fermentations (DF and AD were carried out in glass reactors with capacities of 1 L (working volume 0.7 L), which were tightly connected to the biogas collection tanks. Daily biogas yield was measured on the basis of the amount of water displaced from the 1 L gas tank. Prior to initiating the fermentation process, to create anaerobic conditions air from the reactors was purged by flushing them with nitrogen gas for about 5 min. The glass reactors were incubated in a thermostat under mesophilic conditions, keeping the temperature constant at 35 °C. The reactors were mixed by hand twice a day. The fermentations were continued for 30 days or stopped when there was only residual biogas production.

The biogas efficiency was monitored qualitatively and quantitatively on a daily basis. The content of CH_4 and H_2 was measured with a Madur GA-21 plus portable gas analyzer (Madur Polska sp. z o.o., Zgierz, Poland). The yields of methane and hydrogen were converted into the amount of gas produced per kg of organic dry matter under normal conditions so that it was possible to compare the results with the literature data.

3. Results and Discussion

Cotton fibers consist almost entirely of cellulose (94–100%). Cotton hydrolysis, therefore, leads to glucose as the main product. Acid hydrolysis of the cotton yarn was performed in a Parr Instrument Company Series Mini Reactor 4592 with a volume of 50 cm^3. Hydrolysis was carried out in the temperature range of 80–200 °C for 1–8 h with the use of a H_2SO_4 catalyst at concentrations of 0.5–10%. Prior to HPLC analysis, to determine the glucose concentration samples of the hydrolysates were neutralized with NH_4OH to pH 7.5. The results are summarized in Figure 2, as the arithmetic means of the glucose concentrations of the hydrolysates from the three hydrolysis processes.

The highest concentration of glucose was obtained in the hydrolysates of cotton yarn hydrolyzed in a 2% solution of sulfuric acid in the temperature range of 140–160 °C for 2 h. Our previous studies of cotton hydrolysates obtained via the action of sulfuric acid revealed that only above the temperature of 160 °C, trace amounts of (4.970 min) furfural and (8.558 min) 5-methyl-2-furanocarboxaldehyde were formed in the reaction and a temperature increase led to the creation of other products, such as (7.885 min) 2(5H)-furanone, 5-methyl- and (12.438 min) levulinic acid [42]. What is more, further increasing the temperature of the process of cotton fiber hydrolysis to 200 °C reduced the glucose concentration in the hydrolysates as a result of caramelization and Maillard reactions [43], the products of which may be fermentation inhibitors.

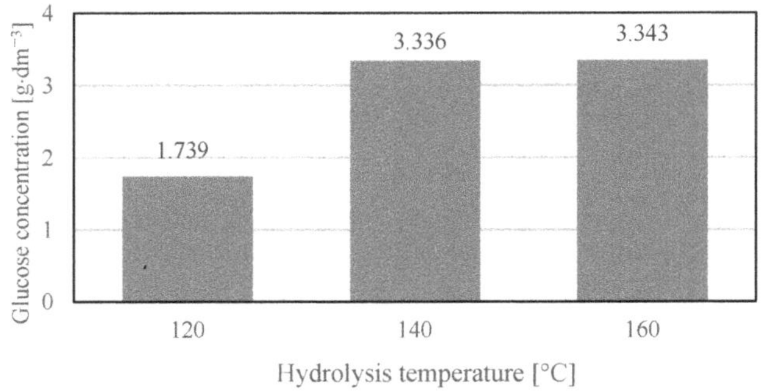

Figure 2. Glucose concentration in hydrolysates depending on the temperature of hydrolysis. Reaction conditions: 2 h, 2% H_2SO_4. Glucose concentrations were the arithmetic mean of the results obtained in the three hydrolysis processes with the standard deviations: 1.739 ± 0.157; 3.336 ± 0.167; 3.343 ± 0.137.

SEM images (Figure 3) of the solid residue after hydrolysis of the cotton fibers clearly showed that the most significant changes in the morphology of the fibers occurred when the process was carried out at high temperatures (≥160 °C). The sample hydrolyzed at 160 °C contained the highest concentration of glucose (3.34 g·dm^{-3}). However, when the temperature of hydrolysis was increased to 200 °C, glucose was not detected and the only products of the reaction were the products of its further transformation. Although no significant changes in the morphology of the cotton fibers were observed in the SEM images after hydrolysis at 140 °C, the amounts of glucose released into the hydrolysate were also very high (3.34 g·dm^{-3}). To study the changes in the structure of cotton fibers, IR and Raman spectroscopy techniques were used.

Both in the IR spectra (Figure 4, Table 1) and Raman spectra (Figure 5, Table 1), changes in the cotton were observed at temperatures above 120 °C. However, the sample treated at 140 °C (Figure 4c) mainly showed changes in the IR band intensities. In the Raman spectra (Figure 5c), these changes were more pronounced, as the bands of the ring in the region 300–600 cm^{-1} were more intense. Glycosidic COC ring breathing and stretching vibrations (1101 cm^{-1}, 1122 cm^{-1}) decreased. In the region of 1200–1600 cm^{-1}, new overlapping bands appeared. The CH stretching band at 2900 cm^{-1} disappeared and new bands of the decomposition products were visible at 1182 cm^{-1}, 1278 cm^{-1}, 1496 cm^{-1}, 1590 cm^{-1}, and 1707 cm^{-1}. These changes were more evident in the Raman spectra for cotton treated at 160 °C (Figure 5d) and 200 °C (Figure 5e). The bands characteristic of cotton completely disappeared. In the IR spectra, C-OH and COC ring bands, as well as CH bending bands, became less visible when cotton was treated at 160 °C (Figure 4d) and finally disappeared when the sample was treated at 200 °C (Figure 4e). At the same time, new bands with maxima at 1701 cm^{-1} and 1608 cm^{-1} appeared and bands in the 2900–3600 cm^{-1} region differed significantly in shape and intensity. Characteristic cotton IR bands in the region 900–1200 cm^{-1} completely disappeared and a group of new overlapping bands in the region 1200–1500 cm^{-1} became more intense. These bands were derived from solid cellulose decomposition products after the release of gases, mainly hydrocarbons [44].

Another important parameter influencing the amount of glucose obtained in cotton hydrolysates is the concentration of H_2SO_4 used as the catalyst. Hydrolysis of glucose samples was carried out at 120 °C for 2 h. Each time, a different concentration of sulfuric acid was used in the range of 0.5–5%. Figure 6 presents the results of the HPLC analysis of the sugar composition of the hydrolysates.

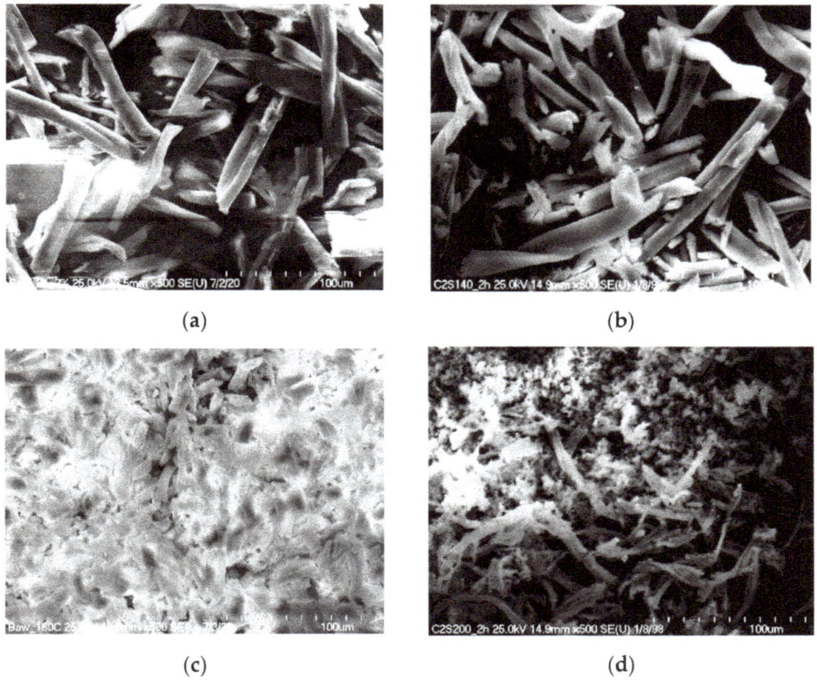

Figure 3. SEM pictures of the solid residue after hydrolysis of cotton, depending on the temperature of the process (2 h, 2% H_2SO_4): cotton treated at (**a**) 120 °C, (**b**) 140 °C, (**c**) 160 °C, and (**d**) 200 °C.

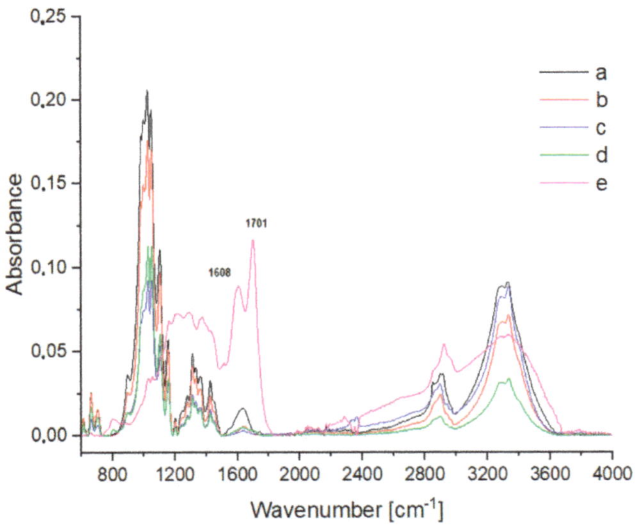

Figure 4. IR spectra of (**a**) washed cotton and cotton treated for 2 h with 2% H_2SO_4 at various temperatures: (**b**) 120 °C; (**c**) 140 °C; (**d**) 160 °C; (**e**) 200 °C.

Table 1. IR and Raman characteristic bands of cotton.

IR [cm^{-1}]	Description [44]	Raman [cm^{-1}]	Description [45,46]
		330, 383	CCC, CCO ring deformation
1029	C-OH alcohol stretching	437, 461	CCC ring deformation
1053	C-OH alcohol stretching	523	COC glycosidic linkage deformation
1106	COC in glycosodic linkage stretching	902, 1001	CH skeletal rotation
1160	CCC ring asymmetric stretching	1101	COC glycosidic ring breathing symmetric stretching
1314	CH bending	1122	COC glycosodic ring breathing asymmetric stretching
1426	CH bending	1344	CH_2 vibration
1600	H_2O adsorbed	1385	CH_2 vibration
2916	CH stretching	1484	CH_2 vibration
3333	OH stretching	2900	CH stretching

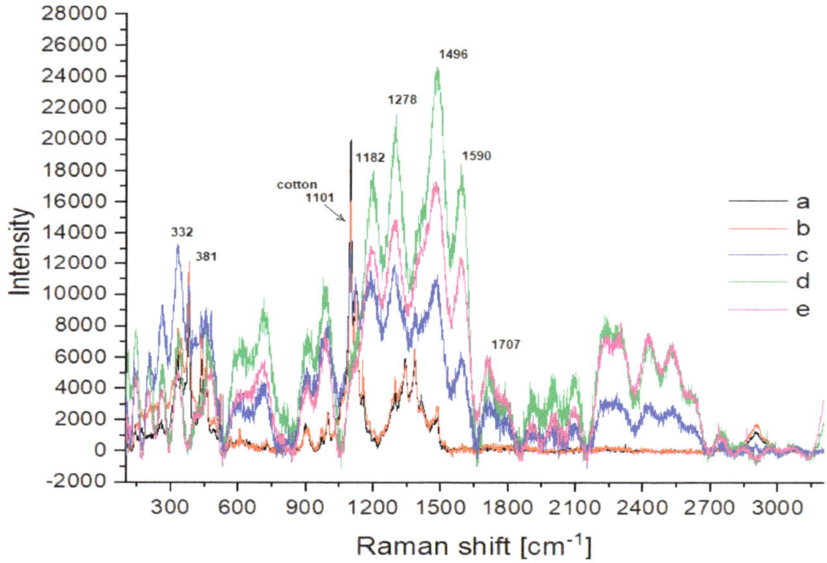

Figure 5. Raman spectra of (**a**) washed cotton and cotton treated for 2 h with 2% H_2SO_4 at various temperatures: (**b**) 120 °C; (**c**) 140 °C; (**d**) 160 °C; (**e**) 200 °C.

High glucose concentrations were also noted in the hydrolysates when cotton yarn hydrolysis was performed for 2 h in concentrated H_2SO_4 solutions (5 and 10% v/v) at a temperature of 120 °C. The use of concentrated acid solutions in industrial reactors is undesirable due to their high corrosivity, as well as the low-quality fermentation media obtained. From the industrial point of view, it is therefore preferable that the hydrolysis process is carried out at the highest possible temperatures, namely, 140–160 °C, with the concentration of the acid catalyst as low as possible.

Figure 7 shows SEM images of the solid residue after hydrolysis. As can be seen, changes in the morphology of the fibers occurred when the process was conducted at 120 °C for 2 h with concentrations of 5% and 10% sulfuric acid. High glucose concentrations (5%—3.216 g·dm^{-3} and 10%—3.323 g·dm^{-3}) were also recorded for both of these samples. The morphologies of the fibers in the solid residues after hydrolysis with a low concentration of sulfuric acid (0.5%) were not significantly altered. This was in agreement with the glucose concentration in the hydrolysate, which was also very low (0.277 g·dm^{-3}). IR and Raman spectroscopy studies were additionally performed to confirm changes in the fiber structure.

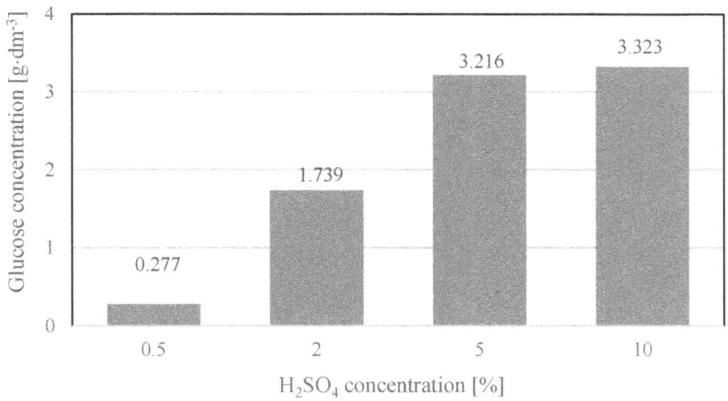

Figure 6. Glucose concentration in hydrolysates, depending on the concentration of the acid used. Reaction conditions 2 h, 120 °C. Glucose concentrations were the arithmetic means of the results obtained in three hydrolysis processes with the standard deviations: 0.277 ± 0.055; 1.739 ± 0.157; 3.216 ± 0.154; 3.323 ± 0.263.

Figure 7. SEM photos of the solid residue after hydrolysis (120 °C, 2 h) of cotton with various concentrations of H_2SO_4: (**a**) 0.5%; (**b**) 2%; (**c**) 5%; (**d**) 10%.

No significant changes were noted in the IR and Raman spectra of the samples exposed to sulfuric acid at various concentrations (Figures 8 and 9). The IR spectrum of the sample treated with 0.5% H_2SO_4 (Figure 8b) differs from the other spectra, but the changes concern only the band intensity. The higher intensity of the OH band (region 3200–3500 cm^{-1}) indicates the influence of water. The Raman spectrum of cotton treated with 5% H_2SO_4 (Figure 9d) differed slightly from the others, but no new bands could be distinguished.

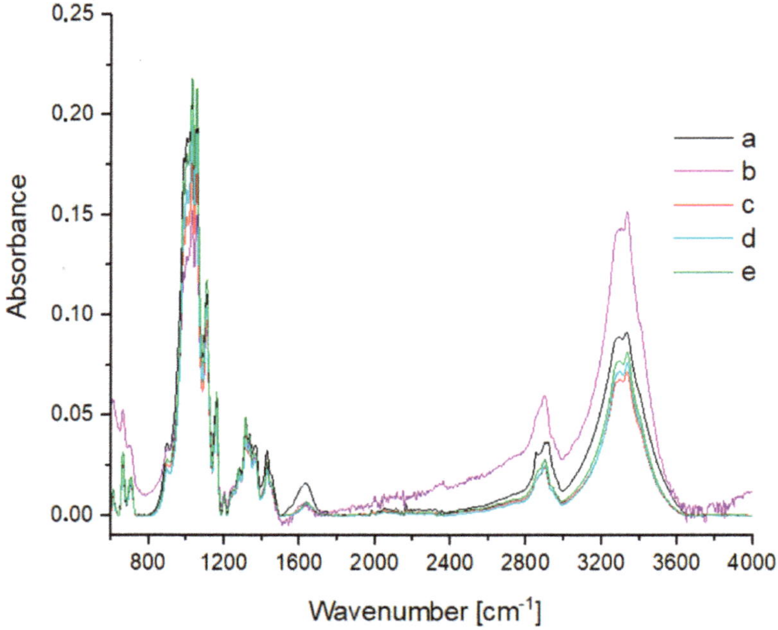

Figure 8. IR spectra of (**a**) washed cotton and cotton treated at 120 °C for 2 h with various concentrations of H_2SO_4: (**b**) 0.5%; (**c**) 2%; (**d**) 5%; (**e**) 10%.

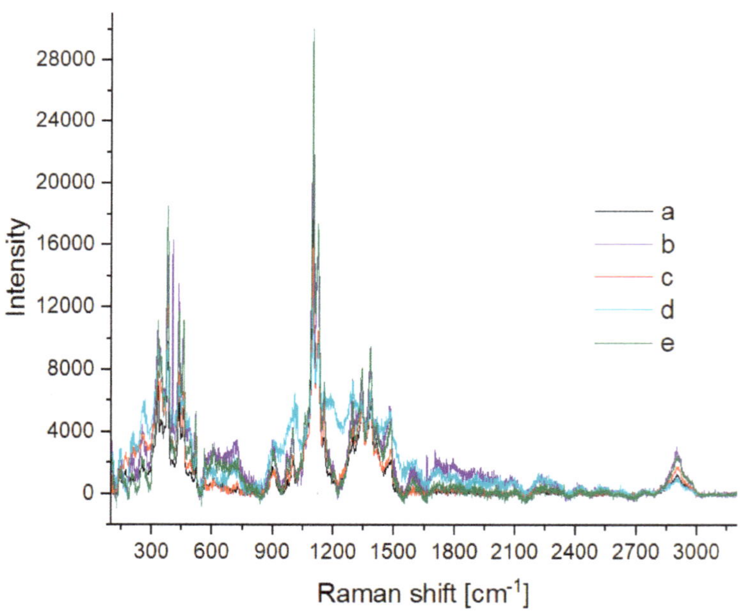

Figure 9. Raman spectra of (**a**) washed cotton and cotton treated at 120 °C for 2 h with various concentrations of H_2SO_4: (**b**) 0.5%; (**c**) 2%; (**d**) 5%; (**e**) 10%.

Another important parameter that influences the amount of glucose in cotton hydrolysates is the time of the hydrolysis process. The data presented in Figure 10 show that increasing the time of the hydrolysis process from 1 h to 8 h promoted the depolymerization of cellulose and the formation of more glucose.

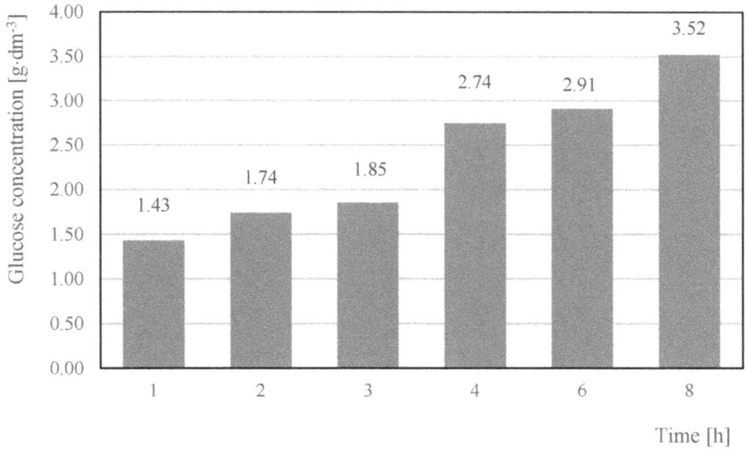

Figure 10. Glucose concentration in hydrolysates, depending on the time of the process. Reaction conditions: 120 °C, 2% H_2SO_4. Glucose concentrations were the arithmetic means of the results obtained in the three hydrolysis processes with the standard deviations: 1.43 ± 0.16; 1.739 ± 0.16; 1.85 ± 0.15; 2.74 ± 0.19; 2.91 ± 0.24; 3.52 ± 0.33.

Figure 11 shows SEM images of the solid residue after hydrolysis was performed for various reaction times. As can be seen, there were no visible changes in the morphologies of the fibers. To examine the changes in the cotton fibers depending on the hydrolysis time, IR and Raman tests were performed.

Figure 11. SEM pictures of the solid residue after various hydrolysis times (120 °C, 2% H_2SO_4): cotton treated for **(a)** 1 h, **(b)** 2 h, **(c)** 3 h, **(d)** 4 h, **(e)** 6 h, and **(f)** 8 h.

Differences can be observed in the IR and Raman spectra of the cotton treated at 120 °C with 2% H_2SO_4 depending on the duration of hydrolysis (Figures 12 and 13). In the IR spectra for samples exposed for 6 h and 8 h to acid hydrolysis, bands for the cellulose ring C−OH and COC bands decreased (Figure 12f,g). The same time bands for OH at 3333 cm^{-1} increased in intensity. In the Raman spectrum, a new band at 1707 cm^{-1} was observed (Figure 13f,g). When cotton was exposed for 6 h to 2% H_2SO_4 at 120 °C (Figure 13f), bands characteristic of cotton disappeared, but a broad band between 600 cm^{-1} and 1000 cm^{-1} appeared. After 8 h of decomposition (Figure 12), new bands formed derived from solid cellulose decomposition products after the release of gases. The new bands were mainly for hydrocarbons [40], similarly to those presented in Figure 5.

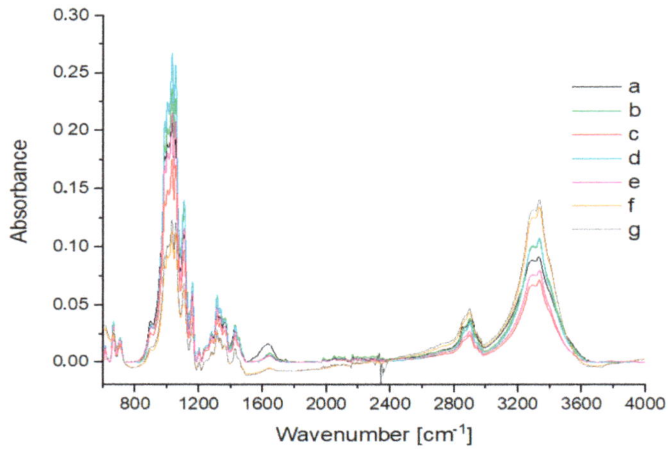

Figure 12. IR spectra of (**a**) washed cotton and cotton treated at 120 °C with 2% H_2SO_4 for various reaction times: (**b**) 1 h; (**c**) 2 h; (**d**) 3 h; (**e**) 4 h; (**f**) 6 h; (**g**) 8 h.

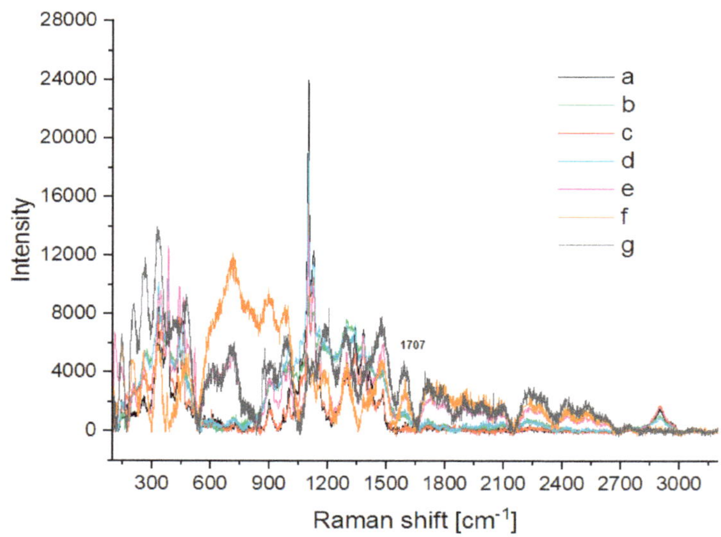

Figure 13. Raman spectra of (**a**) washed cotton and cotton treated at 120 °C with 2% H_2SO_4 for various reaction times: (**b**) 1 h; (**c**) 2 h; (**d**) 3 h; (**e**) 4 h; (**f**) 6 h; (**g**) 8 h.

Based on the results of acid hydrolysis carried out in the microreactor (V = 50 cm^3), the cotton yarn was hydrolyzed at a larger scale under optimized conditions in a Parr Instrument Company Reactor 4577, which is a pressure reactor with a working capacity of 1 L. Cotton hydrolysis with sulfuric acid and phosphorus acid at a concentration of 2% was performed in the temperature range of 120–160 °C. The reactions were continued for 2 h. Prior to the HPLC analysis to determine glucose concentration (Figures 14 and 15), samples of the hydrolysates were neutralized with NH_4OH to pH 7.5. Samples of the hydrolysates were used for the preparation of fermentation media.

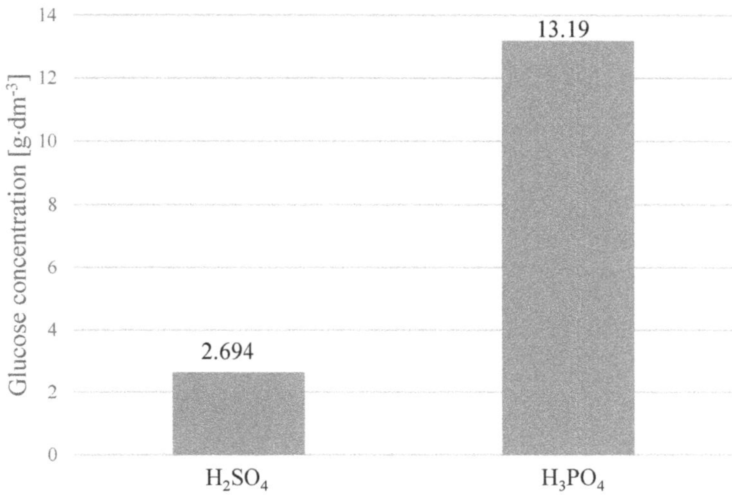

Figure 14. Glucose concentration in hydrolysates, depending on the type of acid used. Reaction conditions: 2 h, 140 °C, 2% H_2SO_4 or 2% H_3PO_4. Glucose concentrations were the arithmetic means of the results obtained in the three hydrolysis processes with the standard deviations: 2.694 ± 0.172; 13.19 ± 0.475.

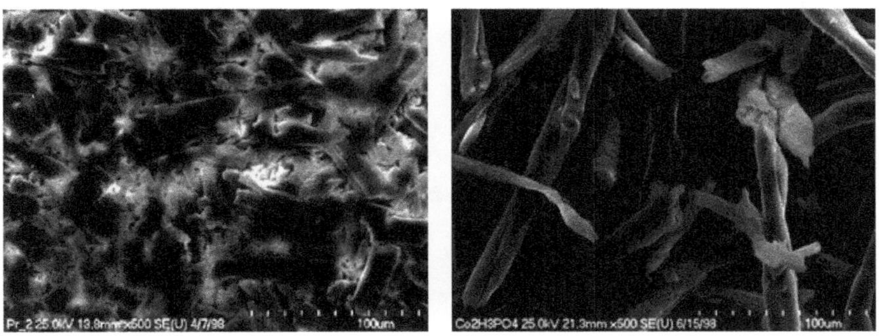

(**a**) (**b**)

Figure 15. SEM pictures of the solid residue after hydrolysis (140 °C; 2 h), depending on the type of acid used: (**a**) cotton treated 2% H_2SO_4; (**b**) cotton treated 2% H_3PO_4.

The use of 2% phosphoric acid as a catalyst for hydrolysis of the β-1,4-glycosidic bonds in cellulose present in cotton fibers had a positive effect on the amount of glucose formed. By continuing the cotton yarn hydrolysis at 140 °C with 2% H_3PO_4 for 2 h, the concentration of glucose in the hydrolysate was five times higher than that obtained under the same conditions with H_2SO_4. The use of phosphoric acid instead of sulfuric acid was also beneficial from the point of view of using hydrolysates as components of the fermentation medium because the ammonium phosphate formed as a result of neutralization was a source of nutrients, which were necessary for the growth of microorganisms.

The structure of the cotton fibers in the solid residue after hydrolysis with sulfuric acid showed much greater destruction compared to the structure of the solid residue after hydrolysis catalyzed with phosphoric acid. However, higher glucose yields were obtained after hydrolysis with phosphoric acid.

Hydrolysis was also carried out under optimized conditions in a Parr Instrument Company Reactor 4551 with a volume of 7.99 L. The results were very similar. The hydrolysates were used in biological materials as a nutrient.

3.1. Yeast Cultivation on Cotton Substrates

A series of biological tests were conducted, which confirmed that the obtained hydrolysates were suitable for biotechnological applications. Cotton yarn hydrolysates neutralized with NH_4OH to pH 6.5, mainly containing glucose, could be used as a carbon source for the biosynthesis of yeast biomass. The nutritional components of the hydrolysates were assimilated and metabolized, resulting in the multiplication of yeast cells. Both supplemented hydrolysates obtained via the action of H_2SO_4 and H_3PO_4 at 140 °C constituted suitable media for the cultivation of the selected yeast strains (Figures 16 and 17). Both *Saccharomyces cerevisiae* Ethanol Red and *Saccharomyces cerevisiae* Tokay LOCK0204 showed greater increases in yeast biomass when cotton hydrolysates obtained via the decomposition of cotton yarn at 140 °C for 2 h with 2% H_3PO_4 were used as sugar nutrients. The highest absorbance was recorded for the yeast *Saccharomyces cerevisiae* Ethanol Red (Figure 16) in a fermentation medium containing hydrolysate from the process using phosphoric acid on cotton yarn at 140 °C for 2 h.

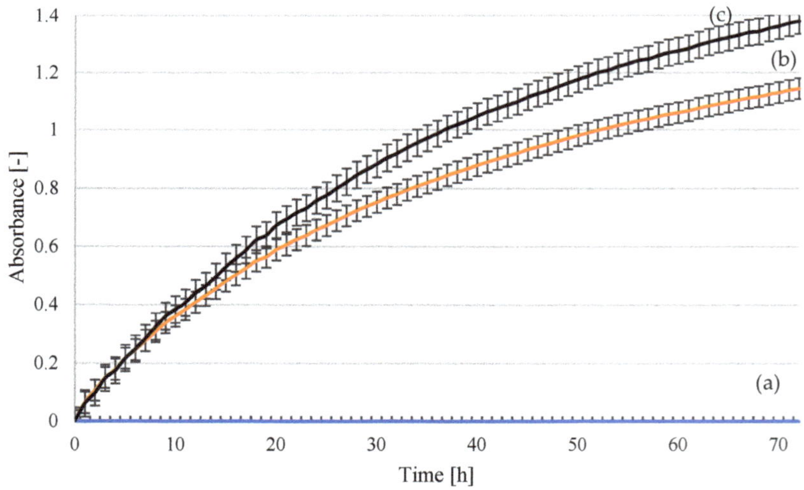

Figure 16. Yeast growth curves for *Saccharomyces cerevisiae* Ethanol Red strain in fermentation media: (**a**) uninoculated medium–reference sample; (**b**) hydrolysate following treatment at 140 °C, 2 h, 2% H_2SO_4; (**c**) hydrolysate following treatment at 140 °C, 2 h, 2% H_3PO_4. The curves show the mean values obtained in three independent measurements.

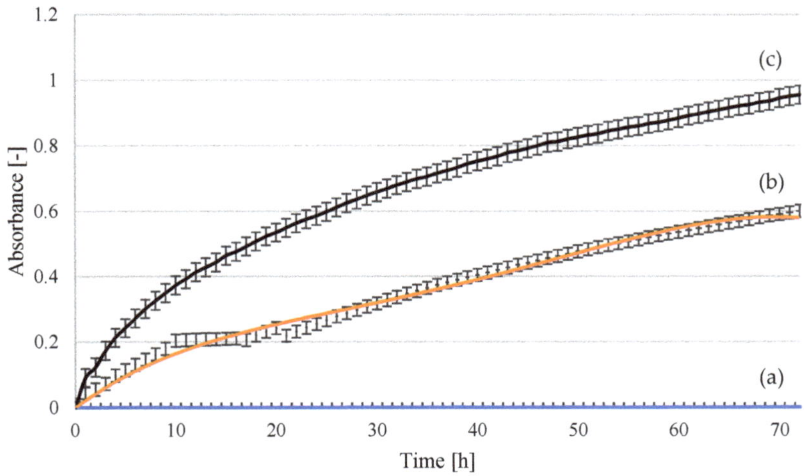

Figure 17. Yeast growth curves for the *Saccharomyces cerevisiae* Tokay LOCK0204 strain in fermentation media: (**a**) uninoculated medium—reference sample; (**b**) hydrolysate following treatment at 140 °C, 2 h, 2% H_2SO_4; (**c**) hydrolysate following treatment at 140 °C, 2 h, 2% H_3PO_4. The curves show the mean values obtained in three independent measurements.

Lower optical density values and a lower number of colony-forming units (Table 2) were observed for the yeast strain *Saccharomyces cerevisiae* Tokay ŁOCK0204 than for *Saccharomyces cerevisiae* Ethanol Red, regardless of the hydrolysate used in the medium. Better growth was observed in the medium containing phosphoric acid hydrolysate. However, in all cases, the stationary phase of growth was not reached during the first 48 h, as happens in model media. The environment of the tested hydrolysates may be conducive to extending the growth time or the rate of utilization of nutrient compounds. This may be related to the release of biologically active compounds during the hydrolysis processes. As reported in previous research [47], this group does not include furfural and levulinic acid.

Table 2. Yeast cell multiplication.

	Strain	Cultivation Time	Medium: Hydrolysate following Treatment at 140 °C, 2 h, 2% H_2SO_4	Medium: Hydrolysate following Treatment at 140 °C, 2 h, 2% H_3PO_4
Colony forming units [cfu/cm^3]	*Saccharomyces cerevisiae* Ethanol Red	0 h	$1.8 \cdot 10^6$	$1.7 \cdot 10^6$
		48 h	$6.8 \cdot 10^7$	$8.6 \cdot 10^7$
		72 h	$9.2 \cdot 10^7$	$1.0 \cdot 10^8$
	Saccharomyces cerevisiae Tokay ŁOCK0204	0 h	$1.1 \cdot 10^6$	$1.4 \cdot 10^6$
		48 h	$3.5 \cdot 10^7$	$5.5 \cdot 10^7$
		72 h	$5.2 \cdot 10^7$	$8.7 \cdot 10^7$

Our results showed that cotton yarn hydrolysate has great potential for replacing simple sugar solutions in fermentation media. The developed processes did not generate fermentation inhibitors. Regardless of the acid used, a systematic increase in yeast biomass was observed. In future work, processes conducted in glass reactors on a laboratory scale will enable full qualitative and quantitative analysis of fermentation products, with particular emphasis on ethanol.

3.2. Cultivation of a Consortium of Microorganisms for the Production of Methane and Hydrogen on Substrates from Cotton Yarn Hydrolysates

Natural cotton fiber is composed of 88–95% cellulose, along with proteins (1.0–1.9%), waxes (0.4–1.2%), pectins (0.4–1.2%), and inorganic compounds (0.7–1.6%). Its composition suggests that it could be an appropriate substrate for methane and hydrogen production [48]. However, the complexity of its structure inhibits the process of fermentation and makes pretreatment necessary. Hydrolysis allows for the release of glucose, but also slightly increases the amount of nitrogen and phosphorus available to microorganisms. Among the numerous hydrolysis treatments available, dilute acid was shown to be the most suitable method for industrial applications and is the most commonly applied for a variety of lignocellulosic biomasses. Our previous research [49] also indicated that acid hydrolysis would be the most appropriate form of pretreatment.

In the present study, to improve the anaerobic digestion (AD) and dark fermentation (DF) efficiency, two acid pretreatment methods were used, with 2% H_2SO_4 (AD-1, AD-2, DF-1, DF2) and 2% H_3PO_4 (AD-3, AD-4, DF-3, DF-4). We also investigated the influence of pH adjustment using NaOH (AD-1, AD-3, DF-1, DF-3) and KOH (AD-2, AD-4, DF-2, DF-4) on the fermentation in batch tests under mesophilic conditions. The AD process was carried out at pH 7 under mesophilic conditions. The DF process was conducted at approximately pH 5.5, followed by heating (1.5 h, 80 °C) to avoid methanation. The characteristics of the applied inoculum and hydrolyzed cotton fibers are presented in Table 3. The inoculum showed greater levels of nitrogen and phosphorous and very low levels of COD, as well as total and volatile solids. Compared with the representative values of nitrogen and phosphorous found in cotton hydrolysates, these concentrations were typically 200 times higher or 300 times higher. The addition of anaerobic sludge not only inoculated the substrate with microorganisms but also provided missing micro- and macroelements, as well as dilution of toxic and inhibitory substances. The lack of elements such as nitrogen or phosphorus significantly reduces the efficiency of the digestion process, even if the substrate contains high amounts of sugars. This was confirmed by our recent study, in which the addition of nitrogen and phosphate compounds was found to have a positive effect on the process of fermentation [50].

Table 3. Characteristics of undiluted substrates used for batch tests.

Indicator	Unit	Inoculum	Cotton Hydrolysate (H_2SO_4)	Cotton Hydrolysate (H_3PO_4)
pH	-	7.47 ± 0.12	5.51 ± 0.15	5.83 ± 0.09
Total solids	g/kg	17.81 ± 0.68	29.04 ± 1.08	27.64 ± 1.21
Volatile solids	g/kg	11.85 ± 0.39	26.48 ± 1.11	25.03 ± 1.36
COD	gO_2/kg	3.14 ± 0.04	7.18 ± 0.06	6.45 ± 0.02
Glucose	g/dm^3	-	2.48 ± 0.05	10.56 ± 0.02
TAN	mgN/dm^3	233.82 ± 0.12	2.13 ± 0.01	1.79 ± 0.02
$P\text{-}PO_4^{3-}$	mgP/dm^3	308.25 ± 3.77	0.76 ± 0.01	4.89 ± 0.01
Iron	$mgFe/dm^3$	0.30 ± 0.01	0.28 ± 0.01	0.35 ± 0.01

± standard deviation.

Figure 18 shows the cumulative CH_4 and H_2 production from four different runs (AD-1–AD-4, average pH = 7, without thermal treatment). Figure 19 shows the cumulative production from four DF experimental runs (DF-1–DF-4, average pH 5.5, with thermal pretreatment for 1.5 h at 80 °C). The final pH in the DF runs increased significantly from approximately 5.50 to 7.25 (DF-1) (Table 3). After the end of the lag phase and with rising pH, methane was also generated with the appearance of hydrogen. The process of dark fermentation did not take place, despite the application of optimal conditions [51] for the production of hydrogen. In the test with H_3PO_4 (DF-3), half the hydrogen production compared to the yield obtained in the test with H_2SO_4 (DF-1) was noted. Adjusting the pH with KOH was found to completely inhibit the production of hydrogen in contrast with using NaOH. At high concentrations, potassium was found to be toxic to anaerobic bacteria. Searmsirimongkol et al. [52] reported that potassium concentrations in the range of 2500–4500 mg·dm^{-3} have a strong inhibitory influence on hydrogen production, with a toxicity level of 12000 mg·dm^{-3}. According to Zou et al. [53], acid hydrolysis of cellulose produces not only glucose but also fructose (an isomer of glucose), 5-(hydroxylmethyl)furfural (HMF, dehydration from fructose), and other carbohydrates. In particular, HMF has a negative impact on the hydrogen formation pathway. At the highest concentration, the inoculum was unable to remove the HMF, resulting in strong inhibition of the hydrogen production. Interestingly, DF had a longer lag phase (approximately 6 days) than that of AD (2 days). The lag phase of dark fermentation usually lasts only a few hours [49]. Muñoz-Páez et al. [54] suggested that lower HMF concentrations improve the production of hydrogen. However, the higher HMF content affected the adaptation time and resulted in a sixfold increase in the lag time compared with the control treatment. The HMF content in our hydrolysates was probably too high and inhibited the process. Only after microbial adaptation and partial degradation of HMF did the pH in the reactor rise and the process shifted to AD, in which the methane production (120 dm^3/kgVS) was much higher than the hydrogen yield (5 dm^3/kgVS) (DF-3, Table 4).

The highest methane and hydrogen production, at 278 dm^3/kgVS and 42 dm^3/kgVS, respectively, was recorded in the AD trial (AD-3) with pH adjustment using NaOH. This was possibly due to the pretreatment with H_3PO_4, which is a much weaker inhibitor than sulfuric acid. Based on our previous research, the addition of phosphorous improves the digestion process efficiency [50]. Biogas from the AD-3 run consisted of 27% CH_4 and 6.5% H_2. The gas released in this experimental run was qualitatively worse than in the AD-1 and AD-2 runs (46% CH_4, 5% H_2) (Table 4). However, the highest cumulative biohydrogen production was observed in the AF-3 run, with an increase of over 50% compared to the other experimental runs. These results were better than those obtained by Yoruklu et al. [55] from the fermentation of cotton straw. In their study, methane and hydrogen production amounted to 83 dm^3/kgVS and 33 dm^3/kgVS, respectively (Table 4). The lower efficiency was due to the slightly different structure of the biomass. First of all, straw contains lignin, which makes hydrolysis more difficult. In the case of AD-1, AD-2, and DF-2, an additional negative effect was caused by sulfate ions. The lowest methane production (87 dm^3/kgVS) was observed in AD-2 with overlapping inhibition (potassium and sulfate ions). Moreover, hydrogen was not produced in DF-2 with the addition of sulfuric acid and potassium hydroxide and in DF-4 with the addition of KOH. This showed that the addition of potassium had a much stronger effect on the bacteria that produce hydrogen at a lower pH (approx. 5.5). Several inhibition mechanisms were reported for sulfate ions in the literature. At very high concentrations (1000 mg·dm^{-3}), sulfate ions were found to diffuse into the microbial cell [52].

Table 4. Operating parameters and performance of batch tests to establish the optimal inoculum/SBP ratio.

		Experimental Run							
Indicator	**Unit**	**AD-1**	**AD-2**	**AD-3**	**AD-4**	**DF-1**	**DF-2**	**DF-3**	**DF-4**
TS	g/kg	67.16 ± 1.08	70.06 ± 0.52	50.90 ± 0.67	52.99 ± 1.25	67.16 ± 1.08	70.06 ± 0.52	50.90 ± 0.67	52.99 ± 1.25
VS	g/kg	36.77 ± 1.11	35.65 ± 0.87	32.07 ± 0.97	32.64 ± 1.35	36.77 ± 1.11	35.65 ± 0.87	32.07 ± 0.97	32.64 ± 1.35
Hydrolysis		2% H_2SO_4 NaOH	2% H_2SO_4 KOH	2% H_3PO_4 NaOH	2% H_3PO_4 KOH	2% H_2SO_4 NaOH	2% H_2SO_4 KOH	2% H_3PO_4 NaOH	2% H_3PO_4 KOH
X/S *	gVS/gVS	2:1	2:1	2:1	2:1	2:1	2:1	2:1	2:1
Inoculum	g	300 ± 0.25	300 ± 0.25	300 ± 0.25	300 ± 0.25	200 ± 0.25	200 ± 0.25	200 ± 0.25	200 ± 0.25
Substrate	g	48.34 ± 1.25	49.86 ± 2.47	55.43 ± 1.82	54.46 ± 2.15	48.34 ± 1.25	49.86 ± 2.47	55.43 ± 1.82	54.46 ± 2.15
$pH_{initial}$	-	7.11 ± 0.09	7.02 ± 0.12	7.09 ± 0.17	7.12 ± 0.08	5.51 ± 0.15	5.48 ± 0.06	5.53 ± 0.18	5.47 ± 0.15
pH_{end}	-	7.61 ± 0.05	6.33 ± 0.10	6.52 ± 0.08	67.32 ± 0.15	7.25 ± 0.02	6.50 ± 0.13	6.22 ± 0.24	7.10 ± 0.23
Duration	days	30	30	30	30	30	30	30	30
Cumulative biogas volume	cm^3/d	785.00 ± 8.05	694.00 ± 6.02	1940.00 ± 10.54	1160.00 ± 9.25	500.00 ± 4.12	249.00 ± 1.08	100.00 ± 0.56	630.00 ± 2.68
SGP	$Ndm^3/kgVS_{feed}$	435.69 ± 3.45	198.67 ± 3.15	687.61 ± 7.28	407.18 ± 2.70	316.36 ± 4.62	217.47 ± 2.49	318.99 ± 1.47	138.20 ± 3.58
SHP	$Ndm^3H_2/kgVS$	20.64 ± 1.52	10.53 ± 0.98	42.47 ± 2.45	9.15 ± 0.96	18.18 ± 0.92	0.00 ± 0.00	5.08 ± 0.58	0.00 ± 0.00
SMP	$Ndm^3CH_4/kgVS$	195.56 ± 6.48	86.59 ± 9.56	277.61 ± 15.04	94.97 ± 1.20	56.86 ± 1.00	0.00 ± 0.00	120.22 ± 8.44	0.00 ± 0.00
H_2 content	%	4.93 ± 0.25	5.31 ± 0.48	6.55 ± 0.78	3.92 ± 1.28	9.15 ± 1.06	0.00 ± 0.00	2.59 ± 0.54	0.00 ± 0.00
CH_4 content	%	45.83 ± 2.02	45.76 ± 3.25	27.01 ± 2.01	35.10 ± 2.00	21.04 ± 2.00	0.00 ± 0.00	44.00 ± 1.77	0.00 ± 0.00
CO_2 content	%	16.00 ± 0.62	12.00 ± 0.14	11.00 ± 0.68	9.00 ± 0.20	5.00 ± 0.54	0.00 ± 0.00	12.00 ± 1.08	5.00 ± 0.04
H_2S content	ppm	986.00 ± 0.00	640.00 ± 0.09	21.00 ± 0.08	10.00 ± 0.14	26.00 ± 0.05	21.00 ± 0.00	39.00 ± 2.56	13.00 ± 0.53
TAN	mgN/dm^3	120.24 ± 0.02	138.35 ± 0.02	159.76 ± 0.02	85.65 ± 0.02	230.59 ± 0.02	293.18 ± 0.02	224.00 ± 0.02	105.41 ± 0.02
$P\text{-}PO_4^{3-}$	mgP/dm^3	156.63 ± 0.02	132.48 ± 0.02	206.88 ± 0.02	301.52 ± 0.02	229.73 ± 0.02	168.38 ± 0.02	194.48 ± 0.02	246.04 ± 0.02

* inoculum/substrate ratio, ± standard deviation.

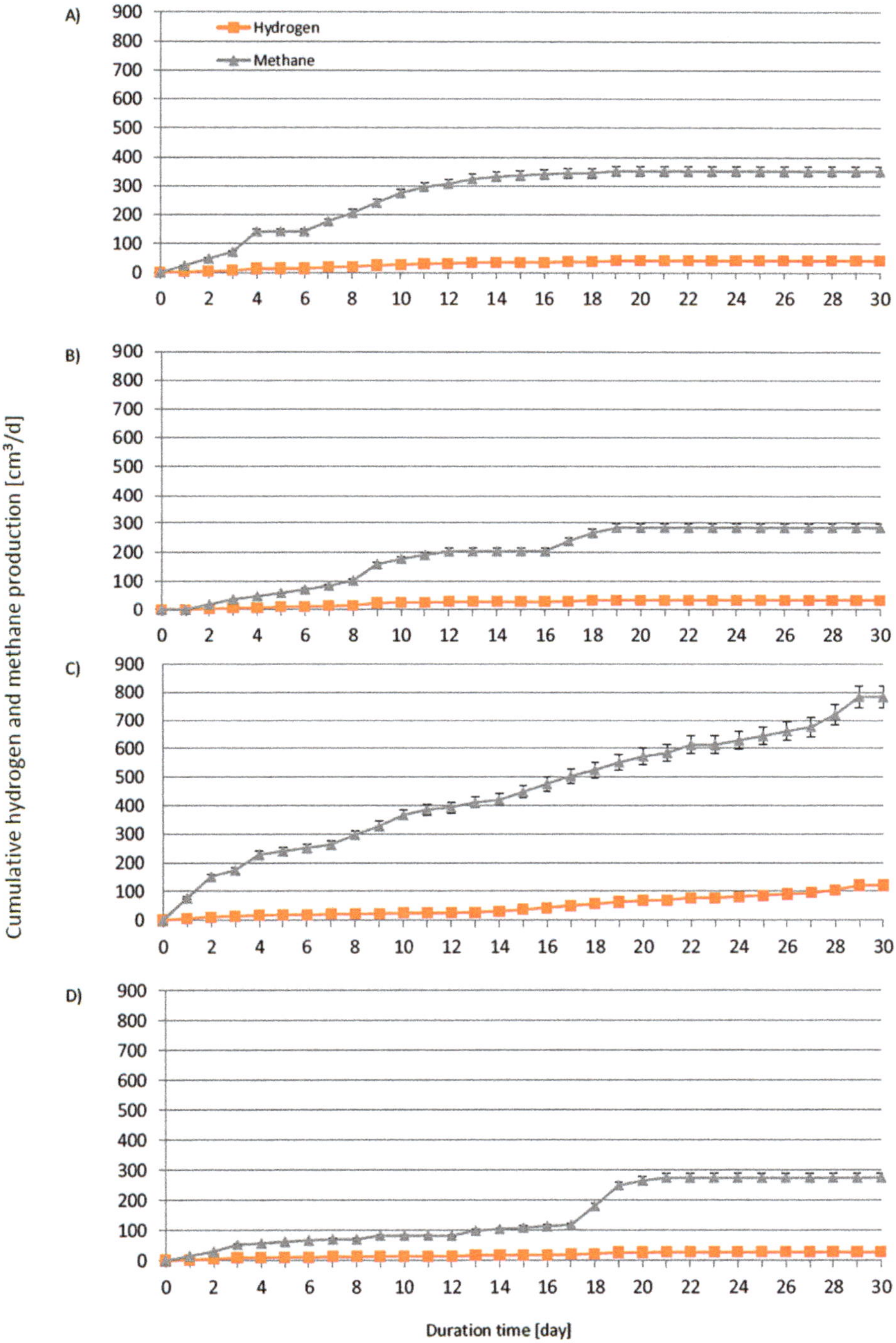

Figure 18. Daily average specific hydrogen and methane production from cotton hydrolyzed in AD systems (mesophilic conditions) with (**A**) H_2SO_4 neutralized with NaOH, (**B**) H_2SO_4 neutralized with KOH, (**C**) H_3PO_4 neutralized with NaOH, and (**D**) H_3PO_4 neutralized with KOH.

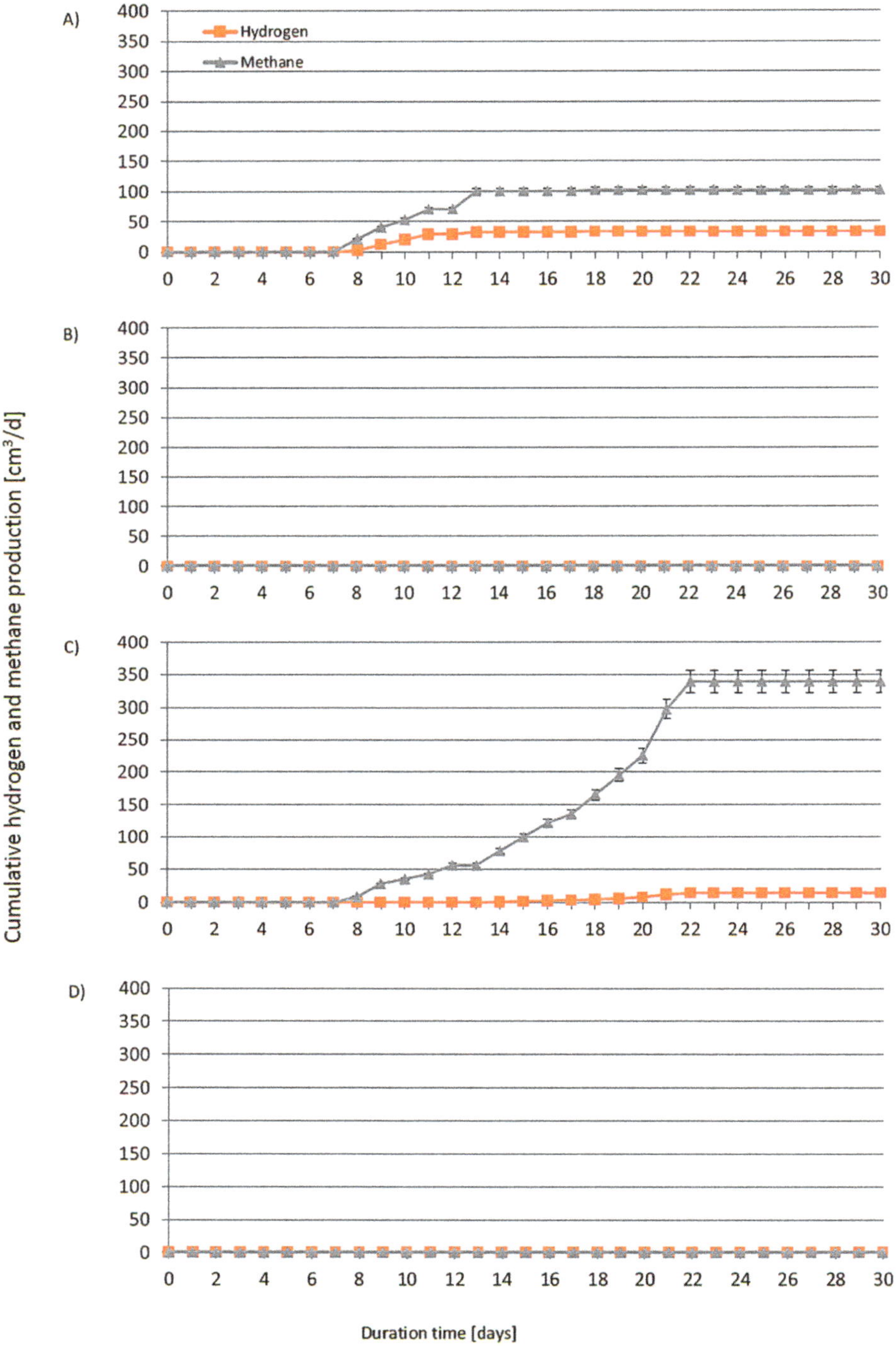

Figure 19. Daily average specific hydrogen and methane production from cotton hydrolyzed in DF systems (mesophilic conditions) with (**A**) H_2SO_4 neutralized with NaOH, (**B**) H_2SO_4 neutralized with KOH, (**C**) H_3PO_4 neutralized with NaOH, and (**D**) H_3PO_4 neutralized with KOH.

4. Conclusions

In this study, we investigated the possibility of using acid hydrolysates from cotton waste as components in fermentation broths for the production of bioethanol and biogas. Currently, most cotton textile waste is sent to landfill. Cotton textile waste may also be subjected to material recycling, i.e., converted into useable yarn or non-woven fabrics. It can also be used to produce energy via incineration/gasification. The cotton yarn hydrolysates showed great potential for replacing simple sugar solutions in fermentation media. When the cotton yarn was treated with 2% H_3PO_4 at 140 °C for 2 h, the concentration of glucose in the hydrolysate (13.19 $g{\cdot}dm^{-3}$) increased fivefold compared to the same conditions with H_2SO_4 (2.65 $g{\cdot}dm^{-3}$). The effectiveness of the process of cotton fiber hydrolysis was confirmed by SEM-EDS, FTIR-ATR, and Raman spectroscopy. The acid hydrolysis of cotton fibers led to the formation of glucose in the hydrolysate, which could therefore be a suitable medium for microorganisms and a source of bio-based products, such as bioethanol or biogas. The production of hydrogen-rich biogas is the most interesting approach. However, a relatively high methane yield of nearly 280 dm^3/kgVS can also be achieved by subjecting cotton textile waste to acid hydrolysis followed by anaerobic digestion.

Most textile products contain a mixture of cotton and synthetic fibers, such as poly (ethylene terephthalate) (PET) or polyamide (PA). Polymer additives and chemicals used in the textile processes could act as fermentation inhibitors and impede the production of biofuels. Therefore, future work will focus on processing cotton textiles with synthetic fibers and textiles modified with various chemical agents. The process of biogas production requires further optimization to increase the yield per unit weight of cotton waste. Biogas could then become an additional product of the textile waste treatment process, adding economic value while also contributing to energy sustainability. The presented solution for using cotton waste to generate bio-energy is, therefore, a step forward in the creation of systems for managing the large amounts of textile waste produced each year, in line with the concept of a circular economy.

Author Contributions: Conceptualization, M.J.B., J.Z.M., M.C. and I.A.W.; methodology, M.J.B., J.Z.M., J.B., W.C.-W., M.C. and I.A.W.; formal analysis, M.J.B., J.Z.M., M.C. and I.A.W.; investigation, M.J.B., J.Z.M., J.B., M.C., D.P., W.C.-W., S.B. and I.A.W.; writing—original draft preparation, M.J.B., J.Z.M., J.B., W.C.-W., M.C. and I.A.W.; writing—review and editing, M.J.B., J.Z.M., J.B., M.C. and I.A.W.; visualization, M.J.B., J.Z.M., D.P. and W.C.-W.; supervision, I.A.W.; project administration, I.A.W. All authors have read and agreed to the published version of the manuscript.

Funding: This research was funded by the National Science Center (Poland), grant number OPUS 2019/33/B/ST8/02005.

Institutional Review Board Statement: Not applicable.

Informed Consent Statement: Not applicable.

Data Availability Statement: The data presented in this study are available on request from the corresponding author.

Acknowledgments: This article was completed while the second author was a doctoral candidate in the Interdisciplinary Doctoral School at Lodz University of Technology, Poland.

Conflicts of Interest: The authors declare no conflict of interest.

Abbreviations

AD	anaerobic digestion
BT	batch test
COD	chemical oxygen demand
DF	dark fermentation
SGP	specific gas production
SHP	specific hydrogen production
SMP	specific methane production
TAN	total ammonium nitrogen
TS	total solids
VFA	volatile fatty acids
VS	volatile solids

References

1. Tiseo, I. Global Waste Generation—Statistics & Facts. Available online: https://www.statista.com/topics/4983/waste-generation-worldwide/ (accessed on 18 January 2022).
2. Ruiz, L. Global Textile Fibre Demand: Trends and Forecast. Available online: www.icac.org (accessed on 18 January 2022).
3. Sanchis-Sebastiá, M.; Ruuth, E.; Stigsson, L.; Galbe, M.; Wallberg, O. Novel sustainable alternatives for the fashion industry: A method of chemically recycling waste textiles via acid hydrolysis. *Waste Manag.* **2021**, *121*, 248–254. [CrossRef] [PubMed]
4. Apparel and Textiles: Trade Statistics. Available online: https://globaledge.msu.edu/industries/apparel-and-textiles/tradestats (accessed on 18 January 2022).
5. Sandin, G.; Peters, G.M. Environmental impact of textile reuse and recycling—A review. *J. Clean. Prod.* **2018**, *184*, 353–365. [CrossRef]
6. The Fiber Year 2009/10 A World Survey on Textile and Nonwovens Industry. *Oerlikon May*. 2010. Available online: https://www.oerlikon.com/ecomaXL/get_blob.php?name=The_Fibre_Year_2010_en_0607.pdf (accessed on 18 January 2022).
7. Pensupa, N.; Leu, S.Y.; Hu, Y.; Du, C.; Liu, H.; Jing, H.; Wang, H.; Lin, C.S.K. Recent Trends in Sustainable Textile Waste Recycling Methods: Current Situation and Future Prospects. In *Chemistry and Chemical Technologies in Waste Valorization*; Springer, Ed.; Springer International Publishing: Cham, Switzerland, 2017; Volume 375, pp. 189–228.
8. Ghaly, A.E.; Ananthashankar, R.; Alhattab, M.V.V.R.; Ramakrishnan, V.V. Production, Characterization and Treatment of Textile Effluents: A Critical Review. *J. Chem. Eng. Process Technol.* **2014**, *5*, 1–19. [CrossRef]
9. The Lenzing Group. The Global Fiber Market. Available online: http://www.lenzing.com/en/investors/equity-story/global-fiber-market.html (accessed on 18 January 2022).
10. Paul, R. *Functional Finishes for Textiles: Improving Comfort, Performance and Protection*; Woodhead Publishing: Cambridge, UK, 2015.
11. Giesz, P.; Mackiewicz, E.; Grobelny, J.; Celichowski, G.; Cieślak, M. Multifunctional hybrid functionalization of cellulose fabrics with AgNWs and TiO_2. *Carbohydr. Polym.* **2017**, *177*, 397–405. [CrossRef]
12. Giesz, P.; Celichowski, G.; Puchowicz, D.; Kamińska, I.; Grobelny, J.; Batory, D.; Cieślak, M. Microwave-assisted TiO_2: Anatase formation on cotton and viscose fabric surfaces. *Cellulose* **2016**, *23*, 2143–2159. [CrossRef]
13. Giesz, P.; Mackiewicz, E.; Nejman, A.; Celichowski, G.; Cieślak, M. Investigation on functionalization of cotton and viscose fabrics with AgNWs. *Cellulose* **2017**, *24*, 409–422. [CrossRef]
14. Nørup, N.; Pihl, K.; Damgaard, A.; Scheutz, C. Evaluation of a European textile sorting centre: Material flow analysis and life cycle inventory. *Resour. Conserv. Recycl.* **2019**, *143*, 310–319. [CrossRef]
15. Subramanian, K.; Chopra, S.S.; Cakin, E.; Li, X.; Lin, C.S.K. Environmental life cycle assessment of textile bio-recycling—Valorizing cotton-polyester textile waste to pet fiber and glucose syrup. *Resour. Conserv. Recycl.* **2020**, *161*, 104989. [CrossRef]
16. Leal Filho, W.; Ellams, D.; Han, S.; Tyler, D.; Boiten, V.J.; Paco, A.; Moora, H.; Balogun, A.L. A review of the socio-economic advantages of textile recycling. *J. Clean. Prod.* **2019**, *218*, 10–20. [CrossRef]
17. Xu, Z.; Qi, R.; Xiong, M.; Zhang, D.; Gu, H.; Chen, W. Conversion of cotton textile waste to clean solid fuel via surfactant-assisted hydrothermal carbonization: Mechanisms and combustion behaviors. *Bioresour. Technol.* **2021**, *321*, 124450. [CrossRef]
18. Alay, E.; Duran, K.; Korlu, A. A sample work on green manufacturing in textile industry. *Sustain. Chem. Pharm.* **2016**, *3*, 39–46. [CrossRef]
19. Campbell, B.T.; Saha, S.; Percy, R.; Frelichowski, J. Status of the global cotton germplasm resources (Crop Science, (2010), 10, (1161-1179)). *Crop Sci.* **2010**, *50*, 1161–1179. [CrossRef]
20. Leading Cotton Producing Countries Worldwide in 2019/2020. Available online: https://www.statista.com/statistics/263055/cotton-production-worldwide-by-top-countries/ (accessed on 18 January 2022).
21. Ward, G.D.; Hewitt, A.D.; Russell, S.J. Fibre composition of donated post-consumer clothing in theUK. *Proc. Inst. Civ. Eng. Waste Resour. Manag.* **2013**, *166*, 29–37. [CrossRef]
22. Patti, A.; Cicala, G.; Acierno, D. Eco-sustainability of the textile production: Waste recovery and current recycling in the composites world. *Polymers* **2021**, *13*, 134. [CrossRef]

23. Wanassi, B.; Azzouz, B.; Hassen, M. Ben Value-added waste cotton yarn: Optimization of recycling process and spinning of reclaimed fibers. *Ind. Crops Prod.* **2016**, *87*, 27–32. [CrossRef]
24. Cuiffo, M.; Jung, H.J.; Skocir, A.; Schiros, T.; Evans, E.; Orlando, E.; Lin, Y.C.; Fang, Y.; Rafailovich, M.; Kim, T.; et al. Thermochemical degradation of cotton fabric under mild conditions. *Fash. Text.* **2021**, *8*, 1–14. [CrossRef]
25. Schmidt, H.; Cieślak, M. Concrete with carpet recyclates: Suitability assessment by surface energy evaluation. *Waste Manag.* **2008**, *28*, 1182–1187. [CrossRef]
26. Wagaye, B.T.; Adamu, B.F.; Jhatial, A.K. Recycled Cotton Fibers for Melange Yarn Manufacturing. In *Cotton Science and Processing Technology. Textile Science and Clothing Technology*; Springer: Singapore, 2020; pp. 529–546.
27. Bedez Ute, T.; Celik, P.; Bunyamin Uzumcu, M. Utilization of Cotton Spinning Mill Wastes in Yarn Production. In *Textile Industry and Environment*; IntechOpen: London, UK, 2019; pp. 1–13.
28. Béchir, W.; Béchir, A.; Mohamed, B.H. Industrial cotton waste: Recycling, Reclaimed fiber behavior and quality prediction of its blend. *Text. Appar.* **2018**, *28*, 14–20.
29. Wang, S.; Yu, X.; Chen, X.; Hou, W.; Niu, M. Recycling of Cotton Fibers Separated from the Waste Blend Fabric. *J. Nat. Fibers* **2020**, *17*, 520–531. [CrossRef]
30. Sun, X.; Lu, C.; Liu, Y.; Zhang, W.; Zhang, X. Melt-processed poly(vinyl alcohol) composites filled with microcrystalline cellulose from waste cotton fabrics. *Carbohydr. Polym.* **2014**, *101*, 642–649. [CrossRef]
31. Low, J.H.; Rahman, W.A.W.A. Plant fibers: Renewable reinforcing filler in polyolefin bio-composites. In *Natural Fibers*; Kozlowski, R.M., Muzyczek, M., Eds.; Nova Science Publishers, Inc.: Hauppauge, NY, USA, 2017; pp. 303–324. ISBN 978-1-53612-071-4.
32. Hsieh, Y.L. Chemical structure and properties of cotton. *Cott. Sci. Technol.* **2006**, 3–34. [CrossRef]
33. Degani, O. Synergism between Cutinase and Pectinase in the Hydrolysis of Cotton Fibers' Cuticle. *Catalysts* **2021**, *11*, 84. [CrossRef]
34. Szymański, Ł.; Grabowska, B.; Kurleto, Ż.; Kaczmarska, K. Celuloza i jej pochodne—Zastosowanie w przemyśle. *Arch. Foundry Eng.* **2015**, *15*, 129–132.
35. Huang, Y.B.; Fu, Y. Hydrolysis of cellulose to glucose by solid acid catalysts. *Green Chem.* **2013**, *15*, 1095–1111. [CrossRef]
36. Håkansson, H.; Ahlgren, P. Acid hydrolysis of some industrial pulps: Effect of hydrolysis conditions and raw material. *Cellulose* **2005**, *12*, 177–183. [CrossRef]
37. Sasaki, C.; Kiyokawa, A.; Asada, C.; Nakamura, Y. Glucose and Valuable Chemicals Production from Cotton Waste Using Hydrothermal Method. *Waste Biomass Valorization* **2019**, *10*, 599–607. [CrossRef]
38. Dahlbo, H.; Aalto, K.; Eskelinen, H.; Salmenperä, H. Increasing textile circulation—Consequences and requirements. *Sustain. Prod. Consum.* **2017**, *9*, 44–57. [CrossRef]
39. Ioelovich, M. Study of Cellulose Interaction with Concentrated Solutions of Sulfuric Acid. *ISRN Chem. Eng.* **2012**, *2012*, 428974. [CrossRef]
40. Nguyen, T.M.; Chang, S.C.; Condon, B.; Thomas, T.P.; Azadi, P. Thermal decomposition reactions of cotton fabric treated with piperazine-phosphonates derivatives as a flame retardant. *J. Anal. Appl. Pyrolysis* **2014**, *110*, 122–129. [CrossRef]
41. Puchowicz, D.; Nejman, A.; Kamińska, I.; Cieślak, M. Effect of Reactive Dyeing on Fabrics Modification with Silver Nanowires (AgNWs). *ACS Omega* **2021**, *6*, 26077–26085. [CrossRef]
42. Binczarski, M.J.; Malinowska, J.; Stanishevsky, A.; Severino, C.J.; Yager, R.; Cieslak, M.; Witonska, I.A. A Model Procedure for Catalytic Conversion of Waste Cotton into Useful Chemicals. *Materials* **2021**, *14*, 1981. [CrossRef] [PubMed]
43. Woo, K.S.; Kim, H.Y.; Hwang, I.G.; Lee, S.H.; Jeong, H.S. Characteristics of the Thermal Degradation of Glucose and Maltose Solutions. *Prev Nutr Food Sci.* **2015**, *20*, 102–109. [CrossRef] [PubMed]
44. Garside, P.; Wyeth, P. Identification of cellulosic fibres by FTIR spectroscopy: Differentiation of flax and hemp by polarized ATR FTIR. *Stud. Conserv.* **2006**, *51*, 205–211. [CrossRef]
45. Kavkler, K.; Demšar, A. Examination of cellulose textile fibres in historical objects by micro-Raman spectroscopy. *Spectrochim. Acta—Part A Mol. Biomol. Spectrosc.* **2011**, *78*, 740–746. [CrossRef]
46. Puchowicz, D.; Cieslak, M. Raman Spectroscopy in the Analysis of Textile Structures. In *Recent Developments in Atomic Force Microscopy and Raman Spectroscopy for Materials Characterization*; IntechOpen: London, UK, 2022.
47. Modelska, M.; Berlowska, J.; Kregiel, D.; Cieciura, W.; Antolak, H.; Tomaszewska, J.; Binczarski, M.; Szubiakiewicz, E.; Witonska, I.A. Concept for Recycling Waste Biomass from the Sugar Industry for Chemical and Biotechnological Purposes. *Molecules* **2017**, *22*, 1544. [CrossRef]
48. Wakelyn, P.J.; Bertoniere, N.R.; French, A.D.; Thibodeaux, D.P.; Triplett, B.A.; Rousselle, M.A.; Goynes, W.R., Jr.; Edwards, J.V.; Hunter, L.; McAlister, D.D. *Cotton Fiber Chemistry and Technology*; CRC Press: Boca Raton, 2007; ISBN 9780429141263.
49. Cieciura-Włoch, W.; Binczarski, M.; Tomaszewska, J.; Borowski, S.; Domański, J.; Dziugan, P.; Witońska, I. The use of acid hydrolysates after furfural production from sugar waste biomass as a fermentation medium in the biotechnological production of hydrogen. *Energies* **2019**, *12*, 3222. [CrossRef]
50. Cieciura-Włoch, W.; Borowski, S.; Domański, J. Dark fermentative hydrogen production from hydrolyzed sugar beet pulp improved by nitrogen and phosphorus supplementation. *Bioresour. Technol.* **2021**, *340*, 125622. [CrossRef]
51. Cieciura-Włoch, W.; Borowski, S. Biohydrogen production from wastes of plant and animal origin via dark fermentation. *J. Environ. Eng. Landsc. Manag.* **2019**, *27*, 101–113. [CrossRef]

52. Searmsirimongkol, P.; Rangsunvigit, P.; Leethochawalit, M.; Chavadej, S. Hydrogen production from alcohol distillery wastewater containing high potassium and sulfate using an anaerobic sequencing batch reactor. *Int. J. Hydrogen Energy* **2011**, *36*, 12810–12821. [CrossRef]
53. Zou, J.; Zhang, G.; Xu, X. One-pot photoreforming of cellulosic biomass waste to hydrogen by merging photocatalysis with acid hydrolysis. *Appl. Catal. A Gen.* **2018**, *563*, 73–79. [CrossRef]
54. Muñoz-Páez, K.M.; Alvarado-Michi, E.L.; Buitrón, G.; Valdez-Vazquez, I. Distinct effects of furfural, hydroxymethylfurfural and its mixtures on dark fermentation hydrogen production and microbial structure of a mixed culture. *Int. J. Hydrogen Energy* **2019**, *44*, 2289–2297. [CrossRef]
55. Civelek Yoruklu, H.; Koroglu, E.O.; Ozdemir, O.K.; Demir, A.; Ozkaya, B. Bioenergy production from cotton straws using different pretreatment methods. *Int. J. Hydrogen Energy* **2020**, *45*, 34720–34729. [CrossRef]

energies

MDPI

Article

A Production and Delivery Model of Hydrogen from Solar Thermal Energy in the United Arab Emirates

Abdulrahman Joubi *, Yutaro Akimoto and Keiichi Okajima

Graduate School of Science and Technology, University of Tsukuba, Tsukuba 3058573, Japan; akimoto@risk.tsukuba.ac.jp (Y.A.); okajima@risk.tsukuba.ac.jp (K.O.)
* Correspondence: abdulrahman.joubi@gmail.com; Tel.: +81-29-853-5756

Abstract: Hydrogen production from surplus solar electricity as energy storage for export purposes can push towards large-scale application of solar energy in the United Arab Emirates and the Middle East region; this region's properties of high solar irradiance and vast empty lands provide a good fit for solar technologies such as concentrated solar power and photovoltaics. However, a thorough comparison between the two solar technologies, as well as investigating the infrastructure of the United Arab Emirates for a well-to-ship hydrogen pathway, is yet to be fully carried out. Therefore, in this study we aim to provide a full model for solar hydrogen production and delivery by evaluating the potential of concentrated solar power and photovoltaics in the UAE, then comparing two different pathways for hydrogen delivery based on the location of hydrogen production sites. A Solid Oxide Cell Electrolyzer (SOEC) is used for technical comparison, while the shortest routes for hydrogen transport were analyzed using Geographical Information System (GIS). The results show that CSP technology coupled with SOEC is the most favorable pathway for large-scale hydrogen from solar energy production in the UAE for export purposes. Although PV has a slightly higher electricity potential compared to CSP, around 42 GWh/km^2 to 41.1 GWh/km^2, respectively, CSP show the highest productions rates of over 6 megatons of hydrogen when the electrolyzer is placed at the same site as the CSP plant, while PV generates 5.15 megatons when hydrogen is produced at the same site with PV plants; meanwhile, hydrogen from PV and CSP shows similar levels of 4.8 and 4.6 megatons of hydrogen, respectively, when electrolyzers are placed at port sites. Even considering the constraints in the UAE's infrastructure and suggesting new shorter electrical transmission lines that could save up to 0.1 megatons of hydrogen in the second pathway, production at the same site with CSP is still the most advantageous scenario.

Keywords: concentrated solar power; photovoltaics; GIS; hydrogen supply chain; renewable energy; the Middle East region; United Arab Emirates

Citation: Joubi, A.; Akimoto, Y.; Okajima, K. A Production and Delivery Model of Hydrogen from Solar Thermal Energy in the United Arab Emirates. *Energies* **2022**, *15*, 4000. https://doi.org/10.3390/en15114000

Academic Editors: Wei-Hsin Chen, Aristotle T. Ubando, Chih-Che Chueh and Liwen Jin

Received: 30 April 2022
Accepted: 26 May 2022
Published: 29 May 2022

1. Introduction

The Middle East region, including the United Arab Emirates, is facing a growing energy demand and a surge in CO_2 emissions [1,2], leading the region into looking for new sources of energy with low carbon footprint.

The region's climatic advantages provide a suitable fit for solar energy [1–7], from sunlight abundance to the vast open areas of deserts [8]. A few projects and plans to exploit the UAE's solar potential, such as concentrated solar power (CSP) and photovoltaics (PV), are already afoot, but addressing its full capabilities is still a distant milestone. Only 2% of the UAE's electricity came from renewable resources by the year 2018 [6], and that comes as a result of the large economic investment renewables require compared to the current traditionally applied technologies [1].

Hydrogen's ability to store surplus energy from renewable sources, such as solar energy, makes it a key player in the future of renewables and global decarbonization efforts by guaranteeing long term sustainability [9–12]. The global demand for hydrogen is on the rise, especially in the mobility sector [13]. Japan already has plans to integrate hydrogen into their future energy mix [14], but as domestic supply of hydrogen fails to meet its targets in CO_2 emissions reduction, importing zero-carbon produced H_2 from overseas seems like a viable option [15–17]. For instance, to realize its goal of a hydrogen-based society, Japan aims to establish an international hydrogen supply chain [18]. A deal to co-operate on the development of such a supply chain has already been established between Japan and the United Arab Emirates, as the UAE and other countries in the Middle East region aspire to becoming major international suppliers of low-carbon hydrogen in the future [19]. The profits from exporting green hydrogen from excess energy in the UAE can provide an incentive to decision makers to seek a higher renewable energy target.

Concentrated solar power plants are deemed as one of the favorable pathways for large-scale hydrogen production from solar power [20]. Its compatibility with thermal energy storages (TES) that can store energy in the form of heat can overcome the challenge of fluctuating energy supply from solar technologies. It can also prevent the intermittent functioning of water electrolyzers and provide more stable hydrogen production [20]. Another advantage of CSP and TES is that the electricity supply can be adapted to the demand, providing more flexible hydrogen production for export, and abating the need for H_2 storage at production site.

Although the UAE's potential for renewable energy and hydrogen production was addressed in a few previous studies, none addresses the possible use of hydrogen as a means for maximum solar energy potential fulfilment. While some studies consider a hydrogen future for the UAE, the focus is only on official targets and cost analysis, or the use of energy from photovoltaics and nuclear plants only for hydrogen production [21–23]. In addition, hydrogen production and utilization are confronted with a spatial gap which infrastructure can solve [9], but the United Arab Emirates' infrastructure is yet to be analyzed with a full plan for hydrogen delivery and transport. Thus, a clear gap between renewable energy application and hydrogen supply chain is present in the case of the United Arab Emirates.

Therefore, this study aims to bridge this gap by using the excess energy from two different scenarios of large-scale solar energy application in the UAE for hydrogen production for the purpose of export, by comparing electricity and hydrogen production from PV and CSP and analyzing two different delivery pathways while examining the UAE's current infrastructure and highlighting the constraints. This will provide an adequate and detailed model for hydrogen production from solar technologies in the UAE with a well-to-ship supply chain for the delivery of hydrogen from plants to ports. The resultant predicted income from hydrogen sales would make large-scale solar energy application feasible, while keeping the UAE as a major energy exporter in the market. This paper also proposes some solutions to the limitations in the UAE's infrastructure capabilities to deliver the produced hydrogen, and analyzes how this will affect the hydrogen supply chains.

2. Materials and Methods

This study investigates the technical potential of hydrogen production from excess electricity of CSP and PV application in the UAE based on our previous study [24], where the potential of solar energy in the United Arab Emirates was estimated by analyzing the suitable areas for solar energy application, satisfying a solar irradiance of 5 $kWh/m^2/day$ and higher and a slope lower than 3% based on the National Renewable Energy Laboratory's (NREL) model for the estimation of rural utility-scale PV and CSP [25]. Depending on the distance from roads, electrical grid, and water supply, these areas were then divided into three different scenarios after the exclusion of populated and protected areas with 10, 25, and 40 km distances, as 40 km is the furthest distance for a buffer zone from main roads and electrical grid before it crosses the UAE borders, and then this distance was divided into three different scenarios for more simple representation. The results show that the two scenarios with distances of 25 km and 40 km from the mentioned parameters can leave extra electricity for hydrogen production, after satisfying the future electricity demand of the UAE and the 27% green energy obligations for the year 2023 [6,26]. Figure 1 shows the GIS results of the areas of the UAE that are suitable for solar energy application in the 25 km and 40 km scenarios.

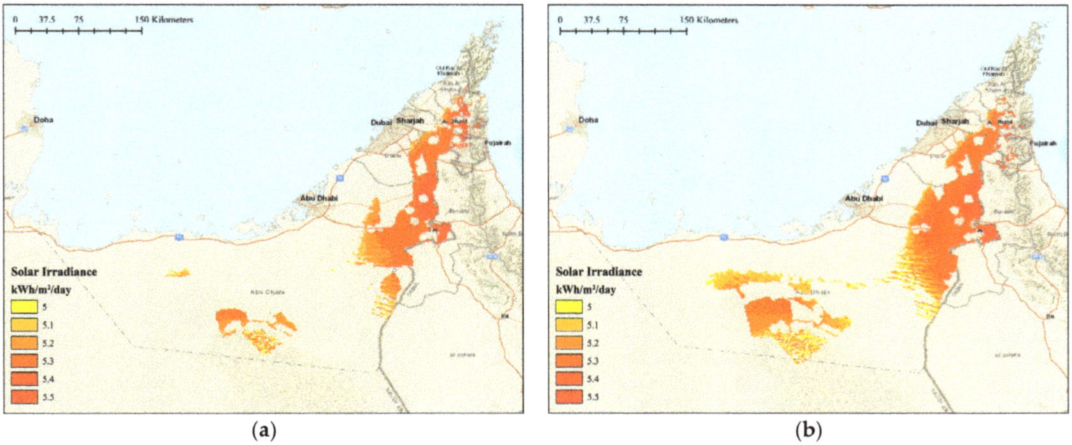

Figure 1. Areas with high solar potential in the UAE (**a**) 25 km and (**b**) 40 km from roads, electrical grid, and water areas.

In Figure 2 we explain the flow of the study by using the areas suitable for solar energy production in the UAE to calculate PV and CSP potentials to having the final amounts of liquid hydrogen ready for export.

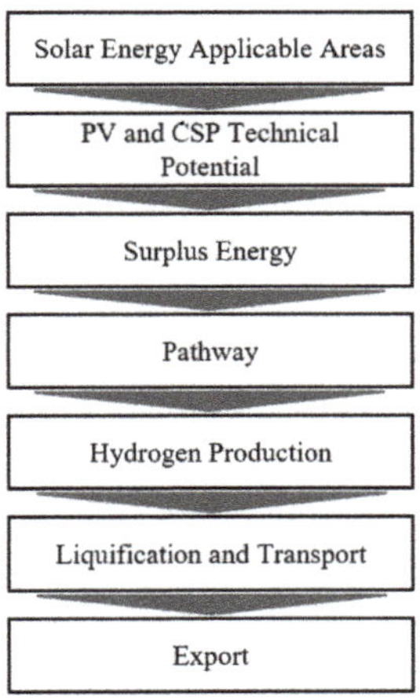

Figure 2. Study flow chart.

The equation used to evaluate the UAE's technical potential of solar energy in MWh is derived from the NREL's technical report as follows [25]:

$$\text{Available land } [\text{km}^2] \times \text{Power density } [\text{MW/km}] \times \text{Region capacity factor } [\%] \times 8760 \text{ [hours per year]} \quad (1)$$

The CSP system assumed in our study is a tower system with dry cooling, with salt as a heat fluid, coupled with a 10-h thermal storage and a solar multiple of 2.4. The power density of this system is 14.9 MW/km^2 according to NREL's module [27]. Following the assessments of NREL, a region capacity factor of 31.5% is used for areas with 5 kWh/m^2/day irradiance or higher [25]. As for the PV system, a one-axis tracking collector with the axis of rotation aligned north-south with zero-degree tilt from horizon as per the NREL model for utility-scale photovoltaics was assumed [25], power density of 30 MW/km^2 is assumed [28], and the UAE's capacity factor of 16% is used as per Al Ali et al.'s analysis for PV application in Abu Dhabi [29]. Table 1 presents the area and expected energy production results after applying all assumptions of both systems in Equation (1).

Table 1. UAE's results for solar energy applicable areas for 25 km and 40 km scenarios.

Scenario	25 km	40 km
Area km^2	10,180	16,207

The hydrogen supply chain model is similar to those of Strachan et al. [30], Balta-Ozkan et al. [31], Reuß et al. [9,32], and Tlili et al. [33], where different delivery pathways are compared and travel distances to demand areas are assessed.

This study proposes two delivery pathways based on the location of H_2 electrolyzers. The first pathway allocates hydrogen plants at the same site as solar energy plants; six main locations are chosen for the plants, covering all areas of the UAE suitable for solar energy application, with direct access to the main road grid and electricity transmission lines in the UAE. In this pathway, trucks are assumed to carry the produced hydrogen to the six main seaports for export in the UAE [34]. Using Geographical Information System (GIS), the shortest distances from plants to ports are analyzed to investigate the applicability of this scenario (road network data from OpenStreetMap [35] is used).

In the second pathway we place hydrogen plants at the port sites, assuming that the excess electricity from solar plants is transmitted through the electrical transmission lines of the UAE towards export ports, and assuming that 3% of the electricity is lost in the grid for every 1000 km traveled in the selected electricity grid lines of 220 kV and above of the UAE [36]. To ensure the least loss of electricity possible in the network, GIS is used to optimize the shortest routes that electricity can travel in this scenario. The second pathway also uses the same CSP plants and ports assumed in the first pathway. The layout of the pathway study methodology is shown in Figure 3.

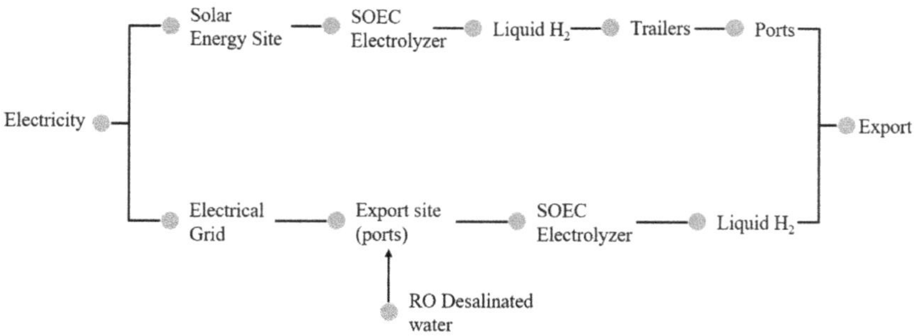

Figure 3. Pathways' layout.

The network analyst function of GIS is used to carry out the spatial analysis after uploading the road network of the UAE and specifying the use of only main roads and highways where hydrogen trailers are allowed to drive. Similarly, the data of the UAE's electric grid are inserted, and lines with 220 voltage and above are chosen for the analysis. Then, the six plants and chosen ports are fixed on the map. Finding the closest facility tool is then used to estimate the shortest routes between each plant and each port for both pathways.

As for H_2 production, solid oxide electrolyzers (SOECs) are used where the electrolyte is a solid ceramic membrane. Although it is a less-mature technology than other electrolyzers, such as alkaline and proton exchange membrane (PEM), and has higher investment cost, SOEC promises greater efficiencies [37]. This electrolyzer is assumed as a higher heating value (HHV), as high-temperature electrolyzers use heat as part of their energy demand [38], making it a perfect fit for thermal solar technologies such as CSP. SOEC electrolyzers use high-temperature steam electrolysis (HTE) at temperatures in the range of 700 °C to 800 °C [39]. This is why, when coupled to high-temperature heat sources such as solar thermal plants, this type of steam electrolyzer offers great advantages in both power requirements and heat provided, and the thermal energy storage integrated into the CSP plants ensures that fast load variations over the electrolyzer could be avoided [40]. The biggest disadvantage of the SOEC electrolyzer is that it is still in the development phase [41]. Chosen efficiencies for the electrolyzer are explained in Table 2. When coupled with PV, a heater is required to increase the temperature of water entering the electrolyzer to the needed levels, and an 11.5 kWh [41] extra energy consumption will be conducted from the PV electricity production for each kg of hydrogen.

For the second pathway, where the hydrogen plants are located at the ports, the SOEC would require extra energy to heat the water at the ports in the amount of 11.5 kWh/kg-H_2 [41]. Desalinated seawater using reverse osmosis (RO) desalination is assumed as the water source for electrolysis in this scenario, and electrical expenses for water desalination are added to the energy demand of electrolyzers. The energy needed can be determined by multiplying the energy need per cubic meter with the amount of water that is necessary, which is [42]:

$$m_{\dot{H}_2O} = \frac{\dot{m}}{M \times \eta} \times M_{H_2O} \tag{2}$$

where $\dot{m}$ (kg/s) is the production rate of hydrogen, M is the molar mass of hydrogen, η is the efficiency of water use, and M_{H_2O} is the molar mass of water.

We assume that hydrogen will be transported and exported in liquid form; liquification will be conducted at solar energy sites in the first pathway, and then transported to ports to be shipped overseas, while in the second pathway hydrogen is to be liquified at port sites to then board the H_2 carrier ships. Liquification is assumed to consume 6.78 kWh for each kg of H_2 [32] that is added to the electricity expenses of the electrolyzer. The Boil Off Gas (BOG) during liquification is neglected in this study.

Table 2. Electrolyzer efficiencies [43].

Electrolyzer	Efficiency	Efficiency
SOEC	39.4 kWh/kg-H_2	82%

3. Results and Discussions

3.1. CSP Technical Potential

In this study we estimate the potential of CSP application in the UAE. Table 3 represents the expected energy production results after applying all assumptions in Equation (1) and the surplus energy that can be used for hydrogen production after deducting the year 2023 energy demand additions and 27% green energy target of the UAE. To simplify this section, only the results of the 25 km energy scenario is shown. The numbers show that the UAE has a high potential for CSP with a large amount of excess energy even after realizing future energy targets.

Table 3. UAE's results for CSP energy potential and excess energy in GWh/km^2.

Scenario	Energy Potential	Excess Energy
25 km	41.115	28.3

3.2. PV Technical Potential

Photovoltaics technical potential is calculated by applying the assumptions into Equation (1) and, similarly to CSP surplus energy, is calculated as shown in Table 4.

Table 4. UAE's results for PV energy potential and excess energy in GWh.

Scenario	Energy Potential	Excess Energy
25 km	42.048	29.223

Figure 4 presents a comparison between the amounts of electricity produced by PV and CSP in the 25 km energy scenario in the UAE from all plants. The energy share produced by PV in the areas suitable for solar energy production in the UAE is larger than that of CSP due to the higher power density of PV technology.

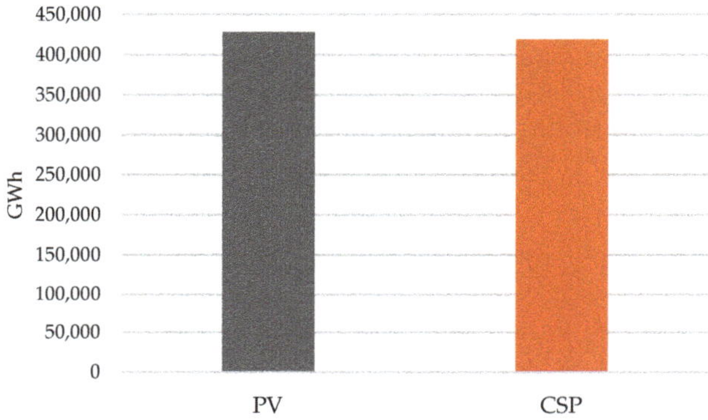

Figure 4. Comparison of the energy potential of PV and CSP in the UAE in the 25 km scenario.

3.3. Hydrogen Plants at the Same Site as Solar Plant Sites Pathway

In this pathway, we first calculate the liquid hydrogen production for all different technologies after applying all assumptions. CSP has the largest amount of hydrogen produced, as shown in Table 5, as the SOEC electrolyzer shows higher efficiencies when directly coupled to a direct source of thermal energy such as the CSP plants. The produced hydrogen quantities from each technology are compared in Figure 5.

Table 5. Produced H_2 in the first pathway from PV and CSP in (kg $\times$ 10^6).

Energy Scenario	PV	CSP
25 km	5157.412	6236.081

Next, we carry a geospatial network analysis to define the shortest routes for the hydrogen delivery from production sites to closest ports. Figure 6 illustrates the four closest ports along with the shortest roads trucks can take to deliver the liquid hydrogen, where two plants can deliver its production to Khalifa port, two to Jebel Ali port, one to port Rashid, and one to the Fujairah port. The distances calculated are to be used for feasibility studies in the future. Table 6 shows the estimated amounts of liquid hydrogen reaching each port in this pathway.

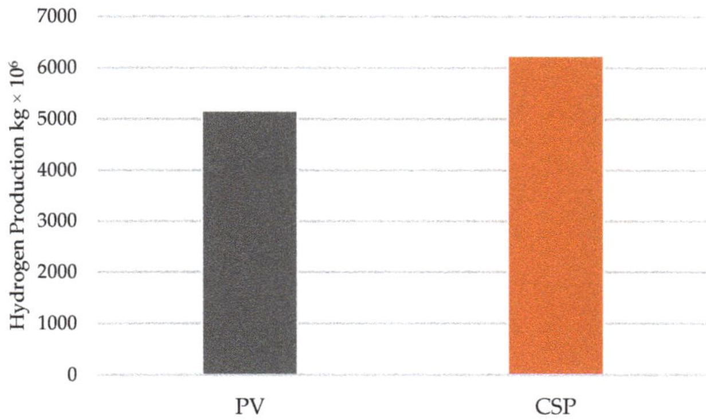

Figure 5. Hydrogen levels produced from PV and CSP comparison, in plants at the same location as solar energy sites pathway.

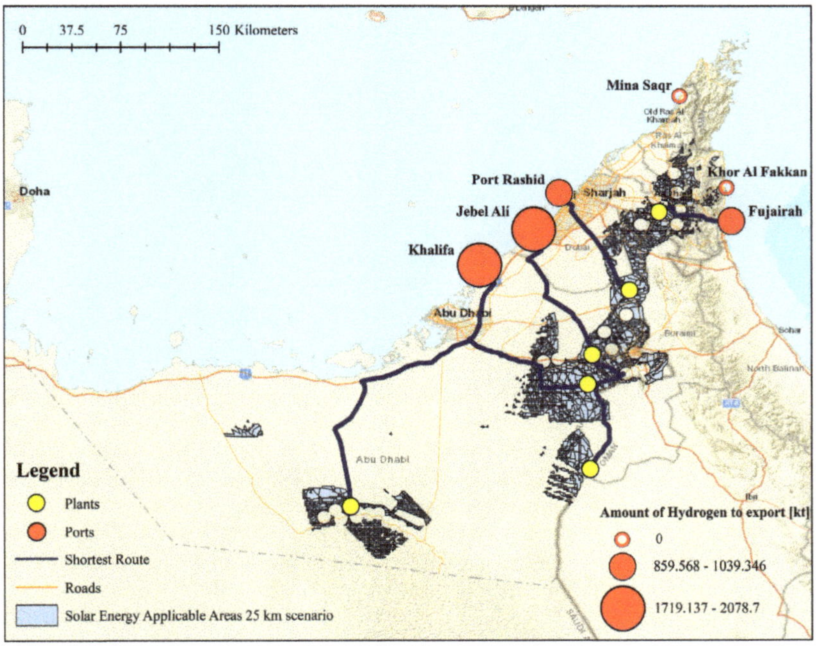

Figure 6. First pathway GIS analysis results for closest ports and shortest routes.

Table 6. Amount of H_2 reaching each port in the first pathway from PV and CSP in (kg $\times$ 10^6).

Port Name	PV	CSP
Khalifa	1719.137	2078.7
Jebel Ali	1719.137	2078.7
Port Rashid	859.568	1039.346
Fujairah	859.568	1039.346

3.4. Hydrogen Plants at Port Sites Pathway

GIS's network analysis is used in this pathway to define the shortest distances of the UAE's electrical grid to deliver the surplus electricity from solar plants to port sites. In

Figure 7 we can see the GIS results, where only port Khalifa and Fujairah are chosen as the closest ports for electricity transmission, with five solar plants transmitting electricity to the Khalifa port while only one solar plant is close enough to use the electrical grid to deliver its electricity to Fujairah port.

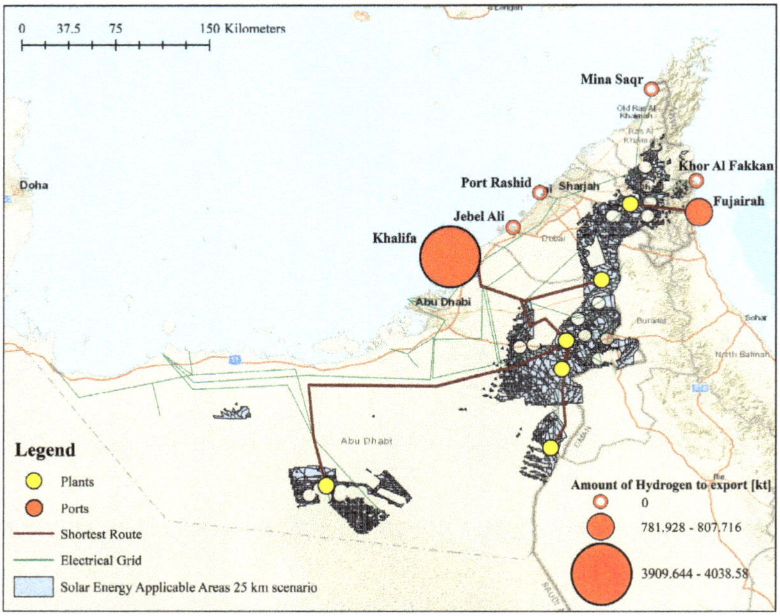

Figure 7. Second pathway GIS analysis results for closest ports and shortest electrical grid routes for electricity transmission from solar plants to port sites.

The sum of distances traveled in the electrical grid is 1031.24 km, which translates to a 3.1% loss of total transmitted energy. The remaining energy is used to desalinate water, provide the necessary energy for heating the water to needed levels before entering the HTE solid oxide electrolyzer and then produce hydrogen at the two chosen ports; the electricity consumption for desalinating the water needed for one kg of H_2 from Equation (2) is 1.8 kWh, which is added to the electrolyzer's consumption along with external steam supply and liquification electricity needs. Table 7 displays the numbers for the amounts of hydrogen produced in this pathway for the 25 km energy supply scenario.

Table 7. Produced H_2 in the second pathway from PV and CSP in (kg $\times$ 10^6).

Energy Scenario	PV	CSP
25 km	4846.296	4691.573

In Figure 8, we compare these results to shows the differences between the amounts of produced H_2 from PV and CSP technologies. The hydrogen levels produced by CSP fall slightly behind those of PV due to the extra energy expenses for providing high-temperature steam to ports in this pathway. The estimated amounts of hydrogen to be produced at each port in this pathway is explained in Table 8.

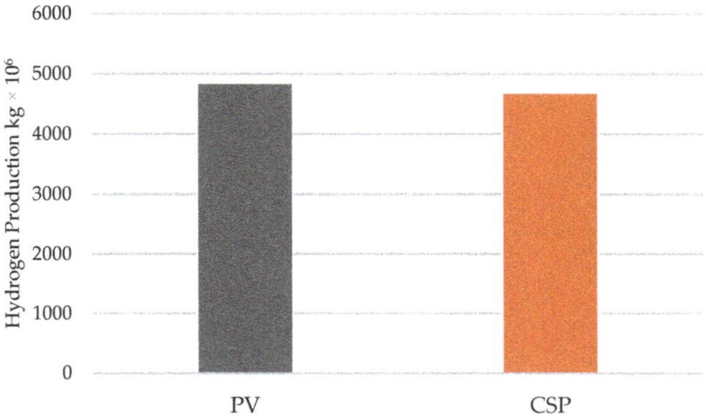

Figure 8. Comparison of hydrogen production in all technologies for 25 km energy scenario, in plants at port sites pathway.

Table 8. Amount of H_2 produced at each port in the second pathway from PV and CSP in (kg $\times$ 10^6).

Port Name	PV	CSP
Khalifa	4038.58	3909.644
Fujairah	807.716	781.928

3.5. Pathways Comparison

Here, we compare hydrogen produced for the 25 km energy scenario to have a better understanding of the results, as shown in Figure 9. The second pathway shows lower levels of H_2 at port sites, as the losses in the electrical grid added to desalination and external heater energy expenses leave less available electricity for hydrogen production than the solar plant sites pathway.

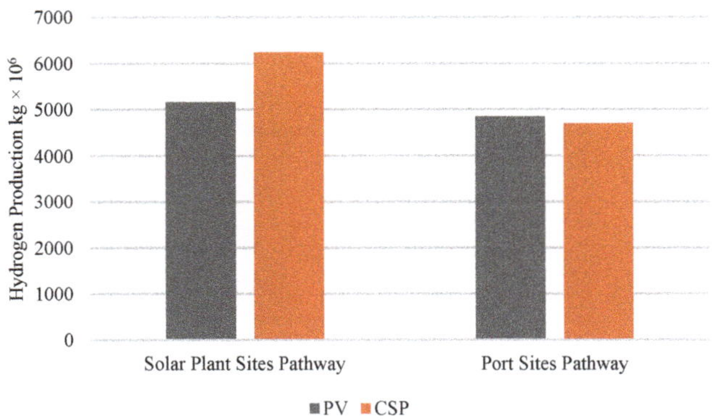

Figure 9. Hydrogen produced for all location scenarios in the 25 km scenario.

3.6. Proposed Shortest Routes

The current infrastructure of both road and electrical grids of the UAE is not ideal for some of the suggested solar plant sites. A few new routes are proposed in both location pathways to cut down the distance of hydrogen delivery for the first pathway, and electricity transmission losses in the second one. In Figure 10, three new routes are built with GIS to bring the distances of plants 4, 5, and 6 closer to ports. Figure 11 shows four suggested electrical transmission lines: the first connects the second plant to the geographically closest port, which is Dubai's Jebel Ali, and three other lines for plants 4, 5 and 6 to port Khalifa.

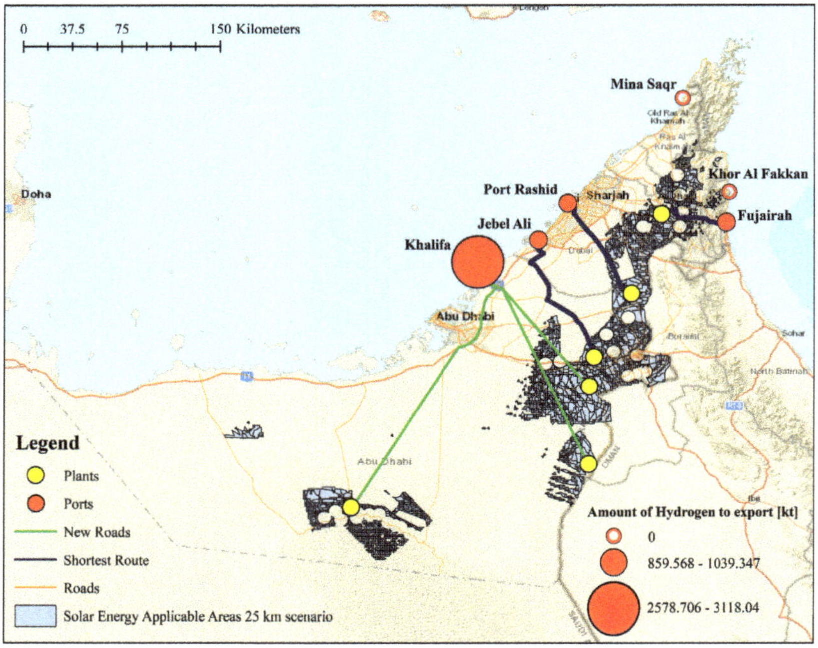

Figure 10. Proposed shortest routes from hydrogen plants to export ports analysis with GIS.

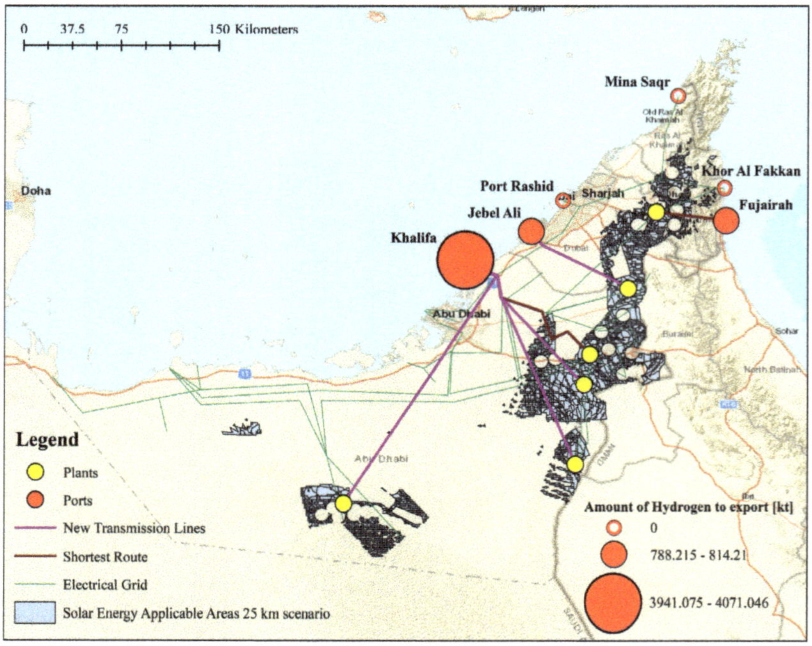

Figure 11. Proposed shortest electrical grid routes from hydrogen plants to export ports.

The new proposed routes for the first pathway will shorten the accumulated distances between all solar plants and ports from 926 km to 781 km. This will reduce time consumed for delivery and the number of trucks used to transport the hydrogen. Table 9 displays the quantities of H_2 to reach UAE's ports after using the proposed new roads in the first pathway.

Table 9. Amount of H_2 reaching each port using the proposed shortest routes in the first pathway from PV and CSP in (kg $\times$ 10^6).

Port Name	PV	CSP
Khalifa	2578.706	3118.04
Jebel Ali	859.568	1039.347
Port Rashid	859.568	1039.346
Fujairah	859.568	1039.346

As for the proposed transmission lines, the shorter traveled distance from 1031.240 km to 773.76 km will reduce the electricity loss in the grid to 2.3% based on a presumed 3% loss for every 1000 km, thus leaving more surplus electricity for hydrogen production at port sites.

Figure 12 shows the increases in hydrogen production if the proposed new electrical grid line is applied compared to the current electrical grid of the UAE, while Table 10 lists the new amounts of liquid hydrogen to be produced at each port in this pathway.

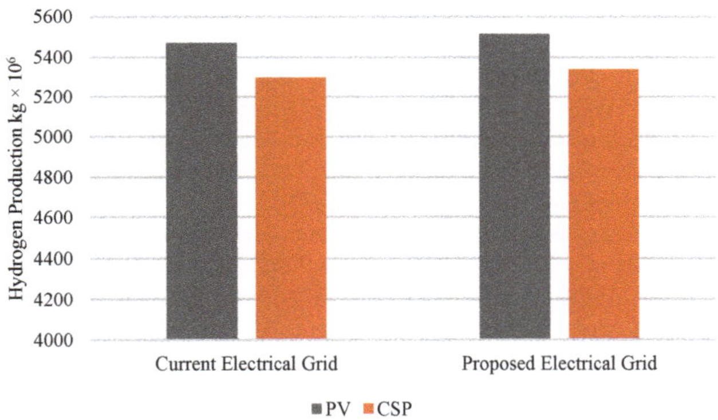

Figure 12. Comparison between the produced H_2 from the current and proposed electrical grid from PV and CSP.

Table 10. Amount of H_2 produced at each port using the proposed shortest electric grid in the second pathway from PV and CSP in (kg $\times$ 10^6).

Port Name	PV	CSP
Khalifa	4071.046	3941.075
Jebel Ali	814.21	788.215
Fujairah	814.21	788.215

The proposed new roads and transmission lines aim to provide a clearer image on which hydrogen production scenario is better for the case of the UAE. However, from a technical point of view, the amount of produced hydrogen is the main parameter. Figure 13 compares the amounts of liquid H_2 produced in each location pathway with the new proposed routes for the 25 km energy scenario. While the new roads do not provide any change in the amounts of hydrogen produced, the shorter transmission lines lead to an increase of hydrogen production at the ports, as explained in the previous section.

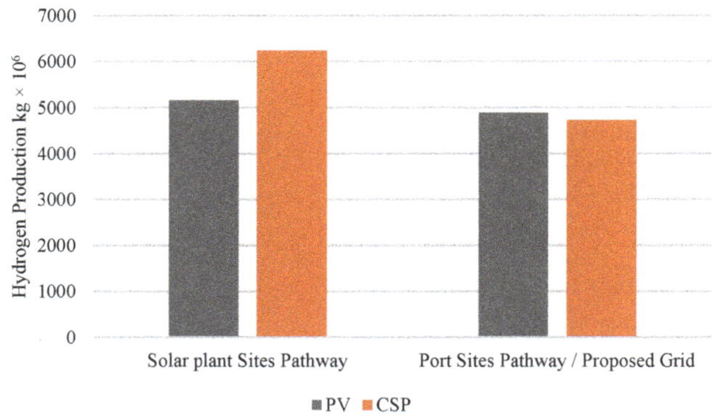

Figure 13. Comparison between the amounts of produced H_2 in the two location pathways with new proposed routes.

The suggested transmission lines close the gap between the amounts of hydrogen produced in each location scenario, separately around 0.1 megatons of hydrogen, while the SOEC electrolyzer still manages to provide higher levels of liquid H_2 when coupled to the CSP plants in the first pathway.

4. Conclusions

In this study we develop a model to evaluate solar power's potential for hydrogen production in the United Arab Emirates. We compare hydrogen output from PV and CSP technologies using solid oxide electric cell water electrolyzer technology, and then investigate the UAE's infrastructure for a H_2 supply chain from well to ship in two different pathways depending on the location of the electrolyzers by estimating the shortest routes using Geographical Information System (GIS), while highlighting the constraints in the UAE's infrastructure and then proposing solutions for these limitations. Produced hydrogen is to be exported as a means of profit to make the investment in large-scale solar energy application in the UAE feasible and set the UAE to become a main green hydrogen exporter in the future.

We have found that the integration of concentrated solar power plants coupled with SOEC electrolyzers for hydrogen production at solar plant sites is the most favorable way to utilize areas suitable for solar energy application in the UAE for large-scale solar hydrogen production. Although PV technology shows an advantage in energy potential, with 42 GWh/km^2 compared to 41.1 GWh/km^2 for CSP, the production of hydrogen from SOEC coupled to CSP plants at the same sites brings about over 1.1 extra megatons of liquid hydrogen to those produced from PV in the same pathway. The second pathway, with electrolyzers positioned at port sites, shows lower levels of hydrogen from both PV and CSP technologies due to the electricity lost in the UAE's electrical grid and the extra energy demands for water heating.

The results from the two delivery paths also show that four ports can be used for export after delivering the liquid hydrogen by trucks from various solar plants in the first one, providing more flexibility to delivery options. In the second pathway, only the Khalifa and Fujairah ports are chosen as the efficient options for electricity transmission and hydrogen production sites.

The proposed shorter routes made the gap of hydrogen produced in both pathways smaller, to around 0.1 megatons of hydrogen. In addition, a third port is added as a suitable option for hydrogen production in the electrolyzers-at-ports scenario. Even with addressing the UAE's infrastructure constraints, the first pathway is still more favorable when it comes to hydrogen generation, while the second pathway addresses the water shortage issue of the UAE and results in hydrogen production levels comparable to those in the first pathway.

A future cost analysis to estimate the levelized cost of energy and hydrogen in each scenario and pathway, along with a life cycle assessment of the environmental impacts, is needed to provide a wider perspective on which pathway and technology is a better fit for the UAE and assures higher hydrogen export profits and the lowest harmful emissions.

Author Contributions: Conceptualization, A.J. and K.O.; methodology, A.J.; software, A.J.; validation, A.J., Y.A. and K.O.; formal analysis, A.J.; investigation, A.J.; resources, A.J.; data curation, A.J.; writing—original draft preparation, A.J.; writing—review and editing, A.J.; visualization, A.J.; supervision, Y.A. and K.O.; project administration, A.J. and K.O.; funding acquisition, K.O. All authors have read and agreed to the published version of the manuscript.

Funding: This research was supported by JSPS Grant-in-Aid for Scientific Research 20H02677.

Informed Consent Statement: Not applicable.

Data Availability Statement: Not applicable.

Acknowledgments: The first author acknowledges funding support from the Japanese government, Ministry of Education, Culture, Sports, Science and Technology, under their international program (MEXT) scholarship.

Conflicts of Interest: The authors declare no conflict of interest.

References

1. El-Katiri, L. *A Roadmap for Renewable Energy in the Middle East and North Africa*; Oxford Institute for Energy Studies: Oxford, UK, 2014; ISBN 9781907555909. Available online: https://a9w7k6q9.stackpathcdn.com/wpcms/wp-content/uploads/2014/01/MEP-6.pdf (accessed on 15 June 2019).
2. Alnaser, W.E.; Alnaser, N.W. The Status of Renewable Energy in the GCC Countries. *Renew. Sustain. Energy Rev.* **2011**, *15*, 3074–3098. [CrossRef]
3. Griffiths, S. A Review and Assessment of Energy Policy in the Middle East and North Africa Region. *Energy Policy* **2017**, *102*, 249–269. [CrossRef]
4. Bhutto, A.W.; Bazmi, A.A.; Zahedi, G.; Klemeš, J.J. A Review of Progress in Renewable Energy Implementation in the Gulf Cooperation Council Countries. *J. Clean. Prod.* **2014**, *71*, 168–180. [CrossRef]
5. IRENA. *Renwable Energy Market Analysis: The GCC Region*; International Renewable Energy Agency: Abu Dhabi, United Arab Emirates, 2016.
6. IRENA. *Renewable Energy Market Analysis: GCC 2019*; International Renewable Energy Agency: Abu Dhabi, United Arab Emirates, 2019.
7. Nematollahi, O.; Hoghooghi, H.; Rasti, M.; Sedaghat, A. Energy Demands and Renewable Energy Resources in the Middle East. *Renew. Sustain. Energy Rev.* **2016**, *54*, 1172–1181. [CrossRef]
8. Bouhal, T.; Agrouaz, Y.; Kousksou, T.; Allouhi, A.; el Rhafiki, T.; Jamil, A.; Bakkas, M. Technical Feasibility of a Sustainable Concentrated Solar Power in Morocco through an Energy Analysis. *Renew. Sustain. Energy Rev.* **2018**, *81*, 1087–1095. [CrossRef]
9. Reuß, M.; Grube, T.; Robinius, M.; Stolten, D. A Hydrogen Supply Chain with Spatial Resolution: Comparative Analysis of Infrastructure Technologies in Germany. *Appl. Energy* **2019**, *247*, 438–453. [CrossRef]
10. Marchenko, O.V.; Solomin, S.V. The Future Energy: Hydrogen versus Electricity. *Int. J. Hydrog. Energy* **2015**, *40*, 3801–3805. [CrossRef]
11. Loisel, R.; Baranger, L.; Chemouri, N.; Spinu, S.; Pardo, S. Economic Evaluation of Hybrid Off-Shore Wind Power and Hydrogen Storage System. *Int. J. Hydrog. Energy* **2015**, *40*, 6727–6739. [CrossRef]
12. Franco, B.A.; Baptista, P.; Neto, R.C.; Ganilha, S. Assessment of Offloading Pathways for Wind-Powered Offshore Hydrogen Production: Energy and Economic Analysis. *Appl. Energy* **2021**, *286*, 116553. [CrossRef]
13. IEA. *World Energy Outlook 2020*; International Energy Agency: Paris, France, 2020; Available online: https://www.iea.org/reports/world-energy-outlook-2020 (accessed on 16 May 2020).
14. Behling, N.; Williams, M.C.; Managi, S. Fuel Cells and the Hydrogen Revolution: Analysis of a Strategic Plan in Japan. *Econ. Anal. Policy* **2015**, *48*, 204–221. [CrossRef]
15. Kamiya, S.; Nishimura, M.; Harada, E. Study on Introduction of CO_2 Free Energy to Japan with Liquid Hydrogen. *Phys. Procedia* **2015**, *67*, 11–19. [CrossRef]
16. Nagashima, M. *Japan's Hydrogen Strategy and Its Economic and Geopolitical Implications*; IFRI: Paris, France, 2018; ISBN 9782365679183.
17. Kim, A.; Kim, H.; Lee, H.; Lee, B.; Lim, H. Comparative Economic Optimization for an Overseas Hydrogen Supply Chain Using Mixed-Integer Linear Programming. *ACS Sustain. Chem. Eng.* **2021**, *9*, 42. [CrossRef]
18. Basic Hydrogen Strategy Determined (METI). Available online: https://www.meti.go.jp/english/press/2017/1226_003.html (accessed on 28 April 2022).
19. IEA. *Global Hydrogen Review 2021*; International Energy Agency: Paris, France, 2021; Available online: https://www.iea.org/reports/global-hydrogen-review-2021 (accessed on 15 November 2021).

20. Rosenstiel, A.; Monnerie, N.; Dersch, J.; Roeb, M.; Pitz-Paal, R.; Sattler, C. Electrochemical Hydrogen Production Powered by PV/CSP Hybrid Power Plants: A Modelling Approach for Cost Optimal System Design. *Energies* **2021**, *14*, 3437. [CrossRef]
21. Kazim, A.; Veziroglu, T.N. Utilization of Solar–Hydrogen Energy in the UAE to Maintain Its Share in the World Energy Market for the 21st Century. *Renew. Energy* **2001**, *24*, 259–274. [CrossRef]
22. Orhan, M.F.; Kahraman, H.; Babu, B.S. Approaches for Integrated Hydrogen Production Based on Nuclear and Renewable Energy Sources: Energy and Exergy Assessments of Nuclear and Solar Energy Sources in the United Arab Emirates. *Int. J. Hydrog. Energy* **2017**, *42*, 2601–2616. [CrossRef]
23. Kazim, A. Strategy for a Sustainable Development in the UAE through Hydrogen Energy. *Renew. Energy* **2010**, *35*, 2257–2269. [CrossRef]
24. Joubi, A.; Okajima, K. Development of an Energy Potential Estimation Model for Concentrated Solar Plants Penetration in the GCC Region. *Int. J. Smart Grid Clean Energy* **2020**, *9*, 1000–1010. [CrossRef]
25. Lopez, A.; Roberts, B.; Heimiller, D.; Blair, N.; Porro, G. *U.S. Renewable Energy Technical Potentials. A GIS-Based Analysis*; U.S. Department of Energy: Golden, CO, USA, 2012. [CrossRef]
26. APICORP. *APICORP's MENA Annual Energy Investment Outlook 2019–2023*; Arab Petroluem Investments Corporation: Dammam, Saudi Arabia, 2019; Available online: https://www.apicorp.org/media-centre/publications/?page_num=2&category_type=&date_year= (accessed on 1 March 2020).
27. Murphy, C.; Sun, Y.; Cole, W.; Maclaurin, G.; Turchi, C.; Mehos, M. *The Potential Role of Concentrating Solar Power within the Context of DOE's 2030 Solar Cost Targets*; U.S. Department of Energy: Golden, CO, USA, 2019.
28. Tran, J.; Grue, N.; Cox, S. *Renewable Energy Data Explorer User Guide*; U.S. Department of Energy: Golden, CO, USA, 2018.
29. Al Ali, M.; Emziane, M. Performance Analysis of Rooftop PV Systems in Abu Dhabi. *Energy Procedia* **2013**, *42*, 689–697. [CrossRef]
30. Strachan, N.; Balta-Ozkan, N.; Joffe, D.; McGeevor, K.; Hughes, N. Soft-Linking Energy Systems and GIS Models to Investigate Spatial Hydrogen Infrastructure Development in a Low-Carbon UK Energy System. *Int. J. Hydrog. Energy* **2009**, *34*, 642–657. [CrossRef]
31. Balta-Ozkan, N.; Baldwin, E. Spatial Development of Hydrogen Economy in a Low-Carbon UK Energy System. *Int. J. Hydrog. Energy* **2013**, *38*, 1209–1224. [CrossRef]
32. Reuß, M.; Grube, T.; Robinius, M.; Preuster, P.; Wasserscheid, P.; Stolten, D. Seasonal Storage and Alternative Carriers: A Flexible Hydrogen Supply Chain Model. *Appl. Energy* **2017**, *200*, 290–302. [CrossRef]
33. Tlili, O.; Mansilla, C.; Linβen, J.; Reuß, M.; Grube, T.; Robinius, M.; André, J.; Perez, Y.; le Duigou, A.; Stolten, D. Geospatial Modelling of the Hydrogen Infrastructure in France in Order to Identify the Most Suited Supply Chains. *Int. J. Hydrog. Energy* **2020**, *45*, 3053–3072. [CrossRef]
34. Seaports—The Official Portal of the UAE Government. Available online: https://u.ae/en/information-and-services/infrastructure/civic-facilities/seaports (accessed on 28 April 2022).
35. GEOFABRIK. OpenStreetMap. Available online: https://www.geofabrik.de/geofabrik/openstreetmap.html (accessed on 28 April 2022).
36. IEA ETSAP. Electricity Transmission and Distribution; International Energy Agency Energy Technology Systems Analysis Programme. 2014. Available online: https://iea-etsap.org/E-TechDS/PDF/E12_el-t&d_KV_Apr2014_GSOK.pdf (accessed on 7 October 2021).
37. Taibi, E.; Miranda, R.; Vanhoudt, W.; Winkel, T.; Lanoix, J.-C.; Barth, F. *Hydrogen from Renewable Power: Technology Outlook for the Energy Transition*; IRENA: Abu Dhabi, United Arab Emirates, 2018; ISBN 978-92-9260-077-8.
38. Houaijia, A.; Roeb, M.; Monnerie, N.; Sattler, C. Solar Power Tower as Heat and Electricity Source for a Solid Oxide Electrolyzer: A Case Study. *Int. J. Energy Res.* **2015**, *39*, 1120–1130. [CrossRef]
39. IEA. *Technology Roadmap Hydrogen and Fuel Cells Technical Annex*; International Energy Agency: Paris, France, 2018; Available online: https://www.iea.org/reports/technology-roadmap-hydrogen-and-fuel-cells (accessed on 30 April 2020).
40. Sanz-Bermejo, J.; Muñoz-Antón, J.; Gonzalez-Aguilar, J.; Romero, M. Optimal Integration of a Solid-Oxide Electrolyser Cell into a Direct Steam Generation Solar Tower Plant for Zero-Emission Hydrogen Production. *Appl. Energy* **2014**, *131*, 238–247. [CrossRef]
41. Hinkley, J.; Hayward, J.; Mcnaughton, R.; Gillespie, R.; Watt, M.; Lovegrove, K.; Matsumoto, A. *Cost Assessment of Hydrogen Production from PV and Electrolysis*; CSIRO: Newcastle, Australia, 2016.
42. Meier, K. Hydrogen Production with Sea Water Electrolysis Using Norwegian Offshore Wind Energy Potentials: Techno-Economic Assessment for an Offshore-Based Hydrogen Production Approach with State-of-the-Art Technology. *Int. J. Energy Environ. Eng.* **2014**, *5*, 104. [CrossRef]
43. Monnerie, N.; von Storch, H.; Houaijia, A.; Roeb, M.; Sattler, C. Hydrogen Production by Coupling Pressurized High Temperature Electrolyser with Solar Tower Technology. *Int. J. Hydrog. Energy* **2017**, *42*, 13498–13509. [CrossRef]

energies

MDPI

Article

Modular Multilevel Converter for a Linear Generator for Wave Energy Converter

Henry M. Zapata * and Marcelo A. Pérez

Department of Electronics, Universidad Técnica Federico Santa María, Valparaiso 2390123, Chile
* Correspondence: henry.zapata@sansano.usm.cl

Abstract: In this paper we propose a modular multilevel converter for a linear generator for a wave energy converter. The coils of this generator are individually controlled to improve energy harvesting performance. This topology involves two stages. The first stage uses a full-bridge to control the harvested current with a reference generated by means of an MPPT method. The second stage uses a half-bridge to control the voltage of the DC-link and the output current. Furthermore, multilevel modular converters allow the generation of a medium DC voltagethat reduces the losses in energy transmission lines from offshore to a coupling common point on the shore.

Keywords: wave energy converter; linear generator; modular multilevel converter

Citation: Zapata, H.M.; Pérez, M.A. Modular Multilevel Converter for a Linear Generator for Wave Energy Converter. *Energies* **2022**, *15*, 6346. https://doi.org/10.3390/en15176346

Academic Editors: Wei-Hsin Chen, Aristotle T. Ubando, Chih-Che Chueh and Liwen Jin

Received: 10 July 2022
Accepted: 24 August 2022
Published: 31 August 2022

Publisher's Note: MDPI stays neutral with regard to jurisdictional claims in published maps and institutional affiliations.

1. Introduction

Concerns about global warming mitigation have increased the development of alternative renewable energy sources, such as solar (photovoltaic panels, PVs), wind (turbines), and ocean energy (wave energy converters). During the last ten years (2011–2021), renewable energies have matured and currently represent 37% of the total operating installed capacity of electrical energy, as shown in Figure 1. Wind power generation is continually improving and increased its power capacity by 93 GW in 2020. Furthermore, PV energy is the renewable energy with the most notable growth, for example, through new technologies such as organic solar cells or polymer solar cells [1–5].

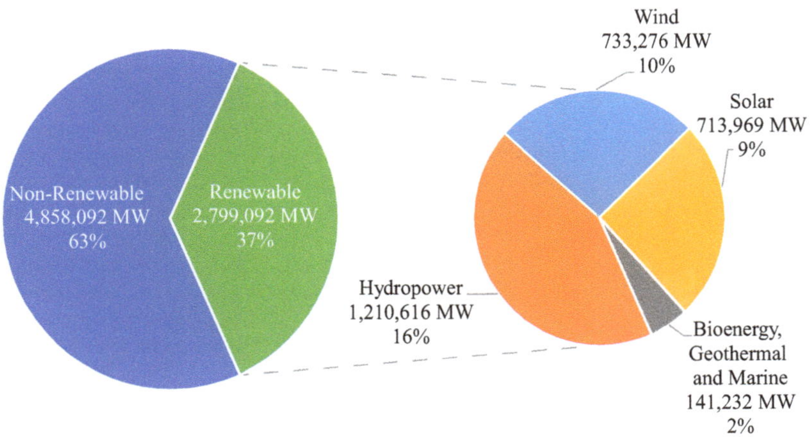

Figure 1. Estimated electricity installed capacity (MW). End 2020. The data were obtained from [6].

Despite the increase in renewable energy generation, some sectors have a low percentage of final energy demand. Therefore, other renewable energy sources are required to meet the targets of gas emissions reductions, and for this purpose ocean energy is an excellent choice. This alternative energy has great potential due to its relatively high power density, but it is the largest untapped energy source [7].

Regarding ocean energy, there are six distinct sources (ocean waves, tidal range, tidal streams, ocean currents, temperature gradients, and salinity gradients), each with different technologies for their conversion. The development of ocean energy technologies has focused on tidal and wave energy, and the installed capacity of ocean energy rose to approximately 527 MW in late 2020 [6–8], as shown in Figure 1.

Ocean waves are generated through the kinetic energy exchange between the wind and the upper surface of the ocean into power. In this paper, we focus on an application using ocean waves to harvest energy. This is a valuable option because the estimated total wave power on the coastlines can reach up to $2TW$, which has a great potential to generate electrical energy [9].

Wave energy converters (WECs) are devices that capture wave energy mechanically and convert it into electrical energy. The development of WECs is still more complex than other types of renewable energies for two reasons. The first corresponds to the diversity of the wave resources in offshore and nearshore locations, which means that the design process involves difficulties in obtaining a high efficiency over the entire range of operations. The other reason involves the instability of energy intensity when capturing maximum energy [10–13].

There are plenty of WECs with different working principles and designs; point absorbers stand out among others because their size is smaller than the wavelengths of ocean waves. These systems harvest energy from the motion of bodies, which are submerged or floating on the surface wave [10,14].

Linear generators are a more efficient option to be used on a point absorber, for example, the Islandberg Project (Sweden), Seacap (France), PowerPod, and PowerPod II (UK), among others [15]. Furthermore, linear generator technology offers a simple and robust structure. In addition, linear generator design presents different properties, such as the use of a planar shape, reducing the detent force through its structure and tubular shape, producing a high energy density [14,16].

A tubular permanent magnet linear generator (TPMLG) can be designed with different topologies and shapes used for the permanent magnets (PMs) to obtain better flux linkage, cogging force reduction, and higher efficiency in WECs [17–19]. The most common topologies are radial, axial, and Halbach configurations for the implementation of TPMLGs [18,20]. Furthermore, topologies have also been developed to avoid demagnetization by modifying the air gap [21].

The development of high-voltage direct current (HVDC) systems based on the use of a voltage source converter (VSC) has increased significantly with the introduction of new semiconductor technologies. HVDC technologies are used in the long-line transmission of high power, the interconnection of different AC systems, the integration of renewable energy sources on a large scale, underground applications, and submarine applications [22,23].

MMC topologies use the same submodules to provide a step in a multilevel waveform and are easily adaptable to high voltage levels. The topology of high-voltage DC/AC conversion is based on the use of MMC converters on cascade submodules [23].

There are two basic submodules (SMs) in MMC topologies—the half-bridge, and the full-bridge submodule. The half-bridge has a voltage output equal to its capacitor voltage ($+v_C$) or zero (0). The full-bridge has a voltage output equal to its positive voltage ($+v_C$), its negative voltage ($-v_C$), or zero. The latter submodule has a higher cost because the number of semiconductors is doubled; however, the full-bridge has the advantage of providing a negative voltage level [24,25].

In this paper, we propose a TPMLG with a radial configuration with a single stator and translator. Furthermore, we simulated this system with five individual coils which harvest energy, each with a full-bridge submodule. Finally, the power converter uses a full-bridge submodule to harvest energy and a half-bridge submodule at the output of the MMC.

2. System Model

A WEC system uses three energy conversion stages to feed the grid. The point absorbers have the advantage of reducing the two first stages in a direct stage. This design uses the action of the incident sea waves to move the translator of a linear generator without additional mechanical systems [26]. This device has a buoy on the surface of the waves and it is connected to an electrical generator system underwater, as shown in Figure 2. The last stage is the interface between the WECs and a coupling common point through a static transmission cable, and this provides a power signal to inject into the grid, as shown in Figure 2. The focus of this paper is on the direct stage.

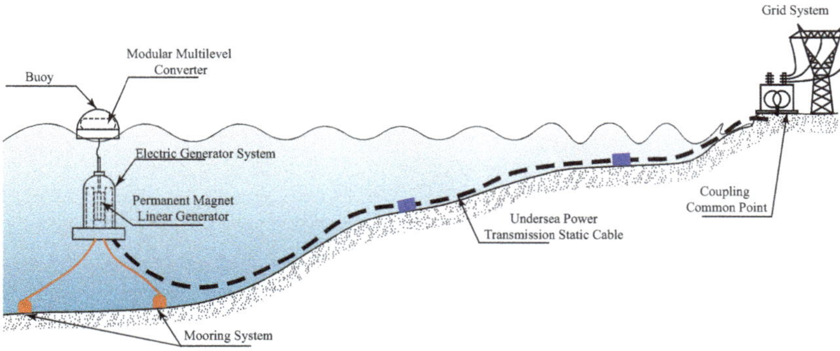

Figure 2. Complete system of a wave energy converter.

2.1. *Wave–Buoy Interaction*

The wave elevation defines the potential energy (E_p) and the water fluid motion defines the kinetic energy (E_k). These energies are stored in the waves, as shown in Equation (1) [27,28].

$$E = E_p + E_k = \rho g \int_0^{\infty} S(f)df = \frac{\rho g H_s^2}{16} \tag{1}$$

where ρ is the fluid density (approximately 1025 $\frac{kg}{m^3}$), g refers to the gravitational acceleration (9.8 $\frac{m}{s^2}$), E is total stored energy, and $S(f)$ is the spectrum distribution of the wave energy of a given location as a wave frequency function (f). The wave spectrum defines the average height (H_s), as shown in Equation (2); this is used to determine the wave power [27,29].

$$\frac{H_s^2}{16} = \int_0^{\infty} S(f)df \tag{2}$$

The integral of the time-average energy transport per unit time and unit area in the direction of wave propagation (the x direction) is called the water-power level J or wave-energy transport, and its expression is shown below in Equation (3) using $v_g = g/2\omega$ for irregular plane waves on deep water [27,28].

$$J = \rho g \int_0^{\infty} S(f)v_g(f)df = \frac{\rho g^2 T_J H_s^2}{64\pi} \tag{3}$$

where T_j is the energy period. For example, an irregular wave of a buoy located in Valparaíso, Chile, where the depth is 4515 m, with energy period T_J = 10.3 s and significant wave height H_s = 2.3 m, the energy transport is J = 26.6 kW/m.

2.2. *WEC Equation of Motion*

The total force acting on a buoy can be decomposed as shown in Figure 3. The excitation force F_e is produced by incident waves and the radiation force F_r is produced by buoy oscillation. The buoyancy force has a part constant F_b when the buoy stands still on the water surface $z_b = 0$; this represents the weight of the buoy and translator, and it also has a variable part $F_h = \rho g S z_b$ related to the displacement of the buoy and the balanced position. There is a gravity force $M_b.g$ and a force line F_{line}. These forces are written in the frequency domain, as shown in Equation (4).

$$-w^2.M_b.z_b(w) = F_e + F_b - F_r - F_h - F_{line} - M_b.g \quad (4)$$

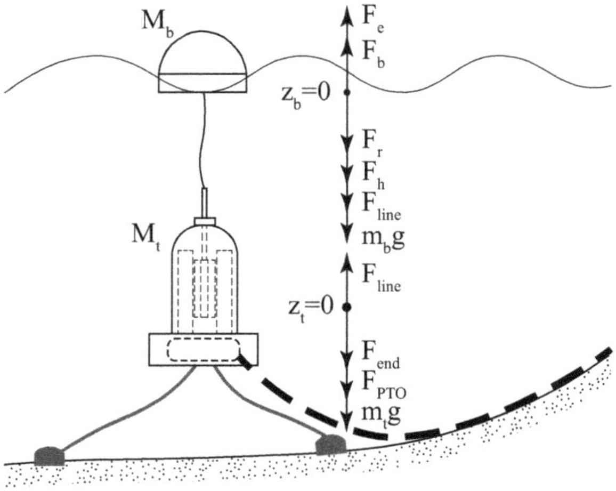

Figure 3. Characteristics of waves. Figure modified to add nomenclature for the vertical reference motion of the buoy and the translator.

The force of the translator is the gravity force $M_t.g$, the line force is F_{line}, and the end-stop force is F_{end}. The latter force is generated when the translator hits the upper-end stop spring or the rubber damper hits the generator bottom. The F_{PTO} is the electromagnetic damping force from the power take-off.

$$-w^2.M_t.z_t(w) = F_{line} - F_{end} - F_{PTO} - M_t.g \quad (5)$$

2.3. Power Take-Off

The PTO uses a linear generator with a tubular shape, divided into two structures. The translator has ring-shaped PMs or PMs arrays with a similar shape, which have an axial configuration (north orientation to outer radial); furthermore, PMs arrays are attached to the PMs' support and a mobile vertical shaft, as shown in Figure 4. The stator has a tubular shape and supports five coils arranged around it.

The magnetic circuit can be analyzed by means of a lumped parameter method related to an electric circuit because they have the same behavior in a stable state [14]. The magnetic circuit is expressed in Equation (6), and it has a net reluctance ($\mathfrak{R}_{xx}$) given by the path of magnetic flux shown in Figure 4, the magnetomotive force ($\mathcal{F}_m$) produced by the PMs, and where Φ indicates the magnetic flux.

$$\mathcal{F}_m = \Phi \mathfrak{R}_n \quad (6)$$

The reluctance ($\mathfrak{R}_{xx}$) is defined by the flux path length (l_x), vacuum permeability (μ_o), the relative permeability of each material (μ_{rx}), and the cross area (A_x), [30], as follows:

$$\mathfrak{R}_{xx} = \frac{l_x}{\mu_o \mu_{rx} A_x} \quad (7)$$

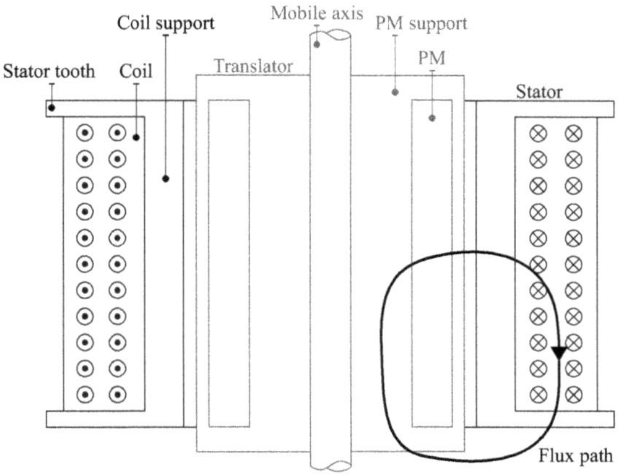

Figure 4. Schematic design of the linear generator. Figure based on [31].

The partial reluctance of the structure describes the net reluctance, using Kirchhoff's circuit laws [32], as shown below.

$$\begin{aligned}\mathfrak{R}_n =& \mathfrak{R}_{PM} + \mathfrak{R}_{oPMs} + \mathfrak{R}_{gap} + \mathfrak{R}_{cs} + \mathfrak{R}_c + \mathfrak{R}_l \\ &+ \mathfrak{R}_{st} + \mathfrak{R}_{cs2} + \mathfrak{R}_{gap2} + \mathfrak{R}_{lPMs} + \mathfrak{R}_{iPMs}\end{aligned} \tag{8}$$

Equation (9) defines the magnetomotive force ($\mathcal{F}_m$) generated by PMs. This is the relation between the product of the remanent magnetization of the PM (B_r) with the width of the PM (w_pm) and the product of the relative permeability of the PM μ_{rpm} with the vacuum permeability (μ_o).

$$\mathcal{F}_m = \frac{B_r\, w_pm}{\mu_o \mu_{rpm}} \tag{9}$$

The flux linkage is considered a sinusoidal function of the vertical displacement (z) to define the magnetic flux for each coil, as shown below.

$$\phi = \Phi_{max} \sin\left(\frac{\pi}{4\,\tau_m} z\right) = \frac{B_r\, w_m}{\mu_o \mu_{rm} \mathfrak{R}_n} \sin\left(\frac{\pi}{4\,\tau_m} z\right) \tag{10}$$

Equation (10) is used to describe the electromotive force (ε) in a coil with many turns (N) [33]; thus,

$$\varepsilon = -N \frac{d\phi}{dh} \frac{dz}{dt} = -\,N \frac{B_r\, w_m}{\mu_o \mu_{rm} \mathfrak{R}_n} \frac{\pi}{4\,\tau_m} \cos\left(\frac{\pi}{4\tau_m} z\right) \frac{dz}{dt} \tag{11}$$

3. MMC Description

An MMC topology was used to harvest energy from each coil of the TPMLG, to control the current harvesting process and generate a medium voltage for transmission. Therefore, each coil was connected to a power converter submodule, and the power converter submodule outputs were connected in series to generate a medium voltage, as shown in Figure 5.

Each submodule was composed of two parts interconnected with a coupled capacitor, as illustrated in Figure 5: an AC-DC converter from a generator coil and a DC-DC converter connected in series to other converters. In the first stage, a full-bridge submodule was used to control the current harvesting with reference to the MPPT, which changed the direction of the torque drive. In the second stage, a half-bridge submodule was used because two states were required to control the capacitor voltage. The MMC connection could require a filter inductor if the transmission line inductance is insufficient to filter the switching harmonics.

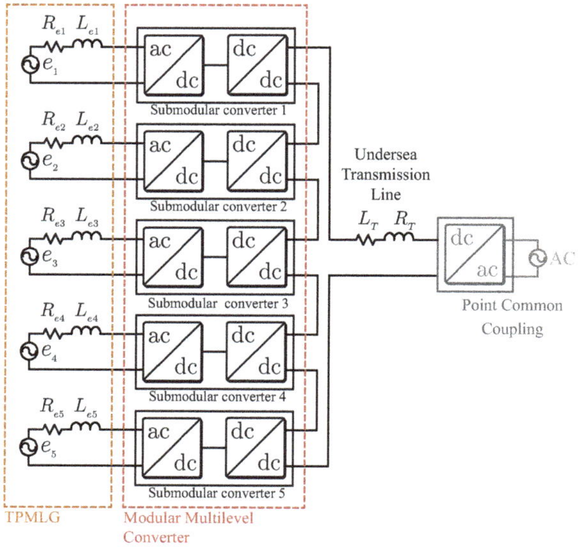

Figure 5. Modular inverter schematic with grid connections.

For the converters shown in Figure 6, we used a controlled decoupled design. The DC-MMC part used a half-bridge submodule to control the voltage of the capacitor.

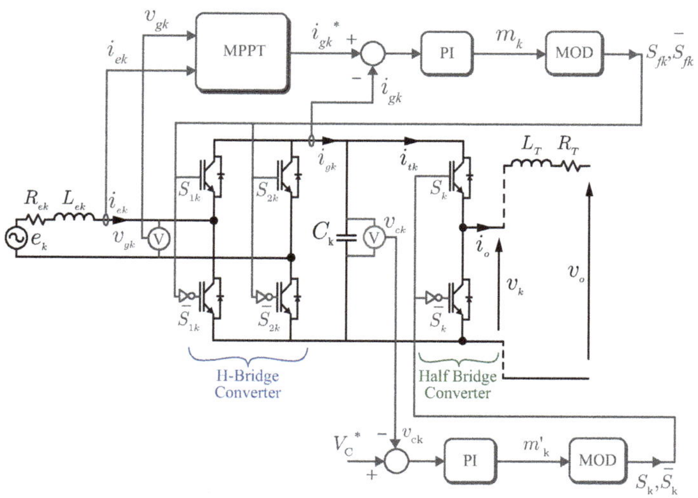

Figure 6. Control scheme and electronic circuit of an MMC submodule.

Through this configuration, one could add the submodules needed as coils have the generator. At the same time, this decreases the frequencies generated by power.

Previous studies on PV systems have presented the same topology to interconnect low-voltage systems (PV systems) to medium- or high-voltage DC systems through one or two voltage transformer steps. These works show the viability of implementing this topology to large-scale connected grids [34].

4. Control Strategy

The control strategy proposed here involves two decoupled control stages, one for an AC-DC converter with the MPPT and one for a DC-DC converter that assembles the MMC, as shown in Figure 6. The first control stage uses the result of the MPPT method as a reference to control the harvesting current of a coil of the generator. The second control stage aims to control each DC-link voltage and the current through all the submodules of the MMC.

Figure 7 shows the simulated voltage of a TPMLG with five coils to show its behavior. Furthermore, the TPMLG has five PM arrays, which are moving at 0.1 Hz.

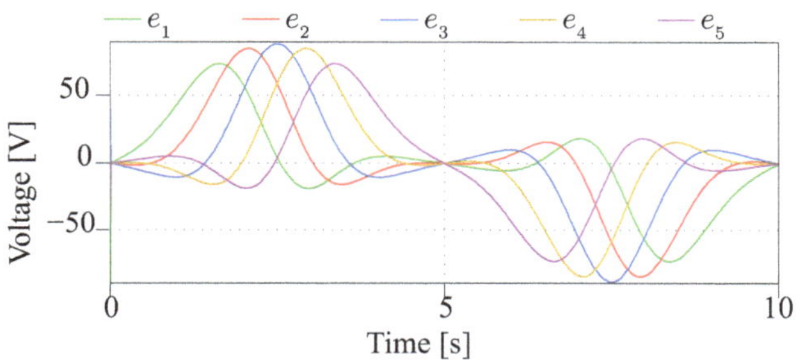

Figure 7. Voltage generated at the terminals' coils.

4.1. Harvesting Control

The MPPT algorithm determines the current reference for the AC-DC stage to maximize the power of each coil. In this paper, we present an algorithm for the MPPT based on the nature of the frequency of the waves and the electrical system measurements. A PI regulator controls the error generated by the MPPT reference and the generated current. Equation (12) shows the transfer function used to design the PI for the modulation index of the full-bridge submodule.

$$\frac{i_{gk}}{m_k} = -\frac{v_{ck}}{R_e + sL_e} \tag{12}$$

where i_{gk} is the input current (coil current), m_k is he modulation index for the converter, R_e is the coil resistance, and L_e is coil inductance. v_{ck} is the capacitor voltage and it is considered a constant parameter with a value of 100 V for the simulation. This control scheme is shown in Figure 6.

Figure 8 shows the input signal and output signal of the PI regulator from the first submodule. It shows the switching signals at the PMW modulator's bipolar output used to control each semiconductor.

The ideal MPPT reference would be the exact behavior of the vertical position of the buoy in relation to the wave, but an absolute reference is estimated to have the following expression:

$$m_{tot}\frac{dv}{dt} + k_s z = 0 \tag{13}$$

$$\beta_g = \beta_w \tag{14}$$

where m_{tot} is a sum of the floater mass and the added mass, k_s is the spring constant point absorber, β_g is the damping coefficient provided by the generator, and β_w hydrodynamic damping coefficient of the point absorber.

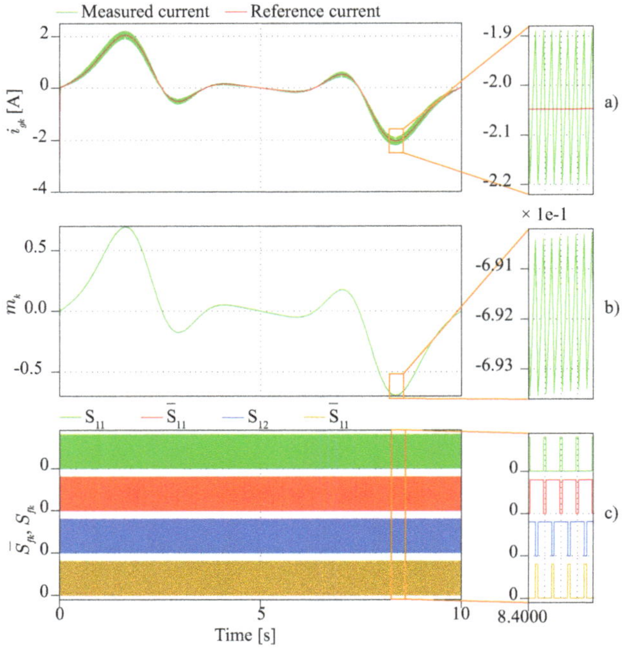

Figure 8. (**a**) Measured current and reference current signals i_{gk}. (**b**) Modulation index m_k. (**c**) Switching signals S_{fk}.

4.2. DC-Link Voltage Control

This control scheme allows one to control the DC-link voltage. The submodules operate at different power levels due to the differing translator positions, and each uses a PI regulator. The PI is designed with the transfer function shown in Equation (15) to generate a modulation index for the half-bridge submodule.

$$\frac{v_{ck}}{m'_k} = -\frac{i_o}{s_c} \tag{15}$$

where v_{ck} is the DC-link voltage, m'_k is the modulation index for the converter, and C is the capacitance of the DC-link. i_o is the output current and is considered a constant parameter. The capacitor current i_c is considered a disturbance of the system, so it does not appear in the control law.

Figure 9 shows the input signal and output signal of the PI regulator to the first submodule. It shows the switching signals at the output of the modulation stage used to control two semiconductors.

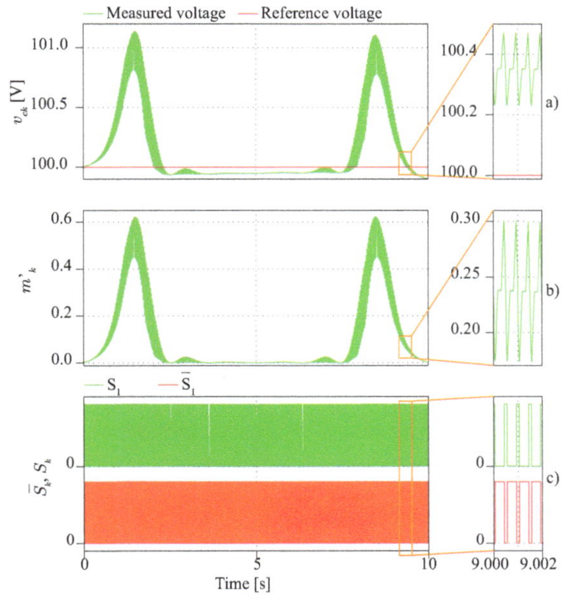

Figure 9. (**a**) Measured voltage and reference voltage signals v_{ck}. (**b**) Modulation index n_k. (**c**) Switching signals S_k.

4.3. Current Control

Loop current control injects a current into the DC output. The topology shown in Figure 10 represents a simplified model of the power circuit. The total current i_T flows through the MMC submodules connected in series, and the total converter voltage v_o is the sum of all submodule voltages, as shown in Figure 5.

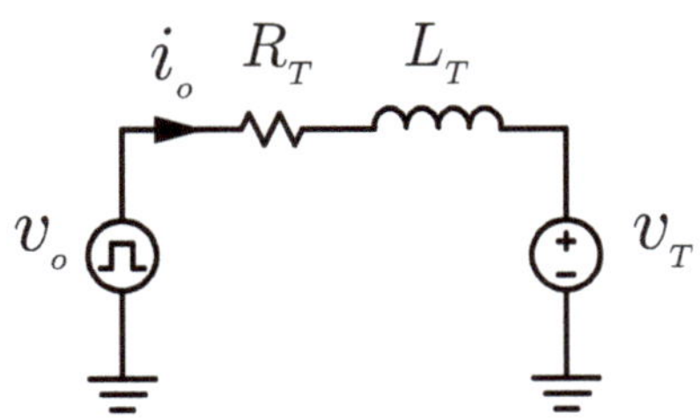

Figure 10. The simplified power circuit model with the grid.

$$v_o = \sum_{k=1}^{N} v_k \tag{16}$$

Equation (16) shows the output voltage of module k (v_k) and the total number of DC submodules (N) in the converter (N=5 in this case). The following equation uses Kirchhoff's voltage law in the topology equivalent of Figure 11.

$$v_T = v_o + i_o R_T + L_T \frac{di_o}{dt} \tag{17}$$

where v_T represent the voltage grid, i_o is the output current, and R_s, L_s are the line parameters. Therefore, an analysis of the power and the sum of all the individual DC-link voltages involves the use of the transfer function to design a PI control to generate the average modulation index and control the grid current, as shown below.

$$\frac{i_o^2}{m_i k} = \frac{N V_{ck}}{\frac{L_L}{2} + R_L} \tag{18}$$

where v_{ck} is the DC-link average of the submodules, N is the number of submodules, and m_i is the modulation index average.

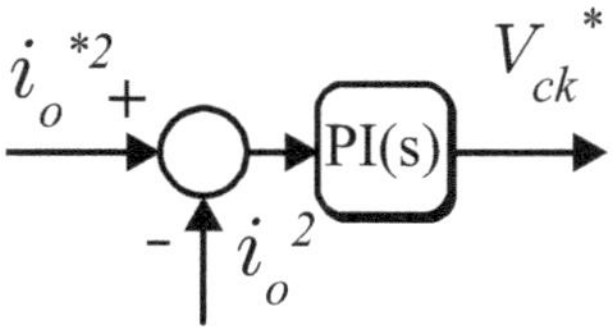

Figure 11. Control scheme for MMC submodules.

5. Conclusions

Wave energy is an under-exploited resource with great potential. Direct absorbers with linear generators present better results because they do not have mechanical connections; they also have the advantage of being placed offshore, avoiding visual contamination and problems with inhabitants near the coast.

In this study, we proposed to improve the energy harvesting performance of WECs by means of a modular multilevel converter to drive a linear generator with independent coils by controlling the magnetic flux and torque depending on the translator position and increasing the efficiency of power transmission. The connection of the windings in series can be undertaken to generate a higher voltage and increase the efficiency of the transmission of power to the coast.

Author Contributions: Conceptualization, M.A.P. and H.M.Z.; methodology, M.A.P.; software, H.M.Z.; validation, H.M.Z. and M.A.P.; formal analysis, H.M.Z.; investigation, H.M.Z.; resources, M.A.P.; data curation, H.M.Z.; writing—original draft preparation, H.M.Z.; writing—review and editing, H.M.Z.; visualization, H.M.Z.; supervision, M.A.P.; project administration, M.A.P.; funding acquisition, M.A.P. All authors have read and agreed to the published version of the manuscript.

Funding: This research was funded by the Advanced Center for Electrical and Electronics Engineering (AC3E) under Grant ANID/FB0008, by the Solar Energy Research Center (SERC) under Grant ANID/FONDAP/15110019, by the ANID/Fondecyt Regular Grant 1211826, DGIIP-PIIC-UTFSM Chile:2019, and by ANID+Doctorado Nacional 2020+21200544.

Institutional Review Board Statement: Not applicable.

Informed Consent Statement: Informed consent was obtained from all subjects involved in the study.

Conflicts of Interest: The authors declare no conflict of interest.

References

1. Ivanova, I.; Agren, O.; Bernhoff, H.; Leijon, M. Simulation of wave-energy converter with octagonal linear generator. *IEEE J. Ocean. Eng.* **2005**, *30*, 619–629. [CrossRef]
2. Gao, Y.; Shen, Z.; Tan, F.; Yue, G.; Liu, R.; Wang, Z.; Qu, S.; Wang, Z.; Zhang, W. Novel benzo[1,2-b:4,5-b']difuran-based copolymer enables efficient polymer solar cells with small energy loss and high VOC. *Nano Energy* **2020**, *76*, 104964. [CrossRef]
3. Gao, Y.; Cui, M.; Qu, S.; Zhao, H.; Shen, Z.; Tan, F.; Dong, Y.; Qin, C.; Wang, Z.; Zhang, W.; et al. Efficient Organic Solar Cells Enabled by Simple Non-Fused Electron Donors with Low Synthetic Complexity. *Small* **2022**, *18*, 2104623. [CrossRef] [PubMed]
4. Huang, J.; He, S.; Zhang, W.; Saparbaev, A.; Wang, Y.; Gao, Y.; Shang, L.; Dong, G.; Nurumbetova, L.; Yue, G.; et al. Efficient and Stable All-Inorganic CsPbIBr2 Perovskite Solar Cells Enabled by Dynamic Vacuum-Assisted Low-Temperature Engineering. *Solar RRL* **2022**, *6*, 2100839. [CrossRef]
5. Nakamura, M.; Yamaguchi, K.; Kimoto, Y.; Yasaki, Y.; Kato, T.; Sugimoto, H. Cd-Free Cu(In,Ga)(Se,S)2 Thin-Film Solar Cell With Record Efficiency of 23.35. *IEEE J. Photovolt.* **2019**, *9*, 1863–1867. [CrossRef]

6. International Renewable Energy Agency (IRENA). *Renewable Capacity Statistics 2021*; IRENA: Masdar, United Arab Emirates, 2021.
7. REN21. Renewables 2021 Global Status Report. In *Technical Report on Renewable Energy Policy Network for the 21st Century*; REN21 Secretariat: Paris, France, 2021.
8. Edenhofer, O.; Pichs-Madruga, R.; Sokona, Y.; Seyboth, K.; Kadner, S.; Zwickel, T.; Eickemeier, P.; Hansen, G.; Schlömer, S.; von Stechow, C.; et al. *Renewable Energy Sources and Climate Change Mitigation: Special Report of the Intergovernmental Panel on Climate Change*; Cambridge University Press: Cambridge, MA, USA, 2011.
9. Gunn, K.; Stock-Williams, C. Quantifying the global wave power resource. *Renew. Energy* **2012**, *44*, 296–304. [CrossRef]
10. Melikoglu, M. Current status and future of ocean energy sources: A global review. *Ocean Eng.* **2018**, *148*, 563–573. [CrossRef]
11. European Commission. *Study on Lessons for Ocean Energy Development*; Technical Report; European Commission (EC): Brussels, Belgium, 2017.
12. Drew, B.; Plummer, A.R.; Sahinkaya, M.N. *A Review of Wave Energy Converter Technology*; Sage Publications: Thousand Oaks, CA, USA, 2009.
13. Czech, B.; Bauer, P. Wave Energy Converter Concepts : Design Challenges and Classification. *IEEE Ind. Electron. Mag.* **2012**, *6*, 4–16. [CrossRef]
14. Pecher, A.; Kofoed, J.P. *Handbook of Ocean Wave Energy*; Springer: London, UK, 2017.
15. *Wave Devices*; European Marine Energy Centre (EMEC): Stromness, UK, 2018.
16. Aamir Hussain Memon, B. Ibrahim, T.; Perumal, N. Portable and pico-scale linear generator for wave energy conversion. In Proceedings of the 2014 5th International Conference on Intelligent and Advanced Systems (ICIAS), Kuala Lumpur, Malaysia, 3–5 June 2014; pp. 1–4.
17. Abdalla, I.I.; Ibrahim, T.; Nor, N.M. Analysis of Tubular Linear Motors for Different Shapes of Magnets. *IEEE Access* **2018**, *6*, 10297–10310. doi: 10.1109/ACCESS.2017.2775863. [CrossRef]
18. Jiabin W.; Jewell, G.W.; Howe, D. A general framework for the analysis and design of tubular linear permanent magnet machines. *IEEE Trans. Magn.* **1999**, *35*, 1986–2000. [CrossRef]
19. Curcic, M.; Quaicoe, J.E.; Bachmayer, R. A novel double-sided linear generator for wave energy conversion. In Proceedings of the OCEANS 2015, Genova, Italy, 8 June 2015; pp. 1–7.
20. Mesantono, L.D.; Danang Wijaya, F.; Haryono, T. Comparison of linear flux permanent magnet generator topologies by using FEMM 2D. In Proceedings of the 2016 8th International Conference on Information Technology and Electrical Engineering (ICITEE), Yogyakarta, Indonesia, 5–6 October 2016; pp. 1–5.
21. Farrok, O.; Islam, M.R.; Islam Sheikh, M.R.; Guo, Y.; Zhu, J.; Lei, G. Oceanic Wave Energy Conversion by a Novel Permanent Magnet Linear Generator Capable of Preventing Demagnetization. *IEEE Trans. Ind. Appl.* **2018**, *54*, 6005–6014. [CrossRef]
22. Páez, J.D.; Frey, D.; Maneiro, J.; Bacha, S.; Dworakowski, P. Overview of DC–DC Converters Dedicated to HVdc Grids. *IEEE Trans. Power Deliv.* **2019**, *34*, 119–128. [CrossRef]
23. Han, X.; Sima, W.; Yang, M.; Li, L.; Yuan, T.; Si, Y. Transient Characteristics Under Ground and Short-Circuit Faults in a ±500 kV MMC-Based HVDC System With Hybrid DC Circuit Breakers. *IEEE Trans. Power Deliv.* **2018**, *33*, 1378–1387. [CrossRef]
24. You, H.; Cai, X. A Three-Level Modular DC/DC Converter Applied in High Voltage DC Grid. *IEEE Access* **2018**, *6*, 25448–25462. [CrossRef]
25. Kish, G.J.; Ranjram, M.; Lehn, P.W. A Modular Multilevel DC/DC Converter With Fault Blocking Capability for HVDC Interconnects. *IEEE Trans. Power Electron.* **2015**, *30*, 148–162. [CrossRef]
26. Binh, P.C.; Tri, N.M.; Dung, D.T.; Ahn, K.K.; Kim, S.; Koo, W. Analysis, design and experiment investigation of a novel wave energy converter. *IET Gener. Transm. Distrib.* **2016**, *10*, 460–469. [CrossRef]
27. Falnes, J.; Kurniawan, A. *Ocean Waves and Oscillating Systems: Linear Interactions Including Wave-Energy Extraction*; Cambridge University Press: Cambridge, MA, USA, 2020; Volume 8.
28. Brooke, J. *Wave Energy Conversion*; Elsevier: Amsterdam, The Netherlands, 2003; Volume 6.
29. Goda, Y. *Random Seas and Design of Maritime Structures*; World Scientific Publishing Company: Singapore, 2010; Volume 33.
30. Hayt, W.H.; Buck, J.A. *Engineering Electromagnetics*, 8th ed.; McGraw-Hill: New York, NY, USA, 2012; p. 593.
31. Zapata, H.M.; Cabrera, F.A.; Perez, M.A.; Silva, C.A.; Jara, W. Model of a permanent magnet linear generator. In Proceedings of the IECON 2019-45th Annual Conference of the IEEE Industrial Electronics Society, Lisbon, Portugal, 14–17 October 2019; Volume 1, pp. 6992–6997.
32. Gieras, J.F.; Piech, Z.J.; Tomczuk, B.Z. *Linear Synchronous Motors*, 2nd ed.; CRC Press: Boca Raton, FL, USA, 2012; p. 520.
33. Zhang, J.; Yu, H.; Chen, Q.; Hu, M.; Huang, L.; Liu, Q. Design and Experimental Analysis of AC Linear Generator with Halbach PM Arrays for Direct-Drive Wave Energy Conversion. *IEEE Trans. Appl. Supercond.* **2014**, *24*, 3–6. [CrossRef]
34. Echeverría, J.; Kouro, S.; Pérez, M.; Abu-rub, H. Multi-modular cascaded DC-DC converter for HVDC grid connection of large-scale photovoltaic power systems. In Proceedings of the IECON 2013-39th Annual Conference of the IEEE Industrial Electronics Society, Vienna, Austria, 10–13 November 2013; pp. 6999–7005. [CrossRef]

energies

Article

Hydrodynamic Response and Tension Leg Failure Performance Analysis of Floating Offshore Wind Turbine with Inclined Tension Legs

Zhaolin Jia [1,2,*], Han Wu [2], Hao Chen [3], Wei Li [4], Xinyi Li [2,*], Jijian Lian [1], Shuaiqi He [2], Xiaoxu Zhang [2] and Qixiang Zhao [2]

1 State Key Laboratory of Hydraulic Engineering Simulation and Safety, Tianjin University, 135 Yaguan Road, Jinnan District, Tianjin 300350, China
2 Hebei Key Laboratory of Intelligent Water Conservancy, Hebei University of Engineering, Handan 056038, China
3 School of Civil and Environmental Engineering, Nanyang Technological University, 50 Nanyang Ave, Singapore 639798, Singapore
4 Power China Huadong Engineering Corporation Limited, 201 Gao Jiao Road, Hangzhou 311122, China
* Correspondence: jzhaolin@126.com (Z.J.); lixinyi0930@tju.edu.cn (X.L.)

Abstract: The tension legs are the essential parts of the tension legs platform-type (TLP-type) floating offshore wind turbine (FOWT) against the extra buoyancy of FOWT. Therefore, the TLP-type FOWT will face the risk of tension leg failure. However, there are seldom analyses on the hydrodynamic response and tension leg failure performance of FOWT with inclined tension legs. In this paper, a hydrodynamic model was established using three-dimensional hydrodynamic theory and applied in the motion response and tension analyses of FOWT with conventional and new tension leg arrangements on Moses. The influence of draft and tension leg arrangement on the performance of FOWT with inclined tension legs were studied. The optimum draft was the height of the column and lower tensions were obtained for the new tension leg arrangement. Moreover, the tension leg failure performance of FOWT with inclined tension legs was evaluated under different failure conditions. The results illustrated that the FOWT with the new tension leg arrangement can still operate safely after one tension leg fails.

Keywords: tension legs platform; floating offshore wind turbine; inclined tension leg; hydrodynamic model; tension analyses; failure performance

Citation: Jia, Z.; Wu, H.; Chen, H.; Li, W.; Li, X.; Lian, J.; He, S.; Zhang, X.; Zhao, Q. Hydrodynamic Response and Tension Leg Failure Performance Analysis of Floating Offshore Wind Turbine with Inclined Tension Legs. *Energies* **2022**, *15*, 8584. https://doi.org/10.3390/en15228584

Academic Editors: Antonio Crespo and Puyang Zhang

Received: 29 September 2022
Accepted: 14 November 2022
Published: 16 November 2022

1. Introduction

With the rapid economic development and growth, the demand for energy is increasing. Due to the depletion of fossil fuel, renewable energy is gaining more and more attention. Wind energy is an alternative because of its environmentally friendly, and it has developed rapidly in recent years [1]. The wind energy is also one of the most efficient renewable energies due to its low operation cost and the availability of a large amount of wind power [2]. Due to visual pollution from shore, routes of coastal ships, etc., the offshore wind farms will inevitably expand from shallow water to deep water. In deep seawater, the construction and installation costs of the fixed foundation for the wind turbine, commonly used in shallow seawater, will sharply increase. Therefore, the floating foundation is an alternative to apply in deep seawater.

The floating foundations mainly include the spar, tension legs platform (TLP), barge, and semi-submersible types. Among them, the TLP-type floating offshore wind turbine is generally composed of an upper wind turbine tower, central column, side column [3,4], side pontoons [5,6] or side spokes [7], supporting components, and a mooring system. The upper wind turbine tower can be assembled on land or in a shipyard to avoid the risk of offshore assembling and installed on the supporting components and central column. Then, the mooring system will be connected to the supporting components and central column. Due to the high stiffness of mooring cables, the natural frequency of TLP-type FOWT is much higher than the wave frequency. Therefore, the TLP-type FOWT should have good hydrodynamic performance [8]. Due to the large buoyancy from the central column, the extra buoyancy needs to be balanced using mooring cables as tension legs [9]. Thus, the mooring system becomes particularly essential for TLP-type FOWT. The failure of certain tension legs may lead to the overturning and sinking of the TLP-type FOWT.

TLP-type structures have been extensively studied and used over the past 40 years as support facility platforms for oil and gas extraction from deep and ultra-deep subsea reservoirs. After serious damages to the floating multi-well units and production platforms in the Gulf of Mexico by hurricanes Katrina and Rita [3], a serious failure with tendons occurred with the A-Typhoon TLP in the Gulf of Mexico during the passage of hurricane Rita [10]. This failure resulted in the capsizing of the platform and led to lessons learned regarding the connection of tendon connectors with subsea support. Many researchers have conducted many numerical simulations and experimental studies for the motion response and the mooring system failure of TLP. Yang and Kim [11] developed a time-domain, nonlinear, global-motion analysis program for floating hulls coupled with risers/mooring line and investigated the transient effects of tendon disconnection on the global performance of TLP during harsh environmental conditions. Jameel et al. [12] and Oyejobi et al. [13,14] studied the hydrodynamic responses of TLP with intact tendons or one tendon missing under moderate and extreme sea conditions. Tabeshpour et al. [15] investigated the effect of damaged tendons on the performance and tension of TLP in the wave frequency range. Yu et al. [16] studied the coupled hydrodynamic response of TLP after the failure of one or multiple tendons under extreme sea conditions. Wu et al. [17] studied the transient effect of one tendon failure on the motion response of Windstar TLP using the FAST software. Cheng et al. [18] analyzed the complex motion of TLP after the mooring system was partially damaged. In conclusion, it is recommended to conduct hydrodynamic analysis of the TLP platform under the combined action of wave, current, and wind loads. These researchers showed that the malfunction of any tendon would cause an increase in the motion response and a surge of the tension response. A study on the tendon failure performance of TLP under complex loads was conducted [19]. Some suggestions were put forward to reduce the risk of tendon failure. Chung et al. [20] proposed a new algorithm for real-time monitoring of the tension and bending moment of tendons. Real-time structural health monitoring can prevent tension legs' failure.

Many scholars studied the hydrodynamic performance and mooring system failure of FOWT in complex marine environments [21–24]. Bae et al. [25] conducted numerical simulation of semi-submersible FOWT with a broken mooring line using CHARM3D and FAST. Li et al. [26] developed a coupled aero-hydro-elastic numerical model to study the transient response and mooring line failure of Spar FOWT. It was found that the drift of semi-submersible FOWT with a broken mooring line was very obvious. This may lead to continuous failure of wind turbines. Le et al. [27] studied the hydrodynamic response of full-submersible FOWT under the action of turbulence and irregular waves using FAST. The results showed that full-submersible FOWT with one mooring line failure maintains good performance. Yang et al. [28] analyzed and predicted the transient behavior and mooring failure of barge FOWT under normal and extreme sea conditions using FAST and AQWA(F2A). Ahmed et al. [29] investigated the hydrodynamic response of Spar FOWT when one or more mooring lines were damaged. Yang et al. [30] analyzed and predicted the hydrodynamic response of a floating offshore wind turbine under different tendon failure scenarios. Sakaris et al. [31] and Sakaris et al. [32] investigated the structural health monitoring on the tendons of FOWT and the tendon damage diagnosis of FOWT, and the results showed that the damage detection can identify the tendon damage.

Wang et al. [33] used AWQA to analyze the "aero-hydro" coupling hydrodynamic response of the TLP-type FOWT. More attention should be paid to the pitch and surge motion of TLP-type FOWT. However, there are seldom studies on the hydrodynamic response and the tension leg failure performance of FOWT with inclined tension legs [34,35]. Therefore, based on the three-dimensional potential flow theory, a hydrodynamic model in deep seawater was established and applied in the hydrodynamic analyses of FOWT with inclined tension legs. The effects of draft and tension leg arrangement on the hydrodynamic and tension responses of FOWT with inclined tension legs are studied. The tension legs' failure performance analyses of FOWT with different tension leg arrangements are also analyzed.

2. Hydrodynamic Theory of Floating Body

The hydrodynamic response of the floating body in deep seawater was decomposed into two parts, as shown in Figure 1. One was the diffraction response and the other was the radiation response. The diffraction velocity potential was generated by the fixed body under the action of an incident wave. The radiation velocity potential was generated by the forced oscillation under still water. The hydrodynamic response was established in 3D Cartesian coordinates, as shown in Figure 1. In Figure 1a,b, the x-y plane is the hydrostatic surface, and the z-axis is vertically upward. In Figure 1c, the six degrees of freedom of FOWT with inclined tension legs are: surge, sway, heave, roll, pitch, and yaw.

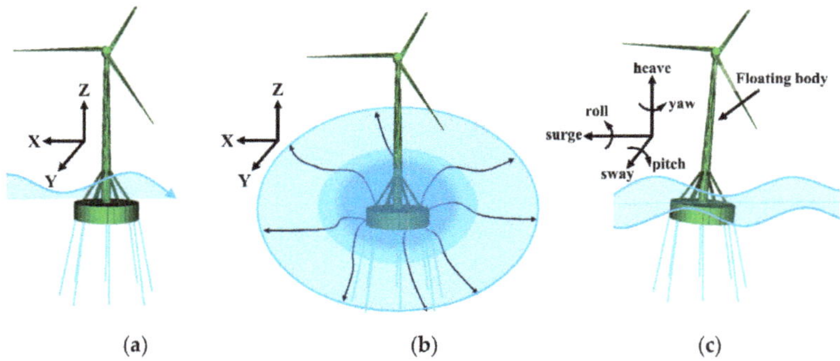

Figure 1. The hydrodynamic response of floating body in waves. (**a**) Diffraction response, (**b**) radiation response, and (**c**) hydrodynamic response.

2.1. Assumptions

The assumptions are listed as follows:

(1) The fluid is uniform, incompressible, and inviscid.
(2) The water flow is irrotational.
(3) The free surface wave is a linear, small-amplitude wave.
(4) The floating body moves in a small-amplitude and simple harmonic motion.

2.2. Load Calculation

The wind load, F_w, can be defined by Equation (1):

$$F_w = q\sum_{1}^{n} C_z C_s A_n \tag{1}$$

where, q is the basic wind pressure, C_z is the height coefficient of the windward structure, C_s is the structure shape coefficient, and A_n is the windward area of the windward component.

The basic wind pressure, q, can be expressed as Equation (2):

$$q = 0.5\rho_a U_{T,Z}^2 \tag{2}$$

where, ρ_a is the air density, and $U_{T,Z}$ is the average wind speed corresponding to the height, Z, at the average time of T.

The current load, F_c, is calculated by Equation (3):

$$F_c = K\frac{\rho}{2}U_c^2 A \tag{3}$$

where, U_c is the current speed, A is the projected area of the floating body component in the plane perpendicular to the current direction, ρ is the mass density of seawater, and K is the coefficient of the current load.

3. Hydrodynamic Modeling of FOWT with Inclined Tension Legs

3.1. Hydrodynamic Model

The hydrodynamic model of FOWT with inclined tension legs in deep seawater was established under a random wave. The random wave was input using the Joint North Sea Wave Project (JONSWAP) spectrum, which is the most systematic observational data of North Sea waves from 1968 to 1969 obtained by the United Kingdom, the Netherlands, the United States, and the Federal Republic of Germany. The JONSWAP spectrum was proposed with the consideration of wind speed and fetch, as Equation (4) [36]:

$$S(\omega) = \frac{\alpha g^2}{\omega^5}\exp\left(-1.25\left(\frac{\omega_p}{\omega}\right)^4\right)\gamma^{\exp\left(-\frac{(\omega-\omega_p)^2}{2(\sigma\omega_p)^2}\right)} \tag{4}$$

where, α is a dimensionless constant, ω_p is the peak angular frequency of the wave, which can be calculated using $\omega_p = 22\left(\frac{g}{U}\right)\left(\frac{gx}{U^2}\right)^{-0.33}$, U (m/s^2) is the wind speed, x (km) is the wind range, γ is the peak enhancement factor, and σ is the peak shape parameter. Here, σ = 0.07 when $\omega \leq \omega_p$ and σ = 0.09 when $\omega > \omega_p$.

In this model, the FOWT with inclined tension legs was subjected to current, wind, and wave. The random wave component was described by the three-parameter JONSWAP spectrum, that included a significant wave height of 1.5 m, a spectrum peak period of 6.67 s, and a peak enhancement factor (γ) of 2.5. The environmental conditions are as shown in Table 1.

Table 1. Environmental conditions.

Wind		Current		Wave		
Speed	**Heading**	**Speed**	**Heading**	**Significant Wave Height**	**Heading**	**Spectrum Peak Period**
15 m/s	90 degrees	1 m/s	90 degrees	1.5 m	90 degrees	6.67 s

3.2. Numerical Model of FOWT with Inclined Tension Legs

The physical model of FOWT with inclined tension legs is shown in Figure 2. The wind turbine tower was installed on the central column and fixed by six hollow steel tubes. The tension legs were connected to the column and the seabed. The numerical model of FOWT with inclined tension legs (shown in Figure 3) was established through the hydrodynamic simulation software Moses [37]. The parameters and material properties of the FOWT foundation with inclined tension legs are presented in Tables 2 and 3.

3.3. RAOs and Motion Responses of FOWT with Inclined Tension Legs

3.3.1. RAOs

By conducting the hydrodynamic analyses of FOWT with inclined tension legs in frequency, the response amplitude operator (RAO) and motion response in each degree of freedom were obtained under the incident wave at 0, 45, 90, 135, and 180 degrees for a FOWT foundation of 15 m (shown in Figure 4).

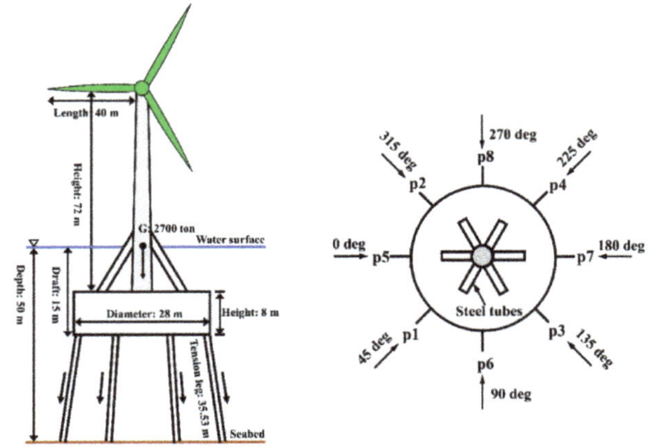

Figure 2. Physical model of FOWT with inclined tension legs.

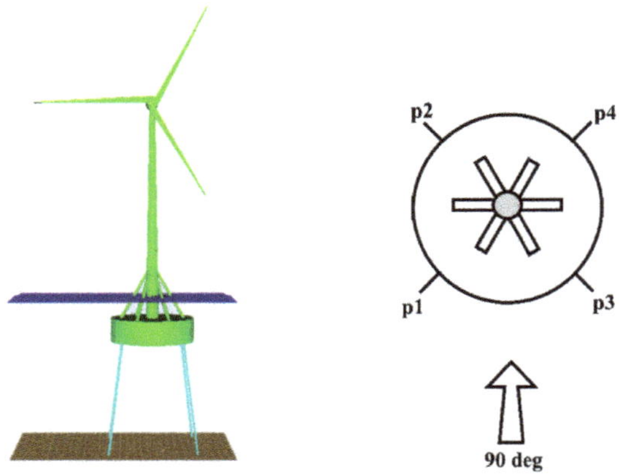

Figure 3. Numerical model of FOWT with four tension legs.

Table 2. Main parameters of FOWT with inclined tension legs.

Parameter (Unit)	Value
Model weight (ton)	2700
Seawater depth (m)	50
Draft (m)	15
Roll Gyradius (m)	35.9
Pitch Gyradius (m)	35.9
Yaw Gyradius (m)	20.1
Column diameter (m)	28
Column height (m)	8
Tension leg outer diameter (m)	0.6
Tension leg length (m)	35.53
Tower diameter (m)	4 m (bottom) to 2 m (top)
Tower length (m)	72
Hollow steel tube diameter (m)	1
Hollow steel tube length (m)	20
Blade length (m)	40

Table 3. Material properties of FOWT with inclined tension legs.

Material Property (Unit)	Value
Yield stress of steel used for FOWT (MPa)	248.04
Coefficient of thermal expansion of steel (°C)	3.6111×10^{-6}
Poisson's ratio of steel	0.3
Steel density (g/cm^3)	7.8492
Modulus of elasticity of steel (MPa)	1.9981×10^{5}
Tension leg weight in water (kg/m)	7884
Tension leg axial stiffness (kN)	3.6×10^{7}

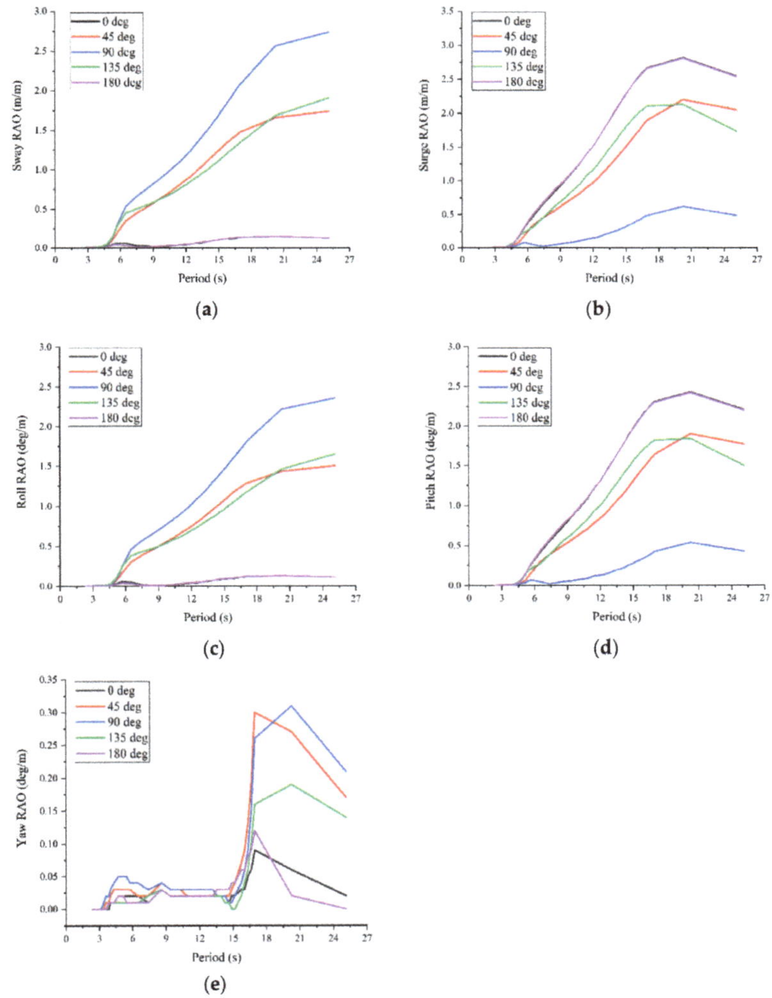

Figure 4. RAO for each degree of freedom: (**a**) Sway RAO, (**b**) Surge RAO, (**c**) Roll RAO, (**d**) Pitch RAO, and (**e**) Yaw RAO.

As in Figure 4a–d, the trendlines of RAOs of four degrees of freedom were basically the same under different angles of incident wave. The RAOs of sway, surge, roll, and pitch were large in 6 to 20 s of the wave period, but the RAO of yaw had the largest response in

15 to 20 s of the wave period. Therefore, it is not recommended to operate in greater than 20 s of the wave period.

3.3.2. Motion Responses

In the time domain, a total of 300 s was simulated, the environmental conditions' input parameters were the same as in Table 1, and the parameters and material properties were the same as in Tables 2 and 3. The simulation results of 100–300 s were selected for analysis, as shown in Figure 5a, and the motion response of sway was greater than that of surge. The maximum value of the motion response of surge was 0.27 m, which occurred at 229 s, and the minimum value was −0.3 m, which occurred at 269 s. The maximum value of the motion response of sway was 0.67 m, which occurred at 208 s, and the minimum value was −1.06 m, which occurred at 212 s. As shown in Figure 5b, the motion response of roll was greater than that of pitch and yaw. The maximum value of the motion response of roll occurred at 208 s, which was 0.47 degrees, and the minimum value occurred at 212 s, which was −0.75 degrees. The maximum value of the motion response of pitch occurred at 268 s, which was 0.21 degrees, and the minimum value occurred at 229 s, which was −0.19 degrees. The maximum value of the motion response of yaw occurred at 277 s, which was 0.17 degrees, and the minimum value occurred a total of 5 times, where the first minimum value occurred at 220 s, which was −0.19 degrees. Since the angle of the incident wave is the same as the direction of roll and sway, these phenomena that the motion responses of roll and sway are larger than those of other degrees of freedom are correct.

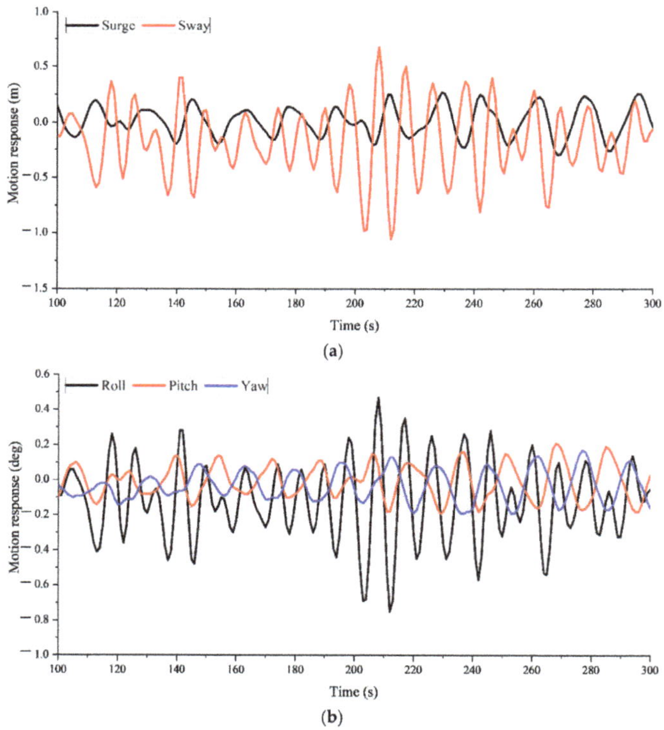

Figure 5. Motion response of each degree of freedom of 4 tension legs: (**a**) surge and sway motion response, and (**b**) roll, pitch, and yaw motion response.

3.4. Parametric Study of FOWT with Inclined Tension Legs

To investigate the effects of draft and tension leg arrangement, a parametric study of FOWT with inclined tension legs was conducted. Four tension legs were used for conventional tension leg arrangement, as shown in Figure 3, while eight tension legs were used for the new tension leg arrangement. One more tension leg was added between the two adjacent tension legs of the conventional tension leg arrangement. The angle between the two tension legs is 45 degrees for the eight tension legs condition. The numerical model for TLP-type FOWT with the new tension leg arrangement is shown in Figure 6.

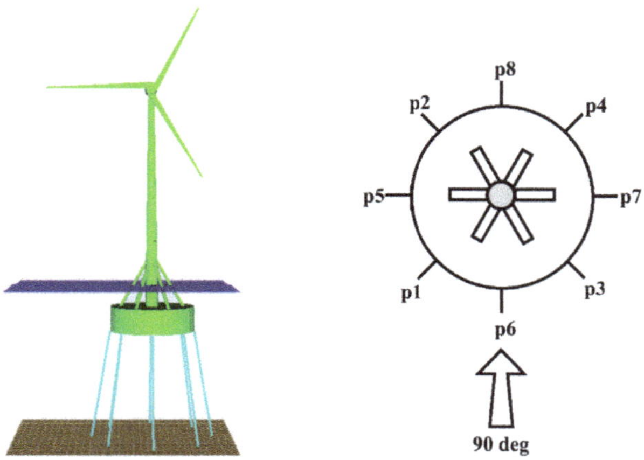

Figure 6. Numerical model with eight tension legs.

3.4.1. Effect of Draft

In this section, the input parameters were the same as in Section 3.3.2 except for the draft. Varying the draft, the maximum value of the motion response of the 5 degrees of freedom in the 300 s time-domain analysis, i.e., the motion amplitude, was selected. Similarly, since the tensions of the four tension legs were not the same, the maximum value of them was selected and the average value was calculated. As shown in Figure 7a, the motion amplitudes of surge and sway gradually decreased when the column draft was between 5.5 and 8 m. When the draft was 8 m, the motion amplitudes of surge and sway reached the minimum value, which was close to 0 m. When the column was below the water surface, the motion amplitudes of surge and sway first increased, then gradually decreased. When the draft was 10 m, the motion amplitudes of surge and sway reached the maximum values, which were 0.64 and 2.7 m, respectively. Moreover, the motion amplitude of surge obtained another minimum value at 12 m of draft. The trendline of Figure 8a is the same as Figure 7a, but the motion amplitude is smaller as the draft increased. As shown in Figure 7b, when the column was above the water surface, the motion amplitudes of roll, pitch, and yaw gradually decreased with the draft. When the draft was 8 m, the motion amplitude reached the minimum values, with roll of 0.93 degrees, pitch of 0 degrees, and yaw of 0.01 degrees, respectively. When the column was below the water surface, the motion amplitude of roll first gradually increased, reached the maximum value of 1.66 degrees at 10 m draft, and then gradually decreased. The motion amplitudes of pitch and yaw increased to the maximum value with pitch of 0.42 degrees and yaw of 0.44 degrees at 10 m draft, then gradually decreased to the minimum value at 12 m draft. However, the motion amplitude of Figure 8b is also smaller than Figure 7b, as the draft increased.

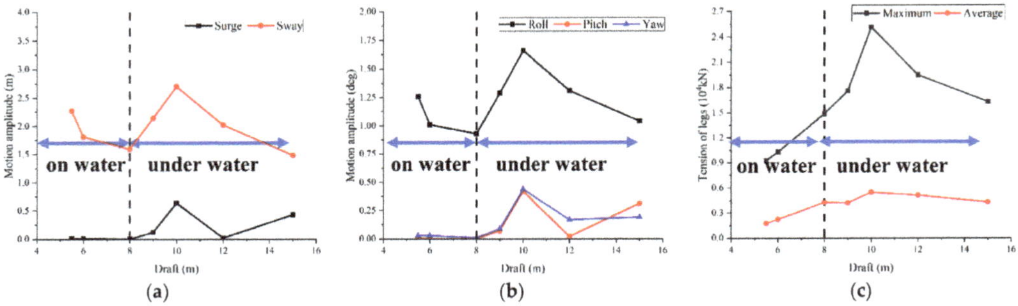

Figure 7. Effect of draft on motion amplitude and tension on four tension legs: (**a**) surge and sway, (**b**) draft on roll, pitch, and yaw, and (**c**) maximum and average tension of tension leg.

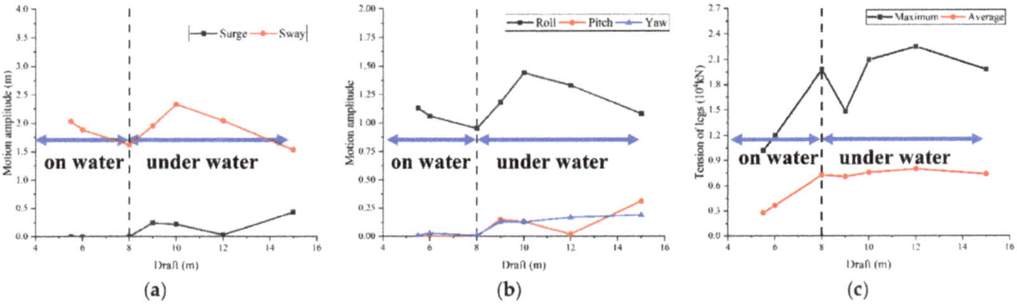

Figure 8. Effect of draft on motion amplitude and tension on eight tension legs: (**a**) surge and sway, (**b**) draft on roll, pitch, and yaw, and (**c**) maximum and average tension of tension leg.

In conclusion, a draft of 8 m, which is the same as the height of the central column, is recommended. At this condition, when the central column is just submerged in seawater, the motion response of each degree of freedom is the minimum. The column with a draft of 10 m, about 1.25 times the column height, should be avoided because the motion response of each degree of freedom is the maximum. Moreover, the hydrodynamic performance of TLP-type FOWT with the new tension leg arrangement was better.

As shown in Figure 7c, when the column was above the water surface, the maximum and average tension gradually increased. With a draft of 8 m, the Moses command was used to extract all the tension of the four tension legs in the 300 s time-domain analysis, where the maximum tension was 1.488×10^4 kN and the average value was 4.26×10^3 kN. When the column was under water, the maximum and average tension increased first then decreased. When the draft was 10 m, the tension reached the maximum values, with the maximum tension of 2.513×10^4 kN and average tension of 5.46×10^3 kN at 10 m draft. As the draft increased, although the maximum tensions in Figure 8c were smaller than those in Figure 7c, the average tensions were all larger than those in Figure 7c. Therefore, the recommended breaking force of the tension leg material should be greater than 2.6×10^4 kN.

3.4.2. Effect of Tension Leg Arrangement

The draft used for the analyses of different tension leg arrangements was 15 m. As shown in Figure 9, the roll of the floating body under 4 and 8 tension legs conditions was the same. As shown in Figure 10, the tension fluctuation of the p2 tension leg under different tension leg arrangements was similar. However, the tension under the 8 tension legs condition was smaller than that under the 4 tension legs condition. This indicated that the tension leg performance of TLP-type FOWT with the new tension leg arrangement was better. The stability of FOWT with inclined tension legs should also be better for the new tension leg arrangement.

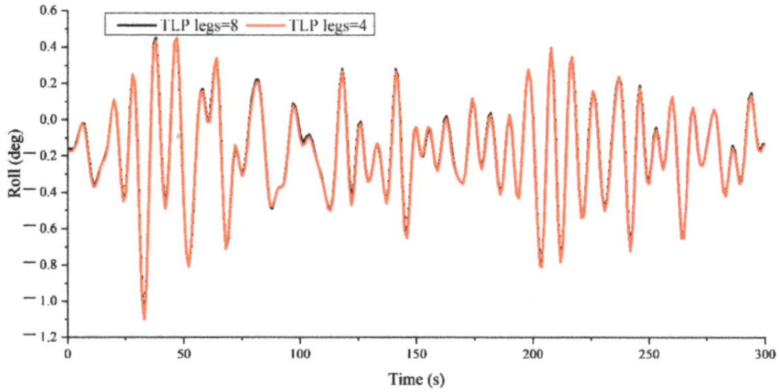

Figure 9. Roll under different tension leg arrangements.

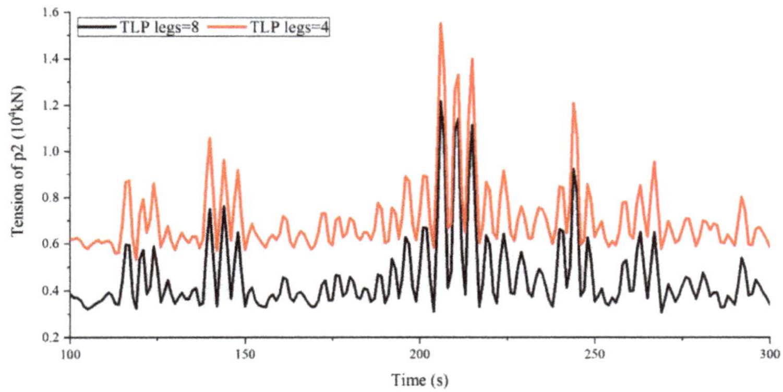

Figure 10. Tension of p2 tension leg under different tension leg arrangements.

4. Failure Performance of Tension Legs

4.1. Failure Conditions of Tension Legs

As the performance and stability under 8 tension legs were better than for 4 tension legs, the TLP-type FOWT with a new tension leg arrangement was proposed. Then, the failure performance of tension legs for the TLP-type FOWT with the new tension leg arrangement was analyzed. The numerical models and input parameters were the same as those of Tables 1–3 in the previous section. The tension of a selected tension leg was monitored throughout the time-domain analysis using the Moses "sensor" command, and as the simulation progressed, the tension leg "action deactivate" command ran when the tension of the monitored tension leg exceeded the set allowable value of 1×10^4 kN, i.e., the tension legs' broken force, and when the tension leg was broken, there was no force.

Figure 11a shows the 8 tension legs arrangement under well operation. Figure 11b,c show the states of FOWT with inclined tension legs after one or two tension legs failed. Under these conditions, the FOWT with inclined tension legs can still maintain well operation. Figure 11d shows the state of FOWT with inclined tension legs after p1, p6, and p3 tension legs failed. Under this condition, the FOWT collapsed. Therefore, the new tension legs arrangement has good failure performance.

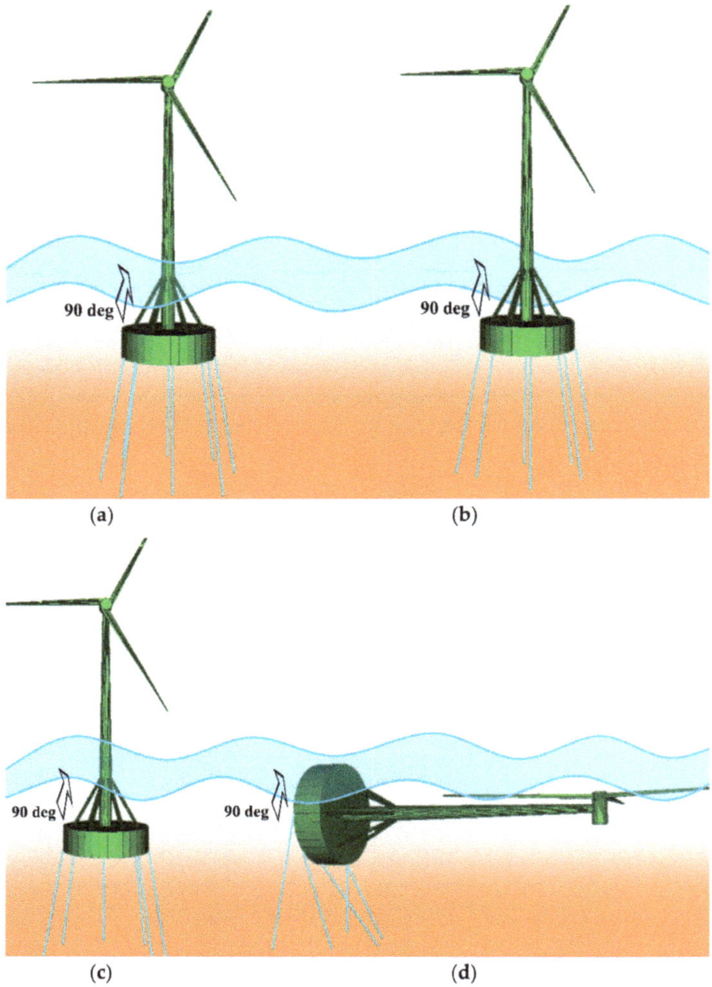

Figure 11. Failure analysis of tension legs under the 8 tension legs condition: (**a**) normal, (**b**) failure of p1, (**c**) failure of p1 and p6, and (**d**) failure of p1, p6, and p3 (collapse).

Figure 12 shows the roll of FOWT with 8 tension legs under different failure modes. The roll after the failure of the p1 tension leg or p1 and p6 tension legs was almost the same as that under the well condition. Therefore, the failure of two tension legs has little effect on the operation of FOWT with inclined tension legs. From Figure 13, an increase of tension for the p2 tension leg was obtained under different failure modes. From Figure 14, a two-times increase of tension for the p3 tension leg was observed after the failure of p1 and p6 tension legs. Then, the more the tension legs failed, the greater the increase of tension for the adjacent tension legs was. Therefore, the operation of FOWT with inclined tension legs is risky after two tension legs fail under this environment condition. However, the FOWT with inclined tension legs still does not collapse.

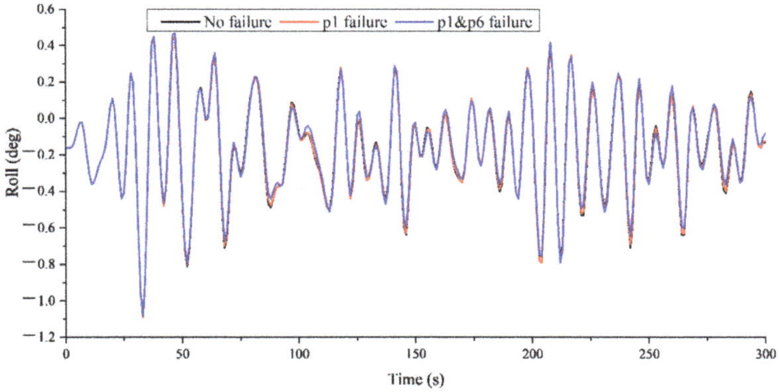

Figure 12. Roll under different failure conditions.

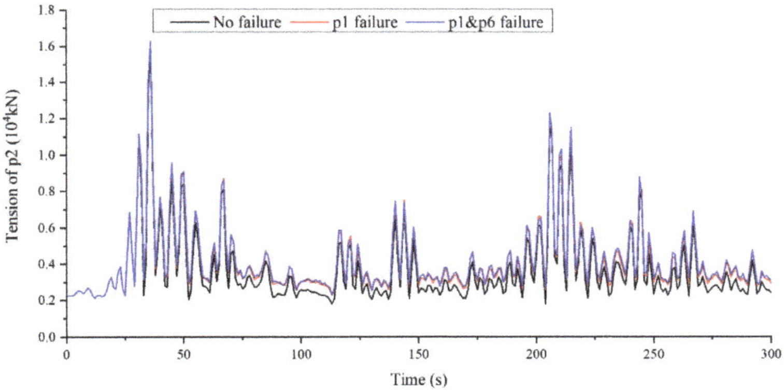

Figure 13. p2 tension time–history curve.

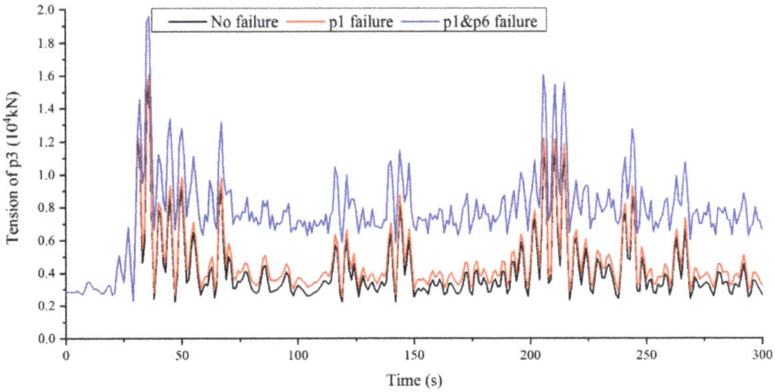

Figure 14. p3 tension time–history curve.

As a comparison, the failure performance of the 4 tension legs arrangement under well operation was conducted. Figure 15b shows that the FOWT with inclined tension legs collapsed after the p1 tension leg failed. Therefore, the FOWT with the conventional tension leg arrangement cannot work well with any failure of tension legs when the wave incidence angle is 90 degrees.

(**a**) (**b**)

Figure 15. Failure analysis of tension leg under the 4 tension legs condition: (**a**) normal and (**b**) p1 failure (collapse).

4.2. Envelope Diagram of Tensions of New Tension Leg Arrangement

Figures 16 and 17 summarize the maximum and average tensions of the 8 tension legs under different failure conditions. Without tension leg failure, the maximum tensions of the 8 tension legs were close. The maximum tension of p7 was the largest, 1.70542×10^4 kN, and that of p4 was the smallest, 1.54101×10^4 kN. In terms of the average tension, the tension of p8 was the largest, 5.0646×10^3 kN, and that of p2 was the smallest, 3.6426×10^3 kN.

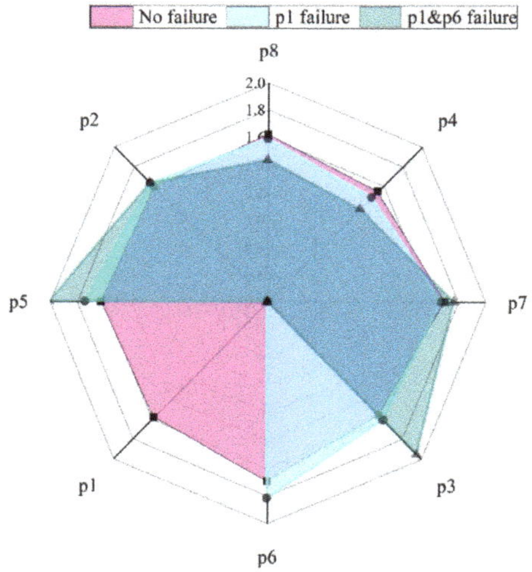

Figure 16. Maximum tension of each tension leg (unit: 10^4 kN).

When p1 failed, the maximum tension of p5 and p6 increased, which were nearest to p1. The maximum tension of p5 increased by 1.1958×10^3 kN, an increment of 7.4%. That of p6 increased by 1.2362×10^3 kN, an increment of 7.3%. At the same time, the maximum tensions of p4, p7, and p8 decreased. The maximum tension of p4, which is opposite to p1, reduced by 3.9%. When p1 and p6 tension legs failed, the maximum tensions of p3 and p5, nearest to p1 and p6, largely increased, by 3.9935×10^3 and 3.8448×10^3 kN, respectively.

The maximum tensions of p4 and p8, opposite to p1 and p6, decreased. A similar trend can be observed for the average tension of tension legs under different failure modes. When p1 and p6 failed, the adjacent tension legs achieved large average tensions.

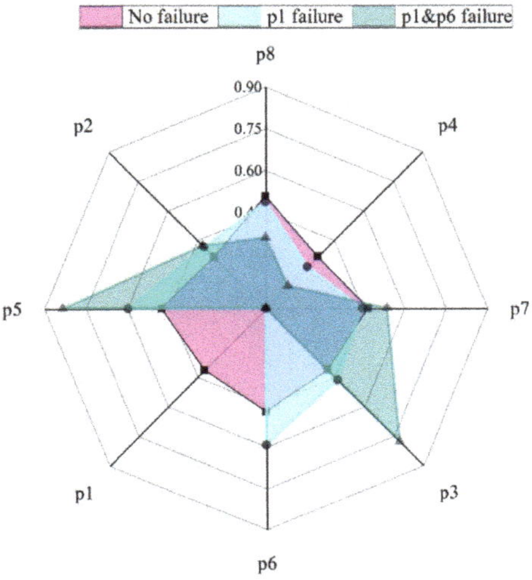

Figure 17. Average tension of each tension leg (unit: 10^4 kN).

When one tension leg failed, the maximum and average tension did not change too much. The FOWT with the new tension leg arrangement can still remain in good operation. When two adjacent tension legs failed, the maximum and average tension of adjacent tension legs greatly increased. However, the FOWT can still operate. There is a risk of collapsing, and therefore some improvement methods need to be performed in time. In conclusion, the FOWT with the new tension leg arrangement has good failure performance and the ability to withstand environmental loads.

5. Conclusions

A hydrodynamic model was applied to analyze the hydrodynamic response of FOWT with inclined tension legs. Failure performance of tension legs was conducted under conventional and new tension leg arrangements. The specific conclusions of this paper can be summarized as follows:

(1) The effects of draft and tension leg arrangement were analyzed. The draft of 8 m, which is the same as the height of the column, is proposed due to the optimum performance of FOWT with inclined tension legs. Compared with the conventional tension leg arrangement, the new tension leg arrangement was subjected to lower tensions. Therefore, the performance of FOWT with the new tension leg arrangement was better.

(2) Based on the tension analyses of FOWT with inclined tension legs under different drafts, it is recommended that the breaking force of the tension leg material should be greater than 2.6×10^4 kN.

(3) The stability of FOWT with the new tension leg arrangement after single or double tension leg failure was largely improved. The FOWT with the new tension leg arrangement had good failure performance and the ability to withstand environmental loads.

(4) Although the FOWT with inclined tension legs could still operate after two tension legs failed, there is a risk of collapsing. Therefore, some improvement methods need to be performed in time.

Author Contributions: Conceptualization, Z.J. and X.L.; methodology, Z.J. and X.L.; software, J.L. and W.L; validation, Z.J., H.W. and H.C.; formal analysis, H.C.; investigation, H.W.; resources, J.L. and S.H.; data curation, Z.J.; writing—original draft preparation, Z.J.; writing—review and editing, X.L.; visualization, X.Z. and W.L.; supervision, X.L.; project administration, Q.Z. All authors have read and agreed to the published version of the manuscript.

Funding: This research was funded by the State Key Laboratory of Hydraulic Engineering Simulation and Safety (Tianjin University) (Grant No. HESS-2115), and the Natural Science Foundation of Hebei Province (Grant No. E2020402110).

Data Availability Statement: Not applicable.

Conflicts of Interest: The authors declare no conflict of interest.

References

1. Leung, D.Y.C.; Yang, Y. Wind energy development and its environmental impact: A review. *Renew. Sust. Energ.* **2012**, *16*, 1031–1039. [CrossRef]
2. Kumar, Y.; Ringenberg, J.; Depuru, S.S.; Devabhaktuni, V.K.; Lee, J.W.; Nikolaidis, E.; Andersen, B.; Afjeh, A. Wind energy: Trends and enabling technologies. *Renew. Sust. Energ.* **2016**, *53*, 209–224. [CrossRef]
3. Ren, Y.; Venugopal, V.; Shi, W. Dynamic analysis of a multi-column TLP floating offshore wind turbine with tendon failure scenarios. *Ocean. Eng.* **2022**, *245*, 110247. [CrossRef]
4. Walia, D.; Schünemann, P.; Hartmann, H.; Adam, F.; Groβmann, J. Numerical and physical modeling of a tension-leg platform for offshore wind turbines. *Energies* **2021**, *14*, 3554. [CrossRef]
5. Ma, Z.; Wang, S.; Wang, Y.; Ren, N.; Zhai, G. Experimental and numerical study on the multi-body coupling dynamic response of a Novel Serbuoys-TLP wind turbine. *Ocean. Eng.* **2019**, *192*, 106570. [CrossRef]
6. Ma, Z.; Ren, N.; Wang, Y.; Wang, S.; Shi, W.; Zhai, G. A Comprehensive study on the serbuoys offshore wind tension leg platform coupling dynamic response under typical operational conditions. *Energies* **2019**, *12*, 2067. [CrossRef]
7. Song, J.; Lim, H.C. Study of Floating Wind Turbine with Modified Tension Leg Platform Placed in Regular Waves. *Energies* **2019**, *12*, 703. [CrossRef]
8. Zhao, Y.; She, X.; He, Y.; Yang, J.; Peng, T.; Kou, Y. Experimental study on new multi-column tension-leg-type floating wind turbine. *China. Ocean. Eng.* **2018**, *32*, 123–131. [CrossRef]
9. Ha, K.; Truong, H.V.A.; Dang, T.D.; Ahn, K.K. Recent control technologies for floating offshore wind energy system: A review. *Int. J. Pr. Eng. Man-Gt.* **2021**, *8*, 281–301. [CrossRef]
10. Cruz, A.M.; Krausmann, E. Damage to offshore oil and gas facilities following hurricanes Katrina and Rita: An overview. *J. Loss Prev. Proc.* **2008**, *21*, 620–626. [CrossRef]
11. Yang, C.K.; Kim, M.H. Transient effects of tendon disconnection of a TLP by hull-tendon-riser coupled dynamic analysis. *Ocean. Eng.* **2010**, *37*, 667–677. [CrossRef]
12. Jameel, M.; Oyejobi, D.O.; Siddiqui, N.A.; Sulong, N.H.R. Nonlinear dynamic response of tension leg platform under environmental loads. *KSCE J. Civ. Eng.* **2017**, *21*, 1022–1030. [CrossRef]
13. Oyejobi, D.O.; Jameel, M.; Sulong, N.H.R. Nonlinear response of tension leg platform subjected to wave, current and wind forces. *Int. J. Civ. Eng.* **2016**, *14*, 521–533. [CrossRef]
14. Oyejobi, D.O.; Jameel, M.; Sulong, N.H.R. Stochastic response of intact and a removed tendon tension leg platform to random wave and current forces. *Arab. J. Sci. Eng.* **2017**, *42*, 1065–1076. [CrossRef]
15. Tabeshpour, M.R.; Ahmadi, A.; Malayjerdi, E. Investigation of TLP behavior under tendon damage. *Ocean. Eng.* **2018**, *156*, 580–595. [CrossRef]
16. Yu, J.; Hao, S.; Yu, Y.; Chen, B.; Cheng, S.; Wu, J. Mooring analysis for a whole TLP with TTRs under tendon one-time failure and progressive failure. *Ocean. Eng.* **2019**, *182*, 360–385. [CrossRef]
17. Wu, H.; Zhao, Y.; He, Y.; Shao, Y.; Mao, W.; Han, Z.; Huang, C.; Gu, X.; Jiang, Z. Transient response of a TLP-type floating offshore wind turbine under tendon failure conditions. *Ocean. Eng.* **2021**, *220*, 108486. [CrossRef]
18. Cheng, S.; Yu, Y.; Yu, J.; Wu, J.; Li, Z.; Huang, Z. Mechanistic research on the complex motion response of a TLP under tendon breakage. *Ocean. Eng.* **2021**, *240*, 109984. [CrossRef]
19. Qi, Y.; Tian, X.; Guo, X.; Lu, H.; Liu, L. The hydrodynamic performance of a tension leg platform with one-tendon failure. *Ships. Offshore. Struc.* **2018**, *14*, 523–533. [CrossRef]
20. Chung, W.C.; Pestana, G.R.; Kim, M. Structural health monitoring for TLP-FOWT (floating offshore wind turbine) tendon using sensors. *Appl. Ocean. Res.* **2021**, *113*, 102740. [CrossRef]
21. Choi, Y.-M.; Nam, B.W.; Hong, S.Y.; Jung, D.W.; Kim, H.J. Coupled motion analysis of a tension leg platform with a tender semi-submersible system. *Ocean. Eng.* **2018**, *156*, 224–239. [CrossRef]
22. Jeon, S.H.; Cho, Y.U.; Seo, M.W.; Cho, J.R.; Jeong, W.B. Dynamic response of floating substructure of spar-type offshore wind turbine with catenary mooring cables. *Ocean. Eng.* **2013**, *72*, 356–364. [CrossRef]

23. Koo, B.J.; Goupee, A.J.; Kimball, R.W.; Lambrakos, K.F. Model tests for a floating wind turbine on three different floaters. *J. Offshore. Mech.* **2014**, *136*, 020907. [CrossRef]
24. Wu, Z.; Li, Y. Platform stabilization of floating offshore wind turbines by artificial muscle based active mooring line force control. *IEEE ASME T. Mech.* **2020**, *25*, 2765–2776. [CrossRef]
25. Bae, Y.H.; Kim, M.H.; Kim, H.C. Performance changes of a floating offshore wind turbine with broken mooring line. *Renew. Energ.* **2017**, *101*, 364–375. [CrossRef]
26. Li, Y.; Zhu, Q.; Liu, L.; Tang, Y. Transient response of a SPAR-type floating offshore wind turbine with fractured mooring lines. *Renew. Energ.* **2018**, *122*, 576–588. [CrossRef]
27. Le, C.; Li, Y.; Ding, H. Study on the coupled dynamic responses of a submerged floating wind turbine under different mooring conditions. *Energies* **2019**, *12*, 418. [CrossRef]
28. Yang, Y.; Bashir, M.; Li, C.; Wang, J. Investigation on mooring breakage effects of a 5 MW barge-type floating offshore wind turbine using F2A. *Ocean. Eng.* **2021**, *233*, 108887. [CrossRef]
29. Ahmed, M.O.; Yenduri, A.; Kurian, V.J. Evaluation of the dynamic responses of truss spar platforms for various mooring configurations with damaged lines. *Ocean. Eng.* **2016**, *123*, 411–421. [CrossRef]
30. Yang, Y.; Bashir, M.; Michailides, C.; Mei, X.; Wang, J.; Li, C. Coupled analysis of a 10 MW multi-body floating offshore wind turbine subjected to tendon failures. *Renew. Energ.* **2021**, *176*, 89–105. [CrossRef]
31. Sakaris, C.S.; Bashir, M.; Yang, Y.; Michailides, C.; Wang, J.; Sakellariou, J.S. Diagnosis of damaged tendons on a 10 MW multibody floating offshore wind turbine platform via a response-only functional model based method. *Eng. Struct.* **2021**, *242*, 112384. [CrossRef]
32. Sakaris, C.S.; Yang, Y.; Bashir, M.; Michailides, C.; Wang, J.; Sakellariou, J.S.; Li, C. Structural health monitoring of tendons in a multibody floating offshore wind turbine under varying environmental and operating conditions. *Renew. Energ.* **2021**, *179*, 1897–1914. [CrossRef]
33. Wang, T.; Jin, H.; Wu, X. Coupled dynamic analysis of a tension leg platform floating offshore wind turbine. *J. Offshore Mech. Arct.* **2020**, *142*, 011901. [CrossRef]
34. Wei, N.; Zhang, Z.; Xu, X.; Yao, W. Stability analysis of a TLP with inclined tension legs under different marine survival conditions. *J. Mar. Sci. Eng.* **2022**, *10*, 1058. [CrossRef]
35. Milano, D.; Peyrard, C.; Capaldo, M.; Ingram, D.; Xiao, Q.; Johanning, L. Impact of high order wave loads on a 10 MW tension-leg platform floating wind turbine at different tendon inclination angles. In Proceedings of the ASME 2019 38th International Conference on Ocean, Offshore and Arctic Engineering: Volume 10 Ocean Renewable Energy, Glasgow, Scotland, UK, 9–14 June 2019; ASME: New York, NY, USA, 2019; p. V010T09A076. [CrossRef]
36. Hasselmann, K.; Barnett, T.P.; Bouws, E.; Carlson, H.; Cartwright, D.E.; Enke, K.; Ewing, J.A.; Gienapp, H.; Hasselmann, D.E.; Kruseman, P.; et al. Measurements of wind-wave growth and swell decay during the Joint North Sea Wave Project (JONSWAP). *Deut. Hydrogr. Z.* **1973**, *8*, 1–95.
37. Manual, M. Moses Pdf. 2009. Available online: http://www.ultramarine.com/hdesk/document/ref_man.htm (accessed on 1 August 2020).

energies

Article

An Optimization Study on the Operating Parameters of Liquid Cold Plate for Battery Thermal Management of Electric Vehicles

Lichuan Wei [1,2], Yanhui Zou [3], Feng Cao [1,*], Zhendi Ma [3], Zhao Lu [1] and Liwen Jin [3,*]

1 School of Energy and Power Engineering, Xi'an Jiaotong University, Xi'an 710049, China
2 Shenzhen Envicool Technology Co. Ltd., Shenzhen 518129, China
3 School of Human Settlements and Civil Engineering, Xi'an Jiaotong University, Xi'an 710049, China
* Correspondence: fcao@xjtu.edu.cn (F.C.); lwjin@xjtu.edu.cn (L.J.)

Abstract: The development of electric vehicles plays an important role in the field of energy conservation and emission reduction. It is necessary to improve the thermal performance of battery modules in electric vehicles and reduce the power consumption of the battery thermal management system (BTMS). In this study, the heat transfer and flow resistance performance of liquid cold plates with serpentine channels were numerically investigated and optimized. Flow rate ($\dot{m}$), inlet temperature (T_{in}), and average heat generation (Q) were selected as key operating parameters, while average temperature (T_{ave}), maximum temperature difference (ΔT_{max}), and pressure drop (ΔP) were chosen as objective functions. The Response Surface Methodology (RSM) with a face-centered central composite design (CCD) was used to construct regression models. Combined with the multi-objective non-dominated sorting genetic algorithm (NSGA-II), the Pareto-optimal solution was obtained to optimize the operation parameters. The results show that the maximum temperature differences of the cold plate can be controlled within 0.29~3.90 °C, 1.11~15.66 °C, 2.17~31.39 °C, and 3.43~50.92 °C for the discharging rates at 1.0 C, 2.0 C, 3.0 C, and 4.0 C, respectively. The average temperature and maximum temperature difference can be simultaneously optimized by maintaining the pressure drop below 1000 Pa. It is expected that the proposed methods and results can provide theoretical guidance for developing an operational strategy for the BTMS.

Keywords: battery thermal management system; response surface methodology; genetic algorithm; multi-objective optimization; serpentine cold plate

Citation: Wei, L.; Zou, Y.; Cao, F.; Ma, Z.; Lu, Z.; Jin, L. An Optimization Study on the Operating Parameters of Liquid Cold Plate for Battery Thermal Management of Electric Vehicles. *Energies* **2022**, *15*, 9180. https://doi.org/10.3390/en15239180

Academic Editor: Fangming Jiang

Received: 3 November 2022
Accepted: 30 November 2022
Published: 3 December 2022

1. Introduction

With the rapid development of global industry, energy shortages and environmental pollution are becoming increasingly serious. The development of a traditional automobile using fossil fuels as a power source is facing huge challenges. On the one hand, the annual oil consumption of traditional fuel vehicles accounts for about 50% of the total global oil consumption [1]. On the other hand, pollution gases and particles such as CO, CO_2, HC, and NO_x in fuel vehicle exhaust are causing global warming and haze [2]. Most countries in the world have formulated strict exhaust emission standards for conventional fuel vehicles. New energy vehicles are being vigorously developed and popularized to reduce the consumption of fossil fuels to achieve energy conservation and emission reduction. In particular, new energy vehicles (electric vehicles), which take electric energy as the main or only power source, have the significant advantages of low energy consumption, low emissions, and near-zero emissions. If renewable energy sources such as solar or wind power are used to provide electricity, greenhouse gas emissions can be reduced by nearly 40% by using new energy vehicles [3]. Therefore, vigorously developing electric vehicles has become one of the most effective ways to achieve energy conservation and emission reduction in the automobile industry.

As the power source of electric vehicles, the battery power pack is one of the core components affecting the overall performance of electric vehicles. Its operating temperature directly affects the charging and discharging efficiency, cycle life, and safety of the battery power pack [4–8]. When the operating temperature is lower than 10 °C, the ionic conductivity and increase in charge transfer resistance, charging and discharging efficiency, output power, and available capacity of lithium batteries are reduced [9]. As a result, the overall performance of electric vehicles will be influenced. In practice, the operating temperature of a lithium battery is not only related to the heat produced by the battery itself but also depends on the performance of the battery thermal management system (BTMS). To rapidly emit the heat generated by the battery pack so that the battery is within the appropriate temperature range and improve the performance of the whole vehicle, a BTMS is indispensable [10–13].

In recent years, commonly used battery thermal management methods in the market include air thermal management technology, liquid thermal management technology, and thermal management technology, combining active and passive methods [14–16]. Saechan et al. [17] conducted numerical research on the air-cooling and heat-management system to optimize the cooling performance during the discharge process. A three-dimensional transient heat transfer model of a cylindrical lithium-ion battery pack was established to study the influence rate of inlet speed, discharge, and other factors on the heat transfer process and the influence of the battery arrangement structure on the cooling performance. The results of the performance optimization show that the maximum temperature and temperature uniformity of the battery are reduced. Liang et al. [18] proposed a cell BTM solution using a new distributed water-cooled component. The multi-physical field model of a BTM system was established, and the influence of three important system parameters (V_{er}, N_{cc}, and T_{in}) on the system performance was studied. The results show that in order to balance the cooling performance, the power consumption and light weight of the system need to be moderately determined. Lyu et al. [19] used experimental methods to develop a new thermal management system for electric vehicle batteries: a combination of thermoelectric cooling, forced air cooling, and liquid cooling. The liquid coolant comes into indirect contact with the battery and acts as a medium to remove the heat generated by the battery during operation. The experimental results show that the system has a good cooling effect under reasonable power consumption.

Due to the limited cooling capacity, Air-BTMS is widely used to solve the heat dissipation problem of low-power and small-capacity lithium battery packs [20–23]. A coupled BTMS based on heat pipes or phase change materials (PCM) usually increases the initial investment and complexity of the system [24–26]. In contrast, a Liquid-BTMS based on a cold plate has the advantages of compact structure and high reliability, which satisfy the requirements of heat dissipation in the limited space [27–30]. Therefore, a Liquid-BTMS has become the main development direction of power battery module thermal management technology. For the cylindrical lithium-ion battery pack, Zhao et al. [31] designed a new type of liquid-cooled cylinder (LCC). The effects of the channel number and the mass flow rate of cooling water on the thermal characteristics of the battery pack were analyzed. It was found that the LCC liquid cooling structure had an excellent cooling performance. Further, the maximum temperature of the battery pack could be controlled under 40 °C when the battery was discharged at 5 C. Gao et al. [32] proposed a novel BTMS design based on flow direction gradient channels and applied it to cylindrical lithium-ion battery modules. Compared with the uniform flow channel design, the gradient flow channel design obviously changes the basic feature of the monotonic rise in temperature along the flow direction. Basu et al. [33], Du et al. [34], and Wang et al. [35] all designed an indirect contact BTMS using a curved, wavy aluminum flat tube combined with a cylindrical lithium battery pack. When the battery pack was charged and discharged at a high rate, the influence of the contact angle between the cooling channel and the battery, channel number, and cooling water flow rate on the maximum temperature and temperature uniformity of the battery was compared and analyzed. It was found that the contact angle had the most significant influence on cooling performance, and its optimal value was 70°.

Although the liquid BTMS has been extensively studied to satisfy the heat dissipation requirements of large-capacity lithium battery packs through the optimization of the liquid cooling structure and the assembly method with battery packs, research on liquid cooling technology for high-energy-density lithium battery modules is still necessary. In the optimal design of liquid-cooled plates, experiments and numerical simulations are mostly used to optimize the flow form and operating conditions of liquid-cooled plates in a characteristic form, which often leads to a huge workload and the insufficient universality of the research. It is difficult to accurately obtain the functional relationship between cold plate performance and operating parameters. To perform the parametric study, the multi-factor analysis-based numerical simulation is necessary and profitable.

A response surface method (RSM) is an approach to obtain the functional relationship between fitting factors and response values by using reasonable experimental design methods [36–39]. The main advantages of an RSM are as follows: (1) It is possible to construct specific functional relations to reflect the complex nonlinear relationship between structural parameters and the performance of liquid cold plates [38,39]; (2) It is convenient to study the interaction of different factors on the performance of a cold plate, and to explore the significance of the influence of each factor [40]; (3) A high precision regression equation can be obtained to guide the design of the structural parameters for a cold plate [41]; (4) Simulation times and complicated calculations can be greatly reduced, and the design cycle can be shortened [42].

The average temperature of lithium-ion batteries at different battery discharging rates, the temperature difference between individual batteries, and the temperature difference of individual batteries themselves all need to be controlled in a practical application. Especially at high discharge rates, it is necessary to further adjust the operating conditions of the cold plate to meet the battery cooling requirements. Referring to the authors' previous work [43], the cold plates constructed with serpentine channels provide better cooling performance compared to those with straight channels. On the basis of the obtained optimal cold plate structure, it is important to study the influence of the cold plate operating parameters on the performance of lithium-ion batteries under different discharging rates.

According to existing studies, a large number of experiments and simulations are usually necessary for the optimal design, which is time-consuming and lacks universality. The analysis of the interaction factor effects based on an elaborate heat transfer analysis and a reliable mathematical method are also required to clarify the significance of the influences. For the operation of the cold plate, a universal optimization method covering a wide range of different operating parameters is essentially needed. Considering the average temperature of the cold plate and the uniformity of the temperature distribution, a design system for operating parameters should be constructed to meet the rapid optimization of cold plate operating parameters under different battery discharging rates and ensure the efficient operation of a lithium-ion battery pack.

Therefore, this paper intends to adopt the RSM to establish a regression model. The flow rate and inlet temperature of the coolant and the average heat generation of the batteries are chosen as independent variables. The heat transfer and flow resistance characteristics are set as objective functions. Based on the response surface analysis, the interaction effects of the operating strategies on the cold plate performance are clarified. The excellent operating conditions of the cold plate are also obtained. Taking the maximum temperature difference and the average temperature of the liquid cold plate surface as constraints, the multi-objective optimization of the operating parameters is carried out using a non-dominated sorting genetic algorithm (NSGA-II). The findings are expected to provide a theoretical basis for the control strategy of battery pack thermal management systems and the optimization of liquid cold plate performance.

2. Numerical Investigation

2.1. Configuration of Liquid Cooling System

As a key component of an indirect liquid cooling system for battery thermal management, the liquid cold plate is responsible for removing the heat generated by the lithium-ion battery during the charging or discharging process. The cooling capability of a liquid cold plate remarkably affects the thermal characteristics of the lithium-ion battery, which has been extensively proven to have a critical relationship with battery charging and discharging performance. Therefore, the development of a liquid cold plate with high efficiency plays an important role for the BTMS in maintaining the battery temperature within a proper and safe range. It should also be noted that both the flow rate and inlet temperature of liquid coolant affect the cooling performance, including the maximum temperature and the temperature difference on the battery module. The required pumping power highly depends on the flow rate and structure of the liquid cold plate. It can be concluded that the optimization study of the liquid cold plate is necessary to obtain the proper operating conditions for the high-efficiency cooling of lithium-ion batteries.

For the reasons stated above, a three-dimensional model was first established using SOLIDWORKS software to explore the effects of the flow rate and the inlet temperature of the coolant on the cooling performance of the liquid cold plate for the lithium-ion battery module. To simplify the model, the heat generation of lithium-ion batteries under different discharging rates was treated as a surface heat source. The average temperature and the maximum temperature difference on the liquid cold plate were used to evaluate the thermal condition of the lithium-ion battery module and were taken as the objectives of the optimization study. The capacity of the battery model is 37.0 A·h, and its dimensions are 148 mm (x) × 26.5 mm (y) × 94 mm (z). As shown in Figure 1, the battery module consists of sixteen cells, two of which are connected in parallel and eight in series. The cold plate is arranged on the side surface of the battery module, as displayed in Figure 2. The length, width, and thickness of the cold plate are 212.0 mm, 110.5 mm, and 15 mm, respectively.

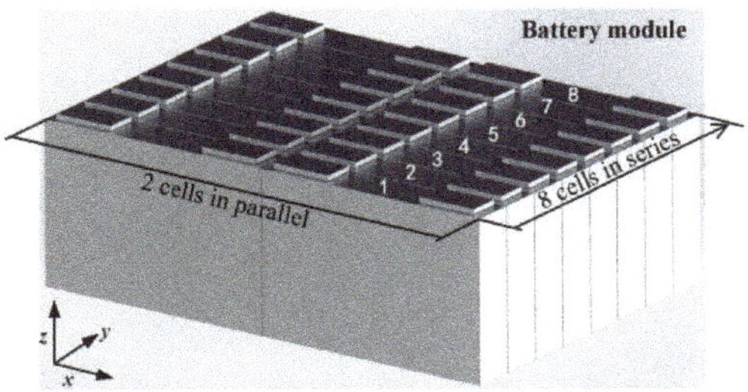

Figure 1. Schematic diagram of lithium-ion battery module.

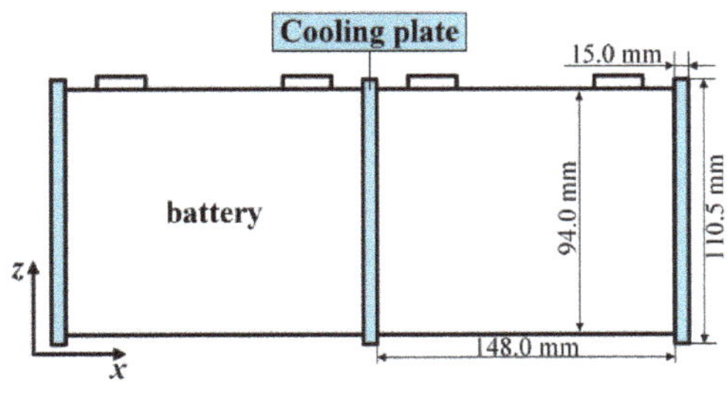

Figure 2. The layout of liquid cold plates for battery module cooling.

As shown in the authors' previous work, the cold plates constructed with serpentine channels provide better cooling performance compared to those with straight channels [43].

Furthermore, the configuration of serpentine channels was optimized considering simultaneously the thermal behavior of the battery module and the pumping power required. The liquid cold plate, consisting of 6 serpentine flow paths with each path including 4 parallel channels, was proven to have better performance, as shown in Figure 3. Each channel was 2.5 mm in width and 9.0 mm in height, while the thickness of each fin was 1.5 mm. The thickness of the substrate and cover plate of the cold plate was 3 mm each. The total thickness of the cold plate was 15 mm. In the present study, the effects of the flow rate and inlet temperature on the thermal characteristics of the BTMS were numerically investigated based on the liquid cold plate described above, and the optimization was carried out involving the pumping power.

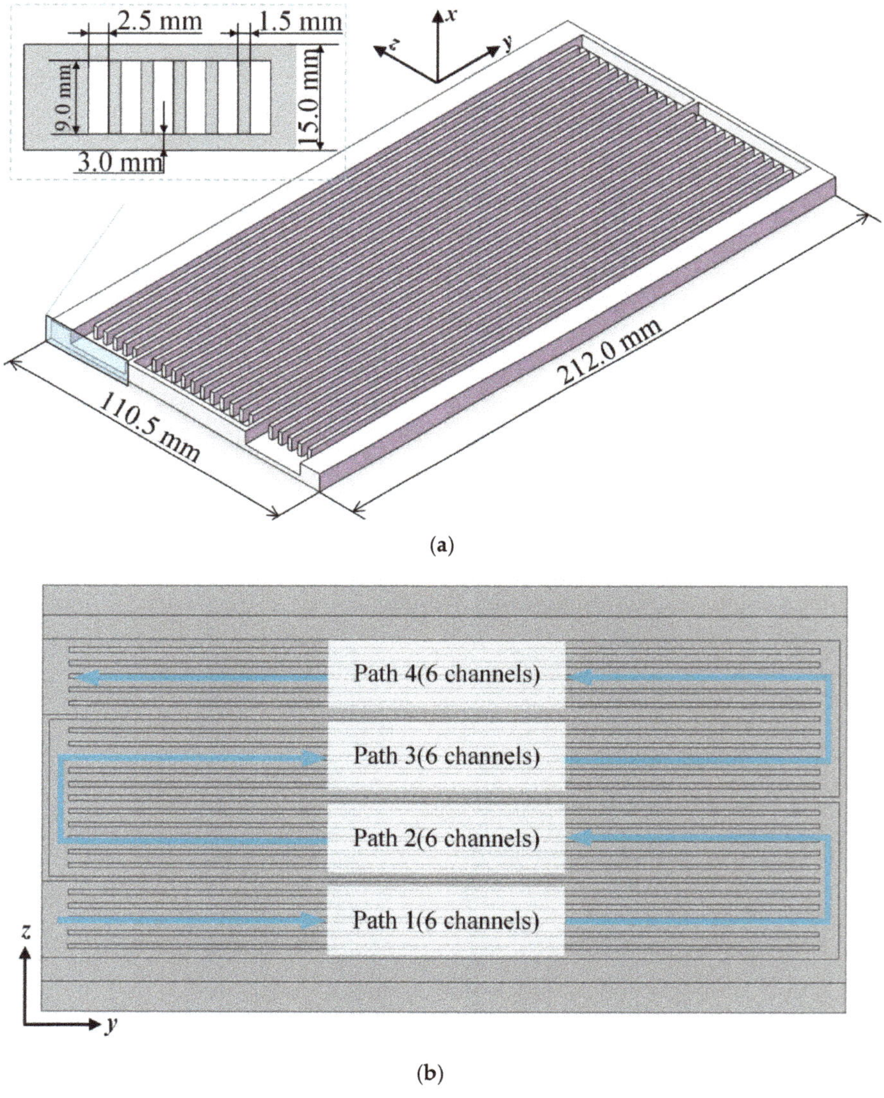

Figure 3. Schematic diagram of liquid cold plates with serpentine flow paths: (**a**) 3D model; (**b**) 2D model of flow paths.

Heat generation from the lithium-ion battery is strongly dependent on the battery charging or discharging rate. In the present study, the discharging rates of 1.0 C, 2.0 C, 3.0 C, and 4.0 C were selected as typical working conditions to represent different heat dissipation demands. In references [44–46], the heat generation rates of a single lithium-ion battery under the various discharging rates were investigated in experiments and electrochemistry simulations based on the pseudo-two-dimensional model. Under the discharging rates of 1.0 C, 2.0 C, 3.0 C, and 4.0 C, the average heat generation rates of a single battery were 5 W, 20 W, 40 W, and 65 W, respectively. The surface heat flux applied on the cold plate was determined accordingly.

2.2. Governing Equations and Boundary Conditions

The material of the liquid cold plate was aluminum, while water was used as the liquid coolant. The following assumptions were made before performing the numerical study:

(1) The total flow rate of the coolant and the heat flux on the cold plate keep constant during the cooling process. Therefore, the flow and heat transfer in the liquid cold plate is incompressible and steady;

(2) Since the temperature variation range of water is less than 10 °C in most cases, the thermal properties of water are considered to be constant;

(3) According to the sizes of the studied liquid cold plate, the effects of gravity and viscous dissipation can be neglected.

The governing equations for the flow and heat transfer in the cold plate can be written as:

$$div(\rho U\phi) = div(\Gamma_\phi grad\phi) + S_\phi \tag{1}$$

where ρ is the density of the coolant, kg·m^{-3}, U is the velocity vector, and ϕ is the unknown to be solved. The left side of the equation is the convective item, whereas the items on the right side are the diffusive item and source item, respectively.

Flow and heat transfer in the cold plates constructed with serpentine channels were numerically investigated by adopting the k-ω turbulent model, and the results were validated using the experiments in Ref [47]. The governing equations to be solved included the continuous equation (ϕ = 1), the momentum equations (ϕ = U_x, U_y, and U_z), the energy equation (ϕ = T), the turbulent kinetic energy equation (ϕ = k), and the specific dissipation rate equation (ϕ = ω).

Uniform velocity and temperature were set as the boundary conditions at the inlet of the cold plate, whereas zero pressure was set at the outlet. Uniform heat flux was applied to the two large surfaces of the cold plate, and adiabatic condition was set for the other surfaces. No-slip boundary condition was set for the contact surfaces between fluid and solid. As stated earlier, the present study aimed to optimize the inlet temperature and flow rate of the coolant of lithium-ion batteries under different discharging rates. Therefore, the inlet temperature and flow rate of coolant and the heat flux were the independent variables in the numerical investigation. The mathematical descriptions of the above boundary conditions are as follows.

The flow and thermal conditions at the Inlet of the liquid cold plate are:

$$u_x = 0, u_y = u_{\text{in}}, u_z = 0, T = T_{\text{in}} \tag{2}$$

The pressure at the outlet of the liquid cold plate is constant and set as:

$$P = 0 \tag{3}$$

The no-slip condition for convective surfaces can be expressed as:

$$u_x = 0, u_y = 0, u_z = 0, T = T_s, -\lambda\frac{dT}{dn}|_{\text{f}} = -\lambda\frac{dT}{dn}|_{\text{s}} \tag{4}$$

The bottom and top walls of the liquid cold plate are heated with uniform heat flux:

$$-\lambda\frac{dT}{dn}|_{\mathrm{s}} = q_{\mathrm{constant}} \tag{5}$$

where u is velocity, T is temperature, P is pressure, λ is thermal conductivity, n indicates the normal direction of the convective surface, and the subscripts f and s refer to fluid and solid, respectively.

2.3. Numerical Scheme

As shown in Figure 4, the calculated domain, including the fluid and solid zones, was discretized with the hexahedral grids using ICEM. The SIMPLE algorithm was used to solve the pressure-velocity coupled term, and the second-order upwind finite scheme was used to solve the momentum and energy equations. The solving process was completed by employing CFD software ANSYS 15.0. The convergence criteria for all the governing equations were set at 10^{-6}. The critical parameters, including flow pressure drop, maximum temperature, and temperature difference, were monitored during the solving process.

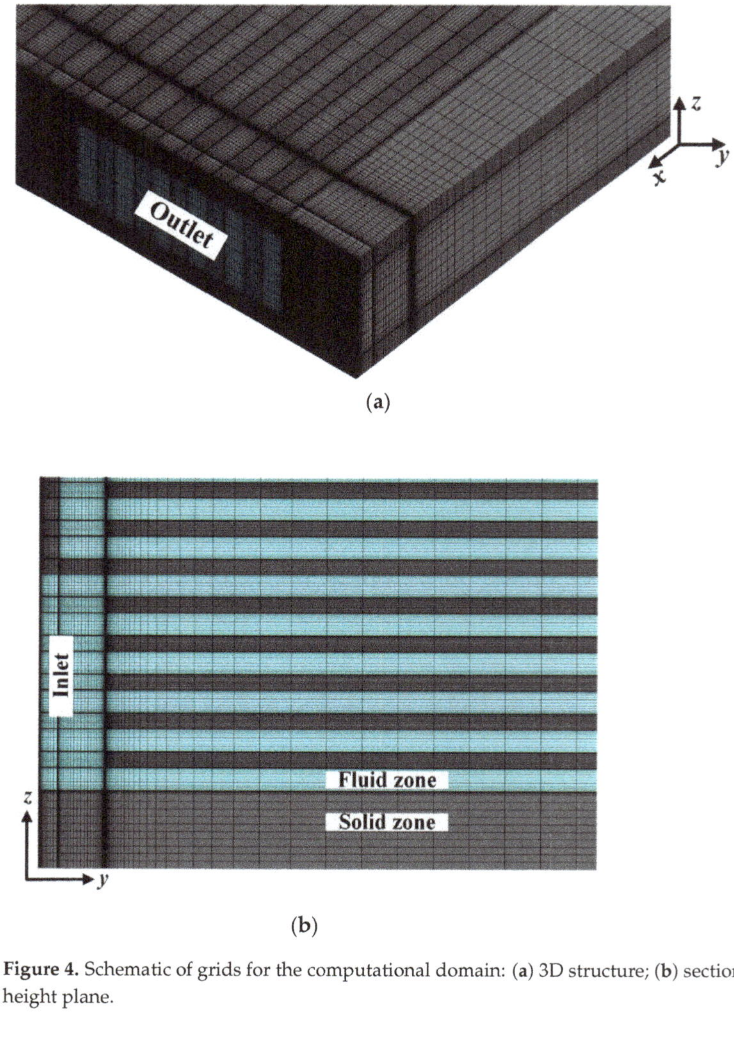

Figure 4. Schematic of grids for the computational domain: (**a**) 3D structure; (**b**) sectional grids at $\frac{1}{2}$2 height plane.

To reduce the effects of grid number on the flow and heat transfer results, the grid independence for the present study was carefully examined by solving the governing equations for the computational domain discretized using different meshing strategies. As shown in Figure 5, the variations of the pressure drop and the maximum temperature difference were less than 0.1 K with the increase in grid number, and the change in the pressure drop became weak. The pressure drop increased by only 0.38% when the number of grids increased from 5,367,900 to 7,047,300. To reduce computational load and ensure solving precision, the meshing, including 5,367,900 grids, was used to solve the flow and heat transfer in the cold plates.

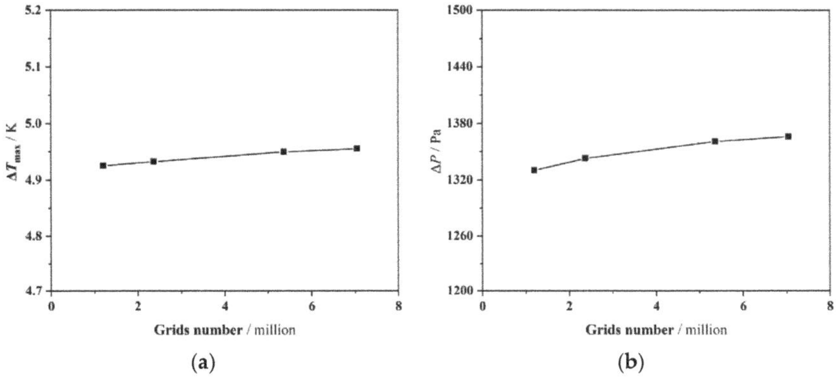

Figure 5. The effects of grid number on the numerically solved results: (**a**) Maximum temperature difference; (**b**) Pressure drop.

3. Optimization Method

It is necessary to change the operating parameters of the liquid cold plate to meet the thermal management demands of the power battery under different discharging rates. The performance of the liquid cold plate changes nonlinearly with the change of key operating parameters, including the flow rate, average heat generation, and inlet temperature. There will be a huge workload when using the variable-controlling approach or orthogonal experiment method. In addition, the improvement of heat transfer and flow resistance is mutually incompatible in practice. The increase in flow pressure drop may happen with the cooling performance enhancement of the liquid cold plate. Therefore, it is difficult to achieve the optimal value for heat transfer and flow resistance at the same time. Compromise solutions considering different design objectives for the liquid cold plates are required.

The RSM is utilized to construct a specific functional relation to reflect the nonlinear relationship between the input parameters and the responses. It can greatly reduce the simulation times and shorten the design cycle. The NSGA -II is invoked to solve complex nonlinear problems and optimize multiple objective functions simultaneously. The optimal solution to multi-objective optimization problems can be visually displayed through the Pareto front.

3.1. Response Surface Methodology

The RSM is a combination of mathematical and statistical methods for modeling and analyzing results, which can effectively show the relationships between the design variables and output responses. The RSM is applied to arrange the experimental design to obtain the relevant experimental data, and then the multiple regression equation is obtained to fit the functional relationship between the variables and the response. The model used to establish the function can be expressed as follows:

$$y = f(X_1, X_2, \cdots, X_k) + \varepsilon \tag{6}$$

where y represents the objective function, X_k and f represent the independent design variables and the approximate response function, respectively, and ε is the residual error.

RSM models are empirical models based on observed data from processes or systems. Therefore, it was necessary to determine the design sample for operating parameters in the process of optimal design. A design of experiment (DOE) based on a popular design method called face-centered central composite design (CCD) was adopted in the design. The numerical simulations were performed according to the set points, including factorial points and center points augmented by axial points in CCD.

As discussed previously, the average temperature and maximum temperature difference in the liquid cold plate and the flow pressure drop of the coolant were selected as the optimization goals. The range of the flow rate was 0.0019–0.0249 kg·s^{-1}, while the inlet temperature was considered within the range of 20–30 °C. The design variables and their levels are summarized in Table 1. Three levels, "−1", "0", and "1", represent the low, middle, and high levels, respectively. The software Design-Expert was utilized to generate the design plans. The DOE and corresponding results are depicted in Table 2. The regression model was fitted according to the sample points obtained using the numerical simulations and the response values.

Table 1. Design variables and levels of the condition.

S/No.	Design Variables	Level		
		−1	0	1
1	Flow rate, kg·s^{-1}	0.0019	0.0134	0.0249
2	Average heat generation rate, W	5	35	65
3	Inlet temperature, °C	20	25	30

Table 2. Simulation design and corresponding results.

	Variables (Code)			Variables (Actual Value)			Response		
No.	$\dot{m}$	Q	T_{in}	$\dot{m}$, kg·s^{-1}	Q, W	T_{in}, °C	T_{ave}, °C	ΔT_{max}, °C	ΔP, Pa
case1	−1	−1	−1	0.0019	5	20	23.89	3.89	30.45
case2	1	−1	−1	0.0249	5	20	20.31	0.38	1377.06
case3	−1	1	−1	0.0019	65	20	70.53	50.52	30.45
case4	1	1	−1	0.0249	65	20	23.98	4.95	1377.06
case5	−1	−1	1	0.0019	5	30	33.83	3.86	25.09
case6	1	−1	1	0.0249	5	30	30.3	0.38	1344.95
case7	−1	1	1	0.0019	65	30	80.46	50.88	25.09
case8	1	1	1	0.0249	65	30	33.93	4.98	1344.93
case9	−1	0	0	0.0019	35	25	52.17	27.29	27.77
case10	1	0	0	0.0249	35	25	27.13	2.67	1360.47
case11	0	−1	0	0.0134	5	25	25.52	0.69	448.10
case12	0	1	0	0.0134	65	25	31.83	9.05	448.10
case13	0	0	−1	0.0134	35	20	23.72	4.88	461.87
case14	0	0	1	0.0134	35	30	33.65	4.87	434.67
case15	0	0	0	0.0134	35	25	28.67	4.87	448.10

3.2. NSGA-II Algorithm

In recent years, various intelligent optimization algorithms have been proposed, such as Multiple Objective Particle Swarm Optimization (MOPSO), Driving Training-Based Optimization (DTBO), and Grey Wolf Optimizer (GWO), which have already been proven to be sufficient for solving specific problems. Compared to them, the genetic algorithm is a simple and mature analysis method. The computation is simple, and its precision was enough for the optimization in the present study. In this study, the multi-objective optimization was executed using a non-dominated sorting genetic algorithm II (NSGA-II), which had the advantages of a fast running speed and good convergence of the solution set [48]. The NSGA -II was performed to acquire the Pareto front and the optimal operating conditions. Compared with the weighting method, a uniformly distributed Pareto-optimal front was obtained to avoid aggregation of solutions in a small interval as well as local convergence.

The flowchart of the optimization procedure is shown in Figure 6. The initial combinations of operating parameters were generated randomly. The objective functions obtained by the RSM were employed in the optimization algorithm to determine the fitness functions in the NSGA-II. Then, the genetic operators, including selection, crossover, and mutation, were used to generate a new population. The non-dominated sorting was conducted to rank the population members and put them into different fronts according to the values of the fitness functions. The crowding distance was used to measure how close an individual was to its neighbors, which was designed to avoid the accumulation of population members over a limited distance. Large crowding distances would result in better diversity in the population. The optimal combinations of operating parameters and the corresponding Pareto optimal solution based on the typical discharging rates could be output when the evolutionary iteration reached the set value. The decision maker could select the appropriate Pareto optimal solution in the optimization domain according to the requirements.

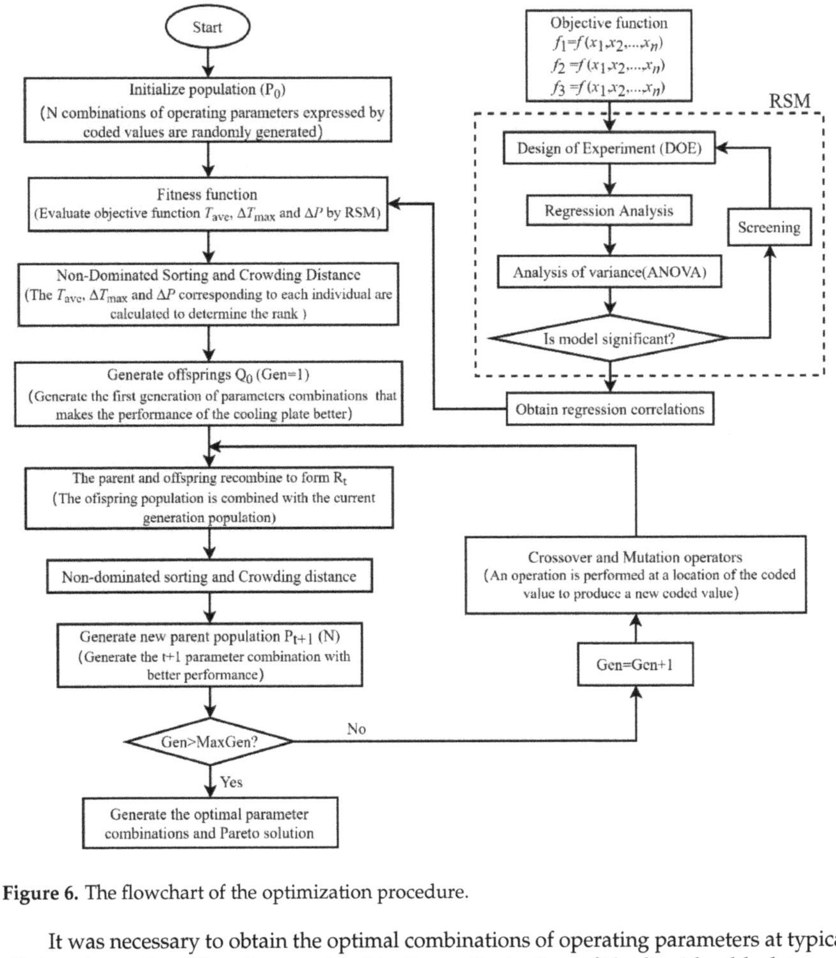

Figure 6. The flowchart of the optimization procedure.

It was necessary to obtain the optimal combinations of operating parameters at typical discharging rates. Therefore, multi-objective optimization of the liquid cold plate was required to obtain Pareto optimal solution at different discharging rates. The multi-objective optimization problem of the cold plate can be described as the following equations:

$$Minimize\begin{cases} T_{\rm ave} = f_1(\dot{m}, T_{in}) \\ \Delta T_{\rm max} = f_2(\dot{m}, T_{in}) & 0.0249 \le \dot{m} \le 0.0498 \\ \Delta P = f_3(\dot{m}, T_{in}) \ldots. & 20 \le T_{in} \le 30 \end{cases} \quad (7)$$

where f_1, f_2, and f_3 are the objective functions of the average temperature ($T_{\rm ave}$), the maximum temperature difference ($\Delta T_{\rm max}$), and the pressure drop (ΔP) at a specific discharging rate, respectively, $\dot{m}$ represents the flow rate of the working medium, and $T_{\rm in}$ is the inlet temperature.

The GAMULTIOBJ function from MATLAB global optimization toolbox, which is a variant of NSGA-II, was utilized. It can effectively solve the problem of premature convergence and avoid falling into local optimum. The tuning parameters used are shown in Table 3.

Table 3. GAMULTIOBJ function parameters.

Parameters	Values
PopulationSize	1000
ParetoFraction	0.3
MaxGenerations	200
MaxStallGenerations	200
FunctionTolerance	10^{-10}
Number of variables	2

4. Results and Discussion

4.1. Numerical Analysis

Figure 7 shows the temperature contours of the heated surface of the liquid cold plate under various flow rates when Q = 35 W and $T_{\rm in}$ = 25 °C. It can be seen that the heat transfer performance and temperature uniformity were improved with the increase in flow rate. The maximum temperature difference and average temperature decreased by 49.57% and 45.05%, respectively, when the flow rate increased from 0.0019 kg·s^{-1} to 0.0134 kg·s^{-1}. However, it is noted that the decreasing trend in the maximum temperature difference slowed down with a further increase in the flow rate. It can be speculated that a further increase in the flow rate had little effect on improving the temperature uniformity when the flow rate exceeded a certain value. Meanwhile, it can be seen from Figure 8 that the flow resistance of the coolant rose significantly with the increase in the flow rate. It is clear that the further increase in flow rate accelerated the rising rate of flow pressure drop. The pressure drop increased by 420.33 Pa when the flow rate varied from 0.0019 kg·s^{-1} to 0.0134 kg·s^{-1}, whereas the rise in pressure drop is up to 912.37 Pa as the flow rate increased from 0.0019 kg·s^{-1} to 0.0249 kg·s^{-1}. Therefore, it is essential to choose the appropriate flow rate in the working medium for the BTMS to achieve the best cooling performance and relatively low flow resistance.

The temperature distributions for the liquid cold plate under different inlet temperatures are shown in Figure 9. It is noted that the maximum temperature difference remained around 4.9 °C and the average temperature increased by 41.86% when the inlet temperature increased from 20 to 30 °C. This indicates that the change of inlet temperature mainly affected the maximum and the average temperatures and had little effect on the temperature uniformity of the cold plate.

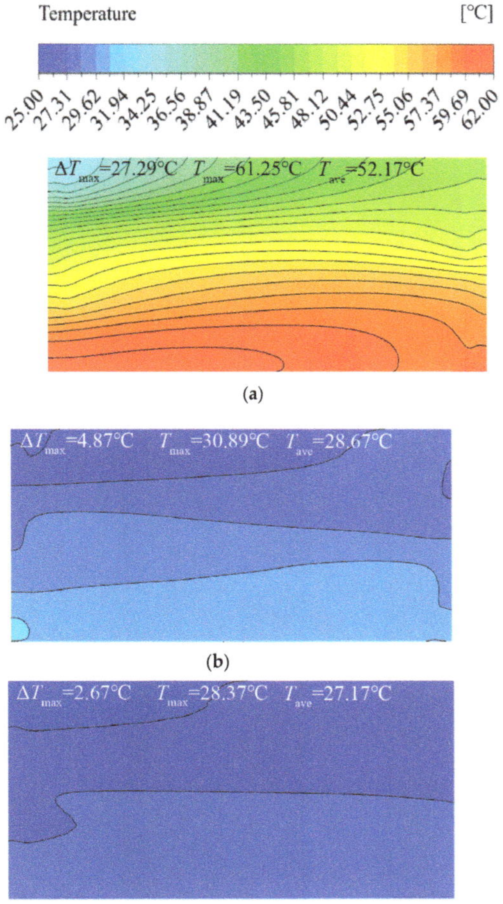

Figure 7. Temperature distribution of the liquid cold plate under different flow rates when $Q = 35$ W and $T_{\text{in}} = 25$ °C: (**a**) $\dot{m} = 0.0019$ kg·s^{-1} (**b**) $\dot{m} = 0.0134$ kg·s^{-1} (**c**) $\dot{m} = 0.0249$ kg·s^{-1}.

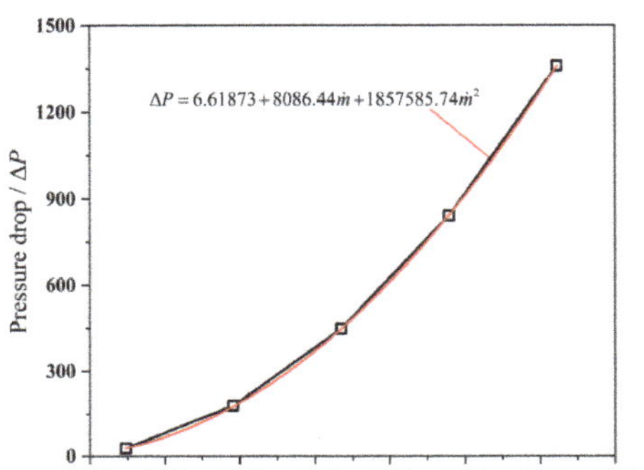

Figure 8. The variation in pressure drop with mass flow rate.

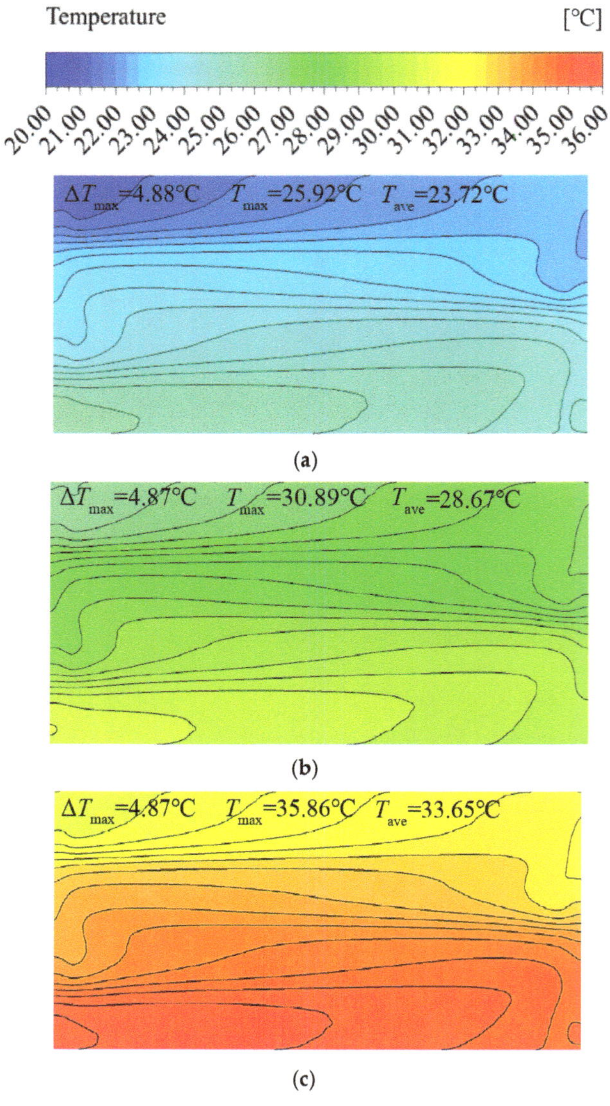

Figure 9. Temperature distribution in the liquid cold plate under different inlet temperatures when $Q = 35$ W and $\dot{m} = 0.0134$ kg·s^{-1}: (**a**) $T_{in} = 20$ °C (**b**) $T_{in} = 25$ °C (**c**) $T_{in} = 30$ °C.

4.2. Response Surface Equations and Validations

This section establishes the regression models to obtain the fitting functions for various optimization objectives. The analysis of variance (ANOVA) aims to examine the significance of each term of the regression models for the prediction of optimization objectives. According to the principle of statistics, *F*-value is calculated as the ratio of the source's mean square to the residual mean square, while the *p*-value is the probability of seeing the observed *F*-value if there are no factor effects. The significance of one item in the regression model can be reflected by a *p*-value. In the present study, one item was considered to have had a significant impact on the response once its *p*-value was less than 0.05; in other words, the terms with a p-value greater than 0.05 were considered insignificant and could be eliminated from the models.

Based on the experimental points (obtained by simulation results) and the response values in Table 2, the ANOVA results of the responses for the average temperature, maximum temperature difference, and pressure drop are summarized in Tables 4–6. For the average and maximum temperature differences, the influences of all factors and their interaction terms were significant. The flow rate of the working medium had the most significant effect, while the effect of the square term of average heat generation was relatively small. For the flow resistance, the flow rate and the inlet temperature had significant effects, and the average heat generation almost did not affect the pressure drop. The term $\dot{m}$ had the maximum F-value and the minimum p-value, demonstrating that the flow rate was the most significant factor affecting the flow resistance of the cold plate. The ANOVA results of the three models, after excluding insignificant terms, are presented in Tables A1–A3.

Table 4. The ANOVA for the average temperature (before elimination).

Source	Sum of Squares	Degrees of Freedom	Mean Square	F-Value	p-Value
Model	4626.93	12	385.58	3.452×10^9	<0.0001
$\dot{m}$	1568.31	1	1568.31	1.404×10^{10}	<0.0001
Q	19.87	1	19.87	1.779×10^8	<0.0001
T_{in}	49.51	1	49.51	4.433×10^8	<0.0001
$\dot{m}\cdot Q$	923.81	1	923.81	8.271×10^9	<0.0001
$\dot{m}\cdot T_{in}$	0.0006	1	0.0006	5484.01	<0.0001
$Q\cdot T_{in}$	0.0002	1	0.0002	1790.70	<0.0001
$\dot{m}\cdot\dot{m}$	331.45	1	331.45	2.968×10^9	<0.0001
$Q\cdot Q$	5.682×10^{-7}	1	5.682×10^{-7}	5.09	0.0587
$T_{in}\cdot T_{in}$	5.682×10^{-9}	1	5.682×10^{-9}	0.0509	0.8280
$\dot{m}\cdot Q\cdot T_{in}$	0.0002	1	0.0002	1450.47	<0.0001
$\dot{m}^2\cdot Q$	141.99	1	141.99	1.271×10^9	<0.0001
$\dot{m}^2\cdot T_{in}$	8.100×10^{-6}	1	8.100×10^{-6}	72.52	<0.0001
$\dot{m}\cdot Q\cdot Q$	1.000×10^{-7}	1	1.000×10^{-7}	0.8800	0.3844
Residual	7.818×10^{-7}	8	1.117×10^{-7}		
Lack of fit	7.818×10^{-7}	3	3.909×10^{-7}		
Pure error	0.0000	5	0.0000		
Cor total	4626.93	19			

Table 5. The ANOVA for the maximum temperature difference (before elimination).

Source	Sum of Squares	Degrees of Freedom	Mean Square	F-Value	p-Value
Model	4273.94	13	328.76	5.904×10^9	<0.0001
$\dot{m}$	302.85	1	302.85	5.439×10^9	<0.0001
Q	34.94	1	34.94	6.276×10^8	<0.0001
T_{in}	0.0000	1	0.0000	323.27	<0.0001
$\dot{m}\cdot Q$	892.30	1	892.30	1.602×10^{10}	<0.0001
$\dot{m}\cdot T_{in}$	0.0122	1	0.0122	2.199×10^5	<0.0001
$Q\cdot T_{in}$	0.0215	1	0.0215	3.866×10^5	<0.0001
$\dot{m}\cdot\dot{m}$	280.85	1	280.85	5.044×10^9	<0.0001
$Q\cdot Q$	2.784×10^{-7}	1	2.784×10^{-7}	5.00	0.0667
$T_{in}\cdot T_{in}$	2.784×10^{-7}	1	2.784×10^{-7}	5.00	0.0667
$\dot{m}\cdot Q\cdot T_{in}$	0.0174	1	0.0174	3.123×10^5	<0.0001
$\dot{m}^2\cdot Q$	120.35	1	120.35	2.161×10^9	<0.0001
$\dot{m}^2\cdot T_{in}$	0.0038	1	0.0038	68,640.45	<0.0001
$\dot{m}\cdot Q\cdot Q$	6.250×10^{-7}	1	6.250×10^{-7}	11.22	0.0154
Residual	3.341×10^{-7}	6	5.568×10^{-8}		
Lack of fit	3.341×10^{-7}	1	3.341×10^{-7}		
Pure error	0.0000	5	0.0000		
Cor total	4273.94	19			

Table 6. The ANOVA for the pressure drop (before elimination).

Source	Sum of Squares	Degrees of Freedom	Mean Square	*F*-Value	*p*-Value
Model	4.748×10^6	9	5.275×10^5	1.833×10^5	<0.0001
$\dot{m}$	4.443×10^6	1	4.443×10^6	1.544×10^6	<0.0001
Q	0.0001	1	0.0001	0.0000	0.9955
T_{in}	1043.63	1	1043.63	362.74	<0.0001
$\dot{m}{\cdot}Q$	0.0001	1	0.0001	0.0000	0.9950
$\dot{m}{\cdot}T_{\text{in}}$	357.98	1	357.98	124.43	<0.0001
$\dot{m}{\cdot}\dot{m}$	1.665×10^5	1	1.665×10^5	57,870.36	<0.0001
$Q{\cdot}Q$	0.0028	1	0.0028	0.0010	0.9756
$T_{\text{in}}{\cdot}T_{\text{in}}$	0.1196	1	0.1196	0.0416	0.8425
Residual	28.77	10	2.88		
Lack of fit	28.77	5	5.75		
Pure error	0.0000	5	0.0000		
Cor total	4.748×10^6	19			

Through the backward elimination process, the ultimate forms of regression equations of average temperature, maximum temperature difference, and pressure drop are expressed as:

$$T_{ave} = 0.12313 - 0.1399\dot{m} + 0.9503Q + 0.9936T_{\text{in}} - 94.9058\dot{m} \cdot Q + 0.1523\dot{m} \cdot T_{\text{in}} + 1.6680 \times 10^{-6}Q \cdot T_{\text{in}} - 177.3068\dot{m}^2 + 4.8611 \times 10^{-7}Q^2 - 0.0026\dot{m} \cdot Q \cdot T_{\text{in}} + 2379.7945\dot{m}^2 \cdot Q + 3.4104\dot{m}^2 \cdot T_{\text{in}} \quad (8)$$

$$\Delta T_{\text{max}} = 0.0451 + 46.8433\dot{m} + 0.9258Q - 0.0036T_{\text{in}} - 88.6916\dot{m} \cdot Q - 1.7198\dot{m} \cdot T_{\text{in}} + 0.0007Q \cdot T_{\text{in}} - 1947.0156\dot{m}^2 + 1.1638 \times 10^{-6}Q^2 + 0.000013T_{\text{in}}^2 - 0.0271\dot{m} \cdot Q \cdot T_{\text{in}} + 2190.9304\dot{m}^2 \cdot Q + 74.0804\dot{m}^2 \cdot T_{\text{in}} - 0.00006\dot{m} \cdot Q^2 \quad (9)$$

$$\Delta P = 17.6338 + 10927.0303\dot{m} - 0.4822T_{\text{in}} - 116.4686\dot{m} \cdot T_{\text{in}} + 1865850\dot{m}^2 \quad (10)$$

where Q represents the average heat generation.

The goodness of fit of the regression models is judged by the coefficient of determination R^2, namely, the ratio of the explained variance to the total variance. The C.V (Coefficient of Variation) is calculated by dividing the standard deviation by the mean. The R^2, $R_{adj}{}^2$, and the C.V of the regression model for the average temperature of the liquid cold plate are 100%, 100%, and 0.0001%, respectively. The R^2, $R_{adj}{}^2$, and the C.V of the regression model for the maximum temperature difference are 100%, 100%, and 0.0024%, respectively. Furthermore, those of the regression models for the pressure drop are 100%, 100%, and 0.2969%, which guarantees the accuracy of the regression models established by the RSM.

Figure 10 exhibits the 3D surface plots of the interaction effects on the average temperature. As can be seen from the three plots, the mass flow rate has the most significant effect on the average temperature. The minimum average temperature corresponds to the smallest average heat generation, the lowest inlet temperature, and the largest flow rate.

The 3D surface plots of interaction effects on the maximum temperature difference are shown in Figure 11. It is seen that the mass flow rate, average heat generation rate, and inlet temperature all had significant influences on the maximum temperature difference. The maximum temperature difference increased with the increase in the average heat generation rate at a low flow rate. With the increase in flow rate, the increase in maximum temperature difference gradually became smaller. The results imply that better temperature uniformity could be achieved with a larger flow rate and lower average heat generation rate.

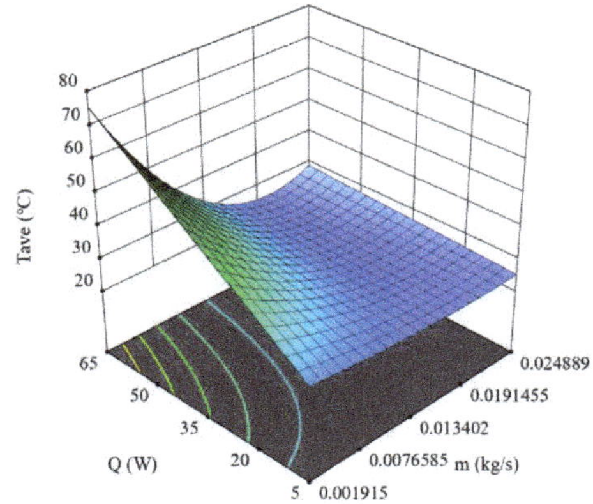

(**a**)

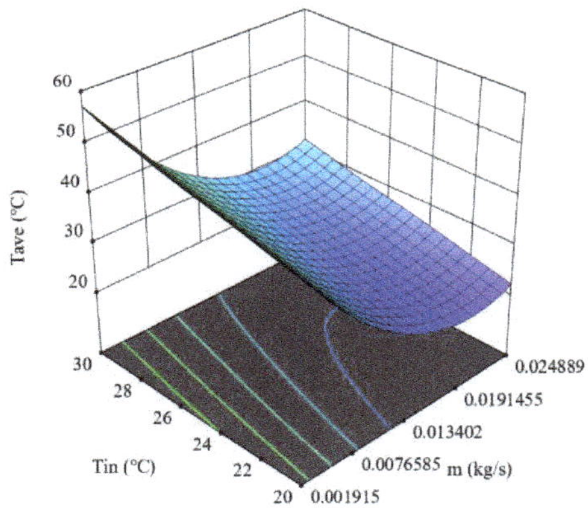

(**b**)

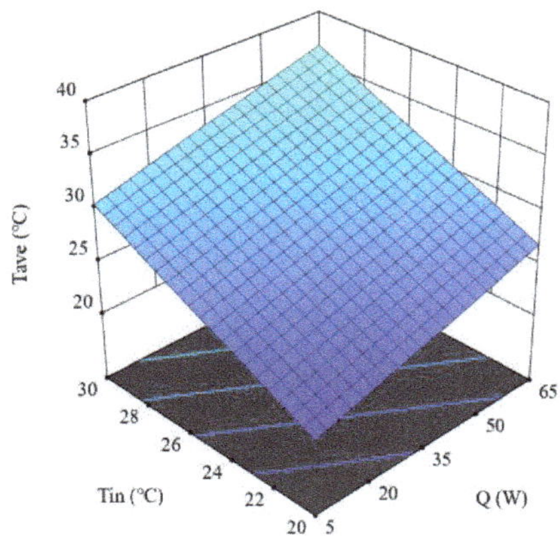

(**c**)

Figure 10. 3D surface plots of the combined effects on the average temperature: (**a**) $Q \cdot \dot{m}$; (**b**) $T_{in} \cdot \dot{m}$ (**c**) $Q \cdot T_{in}$.

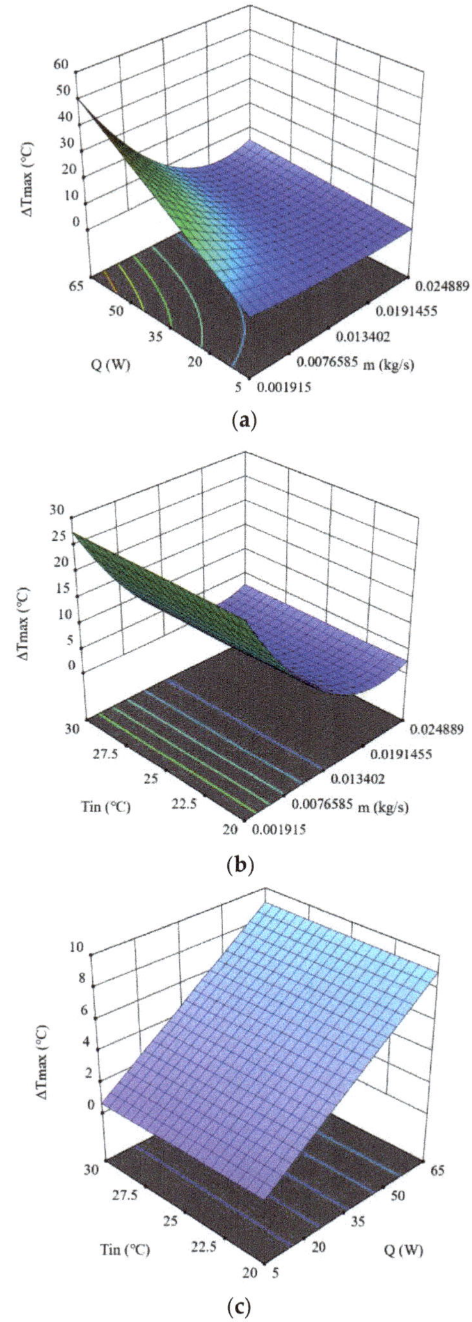

(**a**)

(**b**)

(**c**)

Figure 11. 3D surface plots of the combined effects on the maximum temperature difference: (**a**) $Q{\cdot}\dot{m}$; (**b**) $T_{\text{in}}{\cdot}\dot{m}$ (**c**) $Q{\cdot}T_{\text{in}}$.

Figure 12 shows the combined effects of mass flow rate and inlet temperature on pressure drop. It can be observed that a smaller flow rate reduced the pressure drop. The difference in inlet temperature would lead to a difference in liquid viscosity, resulting in a slight difference in the pressure drop.

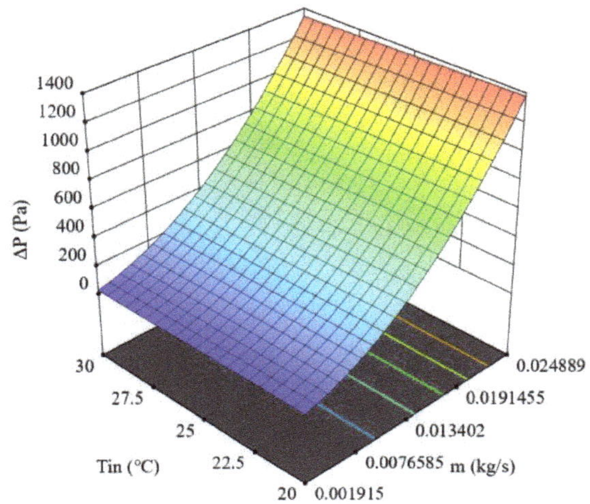

Figure 12. 3D surface plot of the combined effects of inlet temperature and flow rate on the pressure drop.

4.3. Optimization Results

Based on the regression models established in Section 4.2, efforts were made to optimize the operating conditions of the liquid cold plate for the best cooling performance and the lowest flow resistance. It should be noted that there does not exist an operation condition that can simultaneously minimize the maximum temperature and flow resistance. The Pareto optimal solution is essentially a non-inferior solution for multi-objective optimization that can be used to solve the above problem. As shown in Figure 13, the Pareto optimal solutions for the maximum temperature difference, average temperature, and pressure drop of the cold plate were obtained by NSGA-II for different discharging rates of a lithium battery. For any specific required maximum temperature difference and the average temperature, the generated Pareto optimal solution can provide optimal values for the operating parameters that lead to the smallest pressure drop. It can be observed from the figure that the pressure drop presents a downward trend with the increase in the maximum temperature difference and average temperature, which confirms that the improvement of cooling performance of the liquid cold plate is at the expense of an increased flow pressure drop. The operating condition with low inlet temperature and flow rate leads to a large maximum temperature difference and average temperature. Therefore, larger pressure drops need to be overcome when a lower temperature is required for a given liquid cold plate.

Table 7 shows the comparison of two sets of Pareto optimal solutions and the simulation results under different discharging rates. The average temperature and pressure drop in the predicted results obtained based on the RSM and NSGA-II demonstrate a good agreement with CFD results, with relative deviations of less than 11%. Although the values of the maximum temperature differences are small (less than 5 °C), the absolute value of the error is less than 16%.

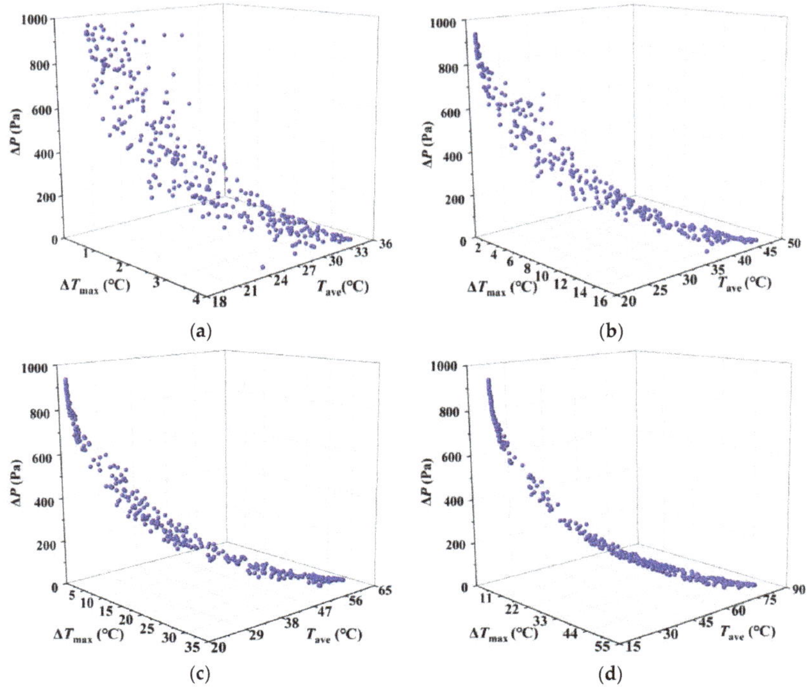

Figure 13. The Pareto solutions for average temperature and pressure drop under different discharging rates: (**a**) 1.0 C (**b**) 2.0 C (**c**) 3.0 C (**d**) 4.0 C.

Table 7. Accuracy verification of Pareto optimal solution.

Discharging Rate	Q kg·s^{-1}	T_{in} °C	Predicted/CFD Results of T_{ave}, °C	Predicted/CFD Results of ΔT_{max}, °C	Predicted/ CFD Results of ΔP, Pa	Error of T_{ave} %	Error of ΔT_{max} %	Error of ΔP %
1.0 C	0.0156	26.86	27.09/ 27.31	0.51/0.58	579.82/ 582.91	−0.81	−12.07	−0.53
	0.0057	28.29	30.68/ 29.68	2.46/2.49	108.95/ 105.79	3.37	−1.20	2.99
2.0 C	0.0163	25.33	26.03/ 27.07	2.14/2.32	629.69/ 628.76	−3.84	−7.88	0.15
	0.0118	28.00	31.18/ 29.85	3.36/3.12	356.27/ 356.58	4.46	7.63	−0.09
3.0 C	0.0174	21.13	21.9/ 24.44	3.86/4.37	718.26/ 722.97	−10.39	−11.67	−0.65
	0.0158	25.54	27.34/ 29.14	4.05/4.79	593.37/ 593.15	−6.18	−15.50	0.04
4.0 C	0.0199	20.00	24.08/ 24.72	3.58/3.97	919.68/ 920.81	−2.59	−9.82	−0.12
	0.0203	20.23	23.60/ 23.25	3.42/3.87	743.05/ 743.85	1.51	−11.63	−0.11

Table 8 shows the variation range of the optimized objectives based on the Pareto optimal solution. It can be seen that the pressure drop is significantly optimized, and the maximum pressure drops under different discharging rates are less than 1000 Pa, which means that the efficiency of the cold plate is essentially improved. The maximum average temperature of the liquid cold plate increases with the discharging rate. The range of the maximum temperature change also expands. Under the same flow rate and inlet temperature, the increased average heat generation results in an upward trend in the cold plate temperature. Furthermore, it reveals that the Pareto optimal solution obtained by multi-objective optimization based on NSGA-II is a solution set that weighs heat transfer performance and flow resistance, which can be used to determine the operating parameters according to the thermal design criteria.

Table 8. The variation range of objectives based on Pareto optimal solution.

Discharging Rate	ΔT_{max}, °C	T_{ave}, °C	ΔP, Pa
1.0 C	0.29~3.90	20.03~33.82	24.39~966.91
2.0 C	1.11~15.66	20.07~45.40	24.44~941.94
3.0 C	2.17~31.39	20.14~60.72	24.56~940.21
4.0 C	3.43~50.92	20.22~80.22	24.53~936.84

5. Conclusions

To ensure the safety and efficiency of electric vehicles, it is necessary to improve the thermal performance of the liquid cold plate in the battery thermal management system. This work presented a method based on the RSM and NSGA-II to achieve the optimization of the operating parameters with relatively low flow resistance. The proposed methods can be used to quickly obtain the optimized operating parameters to balance the maximum temperature, average temperature, and pressure drop of the cold plate in the BTMS. The results can guide the design of the operating strategy for battery thermal management. The main findings are summarized below.

(1) When the flow rate changes from 0.0019 to 0.0249 $kg{\cdot}s^{-1}$, the pressure drop increases from 27.77 Pa to 1360.47 Pa. The average temperature increases by 9.93 °C when the inlet temperature increases from 20 to 30 °C. Although increasing the flow rate and decreasing the inlet temperature can improve the cooling performance of the liquid cold plate, a flow rate greater than 0.0249 $kg{\cdot}s^{-1}$ has no significant effect on the temperature uniformity of the liquid cold plate. Optimization of the BTMS operating parameters is crucial to achieving a better overall performance for the liquid cold plate.

(2) Based on the RSM design, the regression models for the average temperature, maximum temperature difference, and pressure drop are obtained with coefficients of variation (C.V) smaller than 0.3%, which guarantees the accuracy of the regression models. The deviations in the average temperature and pressure drop between the model predictions and CFD simulations are less than 11%.

(3) The Pareto optimal solutions were obtained by NSGA-II for the discharging rates at 1.0 C, 2.0 C, 3.0 C, and 4.0 C. The maximum temperature differences in the cold plate can be controlled within 0.29~3.90 °C, 1.11~15.66 °C, 2.17~31.39 °C, and 3.43~50.92 °C, respectively. Additionally, the maximum average temperatures are within 20.03~33.82 °C, 20.07~45.40 °C, 20.14~60.72 °C, and 20.22~80.22 °C, respectively. The pressure drop is optimized to less than 1000 Pa under various discharging rates. The proposed methods can be used to optimize the operating parameters of the liquid cold plate according to the thermal design criteria.

Author Contributions: Methodology, L.J., L.W. and Y.Z.; Software, Y.Z. and Z.M.; Validation, Z.L.; Formal analysis, Y.Z.; Investigation, L.J. and L.W.; Resources, F.C.; Data curation, Z.M. and Z.L.; Writing—original draft, L.J. and L.W.; Writing—review & editing, L.J.; Supervision, F.C. and L.J.; Project administration, L.J.; Funding acquisition, F.C. All authors have read and agreed to the published version of the manuscript.

Funding: The authors are grateful for the support of the China Post-doctoral Science Foundation Funded Project (Program No. 2021M692534) and the Fundamental Research Funds for the Central Universities (xzy022021007, xzy022021012).

Conflicts of Interest: The authors declare no conflict of interest.

Nomenclature

f	Response function
$\dot{m}$	Mass flow rate ($kg \cdot s^{-1}$)
n	Normal direction
P	Pressure (Pa)
ΔP	Pressure drop (Pa)
Q	Average heat generation (W)
R^2	Dimensionless distance
${R_{adj}}^2$	Thermal resistance ($K \cdot W^{-1}$)
T	Temperature (°C)
T_{ave}	Average temperature (°C)
T_{in}	Inlet temperature (°C)
T_{max}	Maximum temperature (°C)
ΔT_{max}	Maximum temperature difference (°C)
U	velocity vector ($m \cdot s^{-1}$)
u	velocity ($m \cdot s^{-1}$)
X_k	Design variable
x	x-coordinate
y	y-coordinate
z	z-coordinate
Greek:	
λ	Thermal conductivity ($W \cdot m^{-1} K^{-1}$)
ρ	Density ($kg \cdot m^{-3}$)
ε	Residual error
Subscripts:	
k	Number of design variables
in	Inlet fluid
f	Fluid
s	Solid
out	Outlet fluid
x	x direction
y	y direction
z	z direction

Appendix A

Table A1. The ANOVA for the average temperature (after backward elimination).

Source	Sum of Squares	Degrees of Freedom	Mean Square	F-Value	Prob. > F
Model	4626.93	11	420.63	4.273×10^9	<0.0001
$\dot{m}$	1568.31	1	1568.31	1.593×10^{10}	<0.0001
Q	19.87	1	19.87	2.019×10^8	<0.0001
T_{in}	49.51	1	49.51	5.030×10^8	<0.0001
$\dot{m} \cdot Q$	923.81	1	923.81	9.385×10^9	<0.0001
$\dot{m} \cdot T_{\mathrm{in}}$	0.0006	1	0.0006	6222.22	<0.0001
$Q \cdot T_{\mathrm{in}}$	0.0002	1	0.0002	2031.75	<0.0001
$\dot{m} \cdot \dot{m}$	385.68	1	385.68	3.918×10^9	<0.0001
$Q \cdot Q$	6.125×10^{-7}	1	6.125×10^{-7}	6.22	0.0373
$\dot{m} \cdot Q \cdot T_{\mathrm{in}}$	0.0002	1	0.0002	1645.71	<0.0001
$\dot{m}^2 \cdot Q$	141.99	1	141.99	1.442×10^9	<0.0001
$\dot{m}^2 \cdot T_{\mathrm{in}}$	8.100×10^{-6}	1	8.100×10^{-6}	82.29	<0.0001
Residual	7.875×10^{-7}	8	9.844×10^{-8}		
Lack of fit	7.875×10^{-7}	3	2.625×10^{-7}		
Pure error	0.0000	5	0.0000		
total	4626.93	19			

Table A2. The ANOVA for the maximum temperature difference (after backward elimination).

Source	Sum of Squares	Degrees of Freedom	Mean Square	*F*-Value	*p*-Value
Model	4273.94	13	328.76	5.904×10^9	<0.0001
$\dot{m}$	302.85	1	302.85	5.439×10^9	<0.0001
Q	34.94	1	34.94	6.276×10^8	<0.0001
T_{in}	0.0000	1	0.0000	323.27	<0.0001
$\dot{m}{\cdot}Q$	892.30	1	892.30	1.602×10^{10}	<0.0001
$\dot{m}{\cdot}T_{in}$	0.0122	1	0.0122	2.199×10^5	<0.0001
$Q{\cdot}T_{in}$	0.0215	1	0.0215	3.866×10^5	<0.0001
$\dot{m}{\cdot}\dot{m}$	280.85	1	280.85	5.044×10^9	<0.0001
$Q{\cdot}Q$	2.784×10^{-7}	1	2.784×10^{-7}	5.00	0.0667
$T_{in}{\cdot}T_{in}$	2.784×10^{-7}	1	2.784×10^{-7}	5.00	0.0667
$\dot{m}{\cdot}Q{\cdot}T_{in}$	0.0174	1	0.0174	3.123×10^5	<0.0001
$\dot{m}^2{\cdot}Q$	120.35	1	120.35	2.161×10^9	<0.0001
$\dot{m}^2{\cdot}T_{in}$	0.0038	1	0.0038	68,640.45	<0.0001
$\dot{m}{\cdot}Q{\cdot}Q$	6.250×10^{-7}	1	6.250×10^{-7}	11.22	0.0154
Residual	3.341×10^{-7}	6	5.568×10^{-8}		
Lack of fit	3.341×10^{-7}	1	3.341×10^{-7}		
Pure error	0.0000	5	0.0000		
Cor total	4273.94	19			

Table A3. The ANOVA for the pressure drop (after backward elimination).

Source	Sum of Squares	Degrees of Freedom	Mean Square	*F*-Value	Prob. > F
Model	4.748×10^6	4	1.187×10^6	6.154×10^5	<0.0001
$\dot{m}$	4.443×10^6	1	4.443×10^6	2.304×10^6	<0.0001
T_{in}	1043.63	1	1043.63	541.13	<0.0001
$\dot{m}{\cdot}T_{in}$	357.98	1	357.98	185.62	<0.0001
$\dot{m}{\cdot}\dot{m}$	3.031×10^5	1	3.031×10^5	1.571×10^5	<0.0001
Residual	28.93	15	1.93		
Lack of fit	28.93	10	2.89		
Pure error	0.0000	5	0.0000		
total	4.748×10^6	19			

References

1. Amjad, S.; Neelakrishnan, S.; Rudramoorthy, R. Review of design considerations and technological challenges for successful development and deployment of plug-in hybrid electric vehicles. *Renew. Sustain. Energy Rev.* **2020**, *14*, 1104–1110. [CrossRef]
2. Leipzig, I.T.F. Reducing Transport Greenhouse Gas Emissions: Trends & Data. In *Background for the 2010 International Transport Forum, Berlin*; ITF: Paris, France, 2010.
3. Andersen, P.H.; Mathews, J.A.; Rask, M. Integrating private transport into renewable energy policy: The strategy of creating intelligent recharging grids for electric vehicles. *Energy Policy* **2009**, *37*, 2481–2486. [CrossRef]
4. Maja, M.; Morello, G.; Spinelli, P. A model for simulating fast charging of lead/acid batteries. *J. Power Sources* **1992**, *40*, 81–91. [CrossRef]
5. Liu, J.; Gao, D.; Cao, J. Study on the effects of temperature on LiFePO4 battery life. In Proceedings of the 2012 IEEE Vehicle Power and Propulsion Conference, Seoul, Korea, 9–12 October 2012; IEEE: Piscataway, NJ, USA, 2012.
6. Leng, F.; Tan, C.; Pecht, M. Effect of Temperature on the Aging rate of Li Ion Battery Operating above Room Temperature. *Sci. Rep.* **2015**, *5*, 12967. [CrossRef]
7. Cui, S.; Wei, Y.; Liu, T.; Deng, W.; Pan, F. Optimized Temperature Effect of Li-Ion Diffusion with Layer Distance in Li(Nix MnyCoz)O2 Cathode Materials for High Performance Li-Ion Battery. *Adv. Energy Mater.* **2016**, *6*, 1501309. [CrossRef]
8. Liu, X.; Ren, D.; Hsu, H.; Feng, X.; Xu, G.; Zhuang, M.; Han, G.; Lu, L.; Han, X.; Chu, Z. Thermal Runaway of Lithium-Ion Batteries without Internal Short Circuit. *Joule* **2018**, *2*, 2047–2064. [CrossRef]
9. Lei, Z.; Zhang, C.; Li, J.; Fan, G.; Lin, Z. A study on the low-temperature performance of lithium-ion battery for electric vehicles. *Auto. Eng.* **2013**, *35*, 927–933.
10. Ren, D.; Xiang, L.; Feng, X.; Lu, L.; Ouyang, M.; Li, J.; He, X. Model-based thermal runaway prediction of lithium-ion batteries from kinetics analysis of cell components. *Appl. Energ.* **2018**, *228*, 633–644. [CrossRef]
11. Chen, Z.; Xiong, R.; Li, X. Temperature rise prediction of lithium-ion battery suffering external short circuit for all-climate electric vehicles application. *Appl. Energ.* **2018**, *213*, 275–383. [CrossRef]

12. Landini, S.; Leworthy, J.; O'Donovan, T.S. A Review of Phase Change Materials for the Thermal Management and Isothermalisation of Lithium-Ion Cells. *J. Energy Stor.* **2019**, *25*, 100887. [CrossRef]
13. Jiang, Z.; Qu, Z. Lithium–ion battery thermal management using heat pipe and phase change material during discharge–charge cycle: A comprehensive numerical study. *Appl. Energ.* **2019**, *242*, 378–392. [CrossRef]
14. Ling, Z.; Zhang, Z.; Shi, G.; Fang, X.; Wang, L.; Gao, X.; Fang, Y.; Xu, T.; Wang, S.; Liu, X. Review on thermal management systems using phase change materials for electronic components, Li-ion batteries and photovoltaic modules. *Renew. Sustain. Energy Rev.* **2014**, *31*, 427–437. [CrossRef]
15. Xu, X.; He, R. Review on the heat dissipation performance of battery pack with different structures and operation conditions. *Renew. Sustain. Energy Rev.* **2014**, *29*, 301–315. [CrossRef]
16. Liu, X.; Zhang, C.-F.; Zhou, J.-G.; Xiong, X.; Zhang, C.-C.; Wang, Y.-P. Numerical simulation of hybrid battery thermal management system combining of thermoelectric cooler and phase change material. *Energy Rep.* **2022**, *8*, 1094–1102. [CrossRef]
17. Saechan, P.; Dhuchakallaya, I. Numerical study on the air-cooled thermal management of Lithium-ion battery pack for electrical vehicles. *Energy Rep.* **2022**, *8*, 1264–1270. [CrossRef]
18. Liang, G.; Li, J.; He, J.; Tian, J.; Chen, X.; Chen, L. Numerical investigation on a unitization-based thermal management for cylindrical lithium-ion batteries. *Energy Rep.* **2022**, *8*, 4608–4621. [CrossRef]
19. Lyu, Y.; Siddique, A.R.M.; Majid, S.H.; Biglarbegian, M.; Gadsden, S.A.; Mahmud, S. Electric vehicle battery thermal management system with thermoelectric cooling. *Energy Rep.* **2019**, *5*, 822–827. [CrossRef]
20. Intano, W.; Kaewpradap, A.; Hirai, S.; Masomtob, M. Thermal investigation of cell arrangements for cylindrical battery with forced air-cooling strategy. *J. Res. Appl. Mech. Eng.* **2020**, *8*, 11–21.
21. Saw, L.H.; Ye, Y.H.; Tay, A.; Chong, W.T.; Kuan, S.H.; Yew, M.C. Computational fluid dynamic and thermal analysis of Lithium-ion battery pack with air cooling. *Appl. Energ.* **2016**, *177*, 783–792. [CrossRef]
22. Zhou, H.; Zhou, F.; Xu, L.; Kong, J.; Yang, Q. Thermal performance of cylindrical Lithium-ion battery thermal management system based on air distribution pipe. *Int J. Heat Mass Tran.* **2019**, *131*, 984–998. [CrossRef]
23. Lu, Z.; Yu, X.; Wei, L.; Qiu, Y.; Zhang, L.; Meng, X.; Jin, L. Parametric study of forced air cooling strategy for lithium-ion battery pack with staggered arrangement. *Appl. Therm. Eng.* **2018**, *136*, 28–40. [CrossRef]
24. Burban, G.; Ayel, V.; Alexandre, A.; Lagonotte, P.; Bertin, Y.; Romestant, C. Experimental investigation of a pulsating heat pipe for hybrid vehicle applications. *Appl. Therm. Eng.* **2013**, *50*, 94–103. [CrossRef]
25. Tran, T.H.; Harmand, S.; Desmet, B.; Filangi, S. Experimental investigation on the feasibility of heat pipe cooling for HEV/EV lithium-ion battery. *Appl. Therm. Eng.* **2014**, *63*, 551–558. [CrossRef]
26. Tran, T.H.; Harmand, S.; Sahut, B. Experimental investigation on heat pipe cooling for Hybrid Electric Vehicle and Electric Vehicle lithium-ion battery. *J. Power Sources* **2014**, *265*, 262–272. [CrossRef]
27. Pesaran, A.A. Battery Thermal Management in EVs and HEVs: Issues and Solutions. *Battery Man* **2001**, *43*, 34–49.
28. Park, S.; Jung, D. Battery cell arrangement and heat transfer fluid effects on the parasitic power consumption and the cell temperature distribution in a hybrid electric vehicle. *J. Power Sources* **2013**, *227*, 191–198. [CrossRef]
29. Wu, W.; Wang, S.; Kai, C.; Hong, S.; Lai, Y. A critical review of battery thermal performance and liquid based battery thermal management. *Energy Convers. Manag.* **2019**, *182*, 262–281. [CrossRef]
30. Deng, Y.; Feng, C.; Jiaqiang, E.; Zhu, H.; Chen, J.; Wen, M.; Yin, H. Effects of different coolants and cooling strategies on the cooling performance of the power lithium ion battery system: A review. *Appl. Therm. Eng.* **2018**, *142*, 10–29. [CrossRef]
31. Zhao, J.; Rao, Z.; Li, Y. Thermal performance of mini-channel liquid cooled cylinder based battery thermal management for cylindrical lithium-ion power battery. *Energy Convers. Manag.* **2015**, *103*, 157–165. [CrossRef]
32. Gao, R.; Fan, Z.; Liu, S. A gradient channel-based novel design of liquid-cooled battery thermal management system for thermal uniformity improvement. *J. Energy Storage* **2022**, *48*, 104014. [CrossRef]
33. Basu, S.; Hariharan, K.S.; Kolake, S.M.; Song, T.; Sohn, D.K.; Yeo, T. Coupled electrochemical thermal modelling of a novel Li-ion battery pack thermal management system. *Appl. Energ.* **2016**, *181*, 1–13. [CrossRef]
34. Du, X.; Qian, Z.; Chen, Z.; Rao, Z. Experimental investigation on mini-channel cooling-based thermal management for Li-ion battery module under different cooling schemes. *Int. J. Energy Res.* **2018**, *42*, 2781–2788. [CrossRef]
35. Wang, Y.; Zhang, G.; Yang, X. Optimization of liquid cooling technology for cylindrical power battery module. *Appl. Therm. Eng.* **2019**, *162*, 114200. [CrossRef]
36. Zhao, Z.; Bermudez, S.Z.; Ilyas, A.; Muylaert, K.; Vankelecom, I.F.J. Optimization of negatively charged polysulfone membranes for concentration and purification of extracellular polysaccharides from Arthrospira platensis using the response surface methodology. *Sep. Purif. Technol.* **2020**, *252*, 117385. [CrossRef]
37. Cai, M.; Wang, S.; Liang, H.H. Optimization of ultrasound-assisted ultrafiltration of Radix astragalus extracts with hollow fiber membrane using response surface methodology. *Sep. Purif. Technol.* **2012**, *100*, 74–81. [CrossRef]
38. Salahi, A.; Noshadi, I.; Badrnezhad, R.; Kanjilal, B.; Mohammadi, T. Nano-porous membrane process for oily wastewater treatment: Optimization using response surface methodology. *J. Environ. Eng.* **2013**, *1*, 218–225. [CrossRef]
39. Yi, S.; Su, Y.; Qi, B.; Su, Z.; Wan, Y. Application of response surface methodology and central composite rotatable design in optimizing the preparation conditions of vinyltriethoxysilane modified silicalite/polydimethylsiloxane hybrid pervaporation membranes. *Sep. Purif. Technol.* **2010**, *71*, 252–262. [CrossRef]

40. Kumar, A.; Thakur, A.; Panesar, P.S. A comparative study on experimental and response surface optimization of lactic acid synergistic extraction using green emulsion liquid membrane. *Sep. Purif. Technol.* **2019**, *211*, 54–62. [CrossRef]
41. Chanukya, B.S.; Kumar, M.; Rastogi, N.K. Optimization of lactic acid pertraction using liquid emulsion membranes by response surface methodology. *Sep. Purif. Technol.* **2013**, *111*, 1–8. [CrossRef]
42. Song, T.; Yao, Y.; Ni, L. Response surface method to study the effect of conical surface and vortex-finder lengths on de-foulant hydrocyclone with reflux ejector. *Sep. Purif. Technol.* **2020**, *253*, 117511. [CrossRef]
43. Zou, Y.; Wei, L.; Lu, Z.; Qin, S.; Cao, F. An optimal study of serpentine channel with various configurations based on response surface analysis. In *IOP Conference Series: Earth and Environmental Science*; IOP Publishing: Bristol, UK, 2022; Volume 1074, p. 012016.
44. Liebig, G.; Gupta, G.; Kirstein, U.; Schuldt, F.; Agert, C. Parameterization and validation of an electrochemical thermal model of a lithium-ion battery. *Batteries* **2019**, *5*, 62. [CrossRef]
45. Lyu, P.; Huo, Y.; Qu, Z.; Rao, Z. Investigation on the thermal behavior of Ni-rich NMC lithium ion battery for energy storage. *Appl. Therm. Eng.* **2020**, *166*, 114749. [CrossRef]
46. Abdul, Q.Y.; Laurila, T.; Karppinen, J.; Jalkanen, K.; Vuorilehto, K.; Skogström, L.; Paulasto-Kröckel, M. Heat generation in high power prismatic Li-ion battery cell with LiMnNiCoO2 cathode material. *Int. J. Energy Res.* **2014**, *38*, 1424–1437. [CrossRef]
47. Liang, X.; Sun, Z.; Huang, Q. Study on the Channel and Heat Transfer Performance of Liquid Cold Plate Used for LED Module. *China Light Lighting* **2012**, *1*, 6–10.
48. Deb, K.; Agrawal, S.; Pratap, A.; Meyarivan, T. A fast elitist non-dominated sorting genetic algorithm for multi-objective optimization: NSGA-II. *Lect. Notes Comput. Sci.* **2000**, *1917*, 849–858.

Journal of
Marine Science and Engineering

MDPI

Article

Energy-Saving Depth Control of an Autonomous Underwater Vehicle Using an Event-Triggered Sliding Mode Controller

Yu Qi [1], Xinyu Wu [2], Guocheng Zhang [1,*] and Yushan Sun [1]

1 Laboratory of Science and Technology on Underwater Vehicle, Harbin Engineering University, Harbin 150001, China
2 Marine Design & Research Institute of China, Shanghai 200011, China
* Correspondence: zhangguocheng@hrbeu.edu.cn

Abstract: In order to improve the endurance of underwater vehicles and make it possible for the underwater vehicle to inspect long-distance water tunnels, a sliding mode control method based on event triggering is proposed for the depth control of underwater vehicles from the perspective of energy saving. Firstly, the kinematics and dynamics models of underwater vehicle dive surface are established. Secondly, an event-triggered sliding mode controller is designed. According to the Lyapunov function, the stability of the designed controller is proved by theoretical analysis, and Zeno phenomena will not appear in the closed-loop control system. Compared with other controllers, the simulation results show that this controller can effectively realize the depth control of AUV, has strong adaptability and robustness to unmodeled nonlinear dynamics and bounded disturbances, and has the effect of saving computing resources.

Keywords: AUV; depth control; event-triggered; energy saving

Citation: Qi, Y.; Wu, X.; Zhang, G.; Sun, Y. Energy-Saving Depth Control of an Autonomous Underwater Vehicle Using an Event-Triggered Sliding Mode Controller. *J. Mar. Sci. Eng.* **2022**, *10*, 1888. https://doi.org/10.3390/jmse10121888

Academic Editor: Rose Norman

Received: 29 October 2022
Accepted: 17 November 2022
Published: 4 December 2022

1. Introduction

With the wider and wider range of deep-sea research and development activities, various types of underwater vehicles have become essential tools for scientists and engineers to carry out marine exploration and underwater missions [1–7]. Among many vehicles of this kind, autonomous underwater vehicles (AUVs) have the advantages of flexibility, personnel safety, low cost and so on, and have become important equipment for marine exploration and development. AUVs play an important role in hydrological environment surveys, submarine topography surveys, submarine pipeline detection, submarine search and rescue and salvage. For high-quality data collection and long-term tasks, AUVs are expected to be able to possess both satisfactory control ability and low-speed energy-saving cruising [8–14], and so the constant depth navigation of AUVs is essential.

Being equipped for deep-sea travel, remote distances and highly intelligent operation are the development trends for AUVs. Limited onboard energy is one of the key factors that affect the long-term exploration, cruise and operation of AUVs, which restricts the long-distance development of AUVs. If the energy is insufficient during the mission, the AUV needs to return in time and recycle it for charging, which will cause the discontinuity of the mission and destroy the concealment of the mission. Reducing the charging times can effectively improve the operation efficiency and reduce the time cost. Therefore, it is urgent to improve the endurance of AUVs. It is one of the bottlenecks of energy AUV technology development, one which greatly limits the development of AUV endurance. To improve the endurance of AUVs, we usually start with increasing the carrying energy, improving the energy utilization rate and utilizing natural energy such as ocean energy and solar energy. To use natural energy, AUVs need to add additional energy conversion devices, change the existing structure, and generate energy efficiency. This scheme is costly for AUVs. However, the existing batteries have low specific energy, so more energy can only be obtained by increasing the number and volume of batteries. It is relatively costly to improve endurance by increasing the amount of energy carried. Under the existing battery technology, reducing system energy consumption and improving energy utilization rate have more practical application values. To reduce energy consumption, we usually start with the shape of the AUV and energy-saving controls. However, for an existing AUV, changing the shape costs a lot. Therefore, starting from the control level, it is an important research direction of long voyages to study energy-saving strategies in order to improve energy utilization and reduce energy consumption, especially for the detection of long water conveyance tunnels, which is very necessary and meaningful.

2. Research Status

There are several ways to reduce energy consumption of AUVs, such as reducing sailing resistance, installing a variable buoyancy system (VBS) and adopting appropriate control methods. Ren Yitao et al. [15] designed a double-layer flexible skin structure on the surface of underwater vehicles to reduce the resistance of underwater vehicles. Yang Zhuo et al. [16] used a V-shaped microstructure in AUV models, which effectively reduced the resistance and energy consumption. Wang Chao et al. [17] applied drag reduction technology, based on aeration cavitation, to underwater vehicles and verified drag reduction efficiency by performing water tunnel tests. These surface drag reduction technologies are still in the exploratory stage, and the use cost is high. Generally, VBS systems are mounted on submersibles that need large depth changes. Tethy underwater vehicles, developed by the Monterey Bay Institute of Oceanography (MBARI), can last for weeks or even months. Because of their propellers, their speed range is 0.5–1.2 m/s, and they have been successfully used in the investigation and research work in Monterey Bay [18]. However, most AUVs are not equipped with VBS systems. They are usually amongst the most effective means to save energy for AUVs in order to reduce the energy consumption of the system from the point of view of control strategy. This is done to improve the endurance of underwater vehicles. Sarkar et al. [19] designed a suitable switching surface based on sliding mode control, and proposed a suboptimal robust controller based on the Euler–Lagrange optimization algorithm to minimize the total energy and ensure the accuracy of the controller. Feng Yao et al. [20] constructed a cost function based on an AUV state space model, designed an improved predictive control to reduce the energy consumption of the system, and verified the feasibility of the algorithm through depth tracking control simulation. Petar et al. put forward an adaptive oscillator which can learn the frequency, amplitude and phase of external disturbance online, and used this oscillator to process the input signal in order to ignore unnecessary disturbance and achieve the purpose of saving control output energy consumption [21]. Xu Y et al. established a dynamic resistance mathematical model to describe the relationship between the resistance and speed of AUVs, and used the terminal sliding mode control method to reduce the pitch angle and steering amplitude of AUVs. The simulation results show that the pitch angle and steering amplitude are effectively reduced, that the variance of pitch angle is reduced by 90.91%, and that the resistance caused by pitching and the energy consumed by steering are reduced [22]. Niu Xiaoli studied the vertical movement of AUVs near the water surface, analyzed the influence of waves of different depths on the carrier, and used LQR control to control the AUVs' vertical movement posture and energy consumption, which reduced the energy consumption during navigation [23]. However, there was a lack of analysis of and research on calculating resource energy consumption, and the reduction in resource energy consumption could further save energy for AUV control. Ji Zehui, Liu Zhenchuan, Wang Zhenyu et al. [24] of the China Ocean University have developed a submarine pipeline inspection robot to inspect pipeline corrosion. The simulation experiment and underwater experiment, conducted from pool to lake to offshore conditions, have verified the accuracy of the path tracking control system. However, under the complicated environmental conditions of the water conveyance tunnel, the path tracking technology for the outer wall of the submarine pipeline is not applicable. The water conveyance tunnel is usually tens of kilometers long, and so it is necessary and meaningful to increase the endurance of AUVs through energy-saving controls to realize continuous inspection.

In order to achieve AUV depth control, firstly, it is first necessary to establish the AUV diving plane kinematics model and dynamics model. From the AUVs' working environments and their own movement characteristics, their precise mathematical models are hard to establish; even if researchers can design a sufficiently accurate mathematical model, this model is often too complex to be suitable for control system design. Therefore, the model needs to be simplified. The diving behavior of AUVs is often simplified as a multivariable linear system, in which the dynamics characteristics of underwater robot made two major assumptions. First, we assume that the pitch angle of the AUV is very small, and then we assume that the motion dynamics of the pitch angle can be expressed as a linear equation. Based on the above assumptions, the vertical motion model of the AUVs can be linearized [24–27].

Sliding mode control has the advantages of robustness to model inaccuracy and external disturbances. It is very suitable for the control of underwater robots. It has been successfully applied to the dynamic positioning of underwater vehicles. A control algorithm for AUV navigation depth is proposed, which combines the adaptability of adaptive control algorithms and the robustness of sliding mode controllers. The effectiveness of the control algorithm is verified by simulation [28,29]. A high-order sliding mode control (HOSMC) is proposed for deep control of underwater vehicles. The simulation results show that the controller has a good control performance [30,31]. An integrator sliding mode controller is proposed for the depth control of AUVs with positive buoyancy. The simulation results show that the controller has a good control performance [32]. Through the development of depth control systems of ROVs. Adaptive fuzzy algorithms based on sliding mode control are used for uncertainty and disturbance compensation. Two theorems (Lyapunov stability theory and Barbalat lemma) are used to analyze and prove the stability and convergence of the closed-loop system. Numerical results show that the control system has good performance [33]. A model-free high-order sliding mode controller is proposed, and a reasonable transient process is designed, so that the controller has a good performance under any initial error conditions. In a two-degree-of-freedom nonlinear underwater vehicle model, real-time experiments verify the performance of the proposed controller [34].

By designing a reasonable transition process, a model-free high-order sliding mode controller is proposed, which gives the controller have a good performance under any initial error condition. In the verification underwater vehicle model, the performance of the proposed controller is verified by numerical simulation and real-time experiments.

The above-mentioned research uses a sliding mode controller to complete the depth control of an underwater vehicle, and it has achieved good results from the simulation experiment, which proves that the sliding mode controller can be used for the depth control of an underwater vehicle. However, the traditional sliding mode controller has a chattering problem. Frequent changes of the controller increase the energy consumption of the underwater vehicle and reduce its operating time. At the same time, affected by the complex underwater environment, communication may be blocked.

Event-triggered control is a resource-aware sampling strategy that updates control only when a driver condition is met [35]. A state-feedback approach to event-based control method is proposed to complete the control task of the closed-loop system [36]. The event-based control problem of nonlinear stochastic systems is studied, an event-based control method is proposed, and the effectiveness of the method is verified by simulation experiments [37]. The triggering of events can save energy at the control level. When dealing with microprocessors with limited resources and networks with limited bandwidth, the research of event-triggering technology becomes more and more important. The problem of output feedback control based on event-driven observers for linear systems is studied, and the effectiveness and superiority of the event-driven controller are demonstrated through simulation experiments [38].

Event triggering can reduce the energy consumption and communication requirements of underwater vehicles. This article introduces the event-triggered mechanism into the depth control of AUVs. The trigger function and sliding mode control law are designed. Only requiring the AUV to update the control rate at the time of the event trigger, energy-saving control of the AUV depth is realized. It is expected that this method will be applied to the detection of unmanned underwater robots in long-distance water conveyance tunnels.

3. Problem Statements

As shown in Figure 1, in order to establish the motion model of underwater vehicle, it is usually necessary to establish a fixed coordinate system $E - \xi\eta\zeta$ and a satellite coordinate system $O - xyz$.

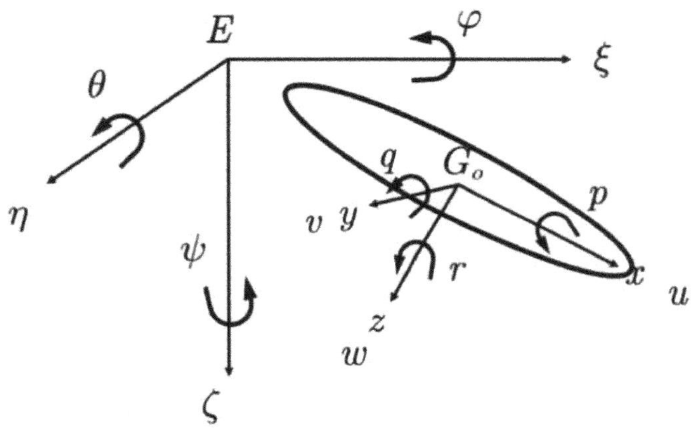

Figure 1. Fixed coordinate system and satellite coordinate system.

The origin of the fixed coordinate system $E - \xi\eta\zeta$ can be any point in space, and it represents the inertial reference system which allows robots to move in space. The coordinate system satisfies the right-hand rule, and its axis $E\zeta$ points to the center of the earth in a positive direction. The $E\xi$ axis and the $E\eta$ axis are perpendicular to each other and at the same time in the horizontal $E\zeta$ plane. Usually, the positive direction of the $E\xi$ axis points to the north of the earth and the positive direction of the $E\eta$ axis points to the east of the earth.

The satellite coordinate system $O - xyz$ is fixed on the robot and the origin is usually taken at the center of gravity of the robot. The coordinate system satisfies the right-hand rule, with the longitudinal axis Ox pointing to the bow parallel to the hull baseline, the horizontal axis Oy pointing to the starboard parallel to the base plane, and the vertical axis Oz pointing to the bottom.

Motion parameters and coordinate components are shown in Table 1:

Table 1. Parameters of Kinematic Model of Underwater Vehicle.

	Variable	X Axis	Y Axis	Z Axis
Fixed coordinate system	displacement	ξ	η	ζ
	angle	ϕ	θ	ψ
Satellite coordinate system	speed	u	v	w
	angular velocity	p	q	r
	force	X	Y	Z
	moment	K	M	N

In this paper, the positive directions of roll angle, pitch angle and yaw angle are all determined from the fixed coordinate system according to the right-hand rule, and only three rotations are needed from the fixed coordinate system for the moving coordinate system.

The first rotation is ψ around the $E\zeta$ axis, $E\xi \to Bx_1, E\eta \to By_1$ then

$$\begin{bmatrix} \xi \\ \eta \\ \zeta \end{bmatrix} = \begin{bmatrix} \cos\psi & -\sin\psi & 0 \\ \sin\psi & \cos\psi & 0 \\ 0 & 0 & 1 \end{bmatrix} \begin{bmatrix} x_1 \\ y_1 \\ \zeta \end{bmatrix} \tag{1}$$

The second rotation is θ around the By_1 axis, $Bx_1 \to Bx, E\zeta \to Bz_1$ then

$$\begin{bmatrix} x_1 \\ y_1 \\ \zeta \end{bmatrix} = \begin{bmatrix} \cos\theta & 0 & \sin\theta \\ 0 & 1 & 0 \\ -\sin\theta & 0 & \cos\theta \end{bmatrix} \begin{bmatrix} x \\ y_1 \\ z_1 \end{bmatrix} \tag{2}$$

The third rotation is φ around the Bx axis, $By_1 \to By, Bz_1 \to Bz$ then

$$\begin{bmatrix} x \\ y_1 \\ z_1 \end{bmatrix} = \begin{bmatrix} 1 & 0 & 0 \\ 0 & \cos\varphi & -\sin\varphi \\ 0 & \sin\varphi & \cos\varphi \end{bmatrix} \begin{bmatrix} x \\ y \\ z \end{bmatrix} \tag{3}$$

These values can be obtained by combining Equations (1)–(3),

$$\begin{bmatrix} \xi \\ \eta \\ \zeta \end{bmatrix} = \begin{bmatrix} \cos\psi\cos\theta & \cos\psi\sin\theta\sin]\varphi - \sin\psi\cos\varphi & \cos\psi\sin\theta\cos\varphi + \sin\psi\sin\varphi \\ \sin\psi\cos\theta & \sin\psi\sin\theta\sin\varphi + \cos\psi\cos\varphi & \sin\psi\sin\theta\cos\varphi - \cos\psi\sin\varphi \\ -\sin\theta & \cos\theta\sin\varphi & \cos\theta\cos\varphi \end{bmatrix} \begin{bmatrix} x \\ y \\ z \end{bmatrix} \tag{4}$$

In order to complete AUV depth control, the AUV diving plane kinematics model is first established. Suppose the surge velocity u is a known constant, and the depth d, pitch angle θ, heave velocity w and pitch angle velocity q are variables of motion state in the diving plane. The vertical kinematics equation of AUV can be expressed as:

$$\begin{cases} \dot{\theta} = q \\ \dot{z} = w\cos\theta - u\sin\theta \end{cases} \tag{5}$$

The dynamic equation can be expressed as:

$$\begin{cases} m[\dot{w} - uq - x_G\dot{q} - z_G] = Z_{\dot{q}}\dot{q} + Z_{\dot{w}}\dot{w} + Z_{uq}uq + Z_{uw}uw + Z_{w|w|}w\left|w\right| + Z_{q|q|}q\left|q\right| + (W-B)\cos\theta + \tau_1 \\ I_{yy}\dot{q} + m[x_G(uq - \dot{w}) + z_G wq] = M_{\dot{q}}q + M_{\dot{w}}\dot{w} + M_{uq}uq + M_{uw}uw + M_{w|w|}w\left|w\right| \\ +M_{q|q|}q\left|q\right| - (x_G W - x_B B)\cos\theta - (z_G W - z_B B)\sin\theta + \tau_2 \end{cases} \tag{6}$$

The hydrodynamic coefficient is the value of the partial derivative of the hydrodynamic component to AUV motion parameters at the expansion point. For example, $Z_{\dot{q}}$ is the influence coefficient of the angular acceleration $\dot{q}$ rotating around the Y axis, which causes the force Z on the Z axis. Then, $M_{w|w|}$ is the second derivative of the influence of w (Z-axis speed) on moment M. The hydrodynamic coefficient is usually obtained by performing an experiment. I_{yy} is the moment of inertia of AUV about the Y axis. x_G, z_G represent the center of gravity coordinates, and x_B, x_B represent the center of gravity coordinates. W is AUV gravity and B is AUV buoyancy. τ_1, τ_2 are control laws. In this paper, they represent force and moment.

Assuming that its second-order viscous damping coefficient is relatively small, it can be regarded as a small external disturbance. Both the center of gravity and the center of floating are at the origin of the coordinate system, $(x_G, y_G, z_G) = (x_B, y_B, z_B) = 0$. The pitch angle is a small amount. Combining Equations (5) and (6), the linearized equations can be obtained as:

$$\begin{bmatrix} m - Z_{\dot{w}} & -Z_{\dot{q}} & 0 & 0 \\ -M_{\dot{w}} & I_{yy} - M_{\dot{q}} & 0 & 0 \\ 0 & 0 & 1 & 0 \\ 0 & 0 & 0 & 1 \end{bmatrix} \begin{bmatrix} \dot{w} \\ \dot{q} \\ \dot{\theta} \\ \dot{z} \end{bmatrix} = \begin{bmatrix} Z_{uw}u & Z_{uq}u + mu & 0 & 0 \\ M_{uw}u & M_{uq}u & 0 & 0 \\ 0 & 1 & 0 & 0 \\ 1 & 0 & -u & 0 \end{bmatrix} \begin{bmatrix} w \\ q \\ \theta \\ z \end{bmatrix} + \begin{bmatrix} \tau_1 \\ \tau_2 \\ 0 \\ 0 \end{bmatrix} + d' \quad (7)$$

where d' represents unmodeled dynamics, external interference and parameter uncertainty. For the convenience of presentation:

$$M = \begin{bmatrix} m - Z_{\dot{w}} & -Z_{\dot{q}} & 0 & 0 \\ -M_{\dot{w}} & I_{yy} - M_{\dot{q}} & 0 & 0 \\ 0 & 0 & 1 & 0 \\ 0 & 0 & 0 & 1 \end{bmatrix}, A' = \begin{bmatrix} Z_{uw}u & Z_{uq}u + mu & 0 & 0 \\ M_{uw}u & M_{uq}u & 0 & 0 \\ 0 & 1 & 0 & 0 \\ 1 & 0 & -u & 0 \end{bmatrix} \quad (8)$$

Assuming that the expected heave speed w_d, expected pitch speed q_d, expected pitch angle θ_d, and expected depth z_d are all constants, namely $\dot{w}_d = \dot{q}_d = \dot{\theta}_d = \dot{z}_d = 0$. Definition: surge velocity error $w_e = w - w_d$, pitch velocity error $q_e = q - q_d$, pitch angle error $\theta_e = \theta - \theta_d$, depth error $z_e = z - z_d$, $x = \begin{bmatrix} w & q & \theta & z \end{bmatrix}^T$, $u(t) = \begin{bmatrix} \tau_1 & \tau_2 & 0 & 0 \end{bmatrix}^T$, $A = M^{-1}A'$, $B = M^{-1}$, $d = M^{-1}d'$, Equation (7) can be displayed as

$$\dot{x}(t) = Ax(t) + Bu(t) + d(t) \quad (9)$$

4. Design of Sliding Mode Control Triggered by Events

The conventional sliding variable for the Equation (9) is defined as

$$s(t) = Cx(t) \quad (10)$$

The control law $u(t)$ is designed to reach the $s(t) = 0$ manifold in a finite time, while always keeping the trajectory on $s(t)$. As shown in the Figure 2.

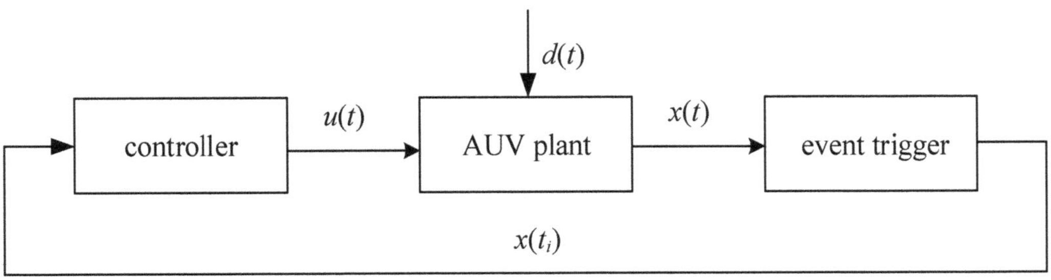

Figure 2. Event-triggered control loop.

Differentiating Equation (10), we can get:

$$\dot{s}(t) = C(Ax(t) + Bu(t) + d(t)) \quad (11)$$

Assume that CB is nonsingular, after that the control law is designed as follows

$$u(t) = -(CB)^{-1}\{CAx(t) + Ksign[s(t)]\} \quad (12)$$

In this paper, the control law adopts a sampling state, and the subsequent update is triggered by events. The control law (12) is given as

$$u(t) = -(CB)^{-1}\{CAx(t_i) + Ksign[s(t_i)]\}, \forall t \in [t_i, t_{i+1}) \quad (13)$$

$s(t) = Cx(t_i)$ denotes the sliding variable at t_i. At the time instant t_{i+1}, the control law is updated and keeps a constant until the next trigger is reached.

The measurement error of th sampling state exists in the system, and it can be defined as $e(t)$. $e(t)$ is given by the following formula:

$$e(t) = x(t_i) - x(t), \forall t \in [t_i, t_{i+1}) \tag{14}$$

In order to achieve stability and accuracy of control, it is necessary to fully consider the control error $e(t)$ and its related problems. For $t = t_i$, $e(t) = x(t_i) - x(t) = 0$, since the control is only updated at this moment in the process, the sliding mode will occur at this moment. Similarly, the deviation from sliding manifold can be obtaind for any $t \in [t_i, t_{i+1})$, $e(t) \neq 0$ and consequently $s(t)$.

In this paper, it is necessary to discuss the problem of robust stabilization of measurement error $e(t)$ by sliding mode control. Simple events can be executed to ensure the stability of the system. In this case, it is critical to consider whether there are Zeno phenomena i.e., in a limited execution time $t \in [t_i, t_{i+1}]$, that is, whether there is the accumulation of execution time between controls. In the paper, we proved that there are no Zeno phenomena, i.e., next, the situation of joining the system (5) is emphatically analyzed, and the results and assumption are as follows

Assumption 1: Firstly, $\alpha \in (0, \infty)$ is determined, and then it is introduced into system (5) and sliding variable (7). Then, the frequency band of the sliding mode near $s(t) = 0$ is defined by the control law (12) as follows

$$\left\{x(t) \in R^n : |s(t_i)| \leq \alpha \|A\|^{-1}\right\} \tag{15}$$

When the system also meets the following conditions

$$\|CAe(t)\| < \alpha \tag{16}$$

Additionally, K also satisfies such conditions

$$K > \sup_{t \geq 0} \|Cd(t)\| + \eta + \alpha \tag{17}$$

where $\eta > 0$.

Proof: We prove the existence and correctness of the sliding mode by considering Lyapunov function at $t \in [t_i, t_{i+1}]$.

$$V = \frac{1}{2}\|s(t)\|^2 \tag{18}$$

Bring (9) into and differentiate V with respect to time to obtain the following formula

$$\dot{V} = s(t)\{C[Ax(t) + Bu(t) + d(t)]\} \tag{19}$$

Sometimes, the control law cannot be updated continuously. The reason for this may be the limitation of the digital processor, and so the above Equation (13) should be used as the control law.

$$\begin{aligned} \dot{V} &= s(t)\{CAx(t) - CAx(t_i) - Ksign[s(t_i)] + d(t)\} \\ &= -s(t)\{CAe(t) + Ksign[s(t_i)] - d(t)\} \end{aligned} \tag{20}$$

It is obvious that, for the Lyapunov function $V < 0$ at $t = t_i$. For $t \in (t_i, t_{i+1})$, under the conditions (16) and (17), we get

$$\begin{aligned} \dot{V} &< -s(t)\{Ksign[s(t_i)] - d(t) - \alpha sign[s(t)]\} < -\eta s(t) sign[s(t_i)] \\ &= -\eta|s(t_i)| + \eta Ce(t) sign[s(t_i)] < -\eta|s(t_i)| + \eta\alpha\|A\|^{-1} \end{aligned} \tag{21}$$

So, the sliding mode will occur in the vicinity of $s(t) = 0$ with a band, as given in (15). The control law is updated whenever

$$\sigma \in [0,1] \; \|C\|\|A\|\|e(t)\| = \sigma\alpha \tag{22}$$

So, $\|C\|\|A\|\|e(t)\| \leq \sigma\alpha$ is met and condition (16) would not be broken.
The triggering instant can be defined as

$$t_{i+1} = \inf\{t \in [t_i, +\infty] : \|C\|\|A\|\|e(t)\| = \sigma\alpha\} \tag{23}$$

In order to avoid event-driven Zeno phenomena, assumption 2 is given. □

Assumption 2: Consider the system (9). Let the control law (13) brings the sliding mode in the system by executing the event (22) for all $t > t_i$ for the increasing time sequence. If t_{i+1} is the triggering instant, then the inter execution time $t_{i+1} - t_i = T_i > 0$ satisfies.

Firstly, the Equation (9) is introduced as a precondition, and the control law (13) executes events (22) through all tt in increasing time series, and the control law makes the system form a sliding mode. When t_{i+1} is the time when the execution event triggers, $t_{i+1} - t_i = T_i > 0$, as the execution interval, should meet the following conditions

$$T_i \geq \frac{1}{\|A\|} \ln(1 + \|CA\|^{-1}\sigma\alpha \frac{\|A\|}{\rho(\|x(t_i)\|) + \beta}) \tag{24}$$

where $\rho(\|x(t_i)\|)$ and β are given as

$$\rho(\|x(t_i)\|) = \|A - B(CB)^{-1}CA\|\|x(t_i)\|, \beta = \|B(CB)^{-1}K\| + d_{\max} \tag{25}$$

Proof: Consider the set $\Gamma = \{t : \|C\|\|A\|\|e(t)\| = 0\}$. Then, for time $t \in [t_i, t_{i+1})\backslash t_i (i = 1, 2, \cdots, n)$, we can get

$$\frac{d}{dt}\|e(t)\| \leq \|\dot{e}(t)\| = \|\dot{x}(t)\| = \|Ax(t) - B(CB)^{-1}CAx(t_i) - B(CB)^{-1}Ksign[s(t_i)] + d(t)\| \tag{26}$$

We substitute $x(t) = x(t_i) - e(t)$ into (26), and then the solution of (26) with the condition $e(t_i) = x(t_i) - x(t_i) = 0$ can be given as

$$\|e(t)\| \leq \frac{\rho(\|x(t_i)\|) + \beta}{\|A\|}(e^{\|A\|(t-t_i)} - 1) \tag{27}$$

$$\sigma\alpha \leq \|CA\|\frac{\rho(\|x(t_i)\|) + \beta}{\|A\|}(e^{\|A\|T_i} - 1) \tag{28}$$

Solve T_i according to Equation (28), and Assumption 2 is proved. □

In order to control the depth faster and better, it is also necessary to observe the heave velocity, pitch angular velocity and surge velocity in time, so it is necessary to establish an observer to improve the control accuracy. Surge velocity in the water conveyance tunnel is regular and fixed, so it can be obtained and input into the system only by prior knowledge and simple acquisition. Therefore, an observer is established for heave velocity and pitch angular velocity.

$$\begin{cases} \dot{x}(t) = \mathrm{A}x(t) + \mathrm{B}u(t) + d(t) \\ s(t) = Cx(t) \end{cases} \tag{29}$$

The basic idea of designing a Luenberger observer is to artificially build a system with the same parameters and inputs as the original system.

$$\begin{cases} \dot{\hat{x}}(t) = A\hat{x}(t) + Bu(t) + d(t) \\ \hat{s}(t) = C\hat{x}(t) \end{cases} \tag{30}$$

The observation error at this time can be defined as

$$\tilde{x} = x - \hat{x} \tag{31}$$

The error differential is

$$\dot{\tilde{x}} = A(x - \hat{x}) \tag{32}$$

As can be seen from the above equation, without introducing feedback, the dynamics of the observed error are completely determined by system matrix A. If the characteristic roots of A are all on the left side of the complex plane, the observed error will approach zero with the passage of time. However, when the system itself is unstable, its zero pole needs to be configured by introducing feedback, and so the design of the Luenberger observer can be expressed as

$$\begin{cases} \dot{\hat{x}}(t) = A\hat{x}(t) + Bu(t) + L[s(t) - \hat{s}(t)] + d(t) \\ \hat{s}(t) = C\hat{x}(t) \end{cases} \tag{33}$$

At this time, the dynamics of the observer error can be expressed as

$$\dot{\tilde{x}} = (A - LC)\hat{x} \tag{34}$$

Then, by selecting L, the characteristic root of the above formula is a proper value, thus ensuring that the error can be attenuated to zero and kept, and it is independent of the control input u(t) and the initial state x(0) of the system. In order to understand the Luenberger observer more vividly, we give its system block diagram. As shown in the Figure 3.

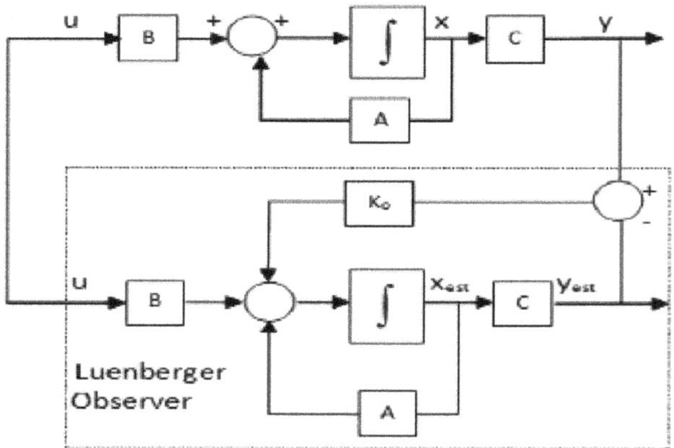

Figure 3. Luenberger observer.

The Luenberger observer controls the observed measurement $x(t)$ by introducing linear feedback. The essence of its control is to make it possible to track $x(t)$.

5. Numerical Studies

REMUS AUVs are used as models for simulation tests. The parameters of the dive plane model of a REMUS are shown in the Appendix A. Choose the expected heave speed $w_d = 0$, expected pitch speed $q_d = 0$, expected pitch angle $\theta_d = 0$, and expected depth $z_d = 0$ m. Suppose the surge velocity $u = 2$ m/s, the initial state $w_0 = 0.02$ m/s, $q_0 = 0.04$ rad/s, $\theta_0 = 0.1$ rad, $z_0 = 10$ m.

Select switch function $C = diag[0.51 \ -0.50 \ -0.73 \ 0.50]$, $K = 0.65$, $\alpha = 0.5$, $\sigma = 0.85$, $d = [0.5 \ \sin\ (10\ t)\ 0.5\ \sin\ (10\ t)\ 0\ 0]^T$. The following simulation results are obtained (Figures 4–8):

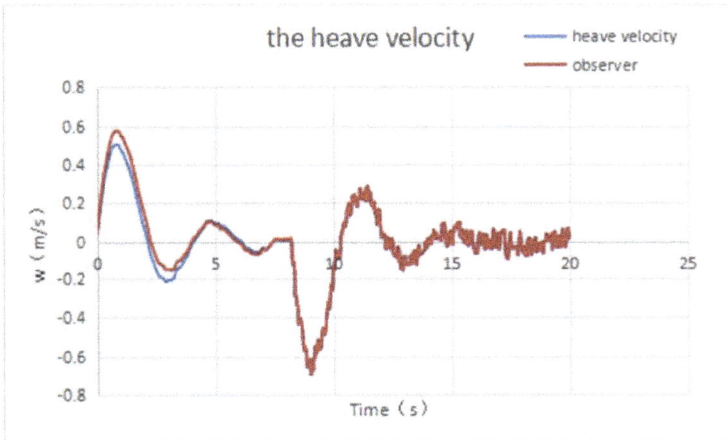

Figure 4. The heave velocity.

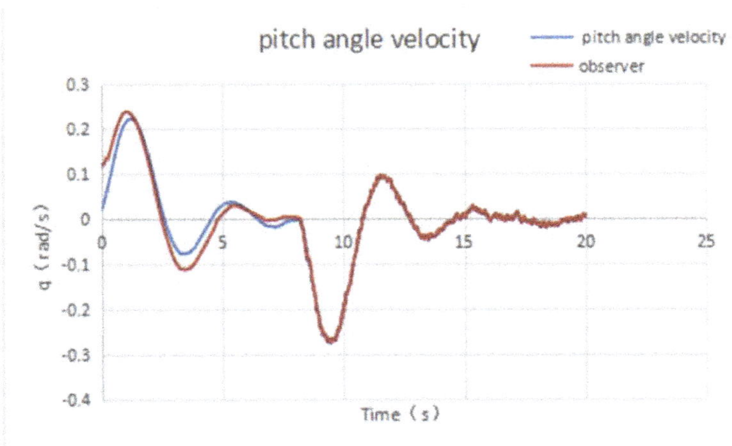

Figure 5. Pitch angle velocity.

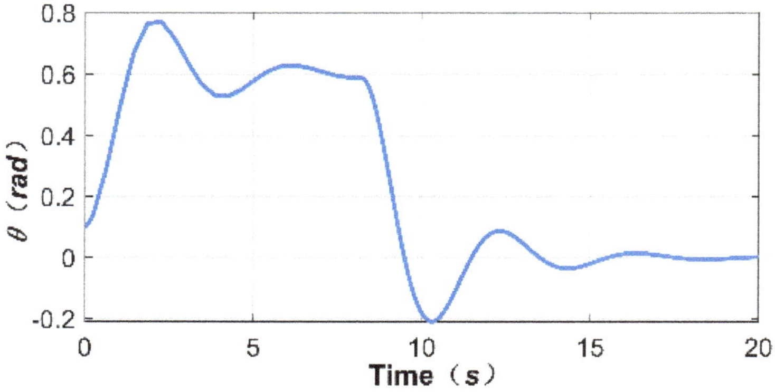

Figure 6. Pitch angle.

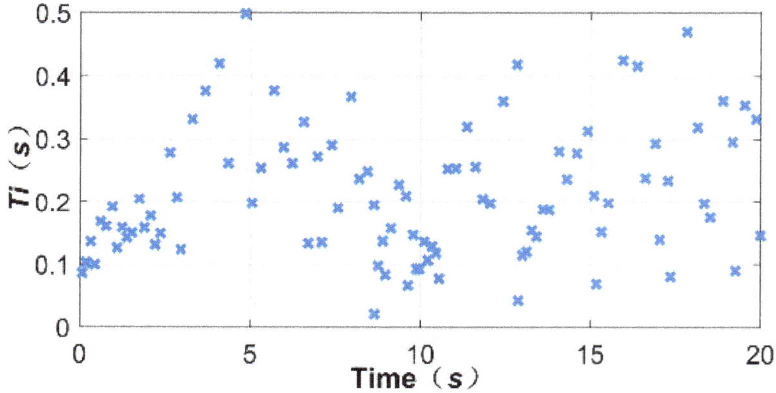

Figure 7. The event-triggered interval.

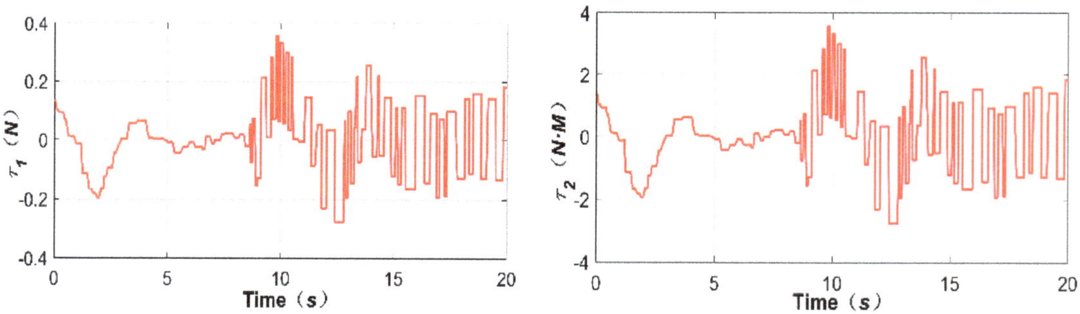

Figure 8. The simulation experiment control law (τ_1 and τ_2).

According to the tracking curve in Figures 4–6 during the simulation process, it can be seen that the event-triggered sliding mode controller in the case of interference works stably and has good robustness. These characteristics mean the control task of AUV sailing depth can be completed. The observer coincides with the actual speed in a short time (within 5 s), which can feed the data back to the controller in time. Figure 8 shows the control input curve of the AUV system. It shown that the frequency of the change in control rate can be effectively reduced based on the action of event triggering. The event-triggered interval is shown in Figure 7. In the adjustment stage (0–10 s), due to a large system error, the event-triggered interval is short, while in the stability stage (10–20 s), the event-triggered interval is long, which can effectively reduce the energy loss caused by the change in control law.

Taking a REMUS AUV as the research object, depth simulation experiments of various control methods are carried out, and the depth control effect of the sliding mode controller, based on event triggering, is compared and verified. As shown in the Figure 9.

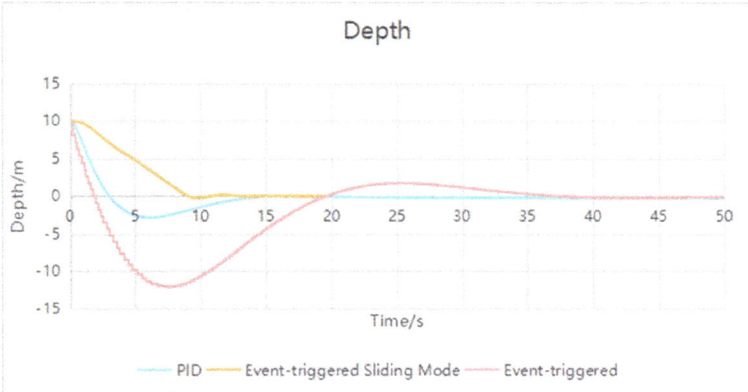

Figure 9. Comparison diagram of three control algorithms.

In the 20 s simulation, there are 513 controls, of which 26 are triggered by events, accounting for about 5% of the total, which can achieve 5.07% energy saving. It can save 5.07% of computing resources. This is shown in the Table 2 below.

Table 2. Event trigger times in simulation.

	Step Number	Percentage
Basic control step number	487	94.93%
Event step number	26	5.07%
Total number of steps	513	100%

In the process of simulation, it can be seen that the PID controller can realize AUV depth control, but there is overshoot. It is stable for about 20 s, but it has a steady-state error of about −0.2 m.

In the simulation process, it can be seen that the event trigger controller can realize AUV depth control, and that it is stable for about 41 s. Although it can reduce the amount of calculation, it takes a long time to stabilize.

It can be seen from the comparison diagram that the sliding mode control based on event trigger can reach the desired depth faster, more accurately and more stably, and can achieve a 5% energy saving in calculation steps.

6. Outfield Experiment

In order to test the feasibility of the control law and ensure the safety of AUVs, the test site was selected in a water area of Shanghai. As shown in the Figure 10.

Figure 10. Experimental site.

The expected depth is set at 0.8 m and 1 m, and the depth variation curve is shown in the following figure. As shown in the Figures 11 and 12 and Table 3.

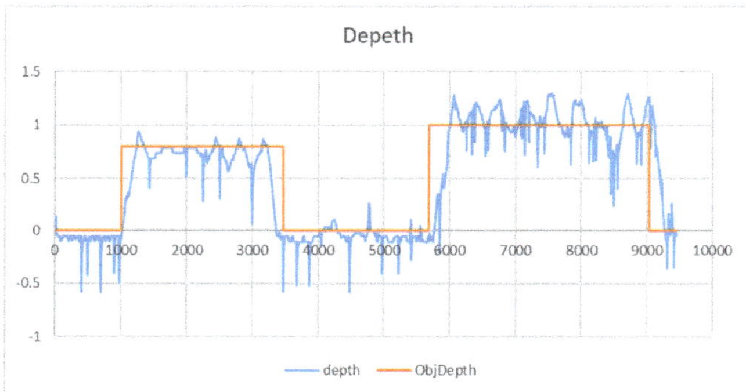

Figure 11. Depth variation curve of fixed-depth experiment.

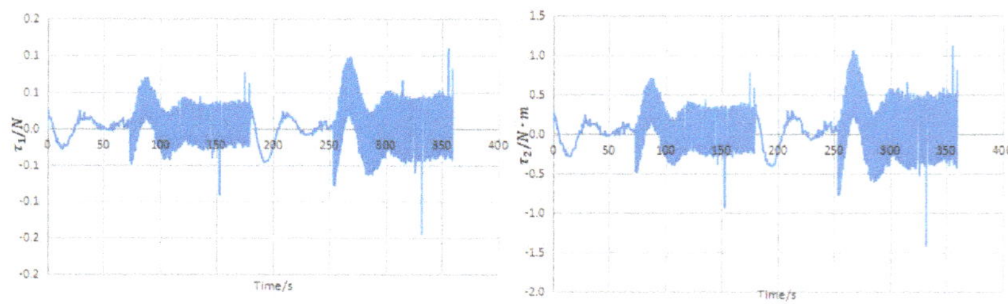

τ_1 τ_2

Figure 12. The outfield experiment control law (τ_1 and τ_2).

Table 3. Event trigger times in outfield experiment.

	Step Number	Percentage
Basic control step number	9352	97.28%
Event step number	261	2.72%
Total number of steps	9613	100%

The outfield experiment proves that the controller can realize the expected depth control of AUV. It can save 2.72% of computing resources.

7. Conclusions

Aiming at the problem of improving the endurance of AUV in the inspection of long water conveyance tunnels, an energy-saving control method is proposed to realize the depth control of AUV, and an event-triggered sliding mode controller is designed. The controller only needs to send instructions to the next module (thrust distribution module) when some discrete events trigger, which can effectively reduce the energy consumption of the controller and achieve the purpose of AUV energy saving. The controller can effectively suppress the interference, realize the convergence of the state control error to a small invariant set containing the origin, and ensure that the whole system will not have Zeno phenomena. Compared with numerical simulation of PID and event-triggered control, this controller has the following advantages:

1. Faster. It can reach the desired depth in a short time (10 s), while other methods can reach a better level in 15–35 s.
2. More accurate. Too much or too little steady-state error will not occur when the desired depth is reached.
3. More stable. There will be no overshoot when the desired depth is reached.
4. Save computing resources. By transmitting only when some discrete events are triggered, the computing resources are saved. The simulation shows that the energy saving effect is 5.07%, and the field experiment shows that the energy saving effect is 2.72%.

The simulation results show that the sliding film controller based on event triggering is faster, more accurate, more stable and energy saving, and the field test proves the feasibility of the controller.

The development of a suitable experimental prototype can directly participate in the inspection of long water conveyance tunnels, save manpower, material resources and energy, and create direct economic value. It can also be used for other tasks, such as underwater terrain scanning and detection, water quality and environmental monitoring, etc. It can bring all kinds of direct or indirect economic benefits. In the research process, the combination of industry–university–research is expected to bring various economic benefits, and especially to bring various benefits to enterprises and scientific research projects. After the technology is patented, it can also be used in the design of other robots such as surface robots and land robots. It can also be used in industrial robots, aviation and other fields.

Author Contributions: Conceptualization, Y.Q., Y.S. and G.Z.; methodology, Y.Q. and X.W.; software, X.W. and G.Z.; validation, X.W. and Y.Q.; formal analysis, Y.Q. and Y.S.; investigation, Y.Q. and Y.S.; resources, G.Z.; data curation, X.W.; writing—original draft preparation, Y.Q. and X.W.; writing—review and editing, Y.Q.; supervision, G.Z. and Y.S. All authors have read and agreed to the published version of the manuscript.

Funding: This research was funded by Key Project of [Heilongjiang Natural Fund: Research on Key Technology of AUV Detection in Water Tunnel] grant number [ZD2020E005] and The APC was funded by [Science and Technology Project Funding of Shaanxi Water Diversion from Han to Wei Engineering Construction Co., Ltd. (SPS-D-15)]. This work was supported in part by the Stable Supporting Fund of Acoustics Science and Technology Laboratory under Grant JCKYS2022604SSJS002.

Institutional Review Board Statement: Not applicable.

Informed Consent Statement: Not applicable.

Conflicts of Interest: The authors declare no conflict of interest.

Appendix A

REMUS AUV parameters [39]

$m = 30.48$ kg, $I_{yy} = 3.45$ kgm^2, $M_{\dot{q}} = -4.88$ kgm^2/rad, $M_{\dot{w}} = -1.93$ kgm, $M_{uq} = -2$ kgm/rad.

$M_{uw} = 24$ kg, $Z_{\dot{q}} = -1.93$ kgm/rad, $Z_{\dot{w}} = -35.5$ kg, $Z_{uq} = -5.22$ kg/rad, $Z_{uw} = -28.6$ kg/m.

References

1. Santana, J.P.; Mathias, N.; Hoveling, R.; Alves, H.; Morais, T. Innovative Benthic Lander for Macroalgae Monitoring in Shallow-Water Environments. *J. Mar. Sci. Appl.* **2022**, *19*, 133–147. [CrossRef]
2. Song, C.; Cui, W. Review of Underwater Ship Hull Cleaning Technologies. *J. Mar. Sci. Appl.* **2020**, *19*, 415–429. [CrossRef]
3. Van Hien, N.; Diem, P.G. A Practical Specialization of MDA/MBSE Approach to Develop AUV Controllers. *J. Marine. Sci. Appl.* **2021**, *20*, 102–116. [CrossRef]
4. Kumar, V.S.; Rajagopal, P. Modelling and Analysis of Turning Motion of a Subsurface Mapping AUV with Split-Hull Design. *J. Marine. Sci. Appl.* **2021**, *20*, 284–301. [CrossRef]
5. Makavita, C.D.; Jayasinghe, S.G.; Nguyen, H.D.; Ranmuthugala, D. Experimental Study of a Modified Command Governor Adaptive Controller for Depth Control of an Unmanned Underwater Vehicle. *J. Marine. Sci. Appl.* **2021**, *20*, 504–523. [CrossRef]
6. Hong, L.; Song, C.; Yang, P.; Cui, W. Two-Layer Path Planner for AUVs Based on the Improved AAF-RRT Algorithm. *J. Marine. Sci. Appl.* **2022**, *21*, 102–115. [CrossRef]
7. Zhang, W.; Wu, L.; Jiang, X.; Feng, X.; Li, Y.; Zeng, J.; Liu, C. Propeller Design for an Autonomous Underwater Vehicle by the Lifting Line Method based on OpenProp and CFD. *J. Marine. Sci. Appl.* **2022**, *21*, 106–114. [CrossRef]
8. Bai, X.; Li, B.; Xu, X.; Jie, X. A Review of Current Research and Advances in Unmanned Surface Vehicles. *J. Marine. Sci. Appl.* **2022**, *21*, 47–58. [CrossRef]
9. Gao, L.; Li, P.; Qin, H.; Deng, Z. Mechatronic Design and Maneuverability Analysis of a Novel Robotic Shark. *J. Marine. Sci. Appl.* **2022**, *21*, 82–91. [CrossRef]
10. Sun, Y.; Jiao, W.; Zhang, G.; Wang, L.; Cheng, J. Research on Stealth Assistant Decision System of Submarine Voyage Stage. *J. Marine. Sci. Appl.* **2020**, *19*, 208–217. [CrossRef]
11. Pais, T.; Gaiotti, M.; Barsotti, B. Evaluation of the Residual Capacity of a Submarine for Different Limit States with Various Initial Imperfection Models. *J. Marine. Sci. Appl.* **2022**, *21*, 59–68. [CrossRef]
12. Shahani, K.; Song, H.; Mehdi, S.R.; Sharma, A.; Tunio, G.; Qureshi, J.; Kalhoro, N.; Khaskheli, N. Design and Testing of an Underwater Microscope with Variable Objective Lens for the Study of Benthic Communities. *J. Marine. Sci. Appl.* **2021**, *20*, 170–178. [CrossRef]
13. Zou, X.; Jiang, G.; Ye, L. Vibration Response Analysis of a New Scientific Research Ship Based on Finite Element Modeling. *J. Marine. Sci. Appl.* **2022**, *21*, 69–81. [CrossRef]
14. Fossen, T.I. *Guidance and Control of Ocean Vehicles*; Wiley: New York, NY, USA, 1994.
15. Ren, J.; Zhao, D.; Chen, H.; Cui, J. Study on drag reduction performance of double-layer flexible skin based on immersion boundary method. *Appl. Sci. Technol.* **2021**, *48*, 1–7.
16. Yang, Z.; Chang, Y.; Li, D.; Wang, H. Design of underwater microstructure functional surface and drag reduction characteristics. *Shipbuild. Eng.* **2020**, *42*, 141–147.
17. Wang, C.; Hu, Z.; Yi, R.; Wen, Y. Experimental study on drag reduction by aeration and cavitation of high-speed underwater vehicle in water tunnel. *Robot* **2018**, *40*, 779–785.
18. Hobson, B.W.; Bellingham, J.G.; Kieft, B.; McEwen, R.; Godin, M.; Zhany, Y. Tethys-Class Long Range AUVs-Extending the Endurance of Propeller-Driven Cruising AUVs from Days to Weeks. In Proceedings of the IEEE 2012 IEEE/OES Autonomous Underwater Vehicles (AUV), Southampton, UK, 24–27 September 2012; pp. 1–8.
19. Sarkar, M.; Nandy, S.; Shome, S.N. Energy Efficient Trajectory Tracking Controller for Underwater Applications: A Robust Approach. *Aquat. Procedia* **2015**, *4*, 571–578. [CrossRef]
20. Yao, F.; Yang, C.; Zhang, M.; Wang, Y. Optimization of the energy consumption of depth tracking control based on model predictive control for autonomous underwater vehicles. *Sensors* **2019**, *19*, 162. [CrossRef]
21. Kormushev, P.; Caldwell, D.G. Improving the Energy Efficiency of Autonomous Underwater Vehicles by Learning to Model Disturbances. In Proceedings of the IEEE 2013 IEEE/RSJ International Conference on Intelligent Robots and Systems, Tokyo, Japan; 2013; pp. 3885–3892.
22. Yao, X.; Meng, L.; Wei, C. On the motion control strategy of AUV to optimize the voyage resistance. In Proceedings of the IEEE 2015 34th Chinese Control Conference (CCC). Hangzhou, China, 28 July 2015; pp. 4274–4279. [CrossRef]
23. Niu, X. Research on Comprehensive Control Strategy of AUV' s Vertical Attitude and Energy Saving Near Water Surface. Master's Thesis, Harbin Engineering University, Qingdao, China, 2015.
24. Ji, Z. Application of Underwater Linear Target Processing Algorithm in Submarine Pipeline Inspection Robot. Master' s Thesis, China Ocean University, Harbin, China, 2015.
25. Marino, R.; Tomei, P. Adaptive output feedback tracking with almost disturbance decoupling for a class ofnonlinear systems. *Automatica* **2000**, *36*, 1871–1877. [CrossRef]

26. Li, J.-H.; Lee, P.-M. Design of an adaptive nonlinear controller for depth control of an autonomous underwater vehicle. *Ocean Eng.* **2005**, *32*, 2165–2181. [CrossRef]
27. Naik, M.S.; Singh, S.N. State-dependent riccati equation-based robust dive plane control of auv with control constraints. *Ocean Eng.* **2007**, *34*, 1711–1723. [CrossRef]
28. Cristi, R.; Papoulias, F.A.; Healey, A.J. Adaptive sliding mode control of autonomous underwater vehicles in the dive plane. *IEEE J. Ocean. Eng.* **1990**, *15*, 152–160. [CrossRef]
29. Song, Y.S.; Arshad, M.R. Sliding Mode Depth Control of a Hovering Autonomous Underwater Vehicle. In Proceedings of the IEEE International Conference on Control System Computing and Engineering, Penang, Malaysia, 27–29 November 2015; pp. 435–440.
30. Salgadojimenez, T.; Jouvencel, B. Using a High Order Sliding Modes for Diving Control a Torpedo Autonomous Underwater Vehicle. In Proceedings of the Oceans Conference, San Diego, CA, USA, 22–26 September 2003; pp. 934–939.
31. García-Valdovinos, L.G.; Salgado-Jiménez, T.; Bandala, M.; Balanzar, L.N.; Hernández-Alvarado, R.; Cruz-Ledesma, J.A. Modelling, Design and Robust Control of a Remotely Operated Underwater Vehicle. *Int. J. Adv. Robot. Syst.* **2014**, *11*, 1–16. [CrossRef]
32. Eng, Y.H.; Soon, H.G.; Chitre, M. Depth Control of an Autonomous Underwater Vehicle, STARFISH//Oceans. In Proceedings of the Oceans' Conference, Sydney, Australia, 14 October 2010.
33. Bessa, W.M.; Dutra, M.S.; Kreuzer, E.J. Depth control of remotely operated underwater vehicles using an adaptive fuzzy sliding mode controller. *Robot. Auton. Syst.* **2008**, *56*, 670–677. [CrossRef]
34. Deng, C.N.; Ge, T. Depth and Heading Control of a Two DOF Underwater System Using a Model-free High Order Sliding Controller with Transient Process. In Proceedings of the International Conference on Measuring Technology and Mechatronics Automation, Hong Kong, China, 16–17 January 2013; pp. 423–426.
35. Åström, K.J. *Event Based Control. Analysis & Design of Nonlinear Control Systems;* Science Press: Beijing, China, 2008; pp. 127–147.
36. Lunze, J.; Lehmann, D. Brief paper: A state-feedback approach to event-based control. *Automatica.* **2010**, *46*, 211–215. [CrossRef]
37. Wang, Y.; Zheng, W.X.; Zhang, H. Dynamic Event-Based Control of Nonlinear Stochastic Systems. *IEEE Trans. Autom. Control* **2017**, *62*, 6544–6551. [CrossRef]
38. Zhang, J.; Feng, G. Event-driven observer-based output feedback control for linear systems. *Automatica* **2014**, *50*, 1852–1859. [CrossRef]
39. Huang, R.; Ding, N. AUV vertical control based on improved PID neural network algorithm. *J. Simul.* **2020**, *32*, 229–235. [CrossRef]

MDPI
St. Alban-Anlage 66
4052 Basel
Switzerland
www.mdpi.com

MDPI Books Editorial Office
E-mail: books@mdpi.com
www.mdpi.com/books

www.ingramcontent.com/pod-product-compliance
Lightning Source LLC
Chambersburg PA
CBHW061455220926
51850CB00004B/97
9783725801978